STARCHES IN SMART PACKAGING

APPLICATIONS OF STARCHES IN FOOD AND PACKAGING

STARCHES IN SMART PACKAGING

Edited by

SWARUP ROY
Department of Food Technology and Nutrition, School of Agriculture, Lovely Professional University, Phagwara, Punjab, India

PRATAP KALITA
Pratiksha Institute of Pharmaceutical Sciences, Guwahati, Assam, India

Series Editor

SNEH PUNIA BANGAR

ELSEVIER

ACADEMIC PRESS
An imprint of Elsevier

Academic Press is an imprint of Elsevier
125 London Wall, London EC2Y 5AS, United Kingdom
50 Hampshire Street, 5th Floor, Cambridge, MA 02139, United States

ISBN: 978-0-443-45396-0

For information on all Academic Press publications visit our website at https://www.elsevier.com/books-and-journals

Publisher: Mara Conner
Acquisitions Editor: Palak Gupta
Editorial Project Manager: Ritika Dagar
Production Project Manager: Aishwarya R.
Cover Designer: Arumugam Kothandan

Typeset by TNQ Tech

Contents

Contributors

Reza Abedi-Firoozjah
Student Research Committee, Kermanshah University of Medical Sciences, Kermanshah, Iran

Abdul Baquee Ahmed
School of Pharmaceutical Sciences, Girijananda Chowdhury University, Tezpur Campus, Guwahati, Assam, India

Tahid Alam
Department of Pharmaceutical Technology, University of North Bengal, Darjeeling, India

Mahmood Alizadeh Sani
Department of Food Science and Technology, School of Nutritional Sciences and Dietetics, Tehran University of Medical Sciences, Tehran, Iran

Saber Amiri
Department of Food Science and Technology, Faculty of Agriculture, Urmia University, Urmia, Iran

Sm Abdul Aziz Barbhuiyan
Department of Pharmaceutical Science, University of Science & Technology Meghalaya, Techno City, Ri-Bhoi, Meghalaya, India

Seyed Alireza Banihashemi
Department of Food Science and Technology, Faculty of Agriculture, Urmia University, Urmia, Iran

Mrinmoy Basak
Assam Down Town University, Guwahati, Assam, India

Deblina Biswas
Department of Instrumentation and Control Engineering, Dr. B R Ambedkar National Institute of Technology, Jalandhar, Punjab, India

Raquel Bonati Moraes Ibsch
Department of Chemical Engineering, University Center of Educational Foundation of Brusque, Brusque, Brazil

Rituparna Borah
Institute of Pharmacy, Assam Don Bosco University, Guwahati, Assam, India; School of Pharmacy, Girijananda Chowdhury University, Guwahati, Assam

Tapoban Bordoloi
NEPEDS College of Pharmaceutical Sciences, Sonapur, Assam, India

Wilson Daniel Caicedo Chacon
Ingeniería de Alimentos, Fundación Universitaria Agraria de Colombia, Bogotá, Colombia

Raja Chakraborty
Institute of Pharmacy, Assam Don Bosco University, Guwahati, Assam, India

Neha Chatterjee
Department of Food Science, University of Wisconsin—Madison, Madison, WI, United States; Glanbia, Twin Falls, ID, United States

Tania Costa
SENAI Santa Catarina University Center, Blumenau, Brazil

Rupak Das
Department of Pharmaceutical Technology, University of North Bengal, Darjeeling, India

Rabin Debnath
Institute of Pharmacy, Assam Don Bosco University, Guwahati, Assam, India

Himangshu Deka
Faculty of Pharmaceutical Science, Assam Down Town University, Guwahati, Assam, India

Satyendra Deka
Pratiksha Institute of Pharmaceutical Sciences, Guwahati, Assam, India

Ganesh Dey
Department of Pharmaceutical Technology, University of North Bengal, Darjeeling, India

Simin Fan
College of Food Science, Sichuan Agricultural University, Ya'an, PR China; Institute of Food Science and Technology, Chinese Academy of Agricultural Sciences, Beijing, PR China

Olaniyi Amos Fawole
University of Johannesburg, South African Chair in Sustainable Postharvest and Agroprocessing Research, Department of Botany and Plant Biotechnology, Johannesburg, South Africa

Ana Caroline Ferreira Carvalho
Federal University of Santa Catarina, Postgraduate Program in Food Engineering, Technology Center, Florianópolis, Santa Catarina, Brazil

Matheus Fiebig
Federal University of Santa Catarina, Postgraduate Program in Food Engineering, Technology Center, Florianópolis, Santa Catarina, Brazil

Valéria Cedeño Fierro
Facultad de Ingeniería de Administración, Universidad Nacional de Colombia, Palmira, Colombia

Abhinab Goswami
Faculty of Pharmaceutical Science, Assam Down Town University, Guwahati, Assam, India

Juan Felipe Grisales Mejía
Chemistry Department, Laboratory of Advanced Analytical Techniques in Natural Products (LATNAP), Universidad de los Andes, Bogotá, Colombia

Anthony Temitope Idowu
University of Limerick, Department of Biological Science, Limerick, Ireland

Nidhi Jaiswal
Department of Food Science and Human Nutrition, University of Illinois Urbana Champaign, Urbana, IL, United States

Ruchi D. Jena
School of Bioengineering and Food Technology, Shoolini University, Solan, Himachal Pradesh, India

Manisha Joshi
School of Bioengineering and Food Technology, Shoolini University, Solan, Himachal Pradesh, India

Pratap Kalita
Pratiksha Institute of Pharmaceutical Sciences, Guwahati, Assam, India

Arezou Khezerlou
Department of Food Safety and Hygiene, School of Health, Gonabad University of Medical Sciences, Gonabad, Iran

Carolina Krebs de Souza
Postgraduate Program in Chemical Engineering, University of Blumenau, Blumenau, Brazil

Susankar Kushari
Department of Pharmaceutical Chemistry, School of Pharmaceutical Sciences, Girijananda Chowdhury University, Guwahati, Assam, India

Gabriel Coelho Leandro
Department of Chemical and Food Engineering, Federal University of Santa Catarina, Florianópolis, Brazil

Łukasz Łopusiewicz
Vizja University, School of Medical & Health Sciences, Warsaw, Poland; University of Greifswald, Department Pharmaceutical Biology, Institute of Pharmacy, Greifswald, Germany

Pooja Sharma Luitel
Department of Pharmaceutical Technology, University of North Bengal, Darjeeling, India

Fatemeh Madanifar
Student Research Committee, Yasuj University of Medical Sciences, Yasuj, Iran

Karine Marafon
Federal University of Santa Catarina, Postgraduate Program in Food Engineering, Technology Center, Florianópolis, Santa Catarina, Brazil

Danila Merino
University of the Basque Country (UPV/EHU), POLYMAT Institute, San Sebastian, Spain; Basque Foundation for Science, Ikerbasque, Bilbao, Spain

Masoud Aman Mohammadi
Student Research Committee, Department of Food Technology, Faculty of Nutrition Science and Food Technology, Nutritional and Food Technology Research Institute, Shahid Beheshti University of Medical Sciences, Tehran, Iran; Food and Beverages Safety Research Center, Urmia University of Medical Sciences, Urmia, Iran

Muryeti
Department of Industrial Technology of Packaging Printing, Politeknik Negeri Jakarta, UI Campus, Depok, Indonesia

Debabrata Nath
Pratiksha Institute of Pharmaceutical Sciences, Guwahati, Assam, India

Rina Ningtyas
Department of Industrial Technology of Packaging Printing, Politeknik Negeri Jakarta, UI Campus, Depok, Indonesia

Abhishek Parasar
Pratiksha Institute of Pharmaceutical Sciences, Guwahati, Assam, India

Sudipta Paul
Department of Pharmaceutical Technology, University of North Bengal, Darjeeling, India

Elane Schwinden Prudencio
Federal University of Santa Catarina, Postgraduate Program in Food Engineering, Technology Center, Florianópolis, Santa Catarina, Brazil; Department of Food Science and Technology, Federal University of Santa Catarina, Florianópolis, Brazil

Bapi Ray Sarkar
Department of Pharmaceutical Technology, University of North Bengal, Darjeeling, India

Syed Nazrin Ruhina Rahman
Department of Pharmaceutics, Royal School of Pharmacy, The Assam Royal Global University, Guwahati, Assam, India

Sadaf Mohajjel Sadeghi
Student Research Committee, Tabriz University of Medical Sciences, Tabriz, Iran; Department of Food Science and Technology, Faculty of Nutrition and Food Sciences, Tabriz University of Medical Sciences, Tabriz, Iran

Oshin Sahni
Department of Food Science, University of Tennessee, Knoxville, TN, United States

Akash Sharma
Institute of Pharmacy, Assam Don Bosco University, Guwahati, Assam, India

Kanika Sharma
Department of Biotechnology, University Institute of Biotechnology, Chandigarh University, Mohali, Punjab, India

Deli Silvia
Department of Industrial Technology of Packaging Printing, Politeknik Negeri Jakarta, UI Campus, Depok, Indonesia

Mansoureh Taghizadeh
Halal Research Center of IRI, Food and Drug Administration, Ministry of Health and Medical Education, Tehran, Iran

Milad Tavassoli
Cellular and Molecular Research Center, Yasuj University of Medical Sciences, Yasuj, Iran; Department of Nutrition, Faculty of Health and Nutrition Sciences, Yasuj University of Medical Sciences, Yasuj, Iran

Hebat-Allah S. Tohamy
National Research Centre, Cellulose and Paper Department, Giza, Dokki, Egypt

Germán Ayala Valencia
Department of Chemical and Food Engineering, Federal University of Santa Catarina, Florianópolis, Brazil

Yiqin Zhang
College of Food Science, Sichuan Agricultural University, Ya'an, PR China

About the editors

Dr. Swarup Roy is currently working as an Associate Professor at the Department of Food Technology and Nutrition, Lovely Professional University, India. Dr. Roy is an eminent scholar at the Department of Food and Nutrition, Kyung Hee University, Republic of Korea. Dr. Roy previously worked as an Assistant Professor at Shoolini University, India (2022–2023). He worked as a Postdoctoral Researcher at the Bio-Nanocomposite Research Center, Kyung Hee University, Seoul, and Inha University, Incheon, Republic of Korea (2017–2021). He also worked as a Research Associate at the Indian Institute of Technology Indore (2016–2017). Dr. Roy completed his PhD degree as a DST INSPIRE Fellow at the University of Kalyani, India (2011–2016). During his doctoral study, he worked on microbial biosynthesis of metallic nanoparticles and the fabrication of bio-nanocomposite materials. During his postdoctoral research, he worked on the fabrication of a bio-composite film for active and intelligent food packaging. His current research work is focused on the preparation and application of biopolymer-based functional composite materials for smart packaging applications. Dr. Roy has already published over 235 research articles, including research papers, reviews, editorials, and book chapters. He has edited 4 books and currently editing 5 books. He is serving as an editor of several reputed journals: *Scientific Reports, Packaging Technology and Science, Journal of Vinyl and Additive Technology, Discover Nano, International Journal of Food Science, International Journal of Biomaterials, Frontiers in Nanotechnology, Frontiers in Food Science and Technology, Current Food Science and Technology Reports.*

Dr. Pratap Kalita is currently working as Professor at Pratiksha Institute of Pharmaceutical Sciences, Guwahati, Assam, India. Dr. Kalita has completed his Ph.D. in Pharmaceutical science from Assam Science and Technology University (ASTU), Guwahati, Assam, India. He worked as Junior Research Fellow (JRF) and Senior Research Fellow (SRF) under a DBT sponsored project awarded to Assam down town University (AdtU), Guwahati, Assam (2017–2020). Dr. Kalita has already published over 85 research articles including research papers, reviews and book chapters. He serves as a Reviewer for several reputed journals and presented several scientific papers, attended a number of national and international conferences. He has received several awards during his professional carrier. He is a member of different professional bodies around the world and involve enthusiastically in scientific research on biopolymers, biomass valorization, bioactives, packaging science, drug delivery, formulation development etc. and to develop active research culture.

Preface

This book discusses smart food packaging based on starches. Starch has the potential to replace the synthetic, non-biodegradable packaging polymers. Starch-based smart packaging is gaining popularity, and this book provides information on smart packaging for food applications. The properties and functions of starch-based packaging can be tuned by adding other polymers, fillers, and functional ingredients. The resulting composite materials provide a barrier to gas and moisture as well as microbes and oxidation, and moreover, they may sense or monitor food conditions in real time. The development of starch-based smart packaging that is both safe and sustainable may be appropriate for enhancing food quality and safety. The 17 chapters that make up the current version are briefly described here.

1. **Introduction to starches in smart packaging**
   Chapter 1 discusses the introduction to starches in smart food packaging. It explores the overview of the starches used in developing smart packaging. This chapter explains recent developments in starch-based packaging.
2. **Starch extraction process and its impact on packaging**
   Chapter 2 discusses the several extraction procedures of starches from different sources. Different principles and procedures to isolate the starches have been discussed in this chapter. This chapter also explains the different extraction methods that impact starch-based packaging applications.
3. **Emerging fabrication process of starch-based smart packaging**
   Chapter 3 concludes the innovative manufacturing techniques involved in the fabrication of the starch-based food packaging. The details of the principles and procedures have been discussed based on the several techniques along with their advantages and disadvantages.
4. **Barrier, thermal, and mechanical properties of starch-based smart packaging**
   Chapter 4 typically deals with the several functional properties of the starch packaging. In addition, this chapter also discusses the different types of bioactive components applied during the fabrication of starch packaging materials.

5. **Antioxidant, antimicrobial, and light barrier properties of starch-based smart packaging**
   Chapter 5 represents the antioxidant, antimicrobial, and light barrier properties of starch packaging materials in an extensive manner. The antioxidant, antimicrobial, and light barrier molecules added to the starch-based packaging are also discussed in this chapter.
6. **Starch nanoparticles-reinforced biopolymer-based active and intelligent packaging**
   Chapter 6 extensively discusses starch nanoparticles along with their fabrication techniques. This chapter also concludes the importance of starch nanoparticles in the development of biopolymer-based active and intelligent packaging. Different physicochemical and characterization attributes of the starch nanoparticles have been discussed in this chapter.
7. **Starch-based smart packaging added with nanomaterials**
   Chapter 7 reveals the recent advancements in starch-based packaging reinforced with nanomaterials. Different types of nanomaterials are used to improve the functional properties of the starch-based films, and the details of the properties have been discussed in this chapter.
8. **Starch-based smart packaging incorporated with plant bioactives**
   Chapter 8 explains the different types of plant bioactives added to the starch packaging. The addition of the plant bioactive or plant extract during the fabrication of the starch packaging improves several functional properties and produces superior antioxidant, antimicrobial, and other relevant properties, and all these aspects are included in this chapter.
9. **Starch-based active packaging added with a cross-linker**
   Chapter 9 presents the discussion of the several cross-linking agents that are added during the fabrication techniques of starch-based packaging materials. Several cross-linking agents have been utilized in the preparation of the starch-based packaging materials, and the characteristics are extensively studied upon the addition of the cross-linker.
10. **Starch-based packaging material added with natural colorant**
    Chapter 10 discusses the natural coloring agents added during the fabrication of the starch-based packaging materials. The chapter also reveals the different properties and characteristics upon the addition of natural coloring agents.

11. **Application of starch film and coatings in improving the shelf life of fresh and processed food**
    Chapter 11 explains the diverse applications of starch-based film and coating for fresh and processed foods. The several fabrication techniques and the different characteristics have been studied in this chapter. The impact of the starch-based film and coating is also studied in this chapter in order to improve the shelf life of the packed food items.
12. **Application of starch-based packaging in monitoring the quality of fresh and processed food**
    Chapter 12 explains the starch-based packaging used in the real-time monitoring and quality maintenance of fresh and processed foods. The chapter also concludes the principles involved in the real-time monitoring application for starch-based packaging.
13. **Industrialization application of starch-based smart packaging**
    Chapter 13 exclusively discusses the recent developments of starch-based packaging and its utilization in industrial fields. Several aspects and considerations have been studied in this chapter.
14. **Biodegradability, composability, environmental fate, and relevant standards in starch-based packaging**
    Chapter 14 represents the details of biodegradability, composability, environmental fate, and relevant standards in starch-based packaging. This chapter also discusses the several tests and standards relevant to starch-based packaging.
15. **Regulatory frameworks and consumer acceptance of starch-based packaging**
    Chapter 15 describes the regulatory concerns of the starch-based packaging. Apart from this, the chapter discusses the consumer acceptance and all the regulatory approvals considered for starch-based packaging.
16. **Life Cycle Assessment and sustainability of starch-based packaging**
    Chapter 16 discusses the life cycle assessment (LCA) and sustainability of the starches in smart packaging. The sustainability aspects and LCA analysis studies of this packaging are important for practical utility in industry.

17. **Safety aspects, challenges, and future outlook in starch-based packaging**
Chapter 17 demonstrates the safety aspects, challenges, and future outlook in starch-based packaging. The chapter also explains in detail the several challenges and safety concerns that need to be focused on to bring the starch-based packaging materials to the market for customers.

Swarup Roy
Pratap Kalita

Acknowledgments

Recently, smart food packaging has received a lot of interest in the food sector. Smart packaging plays a very crucial role in maintaining and monitoring food. The book *Starches in Smart Packaging* has discussed the recent trends and advancements of starch-based sustainable and biodegradable packaging, covering the different aspects from fabrication and characterization to application in food. Moreover, the challenges in industrialization, safety and risk factors, and regulatory and legal aspects are also addressed in this book. The editor would like to acknowledge all the contributors to the book from various countries. The contributions from all the authors are a worthy addition to this book, as they are among the leading researchers in this field.

Dr. Swarup Roy would like to express his gratitude to the Department of Food and Nutrition, School of Agriculture, Lovely Professional University, India. He would like to thank Professor Jong-Whan Rhim for his guidance, inspiration, and support in gaining insights in the field of packaging. Dr. Roy would also like to thank Dr. Deblina Biswas for her continuous support in maintaining a positive and spiritual environment. He would also like to thank all family members and everyone who helped him in this journey.

Dr. Pratap Kalita expresses his gratitude to God for showering immense blessings on him, enabling him to complete this work successfully with determination and confidence. Dr. Kalita would like to thank Professor Abdul Baquee Ahmed, Professor Saikat Sen, and Professor Raja Chakraborty for their invaluable guidance, relentless passion, immense encouragement, and unwavering support. Dr. Kalita expresses his exceptionally sincere, and respectful thanks to his parents, Sri Ganesh Kalita and Mrs. Suchila Kalita. Dr. Kalita also would like to thank Dimpi Kalita for her cooperation, support, constant encouragement, and enlightenment during this journey.

Finally, the editors would like to thank all the staff at Elsevier Publishing Services for their assistance and continuous coordination in publishing the book. Additionally, the editors would like to acknowledge Dr. Sneh Punia Bangar, Dr. Palak Gupta, and Ms. Ritika Dagar for their continuous support and coordination in completing the book.

The blessings of Lord Krishna, the Supreme Soul, are always solicited.

Swarup Roy
Pratap Kalita

CHAPTER ONE

Introduction to starches in smart packaging

Carolina Krebs de Souza[1], Ana Caroline Ferreira Carvalho[2], Karine Marafon[2], Raquel Bonati Moraes Ibsch[3], Tania Costa[4], Matheus Fiebig[2], Elane Schwinden Prudencio[2,5]

[1]Postgraduate Program in Chemical Engineering, University of Blumenau, Blumenau, Brazil
[2]Federal University of Santa Catarina, Postgraduate Program in Food Engineering, Technology Center, Florianópolis, Santa Catarina, Brazil
[3]Department of Chemical Engineering, University Center of Educational Foundation of Brusque, Brusque, Brazil
[4]SENAI Santa Catarina University Center, Blumenau, Brazil
[5]Department of Food Science and Technology, Federal University of Santa Catarina, Florianópolis, Brazil

1.1 Introduction

Considering the continuous increase in the world population, which results in an inescapable rise in food demand and, consequently, in the production of packaging, a more sustainable view of the context is urgent and unavoidable. Growing concern about food safety and changing consumer attitudes toward healthy eating drive the food industry to keep up with trends and meet customer demands, including food packaging as one of the many factors to consider (Madella et al., 2024).

According to the Food and Agriculture Organization of the United Nations (FAO), approximately 1.3 billion tons of food are wasted worldwide every year. This loss not only generates economic damage but also exacerbates environmental impacts through greenhouse gas emissions and the inefficient use of resources, such as water and soil (Durán-Sandoval et al., 2023). While conventional packaging has been designed primarily to contain and protect food, the emerging need is to develop multifunctional systems capable of reducing waste and ensuring food quality throughout the supply chain (Huang & Wang, 2025).

The demand for more sustainable and functional packaging materials has driven the development of smart packaging, which features active and intelligent functions (Carvalho et al., 2025). Unlike conventional packaging, this packaging not only acts on the food or internal atmosphere, reducing its microbial and oxidative activity, but can also monitor, detect, or interact with

Starches in Smart Packaging
ISBN: 978-0-443-45396-0
https://doi.org/10.1016/B978-0-443-45396-0.00009-8

the internal and external environment, providing important information about the quality, safety, and condition of the food. Thus, these materials are effective not only in reducing losses throughout the production chain but also in meeting consumer demands for safer, higher-value products (Zhou et al., 2025).

Active packaging may contain systems that absorb or release substances, such as humidity controllers, oxygen absorbers, carbon dioxide releasers, or ethylene absorbers, as well as components with antimicrobial and antioxidant action, capable of reducing microbial and chemical deterioration (Carvalho et al., 2025; Carvalho, Meinert, et al., 2024).

Intelligent packaging is designed to monitor, detect, and communicate information about the quality, safety, and storage conditions of food throughout the production chain to the end consumer in real-time (Du et al., 2025). This process uses freshness indicators, gas sensors, time-temperature indicators, integrity indicators, as well as RFID/QR tags and biosensors. These categories highlight the broad functional spectrum of smart packaging, extending beyond passive containment to include active and responsive systems (Heo & Lim, 2024; Zhang et al., 2020).

Therefore, the purpose of smart packaging goes beyond helping to extend the shelf life of food, as it also monitors changes that occur in food during transportation and storage, using sensors that indicate the product's conditions in real-time, thereby ensuring consumer safety and satisfaction. Therefore, the information provided by these systems allows consumers, producers, and distributors to make informed decisions, thus preventing the distribution and consumption of unsafe or spoiled food. Furthermore, smart packaging helps reduce the subjective quality assessments traditionally carried out, standardizing assessments across different segments of the food chain (Dodero et al., 2021).

Fig. 1.1 presents a general diagram of a smart packaging system, highlighting its main features, including sensory response, barrier properties, and biodegradability, which collectively ensure food quality, safety, preservation, and environmental compatibility.

Some fruits do not show color changes or other easily noticeable signs of ripeness when ripe, making it difficult to identify their degree of ripeness (Santos et al., 2022). However, they produce volatile compounds during ripening that can be determined using ripeness indicators, which alert consumers through color changes (Alam et al., 2021). Thus, by indicating the color, consumers can choose their preference according to the degree of ripeness (Zhai et al., 2025). Protein-rich foods, such as fish or meat, lose

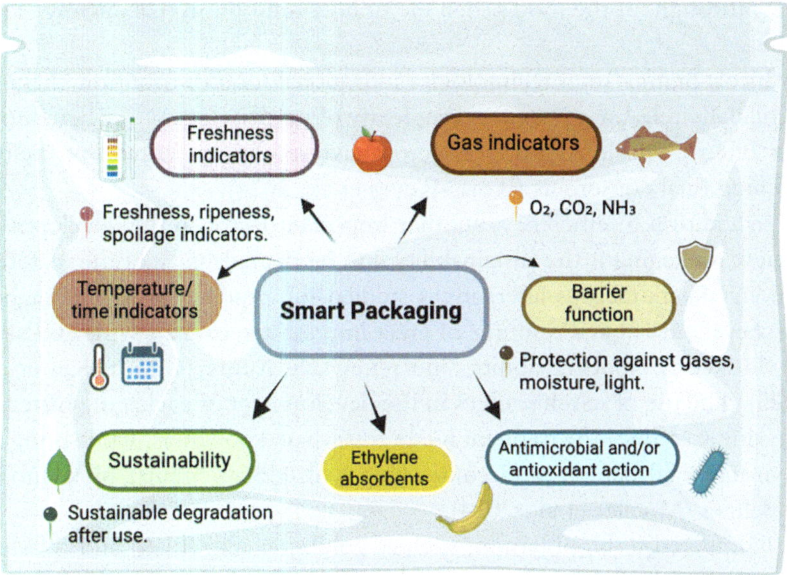

Figure 1.1 General scheme of a smart packaging system.

their freshness during storage because enzymes and bacteria easily break down these products (Wang et al., 2022).

Freshness indicators applied to seafood and meat packaging have enabled retailers and consumers to track spoilage in real-time (Nami et al., 2024). At the same time, biosensor-based RFID tags are being explored to monitor the freshness of beef and poultry during cold chain distribution (Marindra et al., 2023). These innovations illustrate the growing feasibility of integrating smart packaging technologies into commercial supply chains.

This degradation leads to the accumulation of volatile compounds inside the packaging, including ammonia and di- or trimethylamine (organic amines). These compounds can be easily detected by pH and colorimetric indicators, which change color when exposed to volatile compounds, showing consumers the quality of these foods (Bhargava et al., 2020).

Smart packaging, equipped with time and temperature indicators, accurately demonstrates the actual storage conditions inside the packaging. These devices are easy to use because they are integrated into the packaging, allowing consumers to check food quality by observing a color change related to food quality at a specific temperature (Mkhari et al., 2025). Additionally, this system helps maintain ideal food transport conditions, ensuring safety throughout the food's shelf life until consumption (Palanisamy et al., 2025).

Incorporating such technologies into global food distribution chains is particularly relevant given the complexity of modern logistics, where products often travel long distances across diverse climatic conditions before reaching final consumers (Rojas-Reyes et al., 2024).

In addition to efficiency, another important factor in the development of new packaging is the sustainability and biodegradability of the material. Among the numerous alternatives studied for producing this packaging, starches stand out as a resource of great interest due to their wide availability, low cost, biodegradability, and renewable nature (Gurunathan et al., 2025). The use of starch sources in the development of packaging materials also aims to reduce the frequent use of fossil-based polymers, while bringing environmental benefits and consistency with global circular bioeconomy guidelines (Mueller et al., 2024).

Compared to biopolymers, such as polylactic acid (PLA) and cellulose derivatives, starch offers the advantage of being abundant and inexpensive (Carvalho, Bose, et al., 2024). However, it often requires physical or chemical modification, or blending with other polymers, to improve its barrier and mechanical properties (García-Guzmán et al., 2022).

The use of blends and starch modification has expanded the application of starch beyond food processing. Starch-biopolymer composites are now being explored in the manufacture of biodegradable packaging films, smart coatings for food, pharmaceuticals, agricultural cover films, and even 3D-printable biomedical structures (Karnwal et al., 2025; Siddiqui et al., 2024). These applications demonstrate the multidisciplinary application of starch and its important role in materials chemistry, environmental engineering, food technology, and science. The combination of starch from various sources with active compounds demonstrates the excellent potential of this resource in the development of smart food packaging (Huang & Wang, 2025).

Some research has shown that starch-based films containing natural extracts, such as anthocyanins or curcumin, have potential as freshness indicators, as they change color in response to pH changes in the product during food deterioration (da Rocha et al., 2023; Hoffmann et al., 2022). The integration of bioactive compounds into these extracts increases the multifunctionality and applicability of starch-based materials.

Starch is a complex carbohydrate of plant origin that represents the most abundant energy reserve in seeds, tubers, sprouts, and roots of various plants, providing essential carbohydrates that organisms can break down through catabolism (Nsanzabera et al., 2023). This carbohydrate is essential for energy metabolism, and its synthesis plays an important role in the anabolic growth and development of these plants. In addition to its biological importance, starch has proven to be a raw material of crucial industrial application and the subject of academic research (Wu et al., 2024).

The use of economical and sustainable starch modification techniques ensures the production of more stable and accessible biopolymers for the development of biodegradable packaging based on modified starch (Amobonye et al., 2025). These materials represent viable and sustainable alternatives to commonly used plastics and can also be used as bases for smart packaging materials.

Despite their advantages, starch-based materials still have limitations compared to traditional synthetic polymers, particularly in terms of low mechanical strength, high sensitivity to moisture, and thermal instability (Gurunathan et al., 2025; Ribba et al., 2017). On the other hand, they offer significant advantages, such as natural biodegradation, processing versatility, and the possibility of chemical or physical modification to improve their properties (Amobonye et al., 2025).

These materials, in addition to being sustainable, demonstrate efficient multifunctional performance in packaging applications, expanding the applicability of starch-based materials in food preservation, which is particularly evident when applied to differentiated packaging, such as smart packaging (Pawase et al., 2025).

However, challenges in regulatory approval, large-scale production, and consumer acceptance still need to be overcome. Only through collaboration between government, industry, and academia will starch-based smart packaging become a reality, replacing traditional food packaging (Arruda et al., 2025).

1.2 Sources, structure and properties of starches

Starch is one of the primary natural carbon storage polymers in plants and algae, composed of several repeating units of glucose molecules. Its synthesis plays a central role in plant energy metabolism, especially in the anabolic processes of growth and development. In plants, starch is stored in roots, tubers, seeds, and shoots, acting as an energy source for organisms capable of degrading it. Furthermore, this natural polymer is hugely versatile, justifying its application in various industrial products and scientific research (Cifuente et al., 2024).

Its primary botanical sources are cereals (such as corn, rice, wheat, barley, and sorghum), roots and tubers (cassava, potatoes, and sweet potatoes), and legumes (peas and lentils). Its origin directly influences the structure and physicochemical properties of starch, which determine its performance in smart packaging materials (Puri et al., 2025).

Corn starch is the most widely used industrially, featuring a high amylose content (20%–30%) and a gelatinization temperature of 62–72°C, which favors the development of films with good mechanical performance. Cassava starch consists of tiny granules (5–35 μm), high purity, and low lipid and protein content. It is widely used in films that require transparency but have low mechanical strength. Unlike cassava starch, potato starch contains larger granules (15–100 μm) and a high concentration of phosphorus in the amylopectin fraction, which gives it its affinity for water and swelling capacity (Gonçalves et al., 2024). These properties, while increasing the material's flexibility, reduce its water vapor barrier performance (Liu et al., 2009). Rice starch, with even smaller granules (two to eight μm) and a high amylopectin content, is used to produce smooth, homogeneous, and dense films with interesting optical clarity, despite its low mechanical strength (Kowsik & Mazumder, 2018). Wheat starch, on the other hand, because it contains lipids that form amylose-lipid complexes in its composition, exhibits moderate retrogradation and good thermal stability, which favors long-term performance in multilayer or mixed systems (Lee & Choi, 2021).

Defining an ideal starch source for packaging formulations must therefore consider aspects such as the amylose/amylopectin ratio, the morphology of its granules, and the presence/quantity of other components. These intrinsic characteristics will determine processing behavior, as well as compatibility with other biopolymers and sensitivity to environmental factors (such as humidity, pH, and temperature), which are crucial for the development of active and intelligent packaging materials (Pawase et al., 2025).

The integration of starches from underutilized sources, such as banana, yam, or sago, has been considered in scientific studies (Rashwan et al., 2024) as a means to diversify material functionality and increase appreciation for local biodiversity.

Table 1.1 presents some starch sources, their main physicochemical properties, and potential applications as packaging films.

Table 1.1 Comparison of physicochemical properties of starches from different botanical sources.

Botanical source	Granule size (μm)	Amylose content (%)	Gelatinization Temp (°C)	Swelling power	Film transparency	Notable features	References
Corn	5–25	20–30	62–72	Moderate	Medium	Balanced mechanical and barrier properties	Ghasemlou et al. (2013)
Cassava	5–35	17–20	55–65	High	High	High clarity, low tensile strength	Mohammed et al. (2020)
Potato	15–100	20–24	58–68	Very high	Medium	High swelling, poor water barrier	Valencia-Sullca et al. (2018)
Rice	2–8	15–20	68–78	Low	Very high	Smooth films, low strength	Preethi et al. (2021)
Wheat	3–40	25–28	60–70	Moderate	Medium	Good stability due to amylose–lipid complexes	Jia et al. (2007)

The bonding of the polysaccharides amylose and amylopectin forms the starch chemical structure Fig. 1.2. The starch granule is organized into a three-dimensional semicrystalline structure, with alternating amorphous and crystalline lamellae. Crystallinity degree is due to the ordered double-helix arrangement of the amylopectin chains, stabilized by hydrogen bonds (Ai & Jane, 2018).

The three main crystalline polymorphs are type A (typical of cereal starches), type B (root and tuber starches), and type C (a mixture of A and B, typical of legumes). Type A granules are compacted and have a lower hydration capacity, while type B granules have larger cells that retain more water (Zhang et al., 2020).

These structural differences are determined by the proportion of amylase and amylopectin in the starch molecule, which affects the tendency to gelatinize, retrograde, and the mechanical profile of starch-based materials (Jane et al., 2025). These effects also influence the size, shape, and surface morphology of starch granules of different origins Fig. 1.2. Amylose is a polymer formed predominantly by linear chains of D-glucose units linked by α-1,4 bonds, with molecular weights ranging from 10^5 to 10^6 Da, i.e., from 300 to 3000 glucose residues in length. At the same time, amylopectin is a highly branched molecule with α-(1 → 6) linkages occurring every 24–30 glucose residues and a molecular weight in the range of 107–109 Da (Bushra et al., 2023). Branching generally occurs at a low level (~1 branch per 1000 residues) with an α1,4 bond and represents approximately 30% of the starch structure, depending on the plant, its age, and growing conditions (Zhiguang et al., 2025). When extracted from plants or from a solution, amylose forms hydrogen bonds between molecules, resulting in highly rigid gels. However, the gel can crystallize or retrograde, depending on the degree of polymerization and concentration after heating. Due to its linearity, amylose molecules can form strong intermolecular hydrogen bonds, resulting in dense, less permeable films that contribute to greater tensile strength and improved gas barrier properties (Tian et al., 2023).

Amylopectin accounts for 75%–90% of the dry weight of natural starch granules. It is a large, highly branched molecule with an estimated molecular weight of 10^7 to 10^9 Da. This structure is essential in defining the functional and physicochemical properties of starch, featuring chains with approximately 20 glucose residues in α1,3 linkages linked by α1,4 bonds to other branches. A single amylopectin molecule can be 200–400 nm long and approximately 15 nm wide (Sangwongchai et al., 2023).

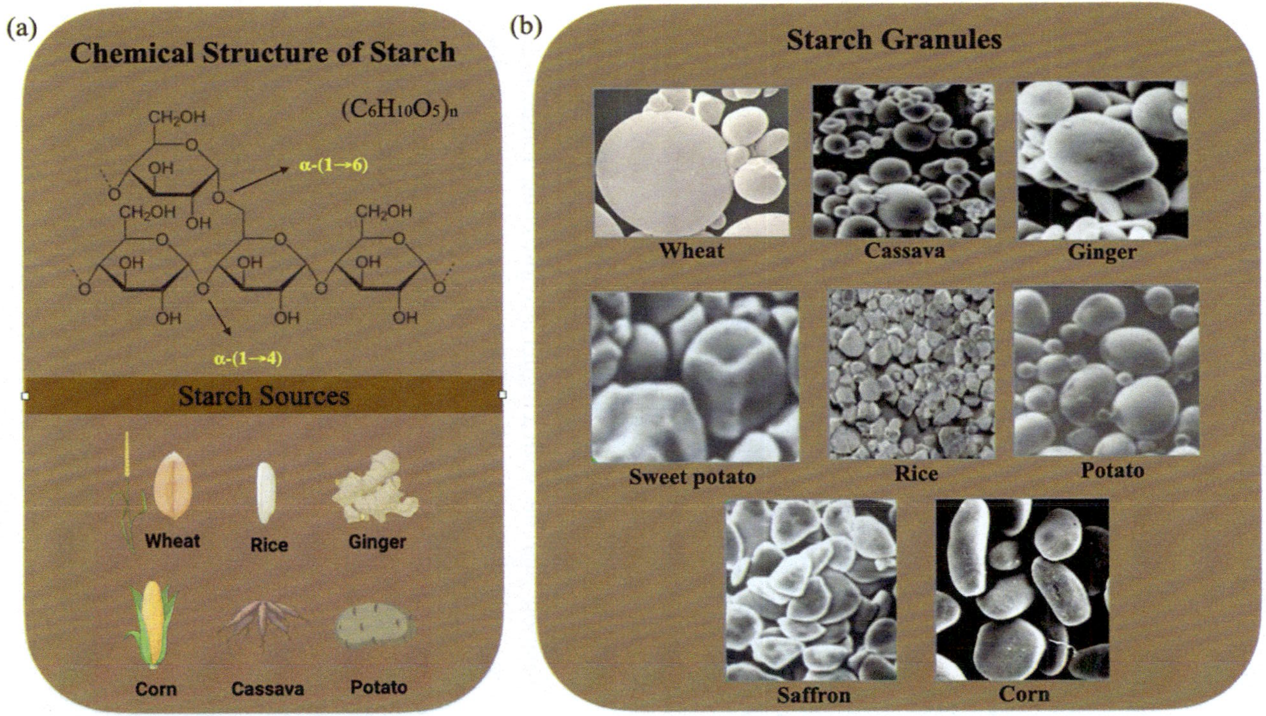

Figure 1.2 Representation of starch structure and granule morphology. (A) simplified chemical structures of amylose and amylopectin, illustrating the linear α-(1 → 4)-glucan chains and the branching α-(1 → 6)-linkages, and some starch sources; (B) micrographs of starch granules from wheat, cassava, ginger, sweet potato, rice, potato, saffron, and corn (Ali et al., 2016; Cereda, 2023).

According to Mauro et al. (2023) amylopectin is more stable than amylose when subjected to extraction, remaining fluid and resulting in high viscosity and elasticity in gels, making it suitable for use in thickeners and pastes (Mauro et al., 2023).

Therefore, starches rich in amylopectin, such as corn and potato, form films with greater flexibility but higher permeability to moisture and gases, making them ideal for use as coatings or multilayer systems where high elasticity is essential (Zhu, 2025). Therefore, matching the amylose/amylopectin ratios through the source or enzymatic modification process is a strategy for modulating starch functionality in packaging applications.

At the nanoscale, starch granules also contain surface pores, channels, and growth rings that affect water diffusion and plasticizer distribution. This structural variation, specific to the starch source, explains why films with distinct morphologies and performances are obtained even under identical processing conditions (Karnwal et al., 2025).

Starch has ideal physicochemical properties for the development and performance of packaging materials, such as gelatinization, retrogradation, crystallinity, and pasting behavior.

Gelatinization is an irreversible thermophysical process that occurs when starch is heated in excess water, resulting in the rupture of the semicrystalline granule structures, leaching of amylose, and a consequent increase in viscosity (Schirmer et al., 2015). This molecular rearrangement is what enables film formation. Starches with a higher amylose content tend to have a higher enthalpy and gelatinization temperature due to the compact double-helix arrangement of the amylose chains. Conversely, starch molecules with a higher amylopectin content and more extensive branching tend to gelatinize at lower temperatures, resulting in films that, despite being more transparent, have less tensile strength (Yang et al., 2024).

After the gelatinization step during heating, retrogradation occurs during cooling of the starch, due to the reorganization of the linear amylose and amylopectin chains through hydrogen bonds (Tao et al., 2024). Controlling the retrogradation step results in improved mechanical strength and barrier properties, whereas excessive retrogradation leads to more brittle and opaque materials, which are undesirable for flexible packaging. The degree of retrogradation can be determined by the starch source, storage conditions, and lipid content, which form amylose-lipid complexes that affect the results of recrystallization (Siva Janika et al., 2025).

Plasticizers are used in the development of flexible materials, especially those made from hydrocolloid materials such as starch and gelatin. Their primary function is to reduce material brittleness by decreasing intermolecular forces and increasing the flexibility of polymer chains. Among the most commonly used plasticizers are glycerol, sorbitol, and citric acid, which, while effective, perform differently depending on the molecular structure of the matrix. For the plasticization of starches rich in amylose, higher concentrations of plasticizers are required to achieve elasticity. In contrast, those with a higher amylopectin fraction tend to absorb more water and separate phases (Sirbu et al., 2024).

The crystalline-to-amorphous ratio determines a material's water vapor permeability (WVP) and oxygen transmission rate (OTR). Amylose-rich starch films exhibit lower WVP and OTR due to their dense molecular packing, whereas those with a higher amylopectin content are more flexible and transparent. These correlations enable the development of biodegradable materials with improved mechanical and optimized barrier properties (Teixeira et al., 2025).

1.3 Smart packaging: Concepts and requirements

Packaging has always played an essential role in the preservation, transportation, and presentation of products, especially in the food industry. Traditionally, its function was limited to protecting and containing products. However, as demand for food safety, traceability, sustainability, and convenience grows, a new generation of packaging systems, called smart packaging, has emerged. These innovative technologies go beyond their traditional functions and begin interacting with the environment, the packaging, or the consumer, providing information and even extending the product's shelf life (Drago et al., 2020). Smart packaging is a packaging system capable of detecting, recording, tracking, communicating, or responding to internal or external stimuli, to improve product safety, quality, traceability, or convenience. They can be divided into two main categories: active packaging and smart or sensory packaging (Soltani Firouz et al., 2021; Vasile & Baican, 2021).

Active packaging interacts directly with the product or its surrounding environment to extend shelf life or maintain food quality. These interactions occur through the release or absorption of substances, including elements such as oxygen scavengers, which prevent oxidation and microbial growth; moisture absorbers, which prevent softening or mold growth; antimicrobial systems, which inhibit the growth of microorganisms directly on the surface of food; and ethylene controllers, used primarily in fruits to delay ripening. These elements are incorporated into the packaging or packaged in sachets to actively modify packaging conditions without negatively affecting the product (Kadirvel et al., 2025).

Smart or sensory packaging, on the other hand, does not directly affect the product itself, but rather monitors the conditions of the food or its environment during storage and transportation, providing information on its quality, safety, or freshness. Examples include time-temperature indicators, which indicate whether the cold chain has been maintained; gas sensors, which detect compounds such as ammonia or carbon dioxide released during food spoilage; Freshness indicators, which change color as food deteriorates; and traceability devices, such as QR codes or RFID tags, which enable logistical monitoring and access to detailed product information (Soltani Firouz et al., 2021). These solutions offer benefits to both the industry and the end consumer, contributing to greater transparency, safety, and trust in the products sold (Azeredo & Correa, 2021).

1.4 Smart packaging performance requirements

For smart packaging to perform its functions effectively, it must meet a set of technical and functional requirements. Among the most important are barrier properties, mechanical strength, biodegradability, sustainability, and responsiveness.

Regarding barrier properties, packaging must be effective against external agents, including oxygen, water vapor, ultraviolet light, and microorganisms. The ability to protect food from these factors is essential to maintain product quality and safety, especially perishable foods. Materials with good barrier properties prevent oxidation, aroma loss, drying out, and excessive moisture in food. These materials also help reduce food waste (González-López et al., 2023; Versino et al., 2023).

Mechanical strength is the packaging's resistance to impacts, compression, vibrations, and other adverse conditions during transportation and handling, while maintaining its structural and functional integrity. This is because mechanical failures can compromise both the protection and the operation of the packaging's smart or active components. Given growing concerns about environmental sustainability, packaging needs to be increasingly biodegradable, recyclable, or compostable. Using renewable materials that decompose naturally in the environment helps reduce the environmental impact of discarded packaging (González-López et al., 2023; Mishra et al., 2024). In this context, replacing conventional plastics with biopolymers represents a promising alternative for both traditional and smart packaging (Versino et al., 2023).

In smart packaging, the ability to accurately and reliably detect and respond to stimuli is essential. This involves the sensors' sensitivity, response speed, and the clarity of the information transmitted to the user. For example, a freshness sensor must display a noticeable color change at the appropriate time, without ambiguity. Furthermore, active systems must be controlled and safe to ensure that released compounds do not compromise consumer health or negatively affect the food (Chaudhary et al., 2022).

1.5 The importance of biopolymers in smart packaging

Biopolymers are polymeric materials obtained from renewable natural sources, such as starch, cellulose, PLA, chitin, and others. Their application in smart packaging has become increasingly relevant, especially given their biodegradability, compostability, and compatibility with other active compounds and sensors. Thus, the advantages of biopolymers include environmental sustainability, compatibility with functional compounds, structural modification capability, and consumer acceptance (Manfroni et al., 2025).

Regarding environmental sustainability, biodegradable biopolymers help reduce the accumulation of plastic waste, promoting a more sustainable life cycle for packaging. Compatibility with functional compounds, such as chitosan, is related to their natural antimicrobial or antioxidant properties, and they are also excellent matrices for incorporating essential oils, natural extracts, or chemical indicators (Marotta et al., 2025; Sinha, 2024).

The structural modification capacity of biopolymers stems from their ability to be tailored to optimize properties such as gas barrier performance, transparency, flexibility, and mechanical strength, enabling their use in various types of smart packaging. Consumer acceptance is linked to packaging materials derived from natural sources, which are generally well-regarded by the public, adding perceived value to the product and favoring brands that invest in environmentally responsible solutions (Babaremu et al., 2023; Dăescu et al., 2024). Thus, smart packaging represents a significant advancement in how products are protected, monitored, and presented to consumers. Whether through active intervention in the packaging's internal environment or through mechanisms that inform about the product's condition, these solutions deliver tangible benefits in terms of safety, quality, and sustainability (Dăescu et al., 2024; Perera et al., 2023).

The technical requirements for the effective performance of these packages range from barrier properties and mechanical strength to sensitivity and biodegradability. In this context, biopolymers play a central role, enabling the development of smart packaging that is both functional and environmentally responsible. The future of packaging undoubtedly depends on the combination of technological innovation and environmental commitment, an equation in which smart packaging based on biopolymers stands out as a promising solution to the challenges of the 21st century (Babaremu et al., 2023). Finally, Fig. 1.3 presents an infographic related to the content described in this section of the present chapter.

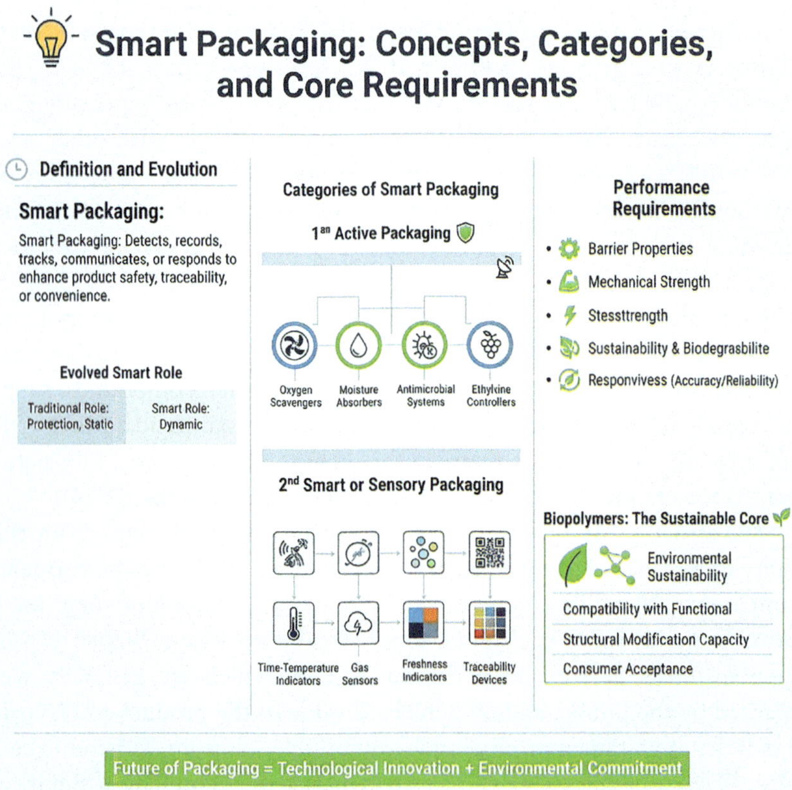

Figure 1.3 Infographic about concepts and requirements of smart packaging.

1.6 The role of starch in smart packaging systems

The evolution of packaging has been driven by the ongoing search for solutions that transcend mere protection, incorporating advanced features that meet the growing demands for food safety, traceability, and sustainability. In this scenario, the concept of smart packaging has emerged as a transformative paradigm, integrating technologies that enable product condition monitoring, interaction with the packaging's internal environment, and even extending food shelf life (Mishra et al., 2024). The development of these innovative packaging is based on the exploration of a variety of materials, with biopolymers playing a prominent role due to their renewable origin, biodegradability, and environmental compatibility. Among biopolymers, starch stands out as a material with great potential: it is widely available, low-cost, and versatile, and its characteristics make it a promising candidate for various applications in smart packaging systems (Fatima et al., 2024).

Starch is an abundant natural polysaccharide derived from various plant sources, including corn, potatoes, rice, wheat, and cassava. Its structure is composed of two main macromolecules: amylose, a linear glucose chain, and amylopectin, a branched chain. The ratio of amylose to amylopectin, along with the morphology of starch granules, varies by botanical source and directly influences the starch's physical and chemical properties. This variability gives starch a range of functionalities that can be exploited in packaging development, whether as a main matrix, an additive, or a modifying agent, paving the way for the creation of systems with characteristics tailored to specific needs (Fatima et al., 2024).

Starch's versatility in smart packaging systems is multifaceted, enabling it to be incorporated in various forms to confer desirable properties. When used as a polymeric matrix, starch forms the structural basis of the packaging material. Starch-based films and coatings can be produced using processes such as solution casting, extrusion, or hot pressing. Because these materials are biodegradable and compostable, they offer a sustainable alternative to conventional petroleum-derived plastics (Fatima et al., 2024). Starch's ability to form transparent, flexible films is particularly attractive for applications where product visibility is important (Amaraweera et al., 2021). However, pure starch has some intrinsic limitations, such as high hydrophilicity, which results in low water resistance and gas permeability, and moderate mechanical properties, which may be insufficient for some demanding applications.

Despite these drawbacks, its abundance and low cost make it a valuable starting point for modifications and combinations (Fatima et al., 2024).

In other applications, starch is incorporated as an additive into formulations of other polymers, whether biopolymers or synthetics. As an additive, starch can act as a filler, reducing the overall material cost and, in some cases, improving the rigidity or biodegradability of the final compound (Fatima et al., 2024). Starch can also function as a nucleating agent, influencing the crystallization of other polymers and, consequently, their mechanical and barrier properties. Furthermore, starch can be used to increase the moisture-absorption capacity of packaging, acting as a natural desiccant in active humidity-control systems (Amaraweera et al., 2021).

Starch functionality can be significantly enhanced through chemical or physical modifications. Gelatinization, for example, improves film-forming ability. Plasticizing starch — typically with glycerol, sorbitol, or other polyalcohols — transforms it into thermoplastic starch (TPS), which can be processed similarly to synthetic plastics (Fatima et al., 2024). Chemical modifications, such as acetylation, etherification, or esterification, can reduce starch's hydrophilicity, improving its water resistance and gas barrier properties. Furthermore, oxidation or cross-linking of starch can increase its mechanical strength and dimensional stability. These modifications are crucial for adapting starch to the specific requirements of smart packaging, allowing it to act as a carrier for active compounds (antimicrobials, antioxidants) or as a component in biosensors (Bajer, 2024; Devi et al., 2025).

To overcome the limitations of pure starch and maximize its potential in smart packaging, blending it with other biopolymers has proven highly effective. Combining starch with polymers that possess complementary characteristics allows for the creation of composite materials with synergistic properties, resulting in superior mechanical, barrier, and functional performance. PLA is a thermoplastic biopolymer derived from renewable sources, known for its good transparency, mechanical strength, and gas barrier properties. However, PLA is relatively expensive and brittle. Starch and PLA blends combine the biodegradability and low cost of starch with the strength and processability of PLA. The starch acts as a low-cost filler, while the PLA improves the blend's moisture resistance and mechanical properties. Challenges such as poor compatibility between the two polymers (due to their different polarities) can be overcome by using plasticizers, compatibilizing agents, or chemical surface modifications of the starch. These blends are particularly promising for packaging films and rigid packaging (Karabanova et al., 2018; Rezghi Rami et al., 2024).

Another alternative would be the use of blends of starch and polyvinyl alcohol (PVA). PVA is a biodegradable synthetic polymer (although not a biopolymer) with excellent film-forming properties, transparency, and a high gas barrier, but it is sensitive to moisture. The combination of starch with PVA results in blends with good processability and improved mechanical properties compared to pure starch. PVA can help improve the starch's water resistance and barrier properties. These blends are often used in water-soluble films or for packaging where an oxygen barrier is crucial (Boonyuen et al., 2017).

It is also possible to use blends of starch with proteins. In this case, proteins such as zein, casein, whey protein, or a combination with soy protein can be used to form packaging materials with additional functional properties. Proteins are known for their excellent gas barrier properties, especially oxygen, in low-humidity conditions, and for their biodegradability. Starch-protein blends can offer greater mechanical strength and a better gas barrier than pure starch films. Furthermore, proteins can confer intrinsic antimicrobial or antioxidant properties, making these blends suitable for active packaging. The compatibility between starch and proteins, which have functional groups that allow interactions, facilitates the formation of homogeneous materials (Deng et al., 2017). Therefore, the choice of blend and compatibilizers depends on the end application and the specific performance requirements of the smart packaging. However, despite the vast potential and improvement strategies, the use of starch in smart packaging still faces significant challenges that must be addressed for widespread industrial adoption. Table 1.2 presents examples of starch-based systems used in smart packaging. However, starch has limitations for use in packaging due to its hydrophilicity and thermal instability.

The main limitation of starch is its high hydrophilic nature. Starch-based materials are very sensitive to moisture and tend to absorb water from the environment, severely compromising their gas barrier properties, especially to oxygen, and their mechanical properties. Water absorption can lead to swelling, softening, loss of structural integrity, and microbial proliferation in the packaging. Although chemical modifications and blends can mitigate this problem, achieving water resistance comparable to that of synthetic plastics remains a challenge. It is crucial to develop more effective approaches to reduce starch's affinity for water while maintaining its biodegradability and other desirable properties (Ye et al., 2021).

Pure starch films, especially starch-based thermoplastics, tend to be relatively brittle and exhibit lower tensile strength and elongation at break than

Table 1.2 Examples of starch-based systems applied to smart packaging.

Smart packaging type	Function/components of starch	Performance/benefits	Limitations	References
Active packaging (starch/PLA blend)	Starch as a low-cost filler and matrix; PLA improves strength.	Sustainable and lower-cost alternative to pure PLA. Improves moisture resistance and mechanical properties (compared to pure starch).	Low compatibility between starch and PLA requires the use of compatibilizers or plasticizers.	Karabanova et al. (2018)
Active packaging (starch/PVA blend)	Starch combined with PVA (polyvinyl alcohol) to form films.	Good processability and excellent gas (oxygen) barrier due to PVA. It can be used in water-soluble films.	PVA is moisture-sensitive, which affects its barrier performance in high humidity.	Boonyuen et al. (2017)
Active packaging (starch/protein blend)	Starch combined with proteins (e.g., zein, casein) or chitosan.	Protein/chitosan provides antimicrobial and antioxidant properties, as well as an improved oxygen barrier in low-humidity conditions.	The effectiveness of the oxygen barrier is highly dependent on humidity. It may require treatment for greater mechanical strength.	Deng et al. (2024)
TPS	Modified/plasticized starch (with glycerol, sorbitol) acting as a matrix.	It is 100% biodegradable and compostable. Processable using traditional methods (extrusion, molding).	High hydrophilicity (sensitivity to water), leading to low gas barrier and mechanical impairment.	Ye et al. (2021)
Reinforced packaging (starch/nanocrystals)	Starch (matrix) reinforced with nanostructures (e.g., cellulose nanocrystals).	Significant improvement in tensile strength and material stiffness.	Increased cost and complexity of formulation/processing.	Ban et al. (2018)

synthetic polymers. This brittleness can lead to cracking and breakage during handling, transportation, and storage, compromising the protection of the packaged product. Blending with more robust polymers and incorporating nanostructures, such as cellulose nanocrystals or nanofibers, are promising strategies for improving starch's mechanical properties. However, there is still a long way to go to achieve the robustness required for all packaging applications. Starch's glass phase transition is also a critical factor, where changes in temperature and humidity can drastically affect its mechanical performance (Ban et al., 2018). Thus, starch can be challenging to process on an industrial scale due to its high viscosity and temperature sensitivity during extrusion or molding. Consistency in starch quality across different batches and sources can also affect packaging reproducibility. The development of additives and processing technologies to enable the efficient, consistent manufacture of starch-based packaging is essential.

1.7 Functionalization and modification of starches for smart packaging

The pursuit of smart and sustainable packaging has driven the development of modified starches capable of acting as biodegradable and functional matrices. The modification and functionalization of starch enable the incorporation of properties such as mechanical strength, antioxidant, antimicrobial, and sensory activities, which are essential for active and smart packaging (Dutta & Sit, 2024; Han et al., 2020).

Starches can be modified by chemical reagents, physical treatments such as thermal processing and radiation, or enzymatic processes, thereby altering their physicochemical and technological behavior, including gelatinization, retrogradation, transparency, and film formation. Modifications are decisive for overcoming the limitations of native starch—particularly its low mechanical strength and high water sensitivity—thereby making it suitable for biodegradable and smart packaging applications (Dutta & Sit, 2024; Han et al., 2020).

Among these approaches, chemical modification remains the most effective approach for improving starch hydrophobicity and expanding its applicability in biodegradable packaging materials. According to Wang et al. (2020), esterification, etherification, oxidation, phosphorylation, and silanization are among the most commonly employed techniques for introducing hydrophobic functional groups onto starch molecules. These reactions reduce water sensitivity, enhance thermal and mechanical stability, and improve the compatibility of starch with hydrophobic polymers, making it more suitable for applications in smart and sustainable packaging (Wang et al., 2020).

On the other hand, physical treatments, including ultrasound, autoclaving, hydrothermal treatment, microwave processing, high-pressure processing, cold plasma, and pulsed electric fields, can modify the granular structure and crystallinity of starch without altering its chemical composition (Table 1.3). These physical processes act on the multi-scale organization of starch, disrupting crystalline regions, altering lamellar ordering, and modifying the amorphous matrix. As a result, they influence key physicochemical properties such as gelatinization, retrogradation, swelling power, and water absorption, which directly affect film formation and stability. Han et al. (2020) demonstrated that emerging physical technologies, such as high hydrostatic pressure and microwave processing, can improve the solubility and mechanical integrity of starch-based materials, while cold plasma increases surface reactivity, enhancing interfacial bonding in composite films (Han et al., 2020). Similarly, Sales et al. (2021) reported that starch gelatinization followed by the casting technique enables homogeneous film formation and improved transparency, thickness uniformity, and mechanical stability. The incorporation of plasticizers, such as glycerol, enhances chain mobility during the gelatinization–casting process, yielding flexible, dense films suitable for active packaging applications (Sales et al., 2021).

Enzymatic modifications, in turn, employ biocatalysts such as pullulanase and isoamylase to cleave glycosidic bonds, thereby selectively enhancing functional and rheological properties. According to Punia Bangar et al. (2022), enzymes such as α-amylase, β-amylase, pullulanase, and isoamylase catalyze the cleavage or rearrangement of glycosidic bonds, resulting in controlled molecular depolymerization and debranching of amylopectin (Punia Bangar et al., 2022). These enzymatic processes enhance freeze–thaw stability, reduce retrogradation, and improve film-forming ability. Ghizdareanu et al. (2023) further demonstrated that controlled α-amylase hydrolysis decreases the molecular weight and gelatinization temperature of starch, thereby improving chain mobility, transparency, and tensile strength of the resulting films (Ghizdareanu et al., 2023). Fatima et al. (2024) reported that enzymatic debranching by pullulanase and isoamylase promotes the formation of type III resistant starch with higher crystallinity, enhancing mechanical performance and stability of starch-

Table 1.3 Starch modification methods and their effects on the final properties of biodegradable films and packaging.

Type of modification	Method	Main effects on the properties of films and packaging	References
Chemical	Oxidation (NaOCl), acetylation (acetic anhydride), phosphorylation (sodium tripolyphosphate/sodium trimetaphosphate), etherification, silanization	Introduces carbonyl, carboxyl, acetyl, or phosphate groups, modifying the molecular structure of starch; increases hydrophobicity, thermal stability, and film flexibility; reduces retrogradation and water sensitivity; enhances compatibility with hydrophobic polymers and film-forming ability	Cheng et al. (2022), Egharevba (2020), Wang et al. (2020)
Physical	Ultrasound, autoclaving, hydrothermal treatment, microwave processing, high-pressure processing, cold plasma, pulsed electric fields, gelatinization followed by casting	Alters granular structure and crystallinity without changing chemical composition; improves solubility, mechanical integrity, transparency, and film uniformity; enhances adhesion in composites; retrogradation and viscosity can be controlled by processing intensity	Han et al. (2020), Egharevba (2020), Sales et al. (2021)
Enzymatic	Hydrolysis by α-amylase, debranching by pullulanase or isoamylase, enzymatic oxidation by laccase/TEMPO systems	Promotes controlled depolymerization, increasing amylose content and crystallinity; improves film flexibility, transparency, and tensile strength; introduces carbonyl and carboxyl groups enhancing hydrophilicity and compatibility with biopolymers	Boccia et al. (2020), Fatima et al. (2024), Ghizdareanu et al. (2023), Punia Bangar et al. (2022)

(Continued)

Table 1.3 Starch modification methods and their effects on the final properties of biodegradable films and packaging.—cont'd

Type of modification	Method	Main effects on the properties of films and packaging	References
Nanoscale modification (nanostructured starches)	Production of starch nanocrystals (SNCs) via acid hydrolysis; starch nanoparticles (SNPs) via ultrasonication, nanoprecipitation, extrusion, or enzymatic–acid hydrolysis; starch-based nanocomposites with nanocellulose, montmorillonite, ZnO, AgNPs	Enhances tensile strength, oxygen and water vapor barrier, thermal stability, and introduces antimicrobial and UV-shielding properties; increases compatibility and reinforcement in biodegradable films	Campelo et al. (2020), Kumari et al. (2020), Mahuwala et al. (2020), Marta et al. (2023)
Functionalization with active molecules	Incorporation of anthocyanins, gallic acid, benzalkonium chloride (BKC), Fe–N/C nanozymes, or other bioactive compounds	Introduces active and intelligent functions, such as pH and gas sensitivity, antioxidant and antimicrobial activity, UV-blocking capacity, and colorimetric freshness indicators; enables diagnostic and preservative functions in biodegradable films	Almeida et al. (2023), Cheng et al. (2022), Jing et al. (2021), Srivastava et al. (2023), Teixeira et al. (2021)

based films (Fatima et al., 2024). In addition, Boccia et al. (2020) reported enzymatic oxidation using laccase/TEMPO systems, which introduce carbonyl and carboxyl groups onto starch chains, thereby increasing hydrophilicity, thermal stability, and interfacial compatibility in biopolymer nanocomposites (Boccia et al., 2020). Moreover, enzymatic modification can be combined with physical or chemical treatments to achieve multifunctional starches with optimized mechanical and rheological properties, making it a promising strategy for sustainable smart packaging applications.

Among these approaches, chemical modification has been the most widely used for producing starches with enhanced performance in smart packaging. Cheng et al. (2022) developed dual-modified cassava starch films by combining oxidation, hydroxypropylation, acetylation, and phosphorylation processes. These chemically modified starches exhibited superior transparency, mechanical strength, water vapor barrier capacity, and pH-sensitive colorimetric behavior when combined with red cabbage extracts, demonstrating their potential as biodegradable matrices for smart and active packaging applications (Cheng et al., 2022).

Building on advances in chemical, physical, and enzymatic modification, researchers have further explored the nanoscale structuring of starch to enhance its mechanical reinforcement, stability, and compatibility with active compounds.

In this way, nanomodified starches represent a new generation of biopolymers designed to enhance the mechanical, thermal, and barrier properties of biodegradable starch-based films. According to Kumari et al. (2020), nanoscale structuring achieved through the production of starch nanocrystals (SNCs), starch nanoparticles (SNPs), or starch-based nanocomposites overcomes the main limitations of native starch, such as low tensile strength and high water sensitivity (Kumari et al., 2020).

Starch nanocrystals (SNCs) are typically produced by controlled acid hydrolysis of the amorphous regions within starch granules, yielding highly crystalline particles (30–900 nm) with increased elastic modulus, thermal stability, and reduced water vapor permeability. These particles exhibit enhanced compatibility with biodegradable polymers and have been successfully applied to reinforce starch-based films for active and intelligent food packaging (Adeyeye et al., 2023).

Starch nanoparticles (SNPs), on the other hand, can be obtained by physical methods, such as ultrasonication, nanoprecipitation, or high-pressure homogenization, or by combined enzymatic–acid hydrolysis, which allows for finer control of particle morphology and crystallinity.

These nanoparticles present a high surface area and reactivity, facilitating the incorporation of bioactive compounds and improving compatibility with other biopolymeric matrices. According to Andrade et al. (2020) acid hydrolysis remains the predominant method for obtaining SNPs, while hybrid techniques that combine ultrasonic or extrusion treatments are gaining relevance for enhancing yield and controlling particle morphology. The authors also highlight that the main industrial applications of SNPs include the reinforcement of biodegradable films and food packaging materials, followed by uses in the pharmaceutical and biomedical sectors, confirming their technological versatility and sustainability potential (Andrade et al., 2020).

In addition, Hassan et al. (2022) reported that ultrasonication and nanoprecipitation are among the most efficient methods for producing monodisperse SNPs, typically ranging in size from 30 to 100 nm. These processes improve dispersion and crystallinity while maintaining starch biodegradability. The authors also emphasize that surface modification of SNPs enhances their compatibility with hydrophobic polymers, enabling their use as reinforcing and carrier agents in functional food packaging (Hassan et al., 2022). Similarly, Marta et al. (2023) reported that SNPs can be produced via top-down or bottom-up strategies, leading to tunable morphology and surface chemistry that determine their effectiveness as emulsifiers, barrier enhancers, or carriers in active packaging (Marta et al., 2023). These nanoparticles have been shown to increase mechanical strength, stability, and antioxidant retention in biodegradable films, expanding their applications beyond packaging to coatings and Pickering emulsions.

Starch-based nanocomposites, in turn, are produced by incorporating nanofillers, such as nanocellulose, montmorillonite, or metal oxide nanoparticles (e.g., ZnO, AgNPs), into starch matrices. The resulting hybrid materials display superior tensile strength, oxygen and moisture barrier properties, and even antimicrobial activity. For instance, Mahuwala et al. (2020) reported that starch/agar nanocomposite films containing ZnO and Ag nanoparticles showed enhanced mechanical integrity, reduced water solubility and vapor permeability, and strong antibacterial activity against *Pseudomonas aeruginosa* and *Staphylococcus aureus* (Mahuwala et al., 2020). Complementarily, Campelo et al. (2020) demonstrated that incorporating metallic or biopolymeric nanofillers into starch matrices improves thermal resistance, elasticity, and functional responsiveness, enabling controlled release of antimicrobial or antioxidant agents in food packaging systems.

This study also highlights that hybrid nanocomposites, particularly those reinforced with nanocellulose or metal oxides, can act as UV barriers and mechanical stabilizers, thereby extending the shelf life of packaged foods (Campelo et al., 2020).

Furthermore, Marta et al. (2023) reported synergistic interactions between starch and nanoscale fillers, resulting in improved gas barrier capacity, flexibility, and active compound retention, which are essential for multifunctional smart packaging. Therefore, nanomodified starches, in the form of SNCs, SNPs, or nanocomposites, represent a promising platform for the development of sustainable, functional, and active food packaging, combining biodegradability, reinforcement, and the potential for smart or antimicrobial functionality (Marta et al., 2023). Given the enhanced structural, mechanical, and barrier properties achieved through nanoscale modification, recent studies have progressed toward combining these materials with functional bioactive compounds, such as antioxidants, antimicrobials, and natural pigments, to create responsive, multifunctional, intelligent packaging systems. The functionalization of starch-based materials with active compounds is a promising approach for developing active and intelligent packaging systems. By incorporating antioxidants, antimicrobial agents, and natural pigments capable of responding to environmental stimuli, starch matrices can be transformed into responsive, bioactive, and eco-friendly materials (Adeyeye et al., 2023; Campelo et al., 2020; Kumari et al., 2020; Mahuwala et al., 2020; Marta et al., 2023).

In this context, Cheng et al. (2022) developed dual-modified cassava starch films containing red cabbage extracts, producing intelligent colorimetric films capable of monitoring food freshness. The anthocyanins derived from red cabbage served as natural pH and ammonia indicators, exhibiting visible color changes over the pH range 2–12. The study demonstrated that dual chemical modification of starch (oxidation and acetylation) not only improved the mechanical strength, transparency, and thermal stability of the films but also stabilized the anthocyanin molecules through electrostatic and hydrogen-bonding interactions. These structural interactions enhanced the pigment's color stability and responsiveness, enabling the films to serve as effective pH-sensitive freshness indicators in intelligent packaging applications (Cheng et al., 2022).

Chemical modification of starch plays a decisive role in tailoring its structural and functional properties for smart packaging applications. Starch oxidation converts the hydroxyl groups of the anhydroglucose units into carbonyl and carboxyl groups, typically using strong oxidizing agents such as sodium hypochlorite (NaOCl) in an alkaline medium. Under these conditions, hypochlorite ions (OCl^-) attack the primary hydroxyls, especially at the C6 position, leading to the formation of carbonyl and carboxyl functionalities. This oxidation process can also cleave some glycosidic bonds, leading to partial depolymerization of starch chains, reduced molecular weight, and significant modification of the physicochemical properties of starch, such as viscosity and film-forming capacity (Cahyana et al., 2023; Egharevba, 2020).

Fig. 1.4 (a) illustrates the oxidation mechanism of starch by sodium hypochlorite, which introduces carbonyl and carboxyl groups into the polymer structure. Starch acetylation, in turn, is an esterification reaction in which part of the hydroxyl groups are replaced by acetyl groups ($-COCH_3$) from acetic anhydride, usually under controlled alkaline conditions. Sodium hydroxide acts as a catalyst, neutralizing the acetic acid formed as a by-product. The incorporation of acetyl groups reduces intermolecular hydrogen bonding, increasing the stability of the starch paste and decreasing

(a) Starch oxidation reaction

$$\text{St–OH} + \text{NaOCl} \rightarrow \text{St–C=O} / \text{St–COOH} + \text{NaCl} + H_2O$$

Starch; Sodium hypochlorite; Oxidized starch; Sodium chloride; Water

(b) Starch acetylation reaction

$$\text{St–OH} + (CH_3CO)_2O \xrightarrow{\text{NaOH}} \text{St–O–COCH}_3 + CH_3COONa + H_2O$$

Starch; Acetic anhydride; Starch acetate or Acetylated starch; Sodium acetate; Water

(c) Starch phosphorylation reaction

$$\text{St–OH} + H_4P_2O_7 \xrightarrow[\Delta]{\text{pH 3 to 6}} \text{St–O–}H_2PO_3 + H_3PO_4$$

Starch; Pyrophosphoric acid; Starch phosphate; Phosphoric acid

Figure 1.4 Most common chemical modification reactions of starch, where (A) represents the oxidation, (B) acetylation, and (C) phosphorylation. *Adapted from Egharevba, H. O. (2020). Chemical properties of starch and its application in the food industry. IntechOpen.*

retrogradation (i.e., hardening during storage), while improving transparency and flexibility of the resulting films. Fig. 1.4 (b) illustrates the acetylation mechanism of starch with acetic anhydride, highlighting the substitution of hydroxyl groups by acetyl moieties that enhance film stability and processability (Cahyana et al., 2023; Egharevba, 2020). Finally, starch phosphorylation consists of the esterification of the hydroxyl groups of glucose residues with phosphate groups ($H_2PO_3^-$) using reagents such as sodium tripolyphosphate (STPP) or sodium trimetaphosphate (STMP). This reaction produces starch phosphate esters ($St–O–PO_3H_2$), which improve the structural stability of starch against temperature and pH fluctuations, while increasing water absorption and gel transparency. The introduction of phosphate groups also enhances starch's ability to form clear, stable gels with greater resistance to syneresis. Fig. 1.4 (c) illustrates the phosphorylation mechanism of starch, showing the ester linkage between starch hydroxyl groups and phosphate groups (Egharevba, 2020).

Teixeira et al. (2021) further emphasized the relevance of anthocyanin-based intelligent packaging, reviewing multiple studies on the incorporation of natural pigments (from sources such as jabuticaba, açaí, purple sweet potato, and grape skin) into biopolymeric matrices. These systems not only provide visual indicators of pH and gaseous metabolites (CO_2, NH_3) but also contribute to antioxidant and antimicrobial activity, supporting both diagnostic and preservative roles in smart packaging technologies (Teixeira et al., 2025).

For antimicrobial functionalization, Srivastava et al. (2023) prepared corn starch-based biocomposite films reinforced with rice husk fibers and functionalized with benzalkonium chloride (BKC). The resulting materials exhibited excellent antibacterial activity against spoilage bacteria, particularly when containing 0.05% BKC and 20% fiber content. This formulation resulted in 61% higher tensile strength, enhanced hydrophobicity, and improved thermal stability compared to native starch films. The combination of natural agro-industrial residues with a quaternary ammonium compound provided a sustainable strategy for extending the shelf life of perishable fruits, such as strawberries, by more than 11 days. The study highlighted that the ionic and hydrophobic interactions between BKC and the starch-fiber matrix enabled controlled antimicrobial release while maintaining biodegradability and environmental compatibility (Srivastava et al., 2023).

In addition to antimicrobial action, the incorporation of antioxidant molecules, such as phenolic acids, has been shown to confer oxidative protection to starch films. Almeida et al. (2023) reported the fabrication of thermoplastic starch–bacterial nanocellulose–gallic acid (TPS–BNC–GA) nanocomposite films. The presence of gallic acid imparted strong antioxidant activity (DPPH inhibition >80%), UV-blocking capacity, and antibacterial properties against *Staphylococcus aureus*, while bacterial nanocellulose reinforcement significantly improved mechanical resistance and oxygen barrier performance. Films containing 10% BNC and 1% GA displayed the best multifunctional performance, effectively delaying browning and weight loss in packaged fresh-cut apples during refrigerated storage. The authors emphasized that gallic acid formed hydrogen bonds within the starch-cellulose matrix, ensuring homogeneous dispersion and controlled release, making these nanocomposites suitable for bioactive and antioxidant packaging applications (Almeida et al., 2023).

Beyond natural extracts, synthetic nanozyme-based systems also offer new avenues for functional packaging design. Jing et al. (2021) developed a Fe–N/C single-atom nanozyme colorimetric sensor array capable of discriminating multiple antioxidants, including glutathione, cysteine, ascorbic acid, uric acid, and melatonin, through a triple-channel peroxidase-mimicking reaction. Although originally designed for biomedical analysis, this concept can be adapted for noninvasive, colorimetric food sensors by integrating similar catalytic nanozymes into starch-based films to detect oxidative degradation or gas evolution during food spoilage. Such hybrid systems exemplify the potential for coupling enzyme-mimetic nanostructures with biopolymers to achieve multifunctional sensing and preservation capabilities (Jing et al., 2021).

Taken together, these studies demonstrate that the functionalization of starch-based films with bioactive and responsive molecules, including natural pigments, phenolic antioxidants, antimicrobial agents, and nanozyme sensors, broadens and strengthens their functional scope, transforming them into materials capable of protection, preservation, and diagnostics within a single biodegradable platform. By integrating chemical modification, nanostructuring, and active compound incorporation, starch-based materials evolve from conventional biodegradable matrices into multifunctional intelligent packaging systems, capable of extending shelf life, preventing microbial proliferation, and providing real-time visual indicators of food freshness (Almeida et al., 2023; Cheng et al., 2022; Jing et al., 2021; Srivastava et al., 2023; Teixeira et al., 2025).

1.8 Starch-based smart packaging applications

1.8.1 Fundamentals of starch functionalization for smart packaging

Starch, an abundant, renewable, and biodegradable polysaccharide, has emerged as a fundamental substrate for the development of sustainable packaging solutions. Traditionally used as a passive barrier material, starch, in its pure form, presents structural limitations, such as low mechanical strength and high susceptibility to moisture, due to its hydrophilic nature (Rhim et al., 2013). However, recent advances in materials and polymer science have enabled the functionalization of starch, generally through blending with other biopolymers (e.g., chitosan or polyvinyl alcohol) or incorporation of nanostructures (e.g., cellulose nanofibers and nanoparticles) and active agents (Ma et al., 2022). The transition from conventional packaging to smart, active systems requires the provision of sensory and responsive capabilities. In the context of starch, this is often achieved by incorporating bioactive compounds that act as colorimetric indicators or as agents with preservative properties. Colorimetric indicators, in particular, enable real-time, visual communication about the quality and freshness of the packaged product, responding to chemical changes resulting from food degradation. Starch's ability to form transparent, non-toxic, and low-cost films positions it as an ideal host polymer for these functional indicators, boosting its relevance in mitigating food waste and improving consumer safety (Singh et al., 2021). Starch's polymeric architecture is thus optimized to transcend its primary barrier function, transforming it into a dynamic component of the value chain.

1.9 Applications in fresh food

Microbial and enzymatic spoilage of fresh foods, such as meat, fish, and fresh produce, is a complex process that invariably leads to the accumulation of volatile metabolites. Early detection of these compounds is crucial. One of the most effective and visually intuitive methods for real-time monitoring is the use of pH indicators incorporated into starch-based films or labels (Hall & Kenny, 2007).

The spoilage of protein products, such as fish and meat, is characterized by the production of biogenic amines and volatile basic nitrogen compounds (TVB-N), such as ammonia and trimethylamine, resulting from microbial activity (Hall & Kenny, 2007). The increased concentration of these volatile compounds raises the pH in the packaging headspace, which can be detected by pH-sensitive dyes.

Starch polysaccharides are ideal matrices for the encapsulation and stabilization of anthocyanins, natural pigments extracted from fruits and vegetables (e.g., red cabbage, purple sweet potato, *Clitoria ternatea* flower), which exhibit distinct color changes in response to pH variations (Hall & Kenny, 2007).

Starch composite films functionalized with anthocyanin extracts demonstrate excellent potential as freshness indicators. For example, in studies of shrimp packaging, a starch film incorporated with anthocyanins from butterfly pea flowers (*Clitoria ternatea*) changed from a pinkish/purple hue (neutral/slightly acidic pH, indicating freshness) to gray or green hues (alkaline pH, indicating advanced spoilage) over 6 days of storage, directly correlating with increased TVB-N levels (Singh et al., 2021). This color change is visible and immediate, allowing the consumer or distributor to assess product quality without the need for laboratory testing or opening the package. Similarly, in monitoring dairy products, such as milk, initial spoilage is marked by lactose fermentation and the production of lactic acid, which lowers the pH. Starch-chitosan films with anthocyanins react to these initial acidic conditions, exhibiting a color change that signals the onset of degradation, long before sensory cues (odor and texture) appear. The accuracy and stability of the film's colorimetric properties are crucial and are optimized by the protective nature of the starch polymer matrix, making it an efficient detection system for extending shelf life and ensuring food safety.

1.10 Applications in meat, dairy, vegetables, and minimally processed products

Starch-based smart packaging offers an essential dynamic monitoring system for highly perishable products such as meat and dairy. For fresh meat and poultry products, the primary concern is the growth of psychrotrophic bacteria, which produce volatile basic metabolites, such as the aforementioned amines. In addition to colorimetric pH indicators (Table 1.4, Fig. 1.5), starch films are often combined with antimicrobial agents to create active and smart packaging systems (Singh et al., 2021).

Table 1.4 Recent practical applications of starch in smart packaging with performance results.

Main starch satrix	Embedded functional agent	Application in food	Primary function	Performance result (example)	References
Corn starch/chitosan	Red cabbage anthocyanin	Fresh chicken fillets	Freshness indicator (pH)	Color change from purple to green/yellow in 72h, correlated with TVB-N.	Qi et al. (2022)
Cassava starch/nanocellulose	Keratin	Chicken	Freshness and barrier indicator	30% increase in tensile strength and visual detection of deterioration in 48h.	Esfahani et al. (2022)
Potato starch	Thyme EO/Ag nanoparticles	Cottage cheese	Antimicrobial and antioxidant	Extends shelf life by 40% (mainly mold/yeast inhibition).	Zhou et al. (2023)
Pea starch	Citric Acid/Gum Arabic	Minimally processed apples	Antioxidant (enzyme inhibition)	Delays enzymatic browning for 5 days under refrigeration.	Chen et al. (2021)

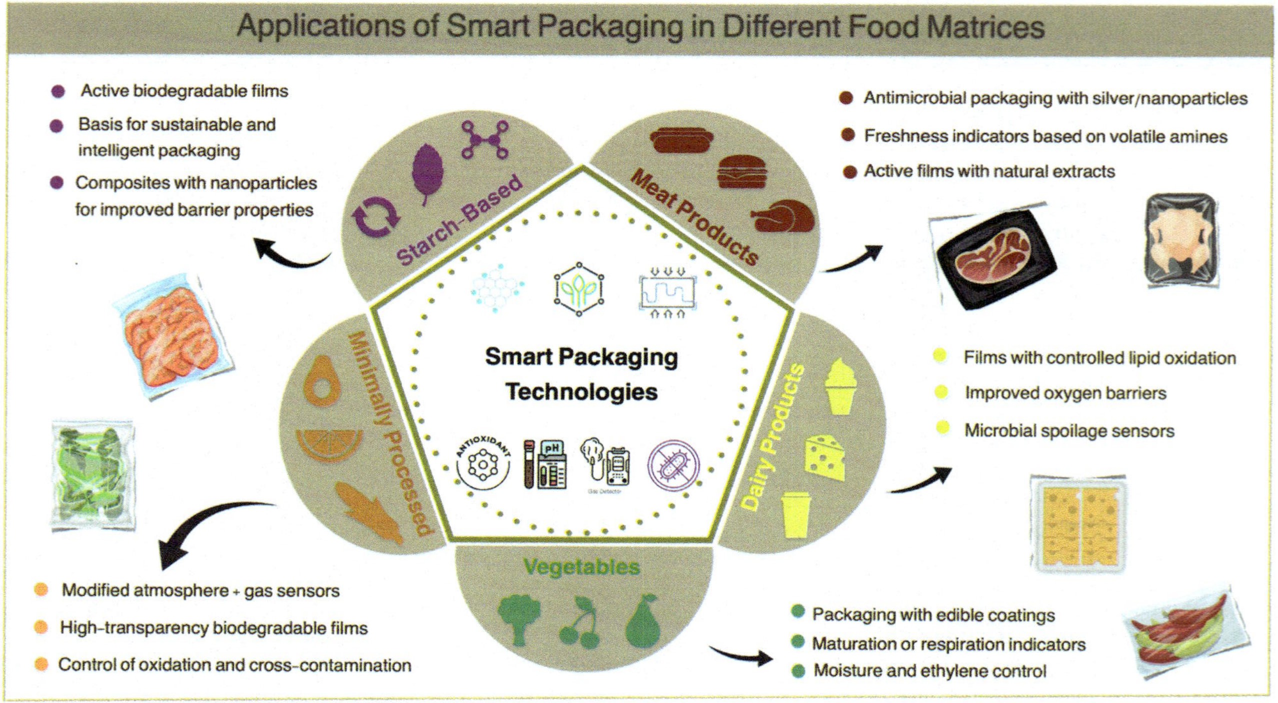

Figure 1.5 Smart packaging technologies can be tailored to different food matrices enhancing safety, quality, and sustainability.

Modified starch, in nanostructured composites, can serve as a vehicle for the controlled release of plant extracts and essential oils (EOs), such as clove, oregano, or cinnamon oil. These EOs contain phenolic compounds that exhibit broad antimicrobial activity against common foodborne pathogens such as *Salmonella* and *Listeria monocytogenes*. In one case study, the incorporation of zinc oxide (ZnO) nanoparticles into starch-chitosan films demonstrated a significant reduction in microbial counts in pork samples, combining the bacteriostatic effect of the metal oxide with visual pH monitoring to indicate freshness (Zhang et al., 2020). The release of the active agents is modulated by the starch matrix, ensuring that the concentration of the antimicrobial agent on the food surface remains effective over time.

In dairy products, the primary challenge is lipid oxidation and the growth of acid-producing bacteria. Starch packaging, in these cases, is functionalized with antioxidant agents, such as tocopherols or green tea extracts, which are slowly released onto the product surface or activated in response to free radicals (Candido et al., 2022). The starch polymer matrix, due to its ability to form hydrogen bonds with these agents, allows for controlled and prolonged release kinetics, overcoming the rapid exhaustion observed with direct surface treatments.

For vegetables and fruits, especially minimally processed ones, starch-based smart packaging focuses on controlling the respiration rate and inhibiting enzymatic browning. Starch is a naturally effective barrier against oxygen in low-humidity conditions, which is beneficial for delaying oxidation and browning of cut surfaces. Active starch packaging applied to fruits, such as strawberries or sliced apples, can be enriched with compounds that absorb ethylene or release enzyme inhibitors. The use of starch-gum arabic films loaded with citric acid has demonstrated the ability to delay browning in fresh vegetable products, maintaining visual and sensory quality for longer periods. The starch acts as a controlled-release matrix, allowing citric acid, a chelating and acidifying agent, to act locally to inhibit the polyphenol oxidase enzyme responsible for enzymatic browning (Chen et al., 2021). Furthermore, micro-hole detection and monitoring of the integrity of the modified atmosphere (MAP) within the package are crucial. Oxygen sensors based on photosensitive dyes, when integrated into starch films, can visually signal the entry of oxygen into the package. This real-time monitoring capability is invaluable for minimally processed products that strictly rely on maintaining a low oxygen concentration to preserve their shelf life and quality.

1.11 Packaging with antioxidant, antimicrobial, and controlled-release functions

The development of multifunctional packaging is the culmination of research into smart starch. Antioxidant functionality, as mentioned, is often achieved by incorporating polyphenols (such as quercetin or catechins), which act by scavenging free radicals and protecting lipids from oxidation. The starch matrix ensures the stability and gradual release of these compounds. This functionality is critical for high-fat foods, such as snacks and oils, where rancidity is the main factor in quality degradation (Chen et al., 2021).

The controlled-release antimicrobial function is based on the incorporation of agents into nanocarriers (nanoparticles or liposomes) that are then dispersed in the starch matrix. The polysaccharide structure modulates the diffusion rate of these agents, ensuring that the inhibitory effect is sustained throughout the product's shelf life. Studies demonstrate that antimicrobial efficacy can be maintained for weeks, depending on the concentration of the active agent and the composition of the starch film (Perdomo & Nguyen, 2022).

The concept of controlled release is key to the success of these active packaging systems. Unlike immediate-release systems, which can have an initial shock effect and rapid exhaustion, the starch matrix is designed to respond to environmental triggers (such as humidity, temperature, or pH) and release the active agent at the optimal concentration, maximizing the preservation effect and minimizing excessive migration into the food, ensuring regulatory compliance (Perdomo & Nguyen, 2022).

Recent studies in the literature confirm the industrial viability and superior performance of starch-based smart packaging. A significant advance involves the development of cassava starch films combined with cellulose nanocrystals and turmeric extract. This material not only significantly improved tensile strength and water vapor barrier performance compared to pure starch film, but also served as an effective colorimetric indicator for monitoring the freshness of tilapia fillets, with visible color changes after just 48 h of refrigerated storage, correlating with bacterial growth. Another notable example is the use of cornstarch films incorporated with silver nanoparticles and thyme essential oil. Tested on cottage cheese, these films demonstrated a synergistic effect: the silver nanoparticles provided a basic antimicrobial effect, while the slowly released thyme oil inhibited the growth of molds and yeasts. The result was an approximately 40% increase in product shelf life compared to conventional packaging. Highlights that starch, when optimized through nanotechnology and bioactive additives, offers a robust, sustainable platform poised to transcend the research stage and impact the global food supply chain. Ultimately, research into starch-based smart packaging attests to the ability of materials science to create sustainable solutions with superior functionality to petrochemical alternatives. Starch functionalization allows for the integration of sensory (pH indicators) and active (antioxidants, controlled-release antimicrobials) capabilities, offering robust tools for food quality and safety management and paving the way for a future with less waste and greater consumer transparency (Candido et al., 2022; Chen et al., 2021; Hall & Kenny, 2007; Perdomo & Nguyen, 2022).

1.12 Challenges and future perspectives

Although the starch-derived smart packaging demonstrates innovative potential and addresses sustainability requirements, its scale-up and large-scale production face considerable challenges in the technical, economic, and regulatory domains.

One of the major barriers to the use of starch is its high moisture sensitivity and hydrophilic nature. This characteristic results in poor moisture barrier properties and reduced mechanical integrity in high-humidity environments. However, the water-resistance properties of the starch packaging material must meet the required packaging application. Although resistant starch-based films exhibit high oxygen barrier properties, reducing respiration and delaying oxidation, laboratory-scale chemical and physical modifications, such as the inclusion of plasticizers and surfactants, and the addition of active ingredients, have mitigated some of these issues. Reliable and economically viable large-scale production methods are still under development (do Val Siqueira et al., 2021; Deng et al., 2024; Farhan et al., 2025; Fatima et al., 2024; Wang et al., 2018).

The mechanical properties of these packaging materials also require improvement for their large-scale production. This fact can be achieved through the use of appropriate additives and a combination of biopolymers. To achieve moderate strength and flexibility, approaches such as nanocomposites, structure modifiers, and crosslinkers have been explored to enhance the film network, thereby overcoming brittleness and low flexibility and improving the overall mechanical strength of films produced from polysaccharides and proteins (Farhan et al., 2025).

The relatively higher cost of bio-based materials used in smart packaging compared to conventional plastics remains a significant barrier to widespread application, especially in cost-sensitive sectors such as food production. Although consumer demand for sustainable solutions is growing, conventional plastics dominate the market due to their affordability and well-established supply chains. Recycling and composting infrastructure for bio-based materials is also limited, especially in regions lacking adequate waste management systems (D'Almeida & de Albuquerque, 2024).

Despite the advantages of smart packaging, food packaging systems with a single freshness-maintenance or monitoring function may not meet all practical needs, necessitating a combination of functions. This fact ultimately hinders their implementation and development at an industrial level. Furthermore, the production costs involved must be considered, as some bio-based materials require process optimization and the discovery of alternative raw materials for their production (Versino et al., 2023).

The performance of biodegradable and smart starch-based packaging materials is influenced by starch type, preparation methods, storage conditions, and other processing variables, often resulting in heterogeneity and poor product stability. Most studies on starch-based food packaging remain at the laboratory scale, with industrial production, regulatory requirements, environmental considerations, and consumer acceptance posing significant challenges to commercialization (Liu et al., 2022). Scaling up nanomodification techniques while maintaining uniform dispersion, preventing aggregation, and ensuring consistent functional performance of embedded active components (such as antioxidants, pH indicators, or antimicrobials) during continuous film extrusion or coating processes requires robust, specialized equipment.

Research involving the preparation of composite materials with natural and synthetic additives using emerging technologies, such as 3D printing and sensor insertion, may lead to improved functionality of the resulting materials. Furthermore, studies dedicated to improving the controlled release kinetics of the compounds to achieve continuous release throughout the shelf life of packaged foods must be better understood (Versino et al., 2023). The shelf-life stability and long-term functional consistency of smart features within the final packaging matrix—especially under varying storage conditions—require further rigorous validation. The goal is to develop materials that retain superior mechanical strength and barrier properties, ideally approaching those of petroleum-based alternatives, without compromising biodegradability; however, using natural and agro-based components (Versino et al., 2023). Moreover, combining starch with other materials still demands extensive investigation to meet practical application needs. Various additives show great promise for developing films with antimicrobial and antioxidant properties; however, the field lacks standardized criteria to evaluate and quantify such functionalities (Liu et al., 2009), although the global market for active and smart packaging materials presents a growing tendency with a forecast of 5.5% from 2021 to 2030 (Versino et al., 2023). Regulatory compliance introduces another layer of complexity, particularly in the context of food safety and consumer confidence. Smart packaging materials—especially those classified as active or intelligent—involve intentionally migrating substances (e.g., essential oils, oxygen scavengers) or indicators that directly interface with the food product or its headspace. Regulatory frameworks for food packaging vary globally, focusing mainly on substance migration limits and toxicological properties. Obtaining regulatory approval, as described in Commission Regulation (EC) n. ° 450/2009 for these novel components requires extensive toxicological data, migration studies, and clarification on acceptable residual levels, while in the EUA, it requires explicit clearance of all materials based on hazardous substance lists. In addition to regulatory compliance, manufacturers must assess potential health impacts from contaminant migration, particularly in smart packaging. Proper labeling of smart devices is also essential to ensure consumer confidence in food safety and in the use of technology. There is a pressing need for harmonized global standards and streamlined protocols for the risk assessment of smart biopolymer systems, and a consumer information compliance statement must be included on all products offered by companies, including those employing active and smart packaging (Thapliyal et al., 2024; Versino et al., 2023). The new development in smart packages includes systems with sensors, indicators, and data carriers that can detect temperature changes, gas concentrations, and microbial activity. Colorimetric indicators—often based on pH-sensitive natural dyes such as anthocyanins—provide visual cues to freshness or spoilage. Time-temperature indicators (TTIs), based on chemical, enzymatic, microbial, or polymeric responses, offer irreversible visual evidence of temperature abuse. Advanced biosensors that integrate antibody-based detection with optical transduction have been developed to identify pathogens, allergens, and contaminants. Additionally, innovations in printed electronics and nanomaterials are enhancing the functionality and integration of smart components in packaging. A prominent example is the use of RFID tags, which offer efficient tracking and environmental monitoring—such as temperature, humidity, pH, and light exposure—throughout the supply chain. Despite their potential, RFID-based systems face challenges related to cost, food safety, read range, recycling, and data privacy. Therefore, future research must focus on low-cost materials and technologies, such as 3D/4D printing with bio-based composites, to enhance functionality while maintaining economic and environmental feasibility. Ultimately, smart packaging can improve supply chain transparency, enabling better inventory control and reducing food waste through accurate demand forecasting (Liu et al., 2022; Thapliyal et al., 2024; Versino et al., 2023). While these advances in smart packaging are promising, challenges remain regarding large-scale commercialization. The high production costs of sensors and the complexity of integrating them into existing packaging systems are barriers to widespread adoption. Furthermore, the need for sophisticated instrumentation limits the use of some advanced solutions. However, like those used in colorimetric detection, optical sensors offer a more affordable and easier-to-use alternative, increasing the likelihood of their adoption in commercial environments (Farhan et al., 2025; Liu et al., 2022; Qi et al., 2022).

1.13 Conclusions

The application of starch in the development of smart packaging is recognized for its innovation, combined with sustainability requirements. With its considerable and comprehensive raw material resource, attractive cost-benefit ratio, and biodegradability, starch presents an attractive alternative to petroleum-derived polymers, aligning the packaging industry with the principles of the circular bioeconomy. However, functionalization and modification techniques for this raw material, such as chemical cross-linking, enzymatic treatment, and the incorporation of nanoparticles (such as starch and clay nanocrystals), leading to the creation of smart packaging, have successfully addressed the main weaknesses of the raw polymer—namely, its high moisture permeability and low mechanical strength. These modifications transform starch from a fragile hydrophilic material into a versatile matrix capable of acting as an effective barrier, a controlled-release system for antimicrobial agents, and a sensitive host for colorimetric and electrochemical sensors. The main successes are the development of starch-based films and coatings that provide real-time responses to food quality indicators, such as pH changes (indicating spoilage) or the presence of specific gases. Applications are already demonstrating tangible benefits across various food sectors, from extending the shelf life of fresh produce and meat through the controlled release of antioxidants to providing clear visual signals of dairy product safety. These developments not only improve food preservation but also critically reduce food waste, offering substantial economic and environmental benefits.

Despite these advantages, challenges related to industrial scale-up, cost competitiveness, and navigating complex regulatory landscapes are the main obstacles for the sector. In this sense, the future focuses on advanced technology integration, combining starch matrices with sophisticated biosensors and leveraging the Internet of Things for autonomous, high-resolution food monitoring. Furthermore, the adoption of 3D printing and additive manufacturing promises to revolutionize packaging customization, enabling distinctive, precise designs and the functional integration of smart features.

Starch has firmly established itself as more than just a sustainable packaging type, but rather as a possibility and support for smart functionality. Continued investment to overcome the remaining technical and economic barriers will solidify its position as the renewable material of choice, ensuring that tomorrow's packaging is not only environmentally friendly but also actively contributes to the safety, quality, and efficiency of the global food supply chain.

Acknowledgments

This work was supported financially by the Conselho Nacional de Desenvolvimento Científico e Tecnológico – CNPq, Brazil (CNPq n° 302867/2023–6; n° 303069/2022–8), and Fundação de Amparo à Pesquisa do Estado de Santa Catarina – FAPESC (2025TR001604) and Coordenação de Aperfeiçoamento de Pessoal de Nível Superior–Brazil (CAPES) Finance Code 001.

AI disclosure

During the preparation of this work the author(s) used Grammarly and ChatGPT in order to during the preparation of this work, the authors used ChatGPT and Grammarly to improve the quality of the written text. After using these tools, the authors reviewed and edited the content as necessary and take full responsibility for the content of the publication. After using this tool/service, the author(s) reviewed and edited the content as needed and take(s) full responsibility for the content of the publication.

References

Adeyeye, S. A. O., Surendra Babu, A., Guruprasath, N., Guruprasath, N., & Sankar Ganesh, P. (2023). Starch nanocrystal and its food packaging applications. *Current Research in Nutrition and Food Science Journal, 11*(1), 01–21. https://doi.org/10.12944/crnfsj.11.1.1

Ai, Y., & Jane, J.-lin (2018). *Understanding starch structure and functionality*. Elsevier BV, pp. 151–178. https://doi.org/10.1016/b978-0-08-100868-3.00003-2

Alam, A. U., Rathi, P., Beshai, H., Sarabha, G. K., & Jamal Deen, M. (2021). Fruit quality monitoring with smart packaging. *Sensors, 21*(4), 1509. https://doi.org/10.3390/s21041509

Ali, A., Wani, T. A., Wani, I. A., & Masoodi, F. A. (2016). Comparative study of the physico-chemical properties of rice and corn starches grown in Indian temperate climate. *Journal of the Saudi Society of Agricultural Sciences, 15*(1), 75–82. https://doi.org/10.1016/j.jssas.2014.04.002

Almeida, T., Karamysheva, A., Valente, B. F. A., Silva, J. M., Braz, M., Almeida, A., Silvestre, A. J. D., Vilela, C., & Freire, C. S. R. (2023). Biobased ternary films of thermoplastic starch, bacterial nanocellulose and gallic acid for active food packaging. *Food Hydrocolloids, 144*. https://doi.org/10.1016/j.foodhyd.2023.108934

Amaraweera, S. M., Gunathilake, C., Gunawardene, O. H. P., Fernando, N. M. L., Wanninayaka, D. B., Dassanayake, R. S., Rajapaksha, S. M., Manamperi, A., Fernando, C. A. N., Kulatunga, A. K., & Manipura, A. (2021). Development of starch-based materials using current modification techniques and their applications: A review. *Molecules, 26*(22). https://doi.org/10.3390/molecules26226880

Amobonye, A., Bendoraitiene, J., Peciulyte, L., & Rutkaite, R. (2025). Review of recent advancements in starch modification: Improving the functionality of starch-based films. *International Journal of Biological Macromolecules, 315*(Pt 2), 144354. https://doi.org/10.1016/j.ijbiomac.2025.144354

Andrade, I. H. P., Camilloto, G. P., & Cruz, R. S. (2020). Estudo prospectivo relativo à nanopartículas de amido/Prospective study on starch nanoparticles. *Brazilian Applied Science Review, 4*(2), 529–546. https://doi.org/10.34115/basrv4n2-010

Arruda, T. R., Machado, G.d. O., Marques, C. S., Souza, A. L.d., Pelissari, F. M., Oliveira, T. V.d., & Silva, R. R. A. (2025). An overview of starch-based materials for sustainable food packaging: Recent advances, limitations, and perspectives. *Macromol, 5*(2). https://doi.org/10.3390/macromol5020019

Azeredo, H. M. C., & Correa, D. S. (2021). Smart choices: Mechanisms of intelligent food packaging. *Current Research in Food Science, 4*, 932–936. https://doi.org/10.1016/j.crfs.2021.11.016

Babaremu, K., Oladijo, O. P., & Akinlabi, E. (2023). Biopolymers: A suitable replacement for plastics in product packaging. *Advanced Industrial and Engineering Polymer Research, 6*(4), 333–340. https://doi.org/10.1016/j.aiepr.2023.01.001

Bajer, D. (2024). Eco-friendly, biodegradable starch-based packaging materials with antioxidant features. *Polymers, 16*(7), 958. https://doi.org/10.3390/polym16070958

Ban, Z., Horev, B., Rutenberg, R., Danay, O., Bilbao, C., McHugh, T., Rodov, V., & Poverenov, E. (2018). Efficient production of fungal chitosan utilizing an advanced freeze-thawing method; quality and activity studies. *Food Hydrocolloids, 81*, 380–388. https://doi.org/10.1016/j.foodhyd.2018.03.010

Bhargava, N., Sharanagat, V. S., Mor, R. S., & Kumar, K. (2020). Active and intelligent biodegradable packaging films using food and food waste-derived bioactive compounds: A review. *Trends in Food Science and Technology, 105*, 385–401. https://doi.org/10.1016/j.tifs.2020.09.015

Boccia, A. C., Scavia, G., Schizzi, I., & Conzatti, L. (2020). Biobased cryogels from enzymatically oxidized starch: Functionalized materials as carriers of active molecules. *Molecules, 25*(11), 2557. https://doi.org/10.3390/molecules25112557

Boonyuen, U., Chamchoy, K., Swangsri, T., Junkree, T., Day, N. P. J., White, N. J., & Imwong, M. (2017). A trade off between catalytic activity and protein stability determines the clinical manifestations of glucose-6-phosphate dehydrogenase (G6PD) deficiency. *International Journal of Biological Macromolecules, 104*, 145–156. https://doi.org/10.1016/j.ijbiomac.2017.06.002

Bushra, R., Ahmad, M., Seidi, F., Qurtulen, J. S., Jin, Y., & Xiao, H. (2023). Polysaccharide-based nanoassemblies: From synthesis methodologies and industrial applications to future prospects. *Advances in Colloid and Interface Science, 318*, 102953. https://doi.org/10.1016/j.cis.2023.102953

Cahyana, Y., Verrell, C., Kriswanda, D., Aulia, G. A., Yusra, N. A., Marta, H., Sukri, N., Esirgapovich, S. J., & Abduvakhitovna, S. S. (2023). Properties comparison of oxidized and heat moisture treated (HMT) starch-based biodegradable films. *Polymers, 15*(9). https://doi.org/10.3390/polym15092046

Campelo, P. H., Sant'Ana, A. S., & Pedrosa Silva Clerici, M. T. (2020). Starch nanoparticles: Production methods, structure, and properties for food applications. *Current Opinion in Food Science, 33*, 136–140. https://doi.org/10.1016/j.cofs.2020.04.007

Candido, G. S., Natarelli, C. V. L., Carvalho, E. E. N., & Oliveira, J. E. (2022). Bionanocomposites of pectin and pracaxi oil nanoemulsion as active packaging for butter. *Food Packaging and Shelf Life, 32*, 100862. https://doi.org/10.1016/j.fpsl.2022.100862

Carvalho, A. C. F., Bose, I., Ghosh, S., Hoffmann, T. G., Bertoli, S. L., Schwinden Prudêncio, E., Roy, S., & De Souza, C. K. (2024b). *Valorization of agrowaste derived materials for food packaging application sustainable materials for Food Packaging and Preservation: Food Security and Sustainability* (pp. 199–221). Brazil: Elsevier. https://doi.org/10.1016/B978-0-443-13567-5.00009-5

Carvalho, A. C. F., Hoffmann, T. G., Helm, C. V., Prudêncio, E. S., Roy, S., & de Souza, C. K. (2025). Smart biopolymer based on rice husk extracts and Pinhão failure applied as an interlayer for sliced mozzarella cheese. *Journal of Vinyl and Additive Technology, 31*(5), 1233–1246. https://doi.org/10.1002/vnl.22222

Carvalho, A. C. F., Meinert, C., Prudêncio, E. S., Fiebig, M., Schmitz, F. W., & de Souza, C. K. (2024a). *Nanoencapsulation and nanocarriers: Edible food packaging Advancements in Nanotechnology for Food and Packaging* (pp. 303–347). Brazil: Elsevier. https://doi.org/10.1016/B978-0-443-21428-8.00011-4

Cereda, M. P. (2023). Application properties of starches extracted from underground starchy crops of South American origin, *Vol. 1. Starchy crops morphology, extraction, properties and applications* (pp. 245–287). Academic Press.

Chaudhary, V., Bangar, S. P., Thakur, N., & Trif, M. (2022). Recent advancements in smart biogenic packaging: Reshaping the future of the food packaging industry. *Polymers, 14*(4), 829. https://doi.org/10.3390/polym14040829

Chen, E., Song, H., Zhao, S., Liu, C., Tang, L., & Zhang, Y. (2021). Comparison of odor compounds of brown sugar, muscovado sugar, and brown granulated sugar using GC-O-MS. *LWT, 142*, 111002. https://doi.org/10.1016/j.lwt.2021.111002

Cheng, M., Cui, Y., Yan, X., Zhang, R., Wang, J., & Wang, X. (2022). Effect of dual-modified cassava starches on intelligent packaging films containing red cabbage extracts. *Food Hydrocolloids, 124*, 107225. https://doi.org/10.1016/j.foodhyd.2021.107225

Cifuente, J. O., Colleoni, C., Kalscheuer, R., & Guerin, M. E. (2024). Architecture, function, regulation, and evolution of α-glucans metabolic enzymes in prokaryotes. *Chemical Reviews, 124*(8), 4863–4934. https://doi.org/10.1021/acs.chemrev.3c00811

Dăescu, D. I., Dreavă, D. M., Todea, A., Peter, F., & Păuşescu, I. (2024). Intelligent biopolymer-based films: Promising new solutions for food packaging applications. *Polymers, 16*(16), 2256. https://doi.org/10.3390/polym16162256

D'Almeida, A. P., & de Albuquerque, T. L. (2024). Innovations in food packaging: From bio-based materials to smart packaging systems. *Processes, 12*(10). https://doi.org/10.3390/pr12102085

da Rocha, J., Mustafa, S. K., Jagnandan, A., Ahmad, M. A., Rebezov, M., Shariati, M. A., & de Souza, C. K. (2023). Development of active and biodegradable film of ternary-based for food application. *Potravinarstvo Slovak Journal of Food Sciences, 17*, 148–158. https://doi.org/10.5219/1853

Deng, H., Su, J., Zhang, W., Khan, A., Sani, M. A., Goksen, G., Kashyap, P., Ezati, P., & Rhim, J. W. (2024). A review of starch/polyvinyl alcohol (PVA) blend film: A potential replacement for traditional plastic-based food packaging film. *International Journal of Biological Macromolecules, 273*. https://doi.org/10.1016/j.ijbiomac.2024.132926

Deng, Y., Wierenga, P. A., Schols, H. A., Sforza, S., & Gruppen, H. (2017). Effect of Maillard induced glycation on protein hydrolysis by lysine/arginine and non-lysine/arginine specific proteases. *Food Hydrocolloids, 69*, 210–219. https://doi.org/10.1016/j.foodhyd.2017.02.007

Devi, N., Geeta, S., Shivani, S. A., Kumar Dubey, S., Sharma, S., & Kumar, S. (2025). Antimicrobial biodegradable packaging films from phosphorylated starch: A sustainable solution for plastic waste. *Carbohydrate Research, 550*, 109404. https://doi.org/10.1016/j.carres.2025.109404

do Val Siqueira, L., Arias, C. I. L. F., Maniglia, B. C., & Tadini, C. C. (2021). Starch-based biodegradable plastics: Methods of production, challenges and future perspectives. *Current Opinion in Food Science, 38*, 122–130. https://doi.org/10.1016/j.cofs.2020.10.020

Dodero, A., Escher, A., Bertucci, S., Castellano, M., & Lova, P. (2021). Intelligent packaging for real-time monitoring of food-quality: Current and future developments. *Applied Sciences, 11*(8), 3532. https://doi.org/10.3390/app11083532

Drago, E., Campardelli, R., Pettinato, M., & Perego, P. (2020). Innovations in smart packaging concepts for food: An extensive review. *Foods, 9*(11), 1628. https://doi.org/10.3390/foods9111628

Du, L., Huang, X., Li, Z., Qin, Z., Zhang, N., Zhai, X., Shi, J., Zhang, J., Shen, T., Zhang, R., & Wang, Y. (2025). Application of smart packaging in fruit and vegetable preservation: A review. *Foods, 14*(3), 447. https://doi.org/10.3390/foods14030447

Durán-Sandoval, D., Durán-Romero, G., & Uleri, F. (2023). How much food loss and waste do countries with problems with food security generate? *Agriculture, 13*(5), 966. https://doi.org/10.3390/agriculture13050966

Dutta, D., & Sit, N. (2024). Comprehensive review on developments in starch-based films along with active ingredients for sustainable food packaging. *Sustainable Chemistry and Pharmacy, 39*, 101534. https://doi.org/10.1016/j.scp.2024.101534

Egharevba, H. O. (2020). *Chemical properties of starch and its application in the food industry*. IntechOpen.

Esfahani, A., Mohammadi Nafchi, A., Baghaei, H., & Nouri, L. (2022). Fabrication and characterization of a smart film based on cassava starch and pomegranate peel powder for monitoring lamb meat freshness. *Food Science and Nutrition, 10*(10), 3293–3301. https://doi.org/10.1002/fsn3.2918

Farhan, A., Fazial, F. F., Azfaralariff, A., Costa, M. J., & Cerqueira, M. A. (2025). Production of polysaccharide and protein edible films: Challenges and strategies to scale-up. *International Journal of Biological Macromolecules, 307*. https://doi.org/10.1016/j.ijbiomac.2025.141909

Fatima, S., Khan, M. R., Ahmad, I., & Sadiq, M. B. (2024). Recent advances in modified starch based biodegradable food packaging: A review. *Heliyon, 10*(6). https://doi.org/10.1016/j.heliyon.2024.e27453

García-Guzmán, L., Cabrera-Barjas, G., Soria-Hernández, C. G., Castaño, J., Guadarrama-Lezama, A. Y., & Rodríguez Llamazares, S. (2022). Progress in starch-based materials for food packaging applications. *Polysaccharides, 3*(1), 136–177. https://doi.org/10.3390/polysaccharides3010007

Ghasemlou, M., Aliheidari, N., Fahmi, R., Shojaee-Aliabadi, S., Keshavarz, B., Cran, M. J., & Khaksar, R. (2013). Physical, mechanical and barrier properties of corn starch films incorporated with plant essential oils. *Carbohydrate Polymers, 98*(1), 1117–1126. https://doi.org/10.1016/j.carbpol.2013.07.026

Ghizdareanu, A.-I., Banu, A., Pasarin, D., Ionita (Afilipoaei), A., Nicolae, C.-A., Raluca Gabor, A., & Pătroi, D. (2023). Enhancing the mechanical properties of corn starch films for sustainable food packaging by optimizing enzymatic hydrolysis. *Polymers, 15*(8), 1899. https://doi.org/10.3390/polym15081899

Gonçalves, E. M., Silva, M., Andrade, L., & Pinheiro, J. (2024). From fields to films: Exploring starch from agriculture raw materials for biopolymers in sustainable food packaging. *Agriculture, 14*(3), 453. https://doi.org/10.3390/agriculture14030453

González-López, M. E., Calva-Estrada, S.d. J., Gradilla-Hernández, M. S., & Barajas-Álvarez, P. (2023). Current trends in biopolymers for food packaging: A review. *Frontiers in Sustainable Food Systems,* 7. https://doi.org/10.3389/fsufs.2023.1225371

Gurunathan, M. K., Navasingh, R. J. H., Selvam, J. D. R., & Čep, R. (2025). Development and characterization of starch bioplastics as a sustainable alternative for packaging. *Scientific Reports, 15*(1). https://doi.org/10.1038/s41598-025-00221-0

Hall, G. J., & Kenny, J. E. (2007). Estuarine water classification using EEM spectroscopy and PARAFAC-SIMCA. *Analytica Chimica Acta, 581*(1), 118–124. https://doi.org/10.1016/j.aca.2006.08.034

Han, Z., Shi, R., & Sun, D. W. (2020). Effects of novel physical processing techniques on the multi-structures of starch. *Trends in Food Science and Technology, 97*, 126–135. https://doi.org/10.1016/j.tifs.2020.01.006

Hassan, N. A., Darwesh, O. M., Smuda, S. S., Altemimi, A. B., Hu, A., Cacciola, F., Haoujar, I., & Abedelmaksoud, T. G. (2022). Recent trends in the preparation of nano-starch particles. *Molecules, 27*(17). https://doi.org/10.3390/molecules27175497

Heo, W., & Lim, S. (2024). A review on gas indicators and sensors for smart food packaging. *Foods, 13*(19), 3047. https://doi.org/10.3390/foods13193047

Hoffmann, T. G., Angioletti, B. L., Bertoli, S. L., & de Souza, C. K. (2022). Intelligent pH-sensing film based on jaboticaba peels extract incorporated on a biopolymeric matrix. *Journal of Food Science and Technology, 59*(3), 1001–1010. https://doi.org/10.1007/s13197-021-05104-6

Huang, K., & Wang, Y. (2025). Advances in bio-based smart food packaging for enhanced food safety. *Trends in Food Science & Technology, 159*, 104960. https://doi.org/10.1016/j.tifs.2025.104960

Jane, K. A., Inamdar, N. N., & Kotagale, N. R. (2025). Effect of gelatinizing temperature and moisture on the retrogradation of coix starch. *International Journal of Biological Macromolecules, 311*. https://doi.org/10.1016/j.ijbiomac.2025.143860

Jia, Y. T., Gong, J., Gu, X. H., Kim, H. Y., Dong, J., & Shen, X. Y. (2007). Fabrication and characterization of poly (vinyl alcohol)/chitosan blend nanofibers produced by electrospinning method. *Carbohydrate Polymers, 67*(3), 403–409. https://doi.org/10.1016/j.carbpol.2006.06.010

Jing, W., Cui, X., Kong, F., Wei, W., Li, Y., Fan, L., & Li, X. (2021). Fe-N/C single-atom nanozyme-based colorimetric sensor array for discriminating multiple biological antioxidants. *Analyst, 146*(1), 207–212. https://doi.org/10.1039/d0an01447h

Kadirvel, V., Palanisamy, Y., & Ganesan, N. (2025). Active packaging system—an overview of recent advances for enhanced food quality and safety. *Packaging Technology and Science, 38*(2), 145–162. https://doi.org/10.1002/pts.2863
Karabanova, L. V., Honcharova, L. A., Babkina, N. V., Sapsay, V. I., & Klymchuk, D. O. (2018). Poss-containing nanocomposites based on polyurethane/poly(hydroxypropyl methacrylate) polymer matrix: Dynamic mechanical properties and morphology. *Polymer Testing, 69*, 556–562. https://doi.org/10.1016/j.polymertesting.2018.06.012
Karnwal, A., Rauf, A., Jassim, A. Y., Selvaraj, M., Al-Tawaha, A. R. M. S., Kashyap, P., Kumar, D., & Malik, T. (2025). Advanced starch-based films for food packaging: Innovations in sustainability and functional properties. *Food Chemistry X, 29*. https://doi.org/10.1016/j.fochx.2025.102662
Kowsik, P. V., & Mazumder, N. (2018). Structural and chemical characterization of rice and potato starch granules using microscopy and spectroscopy. *Microscopy Research and Technique, 81*(12), 1533–1540. https://doi.org/10.1002/jemt.23160
Kumari, S., Yadav, B. S., & Yadav, R. B. (2020). Synthesis and modification approaches for starch nanoparticles for their emerging food industrial applications: A review. *Food Research International, 128*. https://doi.org/10.1016/j.foodres.2019.108765
Lee, B. H., & Choi, H. W. (2021). Improved rheological properties and shelf-life of wheat starch-lipid complex produced by the homogenization process. *Food Science and Biotechnology, 30*(4), 541–544. https://doi.org/10.1007/s10068-021-00889-w
Liu, H., Xie, F., Yu, L., Chen, L., & Lin, L. (2009). Thermal processing of starch-based polymers. *Progress in Polymer Science, 34*(12), 1348–1368. https://doi.org/10.1016/j.progpolymsci.2009.07.001
Liu, D., Zhao, P., Chen, J., Yan, Y., & Wu, Z. (2022). Recent advances and applications in starch for intelligent active food packaging: A review. *Foods, 11*(18), 2879. https://doi.org/10.3390/foods11182879
Ma, Y., Yang, W., Xia, Y., Xue, W., Wu, H., Li, Z., Zhang, F., Qiu, B., & Fu, C. (2022). Properties and applications of intelligent packaging indicators for food spoilage. *Membranes, 12*(5), 477. https://doi.org/10.3390/membranes12050477
Madella, D. K. S. F., de Souza, R. P., & de Melo, N. R. (2024). Consumer perception of the alternative use of "organic packaging" in food products. *Brazilian Archives of Biology and Technology, 67*. https://doi.org/10.1590/1678-4324-2024230126
Mahuwala, A. A., Hemant, V., Meharwade, S. D., Deb, A., Chakravorty, A., Grace, A. N., & Raghavan, V. (2020). Synthesis and characterisation of starch/agar nanocomposite films for food packaging application. *IET Nanobiotechnology, 14*(9), 809–814. https://doi.org/10.1049/iet-nbt.2020.0100
Manfroni, M., Coatti, A., Soccio, M., Siracusa, V., Boanini, E., Salatelli, E., & Lotti, N. (2025). Eco-design of biobased poly(butylene succinate-b-pentamethylene 2,5-furanoate) copolymers with optimized mechanical, thermal and barrier properties for flexible food-packaging. *European Polymer Journal, 225*, 113728. https://doi.org/10.1016/j.eurpolymj.2025.113728
Marindra, A. M. J., Pratama, B. M., & Suroso, D. J. (2023). Non-invasive frozen meat monitoring system using UHF RFID tag antenna-based sensing and RSSI. *International Journal of Advanced Science, Engineering and Information Technology, 13*(1), 1–7. https://doi.org/10.18517/ijaseit.13.1.16919
Marotta, A., Borriello, A., Khan, M. R., Cavella, S., Ambrogi, V., & Torrieri, E. (2025). Boosting food packaging sustainability through the valorization of agri-food waste and by-products. *Polymers, 17*(6), 735. https://doi.org/10.3390/polym17060735
Marta, H., Rizki, D. I., Mardawati, E., Djali, M., Mohammad, M., Cahyana, Y., & Nanoparticles, S. (2023). Preparation, properties and applications. *Polymers, 15*(5), 1167. https://doi.org/10.3390/polym15051167

Mauro, R. R., Vela, A. J., & Ronda, F. (2023). Impact of starch concentration on the pasting and rheological properties of gluten-free gels. effects of amylose content and thermal and hydration properties. *Foods, 12*(12). https://doi.org/10.3390/foods12122281

Mishra, B., Panda, J., Mishra, A. K., Nath, P. C., Nayak, P. K., Mahapatra, U., Sharma, M., Chopra, H., Mohanta, Y. K., & Sridhar, K. (2024). Recent advances in sustainable biopolymer-based nanocomposites for smart food packaging: A review. *International Journal of Biological Macromolecules, 279.* https://doi.org/10.1016/j.ijbiomac.2024.135583

Mkhari, T., Adeyemi, J. O., & Fawole, O. A. (2025). Recent advances in the fabrication of intelligent packaging for food preservation: A review. *Processes, 13*(2). https://doi.org/10.3390/pr13020539

Mohammed, A., Rivers, A., Stuckey, D. C., & Ward, K. (2020). Alginate extraction from sargassum seaweed in the caribbean region: Optimization using response surface methodology. *Carbohydrate Polymers, 245,* 116419. https://doi.org/10.1016/j.carbpol.2020.116419

Mueller, E., Hoffmann, T. G., Schmitz, F. R. W., Helm, C. V., Roy, S., Bertoli, S. L., & de Souza, C. K. (2024). Development of ternary polymeric films based on cassava starch, pea flour and green banana flour for food packaging. *International Journal of Biological Macromolecules, 256.* https://doi.org/10.1016/j.ijbiomac.2023.128436

Nami, M., Taheri, M., Siddiqui, J., Deen, I. A., Packirisamy, M., & Deen, M. J. (2024). Recent progress in intelligent packaging for seafood and meat quality monitoring. *Advanced Materials Technologies, 9*(12). https://doi.org/10.1002/admt.202301347

Nsanzabera, F., Irakoze, E., Manishimwe, A., Mwiseneza, A., Nsengiyumva, J. B., & Nkurikiyimana, F. (2023). Starch metabolism in plant and its applications in food industry. *Advances in Biological Chemistry, 13*(04), 111–127. https://doi.org/10.4236/abc.2023.134009

Palanisamy, Y., Kadirvel, V., & Devi Ganesan, N. (2025). Recent technological advances in food packaging: Sensors, automation, and application. *Sustainable Food Technology, 3*(1), 161–180. https://doi.org/10.1039/d4fb00296b

Pawase, P. A., Pathare, A. M., Bashir, O., Saleem, F., Shrama, E., Mudgal, S., & Ahmad, M. (2025). Physical, chemical, and nano-enabled modifications of starch for sustainable food packaging films: Recent trends, challenges, and prospects. *Carbohydrate Polymer Technologies and Applications, 11.* https://doi.org/10.1016/j.carpta.2025.100986

Perdomo, C., & Nguyen, N. T. (2022). Recent advances in 1D nanostructured catalysts for photothermal and photocatalytic reduction of CO2. *Current Opinion in Colloid & Interface Science, 61,* 101625. https://doi.org/10.1016/j.cocis.2022.101625

Perera, K. Y., Jaiswal, A. K., & Jaiswal, S. (2023). Biopolymer-based sustainable food packaging materials: Challenges, solutions, and applications. *Foods, 12*(12). https://doi.org/10.3390/foods12122422

Preethi, R., Moses, J. A., & Anandharamakrishnan, C. (2021). Development of anacardic acid incorporated biopolymeric film for active packaging applications. *Food Packaging and Shelf Life, 28,* 100656. https://doi.org/10.1016/j.fpsl.2021.100656

Punia Bangar, S., Ashogbon, A. O., Singh, A., Chaudhary, V., & Whiteside, W. S. (2022). Enzymatic modification of starch: A green approach for starch applications. *Carbohydrate Polymers, 287.* https://doi.org/10.1016/j.carbpol.2022.119265

Puri, A., Mohite, P., Ramole, A., Verma, S., Kamble, M., Ranch, K., & Singh, S. (2025). Starch science advancement: Isolation techniques, modification strategies, and multifaceted applications. *Macromol, 5*(3), 40. https://doi.org/10.3390/macromol5030040

Qi, Z., Xue, X., Zhou, H., Yuan, H., Li, W., Yang, G., Xie, P., & Wang, C. (2022). The aqueous assembly preparation of OPs-AgNPs with phenols from olive mill wastewater and its mechanism on antimicrobial function study. *Food Chemistry, 376,* 131924. https://doi.org/10.1016/j.foodchem.2021.131924

Rashwan, A. K., Younis, H. A., Abdelshafy, A. M., Osman, A. I., Eletmany, M. R., Hafouda, M. A., & Chen, W. (2024). Plant starch extraction, modification, and green applications: A review. *Environmental Chemistry Letters, 22*(5), 2483–2530. https://doi.org/10.1007/s10311-024-01753-z

Rezghi Rami, M., Forouzandehdel, S., & Aalizadeh, F. (2024). Enhancing biodegradable smart food packaging: Fungal-synthesized nanoparticles for stabilizing biopolymers. *Heliyon, 10*(18), e37692. https://doi.org/10.1016/j.heliyon.2024.e37692

Rhim, J. W., Park, H. M., & Ha, C. S. (2013). Bio-nanocomposites for food packaging applications. *Progress in Polymer Science, 38*(10–11), 1629–1652. https://doi.org/10.1016/j.progpolymsci.2013.05.008

Ribba, L., Garcia, N. L., D'Accorso, N., & Goyanes, S. (2017). *Disadvantages of starch-based materials, feasible alternatives in order to overcome these limitations starch-based materials in food packaging: Processing, characterization and applications*. Argentina: Elsevier Inc., pp. 37–76. https://doi.org/10.1016/B978-0-12-809439-6.00003-0

Rojas-Reyes, J. J., Rivera-Cadavid, L., & Peña-Orozco, D. L. (2024). Disruptions in the food supply chain: A literature review. *Heliyon, 10*(14). https://doi.org/10.1016/j.heliyon.2024.e34730

Sales, P. F.de, Oliveira Silva, Y. R.de, Silva Santos Lapa, L., & Amaral, F. H. C. (2021). Caracterização e aplicação de filmes biodegradáveis de amido de incorporados de extrato de própolis-verde. *ForScience, 9*(2), e00958. https://doi.org/10.29069/forscience.2021v9n2.e958

Sangwongchai, W., Tananuwong, K., Krusong, K., Natee, S., & Thitisaksakul, M. (2023). Starch chemical composition and molecular structure in relation to physicochemical characteristics and resistant starch content of four thai commercial rice cultivars differing in pasting properties. *Polymers, 15*(3), 574. https://doi.org/10.3390/polym15030574

Santos, S. P. D., Angioletti, B. L., Hoffmann, T. G., Gonçalves, M. J., Bertoli, S. L., Hlebová, M., Khayrullin, M., Gribkova, V., Shariati, M. A., & De Souza, C. K. (2022). Whey based biopolymeric coating as an alternative to improve quality of fresh fruits (malpighia emarginata d.c.) from southern Brazil. *Journal of Microbiology, Biotechnology and Food Sciences, 11*(5). https://doi.org/10.55251/jmbfs.5433

Schirmer, M., Jekle, M., & Becker, T. (2015). Starch gelatinization and its complexity for analysis. *Starch - Stärke, 67*(1–2), 30–41. https://doi.org/10.1002/star.201400071

Siddiqui, M. A. S., Rabbi, M. S., Ahmed, R. U., & Maruf Billah, Md (2024). Biodegradable natural polymers and fibers for 3D printing: A holistic perspective on processing, characterization, and advanced applications. *Cleaner Materials, 14*, 100275. https://doi.org/10.1016/j.clema.2024.100275

Singh, B., Sharma, V., & Kumar, R.and A. (2021). Designing moringa gum-sterculia gum-polyacrylamide hydrogel wound dressings for drug delivery applications. *Carbohydrate Polymer Technologies and Applications, 2*, 100062. https://doi.org/10.1016/j.carpta.2021.100062

Sinha, S. (2024). An overview of biopolymer-derived packaging material. *Polymers from Renewable Resources, 15*(2), 193–209. https://doi.org/10.1177/20412479241226884

Sirbu, E. E., Dinita, A., Tănase, M., Portoacă, A. I., Bondarev, A., Enascuta, C. E., & Calin, C. (2024). Influence of plasticizers concentration on thermal, mechanical, and physicochemical properties on starch films. *Processes, 12*(9). https://doi.org/10.3390/pr12092021

Siva Janika, A. R., Afrin, M. R., Yuvaraj, D., Reddy Gujjula, K., & Rahul, R. (2025). Functionalized starch polymers in packaging: Advances in bioengineering and material science. *Starch - Stärke, 77*(7). https://doi.org/10.1002/star.70039

Soltani Firouz, M., Mohi-Alden, K., & Omid, M. (2021). A critical review on intelligent and active packaging in the food industry: Research and development. *Food Research International, 141*, 110113. https://doi.org/10.1016/j.foodres.2021.110113

Srivastava, V., Singh, S., & Das, D. (2023). Rice husk fiber-reinforced starch antimicrobial biocomposite film for active food packaging. *Journal of Cleaner Production, 421*, 138525. https://doi.org/10.1016/j.jclepro.2023.138525

Tao, H., Fang, X. H., Cai, W. H., Zhang, S., & Wang, H. L. (2024). Retrogradation behaviors of damaged wheat starch with different water contents. *Food Chemistry X, 22*. https://doi.org/10.1016/j.fochx.2024.101258

Teixeira, S. C., Soares, N.de F. F., & Stringheta, P. C. (2021). Desenvolvimento de embalagens inteligentes com alteração colorimétrica incorporadas com antocianinas: uma revisão crítica. *Brazilian Journal of Food Technology, 24*. https://doi.org/10.1590/1981-6723.03321

Teixeira, E.d. M., Stefani, R., Corrêa, A. C., de Campos, A., Luchesi, B. R., Bondancia, T. J., Marconcini, J. M., Corradini, E., Neto, W. P. F., Teodoro, K. B. R., & Correa, D. S. (2025). Assessment of the properties of ginger starch-based films and their potential application as intelligent packaging. *International Journal of Biological Macromolecules, 330*. https://doi.org/10.1016/j.ijbiomac.2025.148120

Thapliyal, D., Karale, M., Diwan, V., Kumra, S., Kumar Arya, R., & Verros, G. D. (2024). Current status of sustainable food packaging regulations: Global perspective. *Sustainability, 16*(13), 5554. https://doi.org/10.3390/su16135554

Tian, J., Qin, L., Zeng, X., Ge, P., Fan, J., & Zhu, Y. (2023). The role of amylose in gel forming of rice flour. *Foods, 12*(6), 1210. https://doi.org/10.3390/foods12061210

Valencia-Sullca, C., Vargas, M., Atarés, L., & Chiralt, A. (2018). Thermoplastic cassava starch-chitosan bilayer films containing essential oils. *Food Hydrocolloids, 75*, 107–115. https://doi.org/10.1016/j.foodhyd.2017.09.008

Vasile, C., & Baican, M. (2021). Progresses in food packaging, food quality, and safety—Controlled-release antioxidant and/or antimicrobial packaging. *Molecules, 26*(5), 1263. https://doi.org/10.3390/molecules26051263

Versino, F., Ortega, F., Monroy, Y., Rivero, S., López, O. V., & García, M. A. (2023). Sustainable and bio-based food packaging: A review on past and current design innovations. *Foods, 12*(5). https://doi.org/10.3390/foods12051057

Wang, D., Cheng, F., Wang, Y., Han, J., Gao, F., Tian, J., Zhang, K., & Jin, Y. (2022). The changes occurring in proteins during processing and storage of fermented meat products and their regulation by lactic acid bacteria. *Foods, 11*(16), 2427. https://doi.org/10.3390/foods11162427

Wang, X., Huang, L., Zhang, C., Deng, Y., Xie, P., Liu, L., & Cheng, J. (2020). Research advances in chemical modifications of starch for hydrophobicity and its applications: A review. *Carbohydrate Polymers, 240*, 116292. https://doi.org/10.1016/j.carbpol.2020.116292

Wang, W., Zhang, H., Jia, R., Dai, Y., Dong, H., Hou, H., & Guo, Q. (2018). High performance extrusion blown starch/polyvinyl alcohol/clay nanocomposite films. *Food Hydrocolloids, 79*, 534–543. https://doi.org/10.1016/j.foodhyd.2017.12.013

Wu, J., Wang, X., Bian, L., Li, Z., Jiang, X., Shi, F., Tang, F., Zhang, Z., & Kumar, R. (2024). Starch and sucrose metabolism plays an important role in the stem development in Medicago sativa. *Functional Plant Biology, 51*(6). https://doi.org/10.1071/fp24073

Yang, Y., Fu, J., Duan, Q., Xie, H., Dong, X., & Yu, L. (2024). Strategies and methodologies for improving toughness of starch films. *Foods, 13*(24), 4036. https://doi.org/10.3390/foods13244036

Ye, L., Ashfaq, M. M., Shi, A., Shah, S. A. R., & Shi, Y. (2021). Performance analysis of indentation punch on high energy lithium pouch cells and simulated model improvement. *Polymers, 13*(12). https://doi.org/10.3390/polym13121971

Zhai, X., Xue, Y., Sun, Y., Ma, X., Ban, W., Marappan, G., Tahir, H. E., Huang, X., Wu, K., Chen, Z., Zou, W., Liu, B., Zhang, L., Yang, Z., & Katona, J. (2025). Colorimetric food freshness indicators for intelligent packaging: Progress, shortcomings, and promising solutions. *Foods, 14*(16), 2813. https://doi.org/10.3390/foods14162813

Zhang, J., Wen, C., Zhang, H., Duan, Y., & Ma, H. (2020). Recent advances in the extraction of bioactive compounds with subcritical water: A review. *Trends in Food Science & Technology, 95*, 183–195. https://doi.org/10.1016/j.tifs.2019.11.018

Zhiguang, C., Zhong, H., Chen, M., Fayong, G., & Jing, L. (2025). The fine structure of starch: A review. *Npj Science of Food, 9*(1). https://doi.org/10.1038/s41538-025-00414-x

Zhou, D., Mohan, L., Li, Q., Geng, F., Li, S., & Wu, Di (2023). Enhancement of liposomal properties of thyme essential oil using lysozyme modification: Physicochemical, storage, and antibacterial properties. *Food Chemistry X, 20*, 101057. https://doi.org/10.1016/j.fochx.2023.101057

Zhou, Z., Wang, J., Li, J., Li, X., wang, G., & Lu, W. (2025). Sustainable biomass materials: Active and intelligent packaging based on cellulose. *Industrial Crops and Products, 233*, 121409. https://doi.org/10.1016/j.indcrop.2025.121409

Zhu, F. (2025). Starch based films and coatings for food packaging: Interactions with phenolic compounds. *Food Research International, 204*, 115758. https://doi.org/10.1016/j.foodres.2025.115758

CHAPTER TWO

Starch extraction process and its impact on packaging

Tahid Alam[1], Pooja Sharma Luitel[1], Ganesh Dey[1], Rupak Das[1], Sudipta Paul[1], Mrinmoy Basak[2], Bapi Ray Sarkar[1]
[1]Department of Pharmaceutical Technology, University of North Bengal, Darjeeling, India
[2]Assam Down Town University, Guwahati, Assam, India

2.1 Introduction

2.1.1 Overview of starch as a biopolymer

Carbohydrates are essential to all living organisms; regardless of whether they are autotrophic or heterotrophic, primary metabolism relies on the conversion of carbon and energy, and therefore remains carbohydrate-centered. It is thus not a surprise that in the biosphere, polysaccharides are the most common polymers. Starch is the quantitatively most common storage carbohydrate on Earth and is produced primarily by plants and in limited amounts by some algae (Dega & Barbhai, 2023). The properties of starch that contribute to its use are primarily based on the botanical origin of the starch. Nevertheless, the methods of extraction also play an important part in the purity of starch, nonstarch constituents, surface morphology, and the extent of degradation of extracted starch. In addition, the nonconventional source of starch is taking a mainstream form because of its peculiar functional characteristics and, in certain instances, valorization of by-products (Fatima et al., 2024). Native starch is also becoming a mainstream product because it lacks the strength to withstand extreme conditions in processing, such as high temperatures, freezing-thawing, exposure to strong acids and alkalis, and high-shear rates. Therefore its use is limited, making it inapplicable in many industrial applications (Garavito et al., 2024).

2.1.2 Sources of starch in food

The top sources of starch are grains (maize, wheat, rice, sorghum, and barley), tubers (potato or sweet potato, cassava, and yam), and unripe fruits (banana and plantain) (Garavito et al., 2024). Traditional sources of starch

Starches in Smart Packaging
ISBN: 978-0-443-45396-0
https://doi.org/10.1016/B978-0-443-45396-0.00012-8

include potato, rice, cassava, wheat, maize, and corn (González-López et al., 2023). Unripe fruits (banana and plantain) and their wastes are needed to generate low-cost starch, raising environmental issues as well as food and agriculture losses (Yadav et al., 2023; Wypij et al., 2023). Food processing industries also produce a lot of agricultural waste every year, such as seeds, kernels, leaves, stems, and roots, and dispose of them. This waste may be one of the possible sources of starch. Pineapple stem starch has a higher amylose content (34.4%) than the most common starches (16%–30%), and during industrial processing, mango kernel is disposed of as waste, and pineapple is reported to have 48.43% starch (Kringel et al., 2020; Akhter et al., 2024). The seeds of jackfruits contain starch as their major composition, consisting of high amylase content of between 22.10% and 38.34%. Potential sources of starch are longan and loquat kernels, with about 59% and 71% of starch, respectively. An estimated 36 million tons of banana peels are produced annually, with significant quantities of starch of high amylose content (25.7%) (González-López et al., 2023). The content of the starch is greater in legumes (21%–27%) than in cereals, and the proportion of starch makes up 45%–51% of the total weight. Potato starch is largely spherical with a size of 5–100 μm (Fatima et al., 2024).

2.1.3 Importance of starch in modern packaging

In developing countries, which do not have recycling systems, biodegradable packaging will be a low-cost and more sustainable solution to the world's pollution. Starch is a very good resource in the manufacturing of biodegradable food packaging because it is widely available and can be easily produced, as opposed to NS films, which are not as good in terms of mechanical and barrier properties (Dega & Barbhai, 2023; Bangar et al., 2023). It has been noted that a combination of RS with other biopolymers, such as guar, konjac glucomannan, carrageenan, chitosan, xanthan gum, and gelatin, is among the effective methods known to amplify the film-forming properties, tensile strength, water resistance, and crystalline structure enhancement. Hybrid RS-based active packaging is also famous in terms of the improved film-forming properties, crystalline structures, increased tensile strength, water resistance, and thermal properties (González-López et al., 2023). The starch is mixed with other biobased polymers and active components, such as nanoparticles (NPs), plant extracts, and essential oils, to produce hybrid biodegradable packaging with low water vapor permeability (WVP), low rates of gas transmission, enhanced antimicrobial activity, and mechanical properties. The carbon footprint of packaged foods would decrease by substituting traditional plastic films with alternatives made of starch, which would lower the amount of CO_2 emitted per ton of plastic used (Dega & Barbhai, 2023).

2.2 Chemical structure and physicochemical properties of starch

2.2.1 Amylose and amylopectin composition of starch

In a polysaccharide chain, the number of glucose units connected together is known as the degree of polymerization. The two primary components of starch are amylose and amylopectin, and they have quite different degrees of polymerization. The linear polymer amylose is made up of 200 to 12,000 glucose units connected by α-1,4 glycosidic linkages (Zhiguang et al., 2025). A branch-like structure derived from the amylose molecule, amylopectin is made up of about 5% branches joined by α-1,6 glycosidic linkages. The three-dimensional arrangement of the atoms or different groups in starch is known as its molecular form, and it is mostly determined by nonbonding interactions, including hydrogen bonds, van der Waals forces, and electrostatic interactions (Nawaz et al., 2020). Fig. 2.1 shows the difference in structure between amylose and amylopectin in that the amylose has a linear structure with α-(1 → 4) glycosidic bonds, whereas amylopectin has a branched structure that is linked by α-(1 → 4) and α-(1 → 6) glycosidic bonds.

2.2.2 Crystallinity, gelatinization, and retrogradation of starch

2.2.2.1 Crystallinity

The polymorphism and crystallinity of starches can have a direct effect on their physicochemical characteristics and operational efficacy. These days, it is commonly used to evaluate procedures including the creation of starch films, hybrid composites, starch and nanocellulose bionanocomposites, and starch nanocrystals (Pinto, 2025). Crystalline A, B, and C types are three distinct polymorphic forms of starch that may be distinguished by their crystal structures, helical packing patterns, and water content. For the A and B type crystallites, these are made up of left-handed parallel-stranded double helices arranged in monoclinic and hexagonal unit cells, respectively. Type C is a different polymorph that coexists in the granules with type A and type B crystals. Another

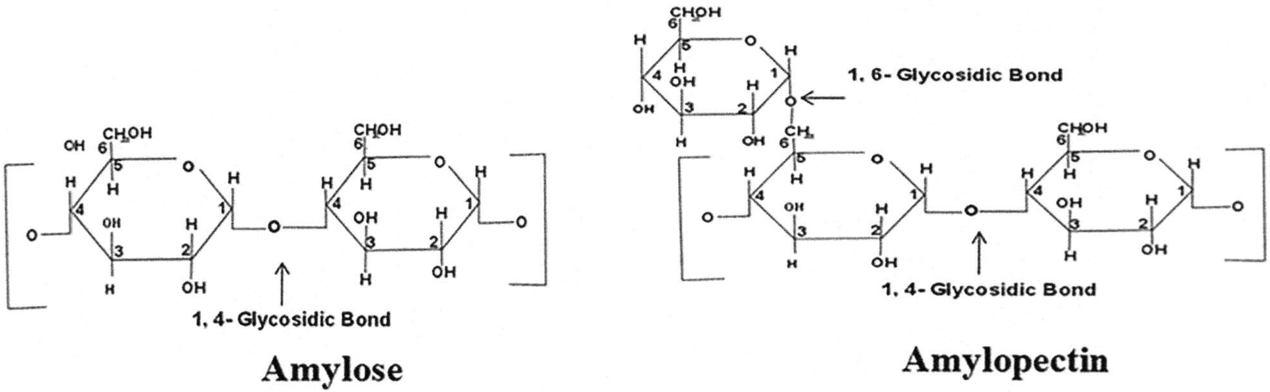

Figure 2.1 The chemical structures of amylose and amylopectin showing α-(1 → 4) and α-(1 → 6) glycosidic linkages, respectively.

crystal polymorph is called V-type, which also originates from single amylose helices, some of which are complexed with endogenous granular lipids. V-type is a different crystal polymorph that also originates from single amylose helices, some of which are complexed with granular lipids (Lopez-Rubio et al., 2008). The technologically useful uses of starches include their capacity to thicken, gel, and create films over food and industrial products.

2.2.2.2 Gelatinization

A physicochemical process known as gelatinization takes place when water is added to heated starch. Natural starch is found in the form of semicrystalline granules that are mostly made up of linear chains called amylose and branched chains called amylopectin. The water molecule (H_2O) has the ability to form four hydrogen bonds, two of which include the two H_2 and the lone pair of electrons from the H_2 and O_2 of two nearby water molecules (Lopes et al., 2025). But because the short side chains of amylopectin molecules have a larger kinetic energy than water molecules, the tetrahedrally captivated bonding of water molecules is engaged in gelatinized starch solutions. In general, this irreversible process occurs when water is heated to cause starch granules to absorb water, swell, lose their crystallinity, and produce a sticky paste (Fan et al., 2024). In this procedure, starch is typically heated in water between 60 and 80 °C.

Disruption of hydrogen bonds → loss of crystalline order → granule swelling → amylose leaching. Gelatinization and consequent retrogradation of starch granules are schematically illustrated in Fig. 2.2. It demonstrates the swelling and amylose leaching upon heating in water (60–80 °C) and the reformation of hydrogen bonds and recrystallization upon cooling.

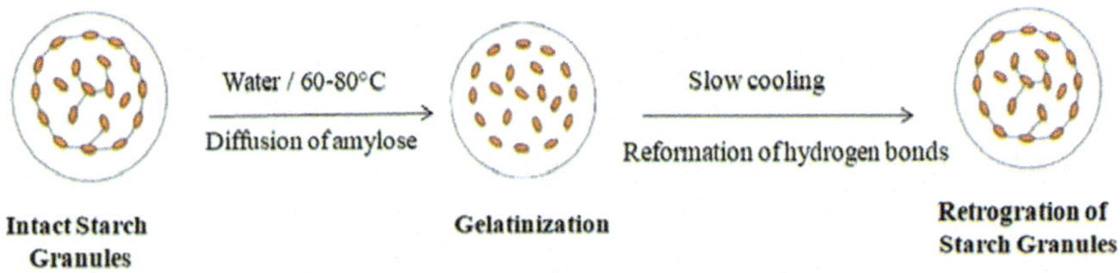

Figure 2.2 Diagrammatic view of starch granule gelatinization and retrogradation.

2.2.2.3 Retrogradation

The process via which gelatinized starch molecules, mostly amylose and amylopectin, reassociate and recrystallize when chilled, leading to enhanced hardness, opacity, and occasionally water separation, is known as retrogradation (Ning et al., 2025). Amylose and amylopectin are two polysaccharides that have distinct functions throughout the retrogradation process. The first kind of junction zone is one in which a double helix is made up of a chain from each polysaccharide. However, from a kinetic perspective, this kind of junction zone would be detrimental because both chains would need to come together, which is more difficult than if they were joined by a link, as in amylopectin (Dega & Barbhai, 2023; Fatima et al., 2024; Garavito et al., 2024; González-López et al., 2023; Yadav et al., 2023; Wypij et al., 2023). The second kind of junction zone is made up of both amylose and amylopectin double helices in the crystal unit cell. It is determined by how close the double helices of the two polysaccharides are to one another, as well as how easily they are attached to their respective molecules (Zhang et al., 2024).

Reformation of hydrogen bonds between starch chains → recrystallization → aggregation.

2.2.3 Functional properties related to packaging

Starch is vibrant in the packaging industries because of its biodegradability, sustainability, versatility, and cost-effectiveness. It assists in the industry transition away from petroleum-based harmful plastics while fulfilling consumer needs, as well as being environmentally friendly.

The crystallinity nature of starch has a great impact on barrier efficacy, processability, mechanical strength, and biodegradability of packing materials. In real applications, the crystallinity nature of starch is frequently changed to get the best flexibility, durability, and strength (Liu et al., 2024).

Gelatinization of starch helps film processing, production, and structural reliability in packing. Starch's gelatinization capability is essential. Manufacturers can customize starch-based packaging materials with desired biodegradability, mechanical strength, and flexibility by modifiable gelatinization conditions (Santhosh et al., 2024).

The biodegradability, storage stability, mechanical strength, and barrier function of packaging made of starch are all significantly impacted by its retrogradation feature. To avoid brittleness while preserving durability and environmental friendliness, retrogradation must be controlled (Fatima et al., 2024).

2.3 Starch extraction process

2.3.1 Conventional extraction methods

Conventionally, starch is extracted from plant sources using physical, chemical, and aqueous procedures. Since they are simple, cost-effective, and require minimal instrumentation, these techniques are mostly used. The primary goal during the starch extraction procedure is to break the cell structure of the plant, release starch granules, and separate starch from fibers, proteins, and lipids, which are nonstarch components. Washing, grinding, filtering, and sedimentation are common techniques, which are generally carried out via wet or dry milling paths (Dorantes-Fuertes et al., 2024).

2.3.1.1 Wet milling

The wet milling technique is one of the most used conventional methods for the extraction of starch, especially from cereals such as rice, wheat, and maize, because it gives optimum recovery of starch with improved purity. Before being stepped in water or a dilute solution (sulfur dioxide or lactic acid), the raw material is washed normally, then peeled and sliced or crushed. Stepping in water or a dilute solution increases the solubilization of protein and promotes the granule release of starch by loosening the structure of the kernel.

Following steeping, the material is mechanically ground, which breaks the cellular structure and releases starch from the matrix. Hydro cyclones or centrifuges are used to separate the resultant slurry, efficiently separating starch from fibrous components and gluten. Finally, the starch fraction is collected and dried for later usage and analysis (Shao et al., 2023; Bidari et al., 2023; Das et al., 2015; Otegbayo et al., 2014).

Though this technique is effective and produces starch with high purity, it is quite water-intensive, require ample amount of water for steeping and cleaning. Moreover, it generates a significant amount of water waste consist sulfur compounds and organic matter, which need to be properly treated before disposal to minimize environmental damage.

2.3.1.2 Dry milling

Unlike wet milling, the dry milling procedure for starch extraction does not need steeping or soaking. Mechanical processes such as grinding, sifting, and air classification techniques are used to separate the fraction that is rich in starch quantity from bran and germ. This method is simple to use, consumes less water, and does not require a high financial budget, especially for small-scale starch production. To proceed with the dry milling technique, the rhizome must first be properly cleaned to remove soil, debris, and other contaminants.

To conduct the dry milling technique, the rhizome first needs to be cleaned properly to remove soil debris and contaminants. The cleaned rhizomes are then split into smaller pieces to make further processing more effective. In some circumstances, a solution preparation phase may be used to partly soften the tissue or as a pretreatment. The pieces are subsequently dried in a hot-air oven to attain a consistent weight, which ensures full moisture elimination and effective grinding. The rhizomes are crushed to a coarse powder to produce finer fractions in starch granules and lower in fibrous debris. The fine fraction is thereafter exposed to starch separation to make starch powder suitable for further product development. Starch extracted using the dry milling technique consists of more protein and fibrous material than starch extracted using the wet milling technique. However, countries with a scarcity of water used the dry milling technique to extract starch from cassava, millet and sorghum (Shittu et al., 2016; Kebakile, 2009).

2.3.1.3 Grating and sedimentation

Extraction of starch from root and tubers such as taro, yam, sweet potato, and cassava, traditional technique grading, and sedimentation are used widely. Before being shredded into a soft pulp by the help of machine or hand graters, tubers are properly washed and peeled. After that pulp is dissolved in water and allowed to settle. During this process, fibers that are light in weight float on the top and heavier starch particles settle to the bottom gradually. To eliminate any unwanted materials, the mixture is rinsed and sieved multiple times.

Finally, the settled starch in the bottom is collected, dried, and crushed to a fine powder. This procedure is easy and inexpensive; however, some starch could be lost throughout the process, and a small amount of fiber may remain mixed with the starch (Dorantes-Fuertes et al., 2024; Moorthy, 2002).

2.3.2 Advanced extraction techniques

Extraction of starch mainly relies on aqueous and mechanical processes. Though recent advancements in bioprocess engineering and food processing have offered fresh methodologies with a goal of improving extraction efficiency, minimal time-consuming, and maintaining the structural and functional stability of starch granules. To increase yields as well as environmental stability, these innovative strategies take advantage of physical, chemical, and biological principles.

2.3.2.1 Enzymatic extraction of starch

Enzymatic extraction is a progressing technique allowing the release of intact starch granules with minimum damage by dissolving the structural polysaccharides in plant cell walls using hydrolytic enzymes. Initially, raw materials such as tubers or root crops are selected and prepared by rinsing to remove soil and debris, followed by peeling to remove the skin. Before being sliced into small pieces, the materials are thoroughly cleaned and then dried to remove the excess moisture present. After that, to increase the surface area available for enzyme action, it was milled into coarse powder. To make a slurry in the enzymatic hydrolysis stage, the powder is dissolved in distilled water.

In the enzymatic hydrolysis stage, before treating with hydrolytic enzymes such as cellulase, pectinase, or hemicellulase, the powder is dissolved in distilled water to make a slurry. pH as well as temperature are maintained, and the mixture is incubated for an hour during this procedure. During this time, enzymatic hydrolysis weakens the cell wall matrix, releasing starch granules into the aqueous medium.

During hydrolysis, fiber and other impurities float or stay in the liquid, which are lighter in weight, and starch granules, which are heavier in weight, sink to the bottom when the mixture is left to stand. After discarding the supernatant part, the starch-rich fraction is washed repeatedly with distilled water to ensure cleanliness. After discarding the supernatant part, to maintain the cleanliness of the starch-rich fraction, it is washed repeatedly. The completed product is dried until it reaches a consistent weight and crushed into a fine powder for storage and further use.

This strategy has the advantage of high recovery efficiency and purity while permitting less damage to granule form than mechanical approaches alone. However, the necessity for controlled conditions and enzymes' relatively higher cost may limit their extensive industrial application (Lumdubwong & Seib, 2000; Perez et al., 1993).

2.3.2.2 Ultrasonication-assisted extraction

Ultrasonication-assisted extraction creates small bubbles called cavitation in the liquid medium by using high-frequency sound waves. The plant cell wall ruptures when these small bubbles collapse and create intense shear and microjets, which lead to the release of granules into the fluid.

The raw materials are carefully cleaned to remove dirt and contaminants for the extraction of starch using the ultrasonication method.

The cleaned samples are then made into little pieces, weighed, and dried until a consistent weight is achieved. The material is crushed into small pieces, sorted, and combined with distilled water to make a slurry after it has dried. The slurry is then treated using ultrasonics. The mixture is filtered and centrifuged to separate the liquid fraction after sonication. The recovered supernatant is concentrated at a lower pressure with a rotary evaporator. Starch is then precipitated from the concentration by adding ethanol in a 3:1 ratio (ethanol: sample). The precipitate is removed for several hours and centrifuged to separate the starch fraction. The purification of precipitate is done by washing it three times with ethanol and acetone, and then evaporating the solvents under a stream of nitrogen gas. Ultimately, the dried extract is collected and stored in a desiccator for further use.

This approach is regarded as efficient since it decreases overall extraction time, needs less solvent, and produces higher purity starch than conventional procedures. Furthermore, ultrasonication can alter physicochemical features such as gelatinization temperature and sticking characteristics, which might be useful in industrial applications.

2.3.2.3 Microwave-assisted extraction

During microwave-assisted extraction of secondary metabolites from plants, energy is quickly delivered to the extractant and the plant matrix. Some compounds found inside vegetative materials, particularly polar molecules such as water, absorb energy effectively. As a result, the interior temperature of plant cells increases dramatically. Interior superheating may produce liquid vaporization within the cells, potentially rupturing the cell walls and/or plasma membranes (Rostagno et al., 2009; Zhang et al., 2011).

Microwave-assisted extraction delivers various advantages, such as reduced extraction time, higher yield, less solvent requirement, and high reproducibility when compared to conventional heating. Structural changes, including surface cracking, deformation, reduced crystallinity, and molecular weight, may change when starch granules are exposed at high power or for prolonged durations since starch granules are sensitive to microwave irradiation. Therefore to increase starch recovery without changing their original crystalline and functional properties, it requires careful optimization of the parameters such as microwave power, irradiation time, ratio of solvent and solid, and moisture content.

During microwave-assisted extraction, the starch-rich material is chopped into small pieces, then ground and suspended in water or ethanol. Following that, the mixture is microwaved at a regulated power (200–600 W) for 2–15 min. The extract is then filtered and centrifuged, followed by washing to eliminate contaminants and preserve the starch's properties when dried at low temperatures.

2.3.2.4 Supercritical fluid extraction

Supercritical fluid extraction (SFE) is a modern extraction technique where fluids are used at critical temperature and pressure conditions, and they behave like the density of a liquid and the diffusivity of a gas, which allows them to penetrate solid materials properly and dissolve target compounds with high selectivity. Carbon dioxide (CO_2) is the most commonly used fluid because it is safe, nontoxic, cost-effective, and leaves no harmful residues after extraction, and it becomes supercritical at mild conditions (31.1°C and 73 atm).

Extraction of starch using the SFE technique starts with preparing raw material such as cereals, tubers, or roots, which are rich in starch. Raw materials are cleaned properly, followed by peeling, drying, and milling into fine powder to increase the area of surface for extraction. Then, this powder material is placed in a high-pressure extraction vessel. Supercritical form of CO_2 is used and sometimes combined with a cosolvent. CO_2 in its supercritical form is utilized, occasionally combining with a cosolvent. The plant material is passed through the supercritical fluid, which dissolves extractable starch fractions or compounds associated with starch. After that, it enters a separator, where the pressure is decreased, transforming CO_2 back into a gas and leaving the extracted product behind. Extract is collected and stored for further use.

This advanced method offers a high extractive value without degrading thermolabile compounds and producing solvent-free extracts, making it an environmentally friendly approach. However, it also has some disadvantages. The technology is so expensive to install and requires a lot of energy and expensive high-pressure equipment. In addition, cosolvents are frequently required for efficient extraction because pure CO_2 on its own is unable to effectively dissolve starch or other hydrophilic compounds (Herrero et al., 2006; Brunner, 2005; Ivanovic et al., 2016).

2.3.3 Factors affecting the extraction process

The extraction of starch from plant sources requires physical, chemical, and enzymatic processes, which are controlled by various factors. Yield, purity, functional qualities, and cost are dependent on these processes. The most important factors are raw material type and maturity, pH, temperature, solvent type, pretreatment techniques, particle size, and the nature of mechanical, chemical, or enzymatic pretreatments.

2.3.3.1 Raw material type and maturity

The source of starch is an important determinant of the process of extraction; cereal grains such as maize, rice, wheat, and sorghum; tubers such as potato, yam, and cassava; legumes such as pea, mung, and bean; and fruits such as banana and breadfruit vary in starch granule morphology, amylose to amylopectin ratio, lipid and protein association, and nonstarch polysaccharide composition. For instance:

Cereal starches (maize, wheat, and rice) are usually blended with high protein (gluten, prolamins) and lipid content, which must be removed

- Tuber starches (potato, cassava) are comparatively cleaner with less nonstarch impurities but are more wet and must be processed gently in order not to cause granule destruction.
- Legume starches (pea and mung bean) are rich in polyphenols and protein, inhibitors of starch recovery

Relatively varying yields from varying material sources: potato gives 15%–20% starch (on a fresh weight basis), cassava 25%–30%, and cereals 60%–70% on a dry weight basis (Moorthy, 2002).

2.3.3.2 Effect of maturity

Plant maturity is of major concern for starch yield and quality. Immature grains or tubers have lower starch levels and more soluble carbohydrates, proteins, and water content, and might cause delayed recovery. As a result, completely developed raw materials have a larger concentration of starch granules, which makes extraction easier. However, overmature or stored materials may degrade starch because of enzymatic activity (amylases), resin-poor extractability, and changed physicochemical qualities, because of which extractability may be poor and physicochemical qualities may be changed.

2.3.3.3 pH, temperature, and solvent type

It is important to maintain the pH under control during starch extraction. Most starches are stable between pH 6.0 and 7.5. Acidic states ($pH < 5.5$) can hydrolyze glycosidic bonds; therefore the breakdown of starch takes place. On the other hand, alkaline conditions (pH > 10) might disrupt granule characteristics or cause partial gelatinization. A gently alkaline environment (pH 8–9) is generally favored for grain starches because it helps in protein solubilization and the removal of pigments attached to starch granules.

Temperature influences both the stability of starch granules and enzyme activity during enzymatic pretreatment. To stop solubilization and loss of granular integrity, most extraction methods are carried out at temperatures lower than gelatinization (50–70°C, depending on source). Excessive heating causes starch gelatinization, which hinders the separation of starch from nonstarch components. Optimal extraction temperatures are generally between 25 and 45°C.

Water is the most used solvent for starch extraction because it is affordable, safe, and nontoxic. However, starch granule-associated colored pigments, proteins, and lipids may not be sufficiently removed by water on its own. Different solvents have been used to solve these problems, such as diluted alkali solutions, such as sodium hydroxide (NaOH) or sodium carbonate (Na_2CO_3), used to dissolve and remove proteins from starch in cereals. Organic solvents isopropanol and ethanol are employed in the removal of lipid content, enhancing starch quality as well as the avoidance of microbiological contamination during processing.

Sulfite solutions such as sodium metabisulfite ($Na_2S_2O_5$) bleach natural colors, inhibit oxidation, and preserve starch in white and stable conditions. When applied appropriately, the solvent solutions improve starch yield and quality and are thus applied in the industrial and food industries.

2.3.3.4 Particle size

Particle size reduction is a significant step in the extraction of starch. Plant material is ground or milled to break down cell walls and release starch granules. Small particle size encourages surface area, solvent penetration, and starch recovery. Excessive grinding leads to starch granule breakage with loss of granular integrity and partial gelatinization, and hence separation becomes difficult and quality decreases. Controlled milling is therefore necessary to balance starch production and granule integrity.

2.3.3.5 Pretreatment methods (mechanical, chemical, enzymatic)

2.3.3.5.1 Mechanical

Mechanical disruption (ultrasonication, grating, mixing, or high-shear homogenization) aids in the physical separation of starch granules from cellular structures. Like, ultrasonication causes cavitation, which tears down cell walls and increases starch release without causing any harm.

2.3.3.6 Chemical

Alkaline and sulfite treatments are commonly used to remove proteins, lipids, and colors from grains. For example, 0.1%–0.2% NaOH may effectively dissolve gluten proteins in wheat and maize starch. Sulfite treatments assist in reducing enzymatic browning while also improving starch whiteness.

2.3.3.7 Enzymatic

Enzyme treatments are increasingly being used in the purification of starch as a result of the capability of enzymic treatments to enable the selective removal of nonstarch components while minimizing chemical damage to the starch granules. Enzymes are used based on the source material and the nature of impurity. Particular enzymes (proteases, cellulases, and hemicellulases) are utilized particularly for the hydrolysis of nonstarch material.

Proteases hydrolyze closely bound protein on starch granules.

Cellulases and hemicellulases break the polysaccharides in the cell wall to produce more release of starch.

Enzymatic pretreatment tends to increase starch yield and quality by restricting chemical degradation.

2.4 Impact of extraction process on starch packaging

A polysaccharide that is utilized as a pharmaceutical excipient (mainly inactive component), serving as a dusting powder for the skin, as a binder, disintegrating, dilution agent, or a filler in tablets as well as capsules forms. The extraction method, which determines quality, granule appearance, chemical structure, and physicochemical properties, starch packaging has a significant impact on its functional characteristics (Benesi et al., 2004). A variety of techniques, including wet milling, enzymatic hydrolysis, ultrasonic treatment, and alkaline extraction, are used to extract starch, which is mostly found in grains, tubers, and roots.

Functional systems of starch can be divided into two groups: Both, intelligent or smart packaging, which alerts the consumer to kinetic changes pertaining to the quality of the food or the environment it contains to reduce losses and guarantee high-quality and safety, intelligent packaging must be simple to use, fast, dependable, sensitive, safe, and most significantly, affordable to visually determine pH variations, which have been effectively used to assess the quality of food (Ceballos et al., 2020), and active packaging, which actively and purposefully alters the product or the environment it contains during storage, transportation, and even usage to improve the safety and quality of food (Faramarzi et al., 2020).

2.4.1 Granule morphology and size distribution

The morphological properties of starches derived from various plants fluctuate depending on the culture and genotype. The biological origin of starch granules is responsible for their variety in size and shape (Svegmark & Hermansson, 1993). Starch granules have different shapes: Spherical, oval, polygonal, and lenticular. Starch granule size and shape are important features, and the granules appear in many different forms. The starch granule size varies from 1 to 110 mm in diameter and exhibits a round, or lenticular to polygonal shape (García-Guzmán et al., 2022).

The biopolymeric chains of native starch are organized in a compact granular structure. It is commonly accepted that the external granule layers have high amylose contents compared to the internal core of the granules (Bello-Pérez et al., 2022).

Laser diffraction analysis was used to determine the starch granules' size (Malvern Instruments Ltd, 2000). The Sirocco 2000 and Hydro 2000 modules were used to evaluate solid and liquid samples. Using 1.5 kHz sonic homogenization, the samples were redispersed to prevent skewed results from particle aggregation. At room temperature, three duplicate measurements were performed, and thepastes were diluted until they reached a 0.001% final concentration. The volume range and median diameter D were used to represent the particle size. Physical-chemical characteristics, including water-binding capacity, swelling power, amylose content, and % light transmittance, were substantially connected with the average granule size of the starches that were isolated from various plant sources (Singh et al., 2006).

When compared to their native counterparts, the granule size distribution of both ultrasound-treated starches showed only minor alterations. This suggests that there was no fragmentation or shrinkage of the starch granules under the treatment conditions. For plantain and taro starch, the particle sizes were roughly 22.4 and 2.3 μm, respectively. The techniques employed in this investigation did not alter the size of the granules, despite the fact that the larger starch granules (plantain) had a greater surface area for capturing ultrasonic waves (Carmona-García et al., 2016) (Fig. 2.3).

Figure 2.3 Different sources of granular shape of starch. *Rashwan AK, Younis HA, Abdelshafy AM, Osman AI, Eletmany MR, Hafouda MA, et al. Plant starch extraction, modification, and green applications: A review. Environmental Chemistry Letters. 2024;22 (5):2483-2530.*

2.4.1.1 Granular shape of starch from different sources

Sorghum starch → Polygonal and irregular shapes
Wheat starch → Lenticular-shaped, small round
Taro starch → Polygonal and irregular shapes
Yam starch → Round, elliptical, or irregular shapes
Cassava starch → Round, irregular
Banana starch → Elongated ovals with ridges
Sweet potato starch → Polygonal, round to cupola form/bell shapes
Potato starch → Oval and spherical shapes
Barley starch →Small spherical granules, Large disc-shaped granules,
Rice starch → Angular, round, and polygonal shapes
Maize starch → Polygonal shapes and round (Rashwan et al., 2024)

2.4.2 Mechanical and barrier properties

Determining the physicochemical characteristics of starch is vital since these parameters directly affect its mechanical (strength, flexibility, and bending properties) and barrier (resistance to moisture, oxygen, and any gases) qualities. These characteristics are essential for use in food packaging, films, and biodegradable materials etc.

The extraction method has an impact on the purity and integrity of the starch granules, which in turn affects their capacity to form films. An Instron 5545 tensile test equipment with a 100 N load cell was used to measure tear resistance in accordance with the ASTM-D1004-13 technique. With an initial grasp separation of 25.4 mm, testing conditions were carried out at a strain rate of 51 mm/min. Instron Bluehill 2.1 software (Instron, Norwood, MA, USA) was used to obtain the results (D, 2013).

Mechanical strength is impacted by extraction techniques that change the amylose/amylopectin ratio, whereas high amylopectin results in softer, more elastic structures, high amylose starch often produces stronger and less flexible films.

Using a TA-XT2i Texture Analyzer (Scarsdale, NY, USA), a 2 mm-diameter flat head stainless probe (TA-52, Texture Technologies Corp., Hamilton, MA, USA), and an extensible platform (TA-108N, Texture Techn. Corp.), puncture resistance was assessed with the ASTM-D7192-18 standard. The films were inserted between two 15 mm-diameter polycarbonate disks. The speed was 33 mm/s. Texture Exponent 32 software (Texture Technologies Corporation, Hamilton, MA, USA) was used to obtain the values (ASTM D, 2018).

2.4.3 Gelatinization and retrogradation properties

In starch granules, the fundamental underlying component affecting functional characteristics is frequently the crystalline organization. Irreversible changes in characteristics, including granule swelling, pasting, loss of optical birefringence, loss of crystalline order, uncoiling and dissociation of the double helices, and starch solubility, are signs of the collapse of crystalline order within the starch granules (Egodage, 2019).

Using DSC, the order–disorder transitions that take place when heating an aqueous suspension of starch granules have been thoroughly studied. The degree of crystallinity and other properties of the starch granule may be connected to the starch transition temperatures and gelatinization enthalpies determined by DSC. This helps determine the starch's thermal and other physical properties and is impacted by its chemical composition. Depending on their composition, starches from various botanical sources showed varying enthalpies of gelatinization and transition temperatures (Singh et al., 2003).

The variations in the degree of crystallinity among the various starches could be the cause of the variations in transition temperatures. A high degree of crystallinity, which confers structural stability and increases the granule's resistance to gelatinization, has been observed to produce high transition temperatures. It has been hypothesized that the gelatinization temperatures represent the degree of crystalline perfection. Amylopectin's molecular structure (unit chain length, extent of branching, molecular weight, and polydispersity), starch composition (amylose to amylopectin ratio and phosphorus content), and granule architecture (crystalline to amorphous ratio) all influence the gelatinization and swelling characteristics (Barichello et al., 1991).

2.4.4 Rheological properties

The starch granule ruptures, swells to multiple times its original size, and amylose concurrently leaks out of the granule during the process of gelatinization. The leached-out amylose forms the three-dimensional system (Singh et al., 2002).

The amylopectin level of starch determines its swelling behavior, and amylose functions as a swelling inhibitor and diluent. The viscosity of starch changes with temperature, concentration, and shear rate in a distinctive way. The Brabender Visco amylograph pasting curves can be used to measure. The processes occurring during the pasting cycle are reflected in the shape of the Brabender peak. The granules' capacity to inflate freely prior to their physical breakdown is indicated by the height of the peak at a certain concentration. High-swelling starches are also less likely to break down when cooked, which is why they show a sharp drop in viscosity once they reach their maximum value. However, the initial concentration of the starch suspension has a significant impact on the peak's shape (Tester & Morrison, 1990; Nurul et al., 1999).

2.4.5 Biodegradability and shelf life

First, these two words, biodegradable and shelf life, are connected with each other; biodegradable refers to how easily a material breaks down naturally by microorganisms into simpler, harmless substances (like water, CO_2, or biomass), while maintaining biodiversity. Example: biodegradable packaging made from starch or cellulose.

And another one, shelf life, refers to how long a product can be stored before it loses its quality, safety, permeability, or usability. Example: food packaged in certain materials, may last longer because the package protects it from moisture or oxygen.

2.4.5.1 Methods of preparation

Plastic trays (10 × 20 × 5 cm) were filled with vegetable compost (soil) until they were about 4 cm high. Each system's 2 × 2 cm samples were weighed before being buried in the ground about 1 cm below the surface. The plastic trays were maintained at room temperature and relative humidity. To keep the compost moist, water was sprayed twice a day. Samples were meticulously removed on the first, sixth, and 12th days.

Purity of starch, granular integrity, chemical modification during extraction, ultrasonic, and physical methods, all these techniques generally improve the biodegradability by maintaining the molecular structure while enhancing surface accessibility for microbial attack (Medina-Jaramillo et al., 2017).

Shelf Life mainly depends on some of the variants, such as:

Moisture sensitivity: Higher water absorption, leading to reduced shelf life.

Crystallinity: crystalline reduces retrogradation to improve storage stability.

Extraction techniques: Extraction can promote rancidity or microbial growth; it ensures the product stability during storage (Bersaneti et al., 2021).

2.4.6 Environmental and economic aspects

Sustainable manufacture of starch-based packaging is supported by extraction techniques that need less energy and fewer chemicals, such as enzymatic extraction. These methods also support the objective of creating biodegradable, environmentally acceptable packaging substitutes for polymers derived from petroleum (Averous & Boquillon, 2004).

2.5 Starch modification for packaging applications

Every starch has drawbacks (such as brittleness, water solubility, and low mechanical strength), so it must be modified to be used in biodegradable packaging materials. Generally, starch modifications are of four types, such as physical modification, chemical modification, thermoplasticization, and biological modification.

2.5.1 Physical modification (heat-moisture treatment, annealing)

For food-contact products, such as biodegradable packaging films, coatings, and composites, physical modification is a common way to modify starch properties without using chemicals. This makes it safe and environmentally most beneficial. Heat-moisture treatment (HMT) and annealing (ANN) are two of the most explored techniques.

Nowadays, nonthermal treatments use less energy and are more environmentally friendly; they have become more popular. Furthermore, these technologies signify significant alterations or shifts in the characteristics of the materials. The physical characteristics of starches are influenced by several treatments (Hu et al., 2017). In this case, starch gelatinization and crystallinity patterns are influenced by high hydrostatic pressure treatments. The disruption of starch granules caused by ozonation affects the weight and size of the particles, raising the crystallinity. There is a lot of interest in nonchemically modified starch that performs similarly to chemically modified starch (La et al., 2019). Chemical qualities are also impacted by nonthermal technologies. It has been discovered that plasma applications show chemical modifications, such as starch molecule crosslinking, altering thermal transitions, such as melting temperature and enthalpy.

Ultrasound treatments are influenced by high temperatures (attained by applying ultrasonic waves) and significant shear forces, which affect the starch's characteristics and the applications themselves. The source of starch, as well as the frequency, intensity, and duration of ultrasound application, all affect the alterations in starch. Typically, the starch granule, pores, fissures, and cracks are damaged by ultrasonic treatments. Modifications in the amylose and amylopectin molecules operate as part of the experimental conditions at the molecular level (Iida et al., 2008).

A few variations and adjustments in treatment duration, pulses, and intensity are documented for starch changes under pulsed electric fields (PEFs) (Han et al., 2009).

2.5.1.1 How heat-moisture treatment and annealing effects on starch properties

HMT raises the temperature at which gelatinization occurs and crystallinity. HMT decreases solubility and swelling power. HMT increases resistance to retrogradation and thermal stability. Annealing produces starch films with a more uniform structure and improved processing stability (Li et al., 2019). Annealing enhances film flexibility and clarity, useful for transparent packaging films.

It produces starch films with better mechanical strength and thermal stability, and reduces water sensitivity and enhances barrier properties (against oxygen and moisture).

2.5.2 Chemical modification (acetylation, oxidation, crosslinking)

One of the most popular methods for modifying the physicochemical, mechanical, and barrier qualities of starch for use in biodegradable packaging is chemical modification. Chemical changes, including acetylation, oxidation, and crosslinking, are used to get beyond the limits of native starch, which include low mechanical strength, poor water resistance, and retrogradation.

By adding functional groups to the starch macromolecules (breaking glycoside bonds), we can chemically modify starches and partially or completely disrupt their physicochemical properties, including structure, composition, gelatinization, retrogradation, and pasting characteristics. Starch undergoes three primary chemical changes: crosslinking (diether, diester), substitution or acetylation (etherification, esterification), and reduction or oxidation (hydrolysis, enzymatic) (Xiong et al., 2014; Ren et al., 2012).

A modification process called starch crosslinking involves the covalent joining of linear or branched strands. The granules' hydration qualities are improved by the crosslinking of starches, which also increases their hydrophobicity and structural stability. Phosphorus chloride ($POCl_3$), sodium trimetaphosphate (STMP), sodium tripolyphosphate (STPP), epichlorohydrin (ECH), and acid have all been utilized in the chemical crosslinking of starch. However, because of the possibility that their poisonous and irritating properties could pose a health concern, glutaraldehyde and epichlorohydrin are not regarded as safe. One of the most popular crosslinking agents for starch granule modifications is citric acid because it is cheap, nontoxic, and may react with the hydroxyl groups of starch molecules to create intermolecular diester-covalent connections (Lauer & Smith, 2020; Bangar et al., 2021).

A modification process called starch reduction involves creating more functional groups. For instance, in starch oxidation, carbonyl and carboxyl are added to the reduced starch, increasing its functionality and reactivity.

When starch was reduced, there were reports of the paste's expanding capacity and viscosity (Vanier et al., 2017).

The process of starch substitution involves transferring some of the hydroxyl groups of the anhydro-glucose units that make up the starch chain to other groups, including acetyl groups, which alters the starch's molecular structure. According to certain research, state starch esterification with organic acids and acid anhydrides improves the physicochemical characteristics of thermoplastic starches (TPS) used in the packaging industry and decreases retrodegradation (Aburto et al., 1999; Kaur et al., 2012).

2.5.3 Biological modification (enzymatic treatments, fermentation)

We know that starch is not a thermoplastic substance; it does exhibit thermoplastic qualities when combined with plasticizers (water and polyols), heat (90–180°C), and shear stress. Gelatinization is necessary for the conversion of starch to a thermoplastic. The granules of starch enlarge, lose their crystalline structure, and become unstructured throughout this irreversible process (Buleon et al., 1998).

Furthermore, as amylose separates from amylopectin and the amylopectin double helix unwinds and dissociates, the starch components partially separate. The primary cause of the crystalline order loss caused by excess water (>63%) is water seeping into the granule, which lowers the interactions between and within the molecules of the starch components. Water heating breaks down the starch granules' crystalline structure and produces amorphous amylopectin (Xie et al., 2013).

A number of gelatinization-related thermal transformations, such as water diffusion, starch granule expansion, melting, and crystallization, take place when TPS are processed (Liu et al., 2009).

2.6 Applications of starch in packaging

Starch, as a natural polysaccharide, has received a lot of interest within the packaging industry as it is biodegradable, renewable, and versatile. Its film-forming capacity, proliferation by food sources such as potato, corn, cassava, and rice, and interference with other biopolymers render it a promising substance in the food and pharmaceutical packaging applications. To satisfy the functional needs of the current packaging, research has concentrated on improving the mechanical, barrier, and bioactive properties of starch-based material, besides making them environmentally sustainable. The subsequent subsections point out the primary use of starch in food packaging, pharmaceutical blister packs, and polymer blends/composites.

2.6.1 Starch films and coatings for food packaging

Starch-containing films and coatings have become particularly crucial in food packaging because they are biodegradable, renewable, and versatile. The packing sector has been identified as the biggest consumer of starch-based bioplastics with an estimated more than 50% of the world bioplastic market (Sobeih et al., 2025)These materials are excellent in the aspect of biodegradation as they demonstrate moderate mechanical strengths and can also serve as a barrier to weathering and they are traditionally used in biodegradable wraps, bags, and food containers (Gamage et al., 2022). Starch-derived edible films and coatings have shown a high level of effectiveness in extending the shelf life of fresh foods by reducing the spoilage of the food through control of moisture and deterioration by exclusion of microbes (Tang & Zhou, 2024; Martins et al., 2024; Khoo et al., 2023; Ebrahimzade et al., 2022). As an example, coating of cherry tomatoes with starch-chitosan NPs impregnated with thymol NPs tremendously reduced weight loss in tomatoes by up to 6% in storage and also significantly inhibited microbial spoilage and increased antioxidant protection (Yao et al., 2024). Although starch does not have any intrinsic antibacterial properties, it can be used as an active packaging material by incorporating bioactive agents, plant extracts, and essential oils. For example, the natural compounds, including polyphenols, flavonoids, tannins, and alkaloids, interfere with bacterial membranes and hinder microbial growth and therefore increase the shelf life of food products (Dai et al., 2024; Zhu et al., 2023; Manzoor et al., 2023), as well as the food industry by-product extracts, which increase antibacterial activity and provide sustainable, low-cost production (Ounkaew et al., 2018; Wang et al., 2021; Ahari et al., 2025). A critical analysis of polysaccharide-based edible coating showed that starch-based composites, particularly in association with essential oils, proteins, or hydrophobic substances, were viable in reducing respiration rates and delaying the change of color in fruit kept at 4°F over a period of 12 days. All these findings highlight the potential of starch-biomacromolecule blends in improving freshness and attaining significant gains in weight retention, firmness, and shelf life in a real storage scenario.

In food packaging, starch-based films are some of the most valuable categories of biobased packaging material in terms of performance, product-adaptability, processing, and cost (Rompothi et al., 2017), since their crystalline regions provide an oil barrier, and amylose films are more protective than amylopectin films (Huang et al., 2019). Also, starch is another polysaccharide, which demonstrated a good film-forming capacity since starch or starch-related products could create an adhesive bond between similar or dissimilar surfaces (Adeyeye et al., 2019, pp. 137–158). Besides its potent film-forming property, starch is an inexpensive polysaccharide raw material to use in food packaging purposes (Mukherjee et al., 2023). Starch-based films have seen use in the coating of fruits and vegetables, providing breathability, lightness, and transparency, despite failure in mechanical strength and sensitivity to moisture (Abdul Khalil et al., 2018; Kocira et al., 2021). In dairy products, starch-based wraps can serve as moisture and mold blocks in cheese, but they must be reinforced with nanocellulose to provide tensile strength (Biegańska, 2022; Gupta et al., 2022; Gürler et al., 2021). However, it is observed that starch-based films or coatings have high hydrophilicity based on the presence of hydroxyl groups as part of the starch-polymer chains. As a result, starch-based packaging systems are susceptible to moisture transfer as well as deterioration of mechanical properties under moist environments. To address this drawback, starch-based films and coatings with the addition of hydrophobic biopolymers, including chitosan, cellulose, and alginate, as well as oils or waxes based emulsifiers, demonstrate promising potential in the field of packaging dry or refrigerated foods (Yu et al., 2021). Particular commercial uses of starch-based films are Chinese yam starch mixed with gallic acid and chitosan used in pork packaging to preserve foods (Rong et al., 2023), cassava starch mixed with PLA through blown film extrusion has been used in modified atmosphere packaging of capsicum. Corn starch nanocrystal films made with chitosan and cellulose nanocrystals (Ochoa-Yepes et al., 2023; Sganzerla et al., 2020; Díaz-Cruz et al., 2022) and mango seed kernel starch films made with micropottery clay (Admase et al., 2022) show better thermal stability, mechanical characteristics, and biodegradability.

2.6.2 Starch in the pharmaceutical blister packaging and capsule coatings

Pharmaceutical packaging is important in ensuring the safety, versatility, stability, efficacy, protection, and usability of the pharmaceuticals during their entire lifecycle, such as manufacturing and administration to the patients. Polymeric excipients have not only a protective effect, but also have a great impact on the efficiency of drug transport, release kinetics, and overall therapy effect due to the choice of polymeric excipients. Because of the recent developments in research and technological advances, starch has been cited as a potential material that can be used in this field and prove to be environmentally sustainable. This is explained by the fact that starch is biodegradable, safe, and versatile (Sarmah et al., 2023), which discusses the swelling-induced mechanically robust starch-based hydrogel as an agent of controlled release of drugs in wound dressing.

Starch-based hydrogel has been studied extensively to determine the potential of its use in medicine and biology. Although subjected to swelling, starch-agar hydrogels do not lose their mechanical integrity. They allow bending and knotting without breaking, and the water absorption rate is approximately 70. The hydrogels had a controlled delivery of ciprofloxacin between 24 and 120 hours, which is in line with nondiffusion trends that provide long-term therapeutic effects. Hydrogel starch, based on the delivery of medication to the colon in harsh digestive conditions, has demonstrated stability, which enables the delivery of medication in a controlled way. This makes them beneficial to be used in the gastrointestinal field. Its ability to resist physiological stress indicates that starch has potential in drug release formulations as a stable medium to be used in controlled drug release formulations (Kobryń et al., 2020). Starch glutaraldehyde cross-linked hydrogel can be used in conjunction with hydrophilic and hydrophobic pharmaceuticals, and the dynamics of their release can be altered depending on the density of crosslinking. This allows starch-based carriers (SBCs) to be used as biodegradable and easily modifiable vehicles, applicable to longer release, targeted delivery, and various sophisticated pharmaceutical applications. This has been because of the delivery systems that are flexible and controllable in nature (Elango et al., 2023). The starch complexes are very effective in the transportation of bioactive components within the gastrointestinal tract. The starch skeleton is broken down, and bioactive compounds trapped on it are liberated during the digestion process. The percentage of release of these bioactive compounds depends on the type and the composition of the complex that is released. Controlled release mechanism shows that starch-based nanogels are effective in encapsulating curcumin and resveratrol, with the structure of the starch matrix showing a great effect on digestibility and release profiles (Tarazi Riess et al., 2022). Studies in which the starch-based nanogels are involved show that the structure of the starch matrix greatly affects the digestibility and release profiles (Yao et al., 2023). In addition, oral films that are made of starch enable rapid dissolution in the mouth cavity, hence enabling the oral use of nutritional supplements like vitamin C. These films are yet another demonstration of the flexibility of starch with regard to pharmaceutical purposes and drug delivery, and an example of patient-friendly dosage forms. Because of its compatibility with proteins, lipids, and hydrophobic compounds, starch has a dual application as a modulator of drug release kinetics and as a protective packaging material. Pharmaceutical packaging can be used as a channel of transferring critical information, including dosage, possible side effects, storage conditions, and specifications by the manufacturer. The introduction of starch-based material in films and blister packs allows packaging to attain its protective purpose and be sustainable in the environment. In addition, it allows a perfect incorporation of visual elements such as branding, logo, and instructions into the packaging. The increase in the need for pharmaceutical packaging with emphasis on sustainability and the safety of the environment corresponds to the use of starch-based materials. Starch is a material that can display a great level of versatility when it comes to modern pharmaceutical packaging processes, and with biodegradable qualities. This material has been known to be safe as an excipient. The properties enable the delivery of drugs into the target, increase the time of release of drugs, and improve pharmaceutical and bioactive compounds (Yang et al., 2024).

2.6.3 Starch-polymer blends and composites

Starch has been extensively investigated with other polymers as a way of addressing its inherent disadvantages, including weakness of moisture resistance and mechanical strength. Its functional properties can be highly improved by mixing starch with biopolymers or using nanomaterials. Similarly, other research revealed biopolymer blends as a means of enhancing moisture resistance, e.g., starch-polyvinyl alcohol composites (Khoo et al., 2023). Nanoclays or graphene can be incorporated in starch-based nanocomposites, and this may greatly enhance thermal stability and mechanical performance, and find use in applications such as automotive parts (Schutz et al., 2024). Moreover, starch composites with biofiber have been established to be used as lightweight automobile parts, minimizing the carbon emission during the use of the vehicle (Thapliyal et al., 2023). Lignocellulosic fiber-reinforced biocomposites are also used in building materials such as window and door frames, as the material has great mechanical strength, hydrophobicity, and biodegradability. The addition of sodium metasilicate and glycerol in potato starch-hemp shive composites increased compressive strength and mechanical performance. Starch films may be used in food packaging together with polymers, essential oils, polyphenols, or nanomaterials to increase antioxidant, UV-blocking, and mechanical properties. This multifunctionality can be used to maintain food in addition to being used in food packaging systems as intelligent packaging systems within starch-based materials. As an example, pH-sensitive pigments and plant extracts can be used as a freshness indicator in food packaging (Yu et al., 2025; Cheng et al., 2024). New developments have also made use of nanotechnology, electrospinning, and 3D printing to create smart packaging materials made out of starch. As an example, starch-based films together with anthocyanin extract can serve as light-responsive freshness indicators in shrimp and other perishable products (Zhang et al., 2020). In a similar manner, starch-based films with chitosan and carbon dots of the green tea extracts have been shown to be useful in the active packaging of meat, increasing its shelf life (Khan et al., 2023). PS film blended with PLA was shown to be useful in reducing the weight loss of cherry tomatoes, besides keeping vitamin C and organic compounds intact, showing it to be a functional composite packaging material (Zhou et al., 2019). The general polysaccharide-based materials are concentrated on the modification of the intrinsic polysaccharide characteristics or the use of such polymers with other biodegradable polymers or natural resources (Singh et al., 2024). Polystarch films supplemented with hydrophobic biopolymers, emulsifiers, or oils demonstrate great prospects in dry/refrigerated food packaging (Yu et al., 2021). Another important issue is environmental sustainability. The results of the soil burial of starch-nanocellulose films revealed that the frequency of biodegradation of films was 3547% in 30 days, and simultaneously, there was an increase in nutrient content of soil. This proves that starch-based composites may compromise between mechanical and environmental degradation performance and make them suitable to sustain packaging, agricultural, and biomedical applications (Malekzadeh et al., 2024). Another quality measure is WVP. Films made using starch tend to be more permeable than films made using PLA, especially when cast using starch pastes that do not contain interchain crosslinks (Agarwal et al., 2023; Hirpara et al., 2021). Starch-based films and foams can be customized to address particular packaging needs, starting with thin film through multilayered materials, through effective structural adaptation and dispersion (Liang et al., 2024).

2.7 Advantages of starch-based packaging

2.7.1 Environmental benefits

Starch is a remarkable, plant-based material that's not only renewable but also biodegradable, making it a great option for creating bioplastics used in food packaging. We can find it in plenty of natural sources such as wheat, corn, rice, and potatoes, which keep it affordable and easy to come by. Thanks to its ability to form films, its eco-friendliness, and its nontoxic properties, starch-based bioplastics are seen as a sustainable choice compared to traditional petroleum-based plastics. Plus, starch can be easily modified or mixed with other biopolymers and plasticizers to boost its strength, barrier qualities, and water resistance, making it even more suitable for a variety of food packaging needs (Onyeaka et al., 2022).

It is a renewable and biodegradable biopolymer that offers a sustainable alternative to traditional petroleum-based plastics. Unlike those synthetic plastics that come from limited fossil resources, starch is sourced from plentiful agricultural crops, making it an affordable and eco-friendly option. One of its standout features is its biodegradability, which means that starch-based materials break down naturally into harmless by-products, helping to tackle plastic pollution and environmental issues. By swapping out petroleum-based plastics for starch-derived bioplastics, we can cut down on our reliance on fossil fuels while also playing a part in reducing waste, lowering our carbon footprint, and supporting a circular economy.

Starch-based packaging materials, made from natural polysaccharides, are naturally biodegradable. When they end up in the environment, these materials can be easily broken down by microorganisms found in soil and water, such as bacteria and fungi. As they biodegrade, starch transforms into simpler compounds, eventually returning to the ecosystem as water, biomass, and carbon dioxide. This natural decomposition process not only helps to prevent the buildup of stubborn plastic waste in landfills and waterways but also plays a role in recycling nutrients back into the environment. So, starch-based packaging is a great eco-friendly alternative to traditional plastics, helping to cut down on pollution while promoting sustainable waste management and maintaining ecological balance (Onyeaka et al., 2022).

The implementation of starch-based materials in packaging adheres to the principles of sustainability, resource efficiency, stimulating circularity, and environmental responsibility throughout the packaging lifecycle (Pei et al., 2024).

2.7.2 Functional advantages

Starch-based composites are a combination of starch and biodegradable polymers such as polylactic acid (PLA) or natural fibers to improve the properties of starch. These composites help to enhance the strength and flexibility as compared to pure TPS. Starch-based composites play a crucial role in keeping the food items fresh for a longer time as they can hinder the moisture and gases from passing through (Pei et al., 2024).

Resistant starch (RS) has a high amount of amylose, which does not easily bond to water. The amylose chains are, on the whole, linear; they pack densely with one another, forming a dense formation bond through hydrogen bonding. Because they pack so densely, it is harder for water to penetrate therefore RS can absorb less water and swell less than natural starch. That same property also makes RS more recalcitrant to digestive enzymes, explaining its behavior in the body (such as dietary fiber) (Fatima et al., 2024).

2.7.3 Economic and industrial benefits

The combination of natural polymers and plant-derived additives into active food packaging is cost-effective, as these are derived from abundant crops or agricultural by-products.

By embracing a circular economy strategy, the food industry has a fantastic opportunity to cut down on the environmental damage caused by plastic waste. The focus here is on creating packaging that can be reused, recycled, or even composted, which helps to reduce the amount of waste piling up in landfills and the pollution that comes with it. When manufacturers adopt these circular economy principles, they can cater to consumers who are looking for eco-friendly products by developing sustainable food packaging solutions. Plus, it boosts resource efficiency by encouraging the use of renewable materials, lessening our dependence on limited fossil resources, and promoting the repurposing of waste. In the end, these practices pave the way for a sustainable future where economic growth, environmental care, and social well-being go hand in hand (Bagri et al., 2025).

Modified starch is extensively used in films, coatings, and laminates for diverse packaging applications. Because of its versatile nature allow the formulation of packaging material optimized according to the product types and environmental conditions (Pei et al., 2024).

2.7.4 Health and safety benefits

Biodegradable packaging materials play an important role for keeping various food products fresh for longer. This includes everything from raw and cooked foods to processed items such as fruits, vegetables, nuts, dried fruits, milk, dairy, poultry, and seafood. These materials typically use natural polymers and bioactive additives that create protective barriers against moisture, oxygen, and harmful microbes. By slowing down spoilage, oxidation, and microbial growth, biodegradable packaging helps preserve the freshness, nutritional value, and overall quality of food. Plus, these eco-friendly options not only reduce our environmental footprint but also ensure that consumers enjoy safe and high-quality food (Bagri et al., 2025).

Starch is a natural carbohydrate that is safe for human consumption (Pei et al., 2024). Starch-based packaging material does not discharge any kinds of harmful chemicals to the foods, such as conventional plastics (Onyeaka et al., 2022). In pharmaceuticals, starch is mostly used as an excipient due to its cost-effectiveness, natural composition, and safety. Being nontoxic, biocompatible, and biodegradable in nature, it has varied applications, including its role as a glidant, binder, disintegrant, diluent, and lubricant in formulations such as capsules, tablets, and granules. Through physical modification techniques such as microwave and ultrasonic treatments, modified starches have shown notable potential as an excipient in controlled drug release formulations. Ultrasonic treatment of tapioca starch shows its improved functionality in drug delivery systems. As compared to native starch, microwave-assisted modification of arrowroot starch has been found to increase its gel-forming capacity, resulting in improved sustained-release properties (Kapoor et al., 2025). The addition of essential nutrients, vitamins, and bioactive phytochemicals to the biopolymer enhances their nutritional value and promotes consumer health and wellness (Pei et al., 2024).

2.8 Future trends, challenges, and constraints

Starch packaging films are cost-effective, biodegradable, abundant, and ecological, and therefore a promising substitute for petroleum-based plastics. They are more attractive and environmentally friendly as alternatives to plastic trash because of their openness, flexibility, and high barrier properties (Karnwal et al., 2025). Despite all of these advantages, it is still hard to use on a large scale. The mechanical weakness, WVP, and instability of the films restrict their use in industries (Pei et al., 2024).

Cellulosic, starch, and plant oil bioplastics are more environmentally friendly compared to the petroleum-based ones. Their compostability, decomposition, and production lower emissions. The advantages are compensated for by manufacturing and cost problems. The cost associated with manufacturing bioplastics is high because of the high cost of agricultural raw materials, complicated processes, and low production volumes. Neither does it have a petroleum-based plastics infrastructure. Certain bioplastics are almost as strong as regular plastics, though others require modification to meet the requirements of industrial performance. The scientists are studying low-cost feedstock such as agricultural residues and biotechnological technologies with modified proteins and polypeptides to increase production efficiency. These techniques can reduce the costs and competitiveness of bioplastics in the world. Bioplastics cause less pollution and carbon emission compared to petroleum plastics, but their production is costly and makes them unhelpful in their usage (Rahardiyan et al., 2023). An example is that thermoplastic starch becomes moist and stale, and hence less stable and strong when stored. The research studies are meant to decrease the extent of retrogradation and water uptake by enhancing starch compositions with plasticizers and NPs. Nevertheless, the cost of production is still high. Because of the high price of bioplastic packaging, consumers might not use it. It should be multidisciplinary in approach. The chemical engineering, environmental science, microbiology, polymer technology, and chemistry specialists should work in cooperation to seek viable and sustainable solutions. The policymakers and the government officials should also be aggressive in commercializing starch-based and other greener products that can be used in day-to-day life (Niranjana Prabhu & Prashantha, 2018).

New trends are geared toward enhancing the mechanical, barrier, and moisture resistance of starch films. With the help of montmorillonite and cellulose nanofibers, these characteristics have been developed. Food packaging films become more flexible and antibacterial by the addition of starch and biopolymers such as polyvinyl alcohol (PVA) or chitosan. These films are versatile due to advanced active and intelligent packaging. Films have been turned into multipurpose by active and intelligent packaging. Film impregnated with clove oil is antimicrobial, and food quality is checked in real time by pH-sensitive indicators.

Film design and use of agricultural waste is possible using 3D printing and reactive extrusion, which are more sustainable (Karnwal et al., 2025). Other new technologies are electrospinning and force spinning, which generate films with distinct physical, chemical, and biological characteristics (García-Guzmán et al., 2022). The inventions also prolong the shelf life of the food, enhance safety, and minimize packaging waste. A combination of active (oxygen quenchers and antimicrobial films) and intelligent (sensors of freshness and QR codes) technologies is a new opportunity for smart packaging. Biodegradability, food safety, antioxidants, and antimicrobials are offered by edible films created out of biopolymers such as starch, proteins, and lipids. On the whole, these innovations are a massive leap into a safer, smarter, and more sustainable packaging.

2.9 Conclusion

Starch material packaging has come up as a sustainable, biodegradable, and environmentally friendly substitute to the petroleum-based plastics. This has increased value due to its dual benefit, having a reduced environmental load and still being functional in a packaging application. Extraction plays a vital role in the determination of starch quality, which determines the morphology of the grains, crystallinity, chemical integrity, and the overall physicochemical behavior. Methods such as enzymatic, alkaline, and ultrasonic extraction have a different impact on purity and structure that subsequently influence the capacity to form films, mechanical strength, and biodegradability.

The morphological and structural characteristics of the starch grains (the size, shape, and crystalline structure) vary according to the botanical source, which directly affect gelatinization, retrogradation, and rheological properties. They are the parameters that define the processing behavior and functional stability of the starch-based films. Besides, the amylose-to-amyloprotein ratio greatly influences mechanical and barrier properties that determine the strength, flexibility, and moisture sensitivity of the film. Their extraction and processing conditions are thus to be optimized so as to offer an improved performance and functionality.

Natural starches, though abundant in nature, have some demerits such as being brittle, sensitive to water, and have low mechanical stability. To overcome these drawbacks, several strategies of modification have been developed: physical, chemical, and biological. Crystallinity, gelatinization temperature, and uniformity of the film are enhanced through physical modification such as heat-moisture treatment and annealing. The biological processes, such as chemical modifications, such as acetylation, oxidation, and crosslinking, can be used to enhance heat stability, tensile strength, and hydrophobicity; in other words, they can be used to provide eco-friendly alternatives to enhance thermoplasticity and filming properties. The combination of the above-mentioned methods makes starch a flexible and versatile packaging material in the contemporary world.

Starch-based packaging has economic and health benefits in addition to environmental sustainability. Starch greenery-wise lessens the use of fossil resources and biodegradability, which mitigates plastic pollution. Its renewable agricultural source helps in supporting its economic viability and scalability. Health and safety-wise, starch is nontoxic, biocompatible, and can be used in food and pharmaceutical industries, which protects the consumer and the quality of the product.

In spite of these advantages, there are still difficulties in reaching industrial-level operation because of the high cost of production, low mechanical strength, and sensitivity to moisture. Nonetheless, the recent developments, including the use of NPs, blending of polymers, and active and intelligent packaging systems, have significantly improved the performance. Additional innovations such as 3D printing, electrospinning, and reactive extrusion also allow the making of tailored and high-performance starch-based films.

To conclude, starch packaging is a potential move toward the development of an eco-friendly material. The interdisciplinary studies combined with materials science, polymer engineering, and biotechnology are crucial to achieve the optimization of extraction, modification, and processing. As the materials based on starch continue to be innovated and developed in cooperation, the materials with the ability to transform the packaging technology include the solution of greater safety, intelligence, and environmental efficiency, which will be grounded in the principles of sustainability and a circular economy.

References

Abdul Khalil, H., Banerjee, A., Saurabh, C. K., Tye, Y., Suriani, A., Mohamed, A., et al. (2018). Biodegradable films for fruits and vegetables packaging application: Preparation and properties. *Food Engineering Reviews, 10*(3), 139–153.

Aburto, J., Hamaili, H., Alric, I., Borredon, E., & Gaset, A. (1999). Advances in the solvent-free synthesis of fatty esters of starch. *Les Colloques de l'INRA*, 81–85.

Adeyeye, O., Sadiku, E. R., Babu Reddy, A., Ndamase, A. S., Makgatho, G., Sellamuthu, P. S., et al. (2019). *The use of biopolymers in food packaging. Green biopolymers and their nanocomposites* (pp. 137–158). Springer.

Admase, A. T., Sendekie, Z. B., & Alene, A. N. (2022). Biodegradable film from mango seed kernel starch using pottery clay as filler. *Journal of Polymers and the Environment, 30*(8), 3431–3446.

Agarwal, S., Singhal, S., Godiya, C. B., & Kumar, S. (2023). Prospects and applications of starch based biopolymers. *International Journal of Environmental Analytical Chemistry, 103*(18), 6907–6926.

Ahari, H., Kalateh-Seifari, F., & Yousefi, S. (2025). Antimicrobial activity of chitosan/corn starch film incorporated with starch nanocrystals/nettle essential oil nanoemulsion for Eleutheronema tetradactylum fillet preservation. *Food Chemistry X, 25*, 102085.

Akhter, M. J., Sarkar, S., Rayhanujjaman, M., Kabir, M. S., & Hosain, M. M. (2024). Characterization of mango seed kernel starch: Extraction and analysis. *Food Chemistry Advances, 5*, 100806.

ASTM D. (2018). *3763–18; Standard Test method for high speed puncture properties of plastics using load and displacement sensors*. West Conshohocken, PA, USA: ASTM international.

Averous, L., & Boquillon, N. (2004). Biocomposites based on plasticized starch: Thermal and mechanical behaviours. *Carbohydrate polymers, 56*(2), 111–122.

Bagri, F., Priyadarshi, R., Pircheraghi, G., Imani, M., & Rhim, J.-W. (2025). Designing entirely biodegradable active food packaging using natural biopolymers and plant-derived additives: A cradle-to-cradle approach. *Materials Today Communications*, 113549.

Bangar, S. P., Scott Whiteside, W., Suri, S., Barua, S., & Phimolsiripol, Y. (2023). Native and modified biodegradable starch-based packaging for shelf-life extension and safety of fruits/vegetables. *International Journal of Food Science and Technology, 58*(2), 862–870.

Bangar, S. P., Whiteside, W. S., Ashogbon, A. O., & Kumar, M. (2021). Recent advances in thermoplastic starches for food packaging: A review. *Food Packaging and Shelf Life, 30*, 100743.

Barichello, V., Yada, R. Y., & Coffin, R. H. (1991). Starch properties of various potato (Solanum tuberosum L) Cultivars susceptible and resistant to low-temperature sweetening. *Journal of the Science of Food and Agriculture, 56*(3), 385–397.

Bello-Pérez, L., Alvarez-Ramirez, J., & Dhital, S. (2022). *Starch and starchy food products: Improving human health*. CRC Press.

Benesi, I. R., Labuschagne, M. T., Dixon, A. G., & Mahungu, N. M. (2004). Stability of native starch quality parameters, starch extraction and root dry matter of cassava genotypes in different environments. *Journal of the Science of Food and Agriculture, 84*(11), 1381–1388.

Bersaneti, G. T., Prudencio, S. H., Mali, S., & Celligoi, M. A. P. C. (2021). Assessment of a new edible film biodegradable based on starch-nystose to increase quality and the shelf life of blackberries. *Food Bioscience, 42*, 101173.

Bidari, R., Abdillah, A. A., Ponce, R. A. B., & Charles, A. L. (2023). Characterization of biodegradable films made from taro peel (Colocasia esculenta) starch. *Polymers, 15*(2), 338.

Biegańska, M. (2022). *Packaging of dairy products: Emerging strategies. Food packaging: The smarter way* (pp. 127–164). Springer.

Brunner, G. (2005). Supercritical fluids: Technology and application to food processing. *Journal of Food Engineering, 67*(1–2), 21–33.

Buleon, A., Colonna, P., Planchot, V., & Ball, S. (1998). Starch granules: Structure and biosynthesis. *International Journal of Biological Macromolecules, 23*(2), 85–112.

Carmona-García, R., Bello-Pérez, L., Aguirre-Cruz, A., Aparicio-Saguilán, A., Hernández-Torres, J., & Alvarez-Ramirez, J. (2016). Effect of ultrasonic treatment on the morphological, physicochemical, functional, and rheological properties of starches with different granule size. *Starch-Stärke, 68*(9–10), 972–979.

Ceballos, R. L., Ochoa-Yepes, O., Goyanes, S., Bernal, C., & Famá, L. (2020). Effect of yerba mate extract on the performance of starch films obtained by extrusion and compression molding as active and smart packaging. *Carbohydrate polymers, 244*, 116495.

Cheng, L., Zhang, M., Bai, H., Yang, F., Zhang, X., Ren, D., et al. (2024). Combined ultraviolet-C radiation and L-cysteine treatment improves the post-harvest quality and volatile compounds of edible Lanzhou lily bulbs (Lilium davidii var. unicolor) by regulating reactive oxygen species metabolism. *Food Chemistry X, 24*, 101813.

D, A. (Ed.). (2013). *Standard test method for tear resistance (graves tear) of plastic film and sheeting.* American Society for Testing and Materials.

Díaz-Cruz, C. A., Caicedo, C., Jiménez-Regalado, E. J., Díaz de León, R., López-González, R., & Aguirre-Loredo, R. Y. (2022). Evaluation of the antimicrobial, thermal, mechanical, and barrier properties of corn starch–chitosan biodegradable films reinforced with cellulose nanocrystals. *Polymers, 14*(11), 2166.

Dai, X., Dong, F., Dong, Z., Bai, Z., & Mao, L. (2024). Enhanced antibacterial and antioxidant activities of chlorogenic acid loaded sweet whey/starch active films for edible food packaging. *Lwt, 199*, 116118.

Das, D., Jha, S., & Kumar, K. J. (2015). Isolation and release characteristics of starch from the rhizome of Indian Palo. *International Journal of Biological Macromolecules, 72*, 341–346.

Dega, V., & Barbhai, M. D. (2023). Exploring the underutilized novel foods and starches for formulation of low glycemic therapeutic foods: A review. *Frontiers in Nutrition, 10*, 1162462.

Dorantes-Fuertes, M.-G., López-Méndez, M. C., Martínez-Castellanos, G., Meléndez-Armenta, R.Á., & Jiménez-Martínez, H.-E. (2024). Starch extraction methods in tubers and roots: A systematic review. *Agronomy, 14*(4), 865.

Ebrahimzade, I., Ebrahimi-Nik, M., Rohani, A., & Tedesco, S. (2022). Towards monitoring biodegradation of starch-based bioplastic in anaerobic condition: Finding a proper kinetic model. *Bioresource Technology, 347*, 126661.

Egodage, R. (2019). *The impact of annealing, acetylation, and dual modification on the structure and physicochemical properties of waxy starches.* Memorial University of Newfoundland.

Elango, B., Shirley, C., Okram, G. S., Ramesh, T., Seralathan, K.-K., & Mathanmohun, M. (2023). Structural diversity, functional versatility and applications in industrial, environmental and biomedical sciences of polysaccharides and its derivatives–A review. *International Journal of Biological Macromolecules, 250*, 126193.

Fan, C., Cheng, L., Hong, Y., Li, Z., Li, C., Ban, X., et al. (2024). Study on the gelatinization and digestive characteristics of wheat starch and potato starch under low moisture conditions. *International Journal of Biological Macromolecules, 269*, 132192.

Faramarzi, S., Anzabi, Y., & Jafarizadeh-Malmiri, H. (2020). Nanobiotechnology approach in intracellular selenium nanoparticle synthesis using Saccharomyces cerevisiae—fabrication and characterization. *Archives of Microbiology, 202*(5), 1203–1209.

Fatima, S., Khan, M. R., Ahmad, I., & Sadiq, M. B. (2024). Recent advances in modified starch based biodegradable food packaging: A review. *Heliyon, 10*(6).

Gürler, N., Paşa, S., & Temel, H. (2021). Silane doped biodegradable starch-PLA bilayer films for food packaging applications: Mechanical, thermal, barrier and biodegradability properties. *Journal of the Taiwan Institute of Chemical Engineers, 123*, 261–271.

Gamage, A., Thiviya, P., Mani, S., Ponnusamy, P. G., Manamperi, A., Evon, P., et al. (2022). Environmental properties and applications of biodegradable starch-based nanocomposites. *Polymers, 14*(21), 4578.

Garavito, J., Peña-Venegas, C. P., & Castellanos, D. A. (2024). Production of starch-based flexible food packaging in developing countries: Analysis of the processes, challenges, and requirements. *Foods, 13*(24), 4096.

García-Guzmán, L., Cabrera-Barjas, G., Soria-Hernández, C. G., Castaño, J., Guadarrama-Lezama, A. Y., & Rodríguez Llamazares, S. (2022). Progress in starch-based materials for food packaging applications. *Polysaccharides, 3*(1), 136–177.

González-López, M. E., Calva-Estrada, S. J., Gradilla-Hernández, M. S., & Barajas-Álvarez, P. (2023). Current trends in biopolymers for food packaging: A review. *Frontiers in Sustainable Food Systems, 7*, 1225371.

Gupta, V., Biswas, D., & Roy, S. (2022). A comprehensive review of biodegradable polymer-based films and coatings and their food packaging applications. *Materials, 15* (17), 5899.

Han, Z., Zeng, X-a, Zhang, B-s, & Yu, S-j (2009). Effects of pulsed electric fields (PEF) treatment on the properties of corn starch. *Journal of Food Engineering, 93*(3), 318–323.

Herrero, M., Cifuentes, A., & Ibañez, E. (2006). Sub-and supercritical fluid extraction of functional ingredients from different natural sources: Plants, food-by-products, algae and microalgae: A review. *Food Chemistry, 98*(1), 136–148.

Hirpara, N. J., Dabhi, M., & Rathod, P. (2021). Development of potato starch based biodegradable packaging film. *Journal of Food Processing & Technology, 12*, 529–541.

Hu, X.-P., Zhang, B., Jin, Z.-Y., Xu, X.-M., & Chen, H.-Q. (2017). Effect of high hydrostatic pressure and retrogradation treatments on structural and physicochemical properties of waxy wheat starch. *Food Chemistry, 232*, 560–565.

Huang, T., Qian, Y., Wei, J., & Zhou, C. (2019). Polymeric antimicrobial food packaging and its applications. *Polymers, 11*(3), 560.

Iida, Y., Tuziuti, T., Yasui, K., Towata, A., & Kozuka, T. (2008). Control of viscosity in starch and polysaccharide solutions with ultrasound after gelatinization. *Innovative Food Science & Emerging Technologies, 9*(2), 140–146.

Ivanovic, J., Milovanovic, S., & Zizovic, I. (2016). Utilization of supercritical CO2 as a processing aid in setting functionality of starch-based materials. *Starch-Stärke, 68*(9–10), 821–833.

Kapoor, D. U., Pareek, A., Sharma, M., Prajapati, B. G., Suttiruengwong, S., & Sriamornsak, P. (2025). Exploring starch-based excipients in pharmaceutical formulations: Versatile applications and future perspectives. *European Journal of Pharmaceutics and Biopharmaceutics*, 114727.

Karnwal, A., Rauf, A., Jassim, A. Y., Selvaraj, M., Al-Tawaha, A. R. M. S., Kashyap, P., et al. (2025). Advanced starch-based films for food packaging: Innovations in sustainability and functional properties. *Food Chemistry X*, 102662.

Kaur, B., Ariffin, F., Bhat, R., & Karim, A. A. (2012). Progress in starch modification in the last decade. *Food Hydrocolloids, 26*(2), 398–404.

Kebakile, M. M. (2009). *Sorghum dry-milling processes and their influence on meal and porridge quality*. University of Pretoria.

Khan, A., Ezati, P., & Rhim, J.-W. (2023). Chitosan/starch-based active packaging film with N, P-doped carbon dots for meat packaging. *ACS Applied Bio Materials, 6*(3), 1294–1305.

Khoo, P. S., Ilyas, R., Uda, M., Hassan, S. A., Nordin, A., Norfarhana, A., et al. (2023). Starch-based polymer materials as advanced adsorbents for sustainable water treatment: Current status, challenges, and future perspectives. *Polymers, 15*(14), 3114.

Kobryń, J., Zięba, T., Sowa, S. K., & Musiał, W. (2020). Influence of acetylated annealed starch on the release of β-escin from the anionic and non-ionic hydrophilic gels. *Pharmaceutics, 12*(1), 84.

Kocira, A., Kozłowicz, K., Panasiewicz, K., Staniak, M., Szpunar-Krok, E., & Hortyńska, P. (2021). Polysaccharides as edible films and coatings: Characteristics and influence on fruit and vegetable quality—A review. *Agronomy, 11*(5), 813.

Kringel, D. H., Dias, A. R. G., Zavareze, E. R., & Gandra, E. A. (2020). Fruit wastes as promising sources of starch: Extraction, properties, and applications. *Starch-Stärke, 72* (3–4), 1900200.

La, F. C. I., de Souza, A. T., Tadini, C. C., & Augusto, P. E. D. (2019). Ozonation of cassava starch to produce biodegradable films. *International Journal of Biological Macromolecules, 141*, 713–720.

Lauer, M. K., & Smith, R. C. (2020). Recent advances in starch-based films toward food packaging applications: Physicochemical, mechanical, and functional properties. *Comprehensive Reviews in Food Science and Food Safety, 19*(6), 3031–3083.

Li, Q., Wu, Q.-Y., Jiang, W., Qian, J.-Y., Zhang, L., Wu, M., et al. (2019). Effect of pulsed electric field on structural properties and digestibility of starches with different crystalline type in solid state. *Carbohydrate Polymers, 207*, 362–370.

Liang, X., Chen, L., McClements, D. J., Zhao, J., Zhou, X., Qiu, C., et al. (2024). Starch-guest complexes interactions: Molecular mechanisms, effects on starch and functionality. *Critical Reviews in Food Science and Nutrition, 64*(21), 7550–7562.

Liu, Z., Chen, L., Qu, L., Zhang, R., Qin, Z., Zhang, H., et al. (2024). Cross-linked poly (ester urethane)/starch composite films with high starch content as sustainable food-packaging materials: Influence of cross-link density. *International Journal of Biological Macromolecules, 256*, 128441.

Liu, D., Wu, Q., Chen, H., & Chang, P. R. (2009). Transitional properties of starch colloid with particle size reduction from micro-to nanometer. *Journal of Colloid and Interface Science, 339*(1), 117–124.

Lopes, H. S. M., da Costa, F. A. T., Komatsu, D., Dufresne, A., & de Menezes, A. J. (2025). Gelatinized cassava starch obtained via low molar ratio hydroxypropylation reaction. *ACS Omega, 10*(12), 12543–12552.

Lopez-Rubio, A., Flanagan, B. M., Gilbert, E. P., & Gidley, M. J. (2008). A novel approach for calculating starch crystallinity and its correlation with double helix content: A combined XRD and NMR study. *Biopolymers: Original Research on Biomolecules, 89*(9), 761–768.

Lumdubwong, N., & Seib, P. (2000). Rice starch isolation by alkaline protease digestion of wet-milled rice flour. *Journal of Cereal Science, 31*(1), 63–74.

Malekzadeh, E., Tatari, A., & Dehghani Firouzabadi, M. (2024). Effects of biodegradation of starch-nanocellulose films incorporated with black tea extract on soil quality. *Scientific Reports, 14*(1), 18817.

Manzoor, A., Yousuf, B., Pandith, J. A., & Ahmad, S. (2023). Plant-derived active substances incorporated as antioxidant, antibacterial or antifungal components in coatings/films for food packaging applications. *Food Bioscience, 53*, 102717.

Martins, P., Brito-Pereira, R., Ribeiro, S., Lanceros-Mendez, S., & Ribeiro, C. (2024). Magnetoelectrics for biomedical applications: 130 years later, bridging materials, energy, and life. *Nano Energy, 126*, 109569.

Medina-Jaramillo, C., Ochoa-Yepes, O., Bernal, C., & Famá, L. (2017). Active and smart biodegradable packaging based on starch and natural extracts. *Carbohydrate polymers, 176*, 187–194.

Moorthy, S. N. (2002). Physicochemical and functional properties of tropical tuber starches: A review. *Starch-Stärke, 54*(12), 559–592.

Mukherjee, C., Varghese, D., Krishna, J., Boominathan, T., Rakeshkumar, R., Dineshkumar, S., et al. (2023). Recent advances in biodegradable polymers–properties, applications and future prospects. *European Polymer Journal, 192*, 112068.

Nawaz, H., Waheed, R., Nawaz, M., & Shahwar, D. (2020). *Physical and chemical modifications in starch structure and reactivity. Chemical properties of starch*. IntechOpen.

Ning, J., Fu, B., Tang, X., Hao, Y., Zhang, Y., & Wang, X. (2025). Starch retrogradation in starch-based foods: Mechanisms, influencing factors, and mitigation strategies. *International Journal of Biological Macromolecules*, 140354.

Niranjana Prabhu, T., & Prashantha, K. (2018). A review on present status and future challenges of starch based polymer films and their composites in food packaging applications. *Polymer Composites, 39*(7), 2499–2522.
Nurul, M., Azemi, B. M., & Manan, D. (1999). Rheological behaviour of sago (Metroxylon sagu) starch paste. *Food Chemistry, 64*(4), 501–505.
Ochoa-Yepes, O., Ceballos, R. L., & Famá, L. (2023). Effect of wheat and oat bran on the physicochemical properties of edible starch-based films obtained by extrusion for food packaging applications. *Starch-Stärke, 75*(1–2), 2200087.
Onyeaka, H., Obileke, K., Makaka, G., & Nwokolo, N. (2022). Current research and applications of starch-based biodegradable films for food packaging. *Polymers, 14*(6), 1126.
Otegbayo, B., Oguniyan, D., & Akinwumi, O. (2014). Physicochemical and functional characterization of yam starch for potential industrial applications. *Starch-Stärke, 66*(3–4), 235–250.
Ounkaew, A., Kasemsiri, P., Kamwilaisak, K., Saengprachatanarug, K., Mongkolthanaruk, W., Souvanh, M., et al. (2018). Polyvinyl alcohol (PVA)/starch bioactive packaging film enriched with antioxidants from spent coffee ground and citric acid. *Journal of Polymers and the Environment, 26*(9), 3762–3772.
Pei, J., Palanisamy, C. P., Srinivasan, G. P., Panagal, M., Kumar, S. S. D., & Mironescu, M. (2024). A comprehensive review on starch-based sustainable edible films loaded with bioactive components for food packaging. *International Journal of Biological Macromolecules, 274*, 133332.
Perez, E., Bahnassey, Y., & Breene, W. (1993). A simple laboratory scale method for isolation of amaranth starch. *Starch-Stärke, 45*(6), 211–214.
Pinto C.C. Unraveling the Semicrystalline Structure of Different Types of Starch (A, B, C and V) using X-Ray Diffraction and Structural Refinement. 2025.
Rahardiyan, D., Moko, E. M., Tan, J. S., & Lee, C. K. (2023). Thermoplastic starch (TPS) bioplastic, the green solution for single-use petroleum plastic food packaging–A review. *Enzyme and Microbial Technology, 168*, 110260.
Rashwan, A. K., Younis, H. A., Abdelshafy, A. M., Osman, A. I., Eletmany, M. R., Hafouda, M. A., et al. (2024). Plant starch extraction, modification, and green applications: A review. *Environmental Chemistry Letters, 22*(5), 2483–2530.
Ren, L., Jiang, M., Wang, L., Zhou, J., & Tong, J. (2012). A method for improving dispersion of starch nanocrystals in water through crosslinking modification with sodium hexametaphosphate. *Carbohydrate Polymers, 87*(2), 1874–1876.
Rompothi, O., Pradipasena, P., Tananuwong, K., Somwangthanaroj, A., & Janjarasskul, T. (2017). Development of non-water soluble, ductile mung bean starch based edible film with oxygen barrier and heat sealability. *Carbohydrate polymers, 157*, 748–756.
Rong, L., Ji, X., Shen, M., Chen, X., Qi, X., Li, Y., et al. (2023). Characterization of gallic acid-chinese yam starch biodegradable film incorporated with chitosan for potential use in pork preservation. *Food Research International, 164*, 112331.
Rostagno, M. A., Villares, A., Guillamón, E., García-Lafuente, A., & Martínez, J. A. (2009). Sample preparation for the analysis of isoflavones from soybeans and soy foods. *Journal of Chromatography A, 1216*(1), 2–29.
Santhosh, R., Ahmed, J., Thakur, R., & Sarkar, P. (2024). Starch-based edible packaging: Rheological, thermal, mechanical, microstructural, and barrier properties–a review. *Sustainable Food Technology, 2*(2), 307–330.
Sarmah, D., Borah, M., Mandal, M., & Karak, N. (2023). Swelling induced mechanically tough starch–agar based hydrogel as a control release drug vehicle for wound dressing applications. *Journal of Materials Chemistry B, 11*(13), 2927–2936.
Schutz, G. F., de Ávila Gonçalves, S., Alves, R. M. V., & Vieira, R. P. (2024). A review of starch-based biocomposites reinforced with plant fibers. *International Journal of Biological Macromolecules, 261*, 129916.

Sganzerla, W. G., Rosa, G. B., Ferreira, A. L. A., da Rosa, C. G., Beling, P. C., Xavier, L. O., et al. (2020). Bioactive food packaging based on starch, citric pectin and functionalized with Acca sellowiana waste by-product: Characterization and application in the postharvest conservation of apple. *International Journal of Biological Macromolecules, 147*, 295–303.

Shao, Y., Jiao, R., Wu, Y., Xu, F., Li, Y., Jiang, Q., et al. (2023). Physicochemical and functional properties of the protein–starch interaction in Chinese yam. *Food Science and Nutrition, 11*(3), 1499–1506.

Shittu, T. A., Alimi, B. A., Wahab, B., Sanni, L. O., & Abass, A. B. (2016). Cassava flour and starch: Processing technology and utilization. *Tropical roots and tubers: Production, processing and technology*, 415–450.

Singh, N., Kaur, L., Sandhu, K. S., Kaur, J., & Nishinari, K. (2006). Relationships between physicochemical, morphological, thermal, rheological properties of rice starches. *Food Hydrocolloids, 20*(4), 532–542.

Singh, P., Pandey, V. K., Singh, R., Singh, K., Dash, K. K., & Malik, S. (2024). Unveiling the potential of starch-blended biodegradable polymers for substantializing the eco-friendly innovations. *Journal of Agriculture and Food Research, 15*, 101065.

Singh, N., Singh, J., Kaur, L., Sodhi, N. S., & Gill, B. S. (2003). Morphological, thermal and rheological properties of starches from different botanical sources. *Food Chemistry, 81*(2), 219–231.

Singh, J., Singh, N., & Saxena, S. (2002). Effect of fatty acids on the rheological properties of corn and potato starch. *Journal of Food Engineering, 52*(1), 9–16.

Sobeih, M. O., Sawalha, S., Hamed, R., Ali, F., & Kim, M. P. (2025). Starch-Derived bioplastics: Pioneering sustainable solutions for industrial use. *Materials, 18*(8), 1762.

Svegmark, K., & Hermansson, A.-M. (1993). Microstructure and rheological properties of composites of potato starch granules and amylose: A comparison of observed and predicted structures. *Food Structure, 12*(2), 6.

Tang, K. H. D., & Zhou, J. (2024). Ecotoxicity of Biodegradable Microplastics and Bio-based Microplastics: A Review of in vitro and in vivo studies. *Environmental Management*, 1–17.

Tarazi Riess, H., Shani Levi, C., & Lesmes, U. (2022). Inclusion of phenolic bioactives in high amylose corn starch for gastro-intestinal delivery. *Frontiers in Nutrition, 9*, 981408.

Tester, R. F., & Morrison, W. R. (1990). Swelling and gelatinization of cereal starches. I. Effects of amylopectin, amylose, and lipids. *Cereal Chemistry, 67*(6), 551–557.

Thapliyal, D., Verma, S., Sen, P., Kumar, R., Thakur, A., Tiwari, A. K., et al. (2023). Natural fibers composites: Origin, importance, consumption pattern, and challenges. *Journal of Composites Science, 7*(12), 506.

Vanier, N. L., El Halal, S. L. M., Dias, A. R. G., & da Rosa Zavareze, E. (2017). Molecular structure, functionality and applications of oxidized starches: A review. *Food Chemistry, 221*, 1546–1559.

Wang, W., Zhang, Y., Yang, Z., & He, Q. (2021). Effects of incorporation with clove (Eugenia caryophyllata) essential oil (CEO) on overall performance of chitosan as active coating. *International Journal of Biological Macromolecules, 166*, 578–586.

Wypij, M., Trzcińska-Wencel, J., Golińska, P., Avila-Quezada, G. D., Ingle, A. P., & Rai, M. (2023). The strategic applications of natural polymer nanocomposites in food packaging and agriculture: Chances, challenges, and consumers' perception. *Frontiers in Chemistry, 10*, 1106230.

Xie, F., Pollet, E., Halley, P. J., & Avérous, L. (2013). Starch-based nano-biocomposites. *Progress in Polymer Science, 38*(10–11), 1590–1628.

Xiong, Z., Ma, S., Fan, L., Tang, Z., Zhang, R., Na, H., et al. (2014). Surface hydrophobic modification of starch with bio-based epoxy resins to fabricate high-performance polylactide composite materials. *Composites Science and Technology, 94*, 16–22.

Yadav, A., Kumar, N., Upadhyay, A., & Pratibha, A. R. K. (2023). Edible packaging from fruit processing waste: A comprehensive review. *Food Reviews International, 39*(4), 2075–2106.

Yang, Z., Xie, F., & Cai, J. (2024). Advances in the delivery of bioactive compounds using starch-based nanogels: Formation mechanism, contemporary challenges, and prospects. *Journal of Agriculture and Food Research, 16*, 101142.

Yao, Y., Cheng, G., & Xie, X. (2024). Design and information interaction study of bio-based materials in the packaging field. *Chemical Engineering Journal Advances, 20*, 100676.

Yao, T., Sui, Z., & Janaswamy, S. (2023). Complexing curcumin and resveratrol in the starch crystalline network alters in vitro starch digestion: Towards developing healthy food materials. *Food Chemistry, 425*, 136471.

Yu, X., Chen, L., Jin, Z., & Jiao, A. (2021). Research progress of starch-based biodegradable materials: A review. *Journal of Materials Science, 56*(19), 11187–11208.

Yu, K., Yang, L., Zhang, S., Zhang, N., Zhu, D., He, Y., et al. (2025). Tough, antibacterial, antioxidant, antifogging and washable chitosan/nanocellulose-based edible coatings for grape preservation. *Food Chemistry, 468*, 142513.

Zhang, K., Huang, T.-S., Yan, H., Hu, X., & Ren, T. (2020). Novel pH-sensitive films based on starch/polyvinyl alcohol and food anthocyanins as a visual indicator of shrimp deterioration. *International Journal of Biological Macromolecules, 145*, 768–776.

Zhang, H.-F., Yang, X.-H., & Wang, Y. (2011). Microwave assisted extraction of secondary metabolites from plants: Current status and future directions. *Trends in Food Science & Technology, 22*(12), 672–688.

Zhang, L., Zhao, J., Li, F., Jiao, X., Zhang, Y., Yang, B., et al. (2024). Insight to starch retrogradation through fine structure models: A review. *International Journal of Biological Macromolecules, 273*, 132765.

Zhiguang, C., Haixia, Z., Min, C., Fayong, G., & Jing, L. (2025). The fine structure of starch: A review. *Npj Science of Food, 9*(1), 50.

Zhou, X., Yang, R., Wang, B., & Chen, K. (2019). Development and characterization of bilayer films based on pea starch/polylactic acid and use in the cherry tomatoes packaging. *Carbohydrate Polymers, 222*, 114912.

Zhu, L., Luo, H., Shi, Z.-W., Lin, C-q, & Chen, J. (2023). Preparation, characterization, and antibacterial effect of bio-based modified starch films. *Food Chemistry X, 17*, 100602.

CHAPTER THREE

Emerging fabrication process of starch-based smart packaging

Valéria Cedeño Fierro[1], Wilson Daniel Caicedo Chacon[2], Juan Felipe Grisales Mejía[3], Gabriel Coelho Leandro[4], Germán Ayala Valencia[4]
[1]Facultad de Ingeniería de Administración, Universidad Nacional de Colombia, Palmira, Colombia
[2]Ingeniería de Alimentos, Fundación Universitaria Agraria de Colombia, Bogotá, Colombia
[3]Chemistry Department, Laboratory of Advanced Analytical Techniques in Natural Products (LATNAP), Universidad de los Andes, Bogotá, Colombia
[4]Department of Chemical and Food Engineering, Federal University of Santa Catarina, Florianópolis, Brazil

3.1 Introduction

Starch is a natural polymer broadly used to manufacture films because it is abundant, cheap, renewable, biocompatible, and biodegradable. Furthermore, this biopolymer has excellent film-forming properties, and the resulting films after drying could be used as packaging materials (Karnwal et al., 2025; Valencia & Sobral, 2018).

Starch becomes in the form of granules in nature, being composed of approximately 60%–85% of amylopectin and 15%–40% amylose, in which amylopectin has a linear (1 → 4)-linked α-glucan with many ramifications by means α-(1 → 6) branch points, while amylose has predominantly a linear (1 → 4)-linked α-glucan (Valencia et al., 2018).

Starch began to be studied in the early 1980s as a potential alternative to manufacture packaging materials that will reduce environmental pollution (Cheng et al., 2021). In this way, several research studied have been carried out to develop starch-based packaging materials by means of conventional techniques at laboratory scales such as casting and tape casting, as well as technologies at industrial scale such as thermo-compression, extrusion, and injection (Khanoonkon et al., 2024; Valencia et al., 2018). In general, starch-based materials are water-sensitive and have poor mechanical and barrier properties to water vapor, oxygen, and carbon dioxide (Thakur et al., 2019). Recent technologies aimed to use electrospinning, 3D-printing, and reactive extrusion to develop starch materials with improved physicochemical properties (Cheng et al., 2021).

Starches in Smart Packaging
ISBN: 978-0-443-45396-0
https://doi.org/10.1016/B978-0-443-45396-0.00007-4

Advanced packaging materials with smart activity are a recent alternative to improve consumer communication. Smart packaging refers to material systems that incorporate indicator, sensing, or communication technologies to monitor and convey information about the condition of the packaged food (Kamboj et al., 2025). In recent years, starch-based films have been used as substrate of smart labels to be used in packaging applications (Gaviria et al., 2021; Koop et al., 2022; Leandro, Laroque, Carciofi, et al., 2024).

The present chapter aims to systematically review the use of starches to manufacture smart materials, focusing in analyze the main chemical compounds used to impart smart properties, as well as to compare conventional and emerging technologies used to manufacture these materials and their food packaging applications.

3.2 Chemical compounds used to manufacture starch-based smart packaging

Among the current strategies to reduce plastic consumption, starch-based packaging materials have emerged as a promising and competitive alternative to conventional plastics. These biopolymeric systems offer multiple advantages that have gained increasing attention in recent years. The continuous reformulation and optimization of these materials represent an active area of research focused on enhancing their physical, chemical, and functional characteristics for potential industrial applications. To achieve this, various chemical compounds have been incorporated (Fig. 3.1), such as plasticizers to improve flexibility, functional additives that impart value-added properties like antioxidant or antimicrobial activity through plant extracts and essential oils, and inorganic particles with antibacterial functionality. Despite their advantages, starch-based materials are inherently brittle, mechanically weak, and highly sensitive to moisture. Therefore, different additives and reinforcing agents have been incorporated to overcome these limitations and develop smart starch-based packaging with improved mechanical performance and multifunctional capabilities.

3.2.1 Mechanical and functional property enhancers

A wide variety of compounds have been incorporated into starch-based packaging formulations to improve their mechanical, optical, and biological properties, including antimicrobial and antioxidant activities. Among these, plasticizers such as glycerol and sorbitol increase flexibility and elongation at

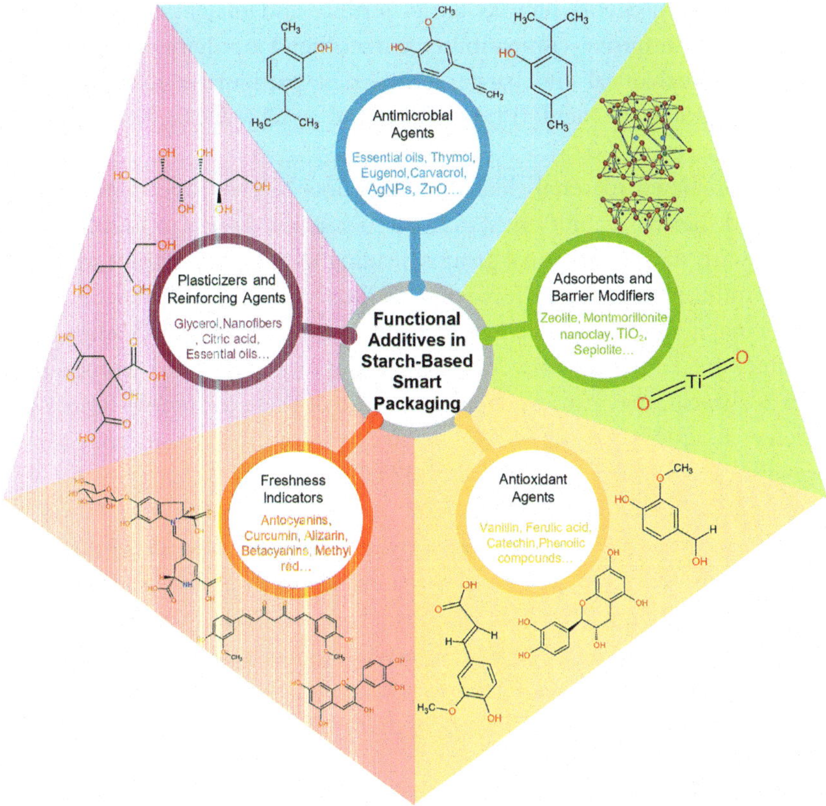

Figure 3.1 Representative molecules and functional roles in starch-based smart packaging systems.

break (EAB). Blending starch with polyvinyl alcohol (PVA) has also proven effective for enhancing film integrity, as this combination promotes hydrogen bonding and a more stable polymeric network (Wang, Zhao, et al., 2024).

Essential oils, besides their well-known biological activities, have been reported to improve elongation and flexibility in starch-based films. Similarly, the incorporation of natural fibers at different scales, from micro to nanostructures, has been extensively studied due to their reinforcing effects. These fibers enhance tensile strength, thermal stability, hydrophobicity, and barrier properties through strong interfacial adhesion and hydrogen bonding between the hydroxyl groups of starch and the fibers (Liu et al., 2025; Nazrin et al., 2025; Nordin et al., 2024).

Crosslinking agents such as citric, ferulic, and maleic acids have also been used to improve compatibility with other biopolymers, resulting in enhanced mechanical and barrier properties, particularly in blends with PVA and polylactic acid (PLA) (Bulatović et al., 2025; Kumar et al., 2024; Wang et al., 2025).

Carbon-based materials like quantum dots (QDs) have demonstrated improved thermal stability in starch-based systems due to molecular interactions with starch functional groups, leading to a more compact structure. Their mild plasticizing effect also facilitates polymer chain mobility, enhancing flexibility and reducing water vapor permeability when small amounts are added. However, the incorporation of certain compounds, such as alizarin, may cause a decline in some of these properties (Qazanfarzadeh et al., 2025). Graphene QDs have also been shown to increase tensile strength (TS) and EAB (Sun et al., 2024).

Inorganic materials including zinc oxide, montmorillonite nanoclay, and titanium dioxide have likewise been incorporated to strengthen mechanical performance and reduce water vapor permeability in starch-based films (Gao et al., 2023; Noviagel et al., 2024; Srinivasan et al., 2025; Torche et al., 2025). These additives can also modify barrier properties by contributing to ethylene adsorption, as observed with sepiolite (Upadhyay et al., 2022), zeolites, and titanium dioxide (Maldonado et al., 2024; Siangyai et al., 2024).

Finally, chemical modifications of starch itself, such as acetylation and oxidation, can enhance compatibility and functionality in blends. Acetylation replaces hydroxyl groups with acetyl groups, improving water resistance (Nasseri et al., 2020), while oxidation produces dialdehyde starch, which significantly increases hydrophobicity (Liu et al., 2023).

3.2.2 Freshness indicators

In recent years, significant attention has been directed toward the incorporation of natural extracts rich in compounds capable of changing color in response to pH variations. These materials have been employed as freshness indicators in smart packaging systems. Their functionality relies on the chemical interactions between the packaging environment and volatile or acidic compounds generated during food storage.

During storage, products can release volatile basic compounds such as ammonia, dimethylamine, and trimethylamine, collectively expressed as total volatile basic nitrogen (TVB-N), as a result of protein and amino acid degradation in meat products. Similarly, organic acids such as lactic acid accumulate in dairy systems. These compounds alter the pH of the headspace surrounding the packaged product, which subsequently induces color changes in the incorporated indicator compounds (Khan & Liu, 2022; Zhang et al., 2025).

The interaction between these chemical indicators and volatile compounds modifies the protonation state or conjugation pattern of the indicator molecules, producing visible color transitions that serve as indicators of product freshness or spoilage. This mechanism allows for the development of pH-responsive packaging materials capable of visually signaling changes in product quality through simple color shifts, without requiring direct contact with the food itself. Several natural and synthetic molecules have been explored for this purpose, including anthocyanins, betacyanins, curcumin, alizarin, and methyl red, whose chemical color-change mechanisms are summarized in Fig. 3.2.

Anthocyanins (Fig. 3.2a) exhibit a well-documented sequence of structural transformations depending on pH. In acidic environments, the molecule predominates in the flavylium cation form, characterized by maximal conjugation and red color. As pH increases, anthocyanins undergo hydration to form the colorless carbinol pseudobase and subsequently the quinonoidal base (violet to blue) via tautomerization and partial hydrolysis of the pyrylium ring. Further deprotonation of hydroxyl groups leads to the blue quinonoidal anion, while ring opening yields the yellow chalcone form (Akhila et al., 2024; Kılıç & Kahyaoglu, 2025; Panwar et al., 2024).

Methyl red (Fig. 3.2b), a synthetic azo dye (–N=N–), exhibits clear protonation-dependent color changes. In its protonated form, present in acidic media, the molecule appears red. At neutral pH, the color shifts to orange, and under basic conditions, the deprotonated form displays a yellow hue (Panwar et al., 2024).

Betacyanins (Fig. 3.2c) are a class of betalain pigments characterized by their red to violet hues under acidic conditions. However, at alkaline pH, oxidative degradation of their phenolic groups disrupts the conjugated chromophore, resulting in yellow coloration. In addition, polymerization of the DOPA moiety can lead to melanin formation, producing brown hues (Khan & Liu, 2022).

Alizarin (Fig. 3.2d), an anthraquinone derivative, changes from yellow in acidic media to reddish and violet hues at higher pH values. This effect is attributed to the deprotonation of phenolic hydroxyl groups, which

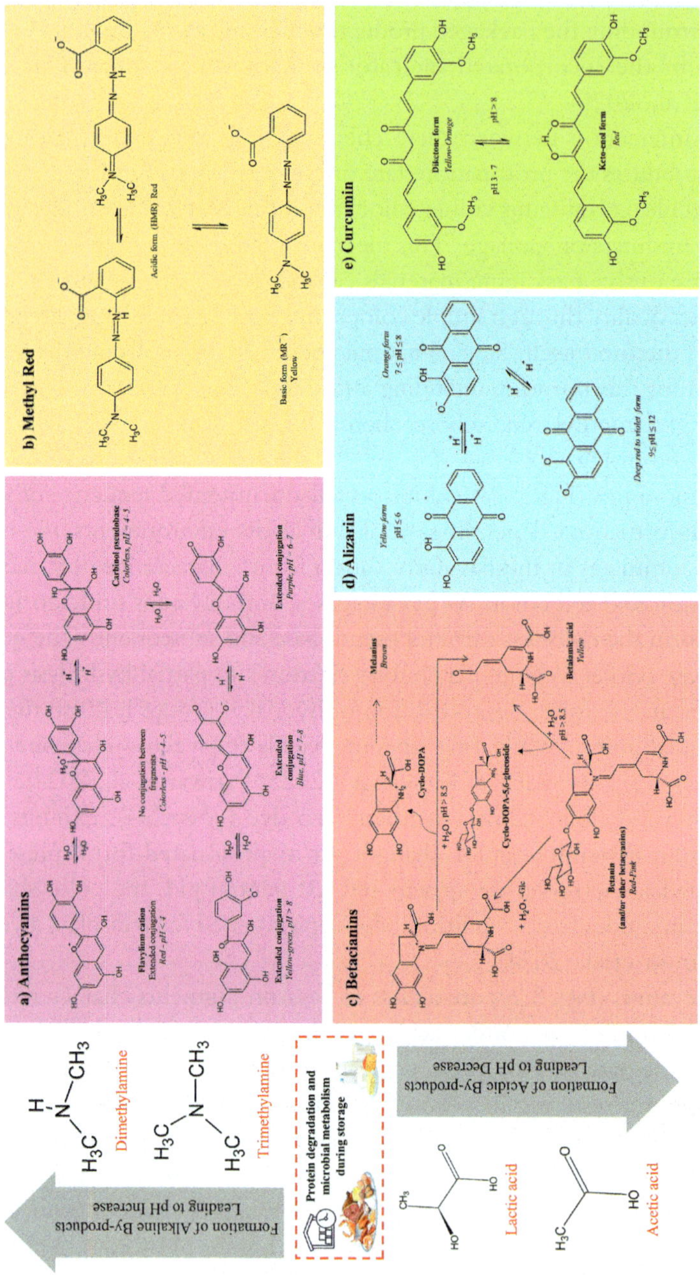

Figure 3.2 Mechanism of color change in pH-sensitive dyes used as freshness indicators in smart packaging.

modifies the electron distribution and the extent of π-conjugation within the molecule (Khan et al., 2023; Zhang et al., 2023).

Curcumin (Fig. 3.2e) undergoes a visible shift from yellow in acidic and neutral media to orange or reddish tones under alkaline conditions. This transition results from the deprotonation of phenolic groups, which increases conjugation and induces a bathochromic shift in the absorption spectrum. Moreover, curcumin exhibits keto–enol tautomerism, a reversible equilibrium that also influences its optical and chemical stability (Channa et al., 2025).

The choice of indicator compound depends primarily on its pH transition range. Although some pigments show limited effectiveness over specific pH intervals, mixtures of dyes, such as combinations of anthocyanins and alizarin, have been reported to extend the detection range and improve color response sensitivity (Zhang et al., 2023).

3.2.3 Antioxidant activity

Antioxidants are compounds that play a crucial role in preventing oxidation in food systems, thereby extending shelf life while preserving product quality (Kurabetta et al., 2025). Oxidation may occur due to the interaction of food with various environmental factors such as exposure to oxygen, light, and temperature, which can act either directly or as catalysts for reactions that generate reactive species. These oxidative reactions are responsible for the loss of nutritional quality through nutrient degradation, fat rancidity, and overall deterioration.

The evaluation of antioxidant capacity can be performed through both in vitro and in vivo methods. Among the in vitro approaches, colorimetric assays are the most widely employed due to their simplicity and reliability. These methods assess the ability of compounds to neutralize synthetic free radicals, such as the 2,2-diphenyl-1-picrylhydrazyl radical (DPPH•) and the 2,2′-azino-bis(3-ethylbenzothiazoline-6-sulfonic acid) cation radical ($ABTS^{+}$•), producing a measurable color change. The reaction mechanism may proceed through single-electron transfer (SET) or hydrogen-atom transfer (HAT), depending on the structural characteristics of the antioxidant compound (Schaich et al., 2015; Wołosiak et al., 2022).

On the other hand, the Oxygen Radical Absorbance Capacity (ORAC) assay is a robust and widely used analytical method that determines the antioxidant potential of a broad range of compounds. This assay measures the ability of compounds to inhibit the oxidation of a fluorescent probe, typically fluorescein, induced by reactive oxygen species. The mechanism is mainly based on hydrogen-atom transfer, reflecting the compound's capacity to neutralize peroxyl radicals and protect the probe from oxidative degradation (Schaich et al., 2015; Wołosiak et al., 2022).

Another frequently applied method is the Thiobarbituric Acid Reactive Substances (TBARS) assay, which provides an indirect measurement of lipid peroxidation. This technique quantifies the malondialdehyde content, a secondary product formed during the reaction of fatty acids with oxygen radicals, offering valuable information about oxidative damage and the protective effect of antioxidants (Chen et al., 2020; Gui et al., 2024).

Among the compounds used for this purpose, plant extracts stand out due to their high content of phenolic compounds, which include molecules such as vanillin (Kurabetta et al., 2025), and essential oils, as well as other phenolic acids and related compounds. In addition, mixtures of phenolic compounds with metal or metal oxide particles have been explored. These systems not only enhance antimicrobial performance but also improve antioxidant activity through synergistic effects between the organic and inorganic components.

3.2.4 Antimicrobial activity

An antimicrobial compound is defined as a substance capable of killing or inhibiting the growth of microorganisms through different mechanisms. Within this classification, various types of compounds have been employed for their antimicrobial potential. Active molecules are commonly found in essential oils rich in constituents such as carvacrol, thymol, and eugenol, as well as in plant extracts abundant in phenolic compounds such as tannins, coumarins, and flavones. These substances can cause cellular disruption, alter membrane permeability, penetrate cell walls, and lead to the loss of intracellular content. Additionally, they can interfere with energy production by affecting ATP synthesis (Calo et al., 2015; Farha et al., 2020; Huang, Zaynab, et al., 2024; Tong et al., 2022).

The antimicrobial effects of these compounds have been extensively studied in packaging systems for meat products, showing strong activity against bacteria such as *E. coli* and *S. aureus* (Adame et al., 2024; Yu et al., 2024). Although their activity can be considerably reduced due to interactions with the complex composition of certain food matrices, several methodologies have been developed to improve their incorporation into formulations, enhance their distribution within the matrix, and even increase their biological activity. Furthermore, these strategies enable controlled compound release, as demonstrated by encapsulation techniques using various wall materials (Castro-Rosas et al., 2017; Morán et al., 2025). It is also important to note that bacterial type significantly influences antimicrobial effectiveness, since morphological and metabolic differences can lead to variable responses to the same antimicrobial agent.

Additionally, metal oxide nanoparticles have been applied due to their ability to interact with microbial cell walls. An example is the incorporation of copper oxide nanoparticles (Suryavanshi et al., 2025), in which positively charged Cu^{2+} ions can modify the charge of the medium through their attraction to the negatively charged bacterial cell wall, leading to cellular disruption, increased permeability, and eventual lysis (George et al., 2020; Gunaki et al., 2024). Similarly, the addition of silver nanoparticles has shown a stronger antibacterial effect than copper nanoparticles during the monitoring of chicken fillet spoilage (Keshtiara & Hadadzadeh, 2025). These nanoparticles act in a manner comparable to copper ions, altering membrane permeability due to their small size, which allows easy penetration of the lipid bilayer, resulting in intracellular fluid loss and/or denaturation of membrane-bound proteins with critical roles in transport systems and ion channels (Lithi et al., 2025).

Likewise, the incorporation of zinc oxide has demonstrated antimicrobial effects, where Zn^{2+} ions can inhibit bacterial growth through mechanisms similar to those previously described (Heryanto et al., 2025). On the other hand, some synthetic compounds have also been used, including quaternary ammonium salts (Gao, Song, et al., 2024) benzalkonium chloride (Srivastava et al., 2023), polyhexamethylene guanidine hydrochloride (Xi et al., 2024), and N-ethylpiperazine (Kanth et al., 2023), among others; however, in recent years, there has been a growing preference for compounds derived from natural sources due to concerns related to cytotoxicity and environmental sustainability.

3.3 Conventional fabrication process of starch-based smart packaging

The most common conventional methods used for the fabrication of starch-based smart packaging are solvent casting, thermocompression molding, and extrusion. These processes have enabled the production of intelligent films or labels capable of monitoring and controlling the preservation state of various foods (Table 3.1).

Table 3.1 Overview of starch-based smart packaging films prepared by conventional fabrication techniques.

Fabrication method	Starch source/type	Other polymers/ biopolymers	Active indicator compound	Indicator concentration	Other bioactive compounds	Monitored food/ material	Reference
Casting	High-amylose corn starch	Chitosan	Red cabbage anthocyanin; curcumin	NA	*Portulaca oleracea* L. flavonoids	Beef	Cao et al. (2025)
Casting	Native cassava starch; oxidized hydroxypropyl starch; acetylated distarch phosphate; oxidized-acetylated starch	NA	RCEs (red cabbage extracts)	885.7 μL RCEs (added to 100 mL solution with 4 g starch)	NA	NA	Cheng et al. (2022)
Casting	Cassava starch	Polyaniline (PANI)	PANI	0.4% of PANI	NA	NA	Chia et al. (2022)
Casting	Cassava starch	NA	Pomegranate peel powder (PPP)	2%, 4%, 6%, and 8% w/w PPP		Lamb meat	Esfahani et al. (2022)
Casting	Starch	Gelatin	RCEs (red cabbage extracts)	NA	NA	Milk	Fotouhi et al. (2023)
Casting	Crosslinked–Acylation cassava starch	NA	Purple sweet potato anthocyanin (PSP)	2 mL; 4 mL; 6 mL; 8 mL; 10 mL (PSP)	NA	Indian Mackerel Fish	Handayani et al. (2025)
Casting	Soluble starch	PVA	Cu-ICA nanocrystal	0.5%; 1.0%; 2.0% (Cu-ICA percentages)	NA	Shrimp	Huang, Wang, et al. (2024)
Casting	Pea starch	κ-carrageenan	Black raspberry extract	0; 2; 3 and 4 mL/g (on PS basis)	NA	Pork	Jiang, Liu, et al. (2024)
Casting	Manihot esculenta starch (MES)	Gum ghatti	Black currant extract (RNE)	RNE (0, 1.5 mL/g, 3 mL/g and 4.5 mL/g, on the basis of MES)	NA	Beef meat	Jiang, Yang, et al. (2024)
Casting	Cassava starch	PVA	Betanina (E−162)	E−162 (1% w/v)	NA	Ammonia vapor (simulates meat degradation)	Kavas et al. (2023)

Casting	Sago starch	NA	Bauhinia Kockiana extract (BKE)	BKE (5, 10, and 15 mg)	NA	Coconut milk	Lau et al. (2024)
Casting	Assam indigenous Joha rice starch	NA	Beetroot extract (BRE)	0.25%, 0.5%, 1.0%, 2.0% w/v (BRE)	NA	Chicken meat; fish; Indian cottage cheese	Lekurwale et al. (2024)
Casting	Potato starch	β-cyclodextrin (β-CD)	Blueberry anthocyanins (BA)	0.2 mg/mL of BA	Cinnamon essential oil	Fatty food	Li et al. (2023)
Casting	Potato starch	Dextrinized starch	Blueberry anthocyanins (BA)	0.2 mg/mL of BA	CA; AA; FA (phenolic acids; copigment to enhance color/pH responsiveness)	Beef	Li et al. (2024)
Casting	Starch	PVA	Anthocyanins of red cabbage extract (CE)	5 mL of CE	Propolis extract	Milk	Mustafa et al. (2021)
Casting	Jackfruit seed starch	PVA	Anthocyanins of purple sweet potato peel extract (PSP)	4% w/v of PSP	NA	Shrimp	Nguyen et al. (2025)
Casting	Corn starch	Chitosan; PVA	Anthocyanin of *Rhododendron arboreum* Sm. flowers extract (RA)	2% w/w of RA	NA	Animal-derived foods (ammonia-sensitive indicator)	Shahi et al. (2025)
Casting	Starch and esterified starch	NA	Black rice anthocyanins	10% w/w relative to starch dry weight	NA	Pork	Song et al. (2024)
Extrusion	Cassava thermoplastic starch	NA	Anthocyanin (ATH)	5, 10 and 20 mg of ATH per 100 g	NA	Beef; fish	Vedove et al. (2021)
Casting	*Dioscorea zingiberensis* starch	NA	Anthocyanin of butterfly pea flower (BPE)	2%, 3%, 4% w/v of BPE	NA	Meat	Wang et al. (2023)

(*Continued*)

Table 3.1 Overview of starch-based smart packaging films prepared by conventional fabrication techniques.—cont'd

Fabrication method	Starch source/type	Other polymers/ biopolymers	Active indicator compound	Indicator concentration	Other bioactive compounds	Monitored food/ material	Reference
Casting	Common maize starch	Ca; CMC; CMS; STMP; SPI	Blueberry anthocyanins (ACN)	100 mg; 0.1 wt% (ACN)	NA	Chicken breast	Wang, McClements, Zhang, et al. (2024)
Casting	Common maize starch	Soy protein isolate	Blueberry anthocyanins (ACN)	100 mg; 0.1 wt% (ACN)	NA	Milk; coconut water	Wang, McClements, Peng, et al. (2024)
Casting	Cassava starch	PVA	*Aronia melanocarpa* anthocyanins	0.025 g; 0.05 g	NA	NA	Wu et al. (2024)
Casting	Acetylated distarch phosphate (ADSP)	NA	Free anthocyanins; microencapsulated anthocyanins	0.1 g	NA	Beef	Wu et al. (2024)
Casting	Carboxymethyl starch	PVA	Cu-Trp nanocrystal	0.5; 1.0; 2.0 (Cu-Trp percentages)	NA	Shrimp	Xu et al. (2024)
Casting	Starch	Gelatin	Purple sweet potato anthocyanins (SPS)	0.1% w/v of SPS	NA	*Flammulina velutipes* mushroom	Zong et al. (2023)
Thermo-compression	Cassava starch	NA	Blueberry residue	2 g of powdered blueberry residue (10% based on starch mass)			Andretta et al. (2019)
Thermo-compression	Cassava starch	Montmorillonite (Mt)	Anthocyanins extracted from the jambolan fruit	ACNs (1%; 3%; 5% basis of starch mass)	NA	Ammonia vapor (simulates meat degradation)	Koop et al. (2022)
Extrusion and thermo-compresion	Corn starch	Poly(L-lactic acid) (PLA)	Cochineal powder (carminic acid) (CP)	0.5 parts per 100 resin (CP)	Clove essential oil	Fresh shrimp	Mohammadi and Fasihi (2025)

3.3.1 Casting

In recent years, the solvent casting method has become a fundamental and simple technique for fabricating intelligent films (Table 3.1). This process typically involves dispersing a biopolymer matrix—such as starch—often mixed with a plasticizer (e.g., glycerol) in a solution, followed by heating until gelatinization occurs (Cheng et al., 2022; Nguyen et al., 2025).

Subsequently, the smart indicator (dye or sensor) is incorporated into the film-forming solution, which is poured into molds or Petri dishes and left to dry under controlled temperature conditions to obtain uniform, flat sheets (Cheng et al., 2022; Nguyen et al., 2025).

Casting is attractive because it requires inexpensive equipment and minimal space (Vedove et al., 2021); however, its main disadvantage lies in its low production capacity, which limits industrial scalability compared to extrusion (Vedove et al., 2021).

These films function as colorimetric sensors that monitor food freshness through visible color transitions induced by pH variations or by the increase of ammonia and volatile amines released during spoilage (Cheng et al., 2022). As illustrated in Fig. 3.3, different active compounds exhibit distinct color responses according to the fabrication method. For solution casting, anthocyanins—extracted from sources such as red cabbage, grapes, or purple sweet potato—are the most widely used pH indicators, showing a gradual color shift from red (<3) to purple (4–6), blue (7–9), and green (>10) as pH increases. Curcumin also acts as a pH-sensitive dye, turning from yellow to orange-reddish above pH seven to eight, while betanin (a betalain pigment from beetroot) transitions from red at $pH < 6$ to yellow at $pH > 8$ (Kavas et al., 2023; Lekurwale et al., 2024). Polyaniline (PANI), a conductive polymer, operates as an ammonia-responsive sensor, exhibiting a reversible color change from green (protonated) to blue (deprotonated) in the presence of increased ammonia (Chia et al., 2022). Similarly, Cu–ICA and Cu–Trp nanocrystals respond to ammonia elevation by shifting from light green or pale blue to deeper blue hues, due to coordination or deprotonation reactions (Huang, Zaynab, et al., 2024; Xu et al., 2024).

More specific chemical systems have also been developed, such as I^-/starch complexes for hydrogen peroxide detection and 1,5-diphenylcarbazide (DPC) for selective Cr(VI) detection in starch/gelatin films (Fotouhi et al., 2023). Furthermore, metal–organic frameworks (MOFs) like Cu-Trp (Cu-L-tryptophan) nanocrystals have been

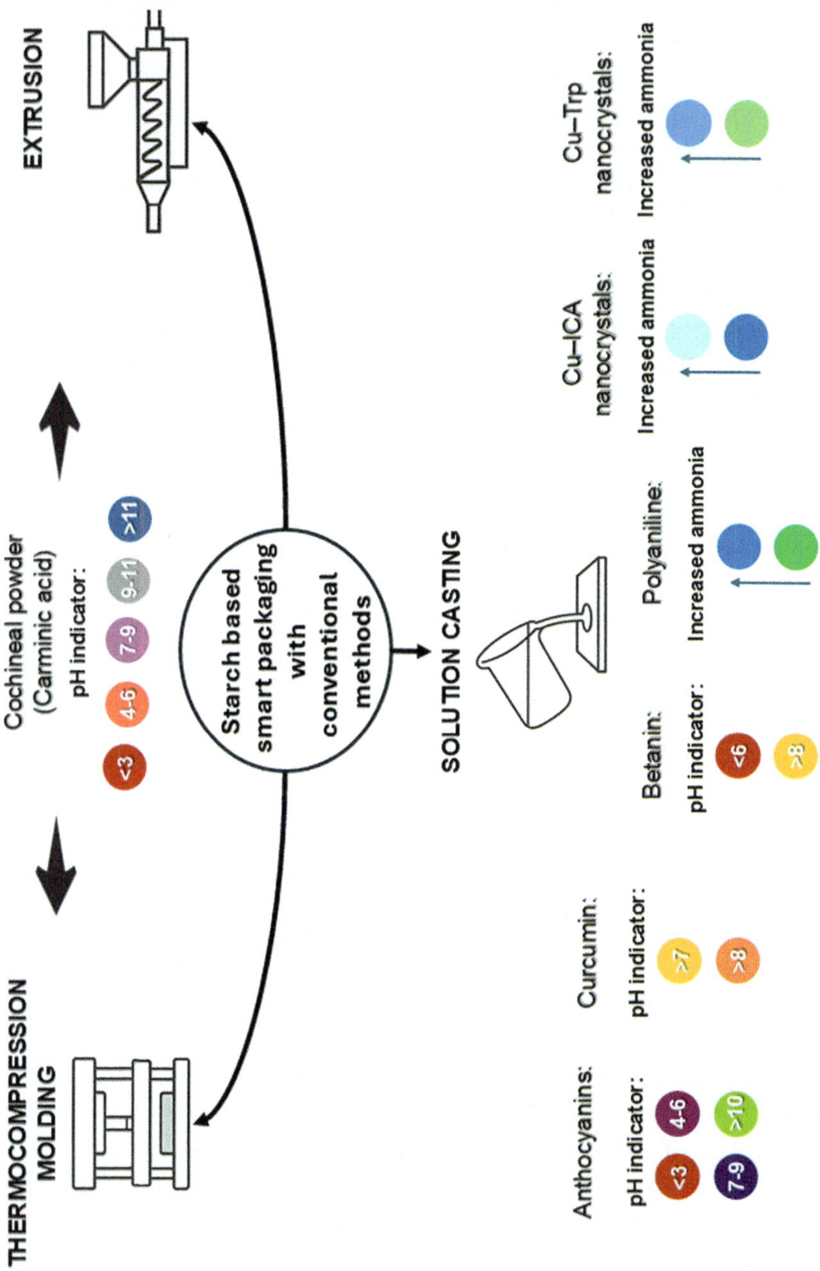

Figure 3.3 Active indicator compounds commonly used in conventional fabrication methods of starch-based smart films.

incorporated into CMS/PVA films as functional compatibilizers and ammonia sensors, showing a color shift from light green to light blue (Xu et al., 2024).

Overall, most recent studies confirm that casting remains the preferred method for the research and development of smart films due to its simplicity and practicality at the laboratory scale, although alternative techniques are being explored to enable industrial-scale production.

3.3.2 Thermocompression molding and extrusion

Thermocompression molding and extrusion represent melt-processing methods that are essential for achieving industrial scalability in smart film production, overcoming the main limitation of the casting technique (Mohammadi & Fasihi, 2025). Extrusion offers a key advantage as a cost-effective and continuous process suitable for high-volume manufacturing (Vedove et al., 2021), whereas thermocompression molding is particularly employed to produce dense, uniform starch-based films at an industrial scale (Andretta et al., 2019). It has been reported that thermocompression can yield films with more intense coloration (higher ΔE values) compared to those obtained by casting, possibly due to faster melting that minimizes pigment degradation (Andretta et al., 2019).

The study by Mohammadi et al. (Mohammadi & Fasihi, 2025) detailed the fabrication of starch-based smart films through extrusion and thermo-compression, focusing on cochineal powder (carminic acid) as the main colorimetric indicator. This work achieved industrial-scale production by blending PLA with TPS and clove essential oil (CEO), employing extrusion to prepare the TPS and hot pressing for final molding. Cochineal powder was specifically selected because, unlike other natural pigments such as anthocyanins, it shows exceptional thermal and light stability, acid–base reversibility, and cost-effectiveness, making it suitable for high-temperature industrial processes such as thermocompression and extrusion. As illustrated in Fig. 3.3, carminic acid acts as a pH-sensitive colorimetric dye whose hue varies markedly across different acidity levels: it appears bright red under highly acidic conditions ($pH < 3$), pink at pH four to six, violet to purple at neutral to slightly alkaline pH (7–9), and bluish or grayish-blue at strongly basic pH (>11). These color changes correspond to structural transitions from protonated to deprotonated anthraquinone species. In food monitoring applications, this translates into a visible shift from orange or pink to purple/pinkish-purple when the film is exposed to ammonia vapors generated by food spoilage (Mohammadi & Fasihi, 2025).

In related research, twin-screw extrusion has been used to produce TPS sheets from cassava starch incorporating anthocyanins as natural pH indicators, pioneering the development of biodegradable smart packaging (Vedove et al., 2021). Extruded TPS is commonly used as feedstock for thermocompression, where it is pressed at $\approx$150 °C and 4 MPa to form uniform films (La Fuente Arias et al., 2023; Mohammadi & Fasihi, 2025). This method has also been employed to prepare pH-indicator films from cassava starch and blueberry residues rich in anthocyanins (Andretta et al., 2019; Esfahani et al., 2022) and to fabricate multilayer laminates combining polar (modified starch) and nonpolar (PLA or PHBV) polymers (La Fuente Arias et al., 2023). Likewise, Koop et al. (2022) successfully fabricated smart labels based on TPS and biohybrid nanocrystals derived from jambolan (*Syzygium cumini*) via thermocompression.

Despite these advances, the number of studies using extrusion and thermocompression is still limited compared to casting (Vedove et al., 2021), mainly due to the low thermal stability and processability of natural pigments such as anthocyanins, which readily degrade at high temperatures (Cheng et al., 2022; Mohammadi & Fasihi, 2025). For this reason, cochineal powder (carminic acid) stands out as a more thermally stable alternative, maintaining its colorimetric response under the elevated temperatures required by industrial processing, making it particularly suitable for extrusion and thermocompression-based smart packaging systems.

3.4 Emerging fabrication process of starch-based smart packaging

Apart from the classical polymeric film production techniques of casting and extrusion, other technologies can be applied to films aiming to modify their properties and/or fabricate films with enhanced physical and chemical properties. Some emerging technologies include cold plasma, 3D printing, layer-by-layer, and electrospinning. By searching in the SCOPUS database using the keywords "starch AND (smart OR intelligent) AND (packaging OR film) AND (layer-by-layer OR plasma OR electrospinning OR 3D)", only 40 documents emerge from 2013 to 2025, demonstrating that this topic is not yet thoroughly studied, and suggesting a promising field of study.

3.4.1 Cold plasma

Cold plasma (CP) is a partially ionized gas containing several reactive special – atomic atoms, excited species, ions, UV radiation, energetic electrons – that is usually formed due to the interaction between a gas and a strong electric field. Several types of reactors can be used – dielectric barrier discharge (DBD) being the most common – and generally are composed of a pair of electrodes. Still, the principle of operation is similar for all reactors: electrons are accelerated by an electric field and collide with particles of a gas, forming reactive species if the electrons were energetic enough at the time of collision, that is, if their energy was equal to the excitation or ionization energy of the atom/molecule with which they interacted (Leandro, Laroque, Monteiro, et al., 2024; Xiao, 2016).

CP treatment outcome depends on many factors, such as plasma formation parameters (voltage, pulse frequency, system pressure, gas used), reactor design (distance between electrodes, material of the electrodes and dielectric barriers, electrode's size), and material treated. Regarding starch films, CP may increase the glass transition temperature and superficial roughness (Pankaj et al., 2017). The increase in glass transition temperature may be due to cross-linking, turning amylose into amylopectin (Leandro, Laroque, Monteiro, et al., 2024). Also, superficial roughness is expected because of plasma etching – bombardment of the film's surface by energetic species, causing the breakage of bonds and removal of material (Leandro, Laroque, Carciofi, et al., 2024; Pankaj et al., 2017). Depending on the gas used, plasma may promote the formation of hydrophilic functional groups on the film's surface, which may lead to increased hydrophilicity (Leandro, Laroque, Carciofi, et al., 2024; Pankaj et al., 2017). Hydrophobicity of starch films may be favored by using, for instance, hexamethyldisiloxane (HMDSO) as ionizing gas (Sifuentes-Nieves et al., 2019).

Few studies have applied CP to develop smart packaging. With LDPE, it has been used to modify its surface, introducing polar functional groups, and thus increasing its hydrophilicity. This allowed for the adhesion of PVA containing anthocyanins from grape skin adsorbed onto a layered silicate. Even though the smart label was sensitive to humidity, there was excellent adhesion to LDPE's surface in dry conditions, which was not possible without CP treatment (Leandro, Laroque, Carciofi, et al., 2024).

As for starch, CP was used to modify smart corn starch-based films with carotenoids and anthocyanins. The treatment results in smart films with reduced amylose content, solubility, and moisture content, and increased water contact angle (Goiana et al., 2025). CP induces depolymerization of amylose, promoting crosslinking between chains, therefore increasing the amylopectin content (Leandro, Laroque, Monteiro, et al., 2024). The oxidation of –OH groups into C=O groups may be the cause of the increase in water contact angle and may also affect the water vapor permeability (WVP), although the latter had no significant change after CP treatments (Goiana et al., 2025). Oxidation may also be the reason for the observed modifications in solubility and moisture content. The low influence on WVP may be due to the superficial nature of CP treatments, affecting just a few nanometers in depth (Leandro, Laroque, Carciofi, et al., 2024). Regarding the films' sensing capabilities, the use of CP enhanced its visual response, facilitating the differentiation between a film in acidic, neutral, and basic pH when the films were treated with CP for 10 min (Goiana et al., 2025).

3.4.2 Electrospinning

Electrospinning is an electro-hydrodynamic processing technique that applies a high intensity electric field to generate polymeric fibers, ranging from nanometers to microns. The process consists of the application of an electric field to a polymeric solution, which forms a Taylor cone – electrically charged conical droplets – at the tip of a spinneret or needle. Electrostatic forces overcome the solution's surface tension, sending a polymer jet toward a grounded electrode, with its solvent evaporating before reaching the electrode, leading to the solidification of the polymer into dry nanofibers. This technique operates under mild conditions, making it ideal for the incorporation of heat-sensitive compounds, and produces nanofibers with a high surface area-to-volume ratio, high porosity, small pore size, flexibility, and absorption capacity, which allows for the controlled release of substances and improves barrier properties for various applications (Amorim et al., 2024; Karnwal et al., 2025; Lv et al., 2024).

This technique produces films with a distinct structure compared to the classical casting method. SEM images reveal a complex arrangement of nanofibers, while the casting method leads to a smoother structure. In addition to these structural differences, these films also vary in optical properties – electrospun starch films may be opaquer, which can enhance their UV-barrier function. Furthermore, the porous nature of electrospun films significantly influences their physical characteristics, increasing flexibility and drastically enhancing hydrophilicity due to the higher porosity of the material (Zhu et al., 2022).

Electrospinning was employed to fabricate films based on modified starch (octenyl succinic anhydride starch, OSA) and PVA with anthocyanins. Mixtures of starch, PVA and anthocyanins with varying anthocyanin content were electrospun. The electrospinning of OSA and the fabrication of nanofibers was only successful when PVA was added to the mixture. Being a non-thermal technique, the resulting films maintained the color changing capacity of anthocyanins, allowing the monitoring of pork and shrimp freshness. The film's fiber networking was well interwoven, with mean fiber diameter ranging from 276 to 490 nm, and hydrophilic, with water contact angle ranging from 62.6 to 50.2 degrees. Higher content of anthocyanins increased the fiber's thickness and decreased the contact angle (Lv et al., 2024).

A similar study was conducted using curcumin in conjunction with anthocyanins. In this case, nanoemulsions of curcumin, clove essential oil, OAS, PVA, and anthocyanins were electrospun to form smart and active films (Lv et al., 2025). Similarly to Lv et al. (2024), the fiber's thickness increased with the addition of anthocyanins, essential oil, and curcumin. However, a clear outer shell formed by the polymers and anthocyanins could be seen, with the oil phase remaining as the core of the nanofibers. As expected, the water contact angle increased with the addition of the oil phase (50–75 degrees, approximately). As for the color changing capabilities, the films containing anthocyanins, essential oil, and curcumin had great color changing capabilities, being much easier to distinguish colors in different pH, especially in the six to eight range, as this is generally the pH of meat stuff (Leandro, Laroque, Carciofi, et al., 2024; Lv et al., 2025).

3.4.2.1 3D printing

Three-dimensional (3D) printing is an additive manufacturing technology that builds complex designs layer by layer, enabling rapid and precise construction. Introduced in the 1980s, this method offers customizability in shapes, patterns, sizes, and pore structures, surpassing the limitations of conventional casting techniques. Beyond traditional sectors like medicine and aerospace, 3D printing has found notable application in the food industry, facilitating the creation of intricate designs, personalized nutrition, and novel textures. For instance, it allows the production of intelligent packaging films and labels, custom tailoring them to specific food requirements and integrating multiple monitoring functions to shorten product development cycles. Starch, with its gel properties, abundance, and cost-effectiveness, is a promising matrix material for 3D printing in food-related applications (Bao et al., 2024; Karnwal et al., 2025; Tang et al., 2024). 3D printed films may present different properties compared to their casted counterparts, having smoother surface, higher tensile strength, but lower elasticity and WVP (Kim et al., 2022).

The colorimetric response of smart labels is influenced by various 3D printing parameters. Starch-based films incorporating anthocyanins and phycocyanins were fabricated via 3D printing, with variables including nozzle size, printing speed, starch concentration, and the incorporation of chondroitin sulfate. Optimal results in terms of printing accuracy, efficiency, and colorimetric response were achieved using a 0.4 mm nozzle, a printing speed of 30 mm/s, and a starch concentration of 15% (w/v) with the addition of 0.6% (w/v) chondroitin sulfate. Use of smaller nozzles (e.g., 0.2 mm) resulted in intermittent printing and films with structural voids, while larger nozzle sizes led to starch aggregation. Notably, both colorimetric stability and response were dependent on nozzle size. Films printed with a 0.2 mm nozzle exhibited greater susceptibility to light due to lower film thickness and pigment content, and the colorimetric reaction to ammonia varied according to nozzle dimension; the 0.4 mm nozzle yielded the most visually discernible color change (Bao et al., 2024).

Compared to other polysaccharides, the use of starch for 3D printing yields good accuracy when used alone. Tang et al. (Tang et al., 2024) compared six different polysaccharides (8% w/v) alone and blended with PVA (1%–7% w/v), showing that pectin displayed the highest accuracy when not blended with PVA (47.2%), followed by xanthan gum (43.04%), konjac glucomannan (35.28%), carboxymethyl cellulose (34.60%), potato starch (34.13%), and sodium alginate (0.71%). In this context, accuracy refers to how closely the printed object matches its intended design. The addition of PVA increased the accuracy of potato starch up to 40% with 3% PVA, and that of sodium alginate and carboxymethyl cellulose up to 15.50% and 53.11%, respectively, with the addition of 7% PVA.

Blends of the films with 3% PVA were prepared and mixed with methyl red and bromothymol blue to monitor CO_2 concentrations. Smart label's color was dependent on the polymers used, and, although all formulations had good colorimetric responses, those with carboxymethyl cellulose and sodium alginate resulted in higher ΔE, which indicates a more noticeable color change (Tang et al., 2024). This translates to improved visibility and effectiveness of smart labels in detecting CO_2 concentrations.

The emerging technologies of cold plasma, electrospinning, and 3D printing represent distinct approaches to the development of starch-based smart packaging. The difference lies in their application: while cold plasma acts as a post-processing technique to modify the surface properties of pre-existing films, electrospinning and 3D printing are fabrication methods that construct the material's structure. This is reflected in their structural impact: plasma induces chemical changes at the surface, such as cross-linking and functionalization; electrospinning creates a porous, high-surface-area network of nanofibers; and 3D printing builds complex, customized geometries layer-by-layer. Consequently, their applications are complementary: cold plasma is ideal for surface functionality, such as enhancing sensor response or adhesion; Electrospinning effectively incorporates sensitive compounds for monitoring; and 3D printing offers great customizability for labels and packaging. Together, these technologies offer new ways to engineer materials with advanced functions.

3.5 Food applications of starch-based smart packaging

Intelligent packaging based on natural polymers, such as starch, represents a development aligned with the Sustainable Development Goals (SDGs), the circular economy, and the bioeconomy. Indeed, this approach not only offers viable alternatives to conventional petroleum-derived plastics but also leverages the functional properties of primary and/or secondary metabolites that can be incorporated, potentially improving the product's shelf life and/or the information conveyed to consumers about the packaged item. Moreover, many of these functional metabolites originate from residues or natural matrices, further enhancing the sustainability of this concept. summarizes several studies that have applied intelligent packaging to food products (Table 3.2).

The deterioration of meat and its derivatives is driven primarily by the action of bacteria and endogenous enzymes on vitamins, lipids, and proteins, with the consequent formation of compounds that modify pH. In this context, intelligent packaging incorporates colorimetric indicators that are sensitive to this parameter, providing an immediate visual signal of a food's freshness and/or condition. Thus, Zou et al. (2025) developed smart materials based on *Euryale ferox* starch reinforced with nano-SiO_2 and incorporated with Chinese rose extract (rich in anthocyanins), and

Table 3.2 Research works on starch that have used intelligent packaging for food applications.

Starch source	Other components	Applied food	Preparation method	Results	References
Potato (*Solanum tuberosum*)	• Potato peel powder • Quince peel powder • Glycerol	Cured cheese	Casting	The films increased hardness without negatively affecting sensory properties during ripening.	Coimbra et al. (2023)
Chayotes (*Sechium edule Sw.*)	• Cinnamon essential oil • Zein–pectin nanoparticles	Ground beef	Casting	The CEO release was controlled into the package and this material had antimicrobial and antioxidant activities, preserving meat.	Wu, Wang, et al. (2023)
Potato (*Solanum tuberosum*)	• Gelatin • Glycerol • Resveratrol	Soybean oil	Thermal extrusion and compression molding	Reduced oxidative rancidity of the oil versus the control during accelerated storage, and decreased cooking loss of the model food.	Wu, Li, et al. (2023)
Maize (*Zea mays*)	• Chitosan • Urea • Graphene oxide • Carboxymethyl cellulose	Cherry tomato	Casting	Improved tomato preservation (better appearance and lower weight loss over 5 days).	Wu, Lv, et al. (2023)
Potato (*Solanum tuberosum*)	• Gelatin • Tween 80 • Anthocyanins extract	Mushroom	Casting	The indicator material changed color with spoilage and, when applied, helped track mushroom freshness.	Zong et al. (2023)

Tapioca	• *Litsea cubeba* essential oil • Pullulan • Glycerol	Beef	Casting	Reduced bacterial load and kept pH, color, texture, and TBARS within acceptable ranges for 10 days at 4 °C versus LDPE.	Adame et al. (2024)
Maize (*Zea mays*)	• Clove essential oil • Sodium caseinate • Arabic gum • Glycerol	Bread	Casting	Increased product shelf life and improved overall quality compared with the control.	Fan et al. (2024)
Not Specified	• PBAT • Glycerol • Blueberry extract	Shrimp	Blown film extrusion	Monitored shrimp spoilage with clear color changes during storage.	Gao, Sun, et al. (2024)
Maize (*Zea mays*)	• Cinnamon essential oil • Glycerin • Ethyl lauroyl arginate • Hydroxypropyl-β-cyclodextrin • Maltodextrin • PBAT	Strawberry	Blown film extrusion	Decreased weight loss and decay rate, preserved firmness and soluble solids, and extended shelf life.	Hu et al. (2024)

monitored in real time the freshness of chicken breast stored at 20 °C for 60 h. In the assay, the pH-sensitive material changed color as TVB-N values increased; at 24 h, TVB-N reached 21.7 mg/100 g, surpassing the regulatory spoilage threshold, which could be clearly distinguished thanks to the smart material. However, color change does not always respond to a decrease in pH, as it may also be associated with an increase. Along these lines, Mali & Pandey (Mali & Pandey, 2024) developed a freshness indicator based on an intelligent film of *Artocarpus lakoocha* seed starch with curcumin, applied to monitor chicken fillets. Leveraging curcumin's chromatic response to pH, the film remained yellow under acidic–neutral conditions and shifted from yellow to orange–red as spoilage progressed. At 8 °C, by day 10, the film turned red when the sample pH rose to 6.94, confirming its utility as a visual freshness indicator.

Fruits are particularly susceptible to the action of microorganisms and enzymes. Although they can exhibit pH variations, their deterioration is chiefly associated with microbial growth and weight (moisture) loss. Likewise, polyphenol oxidase (PPO) induced color change is among the most frequent problems in minimally processed foods. To mitigate this phenomenon, Almeida et al. (2023) developed an active material of thermoplastic starch reinforced with bacterial nanocellulose and incorporated with gallic acid, which they applied as a coating on fresh apple slices stored at 4 °C for 7 days. Owing to its higher antioxidant capacity (provided by gallic acid) and O_2-barrier property, the coating significantly delayed browning: on day 7, the browning index was 52.2 ± 1.0 versus 95.5 ± 1.8 (LDPE), 63.0 ± 1.5 (TPS), and 60.6 ± 1.1 (TPS-BNC10), with a notably improved appearance. In other work, Zhang & Chen (Zhang & Chen, 2023) developed biodegradable starch–alginate films reinforced with montmorillonite and incorporated with star anise oil (S-M-SAO), which were applied to cherry tomatoes. Over 18 days of storage at 18–20 °C, the films reduced weight loss and delayed deterioration: tomatoes with the comparison films began to spoil from day 3, whereas those with S-M-SAO started to deteriorate on day 9.

Intelligent packaging has also been applied to a wide range of other food products. Araújo et al. (2023) prepared, by casting, active corn-starch films with pectin extract and turmeric essential oil (TEO) and used them as wraps for sliced bread (2 × 2 cm), sealed and stored at 25 °C. Thanks to the antimicrobial and antioxidant activity of TEO, the wrapped breads showed no visible contamination for 9 weeks, whereas the cellophane control exhibited browning from week 6. In addition, migration of TEO was observed from week 4 onward, evidenced by the loss of the film's yellow color. On the other hand, Lopes et al. (2024) developed heat-sealable bioplastics from flour derived from carob milling (LBMD) mixed with starch and applied to the packaging of sliced cheese and oat cookies, comparing performance with PA/PE (polyamide/polyethylene). In cheese, molds and yeasts were not detected up to 21 days. Nevertheless, the bioplastic induced 9% dehydration by day 7 (which remained stable through day 21), greater yellowness, and a non-significant trend toward increased hardness. Sensory attributes of odor and taste were acceptable up to 14 days. With respect to volatile compounds, increases of 43% at day 7% and 182% at day 21 were recorded (vs. 130% in PA/PE), dominated by oxidation-derived aldehydes and ketones (benzaldehyde, 3-methylbutanal, nonanal, 3-hydroxy-2-butanone). In cookies, the bioplastic maintained appearance and texture (no loss of crispness), and the panel did not perceive rancidity; moreover, volatiles increased by only 18% at 21 days (vs. 120% in PA/PE), suggesting better containment of oxidation products (pentanal, hexanal, heptanal) and a possible sequestration of hydroxyketones. Taken together, the LBMD/starch packaging preserved cheese quality (no fungal growth and sensory changes after 14 days) and reduced oxidation in cookies compared with PA/PE.

3.6 Future trends and conclusions

The reviewed literature confirmed that starch is an excellent macromolecule to develop smart packaging materials. In general, starch-based films have been used to immobilize biomolecules acting as indicators to communicate food degradation. Among them, anthocyanins have shown potential application in starch-based films to impart color change properties with pH alteration and ammonia production during the food shelf-life. Although the production and application of starch-based smart packaging is proved, the production of these materials at large scale is still a challenge and more research is still required in this area. Furthermore, the water sensitivity of starch-based smart materials must be improved, aiming before industrial applications.

Conflicts of interest

The author declares no conflict of interest.

Acknowledgments

V.C. Fierro gratefully acknowledges the Federal University of Santa Catarina (UFSC) for the collaboration networks established, and CAPES for supporting her doctoral internship and funding these networks through the "MOVE-LA AMERICA" program. J.F. Grisales-Mejía thanks Universidad de los Andes, particularly the Department of Chemistry and the research group *Laboratory of Advanced Analytical Techniques in Natural Products* (LATNAP), led by Dr. Chiara Carazzone, for the support provided during the postdoctoral fellowship. G.C. Leandro gratefully acknowledges CAPES (Coordination for the Improvement of Higher Education Personnel) for his doctoral fellowship. G. A. Valencia would like to thank the CNPq (National Council for Scientific and Technological Development) for the research fellowship (302434/2022–4). All authors would like to thanks their respective universities for the support.

Artificial intelligence statement

This manuscript was not prepared with assistance of Artificial Intelligence.

References

Adame, M. Y., Shi, C., Li, C., Aziz, T., Alharbi, M., Cui, H., & Lin, L. (2024). Fabrication and characterization of pullulan/tapioca starch-based antibacterial films incorporated with Litsea cubeba essential oil for meat preservation. *International Journal of Biological Macromolecules, 268*. https://doi.org/10.1016/j.ijbiomac.2024.131775

Akhila, K., Ramakanth, D., Lakshman Rao, L., Singh, S., & Gaikwad, K. K. (2024). Plant phytochemicals as bio-responsive compounds in smart packaging for food spoilage detection: A comprehensive review. *Sustainable Food Technology, 2*(4), 860–875. https://doi.org/10.1039/d4fb00072b

Almeida, T., Karamysheva, A., Valente, B. F. A., Silva, J. M., Braz, M., Almeida, A., Silvestre, A. J. D., Vilela, C., & Freire, C. S. R. (2023). Biobased ternary films of thermoplastic starch, bacterial nanocellulose and gallic acid for active food packaging. *Food Hydrocolloids, 144*. https://doi.org/10.1016/j.foodhyd.2023.108934

Amorim, L. F. A., Mouro, C., & Gouveia, I. C. (2024). Electrospun fiber materials based on polysaccharides and natural colorants for food packaging applications. *Cellulose, 31*(10), 6043–6069. https://doi.org/10.1007/s10570-024-05956-z

Andretta, R., Luchese, C. L., Tessaro, I. C., & Spada, J. C. (2019). Development and characterization of pH-indicator films based on cassava starch and blueberry residue by thermocompression. *Food Hydrocolloids, 93*, 317–324. https://doi.org/10.1016/j.foodhyd.2019.02.019

Araújo, M. N. P., Grisi, C. V. B., Duarte, C. R., de Almeida, Y. M. B., & Vinhas, G. M. (2023). Active packaging of corn starch with pectin extract and essential oil of Turmeric Longa Linn: Preparation, characterization and application in sliced bread. *International Journal of Biological Macromolecules, 226*, 1352–1359. https://doi.org/10.1016/j.ijbiomac.2022.11.248

Bao, Y., Wang, M., Li, J., Yang, X., Wen, B., Wang, L., He, Y., Si, X., & Li, B. (2024). A starch-based 3D printed intelligent colorimetric film co-loaded natural pigments for visualizing food freshness: Effect of nozzle size on gel structure formation. *Food Hydrocolloids, 155*. https://doi.org/10.1016/j.foodhyd.2024.110218

Bulatović, V. O., Kovač, M., Grgić, D. K., Mandić, V., & Jozinović, A. (2025). Designing sustainable packaging materials: Citric acid-modified TPS/PLA blends with enhanced functional and eco-performance. *Polymers, 17*(19). https://doi.org/10.3390/polym17192571

Calo, J. R., Crandall, P. G., O'Bryan, C. A., & Ricke, S. C. (2015). Essential oils as antimicrobials in food systems - A review. *Food Control, 54*, 111–119. https://doi.org/10.1016/j.foodcont.2014.12.040

Cao, Y., Zhu, L., He, M., Tian, Y., Su, H., & Wang, F. (2025). Characterization of intelligent film based on chitosan/high-amylose starch and Portulaca oleracea L. flavonoids/anthocyanins/curcumin and its application in beef. *International Journal of Biological Macromolecules, 323*. https://doi.org/10.1016/j.ijbiomac.2025.147183

Castro-Rosas, J., Ferreira-Grosso, C. R., Gómez-Aldapa, C. A., Rangel-Vargas, E., Rodríguez-Marín, M. L., Guzmán-Ortiz, F. A., & Falfan-Cortes, R. N. (2017). Recent advances in microencapsulation of natural sources of antimicrobial compounds used in food - A review. *Food Research International, 102*, 575–587. https://doi.org/10.1016/j.foodres.2017.09.054

Channa, I. A., Ashfaq, J., Wasti, M., Memon, A. G., Hasan, M., Shar, M. A., AlHazaa, A., & Tiliakos, A. (2025). pH-responsive halochromic films based on carboxymethyl cellulose infused with anthocyanin or curcumin dye extracts for smart packaging applications. *Hybrid Advances, 10*. https://doi.org/10.1016/j.hybadv.2025.100476

Chen, Y. W., Cai, W. Q., Shi, Y. G., Dong, X. P., Bai, F., Shen, S. K., ... Zhu, X. (2020). Effects of different salt concentrations and vacuum packaging on the shelf-stability of Russian sturgeon (Acipenser gueldenstaedti) stored at 4°C. *Food Control, 109*, 106865. https://doi.org/10.1016/J.FOODCONT.2019.106865.

Cheng, H., Chen, L., McClements, D. J., Yang, T., Zhang, Z., Ren, F., Miao, M., Tian, Y., & Jin, Z. (2021). Starch-based biodegradable packaging materials: A review of their preparation, characterization and diverse applications in the food industry. *Trends in Food Science and Technology, 114*, 70–82. https://doi.org/10.1016/j.tifs.2021.05.017

Cheng, M., Cui, Y., Yan, X., Zhang, R., Wang, J., & Wang, X. (2022). Effect of dual-modified cassava starches on intelligent packaging films containing red cabbage extracts. *Food Hydrocolloids, 124*. https://doi.org/10.1016/j.foodhyd.2021.107225

Chia, M. R., Ahmad, I., & Phang, S. W. (2022). Starch/Polyaniline biopolymer film as potential intelligent food packaging with colourimetric ammonia sensor. *Polymers, 14*(6). https://doi.org/10.3390/polym14061122

Coimbra, P., Marona, B., Henriques, M. H. F., Campos, L., Gomes, D. M. G. S., Vitorino, C., Sousa, J. J. S., Braga, M. E. M., & Gaspar, M. C. (2023). Edible films based on potato and quince peels with potential for the preservation of cured cheese. *Food Packaging and Shelf Life, 40*. https://doi.org/10.1016/j.fpsl.2023.101176

Esfahani, A., Mohammadi Nafchi, A., Baghaei, H., & Nouri, L. (2022). Fabrication and characterization of a smart film based on cassava starch and pomegranate peel powder for monitoring lamb meat freshness. *Food Science and Nutrition, 10*(10), 3293–3301. https://doi.org/10.1002/fsn3.2918

Fan, S., Yin, X., Liu, X., Wang, G., & Qiu, W. (2024). Enhancing bread preservation through non-contact application of starch-based composite film infused with clove essential oil nanoemulsion. *International Journal of Biological Macromolecules, 263*. https://doi.org/10.1016/j.ijbiomac.2024.130297

Farha, A. K., Yang, Q. Q., Kim, G., Li, H. B., Zhu, F., Liu, H. Y., Gan, R. Y., & Corke, H. (2020). Tannins as an alternative to antibiotics. *Food Bioscience, 38*. https://doi.org/10.1016/j.fbio.2020.100751

Fotouhi, M., Nasihatkon, B., Solouki, S., & Seidi, S. (2023). A card instead of a lab: A ligand embedded in a bio-composite of starch/gelatin intelligent film for milk quality test followed by colorimetric analysis. *International Journal of Biological Macromolecules, 228*, 426–434. https://doi.org/10.1016/j.ijbiomac.2022.12.205

Gao, S., Song, H., Wang, Q., Zhang, X., Zhang, H., Wang, W., & Hou, H. (2024). Starch/poly (butylene adipate-co-terephthalate) blown films contained the quaternary ammonium salts with different N-alkyl chain lengths as antimicrobials. *Food Chemistry, 436*, Article 137650. https://doi.org/10.1016/j.foodchem.2023.137650

Gao, S., Sun, S., Zhao, J., Wang, W., & Hou, H. (2024a). A biodegradable pH-response packaging film with blueberry extract: Blown-extrusion fabrication, multifunctional activity, and kinetic investigation. *Food Chemistry, 449*. https://doi.org/10.1016/j.foodchem.2024.139217

Gao, S., Zhang, X., Jiang, J., Wang, W., & Hou, H. (2023). Starch/poly(butylene adipate-co-terephthalate) blown antimicrobial films based on ε-polylysine hydrochloride and different nanomontmorillonites. *International Journal of Biological Macromolecules, 253*(Pt 1), Article 126609. https://doi.org/10.1016/j.ijbiomac.2023.126609

Gaviria, Y. A. R., Palencia, N. S. N., Capello, C., Trevisol, T. C., Monteiro, A. R., & Valencia, G. A. (2021). Nanostructured pH-Indicator films based on Cassava Starch, laponite, and jambolan (Syzygium cumini) fruit manufactured by thermo-compression. *Starch/Staerke, 73*(5–6). https://doi.org/10.1002/star.202000208

George, A., Raj, D. M. A., Raj, A. D., Irudayaraj, A. A., Arumugam, J., Senthil kumar, M., Prabu, H. J., Sundaram, S. J., Al-Dhabi, N. A., Arasu, M. V., Maaza, M., & Kaviyarasu, K. (2020). Temperature effect on CuO nanoparticles: Antimicrobial activity towards bacterial strains. *Surfaces and Interfaces, 21*. https://doi.org/10.1016/j.surfin.2020.100761

Goiana, M. L., Rosa, M.d. F., Mattos, A. L. A., & Fernandes, F. A. N. (2025). Development of plasma-treated corn-starch-based film incorporated with acerola and grape pomace extract possessing pH-Sensing capability. *Polymers, 17*(7). https://doi.org/10.3390/polym17070938

Gui, H., Ma, W., Cao, Y., Chao, H., Fan, M., Dong, Q., & Li, L. (2024). Sustained release, antimicrobial, and antioxidant properties of modified porous starch-based biodegradable polylactic acid/polybutylene adipate-co-terephthalate/thermoplastic starch active packaging film. *International Journal of Biological Macromolecules, 267*, 131657. https://doi.org/10.1016/J.IJBIOMAC.2024.131657.

Gunaki, M. N., Masti, S. P., D'souza, O. J., Eelager, M. P., Kurabetta, L. K., Chougale, R. B., Kadapure, A. J., & Praveen Kumar, S. K. (2024). Fabrication of CuO nanoparticles embedded novel chitosan/hydroxypropyl cellulose bio-nanocomposites for active packaging of jamun fruit. *Food Hydrocolloids, 152*. https://doi.org/10.1016/j.foodhyd.2024.109937

Handayani, N. A., Purbasari, A., Sumardiono, S., Fatikah, A. F., & Nurrudina, F. (2025). The development of intelligent film from crosslinked–acylation cassava starch and purple sweet potato anthocyanin for monitoring Indian mackerel fish freshness. *Journal of Food Process Engineering, 48*(1). https://doi.org/10.1111/jfpe.70043

Heryanto, H., Artasasta, M. A., Tahir, D., Alomari, A., Partini, J., & Akouibaa, A. (2025). Development of a biodegradable smart packaging film with ZnO nanoparticles and anthocyanins for real-time freshness monitoring of shrimps. *Food Bioscience, 69*. https://doi.org/10.1016/j.fbio.2025.106966

Hu, D., Xu, Y., Gao, C., Meng, L., Feng, X., Wang, Z., Shen, X., & Tang, X. (2024). Preparation and characterization of starch/PBAT film containing hydroxypropyl-β-cyclodextrin/ethyl lauroyl arginate/cinnamon essential oil microcapsules and its application in the preservation of strawberry. *International Journal of Biological Macromolecules, 259*. https://doi.org/10.1016/j.ijbiomac.2024.129204

Huang, K., Wang, Y., Xu, Z., Zou, Z., Tang, Q., Li, H., & Peng, D. (2024b). Novel intelligent packaging films based on starch/PVA with Cu-ICA nanocrystal as functional compatibilizer for monitoring food freshness. *International Journal of Biological Macromolecules, 271*. https://doi.org/10.1016/j.ijbiomac.2024.132373

Huang, J., Zaynab, M., Sharif, Y., Khan, J., Al-Yahyai, R., Sadder, M., Ali, M., Alarab, S. R., & Li, S. (2024). Tannins as antimicrobial agents: Understanding toxic effects on pathogens. *Toxicon, 247*, Article 107812. https://doi.org/10.1016/j.toxicon.2024.107812

Jiang, C., Liu, G., Zhang, Q., Wang, S., & Zou, Y. (2024b). Preparation and characterization of intelligent packaging labels based on pea starch, κ-carrageenan and black raspberry extract for monitoring freshness of pork. *Food Chemistry X, 24*. https://doi.org/10.1016/j.fochx.2024.101850

Jiang, C., Yang, H., Liu, T., Zhang, Q., Zou, Y., & Wang, S. (2024a). Fabrication, characterization and evaluation of Manihot esculenta starch based intelligent packaging films containing gum ghatti and black currant (Ribes nigrum) extract for freshness monitoring of beef meat. *Food Chemistry X, 23*. https://doi.org/10.1016/j.fochx.2024.101616

Kılıç, N. N., & Kahyaoglu, L. N. (2025). Smart thermoplastic starch films with dual functionality: Quality preservation of tomatoes and spoilage detection in chicken. *International Journal of Biological Macromolecules, 330*. https://doi.org/10.1016/j.ijbiomac.2025.147898

Kamboj, A., Kaur, G., Jain, H., & Singh, L. (2025). The role of smart packaging and consumer psychology in reducing food waste: A review. *Food and Humanity, 5*, Article 100826. https://doi.org/10.1016/j.foohum.2025.100826

Kanth, S., Malgar Puttaiahgowda, Y., & Kulal, A. (2023). Synthesis, characterization, and antimicrobial activities of a starch-based polymer. *Carbohydrate Research, 532*, Article 108900. https://doi.org/10.1016/j.carres.2023.108900

Karnwal, A., Rauf, A., Jassim, A. Y., Selvaraj, M., Al-Tawaha, A. R. M. S., Kashyap, P., Kumar, D., & Malik, T. (2025). Advanced starch-based films for food packaging: Innovations in sustainability and functional properties. *Food Chemistry X, 29*. https://doi.org/10.1016/j.fochx.2025.102662

Kavas, E., Terzioğlu, P., & Sıcak, Y. (2023). Betanin and nano-calcium carbonate incorporated polyvinyl alcohol/Starch films for active and intelligent packaging applications. *Journal of Polymers and the Environment, 31*(11), 4919–4929. https://doi.org/10.1007/s10924-023-02868-0

Keshtiara, P., & Hadadzadeh, H. (2025). Halochromic smart packaging films based on tragacanth, salep, and bottle gourd cellulose microfibers containing Punica granatum var. pleniflora anthocyanins and antibacterial nanoparticles. *International Journal of Biological Macromolecules, 328*(Pt 2), Article 147591. https://doi.org/10.1016/j.ijbiomac.2025.147591

Khan, A., Ezati, P., & Rhim, J.-W. (2023). Alizarin: Prospects and sustainability for food safety and quality monitoring applications. *Colloids and Surfaces B: Biointerfaces, 223*, Article 113169. https://doi.org/10.1016/j.colsurfb.2023.113169

Khan, M. I., & Liu, J. (2022). Plant betalains: Recent applications in food freshness monitoring films. *Food Packaging and Shelf Life, 34*, Article 100921. https://doi.org/10.1016/j.fpsl.2022.100921

Khanoonkon, N., Dang, K. M., & Yoksan, R. (2024). Injection-molded thermoplastic cassava starch modified with single and mixed polyol plasticizers. *International Journal of Biological Macromolecules, 280*. https://doi.org/10.1016/j.ijbiomac.2024.136335

Kim, Y. H., Priyadarshi, R., Kim, J. W., Kim, J., Alekseev, D. G., & Rhim, J. W. (2022). 3D-Printed Pectin/Carboxymethyl Cellulose/ZnO Bio-Inks: Comparative analysis with the solution casting method. *Polymers, 14*(21). https://doi.org/10.3390/polym14214711

Koop, B. L., Zenin, E., Cesca, K., Valencia, G. A., & Monteiro, A. R. (2022). Intelligent labels manufactured by thermo-compression using starch and natural biohybrid based. *International Journal of Biological Macromolecules, 220*, 964–972. https://doi.org/10.1016/j.ijbiomac.2022.08.127

Kumar, R., Park, K., Ahn, K., Ansari, J. R., Sadeghi, K., & Seo, J. (2024). Maleic acid cross-linked starch/polyvinyl alcohol blend films with improved barrier properties for packaging applications. *International Journal of Biological Macromolecules, 271*(Pt 2), Article 132495. https://doi.org/10.1016/j.ijbiomac.2024.132495

Kurabetta, L. K., Masti, S. P., Gunaki, M. N., Hunashyal, A. A., Chougale, R. B., Dalbanjan, N. P., & Praveen Kumar, S. K. (2025). Vanillin reinforced cationic starch/poly(vinyl alcohol) based antimicrobial and antioxidant bioactive films: Sustainable food packaging materials. *Sustainable Food Technology, 3*(5), 1353–1364. https://doi.org/10.1039/d5fb00132c

La Fuente Arias, C. I., González-Martínez, C., & Chiralt, A. (2023). Lamination of starch/polyesters by thermocompression for food packaging purposes. *Sustainable Food Technology, 1*(2), 296–305. https://doi.org/10.1039/d2fb00038e

Lau, W. N., Mohammadi Nafchi, A., Zargar, M., Rozalli, N. H. M., & Mat Easa, A. (2024). Development and evaluation of Bauhinia Kockiana extract-incorporated sago starch intelligent film strips for real-time freshness monitoring of coconut milk. *International Journal of Biological Macromolecules, 260.* https://doi.org/10.1016/j.ijbiomac.2024.129589

Leandro, G. C., Laroque, D. A., Carciofi, B. A. M., & Valencia, G. A. (2024b). Development of a freshness indicator packaging system using cold atmospheric plasma on LLDPE–PVA with Laponite–anthocyanin biohybrid. *Polymer International, 73*(11), 1001–1012. https://doi.org/10.1002/pi.6679

Leandro, G. C., Laroque, D. A., Monteiro, A. R., Carciofi, B. A. M., & Valencia, G. A. (2024a). Current status and perspectives of starch powders modified by cold plasma: A review. *Journal of Polymers and the Environment, 32*(2), 510–523. https://doi.org/10.1007/s10924-023-03027-1

Lekurwale, S., Mahajan, S., Banerjee, S. K., & Banerjee, S. (2024). Systematic evaluations and integration of Assam indigenous Joha rice starch in intelligent packaging films for monitoring food freshness using beetroot extract. *International Journal of Biological Macromolecules, 277.* https://doi.org/10.1016/j.ijbiomac.2024.134332

Li, J., Bao, Y., Jiang, Q., Wen, B., Wang, L., He, Y., Si, X., & Li, B. (2024). Indicator-enhanced starch-based intelligent film for nondestructive monitoring of beef freshness: Different structural phenolic acids copigment anthocyanin. *Journal of Food Engineering, 383.* https://doi.org/10.1016/j.jfoodeng.2024.112241

Li, J., Bao, Y., Li, Z., Cui, H., Jiang, Q., Hou, C., Wang, Y., Wu, Y., Shang, J., Xiao, Y., Shu, C., Wang, Y., Wen, B., Si, X., & Li, B. (2023). Dual-function β-cyclodextrin/starch-based intelligent film with reversible responsiveness and sustained bacteriostat-releasing for food preservation and monitoring. *International Journal of Biological Macromolecules, 253.* https://doi.org/10.1016/j.ijbiomac.2023.127168

Lithi, I. J., Ahmed Nakib, K. I., Chowdhury, A. M. S., & Sahadat Hossain, M. (2025). A review on the green synthesis of metal (Ag, Cu, and Au) and metal oxide (ZnO, MgO, Co3O4, and TiO2) nanoparticles using plant extracts for developing antimicrobial properties. *Nanoscale Advances,* 7(9), 2446–2473. https://doi.org/10.1039/d5na00037h

Liu, Q., Gao, L., Yang, Q., Ji, N., Lei, D., Xiong, L., & Sun, Q. (2023). Incorporation of oxidized debranched starch/chitosan nanoparticles for enhanced hydrophobicity of corn starch films. *Food Packaging and Shelf Life, 35,* Article 101032. https://doi.org/10.1016/j.fpsl.2023.101032

Liu, Y., Xu, H., Chen, Z., Xie, Z., Wen, H., Chang, X., & Li, G. (2025). Development of lily starch films reinforced with chitosan–honeysuckle essential oil hybrid particles and cellulose nanofibers for enhanced properties. *Foods, 14*(4), 589. https://doi.org/10.3390/foods14040589

Lopes, J., Pettersen, M. K., Grøvlen, M. S., Sharmin, N., Li, K. D., Wetterhus, E., Ferreira, P., Coimbra, M. A., & Gonçalves, I. (2024). Heat-sealable bioplastic films of blended locust bean and potato byproducts for active packaging of fatty foods: Cheese and oat cookies as case studies. *Food Hydrocolloids, 147*. https://doi.org/10.1016/j.foodhyd.2023.109322

Lv, H., He, D., Xu, E., Jin, Z., Zhao, H., Yuan, C., Zhao, M., Gao, W., Li, J., Liu, Y., Wu, Z., Liu, P., & Cui, B. (2025). Starch-based biodegradable active intelligent packaging with color superimposition via emulsion electrospinning nanofiber for meat freshness monitoring and shelf-life extension. *Carbohydrate Polymers, 367*. https://doi.org/10.1016/j.carbpol.2025.123940

Lv, H., Wang, C., He, D., Zhao, H., Zhao, M., Xu, E., Jin, Z., Yuan, C., Guo, L., Wu, Z., Liu, P., & Cui, B. (2024). Intelligent food tag: A starch-anthocyanin-based pH-sensitive electrospun nanofiber mat for real-time food freshness monitoring. *International Journal of Biological Macromolecules, 256*. https://doi.org/10.1016/j.ijbiomac.2023.128384

Maldonado, A., Aguilar, T., Hauser, C., Wehnert, G., Söthje, D., Schlachter, H., Torres, A., Bruna, J., Valenzuela, X., & Rodríguez-Mercado, F. (2024). Ethylene scavenging films based on ecofriendly plastic materials and Nano-TiO2: Preparation, characterization, and in vivo evaluation. *Polymers, 16*(6), 853. https://doi.org/10.3390/polym16060853

Mali, S. N., & Pandey, A. (2024). Development of curcumin integrated smart pH indicator, antibacterial, and antioxidant waste derived Artocarpus lakoocha starch-based packaging film. *International Journal of Biological Macromolecules, 275*. https://doi.org/10.1016/j.ijbiomac.2024.133827

Mohammadi, M., & Fasihi, M. (2025). Eco-friendly polylactic acid/modified thermoplastic starch films enhanced with clove essential oil and cochineal for dual-functional active and intelligent food packaging. *Carbohydrate Polymers, 354*. https://doi.org/10.1016/j.carbpol.2025.123320

Morán, D., Suárez-Iglesias, O., Gutiérrez, G., & Matos, M. (2025). Advances in the use of starch-based matrices for antimicrobials loading: An analysis of natural sources and applications for active packaging for food safety. *Food Bioscience, 69*, Article 106853. https://doi.org/10.1016/j.fbio.2025.106853

Mustafa, P., Niazi, M. B. K., Jahan, Z., Rafiq, S., Ahmad, T., Sikander, U., & Javaid, F. (2021). Improving functional properties of PVA/starch-based films as active and intelligent food packaging by incorporating propolis and anthocyanin. *Polymers and Polymer Composites, 29*(9), 1472–1484. https://doi.org/10.1177/0967391120973503

Nasseri, R., Ngunjiri, R., Moresoli, C., Yu, A., Yuan, Z., & Xu, C. (2020). Poly(lactic acid)/acetylated starch blends: Effect of starch acetylation on the material properties. *Carbohydrate Polymers, 229*, 115453. https://doi.org/10.1016/J.CARBPOL.2019.115453.

Nazrin, A., Ilyas, R. A., Rajeshkumar, L., Hazrati, K. Z., Jamal, T., Mahardika, M., Aisyah, H. A., Atiqah, A., & Radzi, A. M. (2025). Lignocellulosic fiber-reinforced starch thermoplastic composites for food packaging application: A review of properties and food packaging abetted with safety aspects. *Food Packaging and Shelf Life, 47*, Article 101431. https://doi.org/10.1016/j.fpsl.2025.101431

Nguyen, T. T. T., Truong, N. L. Q., Chiem, T. N. Q., Truong, L. D. P., Nguyen, T. T. T., & Tran, T. V. (2025). Development of intelligent films based on jackfruit seed starch/polyvinyl alcohol incorporated with purple sweet potato peel anthocynanin as an indicator. *International Journal of Biological Macromolecules, 319*. https://doi.org/10.1016/j.ijbiomac.2025.145436

Nordin, A. H., Ahmad Ilyas, R., Ngadi, N., Baharuddin, N. H., Nordin, M. L., & Mohd Azami, M. S. (2024). Horse chestnut thermoplastic starch nanocomposite films reinforced with nanocellulose. *Physical Sciences Reviews, 9*(2), 1117–1130. https://doi.org/10.1515/psr-2022-0028

Noviagel, I., Heryanto, H., Putri, S. E., Rauf, I., & Tahir, D. (2024). Tapioca-starch-based bionanocomposites with fructose and titanium dioxide for food packaging and fertilization applications. *International Journal of Biological Macromolecules, 273*. https://doi.org/10.1016/j.ijbiomac.2024.132803

Pankaj, S. K., Wan, Z., De León, J. E., Mosher, C., Colonna, W., & Keener, K. M. (2017). High-voltage atmospheric cold plasma treatment of different types of starch films. *Starch/Staerke, 69*(11–12). https://doi.org/10.1002/star.201700009

Panwar, A., Kumar, S., Dhiman, A., Kumar, V., Gupta, D., & Sharma, A. (2024). pH sensitive indicator film based intelligent packaging systems of perishables: Developments and challenges of last decade. *Microchemical Journal, 207*, Article 111732. https://doi.org/10.1016/j.microc.2024.111732

Qazanfarzadeh, Z., Baptista, H., Nunes, C., & Kumaravel, V. (2025). Development of smart food packaging from bread waste-derived starch and carbon quantum dots. *Industrial Crops and Products, 233*, Article 121426. https://doi.org/10.1016/j.indcrop.2025.121426

Schaich, K. M., Tian, X., & Xie, J. (2015). Hurdles and pitfalls in measuring antioxidant efficacy: A critical evaluation of ABTS, DPPH, and ORAC assays. *Journal of Functional Foods, 14*, 111–125. https://doi.org/10.1016/j.jff.2015.01.043

Shahi, D. K., Awasthi, G. P., Rajendra Bahadur, G. C., Panthi, K. P., Chand, A. B., Shin, M., Kalauni, S. K., Bhattarai, N., Bhatt, L. R., Yu, C., & Joshi, M. K. (2025). Rhododendron arboreum Sm. Anthocyanin-infused starch, chitosan, and polyvinyl alcohol based composite films: Comparative analysis of physical, UV barrier, antioxidant and intelligent behavior. *International Journal of Biological Macromolecules, 302*. https://doi.org/10.1016/j.ijbiomac.2025.140532

Siangyai, R., Sujaridworakun, P., Aht-Ong, D., & Hongrattanavichit, I. (2024). Synthesis and characterization of zeolite – Titanium dioxide composite material for ethylene adsorbing and scavenging dual action in coated paper packaging for postharvest product. *Food Packaging and Shelf Life, 46*, Article 101399. https://doi.org/10.1016/j.fpsl.2024.101399

Sifuentes-Nieves, I., Hernández-Hernández, E., Neira-Velázquez, G., Morales-Sánchez, E., Mendez-Montealvo, G., & Velazquez, G. (2019). Hexamethyldisiloxane cold plasma treatment and amylose content determine the structural, barrier and mechanical properties of starch-based films. *International Journal of Biological Macromolecules, 124*, 651–658. https://doi.org/10.1016/j.ijbiomac.2018.11.211

Song, W., Wu, N., He, Y., Zhao, H., Xu, J., & Ren, L. (2024). Intelligent color changing packaging film based on esterified starch and black rice anthocyanins. *Food Chemistry X, 24*. https://doi.org/10.1016/j.fochx.2024.101930

Srinivasan, S., Ge, C., Lewis, C. L., Begum, S. S., Samui, A. B., & Noyes, H. F. (2025). Effect of chemical modification of corn starch for the development of films for packaging applications: Impact of glutaraldehyde and organically modified montmorillonite incorporation. *Carbohydrate Polymers, 367*. https://doi.org/10.1016/j.carbpol.2025.124019

Srivastava, V., Singh, S., & Das, D. (2023). Rice husk fiber-reinforced starch antimicrobial biocomposite film for active food packaging. *Journal of Cleaner Production, 421*, Article 138525. https://doi.org/10.1016/j.jclepro.2023.138525

Sun, Q., Zhang, L., Huang, M., Ma, M., Zeng, J., & Le, T. (2024). Modification of starch-derived graphene quantum dots as multifunctional nanofillers to produce polymer starch/polyvinyl alcohol composite films for active packaging. *LWT, 198*, Article 115953. https://doi.org/10.1016/j.lwt.2024.115953

Suryavanshi, V. R., Santhosh, R., Singhi, H., Thakur, R., Ahmed, J., Gaikwad, K. K., Kumawat, A. S., Roy, S., & Sarkar, P. (2025). Preparation and characterization of kodo millet starch/gum tragacanth/copper oxide nanoparticles-based antimicrobial food packaging films. *Materials Today Communications, 42*. https://doi.org/10.1016/j.mtcomm.2024.111443

Tang, T., Zhang, M., & Mujumdar, A. S. (2024). Film forming properties of 3D printed polysaccharide based gels and development of CO2 monitoring labels. *Food Packaging and Shelf Life, 46*. https://doi.org/10.1016/j.fpsl.2024.101401

Thakur, R., Pristijono, P., Scarlett, C. J., Bowyer, M., Singh, S. P., & Vuong, Q. V. (2019). Starch-based films: Major factors affecting their properties. *International Journal of Biological Macromolecules, 132*, 1079–1089. https://doi.org/10.1016/j.ijbiomac.2019.03.190

Tong, Z., He, W., Fan, X., & Guo, A. (2022). Biological function of plant tannin and its application in animal health. *Frontiers in Veterinary Science, 8*. https://doi.org/10.3389/fvets.2021.803657

Torche, A., Chouana, T., Akachat, B., Rekbi, F. M. L., Rahmani, Y., El Hadj, M. D. O., Uzun, Ş. A., Parlak, M. E., D'Elia, M., Rastrelli, L., & Sarıcaoğlu, F. T. (2025). Cassava starch films with green-synthesized ZnO nanoparticles from date palm pits for active packaging. *Food Chemistry*, 495. https://doi.org/10.1016/j.foodchem.2025.146392

Upadhyay, A., Kumar, P., Kardam, S. K., & Gaikwad, K. K. (2022). Ethylene scavenging film based on corn starch-gum acacia impregnated with sepiolite clay and its effect on quality of fresh broccoli florets. *Food Bioscience, 46*. https://doi.org/10.1016/j.fbio.2022.101556

Valencia, G. A., Luciano, C. G., Lourenço, R. V., & do Amaral Sobral, P. J. (2018). Microstructure and physical properties of nano-biocomposite films based on cassava starch and laponite. *International Journal of Biological Macromolecules, 107*, 1576–1583. https://doi.org/10.1016/j.ijbiomac.2017.10.031

Valencia, G. A., Sobral, P. J., & do, A. (2018). Recent trends on nano-biocomposite polymers for food packaging. In T. J. Gutiérrez (Ed.), *Polymers for Food Applications* (1st, pp. 101–130). Springer. https://doi.org/10.1007/978-3-319-94625-2_5.

Vedove, T. M. A. R. D., Maniglia, B. C., & Tadini, C. C. (2021). Production of sustainable smart packaging based on cassava starch and anthocyanin by an extrusion process. *Journal of Food Engineering, 289*. https://doi.org/10.1016/j.jfoodeng.2020.110274

Wang, Y., McClements, D. J., Peng, X., Xu, Z., Meng, M., Ji, H., Zhi, C., Ye, L., Zhao, J., Jin, Z., & Chen, L. (2024b). Effects of crosslinking agents on properties of starch-based intelligent labels for food freshness detection. *International Journal of Biological Macromolecules, 261*. https://doi.org/10.1016/j.ijbiomac.2024.129822

Wang, Y., McClements, D. J., Zhang, Z., Zhang, R., He, K., Lin, Z., Peng, X., Xu, Z., Meng, M., Ji, H., Zhao, J., Jin, Z., & Chen, L. (2024c). High water resistance starch based intelligent label for the freshness monitoring of beverages. *Food Chemistry*, 459. https://doi.org/10.1016/j.foodchem.2024.140383

Wang, B., Wang, X., Ma, H., Li, J., Cao, D., Xu, J., Zeng, J., Gao, W., & Chen, K. (2025). Enhancing the mechanical and barrier properties of starch-based green packaging films via acetalization reaction and hydrogen bond dual crosslinking. *Carbohydrate Polymers, 364*, Article 123787. https://doi.org/10.1016/j.carbpol.2025.123787

Wang, L., Yang, C., Deng, X., Peng, J., Zhou, J., Xia, G., Zhou, C., Shen, Y., & Yang, H. (2023). A pH-sensitive intelligent packaging film harnessing Dioscorea zingiberensis starch and anthocyanin for meat freshness monitoring. *International Journal of Biological Macromolecules, 245*. https://doi.org/10.1016/j.ijbiomac.2023.125485

Wang, S., Zhao, B., Hao, Y., & He, Q. (2024a). Development of pH response film based on potato starch/polyvinyl alcohol using Kyoho grape skin extract to monitor pork freshness. *Polymer Engineering and Science, 64*(8), 3431–3443. https://doi.org/10.1002/pen.26744

Wołosiak, R., Drużyńska, B., Derewiaka, D., Piecyk, M., Majewska, E., Ciecierska, M., Worobiej, E., & Pakosz, P. (2022). Verification of the conditions for determination of antioxidant activity by ABTS and DPPH Assays—A practical approach. *Molecules, 27*(1), 50. https://doi.org/10.3390/molecules27010050

Wu, H., Li, T., Peng, L., Wang, J., Lei, Y., Li, S., Li, Q., Yuan, X., Zhou, M., & Zhang, Z. (2023a). Development and characterization of antioxidant composite films based on starch and gelatin incorporating resveratrol fabricated by extrusion compression moulding. *Food Hydrocolloids, 139*. https://doi.org/10.1016/j.foodhyd.2023.108509

Wu, L., Lv, S., Wei, D., Zhang, S., Zhang, S., Li, Z., Liu, L., & He, T. (2023c). Structure and properties of starch/chitosan food packaging film containing ultra-low dosage GO with barrier and antibacterial. *Food Hydrocolloids, 137*. https://doi.org/10.1016/j.foodhyd.2022.108329

Wu, H., Wang, J., Li, T., Lei, Y., Peng, L., Chang, J., Li, S., Yuan, X., Zhou, M., & Zhang, Z. (2023b). Effects of cinnamon essential oil-loaded Pickering emulsion on the structure, properties and application of chayote tuber starch-based composite films. *International Journal of Biological Macromolecules, 240*. https://doi.org/10.1016/j.ijbiomac.2023.124444

Wu, X., Yan, X., Zhang, J., Wu, X., Zhang, Q., & Zhang, B. (2024). Intelligent films based on dual-modified starch and microencapsulated Aronia melanocarpa anthocyanins: Functionality, stability and application. *International Journal of Biological Macromolecules, 275*. https://doi.org/10.1016/j.ijbiomac.2024.134076

Xi, W., Liu, P., Ling, J., Xian, D., Wu, L., Yang, Y., Zhang, J., & Xie, F. (2024). Pregelatinized high-amylose starch enables easy preparation of flexible and antimicrobial composite films for fresh fruit preservation. *International Journal of Biological Macromolecules, 254*(Pt 2), Article 127938. https://doi.org/10.1016/j.ijbiomac.2023.127938

Xiao, D. (2016). Gas discharge and gas insulation. *Energy and Environment Research in China, 6*. https://doi.org/10.1007/978-3-662-48041-0

Xu, R., Xia, L., Tang, Q., Tang, F., Pang, S., Li, H., & Zou, Z. (2024). High-performance carboxymethyl starch/PVA based intelligent packaging films engineered with Cu-Trp nanocrystal as functional compatibilizer. *Food Chemistry, 454*. https://doi.org/10.1016/j.foodchem.2024.139696

Yu, Y., Zhou, J., Chen, Q., Xie, F., Zhang, D., He, Z., Cheng, S., & Cai, J. (2024). Self-reinforced multifunctional starch nanocomposite film for litchi fruit postharvest preservation. *Chemical Engineering Journal, 486*, Article 150262. https://doi.org/10.1016/j.cej.2024.150262

Zhang, M., & Chen, H. (2023). Development and characterization of starch-sodium alginate-montmorillonite biodegradable antibacterial films. *International Journal of Biological Macromolecules, 233*. https://doi.org/10.1016/j.ijbiomac.2023.123462

Zhang, X., Chen, X., Dai, J., Cui, H., & Lin, L. (2023). A pH indicator film based on dragon fruit peel pectin/cassava starch and cyanidin/alizarin for monitoring the freshness of pork. *Food Packaging and Shelf Life, 40*. https://doi.org/10.1016/j.fpsl.2023.101215

Zhang, W., Liu, W., Qi, X., Chen, Y., Yu, Q., Shen, M., Wen, H., & Xie, J. (2025). Natural pigment-based smart packaging films for real-time food spoilage monitoring: Compositions, color-changing mechanisms, applications, and challenges. *Food Packaging and Shelf Life, 49*, Article 101484. https://doi.org/10.1016/j.fpsl.2025.101484

Zhu, W., Zhang, D., Liu, X., Ma, T., He, J., Dong, Q., Din, Z.u., Zhou, J., Chen, L., Hu, Z., & Cai, J. (2022). Improving the hydrophobicity and mechanical properties of starch nanofibrous films by electrospinning and cross-linking for food packaging applications. *LWT*, 169. https://doi.org/10.1016/j.lwt.2022.114005

Zong, Z., Liu, M., Chen, H., Farag, M. A., Wu, W., Fang, X., Niu, B., & Gao, H. (2023). Preparation and characterization of a novel intelligent starch/gelatin binary film containing purple sweet potato anthocyanins for Flammulina velutipes mushroom freshness monitoring. *Food Chemistry*, 405. https://doi.org/10.1016/j.foodchem.2022.134839

Zou, Y., Wang, S., Liu, T., Zhang, Q., Zhang, Y., & Jiang, C. (2025). Development and evaluation of pH-sensitive Euryale ferox starch-based films containing nano-SiO2 and Chinese rose (Rosa chinensis) extract for freshness monitoring of chicken breast meat. *International Journal of Biological Macromolecules, 290.* https://doi.org/10.1016/j.ijbiomac.2024.138959

CHAPTER FOUR

Barrier, thermal and mechanical properties of starch-based smart packaging

Milad Tavassoli[1, 2], Masoud Aman Mohammadi[3], Mansoureh Taghizadeh[4], Fatemeh Madanifar[5]

[1]Cellular and Molecular Research Center, Yasuj University of Medical Sciences, Yasuj, Iran
[2]Department of Nutrition, Faculty of Health and Nutrition Sciences, Yasuj University of Medical Sciences, Yasuj, Iran
[3]Student Research Committee, Department of Food Technology, Faculty of Nutrition Science and Food Technology, Nutritional and Food Technology Research Institute, Shahid Beheshti University of Medical Sciences, Tehran, Iran
[4]Halal Research Center of IRI, Food and Drug Administration, Ministry of Health and Medical Education, Tehran, Iran
[5]Student Research Committee, Yasuj University of Medical Sciences, Yasuj, Iran

4.1 Introduction

Starch-based smart packaging represents a significant advancement in sustainable food systems, offering an eco-friendly alternative to conventional petroleum-based plastics (Arruda et al., 2025; D'Almeida & de Albuquerque, 2024; Kusuma et al., 2024a). The global plastic pollution crisis has intensified the need for biodegradable materials that can reduce environmental impact while maintaining product safety and extending shelf life (Garavito et al., 2024; Upadhyay et al., 2024). Starch, an abundant, renewable, and cost-effective natural polymer, is a promising candidate for such applications due to its inherent biodegradability, biocompatibility, and excellent film-forming capabilities (Deng, Su, et al., 2024; Samsudin & Hani, 2017).

Traditional packaging primarily serves passive roles, but modern innovations integrate active and intelligent functions to enhance food preservation, reduce waste, and improve consumer experience (Rajan & Wani, 2025; Ribeiro et al., 2024). Smart packaging systems integrate indicators and sensors, such as biosensors, time-temperature indicators, and antimicrobial films, to provide real-time monitoring and extend product shelf life (Mussagy et al., 2024; Sani et al., 2024). Starch-based materials are being developed to incorporate these smart functionalities, leading to active and

Starches in Smart Packaging
ISBN: 978-0-443-45396-0
https://doi.org/10.1016/B978-0-443-45396-0.00003-7

intelligent packaging solutions (Murtaza et al., 2024; Rajan & Wani, 2025; Sani et al., 2024).

Despite its potential, starch-based packaging materials inherently possess limitations that hinder their widespread adoption, including low mechanical strength, poor barrier properties against moisture and gases, and high sensitivity to water (Liu et al., 2022; Marichelvam et al., 2019; Yudhistira et al., 2024). To address these challenges, researchers are focusing on enhancing the physicochemical and functional performances of starch films through various modification techniques (Pei et al., 2024; Yudhistira et al., 2024). These innovations involve polymer modifications, the incorporation of bio-nanocomposites, and advanced processing techniques (Karnwal et al., 2025; Xu et al., 2022). For instance, combining starch with other biopolymers like polyvinyl alcohol (PVA) can improve mechanical strength and barrier properties, offering a viable replacement for traditional plastic films (Abedi-Firoozjah et al., 2023; Huang et al., 2024). Intelligent starch-based packaging often incorporates pH indicators, frequently derived from natural colorants like anthocyanins or betalains, to visually signal food freshness or spoilage (Hamzah et al., 2021; Surendhiran et al., 2023; Vedove et al., 2021). These indicators change color in response to changes in the food's environment, such as pH shifts caused by microbial growth (Chen, Zhu, et al., 2024; Mussagy et al., 2024). For instance, a smart packaging film made from sago starch with anthocyanin from red cabbage demonstrated its ability to act as a pH indicator (Hamzah et al., 2021). Similarly, subcritical water-modified oat starch films reinforced with cellulose nanofibrils and beetroot betalain showed promise as pH biosensors (Surendhiran et al., 2023).

Nanomaterials are crucial for enhancing the functional properties of starch-based packaging. The incorporation of various nanofillers, such as starch nanocrystals, cellulose nanofibers, and inorganic carbon materials, improves mechanical strength, barrier properties, and even introduces antimicrobial or antioxidant capabilities (Rout & Pradhan, 2024; Singh et al., 2024). Carbon quantum dots derived from bread waste have been used in starch films to provide UV-barrier, photoluminescent, and antioxidant activities, creating smart films that also address food waste (Rout & Pradhan, 2024). ZnO nanoparticles embedded in starch biofilms also provide antimicrobial properties for food packaging (Kumar & Gautam, 2019).

The ultimate goal of starch-based smart packaging is to extend the shelf life of food products, reduce food waste, and provide real-time information to consumers about product quality and safety, all while being environmentally responsible (Mallick et al., 2020; Murtaza et al., 2024). The global smart packaging market, particularly in the food and drinks industry, is experiencing significant growth, with an expected incremental growth of $37.68 billion by 2023, and a compound annual growth rate (CAGR) of over 5.4% 38. However, challenges remain, including the high production costs and inferior mechanical properties of some bioplastics compared to conventional plastics (Mussagy et al., 2024). Research continues to focus on optimizing material properties, improving processing technologies, and ensuring the safety and sustainability of these innovative packaging solutions (Karnwal et al., 2025; Mussagy et al., 2024; Xu et al., 2022). The overarching goal of starch-based smart packaging is to extend food shelf life, reduce waste, and ensure consumer safety, all while reducing reliance on environmentally harmful plastics. With the global smart packaging market showing rapid growth, particularly in the food and beverage sector, starch-based systems represent a critical area of innovation. However, challenges remain, including production costs and the need for improved material properties. Ongoing research continues to refine these systems, balancing functionality, sustainability, and scalability to meet the demands of future food systems.

4.2 Overview on starch-based smart packaging

The escalating global plastic pollution crisis necessitates the development of such eco-friendly alternatives to petroleum-derived plastics (Arruda et al., 2025; Nath et al., 2023). Starch, a natural biopolymer, offers a promising foundation for these materials due to its inherent properties, though it often requires modification to overcome limitations like low mechanical strength, poor barrier properties, and high moisture sensitivity (Kusuma et al., 2024a; Liu et al., 2022; Xu et al., 2022). The fundamental concept behind starch-based smart packaging involves integrating various functionalities into starch matrices. These functionalities can include active properties like antimicrobial and antioxidant capabilities, as well as intelligent features such as sensing and indicating food freshness or spoilage (Kusuma et al., 2024a; Rout & Pradhan, 2024; Yudhistira et al., 2024). Starch, a natural polymer primarily composed of glucosyl units, is abundant in plants and readily available 1718. Its film-forming properties make it suitable for developing packaging films and coatings (Chen, Zhu, et al., 2024; Xu et al., 2022). A key driver for the development of starch-based smart packaging is the increasing global plastic pollution crisis (Arruda et al., 2025; Forfora et al., 2024). Traditional plastic packaging, despite its desirable mechanical resistance and moldability, poses significant environmental challenges due to its non-biodegradability and waste generation (Nakamoto et al., 2023). Starch-based materials offer a sustainable solution as they are biodegradable, renewable, and have a reduced environmental impact (Deng, Su, et al., 2024; Liu et al., 2022). However, native starch-based films inherently possess certain limitations, including weak mechanical strength, poor barrier properties against moisture and gases, and high moisture sensitivity (Dutta & Sit, 2024; Xu et al., 2022). To overcome these drawbacks and enhance their performance for practical applications, various modification strategies are employed. These strategies often involve blending starch with other biopolymers, incorporating nanofillers, or introducing active ingredients (Dutta & Sit, 2024; Kusuma et al., 2024a). Innovations in starch-based smart packaging include the integration of various functional components. Bioactive compounds, such as antioxidants, can be incorporated into starch films to prevent oxidative rancidity and extend the shelf life of food products (Dutta & Sit, 2024; Gao et al., 2024; Pei et al., 2024). For example Fig. 4.1, propyl gallate has been successfully integrated into starch/poly (butylene adipate-co-terephthalate) (PBAT) blown films, where starch acts as a gatekeeper for controlled release, providing sustained antioxidant protection to packaged foods like peanut butter (Gao et al., 2024).

According to Fig. 4.1, the development of starch-based active packaging is also a growing area, with films engineered to release active compounds or scavenge undesirable substances (Abedi-Firoozjah et al., 2023; Murtaza et al., 2024). For instance, starch/polyvinyl alcohol films containing nano-sized substituted imidazolate material (SIM-1) can monitor food freshness in real-time due to their ammonia-sensitive and antibacterial abilities (Huang et al., 2024). The manufacturing processes for starch-based smart packaging vary, including solvent casting, extrusion, and electrospinning (Qazanfarzadeh et al., 2025; Vedove et al., 2021). Extrusion, for example, has been successfully used to produce sustainable smart packaging based on cassava starch and anthocyanin (Vedove et al., 2021). According to Fig. 4.2, electrospinning allows for the creation of fibers with a high surface area-to-volume ratio, enabling encapsulation of bioactive compounds for controlled release, which can be applied as food additives or in active and intelligent food packaging (Amorim et al., 2024). Electrospinning is a technique used to produce starch fibers with a high surface area-to-volume ratio, allowing for the encapsulation of bioactive compounds. These electrospun fibers can be used as food

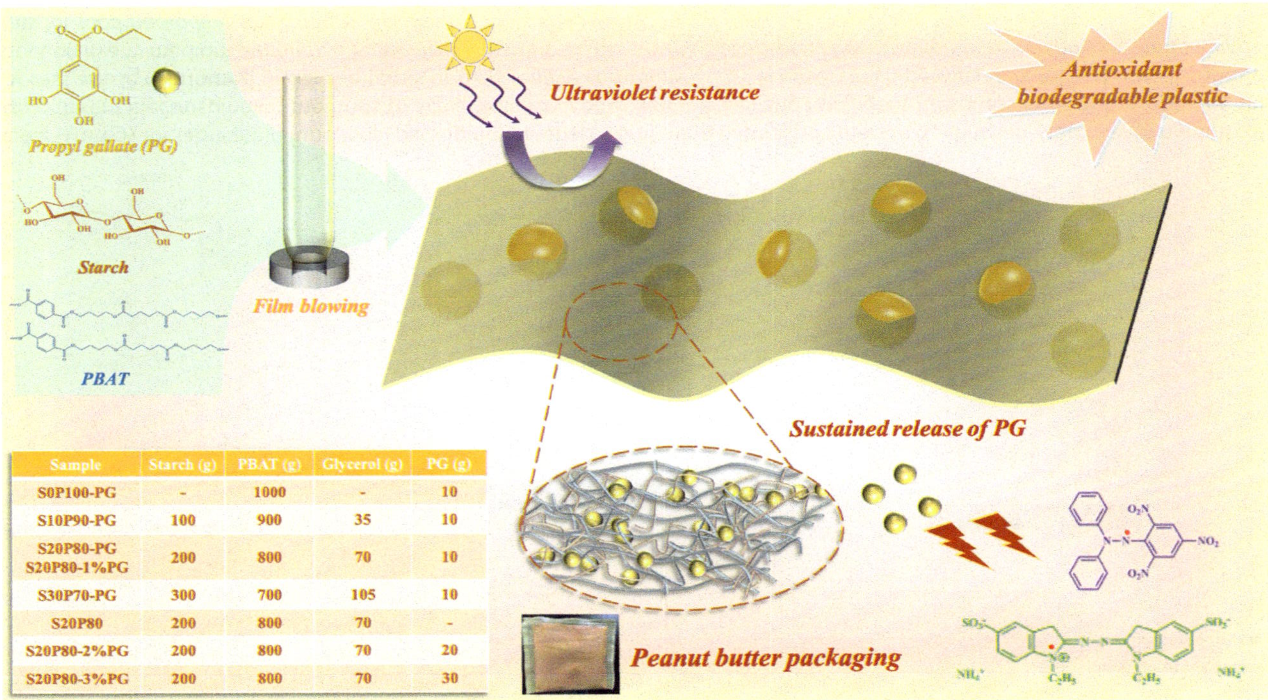

Sample	Starch (g)	PBAT (g)	Glycerol (g)	PG (g)
S0P100-PG	-	1000	-	10
S10P90-PG	100	900	35	10
S20P80-PG S20P80-1%PG	200	800	70	10
S30P70-PG	300	700	105	10
S20P80	200	800	70	-
S20P80-2%PG	200	800	70	20
S20P80-3%PG	200	800	70	30

Figure 4.1 Formulation and film blowing of PBAT/Starch blends plasticized with glycerol and containing Poplar Galate (PG) as a natural antioxidant for ultraviolet-resistant, biodegradable food packaging. *Source:(Gao, S., Li, M., Zhai, X., Wang, W., & Hou, H. (2024). Starch as a smart, cheap, and green gatekeeper for the controlled release of propyl gallate from antioxidant biodegradable packaging films. Food Chemistry, 453, 139627) with reproducing permission from Elsevier under license number 6180081275660.*

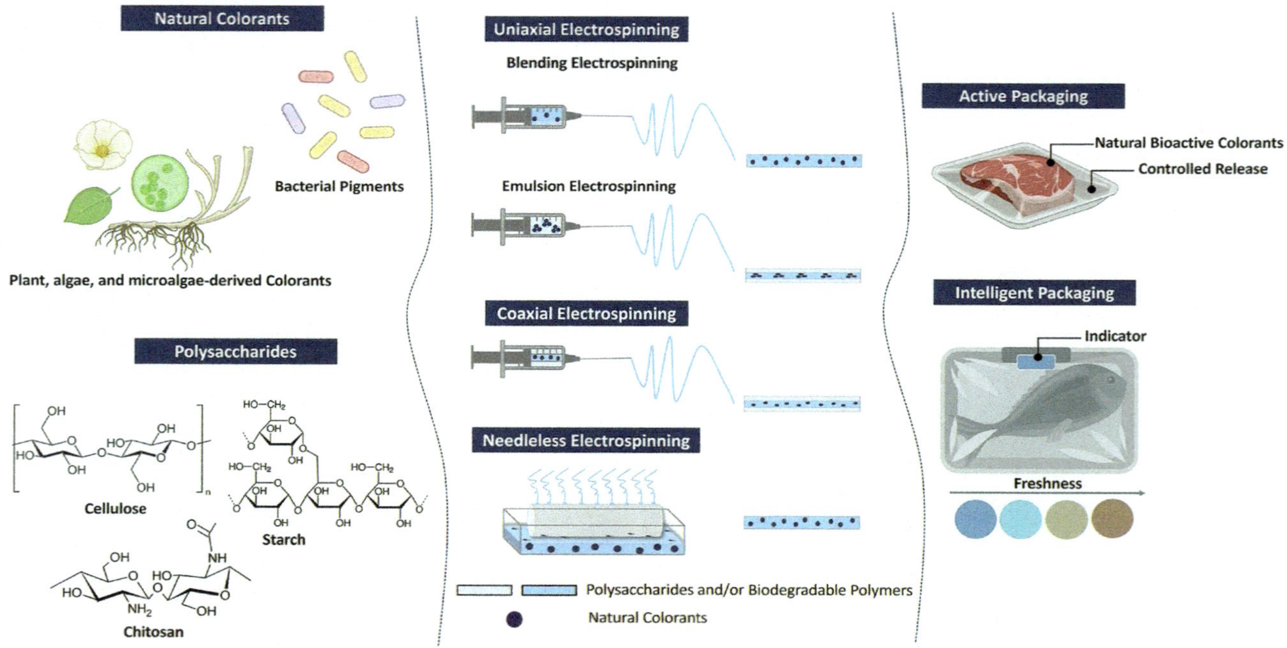

Figure 4.2 Trategy for fabricating functional packaging materials, combining natural polymers (e.g., chitosan, starch) with natural colorants through different electrospinning methods to achieve active (controlled release) and intelligent (freshness indicator) properties, this is an open access article distributed under the terms of the Creative commons CC BY license, which permits unrestricted use, distribution, and reproduction in any medium. *Amorim, L. F., Mouro, C., & Gouveia, I. C. (2024). Electrospun fiber materials based on polysaccharides and natural colorants for food packaging applications. Cellulose, 31(10), 6043–6069.*

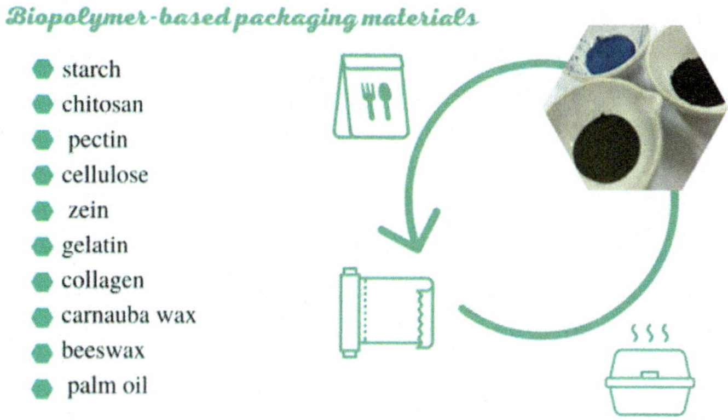

Figure 4.3 Common natural polymers and substances used in the development of biopolymer-based packaging materials. *Source:(Nakamoto, M. M., Assis, M., de Oliveira Filho, J. G., & Braga, A. R. C. (2023). Spirulina application in food packaging: Gaps of knowledge and future trends. Trends in Food Science & Technology, 133, 138–147) with reproducing permission from Elsevier under license number 6180090291727.*

additives or in active and intelligent food packaging. Examples include biofortification with folic acid, anthocyanins as pH indicators in fish meat, and antifungal essential oils in bread (Fonseca et al., 2024).

Also, blending starch with PVA can significantly enhance the mechanical and barrier properties, making them potential replacements for traditional plastic-based food packaging films (Abedi-Firoozjah et al., 2023; Deng, Su, et al., 2024). Chitosan, derived from chitin, is another biopolymer frequently blended with starch. Chitosan offers good mechanical properties and inherent antimicrobial activity, which can be imparted to the composite film (Nakamoto et al., 2023; Rodrigues et al., 2021). Fig. 4.3 depicts a range of biopolymers utilized in packaging, which includes starch, chitosan, pectin, cellulose, zein, gelatin, collagen, carnauba wax, beeswax, and palm oil, along with their respective applications in packaging, such as single-use food packaging and films. The application of these materials contributes to the variety of biopolymer-based packaging options (Nakamoto et al., 2023).

Starch-based smart packaging can integrate indicators and sensors to provide real-time information about food quality and safety (Murtaza et al., 2024; Sani et al., 2024). pH-responsive indicators, often derived from natural pigments like anthocyanins, carotenoids, and betalains, are particularly effective. These natural colorants change hue based on pH shifts, which are often correlated with food spoilage (Chen, Zhu, et al., 2024; Mussagy et al., 2024). For instance, anthocyanins extracted from red cabbage or purple star apple peels can be incorporated into starch films to create pH biosensors that visibly indicate food freshness (Chim-Chi et al., 2024; Hamzah et al., 2021). Fig. 4.4 illustrates the components of smart packaging films, including time/temperature and freshness/pH indicators (natural and synthetic), gas sensors, biosensors, and different bio-based polymers (non-biodegradable and biodegradable, including starch-based). The production process involves biopolymer selection, natural pH indicator integration, and methods like casting, layer-by-layer assembly, or extrusion (Mussagy et al., 2024).

The core function of smart and intelligent packaging: integrating indicators and sensors to monitor internal conditions such as gases, pH, and temperature (Joshi et al., 2024).

Despite significant advancements, challenges remain in the widespread application of starch-based smart packaging. These include issues related to water resistance, mechanical fragility, scalability of production, and cost-effectiveness compared to conventional plastics (Kusuma et al., 2024a; Xu et al., 2022). Research is ongoing to improve these aspects, focusing on modifying starch for better performance, incorporating advanced nanomaterials, and optimizing processing techniques (Liu et al., 2022; Yudhistira et al., 2024). For example, the printability of starch-based films using inkjet printing for QR codes and text has been explored, indicating potential for integrated information systems (Żołek-Tryznowska, Piłczyńska, Murawski, Jeznach, & Niczyporuk, 2024). Efforts are also being made to develop multi-layer active biopolymer systems to reduce gas permeability and improve overall quality (Nowak et al., 2025). The broader context of smart packaging, as shown in various shelf-life extension strategies, includes modified atmosphere packaging, absorbent food trays, and smart stickers. Starch-based solutions are poised to integrate further into these advanced systems, providing sustainable and effective packaging for diverse food products (Saxena et al., 2024). The development of starch-based smart packaging is crucial for achieving a more sustainable and environmentally friendly food industry, mitigating plastic pollution while ensuring food safety and quality for consumers (Arruda et al., 2025; Bangar et al., 2021).

4.3 Barrier properties of starch-based smart packaging

4.3.1 Intelligent packaging

Intelligent packaging is an emerging technology that utilizes communication functions to enhance food quality and safety through monitoring

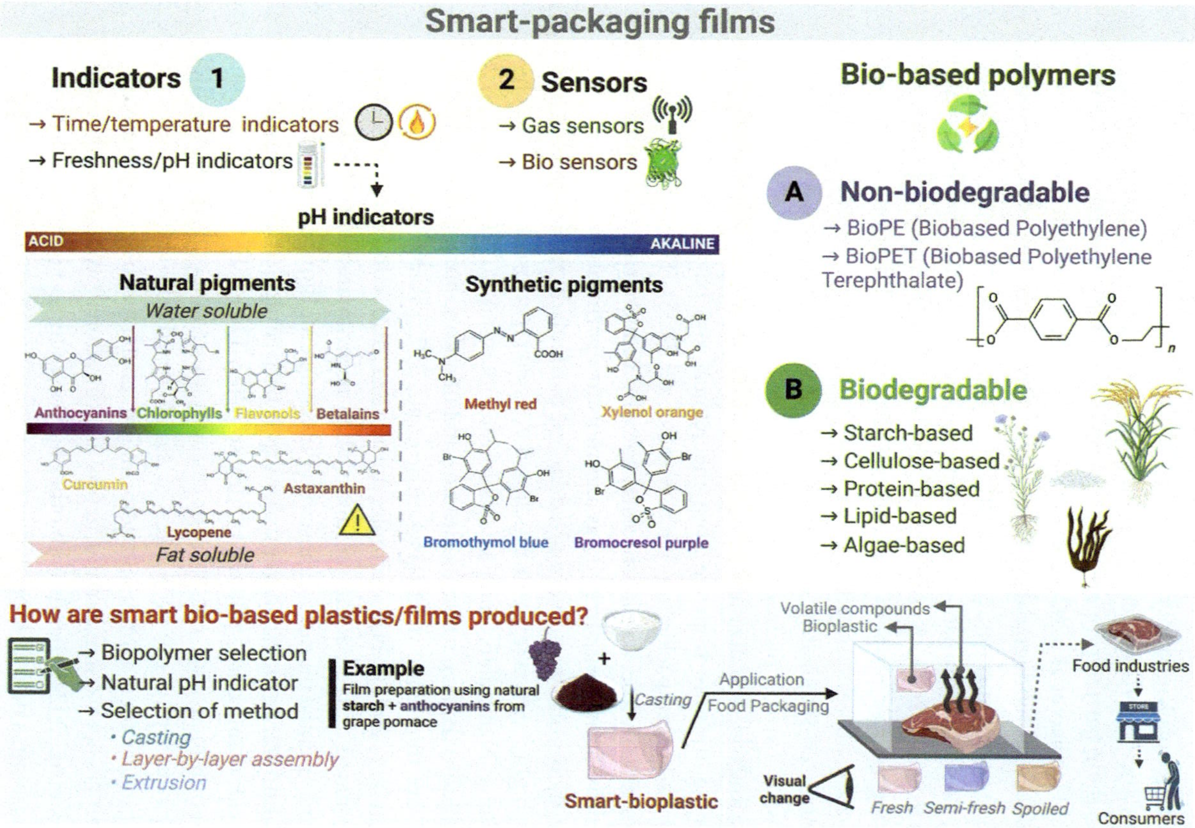

Figure 4.4 Schematic overview of smart-packaging films, showing the integration of pH indicators with bio-based polymers to produce biodegradable plastics that provide a visual freshness indicator for food products. *Source:(Mussagy, C. U., Oliveira, G., Ahmad, M., Mustafa, A., Herculano, R. D., & Farias, F. O. (2024). Halochromic properties of carotenoid-based films for smart food packaging. Food Packaging and Shelf Life, 44, 101325.) with reproducing permission from Elsevier under license number 6180090415530.*

and information provision. Unlike traditional passive packaging, intelligent systems can sense environmental changes and inform users about package conditions, incorporating devices capable of detecting and providing information about packaged food properties (Müller & Schmid, 2019). A critical aspect of their performance, particularly in intelligent packaging applications, is their barrier properties against gases and moisture (Rostamabadi et al., 2024; Shao et al., 2024; Singh et al., 2024). While starch is abundant, renewable, and biodegradable, pure starch films inherently possess limitations such as low mechanical strength, high moisture sensitivity, and inadequate barrier performance, primarily due to their hydrophilic nature (Karnwal et al., 2025; Kusuma et al., 2024a). One primary approach to improving barrier properties involves chemical modifications of starch. For instance, the esterification of starch has been shown to produce films with enhanced intelligent features and colorimetric responses when combined with natural pigments like black rice anthocyanin (Song et al., 2024). Similarly, acetalization reactions and hydrogen bond dual crosslinking can significantly improve both mechanical and barrier properties of starch-based green packaging films (Wang et al., 2025). These chemical alterations modify the starch structure, thereby reducing its hydrophilicity and improving its resistance to gas and moisture permeation.

The incorporation of various additives, particularly nanomaterials, has proven highly effective in enhancing the barrier performance of starch-based films. Nanofillers, such as cellulose nanocrystals (CNC) or bacterial nanocellulose (BNC), montmorillonite, and titanium dioxide nanoparticles (TiO_2), are frequently used (Jayakumar et al., 2022; Karnwal et al., 2025; Singh et al., 2024). For example, lotus seed drill core powder starch (LCPS)-based active packaging films, when incorporated with CNC and grapefruit essential oil-corn nanostarch Pickering emulsion, exhibited lower light transmittance and stronger water vapor barrier properties compared to films made from corn, potato, or rice starches (Deng, Hu, et al., 2024). BNC, combined with red cabbage anthocyanins, not only improves the structural and physical properties of starch-based pH-sensing films but also influences their surface and colorimetric characteristics, demonstrating the interplay between material structure and functional properties (Mahović Poljaček et al., 2024).Blends with other biopolymers also offer a viable strategy. Chitosan, a marine biopolymer known for its non-toxicity, biodegradability, and antimicrobial properties, is often combined with starch to create composite films with improved barrier functions (Chaudhary et al., 2022). Multi-layer systems, such as films comprising furcellaran, chitosan, and gelatin hydrolyzates, can be engineered to reduce gas permeability, especially when active ingredients like silver nanoparticles (AgNPs), montmorillonite, and curcumin-capsaicin mixtures are incorporated into specific layers (Jamróz et al., 2023). Also, ZnO-NPs significantly impact packaging material properties, including mechanical strength, thermal stability, optical characteristics, and barrier properties (Gao et al., 2025). Alkyl ketene dimer (AKD) and corn starch have proven effective as barrier-enhancing additives, with AKD at 7 g/m^2 coating weight showing superior grease resistance compared to starch (Todorova et al., 2022). Polyvinyl alcohol-based coatings combined with AKD and nanoclay significantly improve barrier performance, reducing water vapor transmission rates from 533 to 1.3 $g/m^2 \cdot day$ (Shen et al., 2021). Li, Liu, et al. (2024) developed a bilayer film using polyvinyl alcohol-chitosan/nano-ZnO/sodium alginate combined with anthocyanin extract, demonstrating superior physical properties, lower water vapor permeability, and broader pH sensitivity ($\Delta E > 10$) compared to single-layer films (Li, Liu, et al., 2024). Xu et al. (2024) developed a multifunctional sodium carboxymethyl cellulose film incorporating wheat straw cellulose fibers, cellulose nanofibers, and polydiacetylene/ZnO nanoparticles, achieving superior mechanical strength (>50 folds vs. 10 for ordinary paper), antimicrobial activity, UV protection, and visual spoilage indication for high-protein foods like shrimp (Xu et al., 2024). Similarly, Chen et al. (2020) created pH-responsive cellulose nanofiber films using purple sweet potato anthocyanins and oregano essential oil, demonstrating 99.99% inhibition against *E. coli* and *L. monocytogenes* while providing colorimetric pH indication (Chen et al., 2020). These nanocellulose-based composites offer enhanced barrier properties, thermal stability, and biodegradability compared to petroleum-based packaging (Dong et al., 2023), representing a viable solution for sustainable food packaging applications. Fig. 4.5 depicts a multi-layer packaging structure with components like CHIT (chitosan) and FUR (furcellaran) combined with active ingredients such as CUR-LEO (curcumin-lemongrass essential oil), demonstrating how these layers contribute to active and intelligent packaging functions. The careful selection and arrangement of these layers, along with the incorporation of active substances, lead to enhanced barrier performance and active functionality (Jamróz et al., 2023).

Natural extracts, such as quercetin, elderberry extract, and wild blackberry extract, are also integrated into starch matrices for functional enhancement (Cvetković et al., 2025; Shafi et al., 2025). Starch films containing quercetin exhibit significant barrier properties and contribute to sustainable

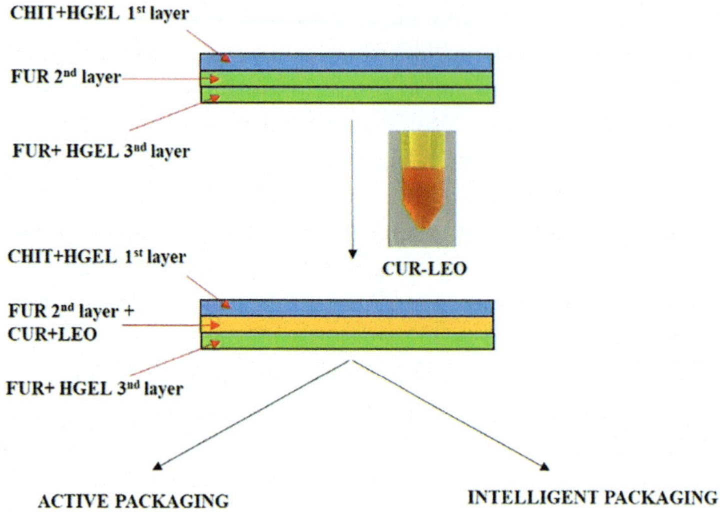

Figure 4.5 Schematic of a multi-functional, layered packaging film. The structure consists of a CHIT + HGEL (chitosan + Hydrogel) base layer, a Fur (furcellaran) middle layer, and a Fur + HGEL top layer, with active compounds (CUR-LEO: Curcumin-Lemon essential oil) incorporated for active packaging and other compounds (CUR + LEO) added for intelligent packaging functionality, this is an open access article distributed under the terms of the Creative commons CC-BY license, which permits unrestricted use, distribution, and reproduction in any medium. *Jamróz, E., Cabaj, A., Tkaczewska, J., Kawecka, A., Krzyściak, P., Szuwarzyński, M., ... Juszczak, L. (2023). Incorporation of curcumin extract with lemongrass essential oil into the middle layer of triple-layered films based on furcellaran/chitosan/gelatin hydrolyzates–In vitro and in vivo studies on active and intelligent properties. Food Chemistry, 402, 134476.*

food packaging solutions (Shafi et al., 2025). Elderberry extract, combined with TiO_2 nanoparticles in starch-based polyvinyl alcohol films, significantly improves tensile strength, moisture barrier, and antimicrobial properties (Jayakumar et al., 2022). Wild blackberry extract, rich in anthocyanins, offers pH-sensitive colorimetric responses and high antioxidant activity, making these films suitable for intelligent packaging applications while maintaining favorable barrier properties (Cvetković et al., 2025). Fig. 4.6 shows how intelligent packaging components, including indicators and sensors, reveal real-time environmental information and facilitate decision-making, emphasizing the importance of effective barrier properties to create a stable microenvironment for accurate sensing. This figure also shows active-releasing and active-scavenging systems, further highlighting the role of barrier properties in regulating gas exchange (Lv et al., 2023).

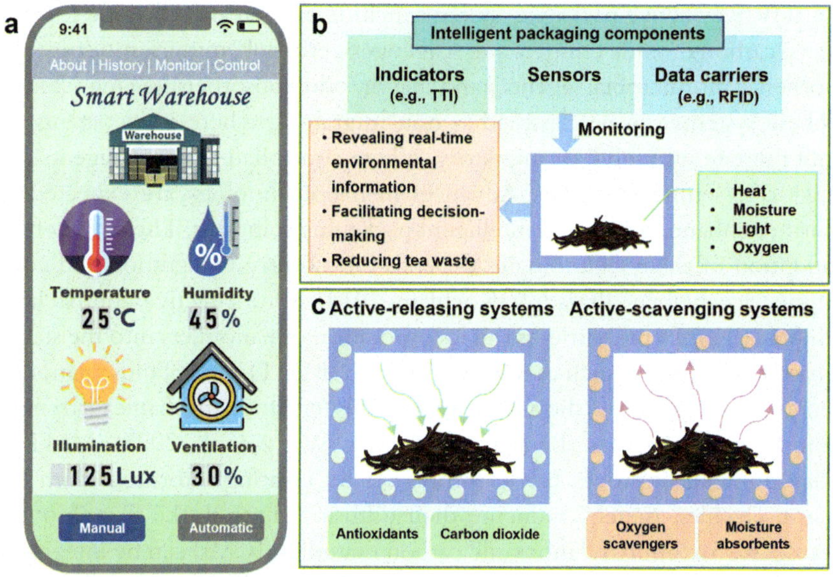

Figure 4.6 Architecture of a smart warehouse system utilizing intelligent packaging components (indicators, sensors, data carriers) to monitor environmental conditions (temperature, humidity, light, oxygen). The system enables real-time decision-making (manual/automatic control) and incorporates active-packaging systems (releasing/scavenging) to reduce tea waste and preserve product quality. *Source: (Lv, H., Feng, X., Song, H., Ma, S., Hao, Z., Hu, H., ... Fan, F. (2023). Tea storage: A not thoroughly recognized and precisely designed process. Trends in Food Science & Technology, 140, 104172.) with reproducing permission from Elsevier under license number 6180090659921.*

Researchers are continuously exploring methods to overcome these limitations, such as tailoring starch properties through various treatments and blending with other materials to create more robust and functional packaging solutions (Shibasaki et al., 2024). The ultimate goal is to develop starch-based intelligent packaging materials that provide superior protection, real-time information, and minimal environmental impact, aligning with the principles of a circular economy and sustainable development (Halder et al., 2024).

4.3.2 Active packaging

Active packaging represents a significant innovation in food packaging, driven by consumer demand for improved food safety, quality, and shelf life (Mane, 2016). Unlike traditional inert packaging, active packaging interacts with food or its environment to inhibit spoilage and maintain quality. Key active packaging systems include oxygen and ethylene scavengers, carbon dioxide emitters and scavengers, ethanol emitters, moisture absorbers, antimicrobial agents, and flavor/odor absorbersn(Mane, 2016). These systems can be classified as non-migratory, where active agents do not migrate into food, or migratory, where controlled release of agents occurs (Hosseinnejad, 2014). Advances in nanotechnology are expected to further enhance active and intelligent packaging solutions. The commercial adoption of these technologies depends on industry recognition and consumer acceptance (Prasad & Kochhar, 2014). One effective approach to improve barrier properties is the incorporation of nanofillers into the starch matrix. Nanofillers such as montmorillonite (MMT) and cellulose nanofibers (CNF) have been shown to significantly enhance mechanical strength and barrier properties (Bangar et al., 2023; Wang et al., 2025). MMT, a clay mineral, improves barrier properties by creating a tortuous path for gas molecules, thereby reducing their diffusion through the film (Nowak et al., 2025). Similarly, inorganic carbon materials (ICMs) can be integrated to overcome the weak mechanical strength, poor barrier properties, and high moisture sensitivity of starch-based films (Xu et al., 2022). Furthermore, combining starch with other biopolymers forms composite films with improved functionalities. For instance, blends of starch with PVA are recognized as promising biodegradable food packaging films due to PVA's excellent film-forming properties and ability to improve the mechanical and barrier performance of starch films (Deng, Su, et al., 2024; Wang et al., 2025). The integration of active compounds further enhances the barrier and protective functions of starch-based packaging. For example, quercetin, when infused into a starch matrix, exhibits antioxidant, antimicrobial, and enhanced mechanical and barrier properties, protecting packaged food from oxidative rancidity and microbial growth (Shafi et al., 2025). Essential oils, such as lemongrass essential oil, and curcumin extract have also been incorporated into multi-layered films to improve UV blocking properties and provide active and intelligent functionalities (Jamróz et al., 2023). Fig. 4.7 highlights that physical improvements in nanostructured food packaging include enhanced mechanical strength, barrier properties, flexibility, stability, and optical properties (Wahab et al., 2024).

Starch-based biocomposites, particularly when used as coatings, also offer promising solutions for improving barrier properties in paper-based food packaging (Sultana, 2025). Starch-based polyelectrolyte complexes (SPECs) can be synthesized to form barrier coatings through industrially scalable processes, with their efficacy influenced by molecular weight and

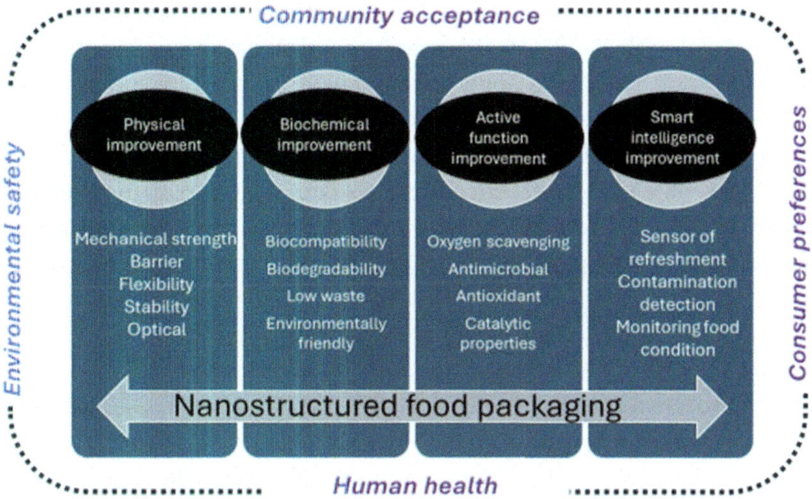

Figure 4.7 A holistic framework for the development of nanostructured food packaging, outlining its key improvement targets: Physical (mechanical, barrier), biochemical (biocompatibility, biodegradability), active (antimicrobial, antioxidant), and smart (sensing, monitoring) properties. The framework connects these material improvements to overarching goals of consumer acceptance, human health, and environmental sustainability. *Source:(Wahab, Y. A., Al-Ani, L. A., Khalil, I., Schmidt, S., Tran, N. N., Escriba-Gelonch, M., ... Hessel, V. (2024). Nanomaterials: A critical review of impact on food quality control and packaging. Food Control, 163, 110466) with reproducing permission from Elsevier under license number 6180090810178.*

charge density of starch polymer chains (Chi et al., 2020). A significant challenge for starch-based materials is their hydrophilic nature, which makes them susceptible to moisture and compromises their barrier effectiveness in humid environments (Deng, Su, et al., 2024). To mitigate this, chemical modifications such as acetalization and dual crosslinking (e.g., hydrogen bonding) can be employed to enhance mechanical and barrier properties. For example, oxidizing native cassava starch to dialdehyde starch (DAS) and then incorporating PVA and CNF significantly improves the film's properties (Wang et al., 2025). Research demonstrates that adding ZnO nanoparticles and nanorods to sago starch-based films enhances multiple properties simultaneously. Mechanical properties improve substantially, with tensile strength increasing from 0.180 to 0.980 MPa when ZnO content increases from 0 to 5 wt%. Water vapor permeability decreases significantly, from 4.992×10^{-10} to 2.723×10^{-10} g m^{-1} s^{-1} Pa^{-1} with 5% ZnO incorporation (Chieh & Siddique, 2019). The enhanced barrier properties, including 40%–55% reduction in oxygen permeability (Nafchi et al.,

2013). Halloysite nanotubes (HNTs) are naturally occurring alumina silicate nanomaterials with unique tubular structures that show significant promise for active food packaging applications. These nanomaterials possess high surface area, thermal stability, and biocompatibility, making them effective functional additives for ethylene scavenging and gas barrier applications (Kumar et al., 2024). Research demonstrates that HNT incorporation into polymer matrices significantly enhances packaging performance. In chitosan/HNT composites, 5 wt% HNT loading resulted in 169.9% increased tensile strength, 52.94% reduced oxygen permeability, and 4.07 times enhanced ethylene scavenging capacity (Wang et al., 2023). Carbon quantum dots (CQDs) incorporated into PVA matrices represent a promising approach for active food packaging applications. These composite films demonstrate exceptional UV-blocking capabilities, with complete blockage of UV-C and UV-B radiation and 20%–60% blockage of UV-A (Patil et al., 2020). Mussagy and colleagues showed that films combining natural β-carotene and astaxanthin with corn starch exhibited positive properties, including increased antioxidant activity and reduced water solubility and water vapor permeability (WVP). This demonstrates the effectiveness of incorporating specific active compounds and biopolymers to improve the barrier functions of starch-based films (Mussagy et al., 2024). Another example involves the development of gelatin-based antibacterial composite films incorporating DAS and AgNPs-immobilized MXene nanosheets, which demonstrate enhanced mechanical strength, reduced water solubility, and improved gas barrier properties, along with antibacterial activity. These modifications are crucial for overcoming the limitations of pure gelatin films, such as poor mechanical strength, high water sensitivity, and low gas barrier properties (Yang et al., 2025). Similarly, biopolymer systems, such as multi-layer films of furcellaran, chitosan, and gelatin, can be modified with active ingredients like AgNPs, montmorillonite, and curcumin-capsaicin mixtures to reduce gas permeability and enhance active properties for food technology applications (Nowak et al., 2025). Overall, while starch offers a highly sustainable base for food packaging, its practical application, especially in active and intelligent systems, heavily relies on advanced modification techniques and the strategic incorporation of various functional components to achieve the necessary barrier properties and overcome its intrinsic limitations (Yudhistira et al., 2024). These advancements are essential for developing packaging solutions that meet both environmental sustainability goals and the growing demand for improved food quality and safety (Konar et al., 2022).

4.4 The thermal properties of starch-based packaging films

The thermal properties of starch-based packaging films play a central role in determining how they can be processed, how stable they remain over time, and how well they perform in real-world applications. These properties affect not only the feasibility of film formation and long-term storage but also the incorporation of bioactive or sensing agents for intelligent and active packaging. In contrast to petroleum-derived plastics, which display relatively stable and predictable glass transition and melting points, starch films are far more complex. Their semi-crystalline structure and extensive hydrogen-bonding networks make them highly sensitive to moisture, leading to thermal transitions that vary widely and are often difficult to control (Nizam et al., 2021; Santhosh et al., 2024). One of the most critical thermal parameters is the glass transition temperature (Tg), which marks the shift of amorphous regions from a rigid, glassy state to a softer, rubber state. For native starch, Tg occurs above the material's degradation point, making unplasticized starch brittle and unsuitable for film formation (Arruda et al., 2025; Karnwal et al., 2025). The addition of plasticizers such as glycerol or sorbitol lowers Tg to a more practical range of 45–65°C, thereby improving flexibility and processability (Che Hamzah et al., 2022). The extent of this reduction depends on both concentration and molecular characteristics of the plasticizer, with small, mobile, hydrophilic molecules such as glycerol being more effective than bulkier counterparts like sorbitol (Xu et al., 2025). These thermal transitions, including Tg values and their relationship with amylose content, are summarized in Table 4.1.

Another defining transition is gelatinization, which usually occurs between 50 and 85°C. During this process, starch granules swell irreversibly, amylose molecules leach out, and the crystalline order collapses into an amorphous matrix. Differential scanning calorimetry (DSC) typically reveals gelatinization endotherms around 62–65°C, although values vary depending on botanical origin and amylose content (Chen et al., 2023; Pooja et al., 2024). After gelatinization and subsequent drying, retrogradation often takes place, as amylose rapidly and amylopectin more gradually reassociate into ordered domains. This recrystallization process increases brittleness, reduces transparency, and alters barrier properties, making its control one of the persistent challenges in starch-based packaging (Arruda et al., 2025). Thermal stability can also be examined using thermogravimetric analysis (TGA), which shows that starch films degrade in multiple stages. The first stage occurs below 100°C and corresponds to the evaporation of free and bound water. At 200–250°C, plasticizers such as glycerol and sorbitol volatilize, while the primary decomposition of the starch backbone takes place between 300 and 400°C, involving glycosidic bond cleavage, dehydration of glucose units, and eventual char formation (Pooja et al., 2024; Todica et al., 2016). Stabilizing additives can shift these thermal events. Cross-linking agents such as citric acid limit chain mobility and strengthen the polymer network, leading to higher degradation onset temperatures; in some cases, Tonset increased from 140°C to 164°C following cross-linking (Shanbhag et al., 2023). Similarly, nanofillers including montmorillonite clay, cellulose nanocrystals (CNC), and Zinc oxide nanoparticles (ZnO) reinforce hydrogen-bond networks and create tortuous diffusion pathways for volatile products, thereby delaying decomposition and enhancing thermal resistance (Chen et al., 2023; Estrada-Monje et al., 2021; Muñoz-Gimena et al., 2023). Thus, starch films exhibit inherent thermal instability compared to commodity plastics, but careful design strategies can significantly improve their performance for packaging applications.

4.4.1 Intelligent packaging

Intelligent packaging films are often designed to incorporate pH-sensitive pigments or other indicator compounds, and their thermal stability is crucial to ensure that the chromogenic functionality of these indicators is preserved throughout the product's shelf life without premature degradation during storage or mild processing (Ahmad et al., 2022; Liu et al., 2022). Thermal analysis of intelligent films based on native cassava starch (NS), oxidized hydroxypropyl starch (OHS), acetylated distarch phosphate (ADSP), and oxidized-acetylated starch (OAS) demonstrates how starch modification influences degradation behavior. Studies on pH-sensitive starch films show that stability can vary considerably depending on the type and concentration of the indicator. For instance, in films where red cabbage extracts (RCEs) were added to native or dual-modified cassava starch including oxidized hydroxypropyl starch, acetylated distarch phosphate (ADSP), and oxidized-acetylated starch the ADSP-based films demonstrated superior stability compared with NS, OHS, or OAS formulations (Cheng, Cui, et al., 2022; Cheng, Yan, et al., 2022). Such enhancements may arise from additional hydrogen bonding and restricted segmental mobility; however, high anthocyanin loadings can also act as plasticizers, so optimization of loading is essential. This improvement is attributed to the cross-linked molecular

Table 4.1 Thermal and mechanical properties starch-based food packaging systems.

Starch source	Amylose content	Thermal stability (Tonset/Tmax) (°C)	Mechanical performance (TS/EB/YM)	Release behavior	Functional application	References
Potato starch	18.1%	–	TS = 3.21 MPa, EB = 61%	Fast release (hydrophilic, amylopectin-rich)	Intelligent films, pH-sensitive indicators	Chavez-Marquez et al. (2023)
Cassava starch	–	–	TS = 1.23-9.86 MPa, EB = 14.83-55.74 % (cross-linked, rigid but less extensible)	Moderate release with balanced kinetics	Intelligent + active (pigments, antioxidants)	Rahmadhia, et al. (2023)
Pumpkin starch	26.6%	Tmax = 312 ± 0.9; onset 290-296; CNF up to 305	TS = 30.32 MPa	Water solubility reduced from 22.98% to 19.6–21.7%	Nanocomposite packaging; CNC for strength, CNF for thermal stability	Zhang et al. (2021)
Cocoyam starch	15.1% amylose, 84.9% amylopectin	Stable ≤250 °C; decomposition heat = 4.64 J/g	High stiffness, compact, creep-resistant	Selective gas permeability (high CO_2 & H_2, low O_2 & N_2)	Modified atmosphere packaging; effective 30 days	Chaudhary et al. (2024)
TPS and MXene	–	–	TS = 36.1 MPa, YM = 1965.3 MPa (at 10 wt % MXene)	Reduced WVP, strong UV shielding	Intelligent packaging: UV-blocking, reinforced	Dong et al. (2025)

(*Continued*)

Table 4.1 Thermal and mechanical properties starch-based food packaging systems.—cont'd

Starch source	Amylose content	Thermal stability (Tonset/Tmax) (°C)	Mechanical performance (TS/EB/YM)	Release behavior	Functional application	References
Esterified corn starch	28% amylose, 72% amylopectin	Moisture/glycerol: 50-280; starch: 280-330; Tmax =318-320	TS = 4.00 MPa, EB = 18.3%	Ethylene degradation up to 12.5% (4% BT)	Active packaging: photocatalytic ethylene scavenger	Wang et al. (2019)
Corn starch TPS (30% glycerol)	25–28%	–	TS = 4.69 MPa, EB = 56%,	CNCs reduce swelling, improve stiffness	Sustainable stiffening fillers for packaging	Arifin et al. (2022)
Corn TPS (cross-linked with zeolite)	18.9%	3-stage decomposition: 30-200, 250-400, >550; peaks 300 & 333	TS ≈ 0.9 MPa, EB = 11%	–	Reinforced antimicrobial TPS films	Gutiérrez (2022)
Corn starch (chitosan)	–	Pure starch 390; with chitosan 371; safe use ≤290-316	TS = 5.19 MPa, EB = 44.6%	Water uptake: control =302%; with chitosan = 251%	Antimicrobial, water-resistant packaging	Tan et al. (2022)

Abbreviations: *BNC*, bacterial nanocellulose; *CNC*, cellulose nanocrystals; *CNF*, cellulose nanofibers; *Cr(Oct)*$_3$, chromium octanoate (antimicrobial catalyst in zeolite); *EB*, elongation at break (%); *EO*, essential oils; *GA*, gallic acid; *GQDs*, graphene quantum dots; *MXene*, 2D Transition metal carbide/nitride (e.g., $Ti_3C_2T_x$); BT, Bi_2WO_6–TiO_2 (photocatalytic filler); *PE*, pickering emulsions (nanocellulose-stabilized droplets); *PUL*, pullulan (polysaccharide); *TPS*, thermoplastic starch; *TS*, tensile strength (MPa); *WE*, grape cane extract (wine extract); *YM*, Young's modulus (MPa or GPa).

structure of ADSP, which forms a robust three-dimensional network requiring greater energy for chain scission, indicating that chemical modification of starch directly enhances the matrix's resilience to heat during indicator integration (Li, Zou, et al., 2024). In another study, sago starch films formulated with red cabbage anthocyanin (RCA) at concentrations between 8%–14% (w/v) showed enhanced thermal stability at the final degradation stage during TGA analysis (Che Hamzah et al., 2022). This suggests that anthocyanins may strengthen the film by forming compact structures that resist heat breakdown. In contrast, polyvinyl alcohol (PVA)/starch films with betanin (E−162) and nano-calcium carbonate ($CaCO_3$) showed only a small improvement, with slightly higher temperatures for residual weight compared to pure PVA/starch films. Here, the first degradation around 200–220°C was linked to moisture and plasticizer loss, followed by starch breakdown at higher temperatures (Kavas et al., 2023). Other studies confirm that the impact of additives can vary. A soybean polysaccharide film with grape skin anthocyanin extract initially showed reduced stability, but adding graphene oxide reversed this effect by forming strong bonds with starch that stabilized the structure (Kafashan et al., 2023). This enhancement is typically linked to the strong hydrogen bonding capacity of GO with starch hydroxyl groups, leading to the formation of a reinforced and thermally stable network (Rapisarda et al., 2021). Similarly, thermoplastic starch (TPS) films with anthocyanins produced by extrusion kept their crystalline structure intact and reached a Tvmax of 327°C, showing that the pigment did not destabilize the material (Vedove et al., 2021). Sodium alginate (SA) films with carboxylated cellulose nanocrystals (C–CNC) and beetroot extract also showed that adding BTE did not significantly change thermal properties, even though strength increased when combined with C–CNC. In another case, oat starch films with betalain and cellulose nanofibrils (CNF) improved the sensitivity of betalain without reducing overall film stability. Cross-linking with CA was also effective; one starch film reached a T80 of 441.7°C compared to 347.7°C for the control (Vedove et al., 2021). This improvement was due to stronger covalent and hydrogen bonds, but excess CA caused acid hydrolysis, weakening stability in some samples (Chen et al., 2021). These findings underscore that the selection of plant-derived extracts and their concentration must be carefully optimized. The thermal degradation of active compounds such as anthocyanins is sensitive to processing conditions, so balancing indicator functionality with overall stability remains a central challenge in designing intelligent packaging systems (Jumaidin et al., 2024; Mahović Poljaček et al., 2024).

4.4.2 Active packaging

Active packaging films are designed to interact with food or its environment to extend shelf life, usually by integrating antimicrobial, antioxidant, or oxygen-scavenging agents. Their thermal stability is essential, as it ensures that active compounds remain effective during heat sealing, sterilization, and long-term storage (Hasanin et al., 2024). Adding bio-fillers and active compounds often improves the stability of starch-based films. For instance, bacterial cellulose (BC) made from tapioca liquid waste, when blended with TPS, significantly increased thermal stability and viscosity (Kavas et al., 2023). Likewise, sugarcane bagasse fiber (SBF) reinforced cassava starch films modified with beeswax (BW) showed that higher SBF levels raised Tg and melting temperatures, confirming better stability (Jumaidin et al., 2024). Starch-silicone oil composites also performed better than native starch alone (Peralta-González et al., 2022). Cellulose nanomaterials are common reinforcements. Films made with CNF showed higher thermal stability than those with CNC, highlighting the importance of particle size and structure (Zhang et al., 2021). In one study, control starch films had a peak endothermic temperature of 104.67°C, but adding CNF shifted this to higher values. The film with 10 g CNF per 100 g starch had the best thermal stability (Pelissari et al., 2017). Other nanoparticles can also help. Talc nanoparticles strengthened thermoplastic corn starch films (López et al., 2015), while kaolin-doped starch nanocomposites achieved optimal stability at 15% kaolin loading (Ashaduzzaman et al., 2020). Starch/TiO_2 nanocomposites showed higher degradation onset (T5%) and mid-point (T50%) temperatures, with increased char residue due to TiO_2's inherent stability (Goudarzi et al., 2017). Ti_3C_2Tx MXene nanoplatelets, even at very low levels ($\leq$1 wt%), enhanced barrier and thermal behavior, especially under humid conditions. Graphene oxide also improved thermal resistance through strong hydrogen bonding with starch hydroxyl groups (Dong et al., 2024). The stability of the bioactive compounds themselves is equally important. Essential oils (EOs), valued for antimicrobial and antioxidant effects, must withstand processing (Ganosi et al., 2023; Mahanta et al., 2021). As shown in Fig. 4.8, TPS films containing nanocellulose-stabilized Pickering emulsions of ho wood, cardamom, and cinnamon oils exhibited good thermal stability under heat, confirmed by TGA and DSC (Souza et al., 2021).

Decomposition followed three stages include moisture and plasticizer loss below 120°C, glycerol and compound volatilization around

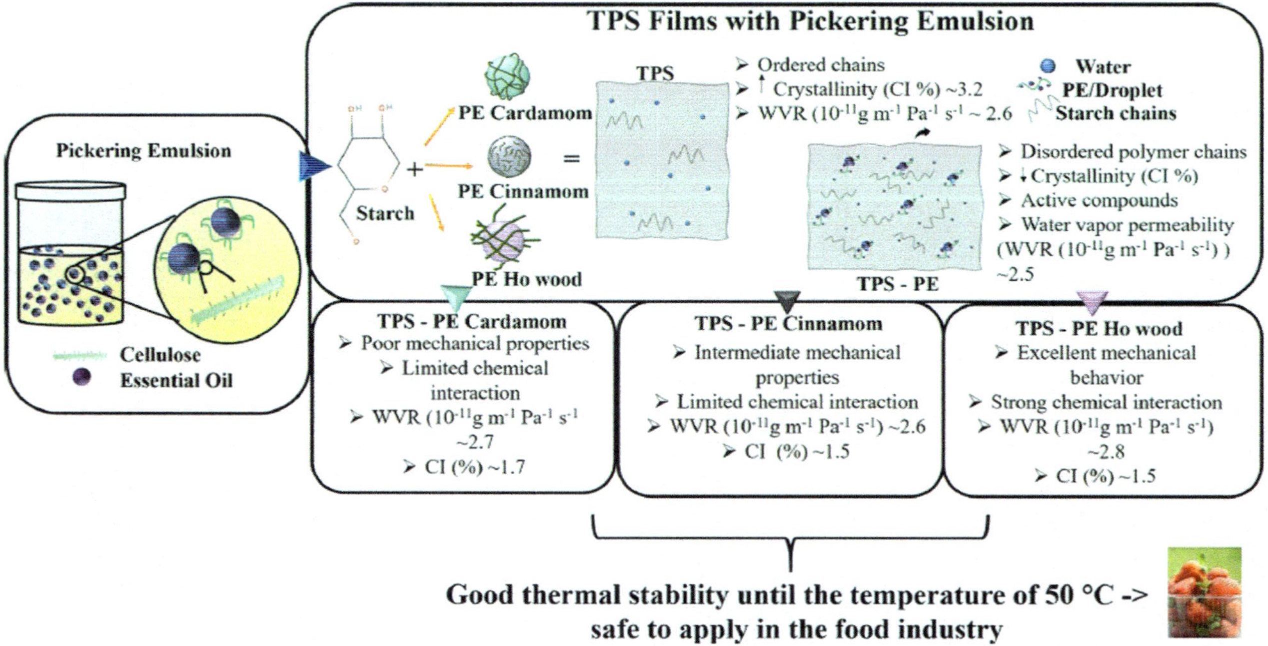

Figure 4.8 Schematic of TPS films with Pickering emulsions of essential oils (cardamom, cinnamon, and ho wood), highlighting structural organization, crystallinity, water vapor permeability, and mechanical performance. *Source:(Souza, A. G., Ferreira, R. R., Paula, L. C., Mitra, S. K., & Rosa, D. S. (2021). Starch-based films enriched with nanocellulose-stabilized Pickering emulsions containing different essential oils for possible applications in food packaging. Food Packaging and Shelf Life, 27, 100615. https://doi.org/10.1016/j.fpsl.2020.100615) with reproducing permission from Elsevier under license number 6180100461796.*

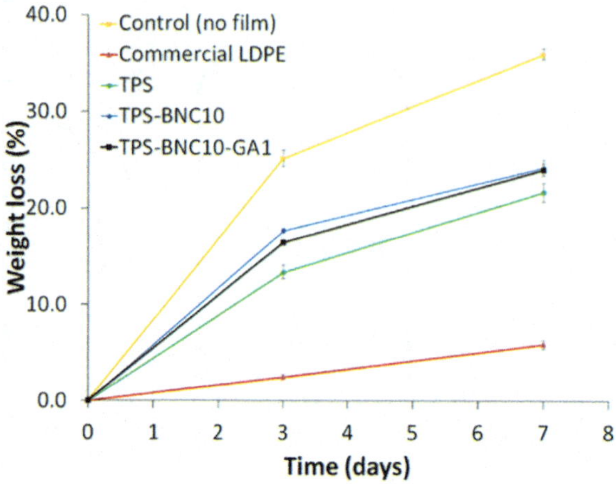

Figure 4.9 Weight loss percentage (%) different films (LDPE commercial film, thermoplastic starch, TPS- bacterial nanocellulose (BNC) 10 and TPS-BNC 10-gallic acid (GA)1). *Source: (Almeida, T., Karamysheva, A., Valente, B. F. A., Silva, J. M., Braz, M., Almeida, A., ... Freire, C. S. R. (2023). Biobased ternary films of thermoplastic starch, bacterial nanocellulose and gallic acid for active food packaging. Food Hydrocolloids, 144, 108934. https://doi.org/10.1016/j.foodhyd.2023.108934) with reproducing permission from Elsevier under license number 6180100656412.*

125–240°C, and starch breakdown at 300–350°C. EOs generally act as secondary plasticizers that can depress Tg and widen the heat-seal window; immobilization via Pickering emulsions/cyclodextrins can mitigate volatility during thermal exposure. In agreement with the trend observed in Fig. 4.9, TPS films with bacterial nanocellulose (BNC) and gallic acid (GA) maintained a maximum degradation temperature near 300–310°C, confirming that the additives did not compromise stability (Almeida et al., 2023).

Blends of starch with other polymers can behave differently. Polybutylene succinate (PBS) mixed with tapioca starch often had lower stability than pure starch, though blends with higher crystallinity resisted heat better. Films from potato peel starch were stable above 100°C and remained flexible, while cassava and potato starch films had glass Tg of about 116–119°C, confirming their suitability for packaging (Ayu et al., 2018). Further modifications also influence behavior. Phosphated TPS and starch nanocrystals made by reactive extrusion retained more char during heating, likely from phosphorus-rich residues, and showed better bonding between glycerol and starch (Gutiérrez & Valencia, 2021). Cross-linking starch oleate

(SO) films with PEG-dilipoic acid altered degradation patterns. For LA-based linkers, degradation began near 220–250°C, with a second phase at 355–450°C due to the SO backbone. The stability increased with longer cross-linkers (Boetje et al., 2023; Matheus et al., 2023). The thermal performance of active starch-based films depends on the starch matrix, plasticizers, and the active or reinforcing agents used. While nanocellulose and certain nanoparticles often strengthen films, the effect of bioactive compounds is less predictable. Careful optimization is therefore essential to ensure that films remain stable without losing their active functions (Torres-Medina & Murillo, 2025).

4.5 Mechanical properties of starch-based smart packaging films

The mechanical performance of starch-based films defined by strength, stiffness, and flexibility is critical for their function in packaging. However, native starch films show poor integrity due to their brittle, semi-crystalline structure formed by dense hydrogen bonding between amylose and amylopectin chains. As a result, elongation at break (EAB) is typically in the single-digit range, while tensile strength (TS) and Young's modulus (YM) are highly sensitive to environmental humidity. Water uptake softens films and increases flexibility, but it simultaneously reduces strength and stiffness, leading to unpredictable performance (Kusuma et al., 2024b; Omotoso et al., 2015; Yang et al., 2024). To overcome these deficiencies, several modification strategies have been employed. Plasticizers such as glycerol or sorbitol disrupt starch-starch hydrogen bonding, increasing chain mobility and EAB but lowering TS and YM (Ghasemlou et al., 2013; Kusuma et al., 2024a; Wang, Xu, Fang, Yan, & Cui, 2022). Reinforcement fillers, including nanoclays, microcellulose fibers, and CNC, counteracts this weakness by enabling efficient stress transfer and significantly improving TS and YM (Menzel et al., 2019; Moraes et al., 2011; Nordin et al., 2018; Wang, Xu, Fang, Yan, & Cui, 2022). Metal oxide nanoparticles such as ZnO and TiO_2 provide similar reinforcement while also imparting antimicrobial or UV-shielding properties, though excessive loading can cause agglomeration (Mirjalili & Ardekani, 2017; Tamimi et al., 2021). Polymer blending offers another route, with chitosan improving strength and water resistance through hydrogen bonding, and PVA enhancing mechanical robustness (Díaz-Cruz et al., 2022; Sun et al., 2019). Crosslinking agents such as citric acid or glutaraldehyde further strengthen films by creating covalent starch networks, though this often reduces flexibility (Rahmah et al., 2017). Mechanical behavior is typically characterized using stress-strain curves. TS reflects the maximum load-bearing capacity, elongation at break quantifies ductility, and YM indicates stiffness. Packaging films require a balanced profile sufficient TS to resist tearing, adequate EAB to accommodate product deformation, and moderate YM to avoid brittleness. Achieving this balance through targeted modification is essential for advancing starch-based materials toward practical smart packaging applications.

4.5.1 Intelligent packaging

Intelligent packaging films integrate colorimetric indicators or other sensing elements into the polymer matrix, but their inclusion often alters the mechanical performance of the films. The central challenge lies in maintaining film strength and flexibility while ensuring indicator functionality (Dong et al., 2020). Several studies have reported reductions in strength and extensibility after indicator incorporation. For example, sago starch films containing red cabbage anthocyanins showed decreases in TS (from 7.15 to 5.08 MPa) and elongation at break (from 16.27% to 12.18%) compared to controls, attributed to disruption of the starch network (Che Hamzah et al., 2022). Similarly, PVA/starch films containing betanin and nano-$CaCO_3$ exhibited a sharp decline in TS (from 27.74 to 14.77 MPa) and reduced elongation at break, likely due to pigment-matrix incompatibility and nanoparticle agglomeration (Kavas et al., 2023). TPS films with anthocyanins also showed compromised integrity, being thicker but significantly weaker and less flexible than controls (Vedove et al., 2021). In contrast, beneficial interactions between indicators and starch matrices have also been observed. Heat moisture-treated lotus seed starch films incorporating red turnip extract and carboxymethyl cellulose showed substantial improvements in TS (up to 21.13 MPa) and elongation (up to 44.31%) compared to native starch films, reflecting synergistic reinforcement from starch modification and extract incorporation (Ali & Dash, 2023). Starch/PVA films with roselle anthocyanins also showed enhanced strength and flexibility due to hydrogen bonding between starch, PVA, and the extract, though excessive pigment loading caused aggregation and reduced compactness (Zhai et al., 2017). Dual-modified cassava starch films with red cabbage extracts further highlighted the trade-off between strength and flexibility of

ADSP films displayed higher strength (13.65 MPa) than native starch (5.73 MPa), but at the cost of lower elongation (47.65% vs. 102.44%) due to increased crosslinking and rigidity (Cheng, Cui, et al., 2022; Cheng, Yan, et al., 2022). Other systems show more favorable outcomes. SA films containing C–CNC and BTE reached TS as high as 55.74 MPa (46), while starch-based optoelectronic nose films with multiple acid-base indicators demonstrated both flexibility and high tensile resistance (Roveda et al., 2024). These findings confirm that the mechanical performance of intelligent starch films depends strongly on the choice of indicator, its concentration, and compatibility with the matrix. Optimal formulations must balance sensitivity with mechanical robustness to ensure packaging viability (Guo et al., 2023).

4.5.2 Active packaging

For starch-based active packaging, improving mechanical properties is critical since films must combine structural integrity with functional performance (Pawase et al., 2025). Various strategies, including the incorporation of reinforcing agents, plasticizers, and chemical modifications, have been extensively employed to improve TS, EAB, and YM (Wang, Gong, et al., 2022). Plasticizers such as glycerol or sorbitol are essential to convert starch into TPS with improved flexibility. For instance, increasing glycerol content in arrowroot starch films lowered TS and stiffness but increased elongation at break due to weakened starch-starch hydrogen bonding (González-Torres et al., 2021; Tarique et al., 2021). However, excess plasticizer can compromise overall strength. Reinforcement with cellulose-based fillers is among the most effective strategies. BNC integrated into TPS films significantly enhanced TS (up to 39 MPa) and modulus (up to 2.0 GPa) without major losses in flexibility, attributed to strong interfacial adhesion (Santos et al., 2025). CNC and CNF derived from plant residues such as banana peels and pumpkin starch also improved TS, although higher fiber contents could restrict chain motion and reduce elongation (Kumar et al., 2023). Other fillers and nanoparticles also reinforce starch matrices. Chitosan nanoparticles increased TS from 1.12 to 10.03 MPa and elongation from 67% to 91% in tapioca starch films (Shapi'i et al., 2022). Kaolin-doped and talc-reinforced starch films similarly improved stiffness and toughness, with talc increasing TS by up to 20% and modulus by 68%–81% (Kwasniewska et al., 2020). Chemical modification offers further improvements. Crosslinking starch chains enhances stiffness and water resistance, while dual modifications such as hydroxypropylation and acetylation, combined with acetylated cellulose nanofibers, yielded biodegradable plastics with strengths comparable to polypropylene (22.89 MPa) (Yao et al., 2024). Organocatalytic crosslinking of starch oleate (SO) films with PEG-dilipoic acid, however, reduced TS with higher crosslinker contents due to soft spacer effects (Boetje et al., 2023). Blending starch with other polymers provides another path. PVA/starch nanocomposites reinforced with cellulose nanofibers from sugarcane bagasse increased TS dramatically from 24 to 85 MPa at 6% CNF loading (Ali et al., 2022). Starch/polybutylene PBS blends also showed improved mechanical behavior, although poor dispersion and weak interfacial bonding at higher starch content reduced performance (Niu et al., 2023). Active compounds themselves may influence mechanics. Grapevine cane extract acted as a mild plasticizer in TPS, lowering strength and modulus but increasing elongation at low concentrations, while higher contents caused aggregation and property loss (WE, 10–15 wt%) (Díaz-Galindo et al., 2020). Ultrasonication has also been shown to improve strength by 20% in potato starch films, particularly in glycerol-containing formulations (Żołek-Tryznowska, Jeznach, Bednarczyk, & Tryznowski, 2024). The mechanical performance of starch-based active films can be significantly enhanced through reinforcement, blending, and controlled chemical modification. Achieving the desired balance of strength, flexibility, and ductility remains central to ensuring both functional activity and structural reliability in sustainable packaging applications (Santhosh et al., 2024).

4.6 Challenges and limitations of starch-based smart packaging

Biodegradable materials derived from starch can significantly contribute to the advancement of eco-friendly food packaging solutions, helping to diminish the environmental impact associated with petroleum-based packaging. The main challenge lies in the inadequate mechanical and barrier properties of starch-based packaging materials, which stem from the inherent characteristics of starch. This issue can be addressed by chemically or physically modifying starch, or by blending it with various biopolymers and functional additives. Once the fundamental requirements for food packaging are satisfied, it is possible to create starch-based active packaging by incorporating antioxidants or antibacterial substances, which can help prolong the shelf life of food and minimize waste (Garavito et al., 2024; Kumar et al., 2023). Additionally, starch-based time and temperature indicator films can provide real-time freshness assessments of food through color changes. Starch-based films have a wide application in food packaging and effectively protect fresh produce. In the future, there is potential for developing active or smart packaging solutions utilizing starch-based materials. Although some advancements have been made in enhancing the properties of starch-based films, further research is necessary to achieve mechanical and barrier properties comparable to those of conventional plastic packaging. The characteristics of starch-based biodegradable materials can be influenced by the type of starch used, preparation methods, storage conditions, and other variables during the manufacturing process, which can lead to inconsistencies and stability issues (Garavito et al., 2024; Pawase et al., 2025). At present, most investigations into food packaging applications are conducted in laboratory settings. The commercialization of starch-based films has also been hindered by factors such as industrial production constraints, safety regulations, environmental concerns, and consumer acceptance. Moreover, combining starch with other materials requires additional research to fulfill practical requirements. Numerous additives hold significant promise for developing antibacterial and antioxidant films. There is still extensive work to be done in formulating antimicrobial agents suitable for starch matrices. Currently, there are no adequate standards for assessing and quantifying antimicrobial and antioxidant effectiveness. The domain of starch-based smart films presents considerable promise, although its functionality has been rather limited thus far. Presently, pH indicators are the most frequently employed in smart packaging applications. Changes in the surrounding environment of the packaged food can also alter the active ingredients, affecting the pH of the system, which may lead to irreversible or time-sensitive reactions in the indicator within the packaging film, potentially causing consumers to misinterpret the quality of the product (Abedi-Firoozjah et al., 2023; Ali et al., 2022).

One of the major barriers to the widespread adoption of starch-based food packaging materials is persuading manufacturers to invest in these alternatives. This challenge is exacerbated by the relatively high production costs associated with creating food packaging compared to traditional synthetic plastics. To reduce these financial barriers, there is a growing trend toward the valorization or recycling of waste and by-products from the agricultural and food industry into valuable food packaging materials. Exploring unconventional sources, especially those that are not widely consumed by the general public, could significantly expand the potential applications of starch-based food packaging materials (Ali & Dash, 2023; Almeida et al., 2023). In addition, research is underway into the use of new plasticizers, particularly deep eutectic solvents, which can enhance the thermal processability of starch-based films. Traditional plasticizers such as glycerol and sorbitol, which are often used in starch film production, often become incorporated into the starch matrix, leading to issues such as retrogradation over time. This retrogradation can compromise the texture and effectiveness of the packaging. In contrast, the use of natural deep eutectic solvents as green plasticizers offers numerous advantages, including improved processability during extrusion, increased starch dissolution and softening, resistance to retrogradation, and greater flexibility in the resulting films. Despite their advantages, the application of these green solvents in starch-based food packaging materials remains unexplored (Arruda et al., 2025; Kusuma et al., 2024a). Consequently, it is essential that future research focuses on investigating different combinations of innovative starch sources and sustainable plasticizers. The goal should be to devise methods that not only reduce production costs and environmental impacts, but also maximize the performance and processability of starch-based packaging materials. Such comprehensive studies will play a pivotal role in advancing the development and application of edible packaging in the food industry, ultimately contributing to increased food safety and sustainability (Liu et al., 2022; Mahović Poljaček et al., 2024).

4.7 Conclusion and future prospectives

Starch is a promising material for biodegradable films due to its biodegradability, edibility, low cost, and thermoprocessable nature. These starch-based films are considered alternatives to conventional packaging that enhance food quality and safety while also serving as carriers for functional ingredients for the development of active, smart, and antioxidant packaging. Although commercial starch-based materials could provide an environmentally friendly solution for the food packaging industry, they currently lack the mechanical, optical, and barrier properties of synthetic alternatives. Research is focused on improving these properties through various methods, including thermoplastic processing, chemical modification, and incorporation with biopolymers and fillers. While starch-based materials are biodegradable, some chemical additives can extend their shelf life, which requires the discovery of new and renewable resources to increase sustainability. Despite the advances, many challenges, including material selection and preparation techniques, hinder commercial viability. Most research is still at the laboratory level, and further studies on the combination of starch with other materials are needed to meet the needs of the industry.

Food safety and human health are important considerations in food product packaging. The emergence of edible packaging, which includes various forms such as coatings and films, offers a compelling and sustainable alternative to extend the shelf life of food products. These edible materials not only provide convenience by allowing direct consumption, but are also biodegradable, thereby reducing the environmental impact compared to traditional plastic packaging. To enhance the performance and protective properties of these edible films, active ingredients such as nanoparticles and natural extracts are often incorporated into them. While bio-based nanoparticles are generally considered safe for use, the incorporation of inorganic nanoparticles raises significant concerns due to the potential health risks associated with their migration into food products. This migration occurs through a diffusion process, which can lead to unintended exposure of consumers. Therefore, extensive research is crucial to assess the cytotoxicity of these inorganic nanoparticles, as existing studies suggest that they may negatively affect cell viability, cause DNA damage, and even penetrate human organs. Accordingly, careful regulation of the size and concentration of nanoparticles used in food packaging is essential to ensure that they remain within established safety thresholds. In conclusion, although starch-based packaging has limitations, its structural properties and advantages make it a strong alternative to traditional petroleum-based packaging and potentially lead to the development of smart food packaging in the future.

References

Abedi-Firoozjah, R., Chabook, N., Rostami, O., Heydari, M., Kolahdouz-Nasiri, A., Javanmardi, F., … Khaneghah, A. M. (2023). PVA/starch films: An updated review of their preparation, characterization, and diverse applications in the food industry. *Polymer Testing, 118*, 107903.

Ahmad, K., Din, Z.u., Ullah, H., Ouyang, Q., Rani, S., Jan, I., … Ali, S. (2022). Preparation and characterization of bio-based nanocomposites packaging films reinforced with cellulose nanofibers from unripe banana peels. *Starch-Stärke, 74*(5–6), 2100283.

Ali, N. A., & Dash, K. K. (2023). Modified lotus seed starch and red turnip peel extract based pH responsive edible films. *Food Packaging and Shelf Life, 40*, 101182. https://doi.org/10.1016/j.fpsl.2023.101182

Ali, M. A. S. S., Jimat, D. N., Nawawi, W. M. F. W., & Sulaiman, S. (2022). Antibacterial, mechanical and thermal properties of PVA/Starch composite film reinforced with cellulose Nanofiber of Sugarcane bagasse. *Arabian Journal for Science and Engineering, 47*(5), 5747–5754. https://doi.org/10.1007/s13369-021-05336-w

Almeida, T., Karamysheva, A., Valente, B. F. A., Silva, J. M., Braz, M., Almeida, A., … Freire, C. S. R. (2023). Biobased ternary films of thermoplastic starch, bacterial nanocellulose and gallic acid for active food packaging. *Food Hydrocolloids, 144*, 108934. https://doi.org/10.1016/j.foodhyd.2023.108934

Amorim, L. F., Mouro, C., & Gouveia, I. C. (2024). Electrospun fiber materials based on polysaccharides and natural colorants for food packaging applications. *Cellulose, 31* (10), 6043–6069.

Arifin, H. R., Djali, M., Nurhadi, B., Hasim, S. A., Hilmi, A., & Puspitasari, A. V. (2022). Improved properties of corn starch-based bio-nanocomposite film with different types of plasticizers reinforced by nanocrystalline cellulose. *International Journal of Food Properties, 25*(1), 509–521.

Arruda, T. R., Machado, G.d. O., Marques, C. S., Souza, A. L.d., Pelissari, F. M., Oliveira, T. V.d., & Silva, R. R. A. (2025). An overview of starch-based materials for sustainable food packaging: Recent advances, limitations, and perspectives. *Macromol, 5*(2), 19.

Ashaduzzaman, M., Saha, D., & Mamunur Rashid, M. (2020). Mechanical and thermal properties of self-assembled Kaolin-Doped starch-based environment-friendly nanocomposite films. *Journal of Composites Science, 4*(2), 38.

Ayu, R. S., Khalina, A., Harmaen, A. S., Zaman, K., Jawaid, M., & Lee, C. H. (2018). Effect of modified tapioca starch on mechanical, thermal, and morphological properties of PBS blends for food packaging. *Polymers, 10*(11), 1187.

Bangar, S. P., Purewal, S. S., Trif, M., Maqsood, S., Kumar, M., Manjunatha, V., & Rusu, A. V. (2021). Functionality and applicability of starch-based films: An eco-friendly approach. *Foods, 10*(9), 2181.

Bangar, S. P., Whiteside, W. S., Chaudhary, V., Akhila, P. P., & Sunooj, K. V. (2023). Recent functionality developments in Montmorillonite as a nanofiller in food packaging. *Trends in Food Science & Technology, 140*, 104148.

Boetje, L., Lan, X., van Dijken, J., Kaastra, G., Polhuis, M., & Loos, K. (2023). Thiol-Ene click cross-linking of starch oleate films for enhanced properties. *Biomacromolecules, 24* (12), 5578–5588. https://doi.org/10.1021/acs.biomac.3c00507

Chaudhary, S., Kumar, V., Sharma, V., Sharma, R., & Kumar, S. (2022). Chitosan nanoemulsion: Gleam into the futuristic approach for preserving the quality of muscle foods. *International Journal of Biological Macromolecules, 199*, 121–137.

Che Hamzah, N. H., Khairuddin, N., Muhamad, I. I., Hassan, M. A., Ngaini, Z., & Sarbini, S. R. (2022). Characterisation and colour response of smart Sago Starch-Based packaging films incorporated with Brassica oleracea anthocyanin. *Membranes, 12*(10). https://doi.org/10.3390/membranes12100913

Chen, N., Gao, H.-X., He, Q., & Zeng, W.-C. (2023). Potato starch-based film incorporated with tea polyphenols and its application in fruit packaging. *Polymers, 15*(3), 588.

Chen, W. C., Judah, S. N. M. S. M., Ghazali, S. K., Munthoub, D. I., Alias, H., Mohamad, Z., & Majid, R. A. (2021). The effects of citric acid on thermal and mechanical properties of crosslinked starch film. *Chemical Engineering Transactions, 83*. https://doi.org/10.3303/CET2183034

Chen, S., Wu, M., Lu, P., Gao, L., Yan, S., & Wang, S. (2020). Development of pH indicator and antimicrobial cellulose nanofibre packaging film based on purple sweet potato anthocyanin and oregano essential oil. *International Journal of Biological Macromolecules, 149*, 271–280.

Chen, Y., Zhu, Z., Ye, Y., Li, Q., Yang, T., Guan, C., & Liu, F. (2024). Comprehensive evaluation of the physicochemical attributes, antioxidant capacity, and pH-Responsive behavior of starch films enhanced by laver incorporation. *Foods, 13*(11), 1600.

Cheng, M., Cui, Y., Yan, X., Zhang, R., Wang, J., & Wang, X. (2022a). Effect of dual-modified cassava starches on intelligent packaging films containing red cabbage extracts. *Food Hydrocolloids, 124,* 107225. https://doi.org/10.1016/j.foodhyd.2021.107225

Cheng, M., Yan, X., Cui, Y., Han, M., Wang, Y., Wang, J., … Wang, X. (2022b). Characterization and release kinetics Study of active packaging films based on modified starch and Red Cabbage Anthocyanin extract. *Polymers, 14*(6). https://doi.org/10.3390/polym14061214

Chi, K., Wang, H., & Catchmark, J. M. (2020). Sustainable starch-based barrier coatings for packaging applications. *Food Hydrocolloids, 103,* 105696.

Chieh, J. S. P., & Siddique, B. M. (2019). Preparation of Degradable Sago Starch film incorporated with zinc oxide (ZnO) nanoparticles for enhanced chemical barrier and mechanical properties. *Key Engineering Materials, 792,* 104–110.

Chim-Chi, Y., Fernández-Méndez, A., Pérez-Pacheco, E., Canto-Pinto, J., Estrada-León, R., Ortiz-Fernández, A., … Pérez-Padilla, Y. (2024). Development of pH indicator smart packaging films with anthocyanins from the purple star apple shell (Chrysophyllum cainito L.). *Journal of Food Measurement and Characterization, 18*(4), 2651–2660.

Chaudhary, S., Kour, M., & Kumar, R. (2024). Bioplastic films from starch of *Colocasia esculenta* and its waste: A smart template for sensing applications. *International Journal of Biological Macromolecules, 281,* Article 136218. https://doi.org/10.1016/j.ijbiomac.2024.136218

Chavez-Marquez, E., Bernedo, M. S. B., de Jara, E. M., Quequezana-Bedregal, M. J., Gutierrez-Oppe, E. E., & Pessôa Filho, P.d. A. (2023). Development of intelligent and active potato starch films based on purple corn cob extract and molle essential oil. *International Journal of Biological Macromolecules, 242,* Article 125080. https://doi.org/10.1016/j.ijbiomac.2023.125080

Cvetković, K., Lazarević, A., Stamenković Stojanović, S., Dordevic, S., Dordevic, D., Karabegović, I., & Danilović, B. (2025). pH-Sensitive starch-based packaging films enhanced with wild blackberry extract. *Processes, 13*(4), 1148.

D'Almeida, A. P., & de Albuquerque, T. L. (2024). Innovations in food packaging: From bio-based materials to smart packaging systems. *Processes, 12*(10), 2085.

Díaz-Cruz, C. A., Caicedo, C., Jiménez-Regalado, E. J., Díaz de León, R., López-González, R., & Aguirre-Loredo, R. Y. (2022). Evaluation of the antimicrobial, thermal, mechanical, and barrier properties of corn Starch–Chitosan biodegradable films reinforced with cellulose nanocrystals. *Polymers, 14*(11), 2166.

Díaz-Galindo, E. P., Nesic, A., Cabrera-Barjas, G., Mardones, C., von Baer, D., Bautista-Baños, S., & Dublan Garcia, O. (2020). Physical-Chemical evaluation of active food packaging material based on thermoplastic starch loaded with grape cane extract. *Molecules, 25*(6), 1306.

Deng, N., Hu, Z., Li, H., Li, C., Xiao, Z., Zhang, B., … Cai, Y. (2024). Physicochemical properties and pork preservation effects of lotus seed drill core powder starch-based active packaging films. *International Journal of Biological Macromolecules, 260,* 129340.

Deng, H., Su, J., Zhang, W., Khan, A., Sani, M. A., Goksen, G., … Rhim, J.-W. (2024). A review of starch/polyvinyl alcohol (PVA) blend film: A potential replacement for traditional plastic-based food packaging film. *International Journal of Biological Macromolecules, 273,* 132926.

Dong, M., Bilotti, E., Zhang, H., & Papageorgiou, D. G. (2024). Multifunctional Ti3C2Tx MXene-reinforced thermoplastic starch nanocomposites for sustainable packaging solutions. *International Journal of Biological Macromolecules, 265*, 130520. https://doi.org/10.1016/j.ijbiomac.2024.130520

Dong, M., Bilotti, E., Zhang, H., & Papageorgiou, D. G. (2025). Recyclable PVA/starch/$Ti_3C_2T_x$ MXene nanocomposite films with superior mechanical and barrier properties. *International Journal of Biological Macromolecules, 295*, Article 139545. https://doi.org/10.1016/j.ijbiomac.2025.139545

Dong, H., Ling, Z., Zhang, X., Zhang, X., Ramaswamy, S., & Xu, F. (2020). Smart colorimetric sensing films with high mechanical strength and hydrophobic properties for visual monitoring of shrimp and pork freshness. *Sensors and Actuators B: Chemical, 309*, 127752. https://doi.org/10.1016/j.snb.2020.127752

Dong, Y., Xie, Y., Ma, X., Yan, L., Yu, H.-Y., Yang, M., ... Jia, B. (2023). Multifunctional nanocellulose based nanocomposites for biodegradable food packaging: Hybridization, fabrication, key properties and application. *Carbohydrate Polymers, 321*, 121325.

Dutta, D., & Sit, N. (2024). Comprehensive review on developments in starch-based films along with active ingredients for sustainable food packaging. *Sustainable Chemistry and Pharmacy, 39*, 101534.

Estrada-Monje, A., Alonso-Romero, S., Zitzumbo-Guzmán, R., Estrada-Moreno, I. A., & Zaragoza-Contreras, E. A. (2021). Thermoplastic starch-based blends with improved thermal and thermomechanical properties. *Polymers, 13*(23). https://doi.org/10.3390/polym13234263

Fonseca, L. M., da Cruz, E. P., Crizel, R. L., Jansen-Alves, C., Dias, A. R. G., & da Rosa Zavareze, E. (2024). New advances of electrospun starch fibers, encapsulation, and food applications: A review. *Trends in Food Science & Technology, 147*, 104467.

Forfora, N., Azuaje, I., Kanipe, T., Gonzalez, J. A., Lendewig, M., Urdaneta, I., ... Argyropoulos, D. (2024). Are starch-based materials more eco-friendly than fossil-based? A critical assessment. *Cleaner Environmental Systems, 13*, 100177.

Ganosi, E., Barda, C., Grafakou, M.-E., Rallis, M. C., & Skaltsa, H. (2023). An In-Depth Stability Study of the Essential Oils from Mentha × piperita, Mentha spicata, Origanum vulgare, and Thymus vulgaris: The Impact of Thermal and Storage Conditions. *Separations, 10*(9), 488.

Gao, Q., Feng, Z., Wang, J., Zhao, F., Li, C., & Ju, J. (2025). Application of nano-ZnO in the food preservation industry: Antibacterial mechanisms, influencing factors, intelligent packaging, preservation film and safety. *Critical Reviews in Food Science and Nutrition, 65*(22), 4327–4353.

Gao, S., Li, M., Zhai, X., Wang, W., & Hou, H. (2024). Starch as a smart, cheap, and green gatekeeper for the controlled release of propyl gallate from antioxidant biodegradable packaging films. *Food Chemistry, 453*, 139627.

Garavito, J., Peña-Venegas, C. P., & Castellanos, D. A. (2024). Production of starch-based flexible food packaging in developing countries: Analysis of the processes, challenges, and requirements. *Foods, 13*(24), 4096.

Ghasemlou, M., Aliheidari, N., Fahmi, R., Shojaee-Aliabadi, S., Keshavarz, B., Cran, M. J., & Khaksar, R. (2013). Physical, mechanical and barrier properties of corn starch films incorporated with plant essential oils. *Carbohydrate Polymers, 98*(1), 1117–1126.

González-Torres, B., Robles-García, M.Á., Gutiérrez-Lomelí, M., Padilla-Frausto, J. J., Navarro-Villarruel, C. L., Del-Toro-Sánchez, C. L., ... Reynoso-Marín, F. J. (2021). Combination of sorbitol and glycerol, as plasticizers, and oxidized starch improves the physicochemical characteristics of films for food preservation. *Polymers, 13*(19), 3356.

Goudarzi, V., Shahabi-Ghahfarrokhi, I., & Babaei-Ghazvini, A. (2017). Preparation of eco-friendly UV-protective food packaging material by starch/TiO2 bio-nanocomposite:

Characterization. *International Journal of Biological Macromolecules, 95*, 306–313. https://doi.org/10.1016/j.ijbiomac.2016.11.065

Guo, Q., Yuan, Y., He, M., Zhang, X., Li, L., Zhang, Y., & Li, B. (2023). Development of a multifunctional food packaging for meat products by incorporating carboxylated cellulose nanocrystal and beetroot extract into sodium alginate films. *Food Chemistry, 415*, 135799. https://doi.org/10.1016/j.foodchem.2023.135799

Gutiérrez, T. J., & Valencia, G. A. (2021). Reactive extrusion-processed native and phosphated starch-based food packaging films governed by the hierarchical structure. *International Journal of Biological Macromolecules, 172*, 439–451. https://doi.org/10.1016/j.ijbiomac.2021.01.048

Gutiérrez, T. J. (2022). Starch-based food packaging films processed by reactive extrusion/thermo-molding using chromium octanoate-loaded zeolite A as a potential triple-action mesoporous material (reinforcing filler/food-grade antimicrobial organocatalytic nanoreactor). *Food Packaging and Shelf Life, 34*, Article 100974. https://doi.org/10.1016/j.fpsl.2022.100974

Halder, S., Shrikrishna, N. S., Sharma, R., Bhat, P., & Gandhi, S. (2024). Raising the bar: Exploring modern technologies and biomaterials for enhancing food safety and quality-a comprehensive review. *Food Control, 159*, 110287.

Hamzah, N., Shaidi, N., Merais, M., & Khairuddin, N. (2021). Smart food packaging from sago starch incorporated with anthocyanin from Brassica oleracea. Paper presented at the IOP Conference Series: Earth and Environmental Science.

Hasanin, M. S., Hassan, Y. R., & Youssef, A. M. (2024). Active packaging films based on the nanoform of chitin, alginate, and layered double hydroxides: Characterization, mechanical properties, permeability, and bioactive properties. *RSC Advances, 14*(50), 37380–37391. https://doi.org/10.1039/d4ra06306f

Hosseinnejad, M. (2014). *Active packaging for food applications-A review.*

Huang, X., Wang, F., Hu, W., Zou, Z., Tang, Q., Li, H., & Xu, L. (2024). Smart packaging films based on corn starch/polyvinyl alcohol containing nano SIM-1 for monitoring food freshness. *International Journal of Biological Macromolecules, 256*, 128373.

Jamróz, E., Cabaj, A., Tkaczewska, J., Kawecka, A., Krzyściak, P., Szuwarzyński, M., … Juszczak, L. (2023). Incorporation of curcumin extract with lemongrass essential oil into the middle layer of triple-layered films based on furcellaran/chitosan/gelatin hydrolysates–In vitro and in vivo studies on active and intelligent properties. *Food Chemistry, 402*, 134476.

Jayakumar, A., Radoor, S., Kim, J. T., Rhim, J. W., Parameswaranpillai, J., Nandi, D., … Siengchin, S. (2022). Titanium dioxide nanoparticles and elderberry extract incorporated starch based polyvinyl alcohol films as active and intelligent food packaging wraps. *Food Packaging and Shelf Life, 34*, 100967.

Joshi, N. C., Negi, P. B., & Gururani, P. (2024). A review on metal/metal oxide nanoparticles in food processing and packaging. *Food Science and Biotechnology, 33*(6), 1307–1322.

Jumaidin, R., Abdul Rahman, A. H., Sapuan, S. M., & Rushdan, A. I. (2024). Effect of sugarcane bagasse on thermal and mechanical properties of thermoplastic cassava starch/beeswax composites. *Physical Sciences Reviews, 9*(1), 1–15.

Kafashan, A., Joze-Majidi, H., Kazemi-Pasarvi, S., Babaei, A., & Jafari, S. M. (2023). Nanocomposites of soluble soybean polysaccharides with grape skin anthocyanins and graphene oxide as an efficient halochromic smart packaging. *Sustainable Materials and Technologies, 38*. https://doi.org/10.1016/j.susmat.2023.e00755

Karnwal, A., Rauf, A., Jassim, A. Y., Selvaraj, M., Al-Tawaha, A., Kashyap, P., … Malik, T. (2025). Advanced starch-based films for food packaging: Innovations in sustainability and functional properties. *Food Chemistry X, 29*, 102662. https://doi.org/10.1016/j.fochx.2025.102662

Kavas, E., Terzioğlu, P., & Sıcak, Y. (2023). Betanin and nano-calcium carbonate incorporated polyvinyl alcohol/Starch films for active and intelligent packaging applications. *Journal of Polymers and the Environment, 31*(11), 4919–4929. https://doi.org/10.1007/s10924-023-02868-0

Konar, N., Gunes, R., Palabiyik, I., & Toker, O. S. (2022). Health conscious consumers and sugar confectionery: Present aspects and projections. *Trends in Food Science & Technology, 123*, 57–68.

Kumar, V., Chakraborty, P., Janghu, P., Umesh, M., Sarojini, S., Pasrija, R., … Sivalingam, A. M. (2023). Potential of banana based cellulose materials for advanced applications: A review on properties and technical challenges. *Carbohydrate Polymer Technologies and Applications, 6*, 100366. https://doi.org/10.1016/j.carpta.2023.100366

Kumar, L., Deshmukh, R. K., Hakim, L., & Gaikwad, K. K. (2024). Halloysite nanotube as a functional material for active food packaging application: A review. *Food and Bioprocess Technology, 17*(1), 33–46.

Kumar, P., & Gautam, S. (2019). Developing ZnO nanoparticle embedded antimicrobial starch biofilm for food packaging. arXiv preprint arXiv:1909.05083.

Kusuma, I., Ma'arif, A., & Rahmadhia, S. N. (2024a). Starch based for intelligent food packaging: A review. Paper presented at the BIO Web of Conferences.

Kusuma, I., Ma'arif, A., & Rahmadhia, S. N. (2024b). Starch based for intelligent food packaging: A review. *BIO Web Conf., 146*, 01099.

Kwasniewska, A., Chocyk, D., Gladyszewski, G., Borc, J., Swietlicki, M., & Gladyszewska, B. (2020). The influence of Kaolin clay on the mechanical properties and structure of thermoplastic starch films. *Polymers, 12*(1). https://doi.org/10.3390/polym12010073

López, O., Castillo, L., Zaritzky, N., Barbosa, S., Villar, M., & García, M. A. (2015). Talc nanoparticles influence on thermoplastic corn starch film properties. *Procedia Materials Science, 8*, 338–345. https://doi.org/10.1016/j.mspro.2015.04.082

Li, H., Liu, G., Ye, K., He, W., Wei, H., & Dang, L. (2024a). A novel pH-sensitive antibacterial bilayer film for intelligent packaging. *Biomass Conversion and Biorefinery, 14*(13), 14303–14316.

Li, X., Zou, F., Kang, X., Gao, W., Cui, B., & Sui, J. (2024b). Effects of acetylated distarch phosphate on the physicochemical characteristics and stability of the oyster sauce system. *Frontiers in Nutrition, 11*, 1412314. https://doi.org/10.3389/fnut.2024.1412314

Liu, D., Zhao, P., Chen, J., Yan, Y., & Wu, Z. (2022). Recent advances and applications in starch for intelligent active food packaging: A review. *Foods, 11*(18), 2879.

Lv, H., Feng, X., Song, H., Ma, S., Hao, Z., Hu, H., … Fan, F. (2023). Tea storage: A not thoroughly recognized and precisely designed process. *Trends in Food Science & Technology, 140*, 104172.

Müller, P., & Schmid, M. (2019). Intelligent packaging in the food sector: A brief overview. *Foods, 8*(1), 16.

Mahanta, B. P., Bora, P. K., Kemprai, P., Borah, G., Lal, M., & Haldar, S. (2021). Thermolabile essential oils, aromas and flavours: Degradation pathways, effect of thermal processing and alteration of sensory quality. *Food Research International, 145*, 110404. https://doi.org/10.1016/j.foodres.2021.110404

Mahović Poljaček, S., Tomašegović, T., Strižić Jakovljević, M., Jamnicki Hanzer, S., Murković Steinberg, I., Žuvić, I., … Karlovits, I. (2024). Starch-Based functional films enhanced with bacterial nanocellulose for smart packaging: Physicochemical properties, pH sensitivity and colorimetric response. *Polymers, 16*(16), 2259.

Mallick, N., Pal, D., Soni, A., Jhariya, D., & Singh, D. (2020). Starch based antimicrobial food packaging film towards a sustainable environment. In *Paper presented at the IOP conference series: Earth and environmental science.*

Mane, K. (2016). A review on active packaging: An innovation in food packaging. *International Journal of Environment, Agriculture and Biotechnology, 1*(3), 238566.
Marichelvam, M., Jawaid, M., & Asim, M. (2019). Corn and rice starch-based bio-plastics as alternative packaging materials. *Fibers,* 7(4), 32.
Matheus, J. R. V., Dalsasso, R. R., Rebelatto, E. A., Andrade, K. S., Andrade, L. M., Andrade, C. J., … Fai, A. E. C. (2023). Biopolymers as green-based food packaging materials: A focus on modified and unmodified starch-based films. *Comprehensive Reviews in Food Science and Food Safety, 22*(2), 1148–1183. https://doi.org/10.1111/1541-4337.13107
Menzel, C., Gonzalez-Martinez, C., Vilaplana, F., Diretto, G., & Chiralt, A. (2019). Incorporation of natural antioxidants from rice straw into renewable starch films. *International Journal of Biological Macromolecules, 146.* https://doi.org/10.1016/j.ijbiomac.2019.09.222
Mirjalili, F., & Ardekani, A. (2017). Preparation and characterization of starch film accompanied with ZnO nanoparticles: Mirjalili and Ardekani Ali. *Journal of Food Process Engineering, 40,* e12561. https://doi.org/10.1111/jfpe.12561
Moraes, J., Müller, C., & Laurindo, J. (2011). Influence of the simultaneous addition of bentonite and cellulose fibers on the mechanical and barrier properties of starch composite-films. *Food science and technology international = Ciencia y tecnología de los alimentos internacional, 18,* 35–45. https://doi.org/10.1177/1082013211427622
Muñoz-Gimena, P. F., Oliver-Cuenca, V., Peponi, L., & López, D. (2023). A review on reinforcements and additives in starch-based composites for food packaging. *Polymers, 15*(13), 2972.
Murtaza, S., Shah, F.-u. H., Shahbaz, M., Murtaza, M. A., Farooq, U., & Ma, Y. (2024). Starch-nanostructured-based active packaging for food applications. In *Starch based nanomaterials for food packaging* (pp. 103–160). Elsevier.
Mussagy, C. U., Oliveira, G., Ahmad, M., Mustafa, A., Herculano, R. D., & Farias, F. O. (2024). Halochromic properties of carotenoid-based films for smart food packaging. *Food Packaging and Shelf Life, 44,* 101325.
Nafchi, A. M., Nassiri, R., Sheibani, S., Ariffin, F., & Karim, A. (2013). Preparation and characterization of bionanocomposite films filled with nanorod-rich zinc oxide. *Carbohydrate Polymers, 96*(1), 233–239.
Nakamoto, M. M., Assis, M., de Oliveira Filho, J. G., & Braga, A. R. C. (2023). Spirulina application in food packaging: Gaps of knowledge and future trends. *Trends in Food Science & Technology, 133,* 138–147.
Nath, P. C., Sharma, R., Debnath, S., Sharma, M., Inbaraj, B. S., Dikkala, P. K., … Sridhar, K. (2023). Recent trends in polysaccharide-based biodegradable polymers for smart food packaging industry. *International Journal of Biological Macromolecules, 253,* 127524.
Niu, Z., Chen, F., Zhang, H., & Liu, C. (2023). High content of thermoplastic starch, Poly (butylenes adipate-co-terephthalate) and Poly(butylene succinate) ternary blends with a good balance in strength and toughness. *Polymers, 15*(9). https://doi.org/10.3390/polym15092040
Nizam, N. M., Rawi, N. F. M., Ramle, S. M., Aziz, A. A., Abdullah, C., Rashedi, A., & Kassim, M. H. M. (2021). Physical, thermal, mechanical, antimicrobial and physicochemical properties of starch based film containing aloe vera: A review. *Journal of Materials Research and Technology.* https://doi.org/10.1016/j.jmrt.2021.08.138
Nordin, N., Othman, S., Kadir, R., & Rashid, S. (2018). Mechanical and thermal properties of starch films reinforced with microcellulose fibres. *Food Research, 2,* 555–563. https://doi.org/10.26656/fr.2017.2(6).110
Nowak, N., Tkaczewska, J., Cholewa-Wójcik, A., Grzebieniarz, W., Piasna-Słupecka, E., Drozdowska, M., & Jamróz, E. (2025). Multi-layer active biopolymer System as a

perspective for improving the quality and scope of applying semi-rigid biodegradable packaging. *Food and Bioprocess Technology, 18*(3), 2682–2696.

Omotoso, M., Adeyefa, O., Animashaun, E., & Osibanjo, O. (2015). Biogradable starch film from cassava, corn, potato and yam. *Chemistry and Materials Research,* 7(12), 15–24.

Patil, A. S., Waghmare, R. D., Pawar, S. P., Salunkhe, S. T., Kolekar, G. B., Sohn, D., & Gore, A. H. (2020). Photophysical insights of highly transparent, flexible and re-emissive PVA@ WTR-CDs composite thin films: A next generation food packaging material for UV blocking applications. *Journal of Photochemistry and Photobiology A: Chemistry, 400*, 112647.

Pawase, P. A., Pathare, A. M., Bashir, O., Saleem, F., Shrama, E., Mudgal, S., & Ahmad, M. (2025). Physical, chemical, and nano-enabled modifications of starch for sustainable food packaging films: Recent trends, challenges, and prospects. *Carbohydrate Polymer Technologies and Applications, 11*, 100986. https://doi.org/10.1016/j.carpta.2025.100986

Pei, J., Palanisamy, C. P., Srinivasan, G. P., Panagal, M., Kumar, S. S. D., & Mironescu, M. (2024). A comprehensive review on starch-based sustainable edible films loaded with bioactive components for food packaging. *International Journal of Biological Macromolecules, 274*, 133332.

Pelissari, F. M., Andrade-Mahecha, M. M., Sobral, P. J.d. A., & Menegalli, F. C. (2017). Nanocomposites based on banana starch reinforced with cellulose nanofibers isolated from banana peels. *Journal of Colloid and Interface Science, 505*, 154–167. https://doi.org/10.1016/j.jcis.2017.05.106

Peralta-González, C., Ramírez-Hernández, A., Rangel-Porras, G., Aparicio-Saguilán, A., Aguirre-Cruz, A., González-García, G., … Páramo-Calderón, D. E. (2022). Synthesis and characterization of the Starch/silicone oil composite and elaboration of its films. *Silicon, 14*(8), 4157–4167.

Pooja, N., Shashank, S., Singh, B. N., & Mazumder, N. (2024). Advancing sustainable bioplastics: Chemical and physical modification of starch films for enhanced thermal and barrier properties. *RSC Advances, 14*(33), 23943–23951. https://doi.org/10.1039/D4RA04263H

Prasad, P., & Kochhar, A. (2014). Active packaging in food industry: A review. *Journal of Environmental Science, Toxicology and Food Technology, 8*(5), 1–7.

Qazanfarzadeh, Z., Baptista, H., Nunes, C., & Kumaravel, V. (2025). Development of smart food packaging from bread waste-derived starch and carbon quantum dots. *Industrial Crops and Products, 233*, 121426.

Rahmah, M., Mohd, N., Norizan, M. N., Aisyah, S., & Fauzi, F. (2017). Swelling and tensile properties of starch glycerol system with various crosslinking agents. *IOP Conference Series: Materials Science and Engineering, 223*, 012059. https://doi.org/10.1088/1757-899X/223/1/012059

Rahmadhia, S. N., Sidqi, A. A., & Saputra, Y. A. (2023). Physical properties of tapioca starch-based film indicators with anthocyanin extract from purple sweet potato (*Ipomea batatas* L.) and response to pH changes. *Sains Malaysiana, 52*(6), 1685–1697.

Rajan, S. S., & Wani, K. M. (2025). A review of smart food and packaging technologies: Revolutionizing nutrition and sustainability. *Food and Humanity*, 100593.

Rapisarda, M., Marken, F., & Meo, M. (2021). Graphene oxide and starch gel as a hybrid binder for environmentally friendly high-performance supercapacitors. *Communications Chemistry, 4*(1), 169. https://doi.org/10.1038/s42004-021-00604-0

Ribeiro, I. S., Maciel, G. M., Bortolini, D. G., Fernandes, I.d. A. A., Maroldi, W. V., Pedro, A. C., … Haminiuk, C. W. I. (2024). Sustainable innovations in edible films and coatings: An overview. *Trends in Food Science & Technology, 143*, 104272.

Rodrigues, P. R., Junior, L. M., de Souza, W. F. C., Sato, H. H., Alves, R. M. V., & Vieira, R. P. (2021). O-ATRP synthesized poly (β-pinene) blended with chitosan for antimicrobial and antioxidant bio-based films production. *International Journal of Biological Macromolecules, 193*, 425–432.

Rostamabadi, H., Demirkesen, I., Colussi, R., Roy, S., Tabassum, N., de Oliveira Filho, J. G., … Galus, S. (2024). Recent trends in the application of films and coatings based on starch, cellulose, chitin, chitosan, xanthan, gellan, pullulan, Arabic gum, alginate, pectin, and carrageenan in food packaging. *Food Frontiers, 5*(2), 350–391.

Rout, S. S., & Pradhan, K. C. (2024). A review on antimicrobial nano-based edible packaging: Sustainable applications and emerging trends in food industry. *Food Control, 163*, 110470.

Roveda, A. C., Dias, B. C., Passini, L. N., Manzani, D., & Petruci, J. F.d. S. (2024). Transparent, flexible, and eco-friendly starch-based films for reversible optoelectronic noses for food spoilage monitoring in smart packaging. *Microchimica Acta, 191*(6), 354. https://doi.org/10.1007/s00604-024-06426-w

Samsudin, H., & Hani, N. M. (2017). Use of starch in food packaging. In *Starch-based materials in food packaging* (pp. 229–256). Elsevier.

Sani, M. A., Zhang, W., Abedini, A., Khezerlou, A., Shariatifar, N., Assadpour, E., … Jafari, S. M. (2024). Intelligent packaging systems for the quality and safety monitoring of meat products: From lab scale to industrialization. *Food Control, 160*, 110359.

Santhosh, R., Ahmed, J., Thakur, R., & Sarkar, P. (2024). Starch-based edible packaging: Rheological, thermal, mechanical, microstructural, and barrier properties – A review. *Sustainable Food Technology, 2*(2), 307–330. https://doi.org/10.1039/d3fb00211j

Santos, T. A., Santos, B. H., Menegassi de Souza, R. F., & Spinacé, M. A. S. (2025). Study on the biodegradation of thermoplastic starch and bacterial cellulose based sandwich panel biocomposites. *International Journal of Biological Macromolecules, 314*, 144191. https://doi.org/10.1016/j.ijbiomac.2025.144191

Saxena, N., Sarkar, M., & Sarkar, B. (2024). Which retail strategy for shelf life products can satisfy consumers under game policy? *Journal of Retailing and Consumer Services, 76*, 103595.

Shafi, Z., Singh, R., Sidiqi, U. S., Bashir, B., Rasool, S., Dash, K. K., … Dar, A. H. (2025). Quercetin infused starch matrix as a sustainable approach to smart packaging: A comprehensive review. *International Journal of Biological Macromolecules, 320*, 145746.

Shanbhag, C., Shenoy, R., Shetty, P., Srinivasulu, M., & Nayak, R. (2023). Formulation and characterization of starch-based novel biodegradable edible films for food packaging. *J Food Sci Technol, 60*(11), 2858–2867. https://doi.org/10.1007/s13197-023-05803-2

Shao, G.-Q., Zhang, H., Xu, D., Wu, F.-F., Jin, Y.-M., Yang, N., … Xu, X.-M. (2024). Insights into starch-based gels: Selection, fabrication, and application. *International Journal of Biological Macromolecules, 258*, 128864.

Shapi'i, R. A., Othman, S. H., Basha, R. K., & Naim, M. N. (2022). Mechanical, thermal, and barrier properties of starch films incorporated with chitosan nanoparticles. *Nanotechnology Reviews, 11*(1), 1464–1477. https://doi.org/10.1515/ntrev-2022-0094

Shen, Z., Rajabi-Abhari, A., Oh, K., Yang, G., Youn, H. J., & Lee, H. L. (2021). Improving the barrier properties of packaging paper by polyvinyl alcohol based polymer coating—effect of the base paper and nanoclay. *Polymers, 13*(8), 1334.

Shibasaki, K., Hsu, Y. I., & Uyama, H. (2024). Facile fabrication of starch-based UV barrier films with remolding ability and reinforcement for water resistance. *Macromolecular Materials and Engineering, 309*(10), 2400137.

Singh, A. K., Lee, M., Jang, D., & Lee, Y. S. (2024). Non-conventional starch nanoparticles: Novel avenues towards improving sustainability of the food packaging sector. *Trends in Food Science & Technology, 143*, 104273.

Song, W., Wu, N., He, Y., Zhao, H., Xu, J., & Ren, L. (2024). Intelligent color changing packaging film based on esterified starch and black rice anthocyanins. *Food Chemistry X, 24*, 101930.

Souza, A. G., Ferreira, R. R., Paula, L. C., Mitra, S. K., & Rosa, D. S. (2021). Starch-based films enriched with nanocellulose-stabilized Pickering emulsions containing different essential oils for possible applications in food packaging. *Food Packaging and Shelf Life, 27*, 100615. https://doi.org/10.1016/j.fpsl.2020.100615

Sultana, A. (2025). Starch-based biocomposite films and coatings in paper-based food packaging: A review. *Environmental Chemistry Letters*, 1–24.

Sun, K.-q., Fangyi, L., Li, J.-y., Li, J.-f., Zhang, C.-w., Chen, S., … Cui, J.-f. (2019). Optimisation of compatibility for improving elongation at break of chitosan/starch films. *RSC Advances, 9*, 24451–24459. https://doi.org/10.1039/C9RA04053F

Surendhiran, D., Roy, V. C., Ho, T. C., Park, J.-S., Lee, S.-C., & Chun, B.-S. (2023). Smart packaging film prepared from subcritical water-modified oat starch and betalain of beetroot extract reinforced with cellulose nanofibrils. *Sustainable Chemistry and Pharmacy, 36*, 101349.

Tamimi, N., Mohammadi Nafchi, A., Hashemi-Moghaddam, H., & Baghaie, H. (2021). The effects of nano-zinc oxide morphology on functional and antibacterial properties of tapioca starch bionanocomposite. *Food Sci Nutr, 9*(8), 4497–4508. https://doi.org/10.1002/fsn3.2426

Tan, S. X., Ong, H. C., Andriyana, A., Lim, S., Pang, Y. L., Kusumo, F., & Ngoh, G. C. (2022). Characterization and parametric study on mechanical properties enhancement in biodegradable chitosan-reinforced starch-based bioplastic film. *Polymers, 14*(2), 278.

Tarique, J., Sapuan, S. M., & Khalina, A. (2021). Effect of glycerol plasticizer loading on the physical, mechanical, thermal, and barrier properties of arrowroot (Maranta arundinacea) starch biopolymers. *Scientific Reports, 11*(1), 13900. https://doi.org/10.1038/s41598-021-93094-y

Todica, M., Nagy, E. M., Niculaescu, C., Stan, O., Cioica, N., & Pop, C. V. (2016). XRD investigation of some thermal degraded starch based materials. *Journal of Spectroscopy, 2016*(1), 9605312. https://doi.org/10.1155/2016/9605312

Todorova, D., Yavorov, N., & Lasheva, V. (2022). Improvement of barrier properties for packaging applications. *Sustainable Chemistry and Pharmacy, 27*, 100685.

Torres-Medina, C. T., & Murillo, E. A. (2025). Polyvinyl Alcohol/Starch films: Improvement of the interactions, migration resistance, thermal stability, gloss, and biodegradability. *Polymer Engineering & Science, 65*(8), 4359–4373. https://doi.org/10.1002/pen.70004

Upadhyay, A., Agbesi, P., Arafat, K. M. Y., Urdaneta, F., Dey, M., Basak, M., … Argyropoulos, D. (2024). Bio-based smart packaging: Fundamentals and functions in sustainable food systems. *Trends in Food Science & Technology, 145*, 104369.

Vedove, T. M. A. R. D., Maniglia, B. C., & Tadini, C. C. (2021). Production of sustainable smart packaging based on cassava starch and anthocyanin by an extrusion process. *Journal of Food Engineering, 289*, 110274. https://doi.org/10.1016/j.jfoodeng.2020.110274

Wahab, Y. A., Al-Ani, L. A., Khalil, I., Schmidt, S., Tran, N. N., Escriba-Gelonch, M., … Hessel, V. (2024). Nanomaterials: A critical review of impact on food quality control and packaging. *Food Control, 163*, 110466.

Wang, C., Gong, C., Qin, Y., Hu, Y., Jiao, A., Jin, Z., … Wang, J. (2022). Bioactive and functional biodegradable packaging films reinforced with nanoparticles. *Journal of Food Engineering, 312*, 110752. https://doi.org/10.1016/j.jfoodeng.2021.110752

Wang, B., Wang, X., Ma, H., Li, J., Cao, D., Xu, J., … Chen, K. (2025). Enhancing the mechanical and barrier properties of starch-based green packaging films via acetalization reaction and hydrogen bond dual crosslinking. *Carbohydrate Polymers*, 123787.

Wang, K., Wu, L., Li, Y., & Li, H. (2023). Preparation and characterization of chitosan/halloysite nanotubes composite film with ethylene scavenging and gas resistance for active food packaging. *Journal of Food Safety, 43*(2), e13027.

Wang, B., Xu, X., Fang, Y., Yan, S., & Cui, B. (2022a). Effect of different ratios of glycerol and erythritol on properties of corn starch-based films. *Frontiers in Nutrition, 9*, 882682. https://doi.org/10.3389/fnut.2022.882682

Wang, H., Wang, L., Ye, S., & Song, X. (2019). Construction of Bi_2WO_6–TiO_2/starch nanocomposite films for visible-light catalytic degradation of ethylene. *Food Hydrocolloids, 88*, 92–100. https://doi.org/10.1016/j.foodhyd.2018.09.021

Xu, H., Cheng, H., McClements, D. J., Chen, L., Long, J., & Jin, Z. (2022). Enhancing the physicochemical properties and functional performance of starch-based films using inorganic carbon materials: A review. *Carbohydrate Polymers, 295*, 119743.

Xu, X., Huang, K., He, D., Huang, B., Tang, Q., Li, H., & Zou, Z. (2025). Biocompatible starch based smart active packaging film engineered with a novel CuNa-MOF colorimetric ammonia sensing agent based on crystal-to-crystal transformation. *Food Packaging and Shelf Life, 49*, 101498. https://doi.org/10.1016/j.fpsl.2025.101498

Xu, Y., Wu, Z., Li, A., Chen, N., Rao, J., & Zeng, Q. (2024). Nanocellulose composite films in food packaging materials: A review. *Polymers, 16*(3), 423.

Yang, Y., Fu, J., Duan, Q., Xie, H., Dong, X., & Yu, L. (2024). Strategies and methodologies for improving toughness of starch films. *Foods, 13*(24), 4036.

Yang, J., Shan, P., Sun, F., Sun, L., Wang, K., Zhuang, Y., … Peng, L. (2025). Development of gelatin-based antibacterial composite films incorporating dialdehyde starch and AgNPs@ MXene for active food packaging. *Food Chemistry*, 145043.

Yao, F., Wu, Z., Gu, Y., Di, Y., Liu, Y., Srinivasan, V., … Li, Y. (2024). Acetylated nanocellulose reinforced hydroxypropyl starch acetate realizing polypropylene replacement for green packaging application. *Carbohydrate Polymers, 331*, 121886. https://doi.org/10.1016/j.carbpol.2024.121886

Yudhistira, B., Husnayain, N., Punthi, F., Gavahian, M., Chang, C.-K., & Hsieh, C.-W. (2024). Progress in the application of emerging technology for the improvement of starch-based active packaging properties: A review. *ACS Food Science & Technology, 4* (9), 1997–2012.

Żołek-Tryznowska, Z., Jeznach, A., Bednarczyk, E., & Tryznowski, M. (2024a). INFLUENCE OF ULTRASONIC TREATMENT ON THE MECHANICAL AND SURFACE PROPERTIES OF POTATO STARCH-BASED FILMS WITH VARIOUS GLYCEROL CONTENT.

Żołek-Tryznowska, Z., Piłczyńska, K., Murawski, T., Jeznach, A., & Niczyporuk, K. (2024b). Study on the printability of starch-based films using ink-jet printing. *Materials, 17*(2), 455.

Zhai, X., Shi, J., Zou, X., Wang, S., Jiang, C., Zhang, J., … Holmes, M. (2017). Novel colorimetric films based on starch/polyvinyl alcohol incorporated with roselle anthocyanins for fish freshness monitoring. *Food Hydrocolloids, 69*, 308–317. https://doi.org/10.1016/j.foodhyd.2017.02.014

Zhang, L., Zhao, J., Zhang, Y., Li, F., Jiao, X., & Li, Q. (2021). The effects of cellulose nanocrystal and cellulose nanofiber on the properties of pumpkin starch-based composite films. *International Journal of Biological Macromolecules, 192*, 444–451. https://doi.org/10.1016/j.ijbiomac.2021.09.187

Further reading

Chen, J., Zhang, J., Wang, N., Xiao, B., Sun, X., Li, J., … Huang, F. (2024a). Critical review and recent advances of emerging real-time and non-destructive strategies for meat spoilage monitoring. *Food Chemistry, 445*, 138755.

Sun, J., Jiang, H., Wu, H., Tong, C., Pang, J., & Wu, C. (2020). Multifunctional bionanocomposite films based on konjac glucomannan/chitosan with nano-ZnO and mulberry anthocyanin extract for active food packaging. *Food Hydrocolloids, 107*, 105942.

CHAPTER FIVE

Antioxidant, antimicrobial, and light barrier properties of starch-based smart packaging

Himangshu Deka[1], Nidhi Jaiswal[2], Debabrata Nath[3], Pratap Kalita[3]

[1]Faculty of Pharmaceutical Science, Assam down town University, Panikhaiti, Guwahati, Assam, India
[2]Department of Food Science and Human Nutrition, University of Illinois Urbana Champaign, Urbana, IL, United States
[3]Pratiksha Institute of Pharmaceutical Sciences, Guwahati, Assam, India

5.1 Introduction

Starch-based smart packaging is a major step forward in sustainable food preservation, which does so by offering a biodegradable and renewable alternative to traditional petroleum-derived plastics while incorporating features such as antioxidant, antimicrobial, and light barrier properties (Arruda et al., 2025; Bajer, 2024; Cui et al., 2021; Kusuma et al., 2024). The development of these types of multifunctional materials is crucial for improving food safety and extending shelf life by reducing different degradation mechanisms (Dang et al., 2025a; Dang et al., 2025b). Starch, which is a naturally abundant polymer, forms the basic matrix for these films, but its innate limitations, such as high-water sensitivity and weak mechanical properties, require modifications and the integration of active compounds to achieve the desired function (Kusuma et al., 2024; Li et al., 2024; Linan et al., 2024).

The integration of antioxidant properties into starch-based packaging is essential for reducing oxidative rancidity in food products, which is a major cause of spoilage (Gao et al., 2024; Mao et al., 2025). These oxidative processes lead to undesirable changes in the quality of food, such as flavor, color, texture, and nutritional value (Mao et al., 2025; Yu et al., 2024). To scavenge the free radicals and prevent lipid oxidation, natural antioxidants, especially phenolic compounds derived from plants, are widely added to starch films (Mao et al., 2025; Shiri et al., 2024). For instance, studies have shown that extracts from honey bee products can enhance the antioxidant activity of octenyl succinylated potato starch films (Pająk et al., 2025).

Starches in Smart Packaging
ISBN: 978-0-443-45396-0
https://doi.org/10.1016/B978-0-443-45396-0.00010-4

Similarly, basil extract has been used to develop cassava starch films with improved antioxidant activity and UV protection, ultimately contributing to the preservation of fresh produce (Iacovone et al., 2025; lacovone et al., 2024). Curcumin-loaded nanocomplexes have also been successfully integrated into corn starch films, significantly enhancing their antioxidant capacity and application in the preservation of blueberries (Li et al., 2024). Propyl gallate, when integrated into starch/poly(butylene adipate-co-terephthalate) (PBAT) films, demonstrates controlled release and effective antioxidant activity with starch acting as a gatekeeper (Gao et al., 2024). Additionally, films incorporating Lycium ruthenicum anthocyanins (LRA) exhibit enhanced antioxidant capacity along with pH-responsive characteristics, making them suitable for intelligent food monitoring (Xie et al., 2025). The use of polyphenols in starch-based films and coatings is also a promising area for preserving fruits and vegetables (Mao et al., 2025).

Antimicrobial properties are also essential for inhibiting the growth of microorganisms, which are major causes for food spoilage and foodborne diseases (Dang et al., 2025a; Dang et al., 2025b). By adding different agents, starch-based films can be made to have antimicrobial properties (Dang et al., 2025a; Dang et al., 2025b; Morán et al., 2025). Chitosan nanoparticles (CNPs) are one of the common additives, imparting both antimicrobial and antioxidant properties to starch films, thereby improving the shelf-life of food (Liu et al., 2023; Othman et al., 2024). Essential oils (EOs) like geraniol, clove essential oil (CEO), and neem extract are also widely used for their broad-spectrum antimicrobial efficacy (Gupta et al., 2025; Jindal et al., 2025; Kraśniewska & Gniewosz, 2025; Zou et al., 2025). A novel starch-based film containing pullulan and geraniol has been shown to enhance hydrophobicity, antioxidant activity, and antimicrobial efficacy for fresh salmon preservation by hydrogen bonding and intermolecular entanglement (Zou et al., 2025). Quaternary ammonium chitosan (QCS) and LRA have also been integrated into starch-based films to produce materials with significant antibacterial activity (Xie et al., 2025). Other methods include the synthesis of novel natural alcohol esters of amino acids, such as L-Trp-Men and L-Trp-Bor, which give excellent antibacterial and antioxidant properties to corn starch/sodium carboxymethyl cellulose films (Jiang et al., 2025). Additionally, many research has also been done to create antibacterial packaging films for meat preservation by combining amylose starch with 2-hydroxypropyl-trimethylammonium chloride chitosan (HTCC) (Deng et al., 2024). Fe-ICA complexes are used in potato starch films to provide improved antioxidant activity, colorimetric ammonia-sensitive functions for freshness monitoring, and enhanced antimicrobial activities (Tian et al., 2025). Similarly, novel Cu/Na metal–organic frameworks (MOFs) have also been integrated into potato starch films to create biocompatible smart active packaging with colorimetric ammonia sensing and antimicrobial functions (Xu et al., 2025).

Light barrier properties are essential for protecting light-sensitive food components such as vitamins, pigments, and lipids from photodegradation, which may result in off-flavors, discoloration, and nutrient loss. Starch has some opacity on its own, but adding UV-absorbing compounds or opaque fillers can greatly improve its ability to block light (Cui et al., 2021). Anthocyanins such as those derived from *Lycium ruthenicum* or butterfly pea can make starch films better at blocking UV rays (Hasan et al., 2025; Xie et al., 2025). For instance, bio-nanocomposite films from sugar palm starch/CNP enhanced with butterfly pea anthocyanins demonstrated excellent UV-blocking (0% UV transmittance for CH/SPS-BP-III film) and improved mechanical strength (Hasan et al., 2025). Also, basil extract-enriched cassava starch films provide UV protection, offering a sustainable solution for food packaging (Iacovone et al., 2024). The incorporation of lignin nanoparticles (LNPs) and cinnamaldehyde (CI) into starch films can also enhance the UV barrier, along with other functions (Firouzjaei et al., 2025). Even UV-C technology has been explored to improve the functional properties of starch-based films, including increasing opacity with prolonged exposure, which can contribute to better light barrier performance (Uyarcan & Güngör, 2024).

5.2 Functional agents/bioactives used in starch packaging

Starch-based packaging systems were earlier used as passive barrier materials. Recently, their application has expanded, and they are now being developed as functional packaging materials that can interact with the packaged food and its surrounding environment. This change is mainly due to the incorporation of functional agents or bioactive compounds into the starch matrix. These functional agents help starch films to provide antimicrobial protection, improve oxidative stability, and respond to environmental conditions such as moisture or oxygen exposure. Some of the functional agents used are shown in Fig. 5.1 Starch has a naturally hydrophilic structure and a heterogeneous molecular arrangement. Because of this, it shows both advantages and limitations when bioactive compounds are added. Even though starch is biodegradable, renewable, and suitable

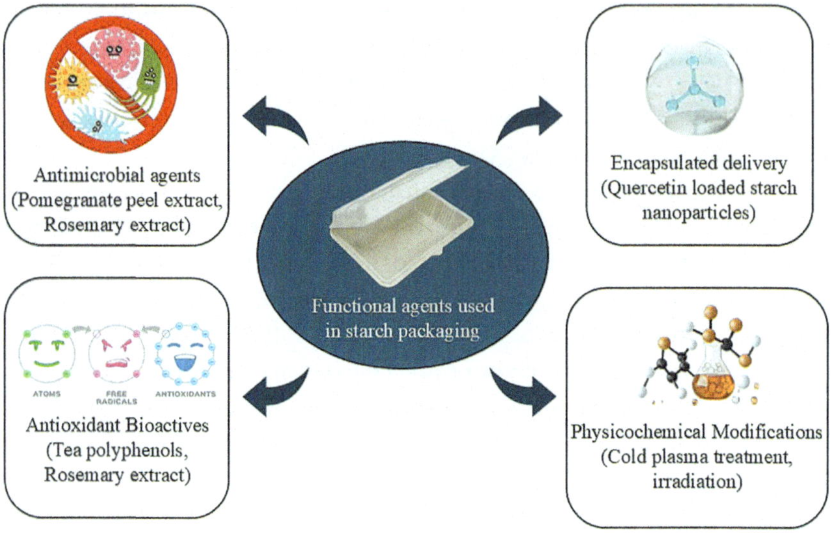

Figure 5.1 Functional agents used in starch packaging.

for food contact applications, its ability to act as a carrier for bioactive compounds depends strongly on its molecular structure, processing history, and the compatibility between starch and the functional agent used (Santhosh et al., 2024; Karnwal, 2025). The selection and addition of functional agents into starch films require careful consideration. This is because factors such as release behavior of the bioactive compound, its stability during processing, and the retention of mechanical and barrier properties of the film are closely related to overall packaging performance. Unlike synthetic polymers, starch films undergo significant structural changes during gelatinization, drying, and storage. These changes affect how bioactive compounds move inside the starch matrix and how long they remain effective during storage and use (Chen et al., 2024; Tosif et al., 2021). Thus, many studies have focused on using natural bioactive compounds along with encapsulation methods and simple physicochemical modification of starch. These methods are mainly used to improve the functional performance of starch-based packaging while still keeping the material biodegradable and safe for food contact applications (Dobson et al., 2019).

5.2.1 Antimicrobial agents in starch-based packaging

Antimicrobial agents are widely used functional additives in starch-based packaging systems because they directly help in improving food safety and extending shelf life. Many studies have focused on using natural antimicrobial compounds instead of synthetic preservatives. These compounds are mainly obtained from plant extracts, phenolic compounds, and agro-industrial by-products. Natural antimicrobial agents are suitable for use in starch-based films. This is because they can interact easily with the starch structure. Starch contains a large number of hydroxyl groups in its molecular structure. Because of this, antimicrobial compounds are able to bind within the starch matrix. These interactions mainly take place through hydrogen bonding and also through hydrophobic interactions (Kechichian et al., 2010; Morán et al., 2025).

Cassava starch films and wheat starch films containing plant-based antimicrobial agents have shown good antimicrobial effects. Antimicrobial agents such as pomegranate peel extract, rosemary extract, and yerba mate extract have been added to these films. These films were effective against different foodborne microorganisms. In one study, starch films incorporated with pomegranate peel extract were able to damage microbial cell membranes and reduce microbial growth (Ali et al., 2019). These films also showed improved protection against UV light. In another study, rosemary extract added to cassava starch films provided antimicrobial activity over a longer period of time. This behavior was mainly linked to the slow and controlled release of phenolic compounds from the starch matrix over time (Piñeros-Hernandez et al., 2017). The method used to prepare starch films also affects antimicrobial performance. Starch films produced using extrusion and compression molding show different behavior when compared to solvent-cast films. Films prepared using mechanical processing usually have a denser internal structure. Because of this dense structure, the movement of antimicrobial compounds within the starch matrix becomes slower. As a result, the release of antimicrobial agents from these films occurs at a slower rate (Ceballos et al., 2020).

In addition, starch films blended with chitosan have shown improved antimicrobial activity. This happens because chitosan itself has antimicrobial properties, which work together with starch in the film structure (Bonilla, Atarés, et al., 2013; Bonilla, Talón, et al., 2013). Even though good antimicrobial results have been reported, some limitations still exist. Adding a high amount of antimicrobial agents can negatively affect the physical properties of starch films. In many cases, excessive addition leads to a softer film structure, lower tensile strength (TS), and higher water vapor permeability (WVP). Because of this, it is important to carefully control the concentration of antimicrobial agents and ensure proper distribution within the starch matrix for food packaging applications (Morán et al., 2025; Smítková et al., 2012).

5.2.2 Antioxidant bioactives and oxidative stability enhancement

Oxidative degradation is a common problem in food products that contain fats and oils. When oxidation occurs, it leads to quality loss such as rancid taste, off-flavors, color changes, and nutrient loss. To reduce these problems, antioxidant bioactive compounds are added to starch-based packaging films. This makes starch films more active in nature and helps in protecting food quality during storage. Natural antioxidants such as polyphenols, flavonoids, and plant extracts are commonly used because they are safe, biodegradable, and suitable for food contact applications. In starch packaging systems, the starch matrix acts as a carrier for these antioxidant compounds and also helps protect them from early degradation. At the same time, starch allows the antioxidants to be released slowly over time (Feng et al., 2018; Silva-Weiss et al., 2013). Tea polyphenols and rosemary extract are widely used antioxidant agents in starch-based films. Feng et al. (2018) reported that starch films incorporated with tea polyphenols showed good free radical scavenging activity. These films did not show a major loss in mechanical strength, which is important for packaging use. In a similar way, cassava starch films containing rosemary extract showed improved oxidative stability when applied to food systems. This effect was mainly related to the gradual release of antioxidant compounds from the starch matrix during storage (Piñeros-Hernandez et al., 2017). The interaction between antioxidant compounds and starch molecules plays an important role in determining film behavior. Phenolic compounds can interact with amylose chains present in starch. These interactions can reduce starch crystallization and slow down retrogradation. They influence how antioxidant compounds are retained and released from the film structure (Dobson et al., 2019). Bonilla, Atarés, et al. (2013) and Bonilla, Talón, et al. (2013) observed that the addition of antioxidants changed film opacity and increased water sensitivity, which showed that improvements in functionality can affect physical properties. Processing conditions impact the performance of antioxidant starch films. Factors like heating conditions during gelatinization, drying rate, and the source of starch influence antioxidant retention and distribution within the film matrix. Poor control of these conditions can lead to uneven distribution and reduced antioxidant effectiveness (Chen et al., 2024; Singh et al., 2022; Tosif et al., 2021). Therefore, antioxidant-loaded starch films need to be designed by considering both the food product being packaged and the processing method used to prepare the films.

5.2.3 Encapsulation and delivery of bioactives within starch matrices

When bioactive compounds are directly added to starch films, several problems can occur. These include fast migration of the bioactive compounds, loss due to volatility, and degradation during processing steps such as heating and drying. Because of these issues, encapsulation techniques are increasingly used in starch-based packaging systems. Encapsulation helps protect bioactive compounds and improves their stability inside the starch matrix. It also helps in controlling the release of bioactive compounds and improves the compatibility between hydrophilic starch and hydrophobic bioactive compounds (Onwulata, 2013; Joye & Julian McClements, 2016).

Starch nanoparticles are commonly used as carriers for bioactive compounds such as quercetin. Jiang et al. (2022) reported that quinoa starch nanoparticles improved the solubility of quercetin and enhanced its antioxidant activity. These nanoparticles also helped in achieving a uniform distribution of quercetin within the starch film matrix. In a similar way, studies have shown that phenolic compounds, when encapsulated, are less sensitive to thermal and oxidative degradation during film preparation compared to directly added bioactives (dos Santos Caetano et al., 2018). Encapsulation also helps in controlling the release behavior of bioactive compounds from starch films. When bioactive compounds are present inside starch-based carriers, they move by diffusion through the film structure. This movement is influenced by factors such as how thick the film is, the humidity around it, and how crystalline the starch matrix is (Onwulata, 2013). Controlled release of bioactive compounds is important because it helps the packaging material remain effective for a longer period during storage. This slow release supports the antimicrobial or antioxidant function of the packaging over time. However, the use of encapsulation also comes with certain limitations. Preparing encapsulated systems makes formulation and processing more complex. Problems such as particle aggregation, stability of the encapsulated bioactive compounds, and difficulty in scaling up the process still need to be addressed before these systems can be widely applied in industrial food packaging (Joye & McClements, 2016; Smítková et al., 2012).

5.2.4 Physicochemical modification of starch to enhance bioactive performance

Apart from adding functional agents directly into starch films, starch itself can also be modified to improve the performance of bioactive compounds. Physicochemical modification methods are used for this purpose. Common methods include cold plasma treatment, moderate electric field processing, and irradiation. These methods change the internal structure of starch. They also affect surface properties and permeability of starch films. Because of these changes, the interaction between starch and bioactive compounds is also affected (Carvalho et al., 2021; Coelho et al., 2017). Cold plasma treatment has been shown to modify the surface of starch films. It increases surface roughness and creates new functional groups on the starch surface. These changes help bioactive compounds attach better to the starch matrix and spread more evenly throughout the film (Carvalho et al., 2021). reported that starch films treated with cold plasma showed better compatibility with functional additives. At the same time, the biodegradable nature of starch was maintained. Moderate electric field treatment has also been applied to starch-based films. This treatment improved the mechanical strength of starch–chitosan films and reduced their moisture sensitivity, which is important for food packaging applications (Coelho et al., 2017).

Radiation-based modification methods are mainly used for polysaccharides other than starch. Even though this is the case, the results from these studies are still useful for starch-based packaging materials. Sabry et al. (2025) reported that controlled irradiation helped improve the cohesion and overall functional performance of biofilms. This shows that similar irradiation-based approaches could also be applied to starch-based packaging systems to improve their performance (Table 5.1). These modification methods help improve the functional performance of starch films without adding large amounts of bioactive compounds. Lower additive levels help maintain the physical strength and structure of the films. This makes the packaging material more stable while still improving its active performance (Chen et al., 2024; Tosif et al., 2021).

5.2.5 Practical considerations

The use of functional agents in starch-based packaging systems helps improve food safety and food quality. These systems are also considered sustainable because starch is biodegradable and obtained from renewable sources. However, using starch-based active packaging in actual food applications is not easy. The type of starch selected, the processing method

Table 5.1 Functional agents incorporated into starch-based packaging systems and their primary effects.

Functional agent	Type	Mechanism of action	Packaging benefit	Reference
Pomegranate peel extract	Antimicrobial	Phenolic compounds disrupt microbial cell membranes	Reduced microbial growth	Ali et al. (2019)
Rosemary extract	Antioxidant	Slow release of phenolic antioxidants from the starch matrix	Improved oxidative stability	Piñeros-Hernandez et al. (2017)
Tea polyphenols	Antioxidant	Free radical scavenging through hydrogen donation	Enhanced antioxidant activity	Feng et al. (2018)
Quercetin (encapsulated)	Antioxidant	Controlled diffusion using starch nanoparticles	Sustained antioxidant delivery	Jiang et al. (2022)
Yerba mate extract	Antimicrobial/ antioxidant	Polyphenol release from compact starch structure	Active and smart packaging behavior	Ceballos et al. (2020)
Encapsulated phenolics	Antioxidant	Protection from thermal and oxidative degradation	Improved stability during processing	dos Santos Caetano et al. (2018)
Quercetin (starch–chitosan film)	Antioxidant	Interaction with polysaccharide matrix and controlled release	Improved functional performance	Pech-Cohuo et al. (2022)
Chitosan (blended with starch)	Antimicrobial	Intrinsic antimicrobial activity combined with film formation	Enhanced antimicrobial efficacy	Bonilla, Atarés, et al. (2013), Bonilla, Talón, et al. (2013)

used, and the functional agent added all affect the final performance of the packaging material. The interaction between starch and bioactive compounds at the molecular level decides how effective the functional agent will be. These interactions also affect the stability of the starch film and its performance during storage and use (Dobson et al., 2019; Morán et al., 2025).

More work is still needed before starch-based active packaging systems can be used at an industrial level. These systems are not yet easy to apply on a large scale. Problems related to large-scale production still exist. Regulatory approval is also a concern. In addition, maintaining consistent performance during processing is difficult. Encapsulation techniques and starch modification methods are useful, but they add extra processing steps. These methods also increase the overall cost of the packaging material. Because of this, starch-based active packaging materials need to be designed carefully. The goal is to improve functionality while still keeping the material environmentally friendly and economically practical.

5.3 Functional properties of starch-based smart packaging

Utilizing the natural abundance, biodegradability, and film-forming ability of starch, starch-based smart packaging is a state-of-the-art, environmentally friendly approach to food preservation. As shown in Fig. 5.2, these

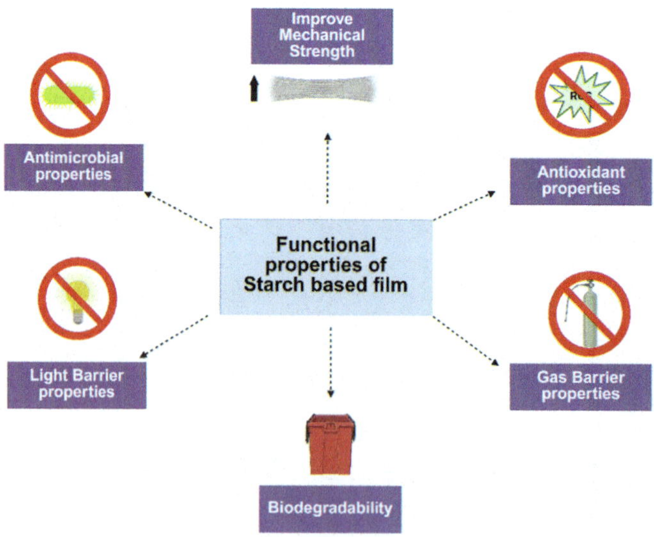

Figure 5.2 Functional properties of starch-based packaging film.

films combine natural indicators, nanoparticles, and bioactive compounds to provide a variety of functional qualities, such as light barrier, pH sensor, antibacterial, and antioxidant. In line with the worldwide demand for environmentally sustainable substitutes for petroleum-based plastics, recent developments highlight their significance in increasing shelf life, cutting waste, and facilitating real-time quality monitoring (Liu et al., 2022).

5.3.1 Antioxidant properties

Starch, which is primarily made up of amylose and amylopectin, has no intrinsic antioxidant capacity; however, it can be incorporated with functional agents like phenolics, flavonoids, EOs, and polyphenols. These substances donate hydrogen atoms or electrons to neutralize reactive oxygen species (ROS), such as superoxide anions and hydroxyl radicals. In packaged foods, especially those high in fats like meats, nuts, and oils, these functional agents can prevent lipid peroxidation, protein denaturation, and the development of an unpleasant flavor.

Multiflower corn EO nanoemulsions greatly outperformed conventional emulsions in a seminal work on maize starch films that showed DPPH (2,2-diphenyl-1-picryhydrazyl) radical scavenging activity up to 85%, due to their increased surface area and regulated volatility. Thiobarbituric acid reactive substances (TBARS) in ground beef were reduced by 65% during refrigeration owing to the nanoemulsions' prolonged release into the food headspace over a period of 15 days, which was made possible by their encapsulation in amylose helices. Similarly, high gallic acid equivalents (55 mg GAE/g extract) in polyvinyl alcohol (PVA)-corn starch blends at 20% (w/w) resulted in 42% DPPH inhibition and 38% ABTS (2,2'-azino-bis(3-ethylbenzothiazoline-6-sulfonic acid) scavenging, with films retaining their integrity under high humidity (Karimi Sani et al., 2025; Liu et al., 2022).

Sago starch films fortified with 1.2 g/100 mL Moringa oleifera black seed extract demonstrated a total phenolic content of 54 mg GAE/100 g, flavonoids at 17 mg QE/100 g, ABTS scavenging of 29 mg EAA/100 g, and DPPH activity of 41 mg EAA/100 g—all of which were statistically better than controls ($P < .05$). This recipe preserved ascorbic acid and sensory qualities while extending the shelf life of mangos by 12 days at 25°C. Ideal for contact with fatty foods, potato starch films containing thymol-loaded MCM-41 mesoporous silica (0.5 g/100 mL) showed strong DPPH scavenging (IC_{50} ~25 μg/mL), and extended thymol release in

50% ethanol simulants. The porous nature of the silica reduced burst effects by promoting Fickian diffusion (Cenobio-Galindo et al., 2019; Cui, 2023).

Through intermolecular hydrogen bonding that improved compatibility and decreased WVP by 30%, cassava starch/gelatin (7:3 ratio) films containing quercetin or TBHQ (Tertiary Butylhydroquinone) minimized pork loin discoloration (a value retention >20%) and delayed lard autoxidation for more than 35 days (peroxide value < 5 meq/kg). In edible starch films, sweet potato peel polyphenols eliminated more than 90% of DPPH and ABTS radicals, provided UV barriers, and increased cherry tomato shelf life to 7 days at 4°C. The antimicrobial synergy also reduced the incidence of mold by 80%. *Aconitum heterophyllum*, *Artemisia annua*, or *Thymus serpyllum* extracts with cornstarch demonstrated dose-dependent DPPH scavenging (70%–92%), corresponded with phenolic profiles via HPLC, and inhibited *E. coli* growth in real-time tests (Ali et al., 2023; Tongdeesoontorn et al., 2021; Zhang et al., 2024).

Other innovations include ascorbic acid increasing the antiradical potency of oxidized potato starch by 20 times at 100 mM concentration; *Zataria multiflora* or black pepper EOs in corn starch increasing DPPH activity proportionately with concentration ($P < 0.05$); and ginger extract in cassava starch achieving 70% ABTS inhibition and 60% TBARS reduction in chicken patties versus synthetic BHT (Butylated Hydroxytoluene). By encouraging matrix amorphization, dialdehyde starch-caffeic acid blends performed better than their quinic acid counterparts, resulting in elastic moduli >2× higher and FRAP (Ferric Reducing Antioxidant Power) values ~15 mmol/g. Tea polyphenol-chitosan-starch films combined antibacterial and antioxidant properties (FRAP 12 mmol/g) (Aminzare et al., 2017; Bajer et al., 2024).

These antioxidants function mechanistically through hydrogen atom transfer (HAT) or single electron transfer (SET), with the hydroxyl groups in starch facilitating dispersion. Acetylation, cyclodextrin complexation, or nanoencapsulation minimize issues such as humidity-accelerated leaching, guaranteeing EU migration restrictions (<10 mg/dm^2. These films are positioned for commercialization in perishable goods packaging since real-world applications verify 20%–50% shelf-life extensions (Farooq et al., 2023; Vianna et al., 2021).

5.3.2 Antimicrobial properties

Starch-based intelligent packaging films have an excellent antimicrobial activity when combined with active agents, such as EOs, bioactive extracts, and inorganic nanoparticles. Antimicrobial efficacy can be considerably enhanced against food-spoiling agents like *Escherichia coli*, *Staphylococcus aureus*, and *Salmonella* spp. by adhering to ZnO nanoparticles and some other nanomaterials to the composition of starch by interfering with the structural integrity of the cell membrane and by the inhibition of microbial metabolic activities (Wu et al., 2025). Furthermore, advanced studies have illustrated that nanocomposite clove EO emulsified with LNPs could exhibit a dramatic boost for antimicrobial efficacy, also protecting fresh produce like cherry tomatoes (Wu et al., 2025). The combination of starch films containing chitosan blends and nanoparticles exhibited enhanced antimicrobial activity, while also improving important mechanical and barrier traits required for food preservation. All these multifunctional films are in line with contemporary trends in biodegradable active packaging by reducing synthetic preservatives and prolonging products' shelf life. Liu et al. have innovatively designed an intelligent starch/PVA film that functions for food packaging and performs two functions in one-go: colorimetric pH monitoring in real time and antimicrobial activity therein that helps in controlling microbial spoilage. Films were prepared employing a solvent casting procedure after priming the sodium trimetaphosphate (STMP) and boric acid double cross-linked starch–PVA as the fundamental biopolymeric matrix. Antimicrobial characteristics were imparted by incorporating natural agents like citric acid or the EO components together with the cross-linking to make a hostile environment to bacteria and fungi through membrane disruption and by enzyme suppression (Liu et al., 2017). Ali et al. developed cornstarch films by using *Aconitum heterophyllum*, *Artemisia annua*, and *Thymus serpyllum* medicinal plant extracts at 5%–15% w/w through a simple solution casting technique, which resulted in substantial improvements for their functional and physical characteristics that meet sustainable food packaging requirements. The extracts demonstrated strong antimicrobial activity, which produced clear inhibition zones of 10–15 mm against important food pathogens such as *S. aureus* and *Salmonella* spp. The extracts from the starch matrix spread through bacterial cell membranes, which created more effective disruption against Gram-positive S. aureus than the lipopolysaccharide-protected Gram-negative *Salmonella*. The system created a potential shelf-life extension for meats, dairy products, and produce (Ali et al., 2023). Another study conducted by Kakimova and colleagues from 2025 found moderate antimicrobial properties against both Gram-positive and Gram-negative pathogens. For instance, compared with controls, the clear zones of inhibition of these films on agar were recorded as 12 mm against *E. coli* and 15 mm against *S. aureus*. The higher relative efficacy against *S. aureus* (Gram-positive) versus *E. coli* (Gram-negative) was supposed to suggest easier chitosan permeation in Gram-positive bacteria through the thick peptidoglycan layer when compared to Gram-negative bacteria with an outer lipopolysaccharide barrier resistant to some extent—this suggests that the film could practically increase a large enough zone of high growth inhibition, suggesting effective diffusion-controlled release from the homogenous matrix of starch and chitosan (Muratkhan et al., 2025). The development of multifunctional intelligent films based on starch fits well with the global trends in biodegradable and sustainable packaging, and along these lines, the development of such advanced systems will decrease dependency on synthetic preservatives by natural and bioactive antimicrobials, thereby leading to a reduced accumulation of chemical residues in food supply systems.

5.3.3 Light barrier properties

One of the most important functions of smart packaging is acting as a protection against light, particularly for foods sensitive to photo-oxidation and the consequent degradation of nutrients. Natural starch-based films have limited capability to block UV light; however, by incorporating core–shell nanocomposites like calcined ZnO–SiO_2 nanoparticles embedded within a silica aerogel matrix, a substantial enhancement in shielding against both ultraviolet and visible light has been demonstrated (Wang et al., 2024). In another study, the effect of the addition of zinc oxide nanoparticles (ZnONPs) on rice starch–gelatin (RS–G) was investigated. ZnONPs were synthesized by a green method using the extract of *Centella asiatica* L. The transmission of UV light (200–280 nm) was found to be 0.06 to 67.80% in the control RS–G composite films; that is, in contrast, no transmission of any UV light was reported at high ZnONP concentrations greater than 0.5%. In the visible wavelengths (400–800 nm) of the control film (without ZnONPs) and RS–G–ZnONPs nanocomposite films, the amount of transmission changed from 75.07% to 83.15% and 0.33% to 18.82%, respectively (Homthawornchoo 2025). In a recent study, nano-SiO_2/potato starch films of various sizes (15, 30, 80, and 100 nm) were synthesized and characterized. The transmission of light and visual clarity was measured through the UV-visible spectrophotometer and has been found to increase with wavelength. Low UV transparency was shown by the additive nano-SiO_2, indicating that nan-SiO_2/potato starch films can protect foods against UV light. In terms of UV barrier resistance, films using 100 nm had higher resistance than the rest. The UV absorbance through 100-nm films at 600 nm was at 30% (Zhang et al., 2018). Nanomaterials such as nano-silica (SiO_2) and other metal oxides effectively reduce the transmission of harmful UV radiation by promoting light scattering and dispersion. This type of improvement in the light barrier properties helps to minimize photodegradation of food while offering protection from external light exposure (Wu et al., 2024). In addition, natural pigments such as anthocyanins, when incorporated into starch-based matrices, exhibit strong light-absorbing properties and can function as intelligent indicators by undergoing color changes in response to light and spoilage-related conditions (Song et al., 2024). Bio-nanocomposite films based on rye starch with 5% nanorod-shaped ZnO nanoparticles offer a viable natural nonmetallic alternative for the packaging of light-sensitive foods, such as dairy products, edible oil, and fresh meat, and are useful against photooxidative rancidity and nutritional deterioration. The films provided complete UV light blockage with 0% transmission over the critical range of 200–400 nm (also ranging into UVA, UVB, and UVC), Furthermore, the nanocomposite showed significant absorption in the near-infrared (NIR) range (700–1100 nm) to avert thermal destruction and light-induced off-flavors by reflecting about 40% of NIR radiation, making it a unique dual-spectrum image hardly seen being exhibited in clear starch films transmitting over 80% NIR (Fallah et al., 2023).

MXene nanoplatelets are increasingly employed as advanced two-dimensional nanofillers in the environmentally friendly high-performance packaging industry, offering prominent gas/moisture barrier values, antimicrobial properties, and mechanical reinforcement. Biodegradable starch-based nanocomposite films have been developed by Dong et al. with the addition of MXene nanoplatelets ($Ti_3C_2T_x$) at 10 wt% results in a dramatic enhancement of the light barrier properties, in which 2D nanosheets served in creating tortuous paths blocking more than 85% of transmitted UV across 200–400 nm thickness due to refractive index contrast ($n \approx 2.2$ vs. starch $n \approx 1.47$) and plasmonic absorption, while at the same time making the film visually transparent (>50% at 600 nm), making them suitable for the protection of oils and dairy from photooxidative degradation (Dong et al., 2024).

Overall, these multifunctional modifications enhance both protective performance and real-time visual quality monitoring in advanced food packaging systems.

5.3.4 Other properties

Starch-based films inherently offer good oxygen barrier performance owing to their strong intermolecule hydrogen bonding networks. However, their hydrophilic nature limits their ability to regulate moisture. Yet, composite formulations blending starch, nanocellulose, nanoclays, or biodegradable polymers have shown a remarkable improvement in water vapor resistance while allowing low oxygen permeability (Santhosh et al., 2024). Nanocellulose and montmorillonite reinforcements cut down water vapor transmission rates and hence improve gas barrier performance, thereby enabling more controlled gas exchange and moisture-mediated growth of starch composites to enhance applicability. Enhanced barrier properties are vital to protect high-moisture foods and allow starch-based packaging to be expanded within real-world food systems.

5.4 Conclusion and future perspectives

Starch-based smart packaging materials provide a promising sustainable alternative to traditional plastics, addressing environmental concerns and enhancing food preservation. The shelf life and safety of packaged foods are comparatively increased by the incorporation of advanced functions such as antioxidant, antimicrobial, and light barrier properties. Despite the innate limitations of starch, which include its poor moisture resistance and weak mechanical properties, ongoing research has made significant progress in overcoming these challenges through the use of active compounds and a variety of modification techniques. Therefore, the current landscape of starch-based smart packaging shows a distinct path toward a highly practical and ecologically friendly solution. Natural extracts, phenolic compounds, and encapsulated active agents have all successfully increased antioxidant capabilities, which are essential for reducing oxidative degradation. These additions maintain the organoleptic properties of food, prevent lipid oxidation, and maintain nutritional value. Similarly, adding CNPs, EOs, and new synthetic compounds to starch matrices has produced antimicrobial properties that are important for inhibiting microbial growth and preventing foodborne illnesses. Additionally, incorporating UV-absorbing substances like anthocyanins and LNPs has improved light barrier properties, which are important for protecting light-sensitive foods from photodegradation. Hence, these developments together transform simple starch films into complex active and intelligent packaging systems. Despite notable progress, additional research and development are still needed in a few areas to fully utilize the potential of starch-based smart packaging.

Improved Mechanical and Barrier Properties: One of the main drawbacks of starch-based films is that their water vapor barrier and mechanical strength are inferior to those of petroleum-based plastics. In order to improve film integrity and water resistance, future research should concentrate on creating innovative cross-linking techniques, optimizing the incorporation of some strengthening agents such as nanocellulose and chitin nanocrystals, and investigating new plasticizers. Strategies involving multilayer structures and composite materials containing other biopolymers may also offer synergistic improvements. Scalability and Cost-Effectiveness: Even though laboratory-scale production of starch-based smart packaging has shown promising results, large-scale production of the same faces difficulties with complexity and cost. To make these sustainable packaging solutions economically feasible for industrial applications, future efforts should focus on a well-organized synthesis process using readily available and inexpensive raw materials and creating effective manufacturing techniques. Advanced Sensing and Smart Functions: The value of starch-based smart packaging will be further increased by incorporating intelligent features like temperature monitors, gas sensors, and real-time freshness indicators. It is essential to conduct research on new pH-sensitive dyes, microbial growth indicators, and volatile compound sensors that can be easily integrated into starch films. Another important area of innovation could be the creation of responsive release mechanisms and self-healing characteristics for active compounds. Biodegradation and End-of-Life Management: Although starch-based materials are naturally biodegradable, it is also important to understand and maximize their degradation kinetics in a variety of environmental settings, such as compost, soil, and marine environments. Aligning with the circular economy, future studies should aim to guarantee complete and rapid biodegradation without leaving harmful residues. Also, the influence of active additives on the biodegradation profiles of these films should be investigated. Regulatory Approval and Consumer Acceptance: It is important to obtain widespread regulatory approval for different innovative active ingredients and clever features in materials that come in contact with food. Also, market penetration will depend on how consumers view and accept these new packaging technologies. To educate consumers about the advantages and security of starch-based smart packaging, research into communication strategies and clear labeling will also be required. Multifunctional Synergies: More effective and efficient packaging solutions may result from exploring synergistic combinations of active compounds that provide multiple benefits at once, such as an antioxidant that also possesses antimicrobial activity. Therefore, by addressing these issues, starch-based smart packaging can provide advanced protection and real-time monitoring capabilities for a variety of food products and can be placed at a position to play an important role in establishing a more sustainable and secure food supply chain.

AI disclosure

During the preparation of this manuscript, the authors used Grammarly to improve the grammar, and all of the information presented in this manuscript was generated by humans.

References

Ali, A., Basit, A., Hussain, A., Sammi, S., Wali, A., Goksen, G., Muhammad, A., Faiz, F., Trif, M., Rusu, A., & Manzoor, M. F. (2023). Starch-based environment friendly, edible and antimicrobial films reinforced with medicinal plants. *Frontiers in Nutrition, 9*, 1066337. https://doi.org/10.3389/fnut.2022.1066337

Ali, A., Chen, Y., Liu, H., Yu, L., Baloch, Z., Khalid, S., Zhu, J., & Chen, L. (2019). Starch-based antimicrobial films functionalized by pomegranate peel. *International Journal of Biological Macromolecules, 129*, 1120–1126. https://doi.org/10.1016/j.ijbiomac.2018.09.068

Aminzare, M., Amiri, E., Abbasi, Z., Hassanzadazar, H., & Hashemi, M. (2017). Evaluation of in vitro antioxidant characteristics of corn starch bioactive films impregnated with Bunium persicum and Zataria multiflora essential oils. *Annual Research & Review in Biology, 15*(5), 1–9.

Arruda, T. R., Machado, G.d. O., Marques, C. S., Souza, A. L.d., Pelissari, F. M., Oliveira, T. V.d., & Silva, R. R. A. (2025). An overview of starch-based materials for sustainable food packaging: Recent advances, limitations, and perspectives. *Macromol, 5*(2). https://doi.org/10.3390/macromol5020019

Bajer, D. (2024). Eco-friendly, biodegradable starch-based packaging materials with antioxidant features. *Polymers, 16*(7), 958. https://doi.org/10.3390/polym16070958

Bonilla, J., Atarés, L., Vargas, M., & Chiralt, A. (2013a). Properties of wheat starch film-forming dispersions and films as affected by chitosan addition. *Journal of Food Engineering, 114*(3), 303–312. https://doi.org/10.1016/j.jfoodeng.2012.08.005

Bonilla, J., Talón, E., Atarés, L., Vargas, M., & Chiralt, A. (2013b). Effect of the incorporation of antioxidants on physicochemical and antioxidant properties of wheat starch--chitosan films. *Journal of Food Engineering, 118*(3), 271–278. https://doi.org/10.1016/j.jfoodeng.2013.04.008

Carvalho, A. P. M. G., Barros, D. R., da Silva, L. S., Sanches, E. A., Pinto, C.d. C., de Souza, S. M., Clerici, M. T. P. S., Rodrigues, S., Fernandes, F. A. N., & Campelo, P. H. (2021). Dielectric barrier atmospheric cold plasma applied to the modification of Ariá (Goeppertia allouia) starch: Effect of plasma generation voltage. *International Journal of Biological Macromolecules, 182*, 1618–1627. https://doi.org/10.1016/j.ijbiomac.2021.05.165

Ceballos, R. L., Ochoa-Yepes, O., Goyanes, S., Bernal, C., & Famá, L. (2020). Effect of yerba mate extract on the performance of starch films obtained by extrusion and compression molding as active and smart packaging. *Carbohydrate Polymers, 244*, 116495. https://doi.org/10.1016/j.carbpol.2020.116495

Cenobio-Galindo, A. J., Pimentel-González, D. J., Del Razo-Rodríguez, O. E., Medina-Pérez, G., Carrillo-Inungaray, M. L., Reyes-Munguía, A., & Campos-Montiel, R. G. (2019). Antioxidant and antibacterial activities of a starch film with bioextracts microencapsulated from cactus fruits (Opuntia oligacantha). *Food Science and Biotechnology, 28*(5), 1553–1561. https://doi.org/10.1007/s10068-019-00586-9

Chen, Y., Zhu, Z., Ye, Y., Li, Q., Yang, T., Guan, C., & Liu, F. (2024). Comprehensive evaluation of the physicochemical attributes, antioxidant capacity, and pH-Responsive behavior of starch films enhanced by laver incorporation. *Foods, 13*(11), 1600.

Coelho, C. C. S., Cerqueira, M. A., Pereira, R. N., Pastrana, L. M., Freitas-Silva, O., Vicente, A. A., Cabral, L. M. C., & Teixeira, J. A. (2017). Effect of moderate electric fields in the properties of starch and chitosan films reinforced with microcrystalline cellulose. *Carbohydrate Polymers, 174*, 1181–1191. https://doi.org/10.1016/j.carbpol.2017.07.007

Cui, C., Ji, N., Wang, Y., Xiong, L., & Sun, Q. (2021). Bioactive and intelligent starch-based films: A review. *Trends in Food Science & Technology, 116*, 854–869. https://doi.org/10.1016/j.tifs.2021.08.024

Cui, Y., Zhang, R., Cheng, M., Guo, Y., & Wang, X. (2023). Sustained release and antioxidant activity of active potato starch packaging films encapsulating thymol with MCM-41. *LWT, 173*, 114342. https://doi.org/10.1016/j.lwt.2022.114342

Dang, X., Guo, B., Du, Y., & Wang, X. (2025b). Advancements in manufacturing and applications of starch-based antimicrobial materials. *Polymer Reviews, 65*(3), 814–853. https://doi.org/10.1080/15583724.2025.2470702

Dang, X., Han, S., & Wang, X. (2025a). Versatile corn starch-based sustainable food packaging with enhanced antimicrobial activity and preservative properties. *Journal of Colloid and Interface Science, 694*, 137698. https://doi.org/10.1016/j.jcis.2025.137698

Deng, B., Chen, J., Li, S., Liu, J., Zhou, Z., Qin, Z., Wang, H., Su, M., Li, L., & Bai, Z. (2024). An antibacterial packaging film based on amylose starch with quaternary ammonium salt chitosan and its application for meat preservation. *International Journal of Biological Macromolecules, 261*. https://doi.org/10.1016/j.ijbiomac.2024.129706

Dobson, C. C., Mottawea, W., Rodrigue, A., Buzati Pereira, B. L., Hammami, R., Power, K. A., & Bordenave, N. (2019). Impact of molecular interactions with phenolic compounds on food polysaccharides functionality. *Advances in Food and Nutrition Research, 90*, 135–181. https://doi.org/10.1016/bs.afnr.2019.02.010

Dong, M., Mastroianni, G., Bilotti, E., Zhang, H., & Papageorgiou, D. G. (2024). Biodegradable starch-based nanocomposite films with exceptional water and oxygen barrier properties. *ACS Sustainable Chemistry & Engineering, 12*(29), 11056–11066.

dos Santos Caetano, K., Almeida Lopes, N., Haas Costa, T. M., Brandelli, A., Rodrigues, E., Hickmann Flôres, S., & Cladera-Olivera, F. (2018). Characterization of active biodegradable films based on cassava starch and natural compounds. *Food Packaging and Shelf Life, 16*, 138–147. https://doi.org/10.1016/j.fpsl.2018.03.006

Fallah, N., Marvizadeh, M. M., Jahangiri, R., Zeinalzadeh, A., & Nafchi, A. M. (2023). High-Barrier and light-protective bionanocomposite film based on Rye Starch/nanorod-ZnO for food packaging applications. *Journal of Chemical Health Risks, 13*(2).

Farooq, S., Mir, S. A., Shah, M. A., Manickavasagan, A., Sunooj, K. V., Siddiqui, M. W., & Mousavi Khaneghah, A. (2023). Incorporation of plant extracts and essential oils into starch-based films: A comprehensive review. *Starch–Stärke, 75*(5–6), 2200098. https://doi.org/10.1002/star.202200098

Feng, M., Yu, L., Zhu, P., Zhou, X., Liu, H., Yang, Y., Zhou, J., Gao, C., Bao, X., & Chen, P. (2018). Development and preparation of active starch films carrying tea polyphenol. *Carbohydrate Polymers, 196*, 162–167. https://doi.org/10.1016/j.carbpol.2018.05.043

Firouzjaei, S. M., Nikzad, M., Kazemi, S., & Pasha, H. Y. (2025). Enhanced functionalities of Starch-Based nanocomposite films incorporating lignin nanoparticles and cinnamaldehyde for sustainable food packaging applications. *Food Biophysics, 20*(2). https://doi.org/10.1007/s11483-025-09968-5

Gao, S., Li, M., Zhai, X., Wang, W., & Hou, H. (2024). Starch as a smart, cheap, and green gatekeeper for the controlled release of propyl gallate from antioxidant biodegradable packaging films. *Food Chemistry, 453*, 139627. https://doi.org/10.1016/j.foodchem.2024.139627

Gupta, R. K., Pipliya, S., Patel, J., Srivastav, P. P., Castro-Muñoz, R., Wang, C. K., & Nemţanu, M. R. (2025). Essential oil-embedded starch based films for active packaging: Insights into migration, stability, and preservation efficiency. *Trends in Food Science and Technology, 163*. https://doi.org/10.1016/j.tifs.2025.105149

Hasan, M., Khaldun, I., Ulva, M., Nisa, K., Gani, A., & Habibati, H. (2025). Bio-nanocomposite films from sugar palm starch/chitosan nanoparticle enriched with butterfly pea anthocyanins. *Case Studies in Chemical and Environmental Engineering, 11*. https://doi.org/10.1016/j.cscee.2025.101174

Homthawornchoo, W., Kaewprachu, P., Pinijsuwan, S., Romruen, O., & Rawdkuen, S. (2022). Enhancing the UV-light barrier, thermal stability, tensile strength, and antimicrobial properties of rice starch–gelatin composite films through the incorporation of zinc oxide nanoparticles. *Polymers, 14*(12), 2505.

Iacovone, C., Fernández, L., Guz, L., & Famá, L. (2025). Multifunctional cassava starch bilayers with basil extract and eco-friendly synthesized silver nanoparticles for scalable and sustainable food packaging. *International Journal of Biological Macromolecules, 309*(Pt 3), 142925. https://doi.org/10.1016/j.ijbiomac.2025.142925

Jiang, F., Du, C., Zhao, N., Jiang, W., Yu, X., & Du, S.-kui (2022). Preparation and characterization of quinoa starch nanoparticles as quercetin carriers. *Food Chemistry, 369*, 130895. https://doi.org/10.1016/j.foodchem.2021.130895

Jiang, Q., Liu, L., Yuan, X., Li, H., Wang, Y., & Li, J. (2025). A completely bio-based corn starch/sodium carboxymethyl cellulose film with antibacterial and antioxidant performance modified by a natural alcohol ester of amino acid and curcumin. *Carbohydrate Polymers, 362*, 123686. https://doi.org/10.1016/j.carbpol.2025.123686

Jindal, Y., Yadav, T., Singh, A., & Mishra, G. (2025). Halloysite nanotube–neem extract reinforced polymeric films: A novel approach for food packaging applications. *Journal of Food Measurement and Characterization*. https://doi.org/10.1007/s11694-025-03755-z

Joye, J. I., & Julian McClements, D. (2016). Biopolymer-based delivery systems: Challenges and opportunities. *Current Topics in Medicinal Chemistry, 16*(9), 1026–1039.

Karimi Sani, I., Hassani, B., Rasul, N. H., Mansouri, E., Eghbaljoo, H., Kaveh, M., Hassani, D., Alizadeh Sani, M., Khezerlou, A., Gholizadeh, H., Mamakani, Z. S., & Jafari, S. M. (2025). Antioxidant packaging films: Application for sustainable food protection. *Current Research in Food Science, 11*, 101222. https://doi.org/10.1016/j.crfs.2025.101222

Karnwal, A., Rauf, A., Jassim, A. Y., Selvaraj, M., Al-Tawaha, A. R. M. S., Kashyap, P., & Malik, T. (2025). Advanced starch-based films for food packaging: Innovations in sustainability and functional properties. *Food Chemistry X*, 102662.

Kechichian, V., Ditchfield, C., Veiga-Santos, P., & Tadini, C. C. (2010). Natural antimicrobial ingredients incorporated in biodegradable films based on cassava starch. *LWT - Food Science and Technology, 43*(7), 1088–1094. https://doi.org/10.1016/j.lwt.2010.02.014

Kraśniewska, K., & Gniewosz, M. (2025). Active packaging based on a PET/PP food-grade film coated with Pullulan and clove essential oil: Physicochemical and antimicrobial properties. *Molecules, 30*(10), 2118. https://doi.org/10.3390/molecules30102118

Kusuma, I., Ma'arif, A., Rahmadhia, S. N., Ma'arif, A., Amzeri, A., Caesarendra, W., & Suwarno, I. (2024). Starch based for intelligent food packaging: A review. *BIO Web of Conferences, 146*, 01099. https://doi.org/10.1051/bioconf/202414601099

Iacovone, C., Guz, L., & Famá, L. (2024). Sustainable innovations in food packaging: Antioxidant basil-enriched cassava starch films with UV protection and enhanced water and mechanical resistance. *Food Packaging and Shelf Life, 45*, 101324. https://doi.org/10.1016/j.fpsl.2024.101324

Li, X., Zhang, X., Lv, J., Zhang, X., Li, Y., Han, X., & Zhang, W. (2024). Development of starch-based films reinforced with curcumin-loaded nanocomplexes: Characterization and application in the preservation of blueberries. *International Journal of Biological Macromolecules, 264*(Pt 1), 130464. https://doi.org/10.1016/j.ijbiomac.2024.130464

Linan, L. Z., Fakhouri, F. M., Nogueira, G. F., Zoppe, J., & Velasco, J. I. (2024). Benefits of incorporating lignin into starch-based films: A brief review. *Polymers, 16*(16). https://doi.org/10.3390/polym16162285

Liu, Q., Gao, L., Yang, Q., Ji, N., Lei, D., Xiong, L., & Sun, Q. (2023). Incorporation of oxidized debranched starch/chitosan nanoparticles for enhanced hydrophobicity of corn starch films. *Food Packaging and Shelf Life, 35*, 101032. https://doi.org/10.1016/j.fpsl.2023.101032

Liu, B., Xu, H., Zhao, H., Liu, W., Zhao, L., & Li, Y. (2017). Preparation and characterization of intelligent starch/PVA films for simultaneous colorimetric indication and antimicrobial activity for food packaging applications. *Carbohydrate Polymers, 157*, 842–849.

Liu, D., Zhao, P., Chen, J., Yan, Y., & Wu, Z. (2022). Recent advances and applications in starch for intelligent active food packaging: A review. *Foods, 11*(18), 2879. https://doi.org/10.3390/foods11182879

Mao, S., Zhang, J., Wu, Q., Xu, Y., Zhang, T., & Lu, C. (2025). Roles of polyphenols incorporated in starch-based films and coatings for fruits and vegetables preservation: A review. *Trends in Food Science & Technology, 163*, 105170. https://doi.org/10.1016/j.tifs.2025.105170

Morán, D., Suárez-Iglesias, O., Gutiérrez, G., & Matos, M. (2025). Advances in the use of starch-based matrices for antimicrobials loading: An analysis of natural sources and applications for active packaging for food safety. *Food Bioscience, 69*, 106853. https://doi.org/10.1016/j.fbio.2025.106853

Muratkhan, M., Zhainagul, K., Svetlana, K., Toimbayeva, D., Temirova, I., Tazhina, S., & Ospankulova, G. (2025). Development and optimization of edible antimicrobial films based on dry heat-modified starches from Kazakhstan. *Foods, 14*(11), 2001.

Onwulata, C. I. (2013). Microencapsulation and functional bioactive foods. *Journal of Food Processing and Preservation, 37*(5), 510–532. https://doi.org/10.1111/j.1745-4549.2012.00680.x

Othman, S. H., Shapi'i, R. A., & Ronzi, N. D. A. (2024). Starch biopolymer films containing chitosan nanoparticles: A review. *Carbohydrate Polymers, 329*, 121735. https://doi.org/10.1016/j.carbpol.2023.121735

Pająk, P., Socha, R., Królikowska, K., Grzyb, J., Hetmańczyk, J., & Zachariasz, P. (2025). Characterization of octenyl succinylated potato-starch based films enriched with extracts from various honey-bee products. *International Journal of Biological Macromolecules, 285*, 138293. https://doi.org/10.1016/j.ijbiomac.2024.138293

Pech-Cohuo, S. C., Martín-López, H., Uribe-Calderón, J., González-Canché, N. G., Salgado-Tránsito, I., May-Pat, A., … Pacheco, N. (2022). Physicochemical, mechanical, and structural properties of bio-active films based on biological-chemical chitosan, a novel ramon (*Brosimum alicastrum*) starch, and quercetin. *Polymers, 14*(7), 1346.

Piñeros-Hernandez, D., Medina-Jaramillo, C., López-Córdoba, A., & Goyanes, S. (2017). Edible cassava starch films carrying rosemary antioxidant extracts for potential use as active food packaging. *Food Hydrocolloids, 63*, 488–495. https://doi.org/10.1016/j.foodhyd.2016.09.034

Sabry, R., Sayed, A., El-Sayed, I. E., & Mahmoud, G. A. (2025). Optimizing pectin-based biofilm properties for food packaging via E-beam irradiation. *Radiation Physics and Chemistry, 229*, 112474.

Santhosh, R., Ahmed, J., Thakur, R., & Sarkar, P. (2024). Starch-based edible packaging: Rheological, thermal, mechanical, microstructural, and barrier properties – A review. *Sustainable Food Technology, 2*(2), 307–330. https://doi.org/10.1039/d3fb00211j

Shiri, Z., Pirsa, S., & Farzi, J. (2024). Eco-friendly biodegradable film based on kombucha mushroom/corn starch/parsley extract: Physicochemical and antioxidant/antibacterial properties. *Food Science and Nutrition, 12*(10), 7924–7937. https://doi.org/10.1002/fsn3.4411

Silva-Weiss, A., Ihl, M., Sobral, P. J. A., Gómez-Guillén, M. C., & Bifani, V. (2013). Natural additives in bioactive edible films and coatings: Functionality and applications in foods. *Food Engineering Reviews, 5*(4), 200–216. https://doi.org/10.1007/s12393-013-9072-5

Singh, G. P., Bangar, S. P., Yang, T., Trif, M., Kumar, V., & Kumar, D. (2022). Effect on the properties of edible starch-based films by the incorporation of additives: A review. *Polymers, 14*(10), 1987.

Smítková, H., Marek, M., & Dobias, J. (2012). Preparation and properties of bioactive packaging materials on the starch base. *Acta Chimica Slovaca, 5*(2), 225. https://doi.org/10.2478/v10188-012-0034-y

Song, W., Wu, N., He, Y., Zhao, H., Xu, J., & Ren, L. (2024). Intelligent color changing packaging film based on esterified starch and black rice anthocyanins. *Food Chemistry X, 24*, 101930. https://doi.org/10.1016/j.fochx.2024.101930

Tian, Z., Li, S., Zou, S., Li, H., Zou, Z., Zhang, X., & Tang, Q. (2025). Colorimetric ammonia-sensitive potato starch-based smart films containing cellulose nanocrystal and Fe-ICA complex for shrimp freshness monitoring. *Industrial Crops and Products, 236*, 121940. https://doi.org/10.1016/j.indcrop.2025.121940

Tongdeesoontorn, W., Mauer, L. J., Wongruong, S., Sriburi, P., Reungsang, A., & Rachtanapun, P. (2021). Antioxidant films from cassava starch/gelatin biocomposite fortified with quercetin and TBHQ and their applications in food models. *Polymers, 13*(7), 1117. https://doi.org/10.3390/polym13071117

Tosif, M. M., Najda, A., Bains, A., Zawiślak, G., Maj, G., & Chawla, P. (2021). Starch–mucilage composite films: An inclusive on physicochemical and biological perspective. *Polymers, 13*(16), 2588.

Uyarcan, M., & Güngör, S. C. (2024). Improving functional properties of starch-based films by ultraviolet (UV-C) technology: Characterization and application on minced meat packaging. *International Journal of Biological Macromolecules, 282*(Pt 3), 137085. https://doi.org/10.1016/j.ijbiomac.2024.137085

Vianna, T. C., Marinho, C. O., Marangoni Júnior, L., Ibrahim, S. A., & Vieira, R. P. (2021). Essential oils as additives in active starch-based food packaging films: A review. *International Journal of Biological Macromolecules, 182*, 1803–1819. https://doi.org/10.1016/j.ijbiomac.2021.05.170

Wang, Y., Li, X., Chen, M., & Zhao, H. (2024). ZnO nanoparticle-reinforced starch films with enhanced UV-barrier and antimicrobial properties for food packaging. *Carbohydrate Polymers, 322*, 121098.

Wu, H., Tong, Y., Jiang, D. J., Qi, M., Xie, X., Yan, G., Li, S., Yuan, X., Wang, W., Liu, Y., & Wu, H. (2025). Lignin nanoparticle-stabilized clove essential oil Pickering emulsion enhanced starch films: Structural properties, antibacterial performance, and preservation effects on cherry tomatoes. *International Journal of Biological Macromolecules, 331*(Pt 1), 148207. https://doi.org/10.1016/j.ijbiomac.2025.148207

Wu, W., Zheng, L., Yu, J., Liu, L., Goksen, G., & Shao, P. (2024). High-sensitivity intelligent packaging films harnessing rose anthocyanins and hydrophilic silica aerogel for visual food freshness monitoring. *Food Quality and Safety, 8*. https://doi.org/10.1093/fqsafe/fyad051

Xie, F., Qin, Z., Luo, Y., He, Z., Chen, Q., & Cai, J. (2025). Synergistically engineered starch-based composite films: Multifunctional platforms integrating quaternary ammonium chitosan and anthocyanins for intelligent food monitoring and sustainable packaging. *Food Chemistry, 478*, 143560. https://doi.org/10.1016/j.foodchem.2025.143560

Xu, X., Huang, K., He, D., Huang, B., Tang, Q., Li, H., & Zou, Z. (2025). Biocompatible starch based smart active packaging film engineered with a novel CuNa-MOF colorimetric ammonia sensing agent based on crystal-to-crystal transformation. *Food Packaging and Shelf Life, 49*, 101498. https://doi.org/10.1016/j.fpsl.2025.101498

Yu, Y., Zhou, J., Chen, Q., Xie, F., Zhang, D., He, Z., Cheng, S., & Cai, J. (2024). Self-reinforced multifunctional starch nanocomposite film for litchi fruit postharvest preservation. *Chemical Engineering Journal, 486*, 150262. https://doi.org/10.1016/j.cej.2024.150262

Zhang, R., Wang, X., & Cheng, M. (2018). Preparation and characterization of potato starch film with various size of nano-SiO2. *Polymers, 10*(10), 1172.

Zhang, Z., Wang, Y., Fang, X., Chen, X., Yin, Z., & Zhang, C. (2024). Preparation of edible film from sweet potato peel polyphenols: Application in fresh fruit preservation. *Frontiers in Sustainable Food Systems, 8*, 1470732. https://doi.org/10.3389/fsufs.2024.1470732

Zou, Y., Ren, J., Fan, Y., Cao, Y., Xie, Y., Xiao, X., & Liu, F. (2025). Enhanced hydrophobicity, antioxidant activity, and antimicrobial efficacy of edible starch-based films incorporating pullulan and geraniol for fresh salmon preservation. *Food Packaging and Shelf Life, 49*, 101530. https://doi.org/10.1016/j.fpsl.2025.101530

Further reading

Liu, F., Ren, J., Yang, Q., Zhang, Q., Zhang, Y., Xiao, X., & Cao, Y. (2024). Improving water resistance and mechanical properties of starch-based films by incorporating microcrystalline cellulose in a dynamic network structure. *International Journal of Biological Macromolecules, 260*, 129404. https://doi.org/10.1016/j.ijbiomac.2024.129404

Liu, Y., Zhang, H., Wang, X., & Chen, L. (2024). Development of starch-based nanocomposite films reinforced with cellulose nanofibers for smart food packaging applications. *Journal of Food Engineering, 398*, 112345.

CHAPTER SIX

Starch nanoparticle-reinforced biopolymer-based intelligent packaging

Abhinab Goswami[1], Syed Nazrin Ruhina Rahman[2], Susankar Kushari[3], Sm Abdul Aziz Barbhuiyan[4], Satyendra Deka[5], Pratap Kalita[5]

[1]Faculty of Pharmaceutical Science, Assam Down Town University, Guwahati, Assam, India
[2]Department of Pharmaceutics, Royal School of Pharmacy, The Assam Royal Global University, Guwahati, Assam, India
[3]Department of Pharmaceutical Chemistry, School of Pharmaceutical Sciences, Girijananda Chowdhury University, Guwahati, Assam, India
[4]Department of Pharmaceutical Science, University of Science & Technology Meghalaya, Techno City, Ri-Bhoi, Meghalaya, India
[5]Pratiksha Institute of Pharmaceutical Sciences, Guwahati, Assam, India

6.1 Introduction

Foods are complex mixtures of various ingredients and exist in solid, liquid, and semisolid forms. Food provides energy and nutrients pertinent to maintaining a healthy life. Thus, it is essential to protect foods from microbial contamination and environmental factors from their subsequent spoilage (Samsudin & Hani, 2017). In this regard, packaging material serves as a comprehensive approach to provide containment and protection for food and to communicate with consumers through its label. Packaging materials are usually prepared using natural or synthetic polymers, which possess diverse properties (such as moisture barrier, gas barrier, oil resistance, mechanical strength, etc.) that are suitable for different foods to ensure quality, safety, and shelf-life maintenance (Kumar et al., 2024; Sharma et al., 2025). Accordingly, packaging can be divided into three categories, viz., primary, secondary, and tertiary packaging. Among the three categories, primary packaging is the one that directly contains and protects food products to warrant their safety and quality (Kumar et al., 2025). Currently, synthetic polymers, such as those derived from petroleum-based sources (polyethylene, propylene, etc.), are widely used to fabricate primary packaging because of their low production costs, excellent functional properties, and relatively easy processing (Ghasemlou et al., 2024). Nevertheless, the

Starches in Smart Packaging
ISBN: 978-0-443-45396-0
https://doi.org/10.1016/B978-0-443-45396-0.00002-5

environmental and health-related issues associated with petroleum-based polymers impede their long-term sustainability (Bai et al., 2024). This led to the exploration of polymers derived from natural sources that are biocompatible and biodegradable for food packaging applications.

Packaging systems fabricated from natural biopolymers, such as cellulose, chitosan, starch, zein, alginate, gluten, etc., often exhibit inadequate functional properties, such as mechanical strength, barrier properties, antimicrobial and antioxidant activities, and thermal stability (Kumari et al., 2024; Mishra et al., 2024). Therefore, to improve the functional properties of biopolymer-based packaging systems, additional components/additives are usually added during the fabrication process (Periyasamy et al., 2025). Based on the nature of the additives, the biopolymer-based packaging systems can be fabricated with improved mechanical strength and barrier properties, along with other functionalities, such as antioxidant, antimicrobial, and sensing capabilities (Ma et al., 2025; Periyasamy et al., 2025). Additives used to improve the properties of biopolymer-based packaging systems can be segregated into two broad categories, viz., synthetic and natural compounds. The synthetic compounds commonly used are gas scavengers, antioxidants, and nanomaterials, while the latter include essential oils, plant extracts, and natural antimicrobials (Baishya et al., 2025; Dutta & Sit, 2023; Qin et al., 2024; Sohrabi et al., 2024).

Among the additives, nanomaterials of both organic and inorganic origin are widely used to improve the functional properties of biopolymer-based packaging systems. Nanomaterials can be ascribed to those materials with at least one dimension in the range of 1–100 nm (Campelo et al., 2020). These materials exhibit properties, such as a high surface area-to-volume ratio and tunable optical, electrical, and mechanical properties (Osman et al., 2023). Incorporating these nanomaterials into the packaging system not only improves barrier properties and mechanical strength but also imparts antimicrobial, antioxidant properties, and sensing capabilities (Khezerlou et al., 2021). Among the nanomaterials, metallic nanoparticles, nanoclays, and metal–organic frameworks are widely incorporated into biopolymer-based active and intelligent packaging systems to enhance their functionalities (Dash et al., 2022). The physicochemical and biological properties of these nanomaterials, such as their small size, high surface area-to-volume ratio, and antimicrobial activity even at low concentrations, have positioned them as widely used additives in active and intelligent packaging systems (Siddiqui et al., 2022). Accordingly, several researchers have incorporated this category of nanomaterials for the fabrication of biopolymer-based active and intelligent systems. For instance, Ezati et al. (2022) developed carboxymethyl cellulose-based functional films containing copper-doped titanium nanoparticles to delay browning in bananas (Ezati et al., 2022). Similarly, Qin et al. (2024) developed ethylene-loaded polycaprolactone nanofiber-based film containing γ-cyclodextrin metal–organic frameworks and titanium nanoparticles to control ethylene levels in banana for possible shelf-life extension (Qin et al., 2024). However, toxicity and environmental concerns, along with the possible risk of migration into the food matrix associated with metal-based nanoparticles, have limited their widespread adoption (Gupta et al., 2024).

In this context, nanomaterials prepared from biopolymers (such as alginate, cellulose, lignin, and starch) can serve as an alternative to inorganic nanomaterials due to their nontoxic, biodegradable, and biocompatible nature. These biopolymer-based nanomaterials can be used as reinforcements to improve the mechanical and barrier properties of biocomposite films (Chen et al., 2022; Sharma et al., 2023, 2024). Moreover, incorporating biopolymer-based nanoparticles like chitosan nanoparticles imparts antimicrobial activity to the packaging system (Dey et al., 2021). These nanomaterials can also serve as carriers for other bioactive compounds, facilitating controlled release of bioactives to extend the shelf-life of foods (Lv et al., 2024; Wang et al., 2022). Thus, integrating biopolymer-based nanoparticles with bioactive compounds into active and intelligent packaging systems possesses the potential to serve as a viable alternative to inorganic nanomaterials.

Starch is a polysaccharide composed of two distinct glucose polymers, viz., amylose and amylopectin. Amylose consists of a linear chain of glucose units linked together through α-1,4-glycosidic bonds, while amylopectin consists of branched glucose units linked linearly through α-1,4-glycosidic bonds with branching points through α-1,6-glycosidic bonds. Starch acts as a major source of energy in plants, usually present in tubers, roots, fruits, grains, and seeds, existing in different shapes and particle sizes varying between 2 and 50 μm. Nanomaterials synthesized by using starch possess a size in the range of 1–100 nm and can be classified as nanocrystals, nanofibers, and nanoparticles. These starch-based nanomaterials can be fabricated by using top-down or bottom-up approaches, such as antisolvent precipitation, acid and enzymatic hydrolysis, high-pressure homogenization, electrospinning, and ultrasonication (Sathianathan et al., 2023; Wang, Chang, et al., 2021; Wang, Qiu, et al., 2021). Apart from their use as filler material in biopolymer-based active and intelligent packaging systems to improve crystallinity and thermal, mechanical, and barrier properties, starch-based nanomaterials also exhibit antimicrobial effects against Gram-positive and Gram-negative bacteria (Fahmy et al., 2020). In this respect, Sharma et al. (2023) studied the influence of starch nanocrystals with chitosan on the functionality of gluten-based nanocomposite films. The resultant films displayed improved mechanical properties with values of 40.82 MPa and 31.09%, respectively, for tensile strength and elongation at break. The nanocomposite films also displayed reduced water vapor transmission rate, opacity, biodegradability within 55 days, and antimicrobial and antioxidant activities (Sharma et al., 2023). Sharma et al. (2024) fabricated montmorillonite, starch nanocrystals, and chitosan-reinforced gluten-based composite films and demonstrated the active film's ability to enhance the shelf-life of litchis. The incorporation of montmorillonite, starch nanocrystals, and chitosan into gluten resulted in films with a uniform surface together with enhanced tensile strength (from 52.44 to 57.00 MPa) and improved water vapor barrier property (from 0.77 to 0.68×10^{-3} $g/m^2/h$) (Sharma et al., 2024). Recently, Yang, Zhang, et al. (2024) and Yang, Zhao, et al. (2024) developed a novel starch-based composite nanofibrous film composed of starch, tannic acid, and Fe^{3+} using a temperature-assisted electrospinning method and demonstrated the potential of the composite nanofibrous film as an active packaging system (Yang, Zhang, et al., 2024; Yang, Zhao, et al., 2024).

Similar to active packaging systems, starch-based nanomaterials are also used to reinforce intelligent packaging systems that can monitor the condition of packaged food and communicate with consumers and producers. These specialized packaging materials utilize components, such as indicators (time, temperature, and pH), gas detectors, and radio frequency identification (RFID) tags, enabling intelligent packaging systems to sense and monitor food products (Qin et al., 2024; Zheng et al., 2023; Zuo et al., 2022). Sohrabi et al. (2024) developed a food freshness colorimetric aerogel consisting of chitosan nanofibers, *Ixiolorion tataricum* anthocyanins, and starch nanoparticles (SNPs) to differentiate minced beef into fresh, medium-fresh, and spoiled categories (Sohrabi et al., 2024). The incorporation of SNPs improved the crystallinity and thermal stability in the aerogel-based intelligent packaging system. In another study, Wang et al. (2022) developed anthocyanin-loaded SNPs and a thermostable, colorimetric waterproof coating with pH-sensing capabilities (Wang et al., 2022). Thus, this chapter aims to provide a comprehensive overview of recent progress and development in starch nanomaterial-reinforced active and intelligent packaging. Moreover, this chapter is anticipated to facilitate further research directed toward applications of starch-based nanomaterials in the development of active and intelligent packaging systems.

6.2 Starch nanoparticles: preparation and characterization

6.2.1 Preparation of starch nanoparticles

SNPs are prepared by using nanotechnology techniques that create particles measuring less than 100 nm in size (Campelo et al., 2020). The characteristics and morphology of these nanoparticles are mainly dependent on their botanical source (Yu et al., 2021). The SNPs are generally prepared via two different types of approaches, i.e., top-down and bottom-up (Qiu et al., 2020). The top-down approach, such as acid hydrolysis, gamma radiation, homogenization, ultrasonication, etc. In this approach, the nanoparticles are formed by fragmenting or breaking down the larger particles into smaller ones by utilizing the chemical or mechanical forces (Yu et al., 2021). In the bottom-up approach, the nanoparticles are prepared utilizing the thermodynamic technique of controlled molecular assembly, including self-assembly or nanoprecipitation methods (Marta et al., 2023). The bottom-up approach is expensive but offers short production times and higher yield of nanoparticles with controlled morphology, i.e., size and shape (Abid et al., 2022). This approach needs specific chemicals and sophisticated equipment for preparing the nanoparticles (Alves et al., 2021). Whereas the top-down approach is simple, cost-effective, and applicable to both laboratory and industrial usage. The top-down approach is less efficient for materials with very small sizes and irregular shapes. Although the top-down approach is user-friendly, it fails to produce particles with the desired morphology. The key factors for selecting a suitable nanoparticle preparation process mainly depend on the production scale and the desired particle size (Abid et al., 2022).

6.2.1.1 Physical method

The physical methods, such as high-pressure homogenization, gamma irradiation, milling, and ultrasonication, are utilized to prepare SNPs. The benefit of the physical method lies in its avoidance of chemicals or solvents, rendering it simpler and environmentally friendly. Moreover, comparatively less time is required for preparing nanoparticles by using this method. However, the drawback of mechanical force usage in this method is likely to degrade the crystalline nature of starch. This method generates less crystalline nanoparticles due to the high energy requirement in the production process, which weakens the crystalline structure of starch (Teodoro et al., 2015). This limitation is not applicable to the gamma irradiation method, as the breakdown of starch structure is done by free radicals, resulting in the formation of highly crystalline nanoparticles. The SNPs prepared via milling methods tend to generate particles with larger sizes. Ultrasonication is considered as most effective method for achieving particles with smaller sizes. In the ultrasonication method, the energy is imparted to the starch granules using a cavitation process, characterized by microbubbles bursting that disperse as sound waves. The covalent bonds present in starch can easily break due to the involvement of high speed and shear force in the process of ultrasonication, thereby decreasing the particle size (Da Silva et al., 2017). The high-pressure homogenizer generates a shear force that can disrupt the hydrogen bonds of starch, resulting in a decrease in particle size and formation of SNPs (Fu et al., 2011).

6.2.1.2 Chemical methods

The chemical method, such as acid hydrolysis, is utilized to prepare SNPs with higher crystallinity. The hydrolyzing solvents (e.g., sulfuric and hydrochloric acids) are used in the acid hydrolysis process (Campelo et al., 2020). Nageli has introduced sulfuric acid as the first hydrolyzing solvent, whereas Lintner used hydrochloric acid for acid hydrolysis of starch conducted under gelatinization temperatures at a specified time (Wang & Copeland, 2015). Both hydrolysis techniques were commercialized by the end of the 19th century. Several factors, such as agitation time, acid concentration, temperature, starch amount, and hydrolysis time, play a crucial role in the acid hydrolysis process for the preparation of SNPs (Hassan et al., 2022). Dularia et al. (2019) prepared SNP-loaded films extracted from water chestnuts via acid hydrolysis. Around 27.5% nanoparticles with 396 nm size were obtained by using 3.16 M sulfuric acid at 40°C for a period of 7 days in a hydrolysis process. Similarly, Roy et al. (2020) developed nanoparticle-loaded composite films from mung bean starch using the solution casting technique. By using sulfuric acid hydrolysis, 32.2% SNPs with 141.77 nm average particle size were yielded.

6.2.1.3 Enzymatic method

The enzymatic method for preparing SNPs through hydrolysis is facilitated by enzymes, such as pullulanase, α-amylase, and glucoamylase. The mechanism of enzyme hydrolysis is that it first attacks the amorphous part, followed by the breakage of the more crystalline part of starch at a slower rate (Marta et al., 2022). This method requires less than 1 day to perform enzymatic hydrolysis of starch granules and produces a higher yield of nanoparticles. Sun et al. (2014) showed that hydrolysis of corn starch using pullulanase enzymes yielded 85% of SNPs. Suriya et al. (2018) obtained 61.33% potato starch-based nanoparticles using the pullulanase enzyme. Kim et al. (2008) on the extraction of around 500-nm-sized, irregularly shaped SNPs from waxy rice using the α-amylase enzyme.

6.2.1.4 Nanoprecipitation method

The nanoprecipitation is recognized as a simple and widely used technique for the preparation of SNPs. In this technique, starch is first gelatinized, followed by precipitation with an antisolvent (e.g., acetone, butanol, ethanol, isopropanol, propanol, etc.). The antisolvent breaks the starch chain during the precipitation step, which leads to the formation of nanoparticles (Kumari et al., 2020). Tan et al. (2009) prepared starch-based nanospheres with controlled size via the nanoprecipitation technique. They have controlled the size of nanoparticles by regulating the concentration of polymer in acetone, the ratio of aqueous and organic phase and the degree of substitution and molecular weight of starch ester. Wu et al. (2016) developed corn starch-based nanoparticles via nanoprecipitation using dimethyl sulfoxide. The increased size of nanoparticles was observed by concentrating the starch solution utilized in the precipitation step. They also reported that nanoparticle size is influenced by the ratio of antisolvent used in this technique; specifically, an increased proportion of antisolvent leads to a decrease in particle size. Caicedo et al. (2019) prepared spherical-shaped potato starch-based nanoparticles with a size ranging from 50 to 150 nm using the nanoprecipitation technique. They found the relative crystallinity values of 25.4% and 23.5% for potato and SNPs, respectively.

6.2.1.5 Combined method

In the combined approach, two or more methods are used simultaneously for preparing the SNPs. This method includes the utilization of acid hydrolysis with ultrasonication, enzymatic hydrolysis with acid hydrolysis, nanoprecipitation with ultrasonication, and others. The utilization of acid hydrolysis with other methods, such as milling, enzymatic hydrolysis, and ultrasonication, as a pretreatment can address the limitations of acid hydrolysis, mainly the extension of treatment duration (Torres et al., 2019). This type of pretreatment can enhance the hydrolysis rate, as it allows the penetration of acid and breaks the amorphous part of starch more rapidly (Kumari et al., 2020). Dai et al. (2018) prepared waxy maize starch nanocrystals by using a planetary ball mill and a sulfuric acid hydrolysis method. The maize starch produces 19.3 and 15.8 wt% yield after performing 3 and 5 days of acid hydrolysis, respectively. The obtained nanoparticles have a size of 31 nm with a round-edge shape. Lin et al. (2020) formulated SNPs from lotus seeds using ultrasonic-assisted pullulanase hydrolysis. An increase in ultrasonication power resulted in particles with smaller sizes and an enhancement in crystallinity values due to the weakening of the bond, followed by the breakage of the amorphous part of starch. The obtained crystallinity value of 74.92% for the prepared nanoparticles. Zhou et al. (2021) prepared SNPs from waxy maize through hydrochloric acid hydrolysis followed by ultrasonication. This combined method produces 80.5% yield, 20–250 nm size, and 29.65% relative crystallinity.

6.2.2 Characterization of starch nanoparticles

6.2.2.1 Amylose content

The amylose content can influence the physicochemical properties, such as adhesion and gel-forming ability, of starch. Based on the botanical source of starch, the amylose content can vary. The analysis of amylose content can be done by using the iodine-binding method. When iodine solution reacts with amylose, it forms a blue-colored solution, whereas amylopectin mixed with iodine solution generates a red-purple-colored solution (Chavan et al., 2021). Minakawa et al. (2019) prepared SNPs from corn, cassava, and yam using ultrasonication. They reported that the nanoparticles prepared from yam starch showed a higher blue value (30%) compared to corn- (25%) and cassava-based (18%) SNPs. The increase in blue value indicates enhanced concentration of starch amylose content. Torres et al. (2019) developed potato SNPs by using three different methods to compare the amylose content. The nanoparticle prepared by nanoprecipitation, acid hydrolysis, and the combined method (i.e., acid hydrolysis with ultrasonication) showed 21.39%, 4.79%, and 3.08% amylose content, respectively. The SNPs prepared by using acid hydrolysis lead to a decrease in amylose content value because the starch amorphous part is acid-labile (Chavan et al., 2021).

6.2.2.2 Morphology analysis

The size and shape of SNPs were analyzed with the help of field emission–scanning electron microscopy (FE-SEM), SEM, or transmission electron microscopy. The SNPs may exhibit various shapes such as ellipsoidal, round, irregular, flat, or platelet. The morphology of nanoparticles can be affected based on the starch source or preparation method (Marta et al., 2023). Minakawa et al. (2019) visualized the prepared SNPs under SEM and found that the nanoparticles formulated via ultrasonication had an eroded and cracked surface. The SNPs are smaller in size (8–32 nm) with different shapes as compared to the conventional starch granules (size 1–3 μm). Similarly, Haaj et al. (2013) reported waxy maize SNPs formulated using the ultrasonication technique, having 30–100 nm size with eroded and cracked particle surfaces. The surface cracks, followed by progressive erosion of starch particles, may be due to the increased duration of sonication time. Hedayati et al. (2020) prepared spherical-shaped SNPs from tapioca through a nanoprecipitation and sonication combined method. An increase in the concentration of starch results in particles with a larger size.

6.2.2.3 *Crystallinity*

The crystallinity of starch can be defined as the ratio of the mass of the crystalline part to the entire mass of SNPs. The crystallinity of starch is attributed to amylopectin content (Kumari et al., 2020). Generally, the starch rich in amylopectin content exhibit higher crystallinity compared to those with a greater amylose content. The crystallinity can be studied by utilizing X-ray diffraction. There are four types (A, B, C, and V) of starch crystallinity patterns. The crystallinity patterns can be influenced based on the variety of starch botanical sources and the preparation method for nanoparticles. Marta et al. (2019) reported that cooking banana starch obtained from different cultivars (such as Nangka, Kapas, and Kepok) had a B-type crystalline pattern, whereas dessert banana starch collected from the Ambon cultivar showed a C-type crystalline pattern. Around 33%–34% crystallinity was obtained for Ambon and Nangka starch, 38.6% crystallinity for Kapas starch, and 39.4% crystallinity for Kepok starch. The waxy maize SNPs prepared by sulfuric acid hydrolysis showed an A-type pattern with 69% crystallinity, while nanoparticles prepared by ultrasonication represented 0% crystallinity. The ultrasonicated nanoparticles completely represent an amorphous nature, as the starch crystallinity is lost due to the ultrasound process (Haaj et al., 2016). The initial crystallinity value of tapioca starch was 25.12%, which reduces to 12.53% after the nanoprecipitation step and further decreases in value to 6.49% due to the nanoparticle formation following the nanoprecipitation and ultrasonication combined method (Hedayati et al., 2020).

6.2.2.4 Thermal properties

With the help of differential scanning calorimetry and thermogravimetric analysis, the thermal properties of SNPs can be evaluated. In comparison to conventional starch granules, nanoparticles exhibit lower thermal stability. This lower stability is attributed to a reduction in the molecular weight of SNPs (Sun et al., 2014). The preparation methods, such as nanoprecipitation, acid hydrolysis, ultrasonication, etc., can further diminish the thermal stability of SNPs. Da Silva et al. (2017) observed a decrease in the degradation temperature of nanoparticles made from cassava starch via ultrasonication. As compared to conventional starch granules, cassava-based nanoparticles exhibited lower thermal stability and a decrease in gelatinization temperature, which could be due to the weakening of the hydrogen bond. Chang et al. (2019) fabricated SNPs from waxy corn and potato starch via a combined method, such as vacuum cold plasma and ultrasonication. A decrease in gelatinization enthalpy from 16.63 ± 0.91 to 9.81 ± 0.19 J/g was found for potato SNPs. Ahmad et al. (2020) reported an increase in degradation temperature of modified sago-based SNPs through high-pressure homogenization, which indicates enhancement of thermal stability of the nanoparticles.

6.2.2.5 Digestibility properties

Starch digestibility is a complex mechanism that involves enzyme diffusion into the substrate, which influences the hydrolysis process, porosity of the substrate, and absorption of the enzyme in nanoparticles. The in vitro studies show higher digestibility of nanoparticles compared to starch granules due to their enhanced surface area (Li et al., 2014). Suriya et al. (2018) reported an enhanced digestibility value of nanoparticles from 41.29% to 43.24% through debranching of starch with pullulanase enzyme hydrolysis, followed by 12–24 h of recrystallization. The increase in starch digestibility resulted in a reduction in retrogradation time. The starch digestibility is influenced by several parameters, including botanical source, amylose–amylopectin ratio, particle size, retrogradation conditions, and crystallinity (Ding et al., 2016).

6.3 Effect of starch nanoparticles in biopolymer-based packaging

In biopolymer-based packaging, SNPs have become a potent addition that greatly enhances performance characteristics that are essential for both environmental sustainability and commercial viability. SNPs improve the mechanical, thermal, and barrier qualities of biopolymer films, which are currently limited by native starch and many biopolymer matrices' low tensile strength, poor moisture resistance, and low thermal stability (Luo et al., 2025). By adding SNPs to starch or starch-containing composites, the microstructure becomes denser, which in turn reduces the permeability of gases and water vapor. Because the nanoparticles make the molecules' paths more complicated, they increase the food's shelf life by decreasing the rate at which they diffuse through the packing film (Adeyeye et al., 2023; Karnwal et al., 2025; Singh et al., 2024). The incorporation of SNPs into biopolymer-based packaging technologies is transforming the environmentally friendly packaging sector, particularly by improving barrier, mechanical, and thermal properties (Fig. 6.1). SNPs provide molecular-level reinforcement to primarily plant-based matrices, such as pure starch or starch blends, effectively addressing long-standing concerns, including low moisture resistance, insufficient tensile strength, and thermal sensitivity in sustainable packaging (Karnwal et al., 2025; Marta et al., 2022; Singh et al., 2024).

6.3.1 Barrier, mechanical, and thermal properties

6.3.1.1 Barrier properties

SNPs significantly increase the barrier characteristics of biopolymer films by producing more convoluted pathways for the passage of water vapor,

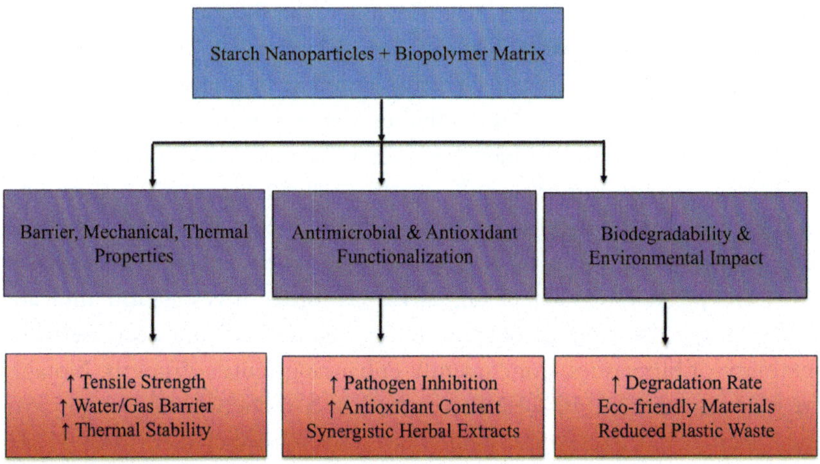

Figure 6.1 Integration of SNPs leads to multifunctional, sustainable biopolymer packaging materials.

oxygen, and other gases. This improvement is critical for maintaining the quality and shelf life of packaged foods because it prevents moisture uptake and oxygen intrusion, which can hasten deterioration (Dong et al., 2024). It is also noteworthy that the water vapor permeability has decreased. For example, research has demonstrated that nanoscale fillers, such as SNPs, can reduce oxygen permeability by 74% and water vapor permeability by 92.9% (Dong et al., 2024). This happens because the nanoparticles have a large surface area, which allows them to interact strongly with the biopolymer matrix and lengthen the diffusion path for molecules that are able to penetrate them (Marta et al., 2022; Santhosh et al., 2024). There is also a correlation between the shape, size, and concentration of the nanoparticles and the performance of the barrier. A uniform dispersion of nanoparticles of a suitable size results in films that are denser and more cohesive, which further restricts the transport of water and gases across the material (Marta et al., 2022; Singh et al., 2024). The practical use of increased barrier qualities is that they lower the risk of food dehydration or contamination, delay the oxidative deterioration process, and enhance overall package effectiveness by reducing the rate at which outside variables can undermine product quality (Adeyeye et al., 2023; Karnwal et al., 2025; Santhosh et al., 2024).

6.3.1.2 Mechanical properties

Mechanical properties of biopolymers, such as tensile strength, Young's modulus, and even elongation at break, are greatly improved when SNPs are incorporated into the material. However, the flexibility of the material could be reduced in specific formulations (Marta et al., 2022; Singh et al., 2024; Siva Janika et al., 2025). The strong interfacial adhesion that exists between the nanoparticles and the polymer matrix is responsible for the gains in mechanical strength that have been seen. Because of this contact, stresses are effectively transmitted, and the structure of the biopolymer is strengthened. For example, the breaking strength of bioplastics may boost from approximately 7.12 MPa to over 12.86 MPa when the right quantities of SNPs are present (Piyada et al., 2013). Another investigation indicated that elevating the content of reinforcing material led to a decrease in the flexibility of bioplastics, concurrently enhancing tensile strength. Notably, the incorporation of 10% SNPs resulted in a reduction of elongation at break from 1550% to 905%, while tensile strength increased from 13.5 to 19.9 MPa (Kong et al., 2017). Despite the fact that different biopolymer matrices, such as polyvinyl alcohol, chitosan, and cellulose derivatives, exhibit a variety of reactions to nanoparticle reinforcement, they all consistently receive the benefits of higher density, decreased porosity, and improved transfer of load through the polymer matrix (Marta et al., 2022; Momtaz et al., 2024). With these modifications, biopolymer films are better able to withstand mechanical strains, splits, and punctures, making them ideal for storing and transporting food products (Marta et al., 2022; Momtaz et al., 2024; Siva Janika et al., 2025).

6.3.1.3 Thermal properties

SNPs additionally affect the thermal properties of biopolymer-based packaging. This feature is crucial for packaging materials that must preserve stability under varying temperatures during production, storage, or utilization (Marta et al., 2022; Santhosh et al., 2024). The incorporation of SNPs significantly improves the thermal stability and durability of the composite films. Nanoparticles enhance interactions between molecules and limit the flexibility of chains of polymers, resulting in elevated temperatures of glass transition (T_g), temperature of melting (T_m), and onset of breaking down temperatures (Santhosh et al., 2024). An increase in ΔH was observed with the inclusion of reinforcing material, attributed to improved interactions between SNPs and the matrix, which necessitate greater energy to disrupt the polymer network (Hakke et al., 2022). In a waterborne polyurethane-based bioplastic reinforced with pea SNPs, the enthalpy change (ΔH) increased from 2.34 J/g at 0% reinforcement to 12.5 J/g at 30% reinforcement (Zou et al., 2011). Another investigation reported a reduction in ΔH, decreasing from 14.32 J/g (unreinforced) to 0.91 J/g in a potato starch bioplastic that included 10% amadumbe SNPs and glycerol, which was attributed to the inhibition of starch crystallization (Mukurumbira et al., 2017). The combined impact of these enhancements, barrier, mechanical, and thermal, establishes SNP-reinforced biopolymers as a viable option for advanced, sustainable packaging materials (Dong et al., 2024; Marta et al., 2022; Santhosh et al., 2024).

6.3.2 Antimicrobial and antioxidant functionalization

Antimicrobial and antioxidant functionalization are important developments in SNP-reinforced biopolymer packaging. These innovations directly address issues regarding the quality and safety of food by limiting the growth of microorganisms and oxidative deterioration. Numerous research papers provide an explanation of how SNP-based films are able to achieve these capabilities by the utilization of a number of different methods and the insertion of a variety of substances, such as nanoparticles of metal oxide, natural plant extracts, and essential oils (Bi et al., 2020; de Souza et al., 2022; Karnwal et al., 2025; Kumar et al., 2024; Ma et al., 2025; Syafiq et al., 2020). Table 6.1 shows various antimicrobial and antioxidant applications of SNP-reinforced films.

6.3.2.1 Antimicrobial functionalization

In biopolymer matrices, SNPs improve the uniformity and controlled release of antimicrobial agents, leading to films with enhanced and broad-spectrum antimicrobial properties. For example, in a study utilizing neem extract to synthesize ZnO nanoparticles in starch-based biofilms, the zones of inhibition toward *Staphylococcus aureus* were up to 22.05% higher and against Escherichia coli up to 15.27% higher than in chemically synthesized equivalents. The reactive oxygen species (ROS) produced by the nanoparticles, when combined with biopolymer carriers, alter the membranes of bacteria and metabolism, thereby boosting the antimicrobial action (Kumar et al., 2024). In line with the findings of many studies, the integration of chitosan or nanocellulose into SNP-based films improves the properties of these films, hence enhancing their physical barrier effect and expanding their antibacterial efficiency. Enhanced antimicrobial activity against a variety of pathogens, such as *Listeria monocytogenes*, *Pseudomonas aeruginosa*, and *Salmonella enteritidis*, can be achieved through the utilization of SNPs in conjunction with herbal extracts, such as pomegranate peel, tea polyphenols, pitanga leaf, or essential oils, including oregano, cinnamon, and bergamot, in order to produce biopolymer-based active packaging (Syafiq et al., 2020).

6.3.2.2 Antioxidant functionalization

It is common practice to functionalize SNP-based films with antioxidants derived from plants or food industry waste in order to prevent oxidative degradation of packaged meals (Bi et al., 2020; Ma et al., 2025; Tambe et al., 2025). Antioxidant tests have shown that starch matrices enhanced with brewer's spent grain extract or onion peel extract can scavenge free radicals and increase the total phenolic content, with IC_{50} values as low as 185.95 μg/mL. Food color, aroma, and nutritional content can be preserved for longer periods of time using starch/antioxidant composite films, which also reduce lipid oxidation (Vieira et al., 2024). In spite of the fact that the nanoparticulate matrix permits progressive diffusion, which ensures that the effect is preserved throughout the duration of the shelf life of the

Table 6.1 Antimicrobial and antioxidant applications of starch nanoparticle-reinforced films.

Sl.No.	Agent type	Agent name/source	Description and mechanism	Effects on packaging	References
1	Antimicrobial/ Antioxidant	Starch–silver hybrid nanoparticles	Synergistic antimicrobial activity of silver and improved film properties of starch	Enhanced antimicrobial performance and packaging durability.	Morán et al. (2025)
2	Smart/colorimetric indicator	Chitosan nanofiber (CHNF)/starch nanoparticles (SNPs)/ *Ixiolirion tataricum* anthocyanins (ITA) aerogel	Controlled porosity, ITA-driven pH color response, hydrogen-bonded stability	Protein-food freshness sensor: Minced beef categorized via reversible color change.	Sohrabi et al. (2024)
3	Film reinforcement/ antimicrobial and antioxidant carrier	SNPs (citrate-esterified, ultrasound-treated)	Ultrasound-assisted citrate esterification yields stable, small, low-polydispersity index A-type crystalline particles	Enhances CH films' strength, crystallinity, and suitability for active packaging.	Yan et al. (2021)
4	Film reinforcement	Taro starch nanoparticles (TSNPs)	Obtained via pullulanase hydrolysis and starch recrystallization; act as reinforcing fillers in starch films	Improves film barrier, mechanical, thermal, and surface properties.	Dai et al. (2015)
5	Film reinforcement	Peanut protein nanoparticles (PNPs)	Antisolvent-fabricated additive (0%–4%) strengthens soy protein–corn starch films	2% loading maximizes soy protein film strength, barrier, and thermal stability.	Li et al. (2015)

(Continued)

Table 6.1 Antimicrobial and antioxidant applications of starch nanoparticle-reinforced films.—cont'd

Sl.No.	Agent type	Agent name/source	Description and mechanism	Effects on packaging	References
6	Smart indicator	Corn starch (CS)/chitosan (CH) film with rose anthocyanin-encapsulated potato amylopectin nanoparticles (ANPs)	ANPs (~193 nm, 89.7% encapsulation) hydrogen-bonded in film, providing a pH-responsive color network.	Real-time shrimp freshness sensor: Durable, pH-responsive, and color-stable for 14 days.	Zheng et al. (2023)
7	Smart/functional coating	SNPs/PDMS/anthocyanin extract	SNP/PDMS hierarchical films: Superhydrophobic and anthocyanin-driven pH colorimetric sensing.	Biodegradable coating with water resistance, self-cleaning, and pH-sensitive food freshness sensing for sustainable packaging.	Wang, Chang, et al. (2021), Wang, Qiu, et al. (2021)
8	Film reinforcement	SNPs (ultrasound and acid hydrolysis)/cellulose nanoparticles (acid hydrolysis from microcrystalline cellulose and sisal)	Higher crystallinity; cellulose nanoparticles (NPs): Superior electrical stability, characterized by size, ζ, X-ray diffraction, and thermogravimetric analysis.	Cellulose NPs enhance starch film strength and barrier, while lowering flexibility.	Santana et al. (2019)
9	Film reinforcement/biocomposite	SNP–CH films	SNP/CH films: enhanced solubility, swelling, hydrophobicity, and uniform morphology.	Optimizes edible film for flexibility, water resistance, and strength.	Restu et al. (2021)

10	Film reinforcement	CS nanocomposite films with SNPs	SNPs (0%–25%) boost film strength, thermal stability, and surface roughness with minor crystallinity increase.	Improves barrier, strength, and thermal stability for biodegradable packaging.	Liu et al. (2016)
11	Film reinforcement/ antioxidant	Potato starch (PS)/ SNP/tea polyphenol (TP) films	SNPs and TP improve mechanical properties; PS–TP forms noncovalent complex with B-type crystallinity; TP absorbed on the PS surface	Enhances tensile strength (up to 9.57 MPa) and elongation at break; improves film quality for starch-based biodegradable packaging.	Yang, Zhang, et al. (2024), Yang, Zhao, et al. (2024)
12	Smart indicator/ functional coating	Starch film with nano-starch/PDMS composite coating and anthocyanins	SNP/PDMS hierarchical films: Superhydrophobic and anthocyanin-driven pH colorimetric sensing.	Repels liquids, self-cleans, and maintains colorimetric freshness monitoring for seafood.	Wang, Chang, et al. (2021), Wang, Qiu, et al. (2021)
13	Film reinforcement/ antioxidant	PS/SNP/TP composite film	SNPs reinforce PS, TPs form B-type crystalline PS–TP complexes with surface adsorption.	5% SNPs/TP maximize PS tensile strength and elongation; higher SNPs reduce performance.	Yang, Zhang, et al. (2024), Yang, Zhao, et al. (2024)
14	Active/biodegradable packaging	Cassava starch bilayer film (TPS-Y/TPS-NP)	Bilayer TP films combine antioxidant/ flexible yerba mate extract and antimicrobial/barrier TiO_2 NPs with compatible starch matrix.	Antibacterial, antioxidant films with strong, flexible, low-water vapour permeability structure, fully compostable in 2 weeks.	Medina-Jaramillo et al. (2024)

(Continued)

Table 6.1 Antimicrobial and antioxidant applications of starch nanoparticle-reinforced films.—cont'd

Sl.No.	Agent type	Agent name/source	Description and mechanism	Effects on packaging	References
15	Film reinforcement	SNPs (3–7 wt%) in Andean PS	Starch–SNP films show optimal reinforcement when the matrix and NPs share the same origin.	Elastic modulus 1.5–3× higher when the starch matrix and SNPs originate from the same source.	Martinez et al. (2016)
16	Film reinforcement	CS films with SNPs (0–25 wt%)	Solution-cast SNPs modify film microstructure, crystallinity, and thermal behavior.	Improved barrier, strength, thermal stability, and biodegradability with slight crystallinity change.	Liu et al. (2016)
17	Film reinforcement/ biodegradable composite	Polybutylene Adipate Terephthalate (PBAT)/ Thermoplastic (TP) films with SNPs	Starch NPs improve gelatinization, compatibility, hydrogen bonding, and PBAT segment structure.	Reinforced, stronger, and more rapidly biodegradable matrix.	González Seligra et al. (2016)

nanocomposites, the effectiveness of antioxidant release from these nanocomposites is enhanced during the initial stages, which offers immediate protection against oxidative stress (Ma et al., 2025; Vieira et al., 2024). In addition to this, in order to obtain high encapsulation efficiency and controlled release of polyphenols and other antioxidants, various techniques have been applied. These techniques include electrospinning and controlled cross-linking, which involves the utilization of substances such as tannic acid. This functionalization not only improves the film's capacity to scavenge free radicals, but it also confers additional features, such as antibacterial activity and protection against ultraviolet radiation. For example, films incorporating SNPs have successfully integrated soy polyphenols, catechins, clove oil, and pomegranate peel extract through physical adsorption or covalent attachment. This method prevents lipid oxidation and maintains organoleptic properties for long periods of time by releasing antioxidants slowly and selectively into food during storage. Films with antioxidant properties not only actively preserve food but also increase its safety by preventing spoiling caused by ROS produced during food deterioration (Karnwal et al., 2025).

6.3.3 Biodegradability of starch nanoparticle biocomposites

Biodegradability is a fundamental advantage of starch-based nanocomposites, establishing them as substitutes for conventional nondegradable petrochemical plastics. The inherent polysaccharide structure of starch renders it particularly vulnerable to enzymatic breakdown by soil microbes. The use of SNPs enhances biodegradability by augmenting the surface area available for microbial degradation and hydrolysis with enzymes (Gamage et al., 2022). Research shows that biopolymer films containing SNPs break down quickly in composting and soil settings, often reaching 80%–90% breakdown in a matter of weeks (Ghasemlou et al., 2022). Native starch bioplastics often have restricted mechanical qualities, despite starch's hydrophilicity and ease of biodegradation. Nanoparticles keep the material biodegradable even after they increase its functional performance by reinforcing the matrix and facilitating efficient hydrolysis. Furthermore, biodegradability is maintained in starch-based nanocomposites that incorporate nanoclay or nanocellulose; in fact, certain formulations exhibit no discernible slowing of degradation rates (Gamage et al., 2022; Ghasemlou et al., 2022; Karnwal et al., 2025). Evaluations of the lifecycle of these starch nanocomposites reveal that they have a lower impact on the environment during the stages of manufacture, usage, and disposal when compared to polymers derived from petroleum. The complete mineralization of these substances into carbon dioxide, water, and biomass protects against the long-term pollution of microplastic resources (Gamage et al., 2022).

6.3.3.1 Environmental impact and sustainability

The renewable characteristics of starch and its biodegradability enhance environmental sustainability. Starch is obtained from plentiful agricultural sources such as corn, potato, cassava, and wheat, rendering it cost-effective and environmentally friendly in comparison to petroleum-based polymers. The degradability of starch nanocomposites diminishes the buildup in landfills and natural environments, aiding in the mitigation of plastic pollution issues. Furthermore, SNP biocomposites often necessitate reduced energy for manufacturing and generate fewer greenhouse gas emissions, since lifecycle assessments reveal diminished carbon footprints and toxicity levels compared to traditional plastics. The application of green production methods for SNPs and the integration of natural additives such as essential oils further mitigate environmental dangers (Gamage et al., 2022; Karnwal et al., 2025). However, the possible environmental risks that could be caused by nanoscale materials need to be controlled with exceptional care. There is a concern over the release of nanoparticles into ecosystems during the process of degradation or disposal. Ongoing research is addressing toxicological implications and safe disposal procedures in order to guarantee that starch nanocomposites are compatible with the environment (Gamage et al., 2022). Despite these advancements, obstacles persist. Technical challenges of scalability, cost-effectiveness, and reliable performance across many climatic conditions must be resolved to enable broad adoption in both developed and emerging countries. Research suggests that while nanoparticle additions often improve biopolymer performance and preserve biodegradability, certain nano-enhancements, specifically some carbon-based compounds or metal oxides, may affect the degradation rate and require meticulous formulation to prevent lasting residues. Moreover, continuous research into the environmental behavior and safety of nanoparticulate substances is essential to ensure ecological compatibility during all phases of the product lifetime (Gamage et al., 2022). In summary, the development of biopolymer packaging that is enabled by SNPs has emerged as an important invention in the field of sustainable materials research. Not only does it provide essential solutions for lowering pollution levels and minimizing waste, but it also promotes the preservation of food in an advanced manner, an increase in agricultural output, and responsible environmental management. As a result of ongoing developments in nanocomposite design, life cycle evaluation, and responsible sourcing, starch-based packaging is well positioned to play a pivotal role in the transition toward environmentally responsible packaging solutions on a worldwide scale (Gamage et al., 2022; Gurunathan et al., 2025; Mafe et al., 2024; Singh et al., 2024).

6.4 Intelligent application of starch nanoparticle-reinforced films

Intelligent packaging is a type of food packaging that exists as labels and sensors that can monitor food quality throughout the supply chain and communicate this information to producers and consumers by virtue of its ability to assess food decomposition, such as oxidation or reduction levels (Mirza Alizadeh et al., 2021). Intelligent food packaging helps address factors such as microbial spoilage, fluctuations in storage and transit temperatures, and packaging integrity, thereby minimizing food waste and the risk of foodborne illness (Mkhari et al., 2025). Fig. 6.2 shows intelligent applications of SNP-reinforced films. Various types of intelligent packaging systems include time-dependent indicators, freshness indicators, leak detectors, RFID tags, etc. (Upadhyay et al., 2024). Among them, freshness or colorimetric indicators utilize pH-sensitive pigments and pigment-immobilization substrates to detect pH changes in spoiled foods and provide a clear indication to consumers of food freshness (Almasi et al., 2022). In this context, Sohrabi et al. (2024) developed a food freshness colorimetric aerogel with controlled porosity based on chitosan nanofibers, *Ixiolirion tataricum* anthocyanin, and SNPs to detect minced beef freshness into three stages, viz., fresh, medium-fresh, and spoiled. SNPs produced by the H_2SO_4 hydrolysis method, obtained with a mean particle size of 86.2 nm, were used to control the porosity and morphological characteristics of the aerogel solid support. SNP-filled aerogel displayed a red color when exposed to a pH 2–5, a claret violet color at pH 6, different intensities of blue color at pH 7–10, bluish-green at pH 11, and greenish-yellow at pH 12, indicating the ability of the colorimetric indicator to display color variation at pH 2–12. Upon exposure to a 0.8 M ammonia solution, the aerogel rapidly changed from red to green with a sensitivity of 54.8% in 5 min, and the color change was reversible. Upon correlating the color change response (ΔE values) of the aerogel, a strong linear correlation and high R^2 values were

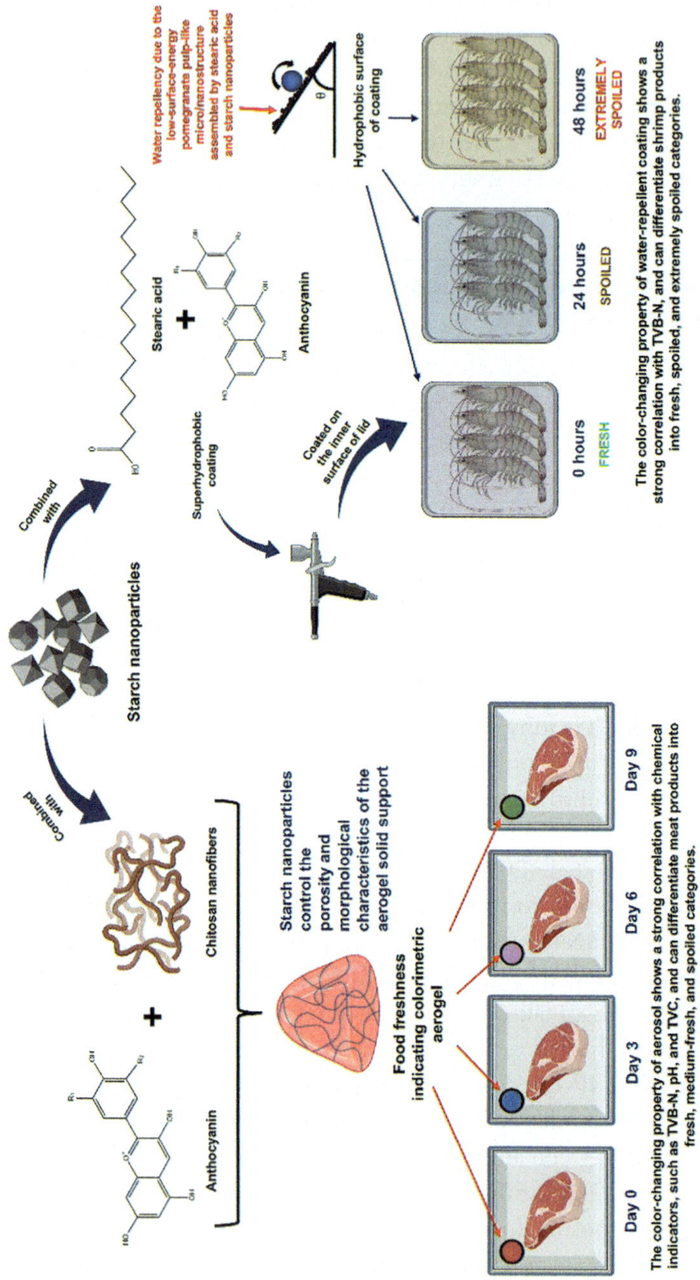

Figure 6.2 Intelligent applications of starch nanoparticle-reinforced films.

attained with pH ($R^2 = 0.95$), total volatile basic nitrogen (TVB-N, $R^2 = 0.94$), and total mesophilic count (TMC, $R^2 = 0.96$), indicating the desirable performance of the aerogel in monitoring minced beef freshness.

In another study, Lv et al. (2024) developed starch-based nanofiber mats as intelligent food tags using roselle anthocyanin as a pH-sensitive indicator and applied the developed intelligent food tags to monitor pork and shrimp freshness. The developed starch-based nanofiber mats displayed dark red, light red, purple, light green, dark green, and yellow, respectively, after exposure to solutions with pH values of 1–2, 3–4, 5–7, 8, 9–10, and 11–12 and color reversibility within 120 s. The potential of nanofiber mats to monitor the freshness of pork and seafood products was evaluated by correlating the color change of the nanofiber mats with freshness expressed by measuring pH, total viable count (TVC), TVB-N, and ΔE values during storage. For pork, the nanofiber mats displayed a series of color changes, viz., red, light red, pink, purple, light green, and dark green, respectively, at 0, 24, 36, 48, 60, and 72 h. The red color of the nanofiber mats indicated the freshness of meat and corresponded to TVB-N, pH, TVC, and ΔE values of 14.5 ± 0.3 mg/100 g, 5.95 ± 0.03, 4.09 ± 0.06, and 17.04 ± 0.63, respectively. The pink color of the nanofiber mats indicated the initiation of the deterioration process and corresponded to TVB-N, pH, TVC, and ΔE values of 19.2 ± 1.0 mg/100 g, 6.12 ± 0.02, 4.54 ± 0.08, and 27.88 ± 0.48, respectively. The pink-to-green color transformation of nanofiber mats indicated that pork had completely spoiled and corresponded to TVB-N and pH values of 55.0 ± 1.0 mg/100 g and 6.93 ± 0.04, respectively. For shrimp products, the nanofiber mats displayed a series of color changes, viz., red, pink, purple, light green, and dark green, respectively, at 0, 6, 12, 24, and 36 h. The transformation from light green (fresh state) to dark green (spoiled) color transformation of nanofiber mats indicated deterioration of shrimps and corresponds to TVB-N, pH, TVC, and ΔE values of 21.9 ± 0.1 to 37.6 ± 0.4 mg/100 g, 6.95 ± 0.13 to 7.57 ± 0.09, 6.70 ± 0.05 to 7.60 ± 0.05, and 54.08 ± 0.91 to 57.43 ± 1.42, respectively. Overall, the obtained results suggest the desirable efficiency of the nanofiber mats to monitor food freshness.

Wang, Chang, et al. (2021) and Wang, Qiu, et al. (2021) fabricated superhydrophobic, eco-friendly, and pH-responsive anthocyanin-rich extract-loaded SNP-based coatings for liquid-food residue reduction and freshness monitoring in shrimps. The developed superhydrophobic coating exhibited a water contact angle of $>152.0°$ and a sliding angle of $<9.0°$, which can be attributed to the formation of hierarchical micro- and nanostructures generated by aggregated SNPs and low surface energy of the polydimethylsiloxane (PDMS) covering. Upon evaluating the freshness monitoring capability of the coating for shrimp freshness, the coating exhibited a color change from purple (at 0 h) to bluish purple (at 36 h) and to bluish gray (at 72 h), with "L" value changing from 53.64 to 61.85 to 59.82, "b" value changing from 1.61 to −6.52 to 2.25 and "a" value changing from 6.10 to 4.33 to −1.52. Similarly, the TVB-N levels of shrimps significantly increase from 5.23 to 20.23–43.67 mg/100 g, indicating fresh, spoiled, and extremely spoiled conditions. Meanwhile, Wang et al. (2022) fabricated a food-grade colorimetric starch-based waterproof coating by using anthocyanin-loaded SNPs and stearic acid with thermal stability and significant water-repelling activity because of the low-surface-energy pomegranate pulp-like micro/nanostructure generated by SNPs and stearic acid. The developed waterproof coating was further evaluated to monitor the freshness of shrimp for 48 h. The coating displayed a distinct color change ($\Delta E > 6.00$), i.e., from purple (L: 39.05, b: 1.02, and a: 13.05) to navy blue (L: 51.13, b: 0.76, and a: 8.08) to grayish yellow (L: 65.22, b: 7.39, and a: 6.72) from 0 to 24–48 h, respectively. Furthermore, the TVB-N levels in shrimps at 0, 24, and 48 h were determined to be 4.12 mg/100 g, 22.63 mg/100 g, and 47.82 mg/100 g, respectively, indicating fresh, spoiled, and extremely spoiled conditions in the shrimps. In a nutshell, the starch-based waterproof coatings developed by Wang, Chang, et al. (2021), Wang, Qiu, et al. (2021), and Wang et al. (2022) possess significant potential to monitor the freshness of foods during transportation and storage without water interference.

In another study, Zheng et al. (2023) fabricated rose anthocyanin-loaded potato amylopectin nanoparticle-based pH indicator films by using corn starch and chitosan to monitor shrimp freshness. The anthocyanin-loaded potato amylopectin nanoparticles displayed mean size and zeta potential values of 192.7 nm and −6.70 nm, respectively. The developed film displayed 28.67% sensitivity to ammonia within 30 min with reversible pH sensing across five alternating cycles of acetic acid and ammonia exposure. Furthermore, the films displayed color changes from pink to colorless to yellow at 4°C to specify the deterioration of shrimps, while indicating the potential of the films in monitoring the freshness of seafood. Fatehi et al. (2025) combined SNPs with bromocresol and utilized them as a pH-sensitive indicator to track the quality of meat for 8 days of storage at refrigerated conditions. The bromocresol and SNP-based indicator changes color from yellow (at 1 day of storage) to blue (at 8 days of storage) due to the increase in the levels of TVB-N (19.6 ± 1.4 to 26.6 ± 1.4 mg/100 g), TVC (3.89 ± 0.08 to 7.17 ± 0.09 log cfu/g), and pH (5.4 ± 0.1 to 6.29 ± 0.35) along with a decreased color score (226.9 ± 3.15 to 104.23 ± 23.29), suggesting a significant correlation between color change and the analyzed chemical indicators in stored meat.

6.5 Challenges and future perspectives

The use of biopolymeric materials as an alternative to conventional plastic packaging has grown in popularity in recent years. Among biopolymers, starch is a plentiful, sustainable, and globally accessible source with intriguing qualities, such as the ability to behave like plastic when it is thermoplastic. Nevertheless, the hydrophilic nature and poor mechanical qualities render the processed material less durable and of lower quality than its synthetic equivalent (Muñoz-Gimena et al., 2023). The active agent, food product, and film matrix characteristics all influence the choice of a suitable nano-enabled delivery system. In addition to enhancing or extending the functional performance of the packaging material, the delivery system should be made to increase the dispersibility, stability, matrix compatibility, and efficacy of these active agents (Khezerlou et al., 2021). Compared to cellulose nanocrystals, starch nanocrystals have a limited capacity for reinforcing, and a greater concentration of the former is required to achieve comparable reinforcing effects. Another concern with starch nanocrystals is that their hydrolysis process takes significantly longer than that of cellulose, which is typically carried out at a lower temperature because starch is more sensitive to acid attack (Le Corre et al., 2010). The primary drawback of SNPs is that their synthesis method is still in its infancy. There is still more research needed to create more innovative techniques that will enable the large-scale production of SNPs with control over the final particle size, even though some pilot plant processes have been developed for SNP preparation. Furthermore, the studies on the production of SNPs with encapsulated bioactive compounds have not yet been commercialized, and more research is required to produce drug-loaded or bioactive SNPs that are satisfactory on a large scale (Morán et al., 2021). The starch content, surfactant addition, and the ratio of starch solution to nonsolvent used in the precipitation are some of the variables that affect the characteristics of SNPs. One simple and manageable way to create nanoparticles is through acid hydrolysis. Nevertheless, selective hydrolysis is a slow process with a low yield.

The high concentration of hydroxyl groups, which aggregate as a result of starch supramolecular interactions, is another drawback. Because of this tendency, SNPs are not appropriate for use in industry. An environmentally acceptable and nonhazardous method is enzymatic hydrolysis. The regeneration parameters were changed to control the SNPs' size and shape. However, when the enzyme activity decreases, the process is not finished. Enzymatic hydrolysis of SNPs is widely used in the food industry. Because no solvent is used, ultrasonic treatment is quicker, easier to handle, and requires fewer rinses than acid hydrolysis. This method's shortcomings include crystalline structure distortion and amorphous or poor crystallinity structure, which vary depending on the X-ray diffraction angle (Hassan et al., 2022). Due to the formation of sulfate groups at the surface, it has been found that when SNPs are prepared using the acid hydrolysis method, their distribution and stability in aqueous solution increase; however, their thermal stability decreases. Furthermore, starch nanocrystals are made to completely remove the amorphous areas while preserving the crystalline structure. It can also be made by separating the nanocrystalline lamellae, which will yield a small amount of platelet starch nanocrystals that have self-aggregated. Poor yield, a long treatment period, and acid involvement are the drawbacks of this technique (Sandhu & Nain, 2017, pp. 213–232). The enzymatic hydrolysis technique was used by Kim et al. (2008) to hydrolyze the native starch and create 500-nm-sized SNPs. The hydrolysis of α-amylase could specifically separate the amorphous regions in the starch granules. Because of their instability and the mechanical energy that ultrasonication provided, the resulting particles swelled quickly, forming starch aggregates. The amorphous structure of starch nanofibers and the intermolecular interactions created during the electrospinning process have an impact on their thermal behavior. Moisture loss happens first, then the carboxyl and hydroxyl groups undergo thermal breakdown. At higher temperatures, the starch polymer chains degrade, which significantly reduces mass (Du et al., 2024). These findings suggest that the structural arrangement and kind of modification made to the polysaccharide of starch nanofibers directly affect their thermal stability, which in turn affects their degradation and potential uses in high-temperature environments. Experiments conducted both in vitro and in vivo have demonstrated that SNPs are an effective vehicle for the controlled release of bioactive compounds. These loaded SNPs are also promising agents for encapsulating medications and antitumor agents that may be used in therapy. Nevertheless, insufficient research has been done on characterizing SNPs for in vivo drug delivery formulations, which is required to investigate potential interactions with the human body and their effects in order to ascertain how SNPs will negatively impact human health over the short and long term. Another primary drawback of SNPs is that their synthesis method has not yet been scaled up. Even though some pilot plant procedures have been developed for the preparation of SNPs, more research is still required to create more innovative techniques that will enable the production of SNPs on a large scale while controlling the final particle size (Morán et al., 2021). Further research is required to obtain satisfactory bioactive or drug-loaded SNPs on a large scale, as the studies found regarding the preparation of SNPs containing encapsulated bioactive compounds have not yet been industrialized.

6.6 Conclusion

Among the different types of starch-based nanomaterials, nanocrystals and nanoparticles have been the most studied for their application as reinforcement material in active and intelligent packaging systems. These nanomaterials possess unique physicochemical properties, opening doors of opportunity for the further development of active and intelligent packaging systems to improve the shelf-life of packaged food products. In general, starch-based nanomaterials have been used to stabilize natural or synthetic compounds with antimicrobial activity or to encapsulate several bioactive compounds, aiming to develop functional packaging systems. Furthermore, nanomaterials based on starch have been used to reduce the water vapour permeability and increase mechanical properties such as tensile strength and elastic modulus in films with potential food applications. The perspectives on the application of starch-based nanomaterials are currently in their nascent stage, and surely the development of new food products and food packaging will be related to these different starch nanoarchitectures.

References

Abid, N., Khan, A. M., Shujait, S., Chaudhary, K., Ikram, M., Imran, M., Haider, J., Khan, M., Khan, Q., & Maqbool, M. (2022). Synthesis of nanomaterials using various top-down and bottom-up approaches, influencing factors, advantages, and disadvantages: A review. *Advances in Colloid and Interface Science, 300*, 102597. https://doi.org/10.1016/j.cis.2021.102597

Adeyeye, S. A. O., Surendra Babu, A., Guruprasath, N., Guruprasath, N., & Sankar Ganesh, P. (2023). Starch nanocrystal and its food packaging applications. *Current Research in Nutrition and Food Science Journal, 11*(1), 01–21. https://doi.org/10.12944/crnfsj.11.1.1

Ahmad, A. N., Lim, S. A., Navaranjan, N., Hsu, Y. I., & Uyama, H. (2020). Green sago starch nanoparticles as reinforcing material for green composites. *Darussalam Polymer, 202*. https://doi.org/10.1016/j.polymer.2020.122646

Almasi, H., Forghani, S., & Moradi, M. (2022). Recent advances on intelligent food freshness indicators; an update on natural colorants and methods of preparation. *Food Packaging and Shelf Life, 32*, 100839.

Alves, M. J.d. S., Chacon, W. D. C., Gagliardi, T. R., Agudelo Henao, A. C., Monteiro, A. R., & Ayala Valencia, G. (2021). Food applications of starch nanomaterials: A review. *Starch/Staerke, 73*(11–12). https://doi.org/10.1002/star.202100046

Bai, F., Chen, G., Hu, Y., Liu, Y., Yang, R., Liu, J., Hou, R., Li, H., Wan, X., & Cai, H. (2024). Understanding the effect of plastic food packaging materials on food flavor: A critical review. *Trends in Food Science & Technology, 148*, 104502. https://doi.org/10.1016/j.tifs.2024.104502

Baishya, H., Dutta, J., & Kumar, S. (2025). Nanofillers reinforcing biopolymer composites for sustainable food packaging applications: A state-of-the-art review. *Advanced Functional Materials*, 2503819.

Bi, F., Zhang, X., Liu, J., Yong, H., Gao, L., & Liu, J. (2020). Development of antioxidant and antimicrobial packaging films based on chitosan, D-α-tocopheryl polyethylene glycol 1000 succinate and silicon dioxide nanoparticles. *Food Packaging and Shelf Life, 24*, 100503. https://doi.org/10.1016/j.fpsl.2020.100503

Caicedo, C. W. D., Ayala, V. G., Aparicio, R. G. M., & Agudelo, H. A. C. (2019). Mathematical models for prediction of water evaporation and thermal degradation kinetics of potato starch nanoparticles obtained by nanoprecipitation. *Starch-Stärke, 71*(1–2), 1800081.

Campelo, P. H., Sant'Ana, A. S., & Clerici, M. T. P. S. (2020). Starch nanoparticles: Production methods, structure, and properties for food applications. *Current Opinion in Food Science, 33*, 136–140.

Chang, R., Ji, N., Li, M., Qiu, L., Sun, C., Bian, X., Qiu, H., Xiong, L., & Sun, Q. (2019). Green preparation and characterization of starch nanoparticles using a vacuum cold plasma process combined with ultrasonication treatment. *Ultrasonics Sonochemistry, 58*, 104660. https://doi.org/10.1016/j.ultsonch.2019.104660

Chavan, P., Sinhmar, A., Nehra, M., Thory, R., Pathera, A. K., Sundarraj, A. A., & Nain, V. (2021). Impact on various properties of native starch after synthesis of starch nanoparticles: A review. *Food Chemistry, 364*. https://doi.org/10.1016/j.foodchem.2021.130416

Chen, Y., Li, Y., Qin, S., Han, S., & Qi, H. (2022). Antimicrobial, UV blocking, water-resistant and degradable coatings and packaging films based on wheat gluten and lignocellulose for food preservation. *Composites Part B: Engineering, 238*, 109868. https://doi.org/10.1016/j.compositesb.2022.109868

Da Silva, N. M. C., Correia, P. R. C., Druzian, J. I., Fakhouri, F. M., Fialho, R. L. L., & De Albuquerque, E. C. M. C. (2017). PBAT/TPS composite films reinforced with starch nanoparticles produced by ultrasound. *International Journal of Polymer Science, 2017*. https://doi.org/10.1155/2017/4308261

Dai, L., Li, C., Zhang, J., & Cheng, F. (2018). Preparation and characterization of starch nanocrystals combining ball milling with acid hydrolysis. *Carbohydrate Polymers, 180*, 122–127. https://doi.org/10.1016/j.carbpol.2017.10.015

Dai, L., Qiu, C., Xiong, L., & Sun, Q. (2015). Characterisation of corn starch-based films reinforced with taro starch nanoparticles. *Food Chemistry, 174*, 82–88. https://doi.org/10.1016/j.foodchem.2014.11.005

Dash, K. K., Deka, P., Bangar, S. P., Chaudhary, V., Trif, M., & Rusu, A. (2022). Applications of inorganic nanoparticles in food packaging: A comprehensive review. *Polymers, 14*(3). https://doi.org/10.3390/polym14030521

de Souza, A. L., Vieira, M. J. A., e Paiva, M. J. D. A., Bittencourt, M. T., Vieira, É. N. R., & Júnior, B. R. D. C. L. (2022). Antimicrobial biodegradable packaging with nanotechnology application. *Research, Society and Development, 11*(8). e3511830406-e3511830406.

Dey, D., Dharini, V., Selvam, S. P., Sadiku, E. R., Kumar, M. M., Jayaramudu, J., & Gupta, U. N. (2021). Physical, antifungal, and biodegradable properties of cellulose nanocrystals and chitosan nanoparticles for food packaging application. *Materials Today: Proceedings, 38*, 860–869. https://doi.org/10.1016/j.matpr.2020.04.885

Ding, Y., Zheng, J., Zhang, F., & Kan, J. (2016). Synthesis and characterization of retrograded starch nanoparticles through homogenization and miniemulsion cross-linking. *Carbohydrate Polymers, 151*, 656–665. https://doi.org/10.1016/j.carbpol.2016.06.007

Dong, M., Mastroianni, G., Bilotti, E., Zhang, H., & Papageorgiou, D. G. (2024). Biodegradable starch-based nanocomposite films with exceptional water and oxygen barrier properties. *ACS Sustainable Chemistry & Engineering, 12*(29), 11056–11066. https://doi.org/10.1021/acssuschemeng.4c04198

Du, C., Jiang, Y., Junejo, S. A., Jia, X., Zhang, B., & Huang, Q. (2024). Metal-anchored oxidized starch-pullulan nanofiber films enhance ethylene adsorption and banana preservation. *International Journal of Biological Macromolecules, 282*(Pt 6), 137399. https://doi.org/10.1016/j.ijbiomac.2024.137399

Dularia, C., Sinhmar, A., Thory, R., Pathera, A. K., & Nain, V. (2019). Development of starch nanoparticles based composite films from non-conventional source - Water chestnut (Trapa bispinosa). *International Journal of Biological Macromolecules, 136*, 1161–1168. https://doi.org/10.1016/j.ijbiomac.2019.06.169

Dutta, D., & Sit, N. (2023). Application of natural extracts as active ingredient in biopolymer based packaging systems. *Journal of Food Science and Technology, 60*(7), 1888–1902. https://doi.org/10.1007/s13197-022-05474-5

Ezati, P., Riahi, Z., & Rhim, J. W. (2022). CMC-based functional film incorporated with copper-doped TiO2 to prevent banana browning. *Food Hydrocolloids, 122*, 107104.

Fahmy, H. M., Salah Eldin, R. E., Abu Serea, E. S., Gomaa, N. M., AboElmagd, G. M., Salem, S. A., Elsayed, Z. A., Edrees, A., Shams-Eldin, E., & Shalan, A. E. (2020). Advances in nanotechnology and antibacterial properties of biodegradable food packaging materials. *RSC Advances, 10*(35), 20467–20484. https://doi.org/10.1039/d0ra02922j

Fatehi, F., Khalili Sadrabad, E., Feilizadeh, M., Derakhshan, Z., Heidari Kochaki, S., Hekmatimoghaddam, S., Jebali, A., & Mohajeri, F. A. (2025). Designing the pH-sensitive indicator based on starch nanoparticle with bromocresol green for monitoring meat spoilage. *International Journal of Environmental Health Research, 35*(5), 1204–1212. https://doi.org/10.1080/09603123.2024.2383427

Fu, Z. Q., Wang, L. J., Li, D., Wei, Q., & Adhikari, B. (2011). Effects of high-pressure homogenization on the properties of starch-plasticizer dispersions and their films. *Carbohydrate Polymers, 86*(1), 202–207. https://doi.org/10.1016/j.carbpol.2011.04.032

Gamage, A., Thiviya, P., Mani, S., Graceraj Ponnusamy, P., Manamperi, A., Evon, P., Merah, O., & Madhujith, T. (2022). Environmental properties and applications of biodegradable starch-based nanocomposites. *Polymers, 14*(21), 4578. https://doi.org/10.3390/polym14214578

Ghasemlou, M., Barrow, C. J., & Adhikari, B. (2024). The future of bioplastics in food packaging: An industrial perspective. *Food Packaging and Shelf Life, 43*, 101279.

Ghasemlou, M., Daver, F., Murdoch, B. J., Ball, A. S., Ivanova, E. P., & Adhikari, B. (2022). Biodegradation of novel bioplastics made of starch, polyhydroxyurethanes and cellulose nanocrystals in soil environment. *Science of the Total Environment*, 815. https://doi.org/10.1016/j.scitotenv.2021.152684

González Seligra, P., Eloy Moura, L., Famá, L., Druzian, J. I., & Goyanes, S. (2016). Influence of incorporation of starch nanoparticles in PBAT/TPS composite films. *Polymer International, 65*(8), 938–945. https://doi.org/10.1002/pi.5127

Gupta, R. K., Guha, P., & Srivastav, P. P. (2024). Investigating the toxicological effects of nanomaterials in food packaging associated with human health and the environment. *Journal of Hazardous Materials Letters, 5*, 100125.

Gurunathan, M. K., Navasingh, R. J. H., Selvam, J. D. R., & Čep, R. (2025). Development and characterization of starch bioplastics as a sustainable alternative for packaging. *Scientific Reports, 15*(1). https://doi.org/10.1038/s41598-025-00221-0

Haaj, S. B., Magnin, A., Pétrier, C., & Boufi, S. (2013). Starch nanoparticles formation via high power ultrasonication. *Carbohydrate Polymers, 92*(2), 1625–1632.

Haaj, S. B., Thielemans, W., Magnin, A., & Boufi, S. (2016). Starch nanocrystals and starch nanoparticles from waxy maize as nanoreinforcement: A comparative study. *Carbohydrate Polymers, 143*, 310–317.

Hakke, V. S., Landge, V. K., Sonawane, S. H., Uday Bhaskar Babu, G., Ashokkumar, M., & Flores, E. M. M. (2022). The physical, mechanical, thermal and barrier properties of starch nanoparticle (SNP)/polyurethane (PU) nanocomposite films synthesised by an ultrasound-assisted process. *Ultrasonics Sonochemistry, 88*. https://doi.org/10.1016/j.ultsonch.2022.106069

Hassan, N. A., Darwesh, O. M., Smuda, S. S., Altemimi, A. B., Hu, A., Cacciola, F., Haoujar, I., & Abedelmaksoud, T. G. (2022). Recent trends in the preparation of nano-starch particles. *Molecules, 27*(17). https://doi.org/10.3390/molecules27175497

Hedayati, S., Niakousari, M., & Mohsenpour, Z. (2020). Production of tapioca starch nanoparticles by nanoprecipitation-sonication treatment. *International Journal of Biological Macromolecules, 143*, 136–142.

Karnwal, A., Rauf, A., Jassim, A. Y., Selvaraj, M., Al-Tawaha, A. R. M. S., Kashyap, P., Kumar, D., & Malik, T. (2025). Advanced starch-based films for food packaging: Innovations in sustainability and functional properties. *Food Chemistry, X*, 102662.

Khezerlou, A., Tavassoli, M., Sani, M. A., Mohammadi, K., Ali, E., & Julian McClements, D. (2021). Application of nanotechnology to improve the performance of biodegradable biopolymer-based packaging materials. *Polymers, 13*(24), 4399. https://doi.org/10.3390/polym13244399

Kim, J. Y., Park, D. J., & Lim, S. T. (2008). Fragmentation of waxy rice starch granules by enzymatic hydrolysis. *Cereal Chemistry, 85*(2), 182–187.

Kong, J., Yu, Y., Pei, X. P., Han, C., Tan, Y., & Dong, L. (2017). Polycaprolactone nanocomposite reinforced by bioresource starch-based nanoparticles. *International Journal of Biological Macromolecules, 102*, 1304–1311. https://doi.org/10.1016/j.ijbiomac.2017.05.019

Kumar, P., Gautam, S., Bansal, D., & Kaur, R. (2024). Starch-based antibacterial food packaging with ZnO nanoparticle. *Journal of Food Science and Technology, 61*(1), 178–191. https://doi.org/10.1007/s13197-023-05834-9

Kumar, P., Kishore, A., Tripathi, S., Lavanya, Chaudhary, V., & Gaikwad, K. K. (2025). Innovative strategies to improve the properties of paper with bio-sourced material and its application in food packaging: A review. *Biomass Conversion and Biorefinery, 15* (9), 13037–13053. https://doi.org/10.1007/s13399-024-06066-3

Kumari, S., Debbarma, R., Nasrin, N., Khan, T., Taj, S., & Bhuyan, T. (2024). Recent advances in packaging materials for food products. *Food Bioengineering, 3*(2), 236–249.

Kumari, S., Yadav, B. S., & Yadav, R. B. (2020). Synthesis and modification approaches for starch nanoparticles for their emerging food industrial applications: A review. *Food Research International, 128*, 108765.

Le Corre, D., Bras, J., & Dufresne, A. (2010). Starch nanoparticles: A review. *Biomacromolecules, 11*(5), 1139–1153.

Li, X., Ji, N., Qiu, C., Xia, M., Xiong, L., & Sun, Q. (2015). The effect of peanut protein nanoparticles on characteristics of protein- and starch-based nanocomposite films: A comparative study. *Industrial Crops and Products, 77*, 565–574. https://doi.org/10.1016/j.indcrop.2015.09.026
Li, F., Zhu, A., Song, X., & Ji, L. (2014). Novel surfactant for preparation of poly(l-lactic acid) nanoparticles with controllable release profile and cytocompatibility for drug delivery. *Colloids and Surfaces B: Biointerfaces, 115*, 377–383. https://doi.org/10.1016/j.colsurfb.2013.12.015
Lin, X., Sun, S., Wang, B., Zheng, B., & Guo, Z. (2020). Structural and physicochemical properties of lotus seed starch nanoparticles prepared using ultrasonic-assisted enzymatic hydrolysis. *Ultrasonics Sonochemistry, 68*, 105199.
Liu, C., Jiang, S., Zhang, S., Xi, T., Sun, Q., & Xiong, L. (2016). Characterization of edible corn starch nanocomposite films: The effect of self-assembled starch nanoparticles. *Starch - Stärke, 68*(3–4), 239–248. https://doi.org/10.1002/star.201500252
Luo, C., Xie, F., Chen, Q., Zhou, J., He, Z., & Cai, J. (2025). Starch-nanoencapsulated polyphenol-induced polysaccharide gel coatings with efficient preservation capability. *Food Chemistry, 467*, 142203. https://doi.org/10.1016/j.foodchem.2024.142203
Lv, H., Wang, C., He, D., Zhao, H., Zhao, M., Xu, E., Jin, Z., Yuan, C., Guo, L., Wu, Z., Liu, P., & Cui, B. (2024). Intelligent food tag: A starch-anthocyanin-based pH-sensitive electrospun nanofiber mat for real-time food freshness monitoring. *International Journal of Biological Macromolecules, 256*(Pt 1), 128384. https://doi.org/10.1016/j.ijbiomac.2023.128384
Ma, S., He, J., Chen, Q., Zhou, J., Xie, F., & Cai, J. (2025). Advancements in starch-based nanomaterials for functional delivery and food packaging: A comprehensive review and future directions. *Journal of Future Foods, 5*(5), 443–454. https://doi.org/10.1016/j.jfutfo.2024.08.002
Mafe, A. N., Edo, G. I., Othuke Akpoghelie, P., Joshua, O. A., Isoje, E. F., Igbuku, U. A., & Essaghah, A. E. A. (2024). Comparative analysis of the environmental impact of biopolymer-based and conventional plastic packaging in food engineering applications. *Al-Mustaqbal Journal of Sustainability in Engineering Sciences, 2*(2). https://doi.org/10.62723/2959-5932.1017
Marta, H., Cahyana, Y., Djali, M., Arcot, J., & Tensiska, T. (2019). A comparative study on the physicochemical and pasting properties of starch and flour from different banana (Musa spp.) cultivars grown in Indonesia. *International Journal of Food Properties, 22*(1), 1562–1575. https://doi.org/10.1080/10942912.2019.1657447
Marta, H., Rizki, D. I., Mardawati, E., Djali, M., Mohammad, M., Cahyana, Y., & Nanoparticles, S. (2023). Preparation, properties and applications. *Polymers, 15*(5), 1167. https://doi.org/10.3390/polym15051167
Marta, H., Wijaya, C., Sukri, N., Cahyana, Y., & Mohammad, M. (2022). A comprehensive Study on starch nanoparticle potential as a reinforcing material in bioplastic. *Polymers, 14*(22), 4875. https://doi.org/10.3390/polym14224875
Martinez, S., Rivon, C., Troncoso, O. P., & Torres, F. G. (2016). Botanical origin as a determinant for the mechanical properties of starch films with nanoparticle reinforcements. *Starch - Stärke, 68*(9–10), 935–942. https://doi.org/10.1002/star.201600143
Medina-Jaramillo, C., Quintero-Borregales, L. M., Goyanes, S., Bernal, C., & Famá, L. (2024). A biodegradable and active bilayer Nanocomposite obtained from starch-yerba mate extract and starch-TiO_2 nanoparticles films for packaging applications. *Starch - Stärke, 76*(3–4). https://doi.org/10.1002/star.202300048
Minakawa, A. F., Faria-Tischer, P. C., & Mali, S. (2019). Simple ultrasound method to obtain starch micro-and nanoparticles from cassava, corn and yam starches. *Food Chemistry, 283*, 11–18.

Mirza Alizadeh, A., Masoomian, M., Shakooie, M., Zabihzadeh Khajavi, M., & Farhoodi, M. (2021). Trends and applications of intelligent packaging in dairy products: A review. *Critical Reviews in Food Science and Nutrition, 62*(2), 383–397. https://doi.org/10.1080/10408398.2020.1817847

Mishra, B., Panda, J., Mishra, A. K., Nath, P. C., Nayak, P. K., Mahapatra, U., Sharma, M., Chopra, H., Mohanta, Y. K., & Sridhar, K. (2024). Recent advances in sustainable biopolymer-based nanocomposites for smart food packaging: A review. *International Journal of Biological Macromolecules, 279.* https://doi.org/10.1016/j.ijbiomac.2024.135583

Mkhari, T., Adeyemi, J. O., & Fawole, O. A. (2025). Recent advances in the fabrication of intelligent packaging for food preservation: A review. *Processes, 13*(2), 539.

Momtaz, F., Momtaz, E., Mehrgardi, M. A., Momtaz, M., Narimani, T., & Poursina, F. (2024). Enhanced antibacterial properties of polyvinyl alcohol/starch/chitosan films with NiO–CuO nanoparticles for food packaging. *Scientific Reports, 14*(1). https://doi.org/10.1038/s41598-024-58210-8

Morán, D., Gutiérrez, G., Carmen Blanco-López, M., Marefati, A., Rayner, M., & Matos, M. (2021). Synthesis of starch nanoparticles and their applications for bioactive compound encapsulation. *Applied Sciences, 11*(10), 4547. https://doi.org/10.3390/app11104547

Morán, D., López-Carballo, G., Gavara, R., Gutiérrez, G., & Matos, M. (2025). Starch-silver hybrid nanoparticles: A novel antimicrobial agent. *Colloids and Surfaces A: Physicochemical and Engineering Aspects, 717,* 136797. https://doi.org/10.1016/j.colsurfa.2025.136797

Muñoz-Gimena, P. F., Oliver-Cuenca, V., Peponi, L., & López, D. (2023). A review on reinforcements and additives in starch-based composites for food packaging. *Polymers, 15*(13), 2972. https://doi.org/10.3390/polym15132972

Mukurumbira, A. R., Mellem, J. J., & Amonsou, E. O. (2017). Effects of amadumbe starch nanocrystals on the physicochemical properties of starch biocomposite films. *Carbohydrate Polymers, 165,* 142–148.

Osman, A. F., Adnan, S. A., Ibrahim, I., Salimi, M. N. A., & Jaafar, M. (2023). Toughening mechanism of thermoplastic starch nano-biocomposite with the hybrid of nanocellulose/nanobentonite. *Polymer, 274,* 125876.

Periyasamy, T., Asrafali, S. P., & Lee, J. (2025). Recent advances in functional biopolymer films with antimicrobial and antioxidant properties for enhanced food packaging. *Polymers, 17*(9), 1257.

Piyada, K., Waranyou, S., & Thawien, W. (2013). Mechanical, thermal and structural properties of rice starch films reinforced with rice starch nanocrystals. *International Food Research Journal, 20*(1), 439.

Qin, Z., Jiang, Q., Chen, M., Li, J., & Zhang, H. (2024). Food packaging for ripening and preserving banana based on ethylene-loaded nanofiber films deposited with nanosized cyclodextrin metal-organic frameworks and TiO2 nanoparticles. *Food Packaging and Shelf Life, 45,* 101332. https://doi.org/10.1016/j.fpsl.2024.101332

Qiu, C., Wang, C., Gong, C., McClements, D. J., Jin, Z., & Wang, J. (2020). Advances in research on preparation, characterization, interaction with proteins, digestion and delivery systems of starch-based nanoparticles. *International Journal of Biological Macromolecules, 152,* 117–125. https://doi.org/10.1016/j.ijbiomac.2020.02.156

Restu, W. K., Sondari, D., Setiasih, G. M., Triwulandari, E., Sampora, Y., Septevani, A. A., Kusumaningrum, W. B., Falah, F., Akbar, F., & Septiyanti, M. (March 31, 2021). Preparation and characterization of edible films from starch nanoparticles and chitosan. *Bio-inspired, Biomimetic and Nanobiomaterials, 10*(1), 1–7.

Roy, K., Thory, R., Sinhmar, A., Pathera, A. K., & Nain, V. (2020). Development and characterization of nano starch-based composite films from mung bean (Vigna radiata).

International Journal of Biological Macromolecules, 144, 242–251. https://doi.org/10.1016/j.ijbiomac.2019.12.113

Samsudin, H., & Hani, N. M. (2017). Use of starch in food packaging. In *Starch-based materials in food packaging* (pp. 229–256). Academic Press.

Sandhu, K. S., & Nain, V. (2017). *Starch nanoparticles: Their preparation and applications Plant Biotechnology: Recent Advancements and developments*. India: Springer Singapore. https://doi.org/10.1007/978-981-10-4732-9_11

Santana, J. S., de Carvalho Costa, É. K., Rodrigues, P. R., Correia, P. R. C., Cruz, R. S., & Druzian, J. I. (2019). Morphological, barrier, and mechanical properties of cassava starch films reinforced with cellulose and starch nanoparticles. *Journal of Applied Polymer Science, 136*(4). https://doi.org/10.1002/app.47001

Santhosh, R., Ahmed, J., Thakur, R., & Sarkar, P. (2024). Starch-based edible packaging: Rheological, thermal, mechanical, microstructural, and barrier properties – A review. *Sustainable Food Technology, 2*(2), 307–330. https://doi.org/10.1039/d3fb00211j

Sathianathan, R. V., Arockia Jayalatha, K., Sundararajan, A., Kannan, S., Paramasivam, N., & Jothipandian, S. (2023). Intelligent food packaging and shelf-life improvement of chapattis using hybrid nanoparticle-based biopolymer electrospin coating. *ACS Food Science & Technology, 3*(4), 753–760. https://doi.org/10.1021/acsfoodscitech.3c00012

Sharma, T., Kaur, G., Singh, A., Kaur, P., Dar, B. N., & Kaur, A. (2023). An emerging sustainable approach for development and characterization of gluten-based nanocomposite films reinforced with starch nanocrystals in conjugation with chitosan. *Sustainable Chemistry and Pharmacy, 36*, 101338. https://doi.org/10.1016/j.scp.2023.101338

Sharma, T., Kaur, G., Singh, A., Kumar, V., & Dar, B. N. (2024). Enhancing litchi shelf life with gluten-based nanocomposite films: A comparative study of montmorillonite and starch nanocrystals reinforced with chitosan. *Scientia Horticulturae, 332*, 113239. https://doi.org/10.1016/j.scienta.2024.113239

Sharma, C., Kundu, S., Singh, S., Saxena, J., Gautam, S., Kumar, A., & Pathak, P. (2025). From concept to shelf: Engineering biopolymer-based food packaging for sustainability. *RSC Sustainability, 3*, 4992–5026.

Siddiqui, J., Taheri, M., Alam, A. U., & Deen, M. J. (2022). Nanomaterials in smart packaging applications: A review. *Small, 18*(1). https://doi.org/10.1002/smll.202101171

Singh, A. K., Lee, M., Jang, D., & Lee, Y. S. (2024). Non-conventional starch nanoparticles: Novel avenues towards improving sustainability of the food packaging sector. *Trends in Food Science and Technology, 143*. https://doi.org/10.1016/j.tifs.2023.104273

Siva Janika, A. R., Afrin, M. R., Yuvaraj, D., Reddy Gujjula, K., & Rahul, R. (2025). Functionalized starch polymers in packaging: Advances in bioengineering and material science. *Starch - Stärke,* 77(7). https://doi.org/10.1002/star.70039

Sohrabi, N., Almasi, H., & Moradi, M. (2024). Controlled porosity intelligent aerogel based on chitosan nanofiber, anthocyanins and starch nanoparticles: Preparation, characterization and application in monitoring food spoilage. *Food Hydrocolloids, 156*, 110261.

Sun, Q., Li, G., Dai, L., Ji, N., & Xiong, L. (2014). Green preparation and characterisation of waxy maize starch nanoparticles through enzymolysis and recrystallisation. *Food Chemistry, 162*, 223–228. https://doi.org/10.1016/j.foodchem.2014.04.068

Suriya, M., Reddy, C. K., Haripriya, S., & Harsha, N. (2018). Influence of debranching and retrogradation time on behavior changes of Amorphophallus paeoniifolius nanostarch. *International Journal of Biological Macromolecules, 120*, 230–236. https://doi.org/10.1016/j.ijbiomac.2018.08.059

Syafiq, R., Sapuan, S. M., Zuhri, M. Y. M., Ilyas, R. A., Nazrin, A., Sherwani, S. F. K., & Khalina, A. (2020). Antimicrobial activities of starch-based biopolymers and biocomposites incorporated with plant essential oils: A review. *Polymers, 12*(10), 2403. https://doi.org/10.3390/polym12102403

Tambe, S. S., Girawale, S. D., Kodam, K. M., Zinjarde, S. S., & Athawale, A. A. (2025). Antioxidant films of starch blends with onion peel extract for packaging coriander. *Journal of Food Science, 90*(7). https://doi.org/10.1111/1750-3841.70383

Tan, Y., Xu, K., Li, L., Liu, C., Song, C., & Wang, P. (2009). Fabrication of size-controlled starch-based nanospheres by nanoprecipitation. *ACS Applied Materials & Interfaces, 1*(4), 956–959. https://doi.org/10.1021/am900054f

Teodoro, A. P., Mali, S., Romero, N., & De Carvalho, G. M. (2015). Cassava starch films containing acetylated starch nanoparticles as reinforcement: Physical and mechanical characterization. *Carbohydrate Polymers, 126*, 9–16. https://doi.org/10.1016/j.carbpol.2015.03.021

Torres, F. G., Arroyo, J., Tineo, C., & Omar, T. (2019). Tailoring the properties of Native Andean Potato Starch nanoparticles using acid and alkaline treatments. *Starch - Stärke, 71* (3–4). https://doi.org/10.1002/star.201800234

Upadhyay, A., Agbesi, P., Arafat, K. M. Y., Urdaneta, F., Dey, M., Basak, M., Hong, S., Umeileka, C., & Argyropoulos, D. (2024). Bio-based smart packaging: Fundamentals and functions in sustainable food systems. *Trends in Food Science and Technology, 145*. https://doi.org/10.1016/j.tifs.2024.104369

Vieira, F. J. A., Ludka, F. R., Diniz, K. M., Klosowski, A. B., & Olivato, J. B. (2024). Biodegradable active packaging based on an antioxidant extract from Brewer's spent grains: Development and potential of application. *ACS Sustainable Resource Management, 1* (11), 2413–2419. https://doi.org/10.1021/acssusresmgt.4c00277

Wang, F., Chang, R., Ma, R., Qiu, H., & Tian, Y. (2021a). Eco-friendly and pH-responsive nano-starch-based superhydrophobic coatings for liquid-food residue reduction and freshness monitoring. *ACS Sustainable Chemistry & Engineering, 9*(30), 10142–10153.

Wang, F., Ma, R., & Tian, Y. (2022). Facile fabrication of thermostable and colorimetric starch-based waterproof coating with edible organic materials. *Food Chemistry, 382*, 132269.

Wang, F., Qiu, L., & Tian, Y. (2021b). Super anti-wetting colorimetric starch-based film modified with poly (dimethylsiloxane) and micro-/nano-starch for aquatic-product freshness monitoring. *Biomacromolecules, 22*(9), 3769–3779.

Wang, S., & Copeland, Les (2015). Effect of acid hydrolysis on starch structure and functionality: A review. *Critical Reviews in Food Science and Nutrition, 55*(8), 1081–1097. https://doi.org/10.1080/10408398.2012.684551

Wu, X., Chang, Y., Fu, Y., Ren, L., Tong, J., & Jiang, Z. (2016). Effects of non-solvent and starch solution on formation of starch nanoparticles by nanoprecipitation. *Starch - Stärke, 68*(3–4), 258–263. https://doi.org/10.1002/star.201500269

Yan, X., Diao, M., Yu, Y., Gao, F., Wang, E., Wang, Z., & Zhang, T. (2021). Influence of esterification and ultrasound treatment on formation and properties of starch nanoparticles and their impact as a filler on chitosan based films characteristics. *International Journal of Biological Macromolecules, 179*, 154–160. https://doi.org/10.1016/j.ijbiomac.2021.03.004

Yang, Z., Zhang, D., & Cai, J. (2024a). Robust design of starch composite nanofibrous films for active food packaging: Towards improved mechanical, antioxidant, and antibacterial properties. *International Journal of Biological Macromolecules, 260*, 129329.

Yang, Q., Zhao, Y., Hu, X., Bao, Q., Zhao, Q., Zhang, F., Guo, K., Li, S., & Li, T. (2024b). Effect of starch nanoparticles and tea polyphenol inclusion on physicochemical and mechanical properties of starch-based films. *Starch-Stärke, 76*(3–4), 2300140.

Yu, M., Ji, N., Wang, Y., Dai, L., Xiong, L., & Sun, Q. (2021). Starch-based nanoparticles: Stimuli responsiveness, toxicity, and interactions with food components. *Comprehensive Reviews in Food Science and Food Safety, 20*(1), 1075–1100. https://doi.org/10.1111/1541-4337.12677

Zheng, L., Liu, L., Yu, J., Farag, M. A., & Shao, P. (2023). Intelligent starch/chitosan-based film incorporated by anthocyanin-encapsulated amylopectin nanoparticles with high stability for food freshness monitoring. *Food Control, 151*. https://doi.org/10.1016/j.foodcont.2023.109798

Zhou, L., He, X., Ji, N., Lei, D., Yang, L., Yang, J., Xiong, L., & Sun, Q. (2021). Preparation and characterization of waxy maize starch nanoparticles via hydrochloric acid vapor hydrolysis combined with ultrasonication treatment. *Ultrasonics Sonochemistry, 80*, 105836. https://doi.org/10.1016/j.ultsonch.2021.105836

Zou, J., Zhang, F., Huang, J., Chang, P. R., Su, Z., & Yu, J. (2011). Effects of starch nanocrystals on structure and properties of waterborne polyurethane-based composites. *Carbohydrate Polymers, 85*(4), 824–831. https://doi.org/10.1016/j.carbpol.2011.04.006

Zuo, J., Feng, J., Gameiro, M. G., Tian, Y., Liang, J., Wang, Y., Ding, J., & He, Q. (2022). RFID-based sensing in smart packaging for food applications: A review. *Future Foods, 6*, 100198.

CHAPTER SEVEN

Starch-based smart packaging added with nanomaterials

Hebat-Allah S. Tohamy
National Research Centre, Cellulose and Paper Department, Giza, Dokki, Egypt

7.1 Introduction

The traditional plastic film that shields our food is simultaneously a ticking environmental time bomb. While effective at preservation, the massive, nondegradable waste stream it creates demands a radical pivot. Every day, countless tons of these plastics, mostly derived from fossil fuels, are manufactured, used for a fleeting moment, and then dumped into landfills or, worse, pollute our oceans, breaking down into microplastics that infiltrate the air we breathe and the food we eat. This unsustainable flood of permanent waste is not something we can recycle our way out of; it mandates an immediate and intelligent redesign toward alternatives that disappear harmlessly after their job is done (Bopp, 2019; Dahiya et al., 2024; Andrady). The necessity for a radical pivot, therefore, is urgent, and the answer lies in smart and intelligent packaging (IntPack). Instead of inert plastic, we have a material that is active and communicative. We define the shift like this: IntPack senses it acts as a sentinel, using simple, visual sensors like a color-changing indicator to tell us, Hey, this is still fresh! or Oops, the food is spoiled! This gives consumers accurate, on-the-spot information, eliminating the guesswork that leads to food waste. On the other hand, Smart packaging acts actively as a helpful helper inside the package, releasing a controlled antimicrobial agent to extend the food's quality (Tohamy, 2025a, 2025b). We need accessible, naturally abundant, and biodegradable sustainable material alternatives, and this is where starch emerges as a leading biopolymer candidate (Compart et al., 2021). As one of the most common and low-cost polymers on Earth, derived readily from crops like corn, potatoes, wheat, and rice, starch possesses an unparalleled natural abundance that makes it a highly scalable and economical solution (Adigwe et al., 2022; Falua et al., 2022). Crucially, starch-based materials are inherently biodegradable and compostable. Unlike petroleum-based plastics that

Starches in Smart Packaging
ISBN: 978-0-443-45396-0
https://doi.org/10.1016/B978-0-443-45396-0.00013-X

persist for centuries, a starch-based package will break down harmlessly into natural components such as water, carbon dioxide, and biomass once its work is over (Jayarathna et al., 2022). By harnessing this simple, yet powerful, natural polymer, we can replace the polluting, permanent nature of traditional plastic packaging with an eco-friendly cycle of life and decay, offering a path to truly smart and green packaging.

While starch offers compelling advantages as a sustainable biopolymer, relying on pure starch films presents significant inherent drawbacks that limit its direct replacement of conventional plastic packaging. The primary issue lies in its poor mechanical strength; films made solely from starch are often brittle, easily torn, and lack the robust tensile strength required for practical handling, sealing, and transport in real-world supply chains (Arruda et al., 2025). For food packaging, pure starch is highly hydrophilic; it readily absorbs moisture from the environment, meaning the film fails to create the essential barrier needed to protect moisture-sensitive products, causing foods to rapidly dry out, become stale, or quickly spoil (Khan et al., 2017). To overcome the inherent weaknesses of pure starch, its fragility and moisture permeability, we introduce nanomaterials inside starch films (Arruda et al., 2025). These nanomaterials will increase the surface-area-to-volume ratio, which allows a small amount of material to interact intensely with and significantly reinforce the entire starch matrix. When dispersed throughout the polymer, these nanomaterials act like a microscopic support, drastically improving the starch's mechanical strength and creating a far more effective barrier against water vapor (Arruda et al., 2025; Gamage et al., 2021). The primary aim of this chapter is to provide a detailed, step-by-step description of the structure, preparation, and comprehensive types of nanomaterials in the starch-nanomaterial nanocomposites. These composites are engineered specifically to overcome the limitations of pure starch and deliver the next generation of smart food packaging applications.

7.2 Starch as a biopolymer matrix

Starch, an important polysaccharide, serves as the primary energy storage carbohydrate in plants and is a major food source for animals. It's constructed from two distinct polymers: amylose, which is linear, and amylopectin, which features highly branched chains (Bertoft, 2017; Giuberti et al., 2014; Seung, 2020). The linear component of starch, amylose, is formed by D-glucopyranosyl units joined primarily by $\alpha(1,4)$ bonds. The other component, amylopectin, is a branched structure where the backbone links are (1,4) bonds, and the crucial (1,6) bonds create the points of branching (Fig. 7.1) (BeMiller, 2025; Li et al., 2023; Falcão et al., 2022). As an abundant, renewably sourced polysaccharide, starch shows great potential. It can be formulated either to replace fossil fuel-based plastics or to be used as an additive in agriculture (Gamage et al, 2022a, 2022b). Starch possesses several distinct environmental safety features, including absorbency, biodegradability, biocompatibility, and nontoxicity (Gamage et al., 2022a, 2022b; Sharma & Bhende, 2024). Consequently, its utility extends far beyond agriculture to the food and biomedical industries. Furthermore, starch is widely utilized in sectors such as food, textiles, chemicals, pharmaceuticals, and paper manufacturing (Phukan & Nongkhlaw, 2022; Kumar et al., 2023). The need for starch is expanding into other industrial uses, such as packaging additives, adhesives, nonfood fillers, and textile stiffening agents (Zarski et al., 2021; García-Guzmán et al., 2022). Starch is traditionally sourced from edible plants such as potato, corn, rice, wheat, and barley (Kumar et al., 2023; Nath & Dutta, 2024). While starch serves as a staple food globally, providing an essential source of carbohydrates and aiding in hunger management, large-scale extraction from food crops poses a challenge: it can reduce food availability and worsen existing hunger issues (Hossain et al., 2021; Majzoobi et al., 2023). Consequently, developing industrial starches from nonedible sources is a commercially sound strategy to replace food-derived starch (Thuppahige et al., 2023). While starch's abundance, renewable nature, and environmental safety features—including biodegradability, biocompatibility, and nontoxicity—make it a highly attractive, versatile material for numerous industrial applications, its widespread utility is fundamentally limited by significant intrinsic drawbacks (Sharma & Bhende, 2024; Zarski et al., 2021). Specifically, starch-based materials typically exhibit poor mechanical strength (e.g., brittleness) and possess a high-moisture permeability, primarily because of their hydrophilic nature (Thakur et al., 2019; Yu et al., 2021). These deficiencies severely restrict its performance in applications requiring robust physical integrity and barrier properties, such as advanced packaging or structural components. Consequently, to unlock starch's full potential as a replacement for conventional materials.

We must use nanostarch materials such as starch nanocrystals (SNCs), starch nanoparticles (SNPs), starch nanofibers (SNF), starch nanomicelle (SNM), and starch nanovesicle (SNV) as potent reinforcing fillers (Adetunji et al., 2024; Chacon et al., 2024; Rayo-Mendez & Rodriguez-

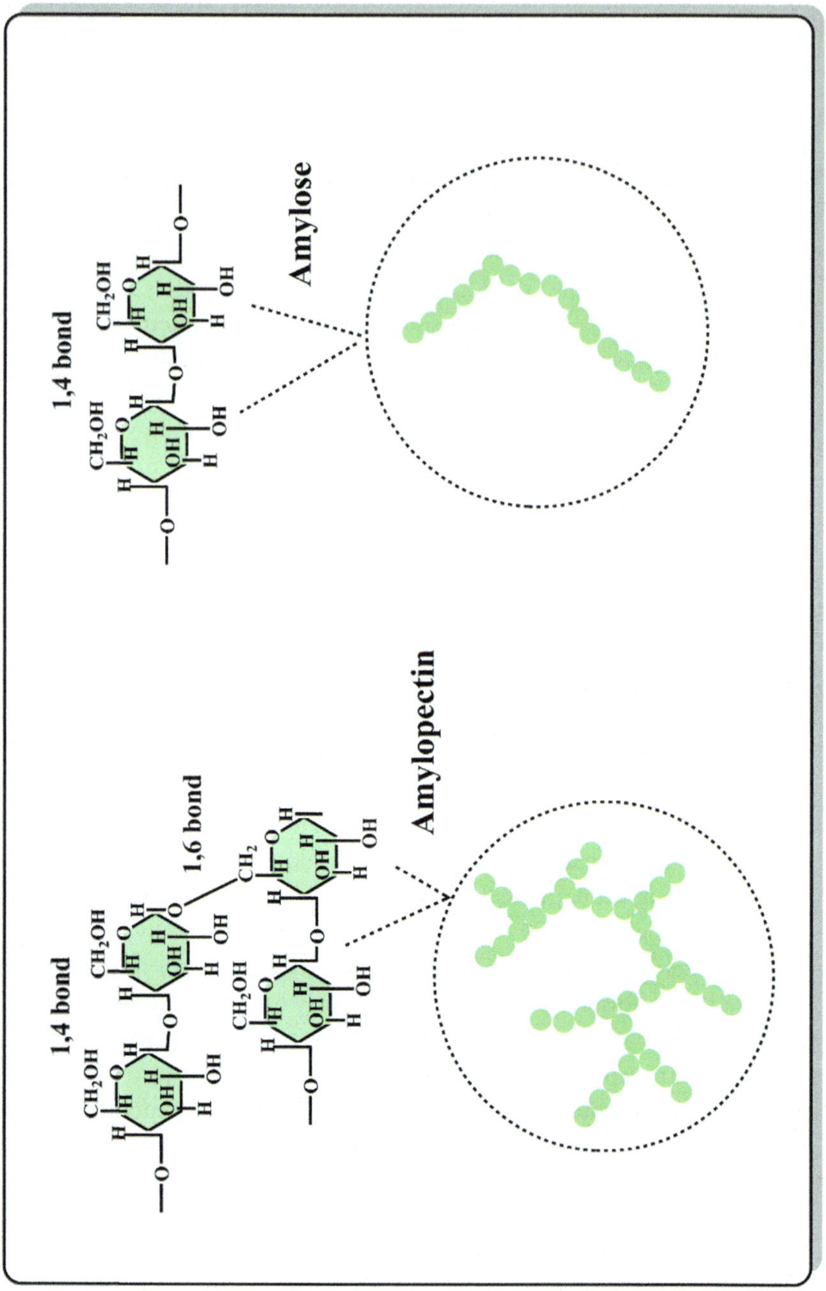

Figure 7.1 Chemical structure of starch.

Llanos, 2024). The nanoscale dimensions and high surface-area-to-volume ratio of these fillers enable the creation of highly efficient starch-based nanocomposites (Thomas et al., 2019). When properly dispersed, these rigid nanocrystalline structures establish a robust, interconnected network within the amorphous starch matrix, which is essential for effective load transfer and dramatically boosting the material's tensile strength and elastic modulus (Yu et al., 2021; Dufresne & Castaño, 2017). Furthermore, the well-packed, crystalline nanostarch acts as an impermeable physical barrier. These forces diffuse water molecules to navigate a tortuous pathway, significantly reducing the material's water vapor permeability, a crucial improvement that expands starch's potential for demanding applications, such as advanced food packaging and biomedical scaffolds (Adetunji et al., 2024; Hubbe et al., 2019). Therefore the transition from macroscale starch to nanoscale reinforcement is a commercially viable and scientifically validated pathway for transforming a low-performance biopolymer into a high-performance, sustainable material.

Starch must be chemically or physically modified; one of the most effective strategies involves reinforcement through the incorporation of various nanomaterials (Dufresne & Castaño, 2017). This approach is crucial for enhancing the material's structural stability and reducing its sensitivity to humidity, thus expanding its use beyond traditional sectors like food and textiles into high-performance engineering applications (Simões, 2024).

7.3 Starch nanostructures

The synthesis of starch nanostructures generally follows one of two approaches: top-down or bottom-up (Fig. 7.2) (Himat et al., 2025). The top-down strategy achieves nanosize by physically or chemically breaking down the larger starch granules. Common examples of this size-reduction process include acid hydrolysis, ultrasonication, homogenization, and gamma irradiation (Himat et al., 2025; Zhou et al., 2023). Conversely, the bottom-up method constructs the nanoscale particles by promoting the assembly of individual starch molecules under precise thermodynamic control, utilizing techniques such as nanoprecipitation and self-assembly

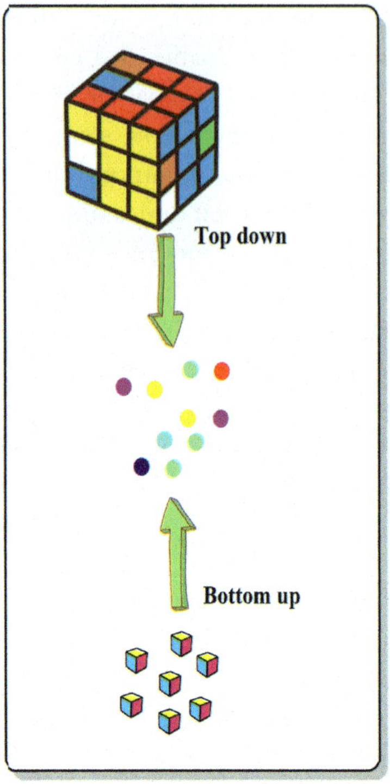

Figure 7.2 Bottom-up and top-down for the preparation of starch nanostructures.

(Himat et al., 2025; Morán et al., 2021). While top-down approaches are simple, cost-effective, and easily scaled for industrial applications, they prove less suitable for starch sources that are already extremely small or irregularly shaped (Jiang et al., 2022; Biswas et al., 2012). Bottom-up techniques, conversely, are typically more expensive but offer shorter production times and result in less structural damage to the starch material (Qin et al., 2024). Fig. 7.2 provides a visual representation illustrating the fundamental difference between these two approaches in starch nanostructures production.

Physical methods for manufacturing SNPs are a popular top-down approach, encompassing treatments such as milling, high-pressure homogenization, ultrasonication, and gamma irradiation (Ribeiro et al., 2025; Kodipyaka & Govindu, 2025). A key benefit of these techniques is their environmental friendliness and simplicity, as they eliminate the need for chemical agents or organic solvents, and generally require a relatively shorter processing time (Sreejit et al., 2025; Hassan et al., 2022). However, a major drawback of mechanical treatments is the high-energy input, which can compromise the structural integrity of the starch. This substantial energy often leads to a weakening and subsequent damage of the starch granule's crystalline structure, resulting in nanoparticles with lower crystallinity (Koekuyt & Marangoni, 2025; Nawaz et al., 2020). The process of milling, for example, achieves nanosize reduction through the constant movement and rotation of a grinding ball, which channels kinetic energy into the sample (Kumar et al., 2023; Yadav et al., 2012). This method typically yields larger starch nanostructures; for instance, corn starch processed by milling has been shown to produce nanoparticles in small size (Campelo et al., 2020). The effect on crystallinity distinguishes certain methods. Most physical processes cause structural damage, but gamma irradiation is an exception (Wi et al., 2007; Reches, 2019). Fragmentation during gamma irradiation is mediated by free radicals that specifically target the more accessible amorphous region of the starch granule, leaving the crystalline domains relatively intact (Kong, 2023; Atrous et al., 2015). Consequently, starch nanostructures produced by gamma irradiation exhibit comparatively high crystallinity. While free radicals are used to break starch chemical bonds, they are considered nonhazardous in this application as they are highly water-soluble and easily dissipate, ensuring the final purified starch nanostructures contain no radical components (Lamanna et al., 2013). Conversely, methods that attack the rigid crystalline region naturally result in starch nanostructures with lower crystallinity. Beyond physical approaches, other preparation routes include chemical methods, such as acid hydrolysis and nanoprecipitation, and enzymatic methods (Hassan et al., 2022; Campelo et al., 2020). Both acid and enzymatic hydrolysis share a similar fragmentation mechanism: they preferentially attack the amorphous regions first, and then proceed to degrade the more resistant crystalline regions at a significantly slower pace (Tsuji, 2022; Shekhar & Mondal, 2024). Nanoprecipitation utilizes a different strategy, first gelatinizing the starch and then using an antisolvent (e.g., ethanol) to cause precipitation, leading to the breakdown and subsequent reassembly of starch chains into smaller particles (Lu & Tian, 2021). Generally, these nanomaterials are categorized into five distinct forms: SNPs, SNCs, SNFs, SNMs, and SNVs (Fig. 7.3).

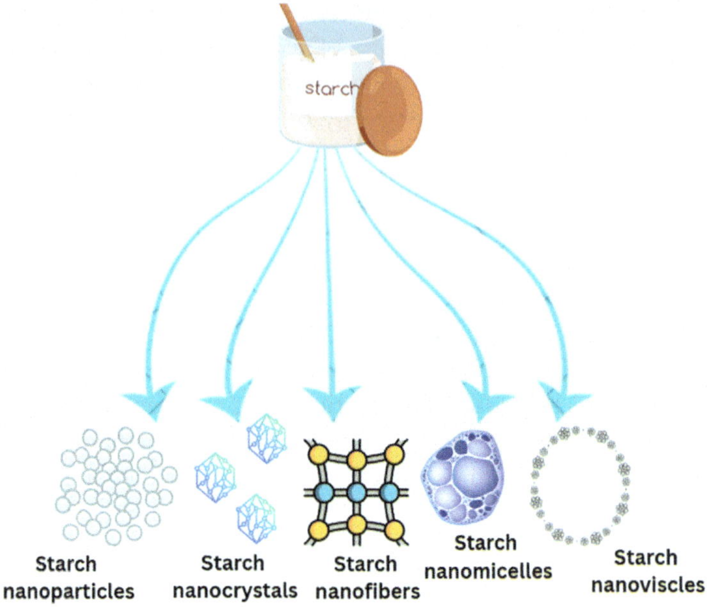

Figure 7.3 Different types of starch nanostructures.

7.3.1 Starch nanoparticles

SNPs are synthesized using both the top-down and bottom-up engineering techniques. A frequently employed method, often classified as a bottom-up approach, is the nanoprecipitation technique. This process works by first dissolving starch in hot water and then swiftly introducing this solution into ethanol, where the starch is insoluble. The sudden shift in solvent polarity triggers the starch chains to precipitate and reorganize themselves into nanoscale particles. Solvents like ethanol are critical as they act as precipitation agents, allowing for precise control over the final size and shape of the SNPs, making this method versatile and widely utilized in starch nanotechnology (Hernández-Giottonini et al., 2024; Chin et al., 2011). The synthesis of SNPs leads to substantial improvements over conventional starch properties, where it exhibits a massive increase in its specific surface area. This enhancement significantly boosts their capacity for both compound absorption and controlled release, making them versatile carriers (Chin et al., 2011).

7.3.2 Starch nanocrystals

The isolation of SNCs is fundamentally driven by the highly organized, semicrystalline structure of native starch granules, which feature alternating amorphous and crystalline regions (Pinto, 2024; Li et al., 2023). The amorphous regions are significantly more vulnerable to breakdown by chemical or enzymatic agents because the crystalline structure remains stable below the starch's gelatinization temperature (Chakraborty et al., 2022). Consequently, methods that selectively attack and remove the amorphous regions effectively release highly crystalline, nanosized platelets. These released platelets, typically ranging in size between 15 and 30 nm in width and 20–40 in width, are defined as SNCs (Lavoine et al., 2019). While waxy starch is often the preferred source for SNC production due to its structure, a wide variety of botanical sources have been successfully studied, including common starches like corn (maize), rice, potato, and wheat, as well as less common ones such as cassava, quinoa, amaranth, starchy seeds, and pines (Sharma et al., 2023; Altaf et al., 2022; Naz et al., 2019). The low recovery yield of SNCs following simple acid hydrolysis has strongly encouraged the search for improved methods (Harun et al., 2022). To address this, various pretreatments, including physical, chemical, or enzymatic modifications, applied to the native starch before isolation are now explored. These pretreatments are critical because they affect not only the final SNC recovery yield but also the intrinsic properties of the nanocrystals (Punia, 2020; BeMiller, 2018).

7.3.3 Starch nanofibers

The formation of SNFs relies on a "bottom-up" assembly technique, predominantly achieved through electrospinning (Alves et al., 2021; Guo et al., 2025). The inherent challenge with starch is its poor solubility, which makes preparing a suitable electrospun solution difficult (Liu et al., 2017). Two principal strategies have emerged to address the solubility hurdle: The first is the electro-wet-spinning setup, where fibers are formed on a grounded metal mesh submerged in a pure ethanol bath. Here, the starch granules are typically dissolved in aqueous dimethyl sulfoxide solution (Kong & Ziegler, 2014). The second approach utilizes a traditional electrospinning setup but prepares the starch by dispersing it in 90% (v/v) aqueous formic acid. The formic acid induces rapid esterification of the starch's glucose units, effectively achieving a "chemical gelatinization" at room temperature (da Cruz et al., 2024; Santos et al., 2024).

7.3.4 Starch nanomicelles

SNMs are tiny, clever delivery trucks made from starch (Rayo-Mendez & Rodriguez-Llanos, 2024). They're excellent at helping less water-friendly (low-polarity) compounds dissolve better. This is a big deal for both medicine and food, where SNMs are being explored as systems to deliver beneficial ingredients (Srinivasan et al., 2025). For instance, they are used to carry compounds such as β-carotene (Ali et al., 2022). Creating these microscopic helpers usually involves a "down-top" strategy. First, we need to modify them through processes like hydrolysis and esterification. This modification gives the long starch chains a dual personality: a hydrophilic part and a hydrophobic part. Once they have this amphiphilic nature, they can act as emulsifiers, naturally wrapping around the hydrophobic "cargo," forming the micelle core and keeping the whole system stable in water. The actual formation process relies on this modified starch's ability to emulsify a hydrophobic substance (Ali et al., 2022). Typically, the hydrophobic compound solution can be encapsulated within the amphiphilic starch in water (Chacon et al., 2024; Kou et al., 2020).

7.3.5 Starch nanovesicles

SNVs are tiny, protective bubbles made from starch (Rayo-Mendez & Rodriguez-Llanos, 2024; Sintov, 2020). SNVs are nanosized containers designed to encapsulate hydrophobic materials, ensuring they remain stable in watery environments, protecting them from degradation, and controlling their release rate (Rayo-Mendez & Rodriguez-Llanos, 2024). Starch is an ideal building block for these nanovesicles because it is highly available, safe, nontoxic, biocompatible, and biodegradable (Petrovic & Barbinta-Patrascu, 2023). While their most studied use is in medicine for drug delivery, they are also highly promising in agriculture (for carrying pesticides and nutrients) and, notably, in food (Rayo-Mendez & Rodriguez-Llanos, 2024; Singh et al., 2023; Wang & Kong, 2025).

The most common way to create SNVs involves a few steps: first, a nanoemulsion of the hydrophobic molecule is prepared (an oil-in-water mix). This nanoemulsion is then added to a gelatinized starch solution, and the whole mixture is dried. The result is oil-charged microcapsules (Petrovic & Barbinta-Patrascu, 2023; Wang et al., 2023; Shaker et al., 2019).

7.4 Nanomaterials for reinforcing and activating starch films

Starch films, which are appealing as biodegradable alternatives to traditional plastics, suffer from two main drawbacks: they are often brittle and do a poor job of blocking moisture and gases. The introduction of nanomaterials has become a powerful strategy to overcome these limitations, effectively reinforcing the films' structure and activating them with new capabilities (Musa et al., 2025; Jamwal & Mittal, 2024). When tiny fillers such as cellulose nanocrystals (CNCs), cellulose nanofibers, or certain inorganic metal oxide nanoparticles are added, they disperse throughout the starch matrix. This network of rigid, nanoscale particles drastically improves the film's mechanics, boosting its tensile strength and stiffness. Furthermore, these tiny additions force gas and water molecules to take a convoluted path to pass through, which significantly improves barrier properties against moisture and oxygen, which are necessary for preserving food quality and extending shelf life (Jagadeesh et al., 2021; Wawrzyńczak et al., 2024).

7.4.1 Cellulosic materials

Cellulose is the planet's most prolific and important construction material. This natural polysaccharide is what gives structure to everything from towering trees to tiny algae and certain bacteria. Chemically, it is a long, linear chain of sugar units (β-1,4-linked D-glucose), and these chains are held together incredibly tightly by strong internal hydrogen bonds, which is why cellulose is tough and has a characteristic crystalline structure (Strachota et al., 2024). In nature, these individual chains bundle together into microscopic protofibrils, which further organize into larger microfibrils, ultimately forming the recognizable cellulose fibers we see in wood or plant stalks (Tohamy et al., 2024). While the crystalline regions are neat and rigid, held fast by those powerful hydrogen bonds and Van der Waals forces, cellulose also has softer, amorphous regions where the chains are tangled or twisted. Because cellulose is so abundant which found everywhere from forest residue to agricultural waste (Tohamy et al., 2024; Tohamy, 2025c). The three most common forms are CNC, cellulose nanofibers (CNF), and bacterial nanocellulose (BC) (Fig. 7.4) (Rostamabadi et al., 2024). They can be used as a reinforcing agent to the pure starch films, which are often brittle and do a poor job of blocking moisture, which limits their usefulness in many applications (Zarski et al., 2021; Mohammed et al., 2024).

7.4.1.1 Cellulose nanocrystals

CNCs are a highly valuable and adaptable type of natural polymer, essentially the super strong, rod-like core of cellulose (Chen et al., 2025). CNCs can increase the strength and stiffness of starch films (Deng et al.,

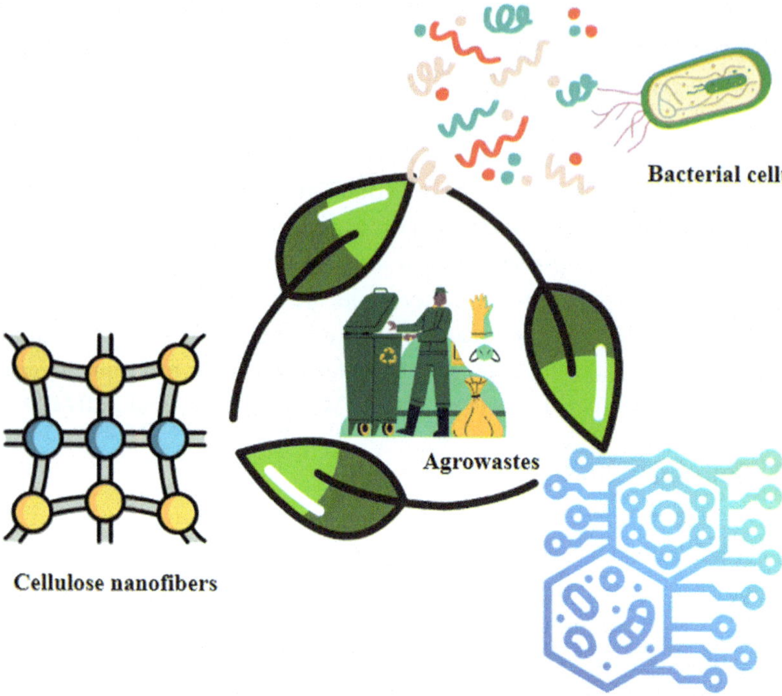

Figure 7.4 Types of cellulosic materials.

2022). Both materials are rich in hydroxyl (–OH) groups, which allows them to form strong hydrogen bonds with each other, creating a dense, highly interactive network. This is why adding a small amount of CNCs to a starch-based material (e.g., potato starch), the CNCs act like a rigid scaffold woven into the softer matrix, dramatically improving the film's mechanical strength and stiffness (Zhong et al., 2022; Zhang et al., 2024). For instance, films made from cornstarch and reinforced with CNCs and beeswax have been shown to significantly boost the film's ability to block unwanted oxygen and water vapor, which is essential for keeping food, such as fresh fruit, preserved longer (Hosseini et al., 2023).

7.4.1.2 Cellulose nanofibers

Cellulose nanofibers (CNFs) are the long, spaghetti-like strands of the material, which still contain both the tough, crystalline sections and the softer, amorphous parts (Hamouda & Abdel-Hamid, 2022). A study by Chen et al. showed that CNF with the highest aspect ratio significantly improved the mechanical and barrier properties of starch-based films (Chen et al., 2019). Similarly, Zhan et al. found that CNF provided better thermal stability to the starch films compared to CNC (Zhang et al., 2021).

7.4.1.3 *Bacterial cellulose*

Cellulose, commonly sourced from plants, can also be produced through enzymatic mechanisms, chemical synthesis from glucose compounds, or biosynthesis using algae, fungi, or bacteria (Mishra et al., 2022; Sohail et al., 2022). Bacterial cellulose (BC) is gaining significant interest as an environmentally friendly nano-biomaterial because of its unique properties. These include a distinct interconnected network structure and strong physical attributes, such as an exceptional ability to hold water, a large specific surface area, excellent chemical stability, inherent environmental friendliness, and remarkable mechanical strength (Sharma et al., 2023). BC is particularly notable because it shares the identical molecular and polymeric structure with plant cellulose but boasts higher crystallinity and is naturally free of lignin, hemicellulose, and pectin (Sohail et al., 2022; Cybulska et al., 2023). This inherent purity makes purification and production easier by eliminating the need for harsh chemical or high-energy treatments (Cybulska et al., 2023). In addition, microorganisms can be genetically modified to achieve desired BC properties, and agro-industrial wastes can be used as a growth medium to reduce costs (Kadier et al., 2021). Febra et al. proved that adding 15% bacterial cellulose to starch films significantly enhanced oxygen ($\approx -95\%$) and water vapor permeability ($\approx -46\%$) barrier properties, maintained transparency, and increased stiffness (Fabra et al., 2016).

7.4.2 Inorganic metal nanoparticles

Incorporating metal nanoparticles (MNPs) into polymeric nanocomposites has become a widespread practice for various applications, with metals such as silver and copper. Silver nanoparticles (AgNPs) are by far the most popular choice, particularly in starch matrices, because of their impressive properties and ease of synthesis (Gamage et al., 2021). In addition, zinc oxide nanoparticles (ZnO NPs) are highly favored for their biocompatibility, nontoxicity, and chemical stability when exposed to high temperatures and UV light (Mohamed et al., 2025; Islam et al., 2022). Its inclusion in starch-based packaging consistently delivers functional improvements. For instance, adding ZnO NPs to buckwheat starch films not only improved their physical strength and water barrier qualities but also conferred effective antimicrobial activity. Specifically, films with 3% ZnO NPs demonstrated antibacterial action against Listeria monocytogenes and successfully preserved the quality of fresh-cut mushrooms during storage (Kim & Song, 2018). Similarly, in another application, ZnO NPs enhanced the tensile strength of starch films and prolonged the shelf life of bread (Anugrahwidya et al., 2022). Copper oxide nanoparticles (CuO NPs) act as powerful multifunctional agents in packaging films, not just structural reinforcements. Researchers incorporating CuO NPs (0.05–2 wt%) into poly(butylene adipate-co-terephthalate) with thermoplastic starch blends via extrusion observed simultaneous improvements across the board, including higher mechanical properties, increased vapor barrier properties, and antimicrobial efficacy (Bumbudsanpharoke et al., 2023).

7.4.3 Carbon-based nanomaterials

The development of next-generation IntPack systems has been fundamentally transformed by the incorporation of carbon nanomaterials (CNMs), offering vital solutions to pressing challenges concerning food safety, preservation of quality, and extending shelf life (Tohamy, 2025a, 2025b). CNMs, such as graphene oxide (GO), carbon nanotubes (CNTs), Fullerenes, and carbon quantum dots (CQDs), are increasingly used to reinforce and add functionality to starch-based films, primarily due to their exceptional mechanical, barrier, and electrical properties (Fig. 7.5) (Singh et al., 2024; Krystyjan et al, 2022a, 2022b; Tohamy, 2025d). These nanoparticles do more than simply boost conventional functions, such as mechanical strength and barrier performance; they also enable intelligent functionalities, including real-time sensing, environmental monitoring, and the controlled release of beneficial active compounds (Tohamy, 2025a, 2025d, 2025e). The crucial role of CNMs in food packaging stems from their capacity to satisfy the increasing demand for both highly efficient and sustainable packaging solutions. The food sector faces continuous requirements to ensure product safety, prolong shelf life, and give consumers instant access to information about product quality (Tohamy, 2025f, 2025g). CNMs provide a modern approach by embedding these smart functions directly into the packaging structure. For example, films based on CDs can signal food freshness with visible color changes, and GO composites can boost moisture barrier properties and structural durability (Anand et al., 2021; Tseng et al., 2012). Moreover, using CNMs in biobased materials supports global sustainability initiatives by decreasing the industry's reliance on traditional plastics (Tohamy, 2024; Tohamy & Greener, 2024).

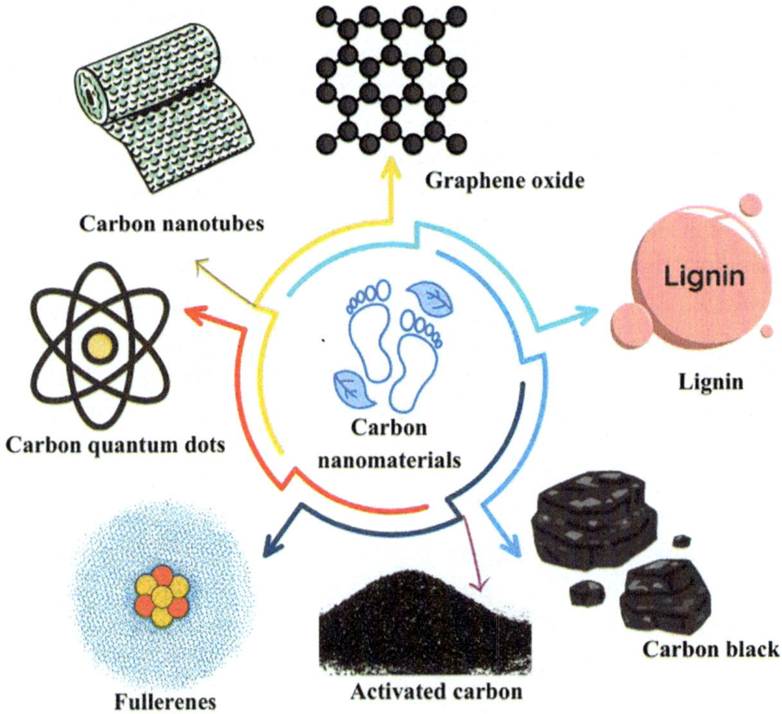

Figure 7.5 Types of carbon nanomaterials.

7.4.3.1 Graphene oxide

GO is a single atomic layer of carbon atoms with surface functional groups (hydroxyl, epoxy, and carboxyl) that enhance its dispersibility in water-based polymer matrices, such as starch (EL-Nashar et al., 2025; Tohamy et al., 2025; Bhadane & Mishra, 2024). The study by Zhang et al. successfully addressed the poor dispersion of CNC in a starch matrix by synthesizing a CNC-GO filler where GO self-assembled on silane-modified CNC. Characterization confirmed the successful modification and uniform dispersion of CNC-GO, leading to a stronger interfacial interaction with the starch matrix. Consequently, the resulting CS/CNC-GO nanocomposite films exhibited significantly enhanced properties compared to those using unmodified CNC. Specifically, the optimal films showed maximum values for tensile strength (53.96 MPa), elongation at break (3.72%), and transmittance (38.76%) at 5 wt% CNC-GO, and a minimum moisture rate (12.13%) at 3 wt% CNC-GO. These improvements demonstrate that the CNC-GO filler is highly effective in reinforcing the strength, improving the clarity, and decreasing the moisture sensitivity of starch-based materials (Zhang et al., 2022). The bionanocomposites developed by Krystyjan et al., utilizing a starch/chitosan matrix enriched with GO nanoparticles via green synthesis, demonstrate superior functional characteristics, particularly as thin films. These composites were selected for their better water vapor barrier properties compared to their thicker counterparts. Furthermore, the materials exhibit desirable sustainability features, showing relatively high biodegradability through susceptibility to enzymatic hydrolysis and acid hydrolysis. Crucially, the composites were confirmed to be nontoxic to blood-stream cells, with cytotoxicity studies showing no effect on cell viability or DNA integrity disruption. The most significant finding relates to their potent antimicrobial activity, showing a strong inhibition or significant slowing of growth for many microorganisms, especially Enterobacteriaceae, *Escherichia coli*, and *Campylobacter*, confirming that both chitosan and GO contribute to the bactericidal activity against Gram-negative and Gram-positive bacteria. Combined with good optical properties, these characteristics position the GO-enriched starch/chitosan bionanocomposites as highly promising, safe, and active elements for food packaging applications (Krystyjan et al, 2022a, 2022b). Ge et al. proposed that starch/GO films showed superior moisture resistance and mechanical strength, which is a significant breakthrough for starch-based packaging materials that are normally hampered by their high-moisture sensitivity (Ge et al., 2017). This methodology is pivotal, as it permits renewable starch resources to be utilized in packaging applications historically dominated by conventional, petroleum-based polymers.

7.4.3.2 Carbon nanotubes

CNTs, which are cylindrical carbon molecules (either single-walled, SWCNTs, or multiwalled, MWCNTs), are recognized as one of the strongest and stiffest materials known, making them excellent reinforcements (Tohamy & Greener, 2024). Because of their distinct physical structure, CNTs are extremely beneficial in food contact applications. Their fundamental structure, graphene sheets rolled into lightweight cylinders, imparts outstanding mechanical reinforcement (Vobornik et al., 2023). The use of starch nanocomposites featuring a synergistic blend of MWCNTs and BC significantly enhances film performance, achieving a 52% increase in TS and a 39% decrease in water vapor permeability (WVP). This finding, supported by related work from Daza-Orsini et al. on starch films reinforced with cellulose nanofibers, demonstrates that combining different nanomaterials is a highly effective strategy for transforming naturally abundant starch into multifunctional, high-performance packaging that addresses multiple material requirements simultaneously (Daza-Orsini et al., 2024).

7.4.3.3 Carbon quantum dots

CQDs are tiny nanoparticles (typically less than 10 nm) with unique optical properties, making them valuable for "smart" applications (Tohamy, 2024, 2025i). Afshar et al. introduced graphene quantum dots (GQDs), synthesized from the accessible and safe precursor citric acid via efficient carbonization, into starch and gelatin nano-biofilms as a viable, decomposable substitute for harmful plastic food packaging. Characterization methods (FT-IR, XRD, FE-SEM, EDX, AFM) confirmed the nano-biofilm formation. The GQDs showed excellent antioxidant (DPPH radical approach) and antimicrobial (disk diffusion method) performance. Incorporating GQDs enhanced the films' physical properties, including moisture contact, swelling index, and solubility. Furthermore, GQDs reduced film transparency, effectively lowering visible light transmission for improved food protection (Bakeshlouy Afshar et al., 2024). The study by Jafarzadeh et al. successfully developed biodegradable sago starch films with enhanced functional properties by incorporating CQDs. The research systematically confirmed that adding CQDs significantly improved the film's characteristics, such as UV-shielding ability, which was greatly enhanced, with the 4% CQDs content reducing UV transmittance across all bands. Furthermore, at a 3% loading, CQDs incorporation reduced moisture content (from 31.25% to 22.95%) and increased the water contact angle (from 37.1° to 67.7°), indicating improved surface hydrophobicity (Jafarzadeh et al., 2025). Yan et al. (2023) developed a fluorescent sensor utilizing CQDs specifically to identify volatile organic compounds (VOCs) linked to food decomposition. Their publication detailed the creation of a CQD-based fluorescent probe designed to detect changes in pH levels. Since pH is closely tied to microbial activity, this functionality allows the probe to serve as a practical indicator of imminent food spoilage (Yan et al., 2023).

7.4.3.4 Carbon black

Carbon black (CB) is a simple, affordable, and versatile carbon additive for starch-based packaging (Si & Zhang, 2024). CB materials are aggregates of nanoparticles of spheroidal shape, typically 10–100 nm in diameter, with an amorphous core surrounded by a shell of stacked graphene-like domains (Khodabakhshi et al., 2020). Although it is less powerful structurally than cutting-edge materials such as graphene or CNTs, it still effectively boosts the film's stiffness and tensile strength. Othman et al. focused on developing sustainable, active food packaging films by blending cassava starch with various concentrations of CB (1%–4% w/v). The study successfully demonstrated that the addition of CBN significantly enhanced the films' functional properties (Si & Zhang, 2024).

7.4.3.5 Activated carbon

Activated carbon (AC) is characterized by its exceptionally porous nature, which includes both micro- and mesopores with diameters typically spanning 2–50 nm. Synthesis usually involves pyrolyzing biomass, followed by a necessary alkali or acid activation step. This extensive porosity makes AC highly effective for diverse applications, such as scavenging ethylene in fruit packaging, managing humidity, and removing undesirable off-flavors (Ramezani et al., 2025). For instance, Kwaśniewska et al. utilized powdered activated carbon (PAC) (1%–15%) to create starch biocomposites, finding that increasing PAC content significantly changed the film color from transparent to opaque black. Although PAC addition decreased overall mechanical strength, the highest concentrations (up to 15%) improved Young's modulus and hardness. Crucially, PAC increased water vapor permeability but conferred hydrophobic properties at moderate concentrations (2%–5%), making these nontoxic, biodegradable films a promising, cheaper alternative for agricultural packaging applications (Kwaśniewska et al., 2021).

7.4.3.6 Fullerenes

Fullerenes are a unique allotrope of carbon, characterized by a complex molecular architecture. Their structure involves carbon atoms bonded together in either a fully closed or partially enclosed network. The best-known example is the C60 molecule, referred to as buckminsterfullerene, which consists of 60 carbon atoms precisely configured in its characteristic soccer ball-like arrangement (Tohamy, 2025d). While fullerenes have been explored across various fields, their practical use in industrial fluid processing remains an emerging research domain. Contemporary investigations have focused on embedding fullerenes within polymer materials to successfully bolster both their barrier characteristics and their mechanical strength (Ghosh et al., 2024).

7.4.3.7 Lignin nanoparticles

Lignin is a vast, complicated, carbon-based polymer, making it the second most abundant natural polymer globally after cellulose. It is the matrix within the cell walls of plants; it is responsible for giving wood and bark their crucial rigidity, mechanical strength, and resistance to compression (Ramesh et al., 2022). Lignin nanoparticles (LNPs) are colloidal particles derived from lignin. LNPs are typically defined as particles of lignin with an average size in the nanometer range (Moreno & Sipponen, 2024). Sun et al. confirmed that developing starch-LNPs composite films yielded a sustainable material with significantly improved properties for food packaging. The optimal composite film showed a remarkable increase, with TS rising 3.04 times to 48.9 MPa and the elastic modulus increasing 2.93 times to 2288.8 MPa, alongside a 5–8 °C improvement in thermal stability (Sun et al., 2023).

7.4.4 Nanoclays

Nanoclays serve as a crucial example of nanomaterials used to dramatically improve the performance of starch-based films, functioning both as a reinforcing agent for mechanical and barrier properties and an activating agent for novel functionalities (Gamage et al, 2022a, 2022b; Muñoz-Gimena et al., 2023). The primary reinforcing mechanism involves the high-aspect-ratio, nanometer-thick clay platelets, such as montmorillonite (MMT). When properly dispersed and exfoliated within the hydrophilic starch matrix, these rigid platelets interact strongly with the starch's hydroxyl (-OH) groups, primarily through hydrogen bonding. This strong interfacial adhesion restricts the movement of the flexible polymer chains, resulting in a significant increase in the film's tensile strength and YM (stiffness), which effectively overcomes the natural brittleness of starch (Wang et al., 2013). Furthermore, the layered structure of the dispersed nanoclays creates a tortuous path within the film (Tabatabaei & Ajji, 2011). Gas molecules (like oxygen) and water vapor must navigate this winding route around the impermeable silicate layers, which dramatically lengthens their diffusion time and subsequently reduces the film's permeability, making the starch a much more effective moisture and oxygen barrier for packaging applications (Schuchardt, 2024; Uddin et al., 2024). Beyond reinforcement, nanoclays act as an activating agent by enabling the films to become active or intelligent (Bandyopadhyay & Ray, 2019). A notable way this is achieved is by using nanoclays as nanocarriers for active compounds. For instance, halloysite nanotubes (HNTs), which possess a unique hollow tubular structure, can be loaded with antimicrobial or antioxidant agents. When incorporated into the starch film, these loaded nanoclays allow for the controlled and sustained release of the active agents. This delayed release extends the film's functional lifespan, allowing it to actively inhibit microbial growth or oxidative decay of the packaged food product, a property not intrinsic to the pure starch film itself (Lvov et al., 2016; Venkatesh et al., 2026). Thus, the introduction of a low concentration of nanoclays transforms the basic, weak, and hydrophilic biopolymer film into a high-performance, multifunctional nanocomposite material (Rhim & Ng, 2007).

7.5 The smart functions of starch nanocomposite

Starch-based materials are increasingly recognized as a sustainable and biodegradable foundation for the next generation of food packaging, offering a compelling alternative to nonrenewable plastics. Derived from an abundant and low-cost natural resource, starch can be easily processed into films and coatings (Falua et al., 2022). Smart packaging represents a revolution in food safety and quality control, moving beyond the simple functions of preservation and mechanical protection offered by traditional packaging (Drago et al., 2020). Instead, it incorporates advanced features to create IntPack systems capable of active monitoring. These systems can provide real-time, easily interpretable signals about the condition of the packaged food, including its quality, safety, and exposure to undesirable temperatures or temperature abuse. This is achieved by embedding specialized components, such as indicators and sensors, directly into the packaging material (Tohamy, 2025b, 2025j). A key material facilitating this shift is the starch nanocomposite, which offers a sustainable, biodegradable platform to integrate cutting-edge nanotechnology (Tohamy, 2025a, 2025b). For example, Tohamy et al. conducted a study focusing on the creation and characterization of a novel nitrogen-doped carbon dots-cellulose sulfate-carboxymethyl cellulose (N–CDs–CS–CMC) composite film, engineered for multifunctional use in tomato packaging and sensing. It exhibited potent antimicrobial activity against common foodborne pathogens, including *Escherichia coli*, *Staphylococcus aureus*, and *Candida albicans*. Furthermore, the film enabled pathogen-specific detection by displaying distinct fluorescence patterns upon interaction with bacteria. By incorporating beetroot during the N–CDs preparation, the film was rendered pH-responsive, showing clear color transitions from a brown/reddish hue in alkaline environments to yellow in acidic conditions. Crucially for food preservation, tomatoes wrapped in the N–CDs–CS–CMC film achieved a 10-day shelf life, a significant improvement over the 4 days recorded for the CS–CMC control.

The film also functioned as an environmental sensor, undergoing rapid decolorization to yellow and causing increased tomato translucency upon exposure to chromium, indicating its potential for heavy metal detection (Tohamy, 2025k). These "smart" functions allow the film to actively monitor the condition of the packaged food, providing real-time, easily interpretable signals about quality, safety, and temperature abuse (Tohamy, 2025d, 2025f, 2025g, 2025l). Table 7.1 shows some examples of starch-based films for food packaging applications.

7.5.1 Colorimetric sensors

Colorimetric sensor technology offers one of the simplest and most effective ways to signal when food has spoiled, achieving this through a clear, visually obvious color change right on the packaging film (Mazur et al., 2024). The process relies on the chemistry of decay. When food, especially items such as meat and fish, begins to spoil due to microbial growth, it releases volatile basic compounds such as ammonia and biogenic amines. These compounds accumulate in the package's enclosed air space, causing the local pH level to change (Tohamy, 2025a, 2025b, 2025e, 2025j, 2025k). To function as a sensor, the packaging film is engineered by incorporating pH-sensitive materials (typically specialized dyes or pigments) into its matrix. When the embedded indicator detects this increase in alkaline spoilage compounds, it undergoes a chromatic shift, or color change. This immediate, visual signal serves as an early warning, instantly alerting the consumer or retailer that the product's freshness is compromised, often long before any traditional signs of decay become apparent (Luo et al., 2022; Zhang et al., 2024; Marappan et al., 2025). In the study by Zhang et al., they aimed to create an IntPack material that could enhance food preservation and offer greater convenience to shoppers. They developed films primarily from a blend of corn starch and polyvinyl alcohol (PVA). To assess the material's ability to act as a food freshness indicator, they incorporated two readily available sources of anthocyanins: purple sweet potato extracts (PSPE) and red cabbage extracts (RCE). The resulting films, when embedded with either PSPE or RCE, demonstrated distinctive color shifts when exposed to various pH buffers. The addition of RCE, or a lower concentration of PSPE, did not significantly affect the WVP of the films ($P > .05$). Conversely, as the amount of either extract was increased, the films exhibited a marked improvement ($P < .05$) in several physical characteristics: the thickness grew (increasing from 64.0 to 97.7 μm with PSPE or

Table 7.1 Starch composites in food packaging applications.

Composition	Results	References
Starch/Chitosan composite sheets with an evenly distributed GO (0.01 wt% GO concentration) and carboxymethyl cellulose.	✓ low water vapor permeability and high water resistance. ✓ sustained at 92% under 50% visible light. ✓ tensile characteristics improved by 91.7% (at 0.01 wt% GO) compared to the control. ✓ increased by 23.8% against Escherichia coli and 25.6% against *Staphylococcus aureus*. ✓ Suggested to be effective in extending the freshness of tomatoes.	Wu et al. (2023)
Cellulose nanofibers with starch.	Enhanced mechanical and high water resistance.	Tibolla et al. (2020)
Banana starch, aloe vera gels, and chitosan.	High water resistance and tensile strength.	Pinzon et al. (2018)
Starch/PVA/purple sweet potato extract films.	Offering antibacterial and colorimetric indication properties.	Wang et al. (2018)
Anthocyanins/sweet potato starch/PVA films.	For real-time shrimp freshness assessment by color sensing ability.	Mahin et al. (2025)
Starch/MWCNTs/BC.	Showed up to a 52% increase in TS and a 39% decrease in WVP.	Daza-Orsini et al. (2024)

85.5 μm with RCE), the mechanical properties strengthened (TS increased from 7.3 to 11.3 MPa or 9.1 MPa, and elongation at break (EB) surged from 92% to 249% or 284% with PSPE or RCE, respectively), and the thermal stability improved. Simultaneously, the film's ability to transmit light was substantially reduced. Notably, at lower concentrations of the extracts, the PS-PSPE film displayed a more vibrant color, superior mechanical performance, and lower light transmission compared to the PS-RCE film, a difference attributed to the varying composition and concentration of anthocyanins in the extracts. The PS-PSPE film, due to its rich anthocyanin content and excellent physical traits, was selected for a practical application: monitoring shrimp freshness. During the spoilage process, the film successfully showed visible color changes that corresponded directly to variations in the concentration of total volatile basic nitrogen (TVB-N), confirming its high potential as a visual indicator for quality monitoring. In essence, the researchers successfully produced a pH-sensitive film by integrating PVA and starch with different amounts of PSPE or RCE. Although this pH-sensitive film shows great promise for ensuring food safety and quality by indicating freshness, the authors acknowledge limitations, such as the anthocyanin's instability, which leads to easy degradation. Therefore further research is essential to overcome these hurdles and facilitate the large-scale production of this innovative packaging material (Zhang et al., 2020). The study by Wu et al. developed a sophisticated pH-sensitive smart packaging film designed to visually signal food spoilage, focusing specifically on milk freshness. The film was created using cassava starch and PVA as the base materials, with Aronia melanocarpa anthocyanins incorporated to act as the color-changing indicator. It was found that a starch/PVA ratio of 3/7 resulted in the best mechanical strength, with a TS of 23.07 MPa and a remarkable elasticity (EAB of 291.22%). Interestingly, these AMA-infused films proved to be highly effective pH sensors, showing clear color changes across a pH range of 2–11. Crucially, they also offered better protection for the packaged food, significantly improving the film's barrier properties against UV light and water vapor. When tested with milk, the films successfully changed color as the milk spoiled, visually indicating the loss of freshness (Wu et al., 2024). Torch et al. introduced a way to monitor food freshness in real-time using a pH-sensitive film that changes color. Their novel approach utilized onion peel powder (OPP), a cheap, polyphenol-rich by-product from agriculture, to create intelligent film labels based on cassava starch. They found that increasing the OPP content significantly boosted the films' total phenolic and flavonoid levels, leading to enhanced antioxidant properties ($P < .0001$). The films also showed a selective antibacterial effect, notably forming clear inhibition zones against *P. aeruginosa* and *E. coli*, two bacteria often associated with food spoilage.

Functionally, the films demonstrated strong pH-sensitivity, displaying a distinct and gradual color change from light to dark brown across the entire pH range of 1–13. This confirmed their suitability as pH indicators. When applied to minced beef packaged and stored at 4 °C, the OPP films successfully tracked freshness over 13 days. The films' color shifts were tightly correlated with both the increasing microbial load and the pH variations as the meat spoiled, effectively signaling the onset of deterioration (Torche et al., 2025). Furthermore, Zhang et al. specifically addressed the creation of IntPack by incorporating anthocyanins extracted from purple sweet potatoes into a carboxymethyl cellulose/starch film matrix. This composite was engineered to serve as a real-time freshness monitor for raw grass carp stored at 20 °C. The key functionality relies on the anthocyanins' natural ability to act as a pH indicator. As the fish degrades, volatile amines accumulate, increasing the packaging's surface pH. This change is visibly signaled by the film undergoing a clear color transition, shifting sequentially from red to blue and then green, thus providing a simple, on-the-spot indication of the product's quality and spoilage level (Zheng et al., 2019). Further advancing this technology, Medina-Jaramillo et al. developed cassava starch-glycerol films enhanced with both green tea and basil extracts. This was a significant step as it showed how readily available natural ingredients could provide the dual benefit needed for smart packaging. In these films, the naturally present chlorophyll and carotenoids were utilized for their pH-responsive color changes. Since food spoilage often results in a change in pH due to the release of volatile compounds, the color transition acts as a simple, visual food quality assessment mechanism (Medina-Jaramillo et al., 2017).

7.5.2 Fluorescent gas sensors

An IntPack film utilizing starch fluorescent sensors is a clever, next-generation food safety tool. It is essentially a special, often edible and biodegradable, wrapper or label made primarily from starch that has been combined with a fluorescent probe. Instead of just changing a visible color, this sensor monitors the food's condition by changing the light it emits when viewed under a UV or specialized light source. This shift in is triggered by microscopic changes, like the presence of spoilage gases or a drop-in oxygen levels inside the package. The sensor provides a highly sensitive, real-time signal about the food's true freshness (by turning on/off the fluorescence), offering a more precise and advanced warning than traditional visual indicators (Song et al., 2022; Idumah et al., 2020; Sharma et al., 2025). The study by Kwaśniewska et al. proposed an exciting way to make simple starch films glow in a unique pattern, opening doors for advanced smart packaging. Their goal was to enhance the light-emitting (fluorescent) properties of the film by embedding a powder called NTA (5-(4-nitrophenyl)-1,3,4-thiadiazol-2-amine). The most compelling discovery involved the NTA's fluorescence. When NTA was dissolved in a liquid, it emitted just one color of light. But when the molecules were trapped and clustered inside the solid starch film, they suddenly emitted two distinct colors of light, a phenomenon called dual fluorescence. It was proven that NTA is a powerful addition to biodegradable starch films. It successfully gives them a sophisticated new light-emitting function (AIE) without sacrificing the film's practical surface qualities, making it a highly valuable component for the next generation of smart packaging materials (Kwaśniewska et al., 2024). Panyong et al. focused on creating a practical and eco-friendly fluorescent sensor to track two substances vital to health and the environment: nitrite (NO_2^-) and sulfite (SO_3^{2-}). They achieved this by synthesizing carbon dots (CDs) using only starch as the raw material, establishing a "green" alternative to traditional probes. This starch-based sensor is highly clever because it displays an easily recognizable "on-off-on" fluorescence response. Initially, the strong natural light emitted by the CDs is rapidly quenched (turned "off") when nitrite is introduced, which happens due to a direct charge transfer. Crucially, if sulfite is then added, the quenched fluorescence is fully recovered (turned "on" again) because the sulfite chemically reacts with the nitrite and reverses the process. This specific sequence allows for the convenient, highly sensitive, and selective detection of both ions. Furthermore, because these starch-derived CDs are nontoxic and compatible with living matter, the sensor was successfully used for intracellular bioimaging, allowing scientists to visualize nitrite and sulfite inside microorganisms' living cells in food (Wang et al., 2021).

7.5.3 Temperature sensors

Starch is a brilliant, natural, and biodegradable base material that's being used to create smart, high-tech temperature labels for food packaging. Instead of acting as the thermometer itself, the starch film serves as a natural, edible platform—like a simple scaffold—where scientists embed tiny, highly responsive indicators. These indicators, which might be fluorescent chemicals or specialized color-changing pigments, are designed to react to temperature changes. For instance, the film might contain a fluorescent dye whose glow dims or brightens in a predictable, measurable way as the temperature rises or falls, providing a real-time signal of thermal abuse. Alternatively, the film might be a time-temperature indicator (TTI) that undergoes a visible, irreversible color change if the food is exposed to unsafe temperatures for too long, alerting shoppers and handlers that the cold chain has been broken. This allows the packaging itself to instantly and safely tell us the true thermal history of the food inside.

7.6 Challenges of starch-based films

Despite the significant promise of starch as a sustainable and abundant resource for biodegradable packaging, its pure form presents substantial limitations that restrict commercial application. The most pronounced drawback is the inherently poor barrier property of starch, particularly its high water vapor permeability (Thakur et al., 2019). As a hydrophilic material, pure starch films readily interact with moisture, leading to rapid water exchange that compromises their barrier function. This high WVP limits their use primarily to packaging low-moisture or dry products, as they are ineffective at preserving the quality of high-moisture foods, which can quickly dry out, or protecting dry foods from becoming soggy (Thakur et al., 2019; Basiak et al., 2018). Furthermore, the strong affinity for water also makes the films mechanically unstable in high-humidity environments; they can become excessively soft, sticky, and lose structural integrity, complicating processing, handling, and transportation (Suh et al., 2020). Another limitation lies in the mechanical performance of pure starch films. While research focuses on achieving high TS and elastic modulus, pure films often lack the necessary toughness and flexibility, exhibiting low EB and poor tear resistance. This means they are prone to cracking or breaking under strain during high-speed packaging operations or normal consumer handling, issues that severely hamper their commercial viability (Thakur et al., 2019). Finally, although often overlooked in early development, achieving optimal characteristics such as impact strength and consistently low oxygen permeability, critical for preventing oxidation and maintaining product freshness, remains a consistent challenge that necessitates the incorporation of complex, multicomponent fillers and polymer blending strategies. Thus while modifications successfully mitigate many of these issues, the limitations of pure starch films mean they cannot be used in most conventional packaging applications without significant, costly compounding (Karnwal et al., 2025). The transition from conventional plastics to sustainable starch-based films faces significant commercialization hurdles, primarily centered on processing and cost-effectiveness. Plastics have become dominant due to their lightweight nature, cost-effectiveness, and exceptional suitability for large-scale commercial processing, particularly through highly efficient extrusion technologies. Starch films, however, struggle to compete (Karnwal et al., 2025; Prabha et al., 2021).

7.7 Future outlook

The first thing we worry about is still how safe and nontoxic these sensors are. When we think about it, we are dealing with starch, which is a complex web of carbohydrates that makes up stuff we eat every day, e.g., bread, rice, and potatoes. Because of this specific food environment, we have to be absolutely sure that the sensor's parts (both the nanoparticles and the material that holds them) would not have any nasty reactions with the starchy food matrix. This means we cannot just do general toxicity testing; we need to develop specific toxicological profiles that account for the environment created by the starch and the stuff it breaks down into. Basically, we have to guarantee that any tiny bits left behind would not hurt consumers over time. The whole point is to design materials that just sit there and do nothing, completely inert inside this very particular chemical context. That challenge leads right into the big headache: migration. A sensor designed for starch has to get up close and personal with the food to check its quality or freshness. By nature, that means the risk of components leaching out goes way up. We need a perfect picture of how those active ingredients, such as metal oxides or carbon nanomaterials, behave when incorporated into the starch composite. Will the signaling elements flake off? We have to design super-tough, self-contained sensor structures that securely lock the active nano-components away from the food while still letting them do their job. It's about building that perfect, nonleaking barrier. Finally, scaling up production is our future outlook. Moving something that works great in a science lab into a billion units for food packaging is a huge step. How do we make millions of these tiny starch sensors cheaply, quickly, and sustainably? The manufacturing process itself must be solid enough to embed the nano-components perfectly every time to stopping the migration problem. To pull this off, we have to find nontoxic, affordable raw materials that stick flawlessly to the packaging and set up quality checks that can quickly verify the safety and function of every sensor made.

Acknowledgments

The author appreciates the National Research Centre and the PRIMA International Project, between the Academy of Scientific Research and Technology (ASRT) and PRIMA, through the Joint Agreement project "Sustainable Antimicrobial Packaging based on a Healthy Intelligent Renewable Approach" with acronym "SAPHIRA".

Author contribution

Hebat-Allah S. Tohamy: Writing – review & editing, Writing – original draft, Visualization, Validation, Supervision, Software, Resources, Project administration, Investigation, Formal analysis, Data curation, Conceptualization.

Data availability

All data is presented in the figures, graphics, and tables in the context of the manuscript.

References

Adetunji, C. O., et al. (2024). Morphology, rheology, properties, and applications of nanostarch-filled polymer blends. In *Nanofillers for binary polymer blends* (pp. 443–464). Elsevier.

Adigwe, O. P., Egharevba, H. O., & Emeje, M. O. (2022). Starch: A veritable natural polymer. *Starch: Evolution and recent advances.*

Ali, A., et al. (2022). Effect of Co-Encapsulated natural antioxidants with modified StarVersatile application of calixarenes and their derivatives: From drug delivery to industrial catalysis and environmental remediationch on the oxidative stability of β-Carotene loaded within nanoemulsions. *Applied Sciences, 12*(3), 1070.

Altaf, A., et al. (2022). A comprehensive review of polysaccharide-based bionanocomposites for food packaging applications. *Discover Food, 2*(1), 10.

Alves, M. J.d. S., et al. (2021). Food applications of starch nanomaterials: A review. *Starch - Stärke, 73*(11–12), Article 2100046.

Anand, A., et al. (2021). Enhanced barrier, mechanical and viscoelastic properties of graphene oxide embedded glass fibre/epoxy composite for marine applications. *Construction and Building Materials, 268*, Article 121784.

Andrady, A.L., Plastics and the Environment.

Anugrahwidya, R., Armynah, B., & Tahir, D. (2022). Composites bioplastic film for various concentration of zinc oxide (ZnO) nanocrystals towards physical properties for high biodegradability in soil and seawater. *Journal of Polymers and the Environment, 30*, 2589–2601.

Arruda, T. R., et al. (2025). An overview of starch-based materials for sustainable food packaging: Recent advances, limitations, and perspectives. *Macromol, 5*(2), 19.

Atrous, H., et al. (2015). Effect of γ-radiation on free radicals formation, structural changes and functional properties of wheat starch. *International Journal of Biological Macromolecules, 80*, 64–76.

Bakeshlouy Afshar, M., Poursattar Marjani, A., & Gozali Balkanloo, P. (2024). Introducing graphene quantum dots in decomposable wheat starch-gelatin based nano-biofilms. *Scientific Reports, 14*(1), 2069.

Bandyopadhyay, J., & Ray, S. S. (2019). Are nanoclay-containing polymer composites safe for food packaging applications?—An overview. *Journal of Applied Polymer Science, 136* (12), Article 47214.
Basiak, E., Lenart, A., & Debeaufort, F. (2018). How glycerol and water contents affect the structural and functional properties of starch-based edible films. *Polymers, 10*(4), 412.
BeMiller, J. N. (2018). Chapter 5 - Physical Modification of Starch. In M. Sjöö, & L. Nilsson (Eds.), *Starch in food* (2nd ed., pp. 223–253). Woodhead Publishing.
BeMiller, J. N. (2025). Overview of food carbohydrates. In *Carbohydrate nutrition* (pp. 1–24). Elsevier.
Bertoft, E. (2017). Understanding starch structure: Recent progress. *Agronomy, 7*(3), 56.
Bhadane, P., & Mishra, A. (2024). Versatile, flexible rice starch–graphene oxide bionanocomposites. *Environmental Science: Water Research & Technology, 10*(10), 2432–2441.
Biswas, A., et al. (2012). Advances in top–down and bottom–up surface nanofabrication: Techniques, applications & future prospects. *Advances in Colloid and Interface Science, 170*(1–2), 2–27.
Bopp, A. F. (2019). The evolution of food preservation and packaging. In *Chemistry's role in food production and sustainability: Past and present* (pp. 211–228). ACS Publications.
Bumbudsanpharoke, N., et al. (2023). Enhanced properties of PBAT/TPS biopolymer blend with CuO nanoparticles for promising active packaging. *Food Packaging and Shelf Life, 37*, Article 101072.
Campelo, P. H., Sant'Ana, A. S., & Clerici, M. T. P. S. (2020). Starch nanoparticles: Production methods, structure, and properties for food applications. *Current Opinion in Food Science, 33*, 136–140.
Chacon, W. D. C., et al. (2024). A brief introduction to starch-based nanomaterials. In G. Ayala Valencia (Ed.), *Starch nanomaterials and food applications* (pp. 1–13). Cham: Springer Nature Switzerland.
Chakraborty, I., et al. (2022). An Insight into the gelatinization properties influencing the modified starches used in food industry: A review. *Food and Bioprocess Technology, 15* (6), 1195–1223.
Chen, L., et al. (2025). Cellulose nanocomposites by supramolecular chemistry engineering. *Nature Reviews Materials, 10*(10), 728–749.
Chen, Q., Liu, Y., & Chen, G. (2019). A comparative study on the starch-based biocomposite films reinforced by nanocellulose prepared from different non-wood fibers. *Cellulose, 26*(4), 2425–2435.
Chin, S. F., Pang, S. C., & Tay, S. H. (2011). Size controlled synthesis of starch nanoparticles by a simple nanoprecipitation method. *Carbohydrate Polymers, 86*(4), 1817–1819.
Compart, J., Li, X., & Fettke, J. (2021). Starch-A complex and undeciphered biopolymer. *Journal of Plant Physiology, 258*, Article 153389.
Cybulska, J., et al. (2023). Influence of pectin and hemicelluloses on physical properties of bacterial cellulose. *Food Chemistry, 429*, Article 136996.
da Cruz, E. P., et al. (2024). Impact of concentration and aging time of pea starch-based polymeric solutions on the fabrication of electrospun nanofibers. *Polymers for Advanced Technologies, 35*(6), Article e6479.
Dahiya, A., et al. (2024). Plastic pollution is a serious menace to ecosystem health with special reference to aquatic ecosystems and its associated challenges, opportunities, and mitigations. *Aquatic Ecosystems Monitoring*, 279–294.
Daza-Orsini, S. M., et al. (2024). Isolation of taro peel cellulose nanofibers and its application in improving functional properties of taro starch nanocomposites films. *International Journal of Biological Macromolecules, 273*, Article 132951.
Deng, Z., et al. (2022). Preparation, characterization and antibacterial property analysis of cellulose nanocrystals (CNC) and Chitosan nanoparticles fine-tuned starch film. *Molecules, 27*(23), 8542.

Drago, E., et al. (2020). Innovations in smart packaging concepts for food: An extensive review. *Foods, 9*(11), 1628.
Dufresne, A., & Castaño, J. (2017). Polysaccharide nanomaterial reinforced starch nanocomposites: A review. *Starch-Stärke, 69*(1–2), Article 1500307.
EL-Nashar, D., Tohamy, H. A. S., & Koriem, A. (2025). Improving acrylonitrile rubber properties by recycled nano graphene oxide prepared from sugarcane bagasse. *Polymer Composites, 46*(10), 9553–9563.
Fabra, M. J., et al. (2016). Improving the barrier properties of thermoplastic corn starch-based films containing bacterial cellulose nanowhiskers by means of PHA electrospun coatings of interest in food packaging. *Food Hydrocolloids, 61*, 261–268.
Falcão, L.d. S., et al. (2022). Starch as a matrix for incorporation and release of bioactive compounds: Fundamentals and applications. *Polymers, 14*(12), 2361.
Falua, K. J., et al. (2022). Valorization of starch to biobased materials: A review. *Polymers, 14* (11), 2215.
Gamage, A., et al. (2022a). Applications of starch biopolymers for a sustainable modern agriculture. *Sustainability, 14*(10), 6085.
Gamage, A., et al. (2022b). Environmental properties and applications of biodegradable starch-based nanocomposites. *Polymers, 14*(21), 4578.
Gamage, A., Punniamoorthy, T., & Madhujith, T. (2021). Starch-based hybrid nanomaterials for environmental remediation. In *Starch-Evolution and recent advances*. IntechOpen.
García-Guzmán, L., et al. (2022). Progress in starch-based materials for food packaging applications. *Polysaccharides, 3*(1), 136–177.
Ge, X., et al. (2017). Improved mechanical and barrier properties of starch film with reduced graphene oxide modified by SDBS. *Journal of Applied Polymer Science, 134*(22).
George, J., et al. (2022). An overview of higher barrier packaging using nanoadditives. *Nanotechnology-Enhanced Food Packaging*, 235–264.
Ghosh, T., et al. (2024). In I. Packaging, S. P. Bangar, & M. Trif (Eds.), *Chapter 4 - Intelligent packaging—sensors* (pp. 101–126). Academic Press.
Giuberti, G., et al. (2014). Factors affecting starch utilization in large animal food production system: A review. *Starch-Stärke, 66*(1–2), 72–90.
Guo, Y., et al. (2025). Biopolymer based fibrous aggregate materials for diagnosis and treatment: Design, manufacturing, and applications. *Advanced Materials, 37*(22), Article 2414877.
Hamouda, R. A., & Abdel-Hamid, M. S. (2022). A comparative Study of cellulose nanocomposite derived from algae and bacteria and its applications. In A. E. Shalan, A. S. Hamdy Makhlouf, S. Lanceros, & Méndez (Eds.), *Advances in nanocomposite materials for environmental and energy harvesting applications* (pp. 151–187). Cham: Springer International Publishing.
Harun, Z., et al. (2022). Acid hydrolysis and optimization techniques for nanoparticles preparation: Current review. *Applied Biochemistry and Biotechnology, 194*(8), 3779–3801.
Hassan, N. A., et al. (2022). Recent trends in the preparation of nano-starch particles. *Molecules, 27*(17), 5497.
Hernández-Giottonini, K. Y., et al. (2024). Starch nanoparticle preparation by the nanoprecipitation technique: Effects of formulation parameters. *Colloids and Surfaces A: Physicochemical and Engineering Aspects, 702*, Article 135022.
Himat, A. S., et al. (2025). Harnessing glucan branching enzymes for amylose-based starch nanoparticle formation: A novel bottom-up approach. *Food Hydrocolloids, 164*, Article 111182.
Hossain, A., et al. (2021). Neglected and underutilized crop species: Are they future smart crops in fighting poverty, hunger and malnutrition under changing climate?. In *Neglected and underutilized crops-towards nutritional security and sustainability* (pp. 1–50). Springer.

Hosseini, S. F., Mousavi, Z., & McClements, D. J. (2023). Beeswax: A review on the recent progress in the development of superhydrophobic films/coatings and their applications in fruits preservation. *Food Chemistry, 424*, Article 136404.

Hubbe, M. A., Tyagi, P., & Pal, L. (2019). Nanopolysaccharides in barrier composites. In *Advanced functional materials from nanopolysaccharides* (pp. 321–366). Springer.

Idumah, C., et al. (2020). A review on innovations in polymeric nanocomposite packaging materials and electrical sensors for food and agriculture. *Composite Interfaces, 27*(1). https://doi.org/10.1080/09276440.2019.1600972

Islam, F., et al. (2022). Exploring the journey of zinc oxide nanoparticles (ZnO-NPs) toward biomedical applications. *Materials, 15*(6), 2160.

Jafarzadeh, S., et al. (2025). Evaluating the influence of carbon quantum dots on starch-based bioplastics: Toward potential food packaging applications. *Food Packaging and Shelf Life, 50*, Article 101555.

Jagadeesh, P., et al. (2021). Influence of nanofillers on biodegradable composites: A comprehensive review. *Polymer Composites, 42*(11), 5691–5711.

Jamwal, V., & Mittal, A. (2024). Recent progresses in nanocomposite films for food-packaging applications: Synthesis strategies, technological advancements, potential risks and challenges. *Food Reviews International, 40*(10), 3634–3665.

Jayarathna, S., Andersson, M., & Andersson, R. (2022). Recent advances in starch-based blends and composites for bioplastics applications. *Polymers, 14*(21), 4557.

Jiang, Z., et al. (2022). Progress in laser ablation and biological synthesis processes:"Top-Down" and "Bottom-Up" approaches for the green synthesis of Au/Ag nanoparticles. *International Journal of Molecular Sciences, 23*(23), Article 14658.

Kadier, A., et al. (2021). Use of industrial wastes as sustainable nutrient sources for bacterial cellulose (BC) production: Mechanism, advances, and future perspectives. *Polymers, 13* (19), 3365.

Karnwal, A., et al. (2025). Advanced starch-based films for food packaging: Innovations in sustainability and functional properties. *Food Chemistry X*, Article 102662.

Khan, B., et al. (2017). Thermoplastic starch: A possible biodegradable food packaging material—A review. *Journal of Food Process Engineering, 40*(3), Article e12447.

Khodabakhshi, S., Fulvio, P. F., & Andreoli, E. (2020). Carbon black reborn: Structure and chemistry for renewable energy harnessing. *Carbon, 162*, 604–649.

Kim, S., & Song, K. B. (2018). Antimicrobial activity of buckwheat starch films containing zinc oxide nanoparticles against Listeria monocytogenes on mushrooms. *International Journal of Food Science and Technology, 53*(6), 1549–1557.

Kodipyaka, R., & Govindu, S. (2025). Recent trends in top-down technologies for the preparation of nanosuspensions. *Indian Journal of Pharmaceutical Education and Research, 59*(2), 493–503.

Koekuyt, H. A., & Marangoni, A. G. (2025). The solid state and nanostructure of starch: Effects on starch functionality. *Critical Reviews in Food Science and Nutrition, 65*(23), 4488–4509.

Kong, X. (2023). Gamma irradiation of starch. In *Physical modifications of starch* (pp. 103–143). Springer.

Kong, L., & Ziegler, G. R. (2014). Fabrication of pure starch fibers by electrospinning. *Food Hydrocolloids, 36*, 20–25.

Kou, Z., et al. (2020). Preparation, characterization, and performance analysis of starch-based nanomicelles. *International Journal of Biological Macromolecules, 145*, 655–662.

Krystyjan, M., et al. (2022a). Polysaccharides composite materials as carbon nanoparticles carrier. *Polymers, 14*(5), 948.

Krystyjan, M., et al. (2022b). The functional and application possibilities of starch/chitosan polymer composites modified by graphene oxide. *International Journal of Molecular Sciences, 23*(11), 5956.

Kumar, Y., et al. (2023). Starch: Current production and consumption trends. In *Starch: Advances in modifications, technologies and applications* (pp. 1–10). Springer.
Kumar, A., Sahu, R., & Tripathy, S. K. (2023). Energy-efficient advanced ultrafine grinding of particles using stirred mills—a review. *Energies, 16*(14), 5277.
Kwaśniewska, A., et al. (2021). Physical properties of Starch/Powdered activated carbon composite films. *Polymers, 13*(24), 4406.
Kwaśniewska, A., et al. (2024). Dual-fluorescent starch biopolymer films containing 5-(4-nitrophenyl)-1,3,4-thiadiazol-2-amine powder as a functional nanofiller. *Scientific Reports, 14*(1), Article 31350.
Lamanna, M., et al. (2013). Development and characterization of starch nanoparticles by gamma radiation: Potential application as starch matrix filler. *Carbohydrate Polymers, 97*(1), 90–97.
Lavoine, N., Durmaz, E., & Trovagunta, R. (2019). Preparation and properties of nanopolysaccharides. In N. Lin, et al. (Eds.), *Advanced functional materials from nanopolysaccharides* (pp. 1–54). Singapore: Springer Singapore.
Li, C., et al. (2023). Recent preparation, modification and application progress of starch nanocrystals: A review. *International Journal of Biological Macromolecules, 250*, Article 126122.
Li, C., Wu, A., & Gilbert, R. G. (2023). Critical examination of the characterization techniques, and the evidence, for the existence of extra-long amylopectin chains. *Comprehensive Reviews in Food Science and Food Safety, 22*(5), 4053–4073.
Liu, G., et al. (2017). Electrospun starch nanofibers: Recent advances, challenges, and strategies for potential pharmaceutical applications. *Journal of Controlled Release, 252*, 95–107.
Lu, H., & Tian, Y. (2021). Nanostarch: Preparation, modification, and application in pickering emulsions. *Journal of Agricultural and Food Chemistry, 69*(25), 6929–6942.
Luo, X., Zaitoon, A., & Lim, L. T. (2022). A review on colorimetric indicators for monitoring product freshness in intelligent food packaging: Indicator dyes, preparation methods, and applications. *Comprehensive Reviews in Food Science and Food Safety, 21* (3), 2489–2519.
Lvov, Y., et al. (2016). Halloysite clay nanotubes for loading and sustained release of functional compounds. *Advanced Materials, 28*(6), 1227–1250.
Mahin, M. I., Rashid, M. H.-A., & Mredul, A. R. (2025). Effects of shrimp chitosan based edible coating on the shelf life of selected vegetables in context of attaining SDGs. *Applied Food Research, 5*(1), Article 100682.
Majzoobi, M., et al. (2023). The role of ancient grains in alleviating hunger and malnutrition. *Foods, 12*(11), 2213.
Marappan, G., et al. (2025). Natural pigment-based pH/gas-sensitive intelligent packaging film for freshness monitoring of meat and seafood: Influencing factors, technological advances, and future perspectives. *Food Reviews International*, 1–38.
Mazur, F., et al. (2024). Digitalization of colorimetric sensor technologies for food safety. *Advanced Materials, 36*(42), Article 2404274.
Medina-Jaramillo, C., et al. (2017). Active and smart biodegradable packaging based on starch and natural extracts. *Carbohydrate polymers, 176*, 187–194.
Mishra, S., et al. (2022). Biochemistry, synthesis, and applications of bacterial cellulose: A review. *Frontiers in Bioengineering and Biotechnology, 10*, Article 780409.
Mohamed, S. A., et al. (2025). Potential application of new antimicrobials biological macromolecule of CNT/ZnO-NPs based on hydroxypropyl methylcellulose/shellac as packaging films. *International Journal of Biological Macromolecules, 306*, Article 141421.
Mohammed, K., et al. (2024). Influence of cellulose viscosity on the physical, mechanical, and barrier properties of the chitosan-based films. *International Journal of Biological Macromolecules, 259*, Article 129383.

Morán, D., et al. (2021). Synthesis of starch nanoparticles and their applications for bioactive compound encapsulation. *Applied Sciences, 11*(10), 4547.

Moreno, A., & Sipponen, M. H. (2024). Overcoming challenges of lignin nanoparticles: Expanding opportunities for scalable and multifunctional nanomaterials. *Accounts of Chemical Research, 57*(14), 1918–1930.

Muñoz-Gimena, P. F., et al. (2023). A review on reinforcements and additives in starch-based composites for food packaging. *Polymers, 15*(13), 2972.

Musa, A. A., et al. (2025). Nano-Enhanced polymer composite materials: A review of Current advancements and challenges. *Polymers, 17*(7), 893.

Nath, R., & Dutta, K. (2024). Sources of starch and its morphology: A review. *Interdisciplinary research in technology and management*, 402–411.

Nawaz, H., et al. (2020). Physical and chemical modifications in starch structure and reactivity. In *Chemical properties of starch*. IntechOpen.

Naz, S., et al. (2019). Agronomic crops: Types and uses. In M. Hasanuzzaman (Ed.), *Agronomic crops: Volume 1: Production technologies* (pp. 1–18). Singapore: Springer Singapore.

Petrovic, S. M., & Barbinta-Patrascu, M.-E. (2023). Organic and biogenic nanocarriers as bio-friendly systems for bioactive compounds' delivery: State-of-the art and challenges. *Materials, 16*(24), 7550.

Phukan, M., & Nongkhlaw, F. T. (2022). Starch: An overview of its conventional and unconventional sources, pharmaceutical and industrial applications. *Curr Trends Pharm Res, 9*(1), 140–161.

Pinto, V. Z. (2024). Starch nanocrystals. In G. Ayala Valencia (Ed.), *Starch nanomaterials and food applications* (pp. 15–35). Cham: Springer Nature Switzerland.

Pinzon, M. I., Garcia, O. R., & Villa, C. C. (2018). The influence of Aloe vera gel incorporation on the physicochemical and mechanical properties of banana starch-chitosan edible films. *Journal of the Science of Food and Agriculture, 98*(11), 4042–4049.

Prabha, K., et al. (2021). Recent development, challenges, and prospects of extrusion technology. *Future Foods, 3*, Article 100019.

Punia, S. (2020). Barley starch modifications: Physical, chemical and enzymatic - A review. *International Journal of Biological Macromolecules, 144*, 578–585.

Qin, Q., et al. (2024). "Bottom-up" and "top-down" strategies toward strong cellulose-based materials. *Chemical Society Reviews, 53*(18), 9306–9343. https://doi.org/10.1039/D4CS00387J

Ramesh, M., et al. (2022). A critical review on wood-based polymer composites: Processing, properties, and prospects. *Polymers, 14*(3), 589.

Ramezani, G., et al. (2025). Carbon nanomaterial-based sensors for smart packaging of food products. *Industrial Crops and Products, 232*, Article 121315.

Rayo-Mendez, L. M., & Rodriguez-Llanos, J. (2024). Starch nanovesicles and nanomicelles. In G. Ayala Valencia (Ed.), *Starch nanomaterials and food applications* (pp. 81–113). Cham: Springer Nature Switzerland.

Reches, Y. (2019). A multi-scale review of the effects of gamma radiation on concrete. *Results in Materials, 2*, Article 100039.

Rhim, J.-W., & Ng, P. K. (2007). Natural biopolymer-based nanocomposite films for packaging applications. *Critical Reviews in Food Science and Nutrition, 47*(4), 411–433.

Ribeiro, A., et al. (2025). Comminution technologies in the pharmaceutical industry: A comprehensive review with recent advances. *Reviews in Chemical Engineering, 41*(1), 69–100.

Rostamabadi, H., et al. (2024). Cellulose nanofibers, nanocrystals, and bacterial nanocellulose: Fabrication, characterization, and their most recent applications. *Future Postharvest and Food, 1*(1), 5–33.

Santos, F. N.d., et al. (2024). Cassava starch esterification with formic acid for fabrication of electrospun fibers. *International Journal of Biological Macromolecules, 266*, Article 131182.

Schuchardt, D., Synthetic layered silicates as a basis for high-performance gas barrier coatings. 2024.

Seung, D. (2020). Amylose in starch: Towards an understanding of biosynthesis, structure and function. *New Phytologist, 228*(5), 1490–1504.

Shaker, D. S., et al. (2019). Nanoemulsion: A review on mechanisms for the transdermal delivery of hydrophobic and hydrophilic drugs. *Scientia Pharmaceutica, 87*(3), 17.

Sharma, S., & Bhende, M. (2024). An overview: Non-toxic and eco-friendly polysaccharides—its classification, properties, and diverse applications. *Polymer Bulletin, 81*(14), 12383–12429.

Sharma, A., et al. (2023). Amaranth & Quinoa Sprouts. In I. Majid, et al. (Eds.), *Advances in plant sprouts: Phytochemistry and biofunctionalities* (pp. 127–151). Cham: Springer International Publishing.

Sharma, P., et al. (2023). Bacterial cellulose: Nano-biomaterial for biodegradable face masks–a greener approach towards environment. *Environmental Nanotechnology, Monitoring & Management, 19*, Article 100759.

Sharma, P., et al. (2025). Nanotechnology and biosensors in food processing, packaging, and safety. In *Food and industry 5.0: Transforming the food System for a sustainable future* (pp. 3–20). Springer.

Shekhar, N., & Mondal, A. (2024). Synthesis, properties, environmental degradation, processing, and applications of Polylactic Acid (PLA): An overview. *Polymer Bulletin, 81* (13), 11421–11457.

Si, W., & Zhang, S. (2024). The green manufacturing of thermoplastic starch for low-carbon and sustainable energy applications: A review on its progress. *Green Chemistry, 26*(3), 1194–1222.

Simões, S. (2024). High-performance advanced composites in multifunctional material design: State of the art, challenges, and future directions. *Materials, 17*(23), 5997.

Singh, R., et al. (2023). Future of nanotechnology in food industry: Challenges in processing, packaging, and food safety. *Global Challenges*, 7(4), Article 2200209.

Singh, A. K., et al. (2024). Synergistic integration of carbon quantum dots in biopolymer matrices: An overview of current advancements in antioxidant and antimicrobial active packaging. *Molecules, 29*(21), 5138.

Sintov, A. C. (2020). AmyloLipid Nanovesicles: A self-assembled lipid-modified starch hybrid system constructed for direct nose-to-brain delivery of curcumin. *International Journal of Pharmaceutics, 588*, Article 119725.

Sohail, M., et al. (2022). Cellulolytic and xylanolytic enzymes from yeasts: Properties and industrial applications. *Molecules, 27*(12), 3783.

Song, T., et al. (2022). Recent advances in bio-based smart active packaging materials. *Foods, 11*(15), 2228.

Sreejit, V., et al. (2025). Green synthesized nano starch: Preparation and emerging avenues in food industry. *Journal of Nanoparticle Research, 27*(4), 1–20.

Srinivasan, G., et al. (2025). Versatile application of calixarenes and their derivatives: From drug delivery to industrial catalysis and environmental remediation. *Critical Reviews in Analytical Chemistry*, 1–58.

Strachota, B., et al. (2024). Potential environmental application of a tough and temperature-responsive nanocomposite hydrogel based on Poly(N-Isopropylacrylamide-co-Sodium methacrylate) and clay. *International Journal of Environmental Research, 19*(1), 7.

Suh, J. H., et al. (2020). Effect of moisture content on the heat-sealing property of starch films from different botanical sources. *Polymer Testing, 89*, Article 106612.

Sun, X., et al. (2023). Sustainable Starch/Lignin nanoparticle composites biofilms for food packaging applications. *Polymers, 15*(8), 1959.

Tabatabaei, S. H., & Ajji, A. (2011). Structure-orientation-properties relationships for polypropylene nanoclay composite films. *Journal of Plastic Film & Sheeting, 27*(1–2), 87–115.

Thakur, R., et al. (2019). Starch-based films: Major factors affecting their properties. *International Journal of Biological Macromolecules, 132*, 1079–1089.

Thomas, M. S., et al. (2019). *Starch, chitin and chitosan based composites and nanocomposites.* Springer.

Thuppahige, V. T. W., et al. (2023). Extraction and characterisation of starch from cassava (Manihot esculenta) agro-industrial wastes. *Lwt, 182*, Article 114787.

Tibolla, H., et al. (2020). Starch-based nanocomposites with cellulose nanofibers obtained from chemical and mechanical treatments. *International Journal of Biological Macromolecules, 161*, 132–146.

Tohamy, H.-A. S. (2024). Cellulosic schiff base hydrogel biosensor for bacterial detection with pH/thermo-responsitivity: DFT calculations and molecular docking. *International Journal of Biological Macromolecules, 283*, Article 137389.

Tohamy, H.-A. S. (2025a). A novel anthocyanins hydroxyethyl cellulose film for intelligent chicken meat packaging with mechanical study, DFT calculations and molecular docking study. *Scientific Reports, 15*(1), Article 27311.

Tohamy, H.-A. S. (2025b). A novel natural chromogenic visual and luminescent sensor platform for Multi-Target analysis in strawberries and shape memory applications. *Foods, 14*(16), 2791.

Tohamy, H.-A. S. (2025c). Microwaved schiff base dialdehyde cellulose-chitosan hydrogels for sustained drug release with DFT calculations. *BMC Chemistry, 19*(1), 114.

Tohamy, H.-A. S. (2025d). Fullerene-functionalized cellulosic hydrogel biosensor with bacterial turn-on fluorescence response derived from carboxymethyl cellulose for intelligent food packaging with DFT calculations and molecular docking. *Gels, 11*(5), 329.

Tohamy, H.-A. S. (2025e). Amylopectin xerogel with onion based sulfur nitrogen doped carbon quantum dots as a chemosensor for chromium and biosensor for microbial spoilage in tomatoes. *Scientific Reports, 15*(1), Article 32667.

Tohamy, H.-A. S. (2025f). Novel colored hydroxypropyl methyl cellulose/magnetite carbon dots films for beef packaging with DFT calculations and molecular docking study. *Scientific Reports, 15*(1), Article 10337.

Tohamy, H.-A. S. (2025g). Novel intelligent naked-eye food packaging pH-sensitive and fluorescent sulfur, nitrogen-carbon dots biosensors for tomato spoilage detection including DFT and molecular docking characterization. *International Journal of Biological Macromolecules*, Article 143330.

Tohamy, H.-A. S. (2025i). Speedy synthesis of magnetite/carbon dots for efficient chromium removal and reduction: A combined experimental and DFT approach. *Emergent Materials, 8*(4), 2425–2437.

Tohamy, H.-A. S. (2025j). Quantum dots as an emerging nanocarrier for drug delivery. *Quantum Dot Nanocarriers for Drug Delivery*, 363–384.

Tohamy, H.-A. S. (2025k). Beet root carbon dots cellulose sulfate film as a novel naked eye pH sensor for chromium and bacterial detection in tomatoes. *Scientific Reports, 15*(1), Article 30235.

Tohamy, H.-A. S. (2025l). Artistic anti-counterfeiting with a pH-responsive fluorescent ink using DFT and molecular electrostatic potential mapping insights. *Scientific Reports, 15*(1), Article 19335.

Tohamy, H.-A. S., et al. (2024). Development and characterization of fluorescent cellulose succinate hydrogels for efficient chromium adsorption. *Journal of Polymer Research, 31*(11), 339.

Tohamy, H.-A. S., et al. (2025). Boosting dielectric properties of polyethersulfone membranes: A strategy using cross-linked graphene oxide/EDTA/Silane. *ECS Journal of Solid State Science and Technology, 14*(3), Article 031007.

Tohamy, H.-A. S., & Greener, S. P. (2024). Carbon Nanotubes/Gelatin-Enhanced recycled paper for fire retardation with DFT calculations. *Journal of Renewable Materials, 12*(12).

Torche, A., et al. (2025). pH-Sensitive cassava Starch/Onion peel powder films as colorimetric indicators for minced beef freshness monitoring. *Foods, 14*(17), 2974.

Tseng, I.-H., et al. (2012). Transparent polyimide/graphene oxide nanocomposite with improved moisture barrier property. *Materials Chemistry and Physics, 136*(1), 247–253.

Tsuji, H., HYDROLYTIC DEGRADATION, in Poly(Lactic Acid). 2022. p. 467-516.

Uddin, M. N., et al. (2024). Research and applications of nanoclays: A review. *SPE Polymers, 5*(4), 507–535.

Venkatesh, P. H., et al. (2026). Tiny tubes with big impact: Exploring the potential of halloysite nanotubes for food packaging applications—recent advancements. In *Smart Halloysite nanotubes* (pp. 275–303). Elsevier.

Vobornik, D., et al. (2023). Measuring the diameter of single-wall carbon nanotubes using AFM. *Nanomaterials, 13*(3), 477.

Wang, Y.-C., et al. (2013). Self-assembled clay films with a platelet–void multilayered nanostructure and flame-blocking properties. *Scientific Reports, 3*(1), 2621.

Wang, W., et al. (2018). High performance extrusion blown starch/polyvinyl alcohol/clay nanocomposite films. *Food Hydrocolloids, 79*, 534–543.

Wang, P., et al. (2021). Starch-Based carbon dots for nitrite and Sulfite detection. *Frontiers in Chemistry, 9*.

Wang, L., et al. (2023). Preparation of starch-based nanoemulsion for sustained release and enhanced bioaccessibility of quercetin. *Colloids and Surfaces A: Physicochemical and Engineering Aspects, 665*, Article 131218.

Wang, J., & Kong, Z. (2025). Plant-derived vesicle-like nanoparticles in food crops: Emerging insights into nutritional biofortification and biomedical applications. *Plant Biotechnology Journal, 23*(8), 3260–3282.

Wawrzyńczak, A., Chudzińska, J., & Feliczak-Guzik, A. (2024). Metal and metal oxides nanoparticles as nanofillers for biodegradable polymers. *ChemPhysChem, 25*(10).

Wi, S. G., et al. (2007). Effects of gamma irradiation on morphological changes and biological responses in plants. *Micron, 38*(6), 553–564.

Wu, L., et al. (2023). Structure and properties of starch/chitosan food packaging film containing ultra-low dosage GO with barrier and antibacterial. *Food Hydrocolloids, 137*, Article 108329.

Wu, X., et al. (2024). Preparation and characterization of pH-sensitive intelligent packaging films based on cassava starch/polyvinyl alcohol matrices containing Aronia melanocarpa anthocyanins. *LWT, 194*, Article 115818.

Yadav, T. P., Yadav, R. M., & Singh, D. P. (2012). Mechanical milling: A top down approach for the synthesis of nanomaterials and nanocomposites. *Nanoscience and Nanotechnology, 2*(3), 22–48.

Yan, M., Wang, B., & Wang, M. (2023). Facilely processable dual-color ratiometric fluorescent sensors for highly sensitive detection of food spoilage. *ACS Food Science & Technology, 3*(6), 1090–1097. https://doi.org/10.1021/acsfoodscitech.3c00086

Yu, X., et al. (2021). Research progress of starch-based biodegradable materials: A review. *Journal of Materials Science, 56*(19), 11187–11208.

Zarski, A., Bajer, K., & Kapuśniak, J. (2021). Review of the most important methods of improving the processing properties of starch toward non-food applications. *Polymers, 13*(5), 832.

Zhang, K., et al. (2020). Novel pH-sensitive films based on starch/polyvinyl alcohol and food anthocyanins as a visual indicator of shrimp deterioration. *International Journal of Biological Macromolecules, 145*, 768–776.

Zhang, L., et al. (2021). The effects of cellulose nanocrystal and cellulose nanofiber on the properties of pumpkin starch-based composite films. *International Journal of Biological Macromolecules, 192*, 444–451.

Zhang, K., et al. (2022). Preparation and characterization of starch-based nanocomposites reinforced by graphene oxide self-assembled on the surface of silanecouplingagent modified cellulose nanocrystals. *International Journal of Biological Macromolecules, 198*, 187–193.

Zhang, J., et al. (2024). Recent advances in pH-sensitive indicator films based on natural colorants for smart monitoring of food freshness: A review. *Critical Reviews in Food Science and Nutrition, 64*(33), 12800–12819.

Zhang, Q., et al. (2024). High-Value utilization of cellulose: Intriguing and important effects of hydrogen bonding interactions—a mini-review. *Biomacromolecules, 25*(10), 6296–6318.

Zheng, L., et al. (2019). The characteristics of starch gels molded by 3D printing. *Journal of Food Processing and Preservation, 43*(7), Article e13993.

Zhong, T., et al. (2022). Interfacial interactions and reinforcing mechanisms of cellulose and chitin nanomaterials and starch derivatives for cement and concrete strength and durability enhancement: A review. *Nanotechnology Reviews, 11*(1), 2673–2713.

Zhou, J., et al. (2023). Advances in starch nanoparticle for emulsion stabilization. *Foods, 12*(12), 2425.

CHAPTER EIGHT

Starch-based smart packaging incorporated with plant bioactives

Rina Ningtyas, Muryeti, Deli Silvia
Department of Industrial Technology of Packaging Printing, Politeknik Negeri Jakarta, UI Campus, Depok, Indonesia

8.1 Introduction

Food packaging is one of the most critical components of the modern supply chain. Its role extends beyond serving as a physical container to protecting food quality, prolonging shelf life, and ensuring food safety (Arruda et al., 2025; Ningtyas, 2021; Onyeaka et al., 2022; Otto et al., 2021; Rodríguez-Rojas et al., 2019). Petrochemical-based plastics have dominated the food packaging industry for several decades due to their low cost, flexibility, and durability. However, their nonbiodegradable nature has created serious environmental challenges (Ahmed et al., 2022; Nayanathara Thathsarani Pilapitiya and Ratnayake, 2024; Siracusa et al., 2008). The accumulation of plastic waste in landfills and aquatic environments has led to the formation of microplastics that threaten ecosystems and human health (Bhatt et al., 2021; Lacourt et al., 2024). These issues have accelerated the search for more environmentally friendly and sustainable packaging alternatives.

Among the various biodegradable biopolymers, starch has emerged as one of the most promising candidates. Starch is abundantly available from plant-based sources such as corn, potato, cassava, and wheat, making it inexpensive and renewable. Its molecular structure, composed of amylose and amylopectin, enables the formation of thin films suitable for packaging applications. Furthermore, its renewable and biodegradable nature contributes to environmental sustainability (Agarwal et al., 2023; Arruda et al., 2025). Nevertheless, starch-based films exhibit several limitations, including brittleness, poor mechanical strength, and high humidity sensitivity (Agarwal et al., 2023). These drawbacks restrict their practical applications, highlighting the need for functional component modification or incorporation to improve their performance. Consequently, considerable research has

Starches in Smart Packaging
ISBN: 978-0-443-45396-0
https://doi.org/10.1016/B978-0-443-45396-0.00017-7

focused on developing functional starch films for smart packaging applications.

Smart packaging has emerged as an innovative strategy to address sustainability challenges while enhancing packaging functionality. Smart packaging can be generally classified into two main categories: active packaging and intelligent packaging. Active packaging interacts with food or its environment to maintain or improve quality, for example, by releasing antimicrobial compounds or absorbing oxygen (Ahari & Soufiani, 2021; Wyrwa & Barska, 2017). Intelligent packaging provides information about the status of the packaged product, such as freshness or temperature indicators (Ahari & Soufiani, 2021). Within the context of starch-based films, the most relevant approach is active packaging, where functional compounds are incorporated into the starch matrix to impart antimicrobial and antioxidant activities. This integration not only enhances the functional performance of starch films but also aligns with consumer trends that favor natural and sustainable preservation methods.

One promising strategy for functionalizing starch-based packaging is the incorporation of plant-derived bioactive compounds. Naturally occurring bioactives from herbs, spices, fruits, and other plant parts are well-known for their antimicrobial, antioxidant, and antifungal properties. Essential oils (e.g., clove, oregano, and thyme) effectively inhibit microbial growth, whereas phenolic compounds and flavonoids act as potent antioxidants capable of scavenging free radicals and preventing lipid oxidation. Incorporating these bioactive compounds into starch films offers two benefits: strengthening the protective function of packaging and reducing the need for synthetic preservatives directly added to food (Nogueira et al., 2020; Pei et al., 2024). From a consumer perspective, the use of natural compounds also supports the "clean label" trend, emphasizing safer and more natural ingredients.

Despite their potential, the application of bioactives in starch films faces several challenges. Many bioactive compounds are volatile, sensitive to environmental conditions (Falcão et al., 2022), and susceptible to uncontrolled migration from the packaging matrix (do Nascimento et al., 2024). Such uncontrolled release can reduce long-term efficacy and raise safety concerns if migration levels exceed regulatory limits. Furthermore, the hydrophobic nature of many bioactives often results in poor compatibility with the hydrophilic–hydrophobic balance of starch matrices, leading to uneven dispersion and adverse effects on mechanical properties. Therefore, it is essential to develop strategies not only for efficient extraction and stabilization of bioactives but also for incorporation techniques that enable controlled release and long-term stability.

This chapter discusses active packaging or biopolymers that address the integration of plant bioactives into starch-based systems, particularly considering recent advances in green extraction methods, incorporation strategies, and regulatory aspects. This chapter aims to provide a comprehensive overview of starch-based smart packaging incorporated with plant-derived bioactive compounds. The scope of this study includes the following: (i) classification and functional properties of bioactive compounds relevant to food packaging, (ii) conventional and novel plant sources and green extraction methods, (iii) incorporation techniques into starch matrices with emphasis on stability and controlled release, (iv) performance evaluation in terms of antimicrobial and antioxidant activity, (v) safety and regulatory aspects, and (vi) industrial challenges and future prospects. By integrating both technical and nontechnical perspectives, this chapter bridges the gap between laboratory research and real-world applications in the food packaging industry.

The novelty of this chapter lies in its focused examination of starch as a biopolymer matrix synergistically combined with plant-derived bioactive within the smart packaging framework. Unlike general reviews on biodegradable packaging, this study emphasizes how plant bioactives can be sustainably extracted, stabilized, and incorporated into starch films to produce functional, safe, and scalable packaging systems. It also highlights current limitations, research gaps, and future opportunities, including integration with circular bioeconomy strategies. Thus, this chapter contributes significantly to the understanding and advancement of starch-based packaging reinforced with plant bioactives as a sustainable and safe solution for modern food systems.

8.2 Overview of plant bioactives in food packaging

Integration of plant bioactives into food packaging systems is a very promising approach to producing multifunctional packaging that is safe, environmentally friendly, and has high added value (Pan et al., 2024). Plant-derived bioactive compounds encompass diverse chemical groups exhibiting antimicrobial, antioxidant, and antifungal activities. In food-packaging contexts, a clear classification is essential to understand their physicochemical characteristics, mechanisms of action, and integration challenges within packaging matrices such as starch. Plant bioactives are generally grouped into five major classes: essential oils, phenolics, flavonoids, terpenoids, and alkaloids. Each class exhibits distinct chemistry and mechanisms, yet all exhibit strong potential in functional food packaging.

8.2.1 Essential oils

Essential oils are complex mixtures of volatile compounds derived from various parts of plants (leaves, flowers, fruits, seeds, bark, and roots) (Żukowska & Durczyńska, 2024). Composition of essential oil is strongly influenced by the plant species, plant part, and extraction method, thus determining their biological effectiveness in food applications (Tongnuanchan & Benjakul, 2014). The various biological activities of essential oils include antiviral, antibacterial, antioxidant, anti-inflammatory, and as a natural alternative to synthetic pesticides (Tongnuanchan & Benjakul, 2014; Żukowska & Durczyńska, 2024).

Tongnuanchan & Benjakul (2014) present a comprehensive review of essential oils as natural bioactive agents for food preservation. This article asserts that essential oils are complex mixtures of volatile compounds, primarily terpenes, terpenoids, and phenolic compounds, that possess potent antimicrobial, antifungal, and antioxidant activities. These activities primarily operate through mechanisms such as disruption of microbial cell membranes, disruption of ion gradients, and inhibition of lipid oxidation in food matrices. In the context of biopolymer-based active packaging such as starch, essential oils are considered highly potential as functional components due to their natural properties, biodegradability, and ability to extend the shelf life of food products. The authors report that essential oils can be incorporated into starch films, edible coatings, or other active packaging systems to inhibit microbial growth without the use of synthetic preservatives. However, major challenges in their application include high volatility, low water solubility, limited thermal stability, and potential changes in the aroma and flavor of food products. Several studies of essential oils as active components in packaging applications are shown in Table 8.1.

Based on Table 8.1, essential oils (EOs) have been reported to be effective as active agents in the development of sustainable food packaging due to their diverse bioactive activities and compatibility with various polymer matrices. In fresh food coating applications, limonene-dominated sweet orange essential oil (SOEO) exhibits high antioxidant activity and is used in the form of WPI/peach gum coacervate microcapsules as an active fresh-

Table 8.1 Representative essential oils used in active and smart food packaging applications.

Essential oil source	Major active compounds	Main bioactivity	Packaging function	Polymer Matrix/ Application	Reference
Rosemary (*Rosmarinus officinalis* L.) essential oil	Phenolic volatile compounds (e.g., 1,8-cineole, camphor, α-pinene, borneol; generally referred to as bioactive volatile compounds)	Strong antimicrobial activity against *E. coli* (Gram-negative) and *S. aureus* (Gram-positive)	Active antimicrobial packaging to inhibit the growth of microorganisms and extend the shelf life of food.	Carrageenan + Carboxymethyl cellulose (CMC) with glycerol plasticizer. Application: Edible film/biopolymer-based food packaging.	Ningtyas et al. (2024)
Sweet orange essential oil (SOEO)	Volatile terpenoid compounds (especially limonene as the dominant component, as well as aldehydes and aromatic esters)	Powerful antioxidant	Active fresh-keeping coating with sustained release to maintain fresh fruit quality	Whey protein isolate (WPI)/Peach gum complex coacervate microcapsules; edible coating application for fresh-cut pear	Xu et al. (2025)

(Continued)

Table 8.1 Representative essential oils used in active and smart food packaging applications.—cont'd

Essential oil source	Major active compounds	Main bioactivity	Packaging function	Polymer Matrix/ Application	Reference
Thyme essential oil	Hymol, accompanied by other monoterpenes (e.g., carvacrol, α-pinene, linalool in smaller amounts)	Potent antifungal against *Alternaria tenuissima, Colletotrichum acutatum, Geotrichum candidum*, and *Cladosporium cladosporioides* through morphological damage and increased permeability of cell membranes	Antifungal active packaging to inhibit fungal growth and reduce post-harvest decay	Polyhydroxybutyrate-co-valerate (PHBV). film modified with thyme essential oil (4–8 wt%) through casting method; application: Post-harvest fresh strawberry packaging	Colín-Chávez et al. (2025)
Eucalyptus essential oil (EEO)	Volatile terpenoid compounds, especially 1,8-cineole (eucalyptol) encapsulated in β-cyclodextrin (EEO@β-CD)	Antibacterial and antioxidant	Active antibacterial food packaging with improved barrier properties (water vapor & oxygen) and biodegradability	Polyvinyl alcohol (PVA)/Konjac glucomannan (KGM) composite film with EEO@ β-CD inclusion complex, application fresh fruit (cherry)	dawam Abdullah et al. (2021)

keeping coating. This system allows for sustained release of active compounds, thereby maintaining the quality and freshness of cut fruit, particularly pears, during storage (Xu et al., 2025). For postharvest fungal microorganism control, thymol-rich thyme essential oil has proven highly effective as an antifungal agent, especially when integrated into a biodegradable PHBV-based film. This active film can disrupt the morphology and increase the permeability of the cell membrane of the fungus that causes strawberry rot, thereby significantly reducing postharvest yield losses (Colín-Chávez et al., 2025). Furthermore, eucalyptus essential oil (EEO) containing 1,8-cineole has been successfully developed as an active antibacterial system through an encapsulation strategy in β-cyclodextrin (EEO@β-CD) and applied to a PVA/konjac glucomannan composite film. This approach not only improves the stability and controlled release of volatile compounds but also provides superior water vapor and oxygen barrier properties, accompanied by high biodegradability, making it effective for packaging fresh fruits such as cherries (Shen et al., 2025).

Several essential oils, namely thyme, clove, and tea tree oil, are very promising as natural alternatives to chemical additives in active food packaging, because they can increase safety and extend the shelf life of food, while also being in line with consumer demands for more natural and sustainable products (Vasile et al., 2017). The addition of essential oils can be applied as natural antimicrobial agents in active food packaging. The extrusion method can reduce tensile strength and increase elongation, while the ionization method has little impact on mechanical properties. In terms of barrier properties, the addition of essential oils can affect oxygen permeability (OTR) and water vapor permeability (WVTR), depending on the method and concentration of essential oil used (Valderrama Solano & de Rojas Gante, 2012). Overall, these studies confirm that the right essential oil–polymer matrix combination is a key strategy in designing active packaging and edible coatings that are environmentally friendly, functional, and applicable to modern food systems.

8.2.2 Phenolic constituents

Phenolic compounds are secondary plant metabolites that have an aromatic structure with one or more hydroxyl groups. Phenolic compounds are generally classified into simple phenolics, such as hydroxybenzoic acids and hydroxycinnamic acids (C6–C1 and C6–C3), and complex polyphenols, including flavonoids, tannins, and lignin, which are formed through the polymerization of phenolic units. Among them, flavonoids (C6–C3–C6) represent the most diverse subgroup and are often present in glycosylated or polymerized forms (Kafkas et al., 2018). These phenolic acids are found abundantly in fruits, vegetables, cereals, and agro-industrial waste, and are known to possess potent antioxidant and antimicrobial activity, primarily through the mechanisms of free radical scavenging, metal ion chelation, and disruption of microbial cell membranes (Ordoñez et al., 2022).

Phenolic compounds have great potential as a substitute for synthetic additives in active food packaging and edible packaging. However, major challenges that still need to be addressed include phenolic stability at temperature and in the environment, controlling migration into food, influencing the mechanical and barrier properties of packaging, and regulatory and sensory issues (Singh et al., 2022). Moreover, selection analytical methods of phenolic are very important to optimize the utilization of bioactive materials ad high value-added materials (Kafkas et al., 2018). Several studies of Phenolic constituents as active components in packaging applications are shown in Table 8.2.

Table 8.2 shows that studies on integrating plant bioactives into starch films consistently demonstrate active, multifunctional, and eco-friendly packaging. Molecules, such as vanillin, quercetin, ferulic acid, fucoidan, anthocyanins, and maltol contribute dual protection, antioxidant and antimicrobial along with value-added sensory roles (natural color/aroma). These innovations starch not only as a biodegradable matrix but also as a platform for smart packaging capable of extending shelf life and preserving quality in a sustainable manner.

Vanillin is a representative simple phenolic that exhibits antioxidant and antimicrobial activity against *Escherichia coli, Bacillus subtilis, Pseudomonas aeruginosa, Staphylococcus aureus,* and *Candida albicans*. Its phenolic group disrupts microbial membranes while scavenging free radicals. Vanillin incorporation into cationic starch/PVA films improves mechanical and barrier properties and prolongs food shelf life (Kurabetta et al., 2025). Quercetin, a polyphenolic flavonoid abundant in fruits, vegetables, and herbs, provides strong antioxidant and antimicrobial effects by inhibiting lipid oxidation, scavenges radicals, and perturbs microbial metabolism. Ramon-starch/chitosan films containing quercetin show enhanced thermal stability and effectively inhibit *S. aureusand S. typhimurium* (Pech-Cohuo et al., 2022). Ferulic acid derived from corn arabinoxylan functions as both an antioxidant and a UV protectant; arabinoxylan-based films

Table 8.2 Several studies of Phenolic constituents as active components in packaging application.

Phenolic constituents	Bioactive category	Bioactive category	Bioactive category	Bioactive category	Ref
Vanillin	*Vanilla planifolia* fruit	Antioxidant, antimicrobial (*E. coli*, *B. subtilis*, *S. aureus*, etc.)	Phenolic groups disrupt microbial membranes and scavenge free radicals	Cationic starch/PVA + vanillin films exhibit improved mechanical and barrier properties	Kurabetta et al. (2025)
Ferulic acid (phenolic)	Corn fiber (arabinoxylan)	Antioxidant and UV protective	Free radical scavenging and oxidation inhibition	Arabinoxylan + ferulic acid films exhibit high antioxidant activity	Serra et al. (2020)
Crab-apple peel extract (polyphenols, flavonoids, and anthocyanins)	*M. asiatica* fruit peel	Antioxidant, antimicrobial	Scavenges free radicals and inhibits lipid oxidation	Potato starch/κ-carrageenan + crab-apple extract films extend fresh meat shelf lifethe shelf life of fresh meat	Lang et al. (2025)
Blackberry pulp (anthocyanins and phenolics)	Blackberry fruit (*Rubus fruticosus*)	Antioxidant, natural colorant	Anthocyanins act as radical scavengers; microencapsulation stabilizes the compounds	Arrowroot starch + blackberry powder films exhibit coloration and controlled bioactive release	Nogueira et al. (2019)
Maltol (3-hydroxy-2-methyl-4-pyrone)	Malt, fruits, and pine leaves	Antimicrobial (*E. coli*, *S. aureus*, and fungi), antioxidant	Disrupts membrane permeability and chelates metal	Acetylated cassava starch + maltol films extend the shelf life of butter cake	Promhuad et al. (2022)

(*Continued*)

Table 8.2 Several studies of Phenolic constituents as active components in packaging application.—cont'd

Phenolic constituents	Bioactive category	Bioactive category	Bioactive category	Bioactive category	Ref
Cranberry powder (anthocyanins, proanthocyanidins, and vitamin C)	Vaccinium macrocarpon	Antioxidant, antibacterial (*E. coli*, *S. aureus*, *L. monocytogenes*)	Anti-adhesive effects of anthocyanins and proanthocyanidins; free radical scavenging	Arrowroot starch/gelatin + cranberry films enhance vitamin C content, color, and sensory quality	Fakhouri et al. (2019)
Grape pomace extract (polyphenols and anthocyanins)	Grape residue (Brs. violeta)	Antioxidant, natural colorant	Polyphenols act as radical scavengers; anthocyanins provide pigmentation	Arrowroot starch + grape pomace films improve solubility and color intensity	Nogueira et al. (2022)

enriched with ferulic acid display high radical-scavenging capacity, underscoring their potential as bioactive packaging (Serra et al., 2020). Maltol is a naturally occurring compound with a mild caramel aroma that shows distinctive promise in starch systems. It exhibits antimicrobial activity against *E. coli, S. aureus, P. aeruginosa, C. albicans,* and *A. brasiliensis*, along with mild antioxidant activity. Mechanistically, maltol perturbs microbial membrane permeability, chelates metal ions, and scavenges radicals. Acetylated cassava starch films containing maltol not only enhance mechanical properties but also extend the shelf life of butter cakes up to 19 days at ambient temperature without the addition of preservatives (Promhuad et al., 2022).

Recent studies have also explored whole-plant extracts. Crab-apple peel extract (rich in polyphenols, flavonoids, and anthocyanins) elevates antioxidants and antimicrobial efficacy in potato starch/κ-carrageenan films and slows lipid oxidation and extends shelf life with strong activity against *S. aureus* and *E. coli* (Lang et al., 2025). Grape pomace extract, a winery by-product rich in polyphenols, anthocyanins, and dietary fiber, provides a valorization route: arrowroot starch films with grape pomace exhibit a natural reddish-purple color, high antioxidant capacity, and improved sensory attributes. Higher extract loadings increase film thickness, water solubility, and WVP, making them suitable for edible films and fruit strips (Nogueira et al., 2022).

In addition to antioxidant and antimicrobial functions, several bioactives act as natural colorants. Rice-starch films formulated from white/red/black rice leverage inherent anthocyanins/phenolics to achieve high antioxidant activity and robust color stability, enabling potential color-indicator roles (da Silva et al., 2022). Microencapsulated blackberry powders combined with arrowroot starch yielded films with high antioxidant capacity, attractive coloration, and controlled active component release (Pei et al., 2024) [17].

8.2.3 Flavonoids

Flavonoids are a large group of naturally occurring polyphenolic compounds derived from plants and are secondary metabolites resulting from the phenylpropanoid–shikimate pathway. Structurally, flavonoids have a C6–C3–C6 carbon skeleton, consisting of two aromatic rings (rings A and B) linked by an oxygenated heterocycle. These compounds play an important role in plants such as natural pigments, UV radiation protectors, defense agents against pathogens, and signaling molecules (Shen et al., 2022). In the context of food and health, flavonoids are widely known for their antioxidant, anti-inflammatory, and antimicrobial activities, thus attracting attention as natural bioactive components in food systems and active packaging.

Flavonoids are classified into several major subclasses based on their chemical structure, oxidation state, and substitutions on the heterocyclic ring. Flavonols are the most abundant group in foods and include compounds such as quercetin, kaempferol, and myricetin, which are found abundantly in berries, apples, onions, and tea and are known to possess potent antioxidant activity. Flavones, such as apigenin and luteolin, are common in green vegetables and herbs and play a role in anti-inflammatory activity and oxidative protection. Flavanones, including naringenin and hesperetin, are predominant in citrus fruits and contribute to their antimicrobial activity and characteristic bitter taste. Flavanols (flavan-3-ols), such as catechin and epicatechin, are found abundantly in tea and cocoa and are precursors to the formation of proanthocyanidins (condensed tannins). Anthocyanidins/anthocyanins function as natural red, purple, and blue pigments in berries and grapes and are pH sensitive. Isoflavones, such as genistein and daidzein, are primarily found in soybeans and legumes and possess phytoestrogenic activity. Chalcones, on the other hand, are open-structured flavonoids that act as intermediates in the biosynthesis pathways of other flavonoids (Shen et al., 2022).

Liu et al. (2025) demonstrated that the incorporation of flavonoids into chitosan-based films significantly enhances the antioxidant and antimicrobial activities of the resulting packaging materials. The effectiveness of this active packaging is highly dependent on the release of flavonoids from the film matrix, which is influenced by various factors, such as the type and form of flavonoids, the characteristics of the film matrix, the presence of reinforcing agents, the manufacturing method and post-production treatment, and the type of food simulation. The stability of chitosan/flavonoid films can be improved through encapsulation, modification of the flavonoid structure, chelation with nano-metals, and mechanical activation. Currently, chitosan/flavonoid films have been widely used for the preservation of fruits, vegetables, fishery products, meat products, and edible oils.

In addition to polyphenols and flavonoids, vitamins A, B, C, D, and E are considered bioactive agents in starch films due to their antioxidant, immunomodulatory, and anti-inflammatory roles protecting cell membranes, enhancing antibody production enhancement, and immune response modulation. Their integration opens opportunities for health-promoting packaging, where the package not only protects but may also contribute supplemental micronutrients (Mironescu et al., 2021).

8.2.4 Terpenoid

Terpenoids are the most abundant and diverse bioactive compounds in nature that have antioxidant, antimicrobial, anti-inflammatory, anticancer, and antidiabetic activities, making them very promising for the functional food industry (Hammed et al., 2025). Terpenoids are composed of isoprene (C_5) units and are classified into monoterpenoids, sesquiterpenoids, diterpenoids, triterpenoids, tetraterpenoids, and polyterpenoids. These compounds play an important role in plant physiology (growth, photosynthesis, and defense) and have high economic value due to their widespread use in the pharmaceutical, food, cosmetics, and agricultural industries (Yang et al., no date)[48]

The selection of extraction methods for terpenoids is very important where conventional extraction methods (such as steam distillation, hydrodistillation, and Soxhlet) have limitations in the form of high energy consumption, long process times, degradation of heat-sensitive compounds, and environmental impacts. Therefore, selecting green extraction technology for terpenoid extraction, combined with advanced characterization, opens significant opportunities for maximizing terpenoid utilization (Hammed et al., 2025).

8.2.5 Alkaloids

Alkaloids are secondary plant metabolites widely known as natural bioactive compounds containing nitrogen (Debnath et al., 2018). Alkaloids are found in bacteria, fungi, animals and plants, with complex and very diverse structures. The wide distribution of alkaloids, accompanied by high structural diversity, often makes the classification process difficult (Valderrama Solano & de Rojas Gante, 2012).

Alkaloids are known to have a variety of biological activities, and some can even be metabolized in the body into active metabolites that enhance their pharmacological effects (Debnath et al., 2018). Traditionally, plant alkaloids have played an important role in traditional medicine since ancient times, including as purgatives, antitussives, sedatives, and as therapies for various diseases. Currently, several alkaloids have been used as models in the development of modern drugs, and some of them have been widely applied in the field of pharmacology, such as codeine, brucine, morphine, ephedrine, and quinine (Valderrama Solano & de Rojas Gante, 2012).

8.3 Sources and extraction of plant bioactives

8.3.1 Common plant sources

Plant bioactives can be obtained from multiple botanical parts, like leaves, flowers, seeds, peels, and roots. Culinary and medicinal herbs remain the most widely used sources of medicinal herbs. Essential oils from oregano, thyme, basil, and rosemary contain key constituents, such as carvacrol, thymol, and rosmarinic acid, that exhibit pronounced antimicrobial activity. In parallel, citrus, grape, and pomegranate peels are rich in flavonoids and phenolic compounds, serving as natural antioxidants for packaging applications.

In addition to the plant species, the plant part plays a crucial role in determining the composition, concentration, and function of the obtained bioactive compounds. A comparative study by (Selvi, 2025) on *Pentanema verbascifolium* showed that the distribution of bioactive compounds is highly dependent on the plant part used. The leaves were reported to be the part richest in phenolic and flavonoid compounds, especially chlorogenic acid, apigenin, and apigenin-7-glucoside, which contributed significantly to high antioxidant activity based on various assays (DPPH, ABTS, FRAP, and CUPRAC) and the highest Relative Antioxidant Capacity Index (RACI) values. In contrast, the root extract, despite having a lower total phenolic content, showed more pronounced enzyme inhibitory activity, including against tyrosinase and α-glucosidase, which was attributed to the presence of specific compounds such as pinoresinol and syringic acid. These findings confirm that aerial parts (leaves and flowers) are generally more suitable as sources of antioxidants, while underground parts (roots) have the potential for more specific functional applications, depending on the target bioactivity. Therefore, the selection of plant parts is a crucial parameter in the design and development of active packaging systems based on natural bioactives, particularly to optimize antioxidant or antimicrobial function in biopolymer matrices such as starch.

The findings regarding the dependence of bioactivity on plant parts are also reinforced by a study on the bael plant (*Aegle marmelos*), which examined volatile bioactive compounds from various plant parts, including roots, stems, leaves, bark, fruit skin, and fruit flesh, using GC–MS. The analysis results showed that each plant part had a significantly different phytochemical profile, both in terms of the type and proportion of bioactive compounds. Some compounds, such as benzene nitro-, benzenepropanoic acid derivatives, and tetradecene, were found in almost all plant parts at varying percentages. However, certain compounds were specific to plant parts, such as phenols, which were only detected in the fruit, especially the skin and were associated with high antioxidant activity. In addition, dibutyl phthalate was identified as the dominant compound in the roots, fruit peel, and fruit flesh, while important bioactive compounds such as coumarin, skimmianine, and cyclobarbital were only found in the roots. This difference further emphasizes that the plant part selection strategy greatly determines the type of bioactivity obtained, whether antioxidant, antimicrobial, or other pharmacological activities. Overall, this study concludes that A. marmelos is a rich source of natural bioactive compounds, with great potential for use in functional food systems and active packaging based on natural compounds (Sharma & Dubey, 2018).

8.3.2 Extraction methods

8.3.2.1 Conventional methods

Conventional techniques are long-established laboratories and industrial practices that typically use heating, organic solvents, and extended processing times without the use of advanced physical aids (e.g., microwaves, ultrasound, high-pressure, or electrical fields). These include liquid–liquid extraction (LLE), solid-phase extraction (SPE), and solid-phase microextraction (SPME), all still widely applied but limited by time efficiency, solvent consumption, and the thermal/oxidative degradation risk for heat-sensitive compounds (Yahya et al., 2018). As summarized by (Barba et al., 2016), the common operations are solid–liquid extraction (SLE), Soxhlet extraction, heating-assisted extraction, maceration, and conventional hot-water extraction.

- Solid–liquid extraction (SLE) Under agitation and moderate heating, SLE dissolves targets from solids into aqueous/organic solvents (e.g., water, ethanol, methanol). While simple and scalable, it is time-consuming, energy-intensive, and may degrade thermolabile species (e.g., anthocyanins and polyphenols).
- Soxhlet. a specific SLE configuration recirculates hot solvent under reflux to approach exhaustive extraction but requires hours of processing and large solvent volumes with thermal-damage risks.

- Heating/maceration can accelerate mass transfer but is suboptimal for phenolics, as high temperatures expedite oxidation and degradation. Before the emergence of pressure-based technologies (e.g., PLE/SWE), hot-water extraction was commonplace, yet it often reduced antioxidant activity and overall recovery relative to modern green techniques (Barba et al., 2016).

8.3.2.2 Green extraction methods

Green extraction aims to reduce environmental impact, minimize the use of hazardous solvents, improve energy efficiency, and deliver food/pharma-grade extracts (Barba et al., 2016). Key techniques include the following:

- Ultrasound-assisted extraction (UAE). Acoustic cavitation (bubble formation/collapse) disrupts cell walls and accelerates mass transfer, shortening extraction time from hours to minutes and increasing polyphenol yields (reports up to ~35% over conventional methods), typically at lower temperatures that protect thermolabile compounds (Barba et al., 2016; Wijngaard et al., 2012). The UAE is readily integrated into existing equipment and is scalable (Barba et al., 2016)·
- High Voltage Electrical Discharges (HVED). HVED (up to 40 kV) generates shock waves, cavitation, reactive species, and electron avalanches that rupture plant tissues, enhancing the release of intracellular compounds. In grape pomace, HVED increased polyphenol yields up to seven-fold compared with untreated material, although careful optimization of energy input control and scale-up is required (Barba et al., 2016)·
- Microwave-assisted extraction (MAE). Microwaves (300 MHz–300 GHz) induce dipole rotation and ionic conduction for rapid volumetric heating, dramatically reducing extraction time (e.g., from 5 h to 5 min) and increasing total phenolics/anthocyanins (Barba et al., 2016, 2017; Wijngaard et al., 2012). Benefits include energy efficiency and compact equipment; however, overheating may form 5-HMF or other undesired products, necessitating precise power/time control (Wijngaard et al., 2012)
- Pulsed electric field (PEF). High-voltage pulses cause electroporation (temporary/permanent pores) that facilitates the diffusion of polyphenols and anthocyanins. Applied to grape pomace, PEF increased polyphenol extraction by ~62% and anthocyanins up to three-fold compared with conventional methods, and operates near ambient temperature, limiting thermal degradation (Barba et al., 2016; Wijngaard et al., 2012)

- Supercritical fluid extraction (SFE). Using CO_2 (critical point 31.1°C) with co-solvents (e.g., ethanol or methanol) enhances the solubility of more polar targets. SFE is well-suited to apolar/lipophilic molecules (essential oils, carotenoids such as β-carotene and lycopene) (Barba et al., 2016; Wijngaard et al., 2012). For instance, CO_2 + 7.5% ethanol at 110 bar and 40°C achieved complete resveratrol recovery from grape skins. Typical variables include temperature, pressure (30–40 MPa), extraction time, particle size, and modifier level (≈10–15% ethanol). It is solvent-residue-free and food-safe, although it is less effective for polar solutes without co-solvent (Wijngaard et al., 2012)·
- Pressurized liquid extraction (PLE)/subcritical water extraction (SWE) Operated at 100–200°C under 4–20 MPa to maintain the liquid phase, these methods exploit reduced water viscosity/surface tension and tunable polarity to extract polar compounds (polyphenols, flavonoids, anthocyanins, and phenolic acids) using water or aqueous ethanol. They offer shorter times and lower solvent use than conventional methods but can cause degradation above 180°C (e.g., gallic, ferulic, chlorogenic acids) and anthocyanin losses above ~100–120°C (Wijngaard et al., 2012)
- High-Pressure Processing (HP). Non-thermal pressures of 200–600 MPa disrupt cellular architecture without high temperatures, enhancing the recovery of heat-sensitive antioxidants (anthocyanins/flavonoids) and improving extract stability, with a reported 40%–80% increase in polyphenol content over conventional processes (Barba et al., 2016)·
- Extraction Assisted by Hydrotropic Solvents. β-Cyclodextrin (β-CD) and related hydrotropes increase the apparent solubility of nonpolar solutes in water via inclusion complexes. Rajha et al. (2015) reported 5.8 mg GAE/g (dry basis) at 37.7 mg/mL β-CD, 66.6°C, and 4.8 h, surpassing conventional ethanolic extraction while remaining safe, selective, and green (Barba et al., 2016).

8.3.3 Bioactivity preservation

The maintenance of bioactive stability is central to defining optimal extraction conditions. Because many plant bioactives are thermolabile and sensitive to oxygen and light, the choice of temperature, pressure, and solvent critically affects the stability during extraction. Key considerations include the following:

- Temperature. Conventional SLE/heating can degrade polyphenols, flavonoids, and anthocyanins during prolonged high-temperature exposure (Barba et al., 2016). In SFE, elevated temperatures accelerate carotenoid degradation (especially lycopene and β-carotene); 60–80°C is often cited for trans-lycopene, whereas >80°C promotes trans→cis isomerization, reducing stability and bioactivity. Light/oxygen also triggers carotenoid oxidation; hence, freeze-drying prior to extraction is recommended (Wijngaard et al., 2012). In PLE/SWE, temperatures above 180°C can degrade gallic, ferulic, and chlorogenic acids, while anthocyanins start deteriorating above 100°C–120°C; hence, non-thermal methods (PEF, UAE) are suitable for preserving these species (Wijngaard et al., 2012).
- Process control. PEF and UAE help maintain integrity by minimizing thermal loads. In contrast, poorly controlled MAE may form HMF and Maillard-type by-products; precise power/time settings are essential (Wijngaard et al., 2012). Selecting water or ethanol–water mixtures (food-grade) across green techniques mitigates unintended side reactions and supports stability (Wijngaard et al., 2012).
- pH and the choice of solvent. Extreme acidic or alkaline extremes can destabilize polyphenols; hence, food-grade solvents are preferred. In SFE, ethanol co-solvent (10%–15%) not only enhances solubility but also helps protect carotenoids from oxidative degradation during SC supercritical processing (Wijngaard et al., 2012)
- Post-processing. Freeze-drying and other gentle dehydration routes preserve activity and improve bioaccessibility by modulating the matrix structure. Non-thermal technologies (HPP, PEF, and ultrasound) can also enhance the release of carotenoids and phenolics from plant matrices (Barba et al., 2017). Notably, the antioxidant capacity does not always correlate with the total phenolic content, as some molecules degrade or react during extraction (Barba et al., 2016). Given these constraints, green extraction is frequently combined with post-extraction encapsulation (e.g., nanoemulsions, cyclodextrin inclusion, micro/nanoencapsulation) to protect bioactives before incorporation into starch films.

8.4 Incorporation of bioactives into starch matrices

Starch has great potential in the development of active packaging by adding bioactive compounds with antioxidant and antimicrobial properties. Starch-based films and coatings enriched with natural extracts or essential oils can extend the shelf life of foods, suppress lipid oxidation, and inhibit the growth of pathogenic microbes (Falcão et al., 2022). In addition, the development of polymer-based food packaging enriches with bioactive compounds from plants as an environmentally friendly alternative to conventional plastic [(Nogueira et al., 2020).

The success of starch-based active packaging development depends not only on the selection of bioactive compounds but also on the incorporation technique into the polymer matrix. The incorporation process determines the dispersion homogeneity, thermal stability, release rate, and functional effectiveness of the bioactive compounds in protecting food products.

Incorporating techniques of bioactives into starch films can be categorized into five major groups: (i) direct mixing, (ii) nanoemulsion, (iii) microencapsulation, (iv) inclusion complexation, and (v) surface coating or lamination. Table 8.3 compares the advantages and disadvantages of various methods for incorporating bioactive compounds into starch matrices. Each technique has its own working principle, advantages, and limitations.

Table 8.3 Comparison of the advantages and disadvantages of various methods for incorporating bioactive compounds into starch matrices.

Method	Advantages	Disadvantages
Blending	Simple and cost-effective	Large particle size and aggregation
Nanoemulsion	Improved bioactive stability and bioavailability of bioactives	Potential instability over time (aggregation and fusion
Microencapsulation	Enhanced physical and chemical stability	Complex production processes
Layer-by-layer Assembly	Allows for multifunctional coatings	Assembly of different types of nano-building blocks is difficult
Coating	Extends shelf life and improves barrier properties	Additional stabilization techniques may be required.

8.4.1 Direct mixing and blending

Direct mixing and blending is the simplest method, in which the bioactive compound is directly mixed into the starch solution before casting or extrusion. It is widely used due to its simple equipment and mild processing conditions. However, because many bioactives are hydrophobic, whereas starch is hydrophilic, nonuniform dispersion and rapid release during storage often occur (Ligarda-Samanez et al., 2025).

For instance, Aloe vera extract was incorporated into a biopolymer matrix through direct blending before casting, followed by oven drying at 50°C for 48 h. This method ensured the homogeneous distribution of the active compound within the film structure, resulting in a biodegradable active film with improved mechanical, hydrophobic, and thermal properties and effectively extended the shelf life of mangoes up to 12 days (Gennaro et al., 2025).

Similarly (Mao et al., 2025), employed the solvent casting technique for starch-based films, consisting of three main steps: preparation of the main matrix solution (starch and other biopolymers), addition of a plasticizer (e.g., glycerol) to improve flexibility, and incorporation of bioactive agents (antimicrobial, antioxidant, or natural extracts) into the homogeneous solution before casting and drying. Research (Amin et al., 2022) resulted starch–κ-carrageenan (PS–κC) films incorporated with crab-apple peel extract (EEC) exhibited smooth and flexible surfaces at low extract concentrations but became slightly rougher at higher concentrations. This technique proved effective for incorporating natural polyphenols into biopolymer films to create antioxidant and antimicrobial active packaging. A similar approach was used to develop smart active bio-nanocomposite films based on chitosan nanoparticles reinforced with purple sweet potato anthocyanin extract (PSPE). Before solvent casting, direct blending successfully produced homogeneous films with enhanced functional performance (Cazón et al., 2025).

8.4.2 Encapsulation

Bioactive compounds derived from secondary metabolites of plants and microorganisms (e.g., polyphenols, flavonoids, terpenoids, essential oils, and vitamins) are known to possess antioxidant, antimicrobial, anti-inflammatory, and therapeutic activities. However, bioactive compounds are generally unstable to light, oxygen, and temperature, making encapsulation in a starch matrix an important strategy to enhance their stability and effectiveness (Falcão et al., 2022)

8.4.2.1 Nanoemulsions

A nanoemulsion is a colloidal system consisting of extremely small oil droplets dispersed within an aqueous phase, where a thin layer of emulsifier molecules surrounds each oil droplet. Nanoemulsion technology is employed to enhance the stability and bioavailability of hydrophobic bioactive compounds, such as oil-soluble vitamins, antimicrobials, and antioxidants (Raddatz & Menezes, 2021). Due to their ultrafine droplet size (<100 nm), nano emulsions offer superior stability, optical clarity, the ability to modulate product texture, and improved bioavailability of lipophilic components (Pédehontaa-Hiaa et al., 2023). Moreover, nanoemulsions protect bioactive compounds from harsh environmental conditions, such as heat, light, and oxygen, while enabling controlled release and targeted delivery.

8.4.2.2 Microencapsulation

Microencapsulation is a process in which a component (the core) is entrapped within a secondary material (the encapsulating matrix). This technique protects bioactive compounds from adverse environmental conditions, such as temperature, pH, humidity, and oxygen, while enhancing their stability and bioavailability. Microencapsulation enables controlled release and targeted delivery of bioactive compounds, and it can also be used to mask undesirable tastes and improve stability during storage and processing (Ligarda-Samanez et al., 2025; Usman et al., 2025) investigated the microencapsulation of bioactive compounds from the Andean wild fruit *Hesperomeles escalloniifolia* (Capachu) through spray drying using quinoa starch (95%) and tara gum (5%). The optimized conditions (inlet temperature of 140°C and encapsulant of 11.6%) produced microcapsules with high phenolic content (9.60 mg GAE/g), strong antioxidant activity, and good physical and chemical stability. FTIR and SEM analyses confirmed physical entrapment without chemical reaction, while release tests showed gradual phenolic diffusion over 24 h, highlighting its potential for natural and eco-friendly applications in food, pharmaceutical, and cosmetic products.

Functional foods such as frozen yogurt have also been enhanced with antioxidant and anticancer compounds from *Salvia officinalis* L. (sage) extract. Microencapsulation via complex coacervation (gum arabic–WPC system) was used to improve the actives' stability and bioavailability. The protein–polysaccharide microcapsule layer uniformly dispersed in the yogurt matrix acts as a carrier system, protecting polyphenols during storage and digestion. During gastrointestinal digestion, the layer gradually degrades, enabling the controlled release of actives in the intestine and thus improving bioaccessibility and bioavailability (Rachmina et al., 2024).

Lamination refers to the process of producing a composite material by bonding multiple layers together. Lamination involves embedding bioactive compounds within or between starch-based films or coatings in the context of starch-based systems. This process can enhance the mechanical strength, barrier performance, and controlled release characteristics of the starch matrix. Laminated films incorporated with antimicrobial agents or UV-blocking compounds have shown great potential for active food packaging applications, extending the shelf life and safety of perishable products (da Conceição Alves de Lima et al., 2025; Iquiapaza et al., 2025).

8.4.2.3 Coating

Edible coatings are thin layers that are applied directly onto the food surface and consumed along with the product. Their main purpose is to protect food from moisture loss, oxidation, respiration, and microbial contamination, thereby extending shelf life and maintaining sensory quality. These coatings are biodegradable, non-toxic, and derived from natural biopolymers, serving as green alternatives to conventional plastics. Moreover, coatings can act as carriers for bioactive compounds, such as antioxidants, antimicrobials, vitamins, and essential oils, enabling their controlled release during storage (Manzoor et al., 2025).

In food applications, four main coating techniques are commonly used: dipping, spraying, spreading, and thin-film hydration. In the dipping method, food products are immersed in the coating solution and then dried to form a protective layer commonly applied to fruits and vegetables (e.g., sodium alginate coating extending carrot shelf life by 7 days). The spraying method uses high pressure (60–80 psi) to form thin, uniform coatings suitable for low-viscosity solutions with essential oils, such as chitosan or alginate. Spreading or brushing involves the manual application of the coating (e.g., on bread or biscuits), allowing for localized control of the thickness. Thin-film hydration is often used in liposome-based systems, where bioactives, such as essential oils, are encapsulated within lipid bilayers for highly efficient controlled release. These techniques provide versatile coating options depending on the food type and material properties (Manzoor et al., 2025).

Films generally act as external physical barrier systems, whereas coatings serve as direct protective layers on the product surface. Both can act as carriers of active compounds such as polyphenols to deliver antioxidants, antimicrobial, and respiration-inhibiting activities; however, their release mechanisms, contact dynamics, and protective areas differ significantly (Motelica et al., 2020).

8.4.3 5. Performance and applications

In active packaging, the focus is the use of plant extracts and essential oils as natural antioxidants and antimicrobials. These bioactives have been shown to inhibit lipid oxidation, prevent the growth of pathogenic microbes, and extend the shelf life of food (Amin et al., 2022). Bioactive compounds derived from fruit are generally rich in antioxidants and antimicrobial agents. By integrating these compounds into packaging, active packaging can be produced, which can extend the shelf life of food, where the packaging has antioxidant and antimicrobial properties, which help inhibit microbial growth and oxidative processes. It also supports the principles of a circular economy by reducing conventional plastic waste and utilizing high-value agricultural by-products (Rodríguez-Rojas et al., 2019). Natural antimicrobial and antioxidant agents (extracted from spices and plants) are preferred as additives because they are considered safe (Generally Recognized as Safe/GRAS) and avoid the side effects, uncontrolled distribution, or undesirable flavors associated with synthetic additives (Otto et al., 2021).

Incorporating plant bioactives into starch matrices not only imparts active functions (antimicrobial and antioxidant) but also measurably alters the resulting films' physical, mechanical, thermal, and barrier performance. This section discusses four key performance aspects, such as antimicrobial performance, antioxidant activity, mechanical and thermal behavior, and barrier properties, and controlled-release mechanisms.

8.4.4 Antimicrobial performance

Incorporating plant bioactives into starch matrices can significantly enhance antimicrobial efficacy, positioning such systems as effective strategies for developing active and sustainable food packaging. Natural compounds, including lignin, curcumin, anthocyanins, essential oils, propolis, and other phenolics, provide microbiological protection while often improving the functional and mechanical properties of starch films.

Essential oils obtained from aromatic plants, including lemon, cinnamon, clove, oregano, tea tree, lemongrass, lavender, and peppermint, have been extensively studied for their antimicrobial efficacy when incorporated into a starch matrix. These oils exhibit broad-spectrum activity against Gram-positive bacteria, Gram-negative bacteria, and fungi. A general trend observed across studies suggests that increasing the concentration of essential oils results in larger zones of microbial inhibition; however, antimicrobial effectiveness is highly dependent on the type of oil, its dominant chemical constituents, and the target microorganism (Syafiq et al., 2020).

For instance, adding lignin to potato-starch films not only increases flexibility and antioxidant capacity but also delivers strong bactericidal effects, reducing the viability of *Escherichia coli* and *Staphylococcus aureus* by up to 99.9% (Ebrahimi et al., 2025). Similarly, the co-incorporation of curcumin and anthocyanins into palm-sugar starch/chitosan bio-nanocomposite films improves antimicrobial interactions at the film surface while modulating release kinetics, thereby enabling sustained antimicrobial activity during storage (Rachmina et al., 2024). Cinnamon essential oil incorporated into starch films exhibits strong activity against *Bacillus cereus, Bacillus subtilis, Aspergillus niger, and Aspergillus brasiliensis,* although excessive oil content may reduce tensile strength, highlighting the need for formulation optimization (Marinopoulou et al., 2022).

Comparable multifunctional effects have been reported for rosemary essential oil (REO), which reduced *S. aureus* and *E. coli* growth by 96.2% and 99.0%, respectively, while increasing film elongation and water solubility without significantly affecting water vapor permeability (Lemos et al., 2025). Extracts from cashew apple bark also improved film performance, delivering significant antimicrobial activity against *S. aureus* and *E. coli,* as well as thermal stability and reduced water vapor permeability (WVP) (da Conceição Alves de Lima et al., 2025). Green propolis added to starch/chitosan films afforded high antioxidant and antimicrobial activities while maintaining mechanical strength, with efficacy against *Candida albicans, E. coli,* and *S. aureus,* a strong candidate for multifunctional antimicrobial packaging (Iquiapaza et al., 2025).

At the mechanistic level, essential oils rich in thymol, carvacrol, eugenol, and limonene exert antimicrobial action by penetrating microbial lipid membranes, disrupting membrane integrity, and inducing leakage of intracellular ions and metabolites. This process collapses the proton gradient and inhibits ATP synthesis. Concurrently, phenolic-rich extracts containing anthocyanins, tannins, and phenolic acids contribute antioxidant effects through radical scavenging, enzyme inhibition, and metal chelation (e.g., Fe and Cu), thereby suppressing oxidative and microbial spoilage pathways. The synergy between these mechanisms enables effective inhibition of bacterial and fungal growth while simultaneously slowing oxidative deterioration, making such bioactives particularly suitable for active food packaging systems (Motelica et al., 2020).

The effectiveness of antimicrobial performance is closely linked to the release behavior of bioactive compounds from the polymer matrix. In hydrophilic matrices such as starch, moisture-induced swelling facilitates diffusion-controlled release, while hydrogen bonding between starch hydroxyl groups and phenolic compounds can retard diffusion and promote sustained release. Encapsulation strategies, including nano- and micro-encapsulation, further enhance stability and regulate release rates of volatile or degradation-sensitive bioactives. Environmental factors such as temperature, humidity, pH, and light exposure, along with polymer structure and bioactive concentration, significantly influence release kinetics. Therefore, effective packaging design must balance bioactive stability, controlled release, and sensory acceptability to ensure long-term antimicrobial efficacy (Siddiqui et al., 2024).

An illustration of the antimicrobial mechanism of a biodegradable film containing an antimicrobial agent derived from bioactive plant compounds is shown in Fig. 8.1. In this system, the antimicrobial agent is embedded in a biopolymer matrix and functions through active release or active capture mechanisms. Upon contact with food, the active compound is gradually released from the packaging layer to the food surface, ensuring a sustained antimicrobial concentration at the food-packaging interface. This controlled release enhances the effectiveness of the antimicrobial agent while minimizing its direct addition to the food matrix (Jafarzadeh et al., 2020)

At the microbial level, the released bioactive compounds interact directly with microbial cells through various mechanisms of action. These compounds disrupt the cytoplasmic membrane by disrupting the phospholipid structure, leading to increased membrane permeability and intracellular electrolyte leakage. Furthermore, the integrity of the outer cell layer is compromised, resulting in structural damage to the cell envelope. Some bioactive agents, particularly phenolic compounds and essential oils, are also capable of chelating essential nutrients and metal ions required for microbial metabolism, thereby inhibiting enzymatic activity and energy production. Furthermore, antimicrobial agents can penetrate cell membranes and disrupt intracellular processes, including enzyme systems involved in cellular metabolism and genetic replication. Some compounds inhibit DNA replication by interacting with bacterial enzymes such as DNA gyrase and topoisomerase, ultimately suppressing cell division and causing

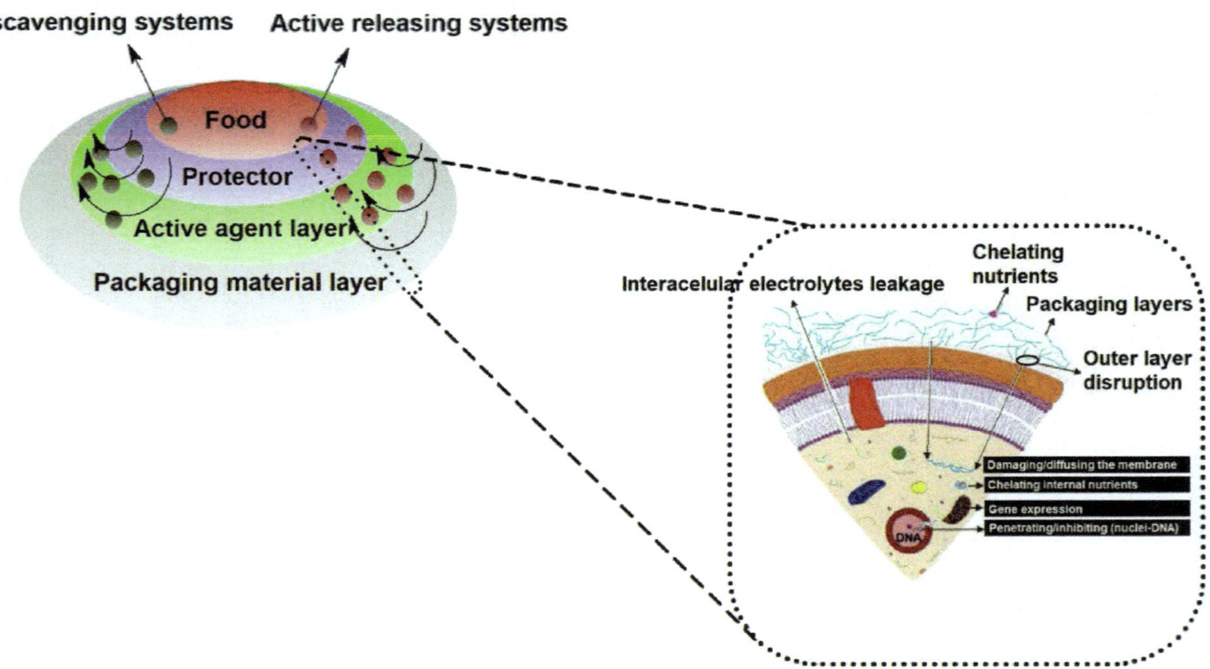

Figure 8.1 ***Antimicrobial mechanism of biodegradable films incorporated with antimicrobial agents.*** This figure illustrates the antimicrobial mechanisms of biodegradable food packaging films containing antimicrobial agents. The active compounds migrate from the polymer matrix to the food packaging interface, where they interact with microbial cells through membrane disruption, enzyme inhibition, and interference with intracellular metabolic processes. Such controlled release and localized activity effectively suppress microbial growth while minimizing direct contact between antimicrobial agents and the food matrix. *From Jafarzadeh, S., Jafari, S. M., Salehabadi, A., Nafchi, A. M., Uthaya Kumar, U. S. & Khalil, H. P. S. A. (2020). Biodegradable green packaging with antimicrobial functions based on the bioactive compounds from tropical plants and their by-products. Trends in Food Science & Technology, 100, 262–277. https://doi.org/10.1016/j.tifs.2020.04.017*

microbial growth stasis. The combined effects of membrane disruption, metabolic inhibition, nutrient deprivation, and genetic disruption lead to irreversible cell damage and microbial cell death. This multi-target antimicrobial mechanism explains the effectiveness of biodegradable active packaging systems in inhibiting spoilage and pathogenic microorganisms and extending the shelf life of food products (Jafarzadeh et al., 2020).

8.4.5 Antioxidant activity

Plant bioactive substantially improves the antioxidant performance of starch-based films, thereby strengthening their protection against oxidative degradation that compromises their quality and shelf life. Natural agents, such as polyphenols, lignin, essential oils, and plant extracts, serve as free-radical scavengers and thermal stabilizers in some systems, yielding functional yet environmentally friendly films.

Phenolic compounds and flavonoids contain hydroxyl groups capable of donating protons/electrons to quench OH and O_2 radicals. These compounds also chelate transition metals (e.g., Fe^{2+}, Cu^{2+}) that catalyze the Fenton reaction, thereby limiting the formation of new radicals. Many antioxidants act as chain-breaking antioxidants, stopping the propagation of radicals in foods. This action helps maintain product quality during storage in packaging systems. Plant bioactives incorporated into starch matrices enhance antioxidant performance through several mechanisms. The addition of lignin significantly increases radical scavenging capacity due to its phenolic structure, up to a 93% increase at 20% lignin content (Ebrahimi et al., 2025). Rosa roxburghii Tratt fruit pulp (RRTP) contains many phenolic compounds that increase the antioxidant activity in corn starch films, reaching 88.59% DPPH radical scavenging at 30% RRTP concentration (Li et al., 2025). The formation of hydrogen bonds between starch and bioactive compounds improves the compatibility and mechanical integrity of starch, thus supporting antioxidant action (Li et al., 2025). In addition, the addition of RRTP or curcumin nanocomplex reduces the permeability of water vapor and oxygen, protecting food from oxidative damage and helping to maintain the antioxidant effectiveness of the film coating over time (Li et al., 2024, 2025).

A representative study showed that co-loading curcumin and anthocyanins into palm-sugar starch/chitosan bio-nanocomposites significantly enhanced their functional properties. The release kinetics differed: curcumin followed the Higuchi model, whereas anthocyanins followed the

Korsmeyer–Peppas model, indicating staged and controlled antioxidant release. Co-loading also lowered the melting enthalpy of the CH/SPS matrix, indicating altered thermal structure and molecular interactions (Rachmina et al., 2024). Similar multifunctional improvements were observed in PVA/starch films containing spent-coffee extract and citric acid, which exhibited enhanced antioxidant release efficiency, antimicrobial activity, water resistance, and mechanical strength (Ounkaew et al., 2018).

Water-soluble lignin derivatives in potato-starch films markedly increased antioxidant capacity, conferred strong bactericidal effects, and delayed retrogradation during storage, indicating increased film stability over time (Ebrahimi et al., 2025). Oxidized starch combined with licorice residue extract also enhanced antioxidant and antimicrobial activities, with a small decline in tensile strength offset by increased elongation at break, i.e., improved flexibility (Zeng et al., 2021). Enriching corn-starch films with Rosa roxburghii Tratt fruit pomace significantly improved the barrier, mechanical, and antioxidant properties of the film, effectively maintaining mushroom freshness for 12 days at 4°C, promising for chilled active packaging (Li et al., 2025). Similarly, black tea extract added to cassava starch films improved antioxidant activity and bolstered UV and water vapor protection while slowing lipid oxidation for up to 35 days of storage (Rajapaksha & Shimizu, 2021).

Diagram of antioxidant action in biodegradable food packaging in Fig. 8.2. This diagram connects three main aspects, namely the source of antioxidants, the method of integration into the polymer, and the final function of the packaging. This diagram shows that environmental factors such as oxygen, humidity, UV radiation, and temperature from the atmosphere can trigger the formation of reactive oxygen species (ROS), which accelerate oxidation reactions in food. To inhibit this process, antioxidants are integrated into a biodegradable polymer matrix. Antioxidants embedded in the polymer matrix gradually migrate from the packaging to the food surface, forming an active protection zone at the packaging-food interface. In this area, antioxidants work by neutralizing free radicals and ROS before these reactive compounds attack sensitive food components such as lipids, proteins, and vitamins (Borkowski & Kępka-Borkowska, 2025).

Figure 8.2 also shows the consequences of this antioxidant mechanism: the elimination of reactive oxygen species, inhibition of oxidation-induced microbial growth, and slowing food quality degradation. The ultimate impact is seen in shelf-life increasing, color and aroma preservation, and product quality stability during storage. Thus, biodegradable packaging

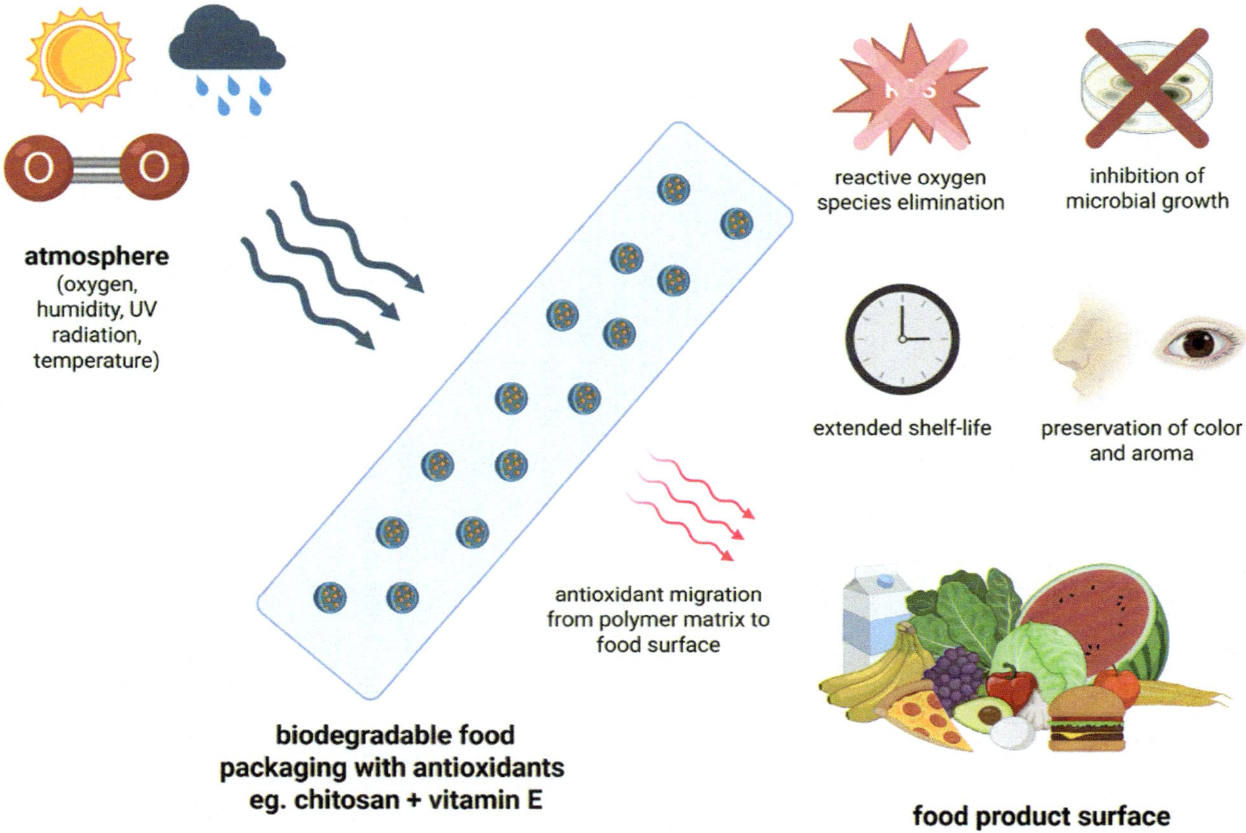

Figure 8.2 ***Antioxidant action in biodegradable food packaging films.*** This figure illustrates the mechanism of antioxidant action in biodegradable food packaging systems. Antioxidant compounds incorporated within the polymer matrix migrate toward the food surface, where they scavenge reactive oxygen species (ROS) and inhibit oxidative reactions. By neutralizing free radicals and limiting oxygen-induced degradation, the antioxidant-enriched packaging effectively delays lipid oxidation and extends the shelf life of packaged food products. *From Borkowski, M. & Kępka-Borkowska, K. (2025). Antioxidants in biodegradable food packaging polymer materials – a review. Polimery/Polymers, 70(7–8), 431–448. https://doi.org/10.14314/polimery.2025.7.1.*

with antioxidants acts as an active protection system that works directly on the food surface, rather than simply acting as a passive barrier (Borkowski & Kępka-Borkowska, 2025).

Overall, plant bioactive embedded in starch films not only boost antioxidant activity but also provide synergistic improvements in mechanical, barrier, and antimicrobial performance. Compounds such as curcumin, anthocyanins, lignin, propolis, and other phenolic-rich extracts slow oxidative reactions via radical scavenging and controlled release, extending shelf life without synthetic additives and advancing sustainable, biodegradable active packaging.

8.4.6 Mechanical properties

The addition of bioactive compounds can improve the active function of the film but often causes a decrease in mechanical strength or an increase in water vapor permeability, depending on the type and concentration (Nogueira et al., 2020). The mechanical effect of films with the addition of bioactive compounds is highly dependent on the type of biopolymer, concentration of bioactive compounds, preparation method, and physicochemical interactions between functional groups (Syafiq et al., 2020). The addition of essential oils tends to decrease tensile strength but increases elongation at break because essential oils act as plasticizers that reduce the cohesion between starch polymer chains and form a more heterogeneous matrix. At low concentrations, some essential oils can even improve film flexibility without significant mechanical degradation, but at high concentrations the mechanical properties generally decrease (Syafiq et al., 2020).

Conversely, certain bioactives promote simultaneous improvements in strength and flexibility. For example, pomegranate rind extract incorporated into citric-acid–crosslinked PVA/starch films increased both tensile strength and elongation, suggesting synergistic reinforcement and controlled plasticization (Dhaundiyal et al., 2025). Specific phenolic compounds, such as quercetin, have been associated with stronger interfacial interactions and enhanced crack resistance (Sunarsono et al., 2025). Anthocyanin extract from purple sweet potato embedded in starch–chitosan matrices increased tensile strength from 25.56 to 43.96 MPa, underscoring how well-dispersed bioactives (or their carriers) can reinforce starch networks (Hasan et al., 2025).

8.4.7 Thermal stability

Bioactive incorporation can alter thermal behavior by modifying hydrogen bonding, crystallinity, and free volume within starch matrices. In starch films containing bioactive proteins such as lactoperoxidase and lysozyme, increases in glass transition temperature (Tg) accompanied by increased brittleness have been observed (Moreno et al., 2015). The addition of citric acid to potato starch reduced the relative crystallinity (from 27.2% to 19.9%) and consistent with chain scission/esterification and modified pasting behavior (Iwasenko Giacomozzi et al., 2021). The gelatinization enthalpy also decreased (from 13.72 to 9.63 J/g), reflecting changes in the double-helix stability (Yadav et al., 2023). Phenolic acids can further tailor the thermal response: caffeic acid altered the thermal characteristics of starch, plausibly via hydrogen bonding and specific starch–phenolic interactions that reorganize local chain packing (Igoumenidis et al., 2018), while other phenolics modified the spatial configuration and noncovalent binding energies of starch, ultimately impacting thermal stability (Chen et al., 2020).

8.4.8 Barrier properties (gas and water vapor)

Concerning barrier properties, particularly water vapor permeability (WVP), summarized research results are varied. In general, hydrophobic essential oils can reduce water absorption and WVP, especially when combined with surfactants or enhancers such as nanocellulose. The addition of nanocellulose has been shown to increase mechanical strength while improving barrier properties through the tortuous path effect and increased hydrogen bonding within the film matrix (Syafiq et al., 2020). Bioactive incorporation affects WVP and gas permeability through competing mechanisms, including network densification, hydrophobization, plasticization, and changes in morphology or crystallinity.

For WVP, both reductions and increases are reported depending on the matrix design. The WVP decreases when the added phase reinforces the film network or raises the hydrophobicity. For instance, adding lentil proteins to starch (via extrusion or thermo-compression) reduced WVP by up to 50% (Ochoa-Yepes et al., 2019). Likewise, poly (vinyl alcohol) (PVOH) blended with potato starch significantly lowered WVP, with a particularly strong effect at 60% PVOH (Gómez-Aldapa et al., 2020). In contrast, acid/oxidative chemical modification of starch can increase WVP, especially under dual modifications that increase water affinity (Biduski et al., 2017).

Increases in WVP were also noted in corn starch films containing jabuticaba-peel flour (FJP) (Ribeiro Sanches et al., 2021) and glycerol, the latter acting as a plasticizer that increases water uptake. Glycerol expands the free volume and chain mobility in thermoplastic starch, elevating water absorption and WVP, effects that can intensify at higher pH (Marinopoulou et al., 2022).

8.4.9 Controlled release and migration studies

Controlled release and migration of plant bioactives in starch-based active packaging are central to long-term functionality. Bioactives are designed not only as passive functional fillers but also as reservoirs that deliver timed release to maintain antioxidant and antimicrobial activity during storage.

Release behavior depends on bioactive matrix interactions and environmental conditions (humidity and temperature). In palm-sugar starch/chitosan bio-nanocomposites, curcumin release followed Higuchi kinetics ($R^2 = 0.9968$) and anthocyanins followed Korsmeyer–Peppas ($R^2 = 0.9808$), indicating the combined roles of diffusion and matrix relaxation (Rachmina et al., 2024). Thermoplastic starch (TPS) films containing thymol, alone or blended with poly(ε-caprolactone) (TPS–PCL), released thymol over 7 days. The TPS–PCL blend showed a higher release rate, which was attributed to the increased molecular mobility imparted by the hydrophobic PCL phase (Lucic Skoric et al., 2022).

The release profiles are also medium-dependent. Starch/PVA films containing phenolics from Clitoria ternatea exhibited release patterns influenced by the food simulant, with aqueous media accelerating release through hydration and swelling of the starch–PVA network (do Nascimento et al., 2024). In starch/PBAT films containing propyl gallate, release correlated with the starch ratio and was faster in high-fat simulants (Gao et al., 2024).

Migration studies provide critical insights into food contact safety and functional persistence. In films with essential oils, migration is tightly linked to the encapsulation method: nanoemulsions and cyclodextrin inclusion complexes improve EO stability and govern the migration of volatile constituents, preserving functional efficacy over longer periods (Gupta et al., 2025).

Multiple factors, including temperature, time, film microstructure, and composition, control release and migration. Higher temperature and prolonged processing generally accelerate diffusion, whereas added phases such as PCL (Lucic Skoric et al., 2022), maltodextrin, cellulose, or bentonite can moderate release by altering pore structure and polarity [(Velloso et al., 2024). Mathematically, the Higuchi and Korsmeyer–Peppas models are widely used to interpret kinetics and guide the design toward product-specific release profiles [(Rachmina et al., 2024; do Nascimento et al., 2024).

From an application and sustainability standpoint, starch-based films with bioactives have been employed to extend shelf life via staged release of antioxidants/antimicrobials, controlling lipid oxidation and microbial growth during storage (do Nascimento et al., 2024; dos Santos et al., 2023; dos Santos et al., 2023). From a safety and regulatory perspective, migration studies are essential to ensure that released amounts remain below international limits (e.g., *European Food Safety Authority* – EFSA) (Wongphan et al., 2025)˙. Thus, a thorough understanding of release and migration mechanisms is fundamental to engineering safe, effective, and environmentally responsible active packaging.

8.5 Safety, regulations, and industrial challenges

8.5.1 Safety aspects

The incorporation of plant-derived bioactive compounds into starch-based food packaging offers substantial benefits in terms of food preservation, safety, and sustainability. Nevertheless, the translation of these systems into real food-contact applications requires careful consideration of safety-related issues to protect consumer health and comply with regulatory frameworks. Key safety concerns include the migration of active substances, toxicity and allergenicity, regulatory compliance, environmental compatibility, and the preservation of film functionality.

The migration of substances from packaging into food remains the most critical safety issue. The use of metal or metal oxide nanoparticles to enhance antimicrobial or barrier properties has raised toxicological and environmental concerns because of their potential accumulation and bioavailability (Singh et al., 2024; Vieira et al., 2022). The concentration of such additives must remain below regulatory limits to mitigate these risks, and migration behavior must be rigorously evaluated. Similarly, the toxicity and allergenicity of natural bioactives, such as essential oils, phenolic antioxidants, and antimicrobial extracts must be assessed to ensure their suitability for food contact (Bhat et al., 2025). Several studies have reported favorable biocompatibility outcomes; for instance, starch and gelatin films incorporated with corn stigma extract maintained cell viability above 90%, indicating a low toxicological risk (Boeira et al., 2022).

Regulatory compliance is another crucial safety dimension. International authorities, such as the European Food Safety Authority and the United States Food and Drug Administration (FDA), require comprehensive toxicological testing, migration analysis, and material characterization before approving new active packaging materials for commercial use (Pawase et al., 2025; Singh et al., 2024). These regulatory procedures ensure that both the active agents and the starch matrix comply with established safety thresholds and do not pose long-term risks to consumers.

From an environmental standpoint, starch-based packaging is inherently advantageous due to its biodegradability and renewable origin. However, the inclusion of bioactive compounds must not compromise biodegradation behavior. Existing evidence suggests that films enriched with natural antioxidants, particularly polyphenols, retain their biodegradability while simultaneously enhancing oxidative stability and antimicrobial protection (Mao et al., 2025; Nahar et al., 2025). Importantly, bioactive incorporation can also reinforce functional properties such as tensile strength, elasticity, and barrier performance, thereby extending shelf life without undermining environmental integrity (Andrade et al., 2025; Kupervaser et al., 2023).

Technological strategies such as encapsulation, polymer blending, and surface modification have been employed to ensure safety and functionality. Encapsulation enables the controlled release of active agents and prevents premature degradation, while blending starch with biopolymers such as polyvinyl alcohol or nanocellulose improves stability, mechanical strength, and migration control (Andrade et al., 2025)' Collectively, these approaches support the development of safe, effective, and environmentally responsible starch-based bioactive packaging.

8.5.2 Consumer perception and acceptance

Beyond technical performance, consumer perception and acceptance strongly influence the successful adoption of starch-based bioactive packaging. Consumers increasingly evaluate packaging not only as a protective material but also as a determinant of food safety, health, and environmental responsibility.

One of the primary concerns of consumers is the migration of substances from packaging into food. The presence of bioactive agents may raise concerns regarding potential toxicity or allergenicity if safety is not clearly demonstrated. Therefore, the use of food-grade, nontoxic bioactives and transparent validation through regulatory approval processes are essential for building consumer trust (Wongphan et al., 2025).

At the same time, there is a growing consumer preference for natural and plant-based ingredients over synthetic additives. Phytochemicals, such as polyphenols, flavonoids, and essential oils, are often associated with health benefits, sustainability, and reduced environmental impact, which positively influences consumer percept (Siddiqui et al., 2024). However, the sensory impact of bioactive compounds is one of the challenges in consumer acceptance. Certain natural compounds can impart strong odors, colors, or tastes to food, which may reduce consumer appeal. Therefore, optimizing the concentration and release profile of bioactive compounds in starch-based films is essential for maintaining sensory neutrality without compromising functionality (Siddiqui et al., 2024).

Environmental sustainability further reinforces consumer acceptance. Biodegradable starch-based films closely align with consumer values related to waste reduction and eco-friendly lifestyles (Garavito et al., 2024; Gonçalves et al., 2024; Pei et al., 2024). In addition, consumers are increasingly recognizing the functional benefits of active packaging, particularly its ability to extend shelf life and reduce food waste, attributes that enhance perceived product value. Therefore, clear labeling, credible sustainability claims, and transparent communication of health and functional benefits play a central role in shaping consumers' willingness to adopt and pay for bioactive starch-based packaging (Andrade et al., 2025; Cui et al., 2021; Pei et al., 2024).

8.5.3 Industrial Scalability

Although the functional benefits of starch-based films with integrated bioactive compounds have been demonstrated in laboratory-scale research, industrial implementation faces several materials, processing, and technological challenges. The first major limitation concerns the stability and solubility of the bioactive compounds. Many plant derived actives, such as phenolic compounds, flavonoids, and essential oils exhibit poor water solubility, thermal instability, or high volatility, which can hinder their performance during film processing and storage. Advanced techniques such as encapsulation, nanoemulsion formation, and polymer blending have been widely explored to overcome these limitations. These methods improve the dispersion, protection, and controlled release of bioactive agents within the starch matrix (Andrade et al., 2025; Tian et al., 2023).

Another critical issue is related to the mechanical properties of starch-based films. Native starch is brittle, hygroscopic, and lacks sufficient flexibility or tensile strength for practical packaging applications. Physical, chemical, and enzymatic modifications of starch, as well as the incorporation of reinforcing nanomaterials such as nanocellulose, bentonite, or montmorillonite, have proven effective in enhancing strength, elasticity, and barrier performance. However, balancing mechanical integrity with the controlled release of bioactives remains a complex task that requires precise formulation design (Kupervaser et al., 2023).

Moisture sensitivity poses yet another significant obstacle for industrial applications. Due to the hydrophilic nature of starch, films readily absorb moisture, resulting in structural deformation and compromised barrier properties. While additives such as hydrophobic lipids or crosslinking agents can mitigate this issue, their optimization at the industrial scale demands further investigation (Arruda et al., 2025; Garavito et al., 2024; Pei et al., 2024).

Scaling up production from laboratory systems to industrial manufacturing introduces additional complexities, including raw material standardization, process reproducibility, extrusion and casting uniformity, and post-production handling. Industrial feasibility also depends on achieving cost-effective production methods without compromising the environmental benefits of starch-based films (Siddiqui et al., 2024; Zainal Arifin et al., 2023). Integration with emerging technologies such as cold plasma treatment, high-pressure processing, ultrasound, and pulsed electric fields (PEF) offers great potential to improve surface activation, antimicrobial performance, and overall film durability. However, incorporating these technologies into existing production lines requires substantial capital investment, technical expertise, and regulatory validation (Yudhistira et al., 2024).

To enable commercialization, the industry must establish robust structure–property relationships linking starch modification, bioactive incorporation, and final film performance. Advanced characterization and standardized testing protocols are essential for ensuring consistent quality, regulatory compliance, and economic viability at scale (Arruda et al., 2025).

8.6 Conclusion

Starch-based smart packaging incorporating plant-derived bioactives has evolved from a laboratory concept to a credible strategy for sustainable and functional food protection. As a renewable and biodegradable polymer, starch provides a versatile matrix capable of hosting natural antioxidants and antimicrobials that address microbial spoilage and oxidative degradation, which are key challenges in modern food systems. When properly engineered, these materials can simultaneously act as physical barriers and controlled-release reservoirs, thereby extending shelf life and reducing the need for synthetic additives.

This chapter synthesized the scientific principles and practical considerations underpinning starch-based bioactive packaging. The major classes of plant bioactives, their mechanisms of action, and their chemical characteristics relevant to packaging design were highlighted. Furthermore, it emphasized the importance of valuing agricultural by-products as bioactive sources to support circular bioeconomy objectives. Advances in green extraction and incorporation strategies, particularly encapsulation and inclusion complexes have been shown to enhance stability, compatibility, and release control compared with direct mixing approaches.

Performance studies have demonstrated that starch-based films can deliver effective antimicrobial and antioxidant activity while maintaining acceptable mechanical and barrier properties, if formulation parameters are carefully optimized. However, safety, regulatory compliance, and consumer acceptance remain central considerations, necessitating rigorous migration studies, toxicological assessments, and transparent benefits communication.

From an industrial perspective, successful commercialization depends on scalable processing methods, consistency of raw materials, cost efficiency, and the integration of advanced surface and structuring technologies. The remaining knowledge gaps include long-term stability under realistic storage conditions, interactions among multi-component bioactives, chronic exposure risks associated with nano-enabled systems, and comprehensive life-cycle assessments. Addressing these challenges through harmonized testing, in situ performance evaluation, and techno-economic analysis will accelerate the transition from laboratory prototypes to market-ready solutions.

In conclusion, starch-based packaging systems enriched with plant-derived bioactives represent a scientifically robust and socially relevant pathway toward safer, more sustainable food packaging. With deliberate design choices and comprehensive validation, these materials hold strong potential to contribute to a more resilient and environmentally responsible food supply chain.

AI disclosure

During the preparation of this work the author(s) used ChatGPT in order to Used ChatGPT to translate Indonesian draft to academic English and to improve grammar/clarity. All content was reviewed and verified by the authors. After using this tool/service, the author(s) reviewed and edited the content as needed and take(s) full responsibility for the content of the publication.

References

Agarwal, S., Singhal, S., Godiya, C. B., & Kumar, S. (2023). Prospects and Applications of Starch based biopolymers. *International Journal of Environmental Analytical Chemistry, 103* (18), 6907–6926. https://doi.org/10.1080/03067319.2021.1963717

Ahari, H., & Soufiani, S. P. (2021). Smart and active food packaging: Insights in novel food packaging. *Frontiers in Microbiology, 12*. https://doi.org/10.3389/fmicb.2021.657233

Ahmed, M. W., Haque, M. A., Mohibbullah, M., Khan, M. S. I., Islam, M. A., Mondal, M. H. T., & Ahmmed, R. (2022). A review on active packaging for quality and safety of foods: Current trends, applications, prospects and challenges. *Food Packaging and Shelf Life, 33*. https://doi.org/10.1016/j.fpsl.2022.100913

Amin, U., Khan, M. K. I., Maan, A. A., Nazir, A., Riaz, S., Khan, M. U., Sultan, M., Munekata, P. E. S., & Lorenzo, J. M. (2022). Biodegradable active, intelligent, and smart packaging materials for food applications. *Food Packaging and Shelf Life, 33*, 100903. https://doi.org/10.1016/j.fpsl.2022.100903

Andrade, M. A., Barbosa, C. H., Ribeiro-Santos, R., Tomé, S., Fernando, A. L., Silva, A. S., & Vilarinho, F. (2025). Emerging trends in active packaging for food: A six-year review. *Foods, 14*(15). https://doi.org/10.3390/foods14152713

Arruda, T. R., Machado, G.d. O., Marques, C. S., Souza, A. L.d., Pelissari, F. M., Oliveira, T. V.d., & Silva, R. R. A. (2025). An overview of starch-based materials for sustainable food packaging: Recent advances, limitations, and perspectives. *Macromol, 5*(2). https://doi.org/10.3390/macromol5020019

Barba, F. J., Mariutti, L. R. B., Bragagnolo, N., Mercadante, A. Z., Barbosa-Cánovas, G. V., & Orlien, V. (2017). Bioaccessibility of bioactive compounds from fruits and vegetables after thermal and nonthermal processing. *Trends in Food Science and Technology, 67*, 195–206. https://doi.org/10.1016/j.tifs.2017.07.006

Barba, F. J., Zhu, Z., Koubaa, M., Sant'Ana, A. S., & Orlien, V. (2016). Green alternative methods for the extraction of antioxidant bioactive compounds from winery wastes and by-products: A review. *Trends in Food Science and Technology, 49*, 96–109. https://doi.org/10.1016/j.tifs.2016.01.006

Bhat, A. A., Ayub, A., & Wani, A. K. (2025). Bioactive compounds in Active Food Packaging (ACP): Health and safety considerations. *Engineering Materials, 21*, 335–357. https://doi.org/10.1007/978-981-96-0369-5_10

Bhatt, P., Pathak, V. M., Bagheri, A. R., & Bilal, M. (2021). Microplastic contaminants in the aqueous environment, fate, toxicity consequences, and remediation strategies. *Environmental Research, 200*. https://doi.org/10.1016/j.envres.2021.111762

Biduski, B., Silva, F. T.d., Silva, W. M.d., Halal, S. L.d. M. E., Pinto, V. Z., Dias, A. R. G., & Zavareze, E.d. R. (2017). Impact of acid and oxidative modifications, single or dual, of sorghum starch on biodegradable films. *Food Chemistry, 214*, 53–60. https://doi.org/10.1016/j.foodchem.2016.07.039

Boeira, C. P., Flores, D. C. B., Alves, J.d. S., Moura, M. R.d., Melo, P. T. S., Rolim, C. M. B., Nogueira-Librelotto, D. R., & Rosa, C. S.d. (2022). Effect of corn stigma extract on physical and antioxidant properties of biodegradable and edible gelatin and corn starch films. *International Journal of Biological Macromolecules, 208*, 698–706. https://doi.org/10.1016/j.ijbiomac.2022.03.164

Borkowski, M., & Kępka-Borkowska, K. (2025). Antioxidants in biodegradable food packaging polymer materials – A review. *Polimery/Polymers, 70*(7–8), 431–448. https://doi.org/10.14314/polimery.2025.7.1

Cazón, P., Mateus, A. R., & Silva, A. S. (2025). Advances in active packaging using natural biopolymers and fruit by-products for enhanced food preservation. *Food Research International, 213*, 116439. https://doi.org/10.1016/j.foodres.2025.116439

Chen, N., Gao, H. X., He, Q., Yu, Z. L., & Zeng, W. C. (2020). Interaction and action mechanism of starch with different phenolic compounds. *International Journal of Food Sciences & Nutrition, 71*(6), 726–737. https://doi.org/10.1080/09637486.2020.1722074

Colín-Chávez, C., Tapia-Esquivel, K. N., Virgen-Ortiz, J. J., Martínez-Téllez, M.Á., Peralta, E., Tejeda-Navarro, D. E., Hernández-Cristobal, O., Talamantes-Soto, R. P., & Hurtado-Macias, A. (2025). In vitro control of fungal phytopathogens from strawberry fruits using active packaging films made with thyme essential oil and polyhydroxybutyrate-co-valerate. *Journal of Natural Pesticide Research, 14*, 100160. https://doi.org/10.1016/j.napere.2025.100160

Cui, C., Ji, N., Wang, Y., Xiong, L., & Sun, Q. (2021). Bioactive and intelligent starch-based films: A review. *Trends in Food Science and Technology, 116*, 854–869. https://doi.org/10.1016/j.tifs.2021.08.024

da Conceição Alves de Lima, A., Monteiro Santana Silva Brito, A., Fonseca Caetano, V., & Vinhas, G. M. (2025). Incorporation of cashew bark extract and gamma irradiation effects on biodegradable bacterial cellulose/kappa-carrageenan film. *Express Polymer Letters, 19*(6), 610–627. https://doi.org/10.3144/expresspolymlett.2025.46

da Silva, L. R., Velasco, J. I., & Fakhouri, F. M. (2022). Bioactive films based on Starch from white, red, and Black rice to food application. *Polymers, 14*(4). https://doi.org/10.3390/polym14040835

Dawam Abdullah, A. H., Firdiana, B., Choerun Nissa, R., Satoto, R., Karina, M., Fransiska, D., Nurhayati, N., Agusman, A., Eko Irianto, H., Priambudi, P., Marliah, S., & Ismadi, I. (2021). Effect OF K-CARRAGEENAN ON mechanical, thermal and biodegradable properties OF starch–CARBOXYMETHYL cellulose (CMC) BIOPLASTIC. *Cellulose Chemistry and Technology, 55*(9–10), 1109–1117. https://doi.org/10.35812/CelluloseChemTechnol.2021.55.95

Debnath, B., Singh, W. S., Das, M., Goswami, S., Singh, M. K., Maiti, D., & Manna, K. (2018). Role of plant alkaloids on human health: A review of biological activities. *Materials Today Chemistry, 9*, 56–72. https://doi.org/10.1016/j.mtchem.2018.05.001

Dhaundiyal, A., Jamwal, V., Mittal, A., Pal, K., Jeevitha, G. C., Tripathi, R., Paul, N., Rameshpathy, M., & Guimarães Junior, M. (2025). Harnessing pomegranate rind waste to develop novel bioactive packaging films using citric acid crosslinked polyvinyl alcohol (PVA)/starch-based polymer matrix. *Biomass Conversion and Biorefinery, 15*(17), 24279–24298. https://doi.org/10.1007/s13399-024-06019-w

do Nascimento, J. V., Silva, K. A., Giuliangeli, V. C., Mendes, A. L. D., Piai, L. P., Michels, R. N., Dal Bosco, T. C., Ströher, G. R., & Shirai, M. A. (2024). Starch-PVA based films with Clitoria ternatea flower extract: Characterization, phenolic compounds release and compostability. *International Journal of Biological Macromolecules, 255.* https://doi.org/10.1016/j.ijbiomac.2023.128232

dos Santos, L. F., Biduski, B., Lopes, S. T., Bertolin, T. E., & dos Santos, L. R. (2023). Brazilian native fruit pomace as a source of bioactive compounds on starch-based films: Antimicrobial activities and food simulator release. *International Journal of Biological Macromolecules, 242.* https://doi.org/10.1016/j.ijbiomac.2023.124900

Ebrahimi, M., José Fabra, M., Martínez-Abad, A., Aussenac, T., Acha, V., & López-Rubio, A. (2025). Water soluble lignin extract from ozonation of Miscanthus giganteus for bioactive starch packaging films. *Food Hydrocolloids, 168*, Article 111564. https://doi.org/10.1016/j.foodhyd.2025.111564

Fakhouri, F. M., Nogueira, G. F., de Oliveira, R. A., & Velasco, J. I. (2019). Bioactive edible films based on arrowroot starch incorporated with cranberry powder: Microstructure, thermal properties, ascorbic acid content and sensory analysis. *Polymers, 11*(10). https://doi.org/10.3390/polym11101650

Falcão, L.d. S., Coelho, D. B., Veggi, P. C., Campelo, P. H., Albuquerque, P. M., & de Moraes, M. A. (2022). Starch as a matrix for incorporation and release of bioactive compounds: Fundamentals and applications. *Polymers, 14*(12). https://doi.org/10.3390/polym14122361

Gao, S., Li, M., Zhai, X., Wang, W., & Hou, H. (2024). Starch as a smart, cheap, and green gatekeeper for the controlled release of propyl gallate from antioxidant biodegradable packaging films. *Food Chemistry, 453*, Article 139627. https://doi.org/10.1016/j.foodchem.2024.139627

Garavito, J., Peña-Venegas, C. P., & Castellanos, D. A. (2024). Production of starch-based flexible food packaging in developing countries: Analysis of the processes, challenges, and requirements. *Foods, 13*(24). https://doi.org/10.3390/foods13244096

Gennaro, M., Büyüktaş, D., Carullo, D., Pinto, A., Dallavalle, S., & Farris, S. (2025). UV-Shielding biopolymer coatings loaded with bioactive compounds for food packaging applications. *Coatings, 15*(7), 741. https://doi.org/10.3390/coatings15070741

Gómez-Aldapa, C. A., Velazquez, G., Gutierrez, M. C., Rangel-Vargas, E., Castro-Rosas, J., & Aguirre-Loredo, R. Y. (2020). Effect of polyvinyl alcohol on the physicochemical properties of biodegradable starch films. *Materials Chemistry and Physics, 239.* https://doi.org/10.1016/j.matchemphys.2019.122027

Gonçalves, E. M., Silva, M., Andrade, L., & Pinheiro, J. (2024). From fields to films: Exploring starch from agriculture raw materials for biopolymers in sustainable food packaging. *Agriculture, 14*(3), 453. https://doi.org/10.3390/agriculture14030453

Gupta, R. K., Pipliya, S., Patel, J., Srivastav, P. P., Castro-Muñoz, R., Wang, C. K., & Nemţanu, M. R. (2025). Essential oil-embedded starch based films for active packaging: Insights into migration, stability, and preservation efficiency. *Trends in Food Science and Technology, 163.* https://doi.org/10.1016/j.tifs.2025.105149

Hammed, A., Yeasmen, N., & Orsat, V. (2025). Eco-Friendly extraction and characterization of terpenoids from plants as functional food ingredients: A review. *Journal of Food Biochemistry, 2025*(1). https://doi.org/10.1155/jfbc/9746960

Hasan, M., Utami, A., Siregar, C., Hanum, L., Khaldun, I., & Nazar, M. (2025). Thermal, structural and mechanical characterization of bio-nanocomposite films from purple sweet potato starch/chitosan nanoparticles integrated with anthocyanin extract. *Results in Engineering, 25*, Article 104448. https://doi.org/10.1016/j.rineng.2025.104448

Igoumenidis, P. E., Zoumpoulakis, P., & Karathanos, V. T. (2018). Physicochemical interactions between rice starch and caffeic acid during boiling. *Food Research International, 109*, 589–595. https://doi.org/10.1016/j.foodres.2018.04.062

Iquiapaza, I. Y. C., Aguilar, G. J., Zitei-Baptista, L. F., da Silva Ferreira, M. E., & Tapia-Blácido, D. R. (2025). Development of active films from Yana Winku Potato Starch/Chitosan blends enriched with alkaline Green Propolis extract. *Journal of Applied Polymer Science, 142*(36). https://doi.org/10.1002/app.57412

Iwasenko Giacomozzi, J., Ruivo Válio Barretti, B., Soltes de Almeida, V., Bet, C. D., da Silva Carvalho Filho, M. A., Lacerda, L. G., Mottin Demiate, I., & Schnitzler, E. (2021). Technological properties of potato starch treated by Heat-Moisture Treatment with addition of organic acids. *Ukrainian Food Journal, 10*(1), 90–99. https://doi.org/10.24263/2304-974x-2021-10-1-8

Jafarzadeh, S., Jafari, S. M., Ali, S., Mohammadi Nafchi, A., Seeta Uthaya Kumar, U., & Abdul Khalil, H. P. S. (2020). Biodegradable green packaging with antimicrobial functions based on the bioactive compounds from tropical plants and their by-products. *Trends in Food Science & Technology, 100*, 262–277. https://doi.org/10.1016/j.tifs.2020.04.017

Kafkas, N. E., Kosar, M., Öz, A. T., & Mitchell, A. E. (2018). Advanced analytical methods for phenolics in fruits. *Journal of Food Quality, 2018*, 1–6. https://doi.org/10.1155/2018/3836064

Kupervaser, M. G., Traffano-Schiffo, M. V., Dellamea, M. L., Flores, S. K., & Sosa, C. A. (2023). Trends in starch-based edible films and coatings enriched with tropical fruits extracts: A review. *Food Hydrocolloids for Health, 4*. https://doi.org/10.1016/j.fhfh.2023.100138

Kurabetta, L. K., Masti, S. P., Gunaki, M. N., Hunashyal, A. A., Chougale, R. B., Dalbanjan, N. P., & Praveen Kumar, S. K. (2025). Vanillin reinforced cationic starch/poly(vinyl alcohol) based antimicrobial and antioxidant bioactive films: Sustainable food packaging materials. *Sustainable Food Technology, 3*(5), 1353–1364. https://doi.org/10.1039/d5fb00132c

Lacourt, C., Mukherjee, K., Garthoff, J., O'Sullivan, A., Meunier, L., & Fattori, V. (2024). Recent and emerging food packaging alternatives: Chemical safety risks, current regulations, and analytical challenges. *Comprehensive Reviews in Food Science and Food Safety, 23*(6). https://doi.org/10.1111/1541-4337.70059

Lang, X., Wang, N., An, X., & Wang, C. (2025). Crab-Apple (Mulus asiatica nakai) Peel extract-enhanced potato Starch/κ-Carrageenan bioactive films: Structural characterization, antioxidant-antimicrobial efficacy, and application in meat preservation. *Polymers, 17*(10), 1328. https://doi.org/10.3390/polym17101328

Lemos, G. S., Pires, J. B., da Silva, F. T., de Souza, E. J. D., Zandoná, G. P., Santos, F. N.d., Hackbart, H. C.d. S., Gandra, E. A., Rodrigues, R.d. S., & Zavareze, E.d. R. (2025). Antimicrobial potential and stability of rosemary essential oil incorporated into Potato Starch-Based films. *Starch/Staerke, 77*(9). https://doi.org/10.1002/star.70082

Li, T., Guan, L., Zeng, M., Li, T., Lei, S., Tang, X., Pan, Y., Li, S., Zhou, M., Yuan, X., Zhang, Z., & Wu, H. (2025). Fabrication of corn starch-based composite films functionalized by Rosa roxburghii Tratt fruit pomace via extrusion compression molding for active food packaging. *International Journal of Biological Macromolecules, 284*(Pt 1), Article 138091. https://doi.org/10.1016/j.ijbiomac.2024.138091

Li, X., Zhang, X., Lv, J., Zhang, X., Li, Y., Han, X., & Zhang, W. (2024). Development of starch-based films reinforced with curcumin-loaded nanocomplexes: Characterization and application in the preservation of blueberries. *International Journal of Biological Macromolecules, 264*(Pt 1), Article 130464. https://doi.org/10.1016/j.ijbiomac.2024.130464

Ligarda-Samanez, C. A., Choque-Quispe, D., Moscoso-Moscoso, E., Ramos-Pacheco, B. S., Palomino-Rincón, H., & Peralta-Guevara, D. E. (2025). Microencapsulation of bioactive compounds from Hesperomeles escalloniifolia Schltdl (Capachu) in quinoa starch and Tara Gum. *CYTA - Journal of Food, 23*(1). https://doi.org/10.1080/19476337.2025.2564354

Liu, X., Xu, F., Yong, H., Chen, D., Tang, C., Kan, J., & Liu, J. (2025). Recent advances in chitosan-based active and intelligent packaging films incorporated with flavonoids. *Food Chemistry X, 25,* 102200. https://doi.org/10.1016/j.fochx.2025.102200

Lucic Skoric, M., Milovanovic, S., Zizovic, I., Ortega-Toro, R., Santagata, G., Malinconico, M., & Kalagasidis Krusic, M. (2022). Supercritical CO2 impregnation of Thymol in thermoplastic starch-based blends: Chemico-physical properties and release kinetics. *Polymers, 14*(20), 4360. https://doi.org/10.3390/polym14204360

Manzoor, A., Khan, T., Lone, A. H., Haque, A., & Banat, F. (2025). Bioactive alginate and carrageenan nanocomposite packaging: A review of recent insights on fabrication techniques, applications and environmental concerns. *Carbohydrate Polymer Technologies and Applications, 12,* Article 101011. https://doi.org/10.1016/j.carpta.2025.101011

Mao, S., Zhang, J., Wu, Q., Xu, Y., Zhang, T., & Lu, C. (2025). Roles of polyphenols incorporated in starch-based films and coatings for fruits and vegetables preservation: A review. *Trends in Food Science & Technology, 163,* Article 105170. https://doi.org/10.1016/j.tifs.2025.105170

Marinopoulou, A., Zoumaki, M., Goulas, A., Raphaelides, S., & Karageorgiou, V. (2022). Biodegradable films from spray dried Starch inclusion complexes with bioactive Compounds—The effect of glycerol and pH. *Starch - Stärke, 74*(9–10). https://doi.org/10.1002/star.202200133

Mironescu, M., Lazea-Stoyanova, A., Barbinta-Patrascu, M. E., Virchea, L. I., Rexhepi, D., Mathe, E., & Georgescu, C. (2021). Green design of novel starch-based packaging materials sustaining human and environmental health. *Polymers, 13*(8). https://doi.org/10.3390/polym13081190

Moreno, O., Atarés, L., & Chiralt, A. (2015). Effect of the incorporation of antimicrobial/antioxidant proteins on the properties of potato starch films. *Carbohydrate Polymers, 133,* 353–364. https://doi.org/10.1016/j.carbpol.2015.07.047

Motelica, L., Ficai, D., Ficai, A., Oprea, O. C., Kaya, D. A., & Andronescu, E. (2020). Biodegradable antimicrobial food packaging: Trends and perspectives. *Foods, 9*(10). https://doi.org/10.3390/foods9101438

Nahar, L., Habibi, E., Gavril, G. L., Abdelfattah, G. M. M., Wrona, M., Nerín, C., Guo, M., & Sarker, S. D. (2025). Towards sustainable food packaging using natural compounds: A review of current research update. *Food and Bioproducts Processing, 150,* 260–274. https://doi.org/10.1016/j.fbp.2025.01.015

Nayanathara Thathsarani Pilapitiya, P. G. C., & Ratnayake, A. S. (2024). , the world of plastic waste: A review. *Cleaner Materials, 11,* Article 100220. https://doi.org/10.1016/j.clema.2024.100220

R. Ningtyas, Tren Teknologi Kemasan Pangan. (2021),

Ningtyas R., Reksa K.D., Muryeti 2024 Production and characterization of active antimicrobial packaging based on carrageenan, carboxymethyl cellulose (CMC) and rosemary essential oil. 060005 060005 doi: 10.1063/5.0229994.

Nogueira, G. F., de Oliveira, R. A., Velasco, J. I., & Fakhouri, F. M. (2020). Methods of incorporating plant-derived bioactive compounds into films made with agro-based polymers for application as food packaging: A brief review. *Polymers, 12*(11), 1–34. https://doi.org/10.3390/polym12112518

Nogueira, G. F., Fakhouri, F. M., & de Oliveira, R. A. (2019). Effect of incorporation of blackberry particles on the physicochemical properties of edible films of arrowroot starch. *Drying Technology, 37*(4), 448–457. https://doi.org/10.1080/07373937.2018.1441153

Nogueira, G. F., Soares, I. H. B. T., Soares, C. T., Fakhouri, F. M., & de Oliveira, R. A. (2022). Development and characterization of arrowroot starch films incorporated with grape Pomace extract. *Polysaccharides, 3*(1), 250–263. https://doi.org/10.3390/polysaccharides3010014

Ochoa-Yepes, O., Di Giogio, L., Goyanes, S., Mauri, A., & Famá, L. (2019). Influence of process (extrusion/thermo-compression, casting) and lentil protein content on physicochemical properties of starch films. *Carbohydrate Polymers, 208*, 221–231. https://doi.org/10.1016/j.carbpol.2018.12.030

Onyeaka, H., Obileke, K., Makaka, G., & Nwokolo, N. (2022). Current research and applications of starch-based biodegradable films for food packaging. *Polymers, 14*(6). https://doi.org/10.3390/polym14061126

Ordoñez, R., Atarés, L., & Chiralt, A. (2022). Biodegradable active materials containing phenolic acids for food packaging applications. *Comprehensive Reviews in Food Science and Food Safety, 21*(5), 3910–3930. https://doi.org/10.1111/1541-4337.13011

Otto, S., Strenger, M., Maier-Nöth, A., & Schmid, M. (2021). Food packaging and sustainability – Consumer perception vs. correlated scientific facts: A review. *Journal of Cleaner Production*, 298. https://doi.org/10.1016/j.jclepro.2021.126733

Ounkaew, A., Kasemsiri, P., Kamwilaisak, K., Saengprachatanarug, K., Mongkolthanaruk, W., Souvanh, M., Pongsa, U., & Chindaprasirt, P. (2018). Polyvinyl alcohol (PVA)/Starch bioactive packaging film enriched with antioxidants from spent coffee ground and citric acid. *Journal of Polymers and the Environment, 26*(9), 3762–3772. https://doi.org/10.1007/s10924-018-1254-z

Pan, J., Li, C., Liu, J., Jiao, Z., Zhang, Q., Lv, Z., Yang, W., Chen, D., & Liu, H. (2024). Polysaccharide-Based packaging coatings and films with phenolic compounds in preservation of fruits and Vegetables—A review. *Foods, 13*(23), 3896. https://doi.org/10.3390/foods13233896

Pawase, P. A., Pathare, A. M., Bashir, O., Saleem, F., Shrama, E., Mudgal, S., & Ahmad, M. (2025). Physical, chemical, and nano-enabled modifications of starch for sustainable food packaging films: Recent trends, challenges, and prospects. *Carbohydrate Polymer Technologies and Applications, 11*. https://doi.org/10.1016/j.carpta.2025.100986

Pech-Cohuo, S. C., Martín-López, H., Uribe-Calderón, J., González-Canché, N. G., Salgado-Tránsito, I., May-Pat, A., Cuevas-Bernardino, J. C., Ayora-Talavera, T., Cervantes-Uc, J. M., & Pacheco, N. (2022). Physicochemical, mechanical, and structural properties of bio-active films based on biological-chemical chitosan, a novel ramon (Brosimum alicastrum) Starch, and quercetin. *Polymers, 14*(7). https://doi.org/10.3390/polym14071346

Pédehontaa-Hiaa, G., Gaudière, F., Khelif, R., Morin-Grognet, S., Labat, B., Lutzweiler, G., Le Derf, F., Atmani, H., Morin, C., & Ladam, G. (2023). Polyvalent incorporation of anionic β-cyclodextrin polymers into layer-by-layer coatings. *Colloids and Surfaces A: Physicochemical and Engineering Aspects, 664*, Article 131154. https://doi.org/10.1016/j.colsurfa.2023.131154

Pei, J., Palanisamy, C. P., Srinivasan, G. P., Panagal, M., Kumar, S. S. D., & Mironescu, M. (2024). A comprehensive review on starch-based sustainable edible films loaded with bioactive components for food packaging. *International Journal of Biological Macromolecules, 274*. https://doi.org/10.1016/j.ijbiomac.2024.133332

Promhuad, K., Bumbudsanpharoke, N., Wadaugsorn, K., Sonchaeng, U., & Harnkarnsujarit, N. (2022). Maltol-Incorporated acetylated cassava starch films for shelf-life-extension packaging of bakery products. *Polymers, 14*(24), 5342. https://doi.org/10.3390/polym14245342

Rachmina, R., Hasan, M., Hasanah, U., & Halim, A. (2024). Sugar palm Starch/Chitosan bionanocomposite films incorporated with anthocyanin and curcumin – Thermal properties and release kinetics. *Journal of Ecological Engineering, 25*(2), 300–308.

Raddatz, G. C., & Menezes, C. R.de (2021). Microencapsulation and co-encapsulation of bioactive compounds for application in food: Challenges and perspectives. *Ciência Rural, 51*(3). https://doi.org/10.1590/0103-8478cr20200616

Rajapaksha, S. W., & Shimizu, N. (2021). Development and characterization of functional starch-based films incorporating free or microencapsulated spent Black tea extract. *Molecules, 26*(13), 3898. https://doi.org/10.3390/molecules26133898

Rajha, H. N., Chacar, S., Afif, C., Vorobiev, E., Louka, N., & Maroun, R. G. (2015). β-cyclodextrin-assisted extraction of polyphenols from vine shoot cultivars. *Journal of Agricultural and Food Chemistry, 63*(13), 3387–3393. https://doi.org/10.1021/acs.jafc.5b00672

Ribeiro Sanches, M. A., Camelo-Silva, C., Tussolini, L., Tussolini, M., Zambiazi, R. C., & Becker Pertuzatti, P. (2021). Development, characterization and optimization of biopolymers films based on starch and flour from jabuticaba (Myrciaria cauliflora) peel. *Food Chemistry, 343*. https://doi.org/10.1016/j.foodchem.2020.128430

Rodríguez-Rojas, A., Arango Ospina, A., Rodríguez-Vélez, P., & Arana-Florez, R. (2019). ¿What is the new about food packaging material? A bibliometric review during 1996–2016. *Trends in Food Science and Technology, 85*, 252–261. https://doi.org/10.1016/j.tifs.2019.01.016

Selvi, B. (2025). Comparative study of Pentanema verbascifolium extracts: Phytochemical composition, antioxidant potential, and enzyme inhibition across plant parts. *Food Bioscience, 68*, 106607. https://doi.org/10.1016/j.fbio.2025.106607

Serra, M., Weng, V., Coelhoso, I. M., Alves, V. D., & Brazinha, C. (2020). Purification of arabinoxylans from corn fiber and preparation of bioactive films for food packaging. *Membranes, 10*(5). https://doi.org/10.3390/membranes10050095

Sharma, N., & Dubey, W. (2018). Determination OF volatile bioactive compounds from extracts OF bael (AEGLE marmelos) plant parts and their comparative analysis. *Asian Journal of Pharmaceutical and Clinical Research, 11*(3), 393. https://doi.org/10.22159/ajpcr.2018.v11i3.23529

Shen, N., Wang, T., Gan, Q., Liu, S., Wang, L., & Jin, B. (2022). Plant flavonoids: Classification, distribution, biosynthesis, and antioxidant activity. *Food Chemistry, 383*, 132531. https://doi.org/10.1016/j.foodchem.2022.132531

Shen, H., Zhou, X., Gao, Y., Lyu, L., Wang, G., & Qian, Y. (2025). Biodegradability and barrier properties of polyvinyl alcohol and Konjac glucomannan composite films loaded with eucalyptus essential oil@β-Cyclodextrin for sustainable food packaging. *International Journal of Biological Macromolecules, 334*, 149053. https://doi.org/10.1016/j.ijbiomac.2025.149053

Siddiqui, S. A., Khan, S., Mehdizadeh, M., Bahmid, N. A., Adli, D. N., Walker, T. R., Perestrelo, R., & Câmara, J. S. (2024). Corrigendum to "Phytochemicals and bioactive constituents in food packaging - A systematic review" [heliyon 9(11) (November 2023) e21196]. *Heliyon, 10*(2), Article e22184. https://doi.org/10.1016/j.heliyon.2023.e22184

Singh, A. K., Kim, J. Y., & Lee, Y. S. (2022). Phenolic compounds in active packaging and edible Films/Coatings: Natural bioactive molecules and novel packaging ingredients. *Molecules, 27*(21), 7513. https://doi.org/10.3390/molecules27217513

Singh, A. K., Lee, M., Jang, D., & Lee, Y. S. (2024). Non-conventional starch nanoparticles: Novel avenues towards improving sustainability of the food packaging sector. *Trends in Food Science and Technology, 143*. https://doi.org/10.1016/j.tifs.2023.104273

Siracusa, V., Rocculi, P., Romani, S., & Rosa, M. D. (2008). Biodegradable polymers for food packaging: A review. *Trends in Food Science and Technology, 19*(12), 634–643. https://doi.org/10.1016/j.tifs.2008.07.003

Sunarsono, H., Abral, H., Mahardika, M., Railis, R. M., Sapuan, S. M., Azka, M. A., Rahmad, & Arrafi, M. R. (2025). Enhancing mechanical properties of durian (Durio ziberthinus murr.) seed starch biopolymer filled with Indonesian bay leaf (Syzygium polyanthum) extract. *Biomass and Bioenergy, 202*. https://doi.org/10.1016/j.biombioe.2025.108159

Syafiq, R., Sapuan, S. M., Zuhri, M. Y. M., Ilyas, R. A., Nazrin, A., Sherwani, S. F. K., & Khalina, A. (2020). Antimicrobial activities of starch-based biopolymers and biocomposites incorporated with plant essential oils: A review. *Polymers, 12*(10), 2403. https://doi.org/10.3390/polym12102403

Tian, B., Liu, J., Liu, Y., & Wan, J. B. (2023). Integrating diverse plant bioactive ingredients with cyclodextrins to fabricate functional films for food application: A critical review. *Critical Reviews in Food Science and Nutrition, 63*(25), 7311–7340. https://doi.org/10.1080/10408398.2022.2045560

Tongnuanchan, P., & Benjakul, S. (2014). Essential oils: Extraction, bioactivities, and their uses for food preservation. *Journal of Food Science, 79*(7). https://doi.org/10.1111/1750-3841.12492

Usman, I., Sana, S., Jaffar, H. M., Munir, M., Afzal, A., Sukhera, S., Boateng, I. D., Afzaal, M., & Urugo, M. M. (2025). Recent progress in edible films and coatings: Toward green and sustainable food packaging technologies. *Applied Food Research, 5*(2). https://doi.org/10.1016/j.afres.2025.101070

Valderrama Solano, A. C., & de Rojas Gante, C. (2012). Two different processes to obtain antimicrobial packaging containing natural oils. *Food and Bioprocess Technology, 5*(6), 2522–2528. https://doi.org/10.1007/s11947-011-0626-3

Vasile, C., Sivertsvik, M., Miteluţ, A., Brebu, M., Stoleru, E., Rosnes, J., Tănase, E., Khan, W., Pamfil, D., Cornea, C., Irimia, A., & Popa, M. (2017). Comparative analysis of the composition and active property evaluation of certain essential oils to assess their potential applications in active food packaging. *Materials, 10*(1), 45. https://doi.org/10.3390/ma10010045

Velloso, C. C. V., Borges, R., Badino, A. C., Oliveira-Paiva, C. A., Ribeiro, C., & Farinas, C. S. (2024). Modulation of starch-based film properties for encapsulation of microbial inoculant. *International Journal of Biological Macromolecules, 283*. https://doi.org/10.1016/j.ijbiomac.2024.137605

Vieira, I. R. S., de Carvalho, A. P. A.d., & Conte-Junior, C. A. (2022). Recent advances in biobased and biodegradable polymer nanocomposites, nanoparticles, and natural antioxidants for antibacterial and antioxidant food packaging applications. *Comprehensive Reviews in Food Science and Food Safety, 21*(4), 3673–3716. https://doi.org/10.1111/1541-4337.12990

Wijngaard, H., Hossain, M. B., Rai, D. K., & Brunton, N. (2012). Techniques to extract bioactive compounds from food by-products of plant origin. *Food Research International, 46*(2), 505–513. https://doi.org/10.1016/j.foodres.2011.09.027

Wongphan, P., Canellas, E., Nerín, C., Estremera, C., Harnkarnsujarit, N., & Vera, P. (2025). Screening and relative quantification of migration from novel thermoplastic Starch and PBAT blend packaging. *Foods, 14*(13), 2171. https://doi.org/10.3390/foods14132171

Wyrwa, J., & Barska, A. (2017). Innovations in the food packaging market: Active packaging. *European Food Research and Technology, 243*(10), 1681–1692. https://doi.org/10.1007/s00217-017-2878-2

Xu, T., Liu, Y., Zhu, T., Zhang, L., Liu, C., & Zheng, L. (2025). Fabrication of whey protein isolate-peach gum complex coacervated microcapsules containing sweet orange essential oil and their application on fresh-cut pear preservation. *Food Chemistry, 495*, 146514. https://doi.org/10.1016/j.foodchem.2025.146514

Yadav, P., Sabith, M., & Bosco, S. J. D. (2023). Effect of citric acid modification on Physio-Chemical, structural, rheological, thermal properties, and in vitro digestibility of Amaranthus paniculatus (Rajgeera) starch. *Starch - Stärke, 75*(9–10). https://doi.org/10.1002/star.202200120

Yahya, N. A., Attan, N., & Wahab, R. A. (2018). An overview of cosmeceutically relevant plant extracts and strategies for extraction of plant-based bioactive compounds. *Food and Bioproducts Processing, 112*, 69–85. https://doi.org/10.1016/j.fbp.2018.09.002

Yang W., Chen X., Li Y., Guo S., Wang Z., Yu X.. Advances in Pharmacological Activities of Terpenoids. Natural Product Communications 15 (3) 1–13.

Yudhistira, B., Husnayain, N., Punthi, F., Gavahian, M., Chang, C. K., & Hsieh, C. W. (2024). Progress in the application of emerging technology for the improvement of starch-based active packaging properties: A review. *ACS Food Science & Technology, 4* (9), 1997–2012. https://doi.org/10.1021/acsfoodscitech.4c00260

Zainal Arifin, M., Adzahan, N. M., Zainal Abedin, N., & Lasik-Kurdyś, M. (2023). Utilization of food waste and By-Products in the fabrication of active and intelligent packaging for seafood and meat products. *Foods, 12*(3), 456. https://doi.org/10.3390/foods12030456

Zeng, F., Weng, Z., Zheng, H., Xu, M., Liang, X., & Duan, J. (2021). Preparation and characterization of active oxidized starch films containing licorice residue extracts and its potential against methicillin-resistant S. aureus. *International Journal of Biological Macromolecules, 187*, 858–866. https://doi.org/10.1016/j.ijbiomac.2021.07.179

Żukowska, G., & Durczyńska, Z. (2024). Properties and applications of essential oils: A review. *Journal of Ecological Engineering, 25*(2), 333–340. https://doi.org/10.12911/22998993/177404

CHAPTER NINE

Starch-based active packaging added with a cross-linker

Simin Fan[1,2], Yiqin Zhang[1]
[1]College of Food Science, Sichuan Agricultural University, Ya'an, PR China
[2]Institute of Food Science and Technology, Chinese Academy of Agricultural Sciences, Beijing, PR China

9.1 Introduction

The global packaging industry is currently at a pivotal juncture, facing mounting pressure to transition from traditional petroleum-based plastics toward sustainable and environmentally benign alternatives (Mendes & Pedersen, 2021). Conventional plastics, derived from non-renewable resources, pose a severe threat to ecosystems due to their recalcitrant nature, contributing significantly to land and marine pollution, and microplastic contamination (Yates et al., 2021). This environmental crisis, coupled with growing consumer awareness and stringent governmental regulations, has catalyzed intensive research into biodegradable polymers (Velasquez et al., 2025). Among the myriads of bio-based materials, starch has emerged as one of the most promising candidates for developing green packaging materials, owing to its abundance, complete biodegradability, non-toxicity, and low cost (Cheng, Chen, McClements, Yang, Zhang, Ren, et al., 2021; Cui et al., 2021).

Starch, a natural polysaccharide stored in plant granules, is composed of two primary macromolecules, linear amylose and highly branched amylopectin. The inherent hydrophilicity of these molecules, due to the plethora of hydroxyl groups, grants starch its biodegradability but also presents the most significant drawback for packaging applications: poor mechanical strength and high sensitivity to moisture (Cheng et al., 2021; Han et al., 2020; Matheus, Dalsasso, Rebelatto, Andrade, Andrade, Andrade, et al., 2023; Zhang, Chen, et al., 2021a). Native starch-based films often exhibit brittleness, low tensile strength, and unsatisfactory water vapor barrier properties, which severely limit their practical utility (Cho et al., 2024; Karmakar et al., 2024). To overcome these limitations, various strategies have been

Starches in Smart Packaging
ISBN: 978-0-443-45396-0
https://doi.org/10.1016/B978-0-443-45396-0.00011-6

employed, including chemical modification, blending with other polymers, and the incorporation of plasticizers (Matheus et al., 2023).

In this context, the use of cross-linking agents has been identified as a highly effective chemical modification technique to enhance the performance of starch-based materials. Cross-linking involves the formation of covalent bonds between starch molecular chains, creating a three-dimensional network structure (Cho et al., 2024). This process fundamentally alters the physicochemical properties of the starch matrix. The introduction of cross-links restricts the mobility of the polymer chains, leading to notable improvements in mechanical strength (Yang et al., 2024), thermal stability (Liu, Wen, et al., 2024), and, most importantly, resistance to water and humidity (Srinivasan et al., 2025; Zhu, Zhang, Liu, Ma, He, Dong, et al., 2022). Additionally, the cross-linking process can influence the release kinetics and efficacy of the encapsulated active compounds in food packaging (Li et al., 2025; Xu, Du, Shan, Wang, Li, Tao, et al., 2025a; Zhang et al., 2025). Therefore, understanding the interplay between the cross-linking density, the network morphology, and the behavior of the active agent is paramount for designing efficient packaging systems.

Therefore, this chapter aims to provide a comprehensive and systematic overview of the cross-linking technology applied to starch-based active packaging. The primary objective is to consolidate the current understanding of various cross-linkers, focusing on their distinct cross-linking mechanisms with starch molecules. Furthermore, this article will critically analyze and summarize the profound effects of these cross-linking reactions on the physicochemical properties of the resulting starch-based films. By synthesizing the existing literature, this review aims to bridge the gap between cross-linker chemistry and the functional performance of starch-based films, comparing the advantages and limitations of various agents, and paving the way for future innovations in high-performance, biodegradable active packaging.

9.2 Types of cross-linking agent for starch-based films

The performance of starch-based films is profoundly influenced by the type of cross-linking agent used (Cheng et al., 2025). These agents facilitate the formation of bridges or bonds between starch polymer chains, creating a robust three-dimensional network (Dai et al., 2021). Based on their chemical nature and reaction mechanisms (shown in Fig. 9.1, cross-linkers for starch can be broadly categorized into several types, each imparting distinct characteristics to the final film (Table 9.1).

9.2.1 Chemical cross-linkers

Chemical cross-linkers represent the most prevalent category for modifying starch-based films, functioning by establishing covalent bridges between starch molecules or starch derivatives. This process creates a robust three-dimensional network that significantly enhances the film's performance (Cho et al., 2024). These cross-linking reactions typically target the hydroxyl or carboxyl groups on starch chains. The cross-linking process itself often requires specific conditions, such as elevated temperature, a particular pH, or the presence of a catalyst, to proceed efficiently. Once formed, the covalent bonds are generally irreversible, resulting in a permanently stabilized polymer structure.

Compounds like glutaraldehyde have been traditionally used due to their high cross-linking efficiency. They form acetal linkages with starch hydroxyl groups (Fig. 9.2a), significantly enhancing mechanical strength and water resistance (Gonenc & Us, 2019; Tavassoli, Abedi-Firoozjah, Bahramian, Hashemi, Noori, Oladzadabbasabadi, et al., 2025). For example, Wang et al. (2016) (Wang et al., 2016) employed glutaraldehyde vapor to cross-link electrospun starch nanofibers. Their results demonstrated that this cross-linking treatment significantly enhanced the mechanical properties and water resistance of the starch membrane without altering its fibrous morphology. Notably, the tensile strength of the cross-linked fibers exhibited an approximately tenfold increase compared to their non-crosslinked counterparts. However, concerns regarding the potential toxicity and residual formaldehyde of some aldehyde-based cross-linkers have limited their application in food packaging, driving the search for greener alternatives. Citric acid (CA) is a prime example of a non-toxic, bio-based chemical cross-linker. Its cross-linking mechanism involves the formation of ester linkages between the carboxyl groups of CA and the hydroxyl groups of starch, often under heat and with the aid of a catalyst like sodium hypophosphite (Fig. 9.2b) (Chi et al., 2024; Skov et al., 2025; Yao et al., 2022). CA cross-linking not only improves mechanical and barrier properties but can also introduce antimicrobial activity to the film matrix, as its low pH can inhibit microbial growth (Lin et al., 2024). In a study aimed at developing a simple and eco-friendly modification method for

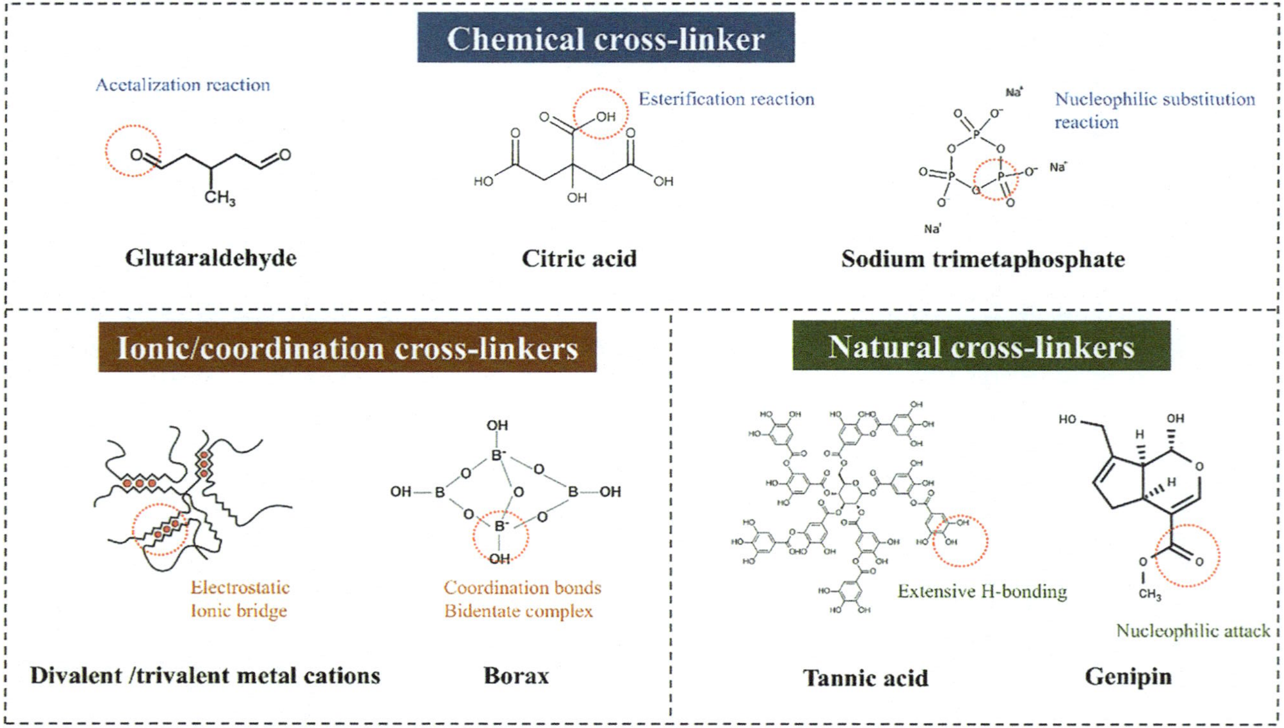

Figure 9.1 Chemical structures and reaction sites of common cross-linkers in starch-based films.

Table 9.1 Advantages and disadvantages of different cross-linker types.

Cross-linker type	Representative	Advantages	Disadvantages	References
Chemical cross-linkers	Glutaraldehyde; citric acid; maleic anhydride; sodium trimetaphosphate	High cross-link density; strong improvement in water resistance and mechanical strength; tunable by dosage	Toxic residues, regulatory restrictions; potential odor/color changes	Li et al. (2025) Yuan, Wei, et al. (2025b) Zhao, Zheng, et al. (2024) (Tavassoli et al.,2025)
Ionic/coordination cross-linkers	Ca^{2+}, Mg^{2+}, Al^{3+}, Fe^{3+} and their salts; boric acid salts	Mild processing; dynamic/reversible cross-links; potential responsive behavior; relatively safe	Migration risk; regulatory limits in some regions; cross-link density and stability depend on pH/temperature	Kumar et al. (2025) Wei et al. (2023).
Natural cross-linkers	Genipin; tannic acid; gallic acid; phytic acid	No chemical residues; safe for food contact; potential edible applications	Generally weaker and less stable under humidity/temperature changes; non-permanent	Cheng et al. (2025) Ahmed et al. (2024) Zhang et al. (2023b)
Enzymatic cross-linkers	Transglutaminase; laccase	Mild processing; food-grade; specific cross-linking; compatible with protein-containing blends	Requires reactive amino groups; cost and enzyme activity considerations; slower reaction	AL-Hassan and Norziah (2017) Rosseto et al. (2025)

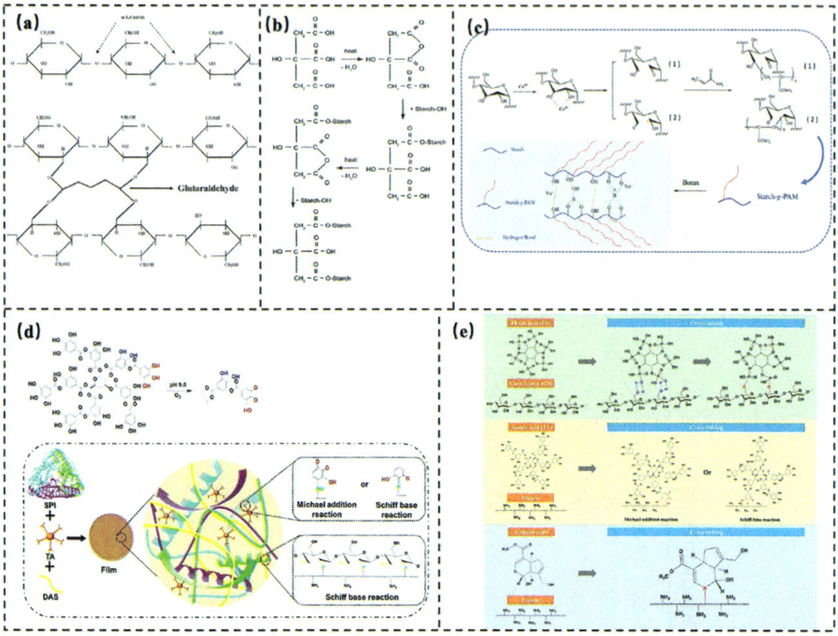

Figure 9.2 Schematic of starch polymer chains before and after cross-linking. (A) Cross-linking reaction of starch with glutaraldehyde (Gonenc & Us, 2019). (B) Crosslinking of starch with citric acid (Chi et al., 2024). (C) Borate anions ($B_4O_7^{2-}$) react with the hydroxyl groups of two adjacent starch chains, forming a complex di-diol ester bridge (Lu et al., 2024). (D) enhancing the properties of soy protein isolate and dialdehyde starch films through tannic acid crosslinking (Zhao, Chen, et al., 2024).

starch-based packaging film, a starch/polyvinyl alcohol/carboxylate cellulose nanofiber film was fabricated through a combination of hydrogen bonding and CA esterification cross-linking (Yuan, Wei, et al., 2025b). The CA-induced three-dimensional network was found to enhance the water resistance of the film. Furthermore, this cross-linked structure, stabilized by strong hydrogen and ester bonds, effectively suppressed the structural retrogradation of the film during storage.

Furthermore, organophosphates, such as sodium trimetaphosphate (STMP), are extensively employed as cross-linkers in starch-based packaging films. As a common phosphorylating agent, STMP introduces distarch phosphate di-ester bridges between the hydroxyl groups of adjacent starch chains. This covalent cross-linking network significantly enhances the mechanical integrity and aqueous stability of the films. Consequently, the modified starch matrix exhibits increased viscosity, improved stability, and superior resistance to swelling and dissolution in aqueous environments (Sharma et al., 2021).

9.2.2 Ionic/coordination cross-linkers

This class of cross-linkers operates via ionic interactions or coordination bonds, rather than stable covalent bonds, to form three-dimensional networks. They are frequently applied to modified starch systems, like oxidized starch or starch nanocrystals. The cross-linked network arises through two primary mechanisms: the formation of ionic bridges or coordination cross-links between polymer chains, or reversible complexation with the rotatable dihydroxy/ortho-hydroxy groups on sugar molecules. Since these interactions are generally mild and reversible, the resulting networks are highly susceptible to environmental conditions such as temperature, pH, and ionic strength (Wei et al., 2023).

Divalent or trivalent metal cations, such as calcium (Ca^{2+}), magnesium (Mg^{2+}), cuprum (Cu^{2+}) or aluminum (Al^{3+}), can act as ionic cross-linkers (Zhang et al., 2023a). They form "egg-box" structures with anionic groups (e.g., carboxylate groups from oxidized starch), bridging adjacent polymer chains. This process effectively strengthens the film network, improving its tensile strength and reducing its water vapor permeability. In some starch systems with carboxylic acid groups or modified ones, metal ions enhance the network through bridging effects, but attention should be paid to metal migration and regulatory restrictions (Kumar et al., 2025).

Borax (sodium tetraborate) is a classic coordination cross-linker for starch (Zhang, Zhao, et al., 2021b). In aqueous solution, borate anions ($B_4O_7^{2-}$) react with the hydroxyl groups of two adjacent starch chains, forming a complex di-diol ester bridge (Fig. 9.2c) (Lu et al., 2024; H. Lu, Ji, Li, Wang, Xiong, Zhou, et al., 2019;). This results in a dramatic improvement in film toughness and flexibility. A study on the cross-linking modification of starch with boric acid revealed that the resulting thermoplastic films exhibited significantly improved mechanical strength, water resistance, and transparency. The film incorporating 0.5% boric acid showed the maximum tensile strength and the optimal water vapor barrier properties (Yoğurtçu & Gürler, 2024). However, the potential toxicity of boron restricts its use in food-contact applications.

9.2.3 Natural cross-linkers

The growing demand for fully bio-based and non-toxic packaging has spurred the exploration of natural compounds as green cross-linkers. These agents typically form covalent networks between starch chains via acid/alkaline or enzymatic/oxidative catalysis. A key advantage of these methods is their superior biocompatibility and safety. Representative cross-linking mechanisms include the oxidative coupling of phenolic compounds, esterification/etherification reactions involving hydroxyl and carboxyl groups, and co-cross-linking within systems containing starch derivatives or proteins.

Polyphenolic compounds, such as tannic acid (TA), gallic acid (GA), and cinnamic acid (CA), serve as effective cross-linkers and modifiers for starch-based film. Their primary mechanisms involve oxidative polymerization to form radical intermediates or the formation of covalent bonds with starch or co-polymers under alkaline/enzymatic conditions. These interactions significantly enhance the mechanical strength and water resistance of the resulting starch-based films. Furthermore, polyphenols are often applied in synergy with emulsifiers or plant extracts to achieve multifunctional performance enhancements (Kurabetta, Masti, Eelager, Gunaki, Madihalli, Hunashyal, et al., 2023). For example, TA acts as a versatile modifier and cross-linker for starch-based films (Zhang et al., 2023b). Its mechanism is primarily based on the formation of numerous strong hydrogen bonds between its many phenolic hydroxyl groups and the hydroxyl groups on starch chains. Additionally, under certain conditions (e.g., alkaline pH or the presence of free radical initiators), TA can undergo oxidation and form quinone structures, which may subsequently react with nucleophilic sites on starch, leading to covalent-like interactions (Fig. 9.2d) (Zhao, Tian, et al., 2024).

Genipin, extracted from the fruits of *Gardenia jasminoides*, genipin is one of the most well-established and efficient natural cross-linkers. It is renowned for its ability to react with primary amine groups (-NH_2), forming intense blue-pigmented and stable heterocyclic compounds (Ahmed et al., 2024). While native starch possesses limited amine groups, genipin demonstrates high efficacy in cross-linking starch-protein blend films or modified starch containing amino moieties. The cross-linking reaction results in a robust three-dimensional network that significantly enhances the mechanical strength, reduces water solubility, and improves the thermal stability of the films (Gonçalves, Hernández, Cruz, Lopes, Barra, Nunes, et al., 2021). Its low cytotoxicity compared to synthetic alternatives like glutaraldehyde makes it a promising candidate for food and biomedical applications. Cheng et al. (2025) (Cheng et al., 2025) investigated the effects of three natural cross-linkers (phytic acid, TA, and genipin) on the properties of corn starch-based/soy protein isolate films (Fig. 9.2e). The results demonstrated that these natural cross-linkers effectively enhanced the mechanical properties, thermal stability, and UV barrier capability of the films. Additionally, TA and genipin further improved the surface hydrophobicity and moisture resistance of the composite films, TA cross-linked film exhibited antioxidant activity.

In summary, natural cross-linkers provide a pathway to develop safer and more sustainable starch-based packaging. While their cross-linking efficiency might sometimes be lower than that of potent synthetic agents, their multifunctionality combining cross-linking with antioxidant, antimicrobial, or pH-indicating properties, makes them highly valuable for next-generation active and intelligent packaging solutions.

9.2.4 Enzymatic cross-linkers

Enzymatic cross-linking offers a highly specific, mild, and environmentally friendly alternative to chemical methods. Transglutaminase (TGase) is a commonly used enzymatic cross-linker (Yang et al., 2025). This enzyme catalyzes the formation of an isopeptide bond between the γ-carboxamide group of a protein-bound glutamine residue and the ε-amino group of a protein-bound lysine residue. Therefore, its primary application in starch films is in starch-protein composite systems, where it cross-links the protein phase, creating a reinforced interpenetrating network that enhances the overall integrity and water resistance of the film (AL-Hassan & Norziah, 2017; Rosseto, Toniciolli Rigueto, Ferreira Menezes, Santos Nunes, Mello da Silveira, Gottschalk de Abreu et al., 2025). In addition, laccase can also be used for the cross-linking of starch-based films (Alizadeh Sani et al., 2024). This oxidase enzyme can catalyze the formation of free radicals in phenolic compounds. In starch films, it is typically used to cross-link phenolic acids (e.g., ferulic acid) that have been grafted onto starch or are present in composite systems. The radical coupling leads to the formation of dimeric and oligomeric cross-links, strengthening the film matrix (Zhu et al., 2021).

9.3 Effects of cross-linking modification on film properties

9.3.1 Mechanical strength

The enhancement of mechanical strength is one of the most critical and widely reported outcomes of cross-linking in starch-based films. Native starch films are often brittle and exhibit poor mechanical integrity, primarily due to the random arrangement of starch chains and the plasticizing effect of water molecules (Xu, Li, et al., 2025). Cross-linking introduces covalent or ionic bridges that effectively connect adjacent starch molecules, creating a robust and cohesive three-dimensional network within the film matrix (Wei et al., 2024a; Yuan, Du, et al., 2025a). The primary mechanism behind this reinforcement is the restriction of molecular chain mobility (Zhu et al., 2022). When a mechanical stress is applied, the cross-linked network distributes the load more uniformly throughout the material, thereby increasing the force required for deformation and fracture. This is quantitatively observed as a significant improvement in tensile strength (TS) (Zhao, Chen, Zhuo, Zhong, Shi, Liu, et al., 2024b). For instance, as highlighted in a previous example, cross-linking electrospun starch nanofibers with glutaraldehyde vapor led to an order-of-magnitude increase in tensile strength (Zhu et al., 2022). Similarly, cross-linking with CA, genipin or STMP consistently results in films with higher TS compared to their native counterparts (Cheng et al., 2025; Sharma et al., 2021; Yao et al., 2022).

The effect on the elongation at break (EAB), which indicates film flexibility, is more complex and depends on the cross-linker type and density (Cho et al., 2024; Duarte et al., 2023; Liu, Chen, Qu, Zhang, Qin, Zhang, et al., 2024a). At moderate levels, cross-linking can sometimes increase or maintain EAB by preventing the premature fracture of the rigid starch matrix (Wei et al., 2024a). However, a high degree of cross-linking typically reduces chain slippage and mobility, leading to a stiffer and more brittle material with decreased EAB (Sarker et al., 2025).Therefore, optimizing the cross-linker concentration is crucial to achieve a balance between high strength and adequate flexibility (Khadsai, Janmanee, Sam-Ang, Nuanchawee, Rakitikul, Mankhong, et al., 2024). As shown in Table 9.2, the formation of this reinforced network also contributes to an increase in the film's storage modulus, as demonstrated by dynamic mechanical analysis (DMA), confirming enhanced stiffness and load-bearing capacity in the glassy state (Cho et al., 2024).

In summary, cross-linking serves as a powerful strategy to overcome the inherent mechanical weaknesses of starch-based films. By forming a firm molecular network, it markedly improves tensile strength and stiffness. The final mechanical performance is a direct result of the cross-linking density and the nature of the bonds formed, necessitating careful control over the modification process to tailor the properties for specific packaging applications, where resistance to stretching, puncture, and tearing is paramount.

Table 9.2 Comparison of different cross-linkers and their effects on starch films.

Category	Cross-linker	Cross-linking Mechanism	Key effects on film properties	Advantages	Disadvantages	Reference
Chemical cross-linker	Glutaraldehyde	Schiff base reaction with $-NH_2$; acetalization with –OH	Significantly enhancing mechanical strength and water resistance	Highly effective, strong mechanical enhancement	Potential cytotoxicity; residual aldehyde concerns limits food use	Cho et al. (2024), Wang et al. (2016)
	Citric acid	Esterification with starch –OH groups	Not only improves mechanical and barrier properties but can also introduce antimicrobial activity to the film matrix	Bio-based, non-toxic, potential antimicrobial effect	Requires high temperature & catalyst; may increase film brittleness	Yuan, Wei, et al. (2025b)
	Sodium Trimeta-phosphate	Forms phosphate di-ester bridges	Significantly enhances the mechanical integrity and aqueous stability of the films	High efficiency, improves aqueous stability	Potential phosphorus residue; regulatory concerns for food contact	Sharma et al. (2021)
Ionic/coordination cross-linkers	Divalent or trivalent metal cations	Ionic bridges with anionic groups (e.g., in oxidized starch)	Effectively strengthens the film network, improving its tensile strength and reducing its water vapor permeability	Simple process, uses divalent cations	Requires modified starch with anionic charges	Kumar et al. (2025)
	Borax (sodium tetraborate)	Forms di-diol complexes with starch –OH groups	Significantly improved mechanical strength, water resistance, and transparency.	Simple process, uses divalent cations	Requires modified starch with anionic charges	Yoğurtçu and Gürler (2024)

(*Continued*)

Table 9.2 Comparison of different cross-linkers and their effects on starch films.—cont'd

Category	Cross-linker	Cross-linking Mechanism	Key effects on film properties	Advantages	Disadvantages	Reference
Natural cross-linkers	Tannic Acid	Extensive H-bonding; possible oxidative coupling	Significantly enhance the mechanical strength and water resistance of the films	Multifunctional (cross-linker & active agent); fully natural	Color change; mechanism relies heavily on physical bonds	Yao et al. (2022)
	Genipin	Nucleophilic attack by $-NH_2$ groups (e.g., in proteins)	Significantly enhances the mechanical strength, reduces water solubility, and improves the thermal stability of the films	Low cyto-toxicity; natural origin	High cost; low reactivity with pure starch; turns blue	Cheng et al. (2025)
Enzymatic cross-linkers	Transglutaminase	Catalyzes isopeptide bond between Lys & Gln	Enhances the overall integrity and water resistance of the film	High specificity; mild reaction conditions	Primarily effective in starch-protein composite systems	Zhu et al. (2021)

9.3.2 Barrier properties

Effective moisture, oxygen, and gas barrier properties are critical for maintaining product quality and prolonging shelf life, particularly for food packaging (Wu et al., 2021). Native starch films commonly show inferior barrier performance because of their inherent hydrophilicity and the presence of free volume in the amorphous polymer regions, enabling relatively facile diffusion of small molecules (Jiang et al., 2023; Santhosh et al., 2024). Cross-linking modifications provide an effective strategy to substantially enhance barrier performance by altering the film's microstructure and physicochemical attributes.

Cross-linking yields a notable enhancement of water vapor barrier properties, driven by two principal effects: (i) a reduction in free hydroxyl (-OH) groups available for water sorption, consumed during network formation, and (ii) the creation of a denser, more tortuous polymer architecture that impedes water-vapor diffusion (Liu, Chen, et al., 2024). CA mediated cross-linking, for instance, forms hydrophobic ester linkages, thereby lowering the matrix's hydrophilicity (Yao et al., 2022). Water vapor permeability (WVP) is commonly used to quantify this barrier. Studies have consistently shown a substantial decrease in WVP following cross-linking treatments with agents such as genipin (Cheng et al., 2025), STMP (Sharma et al., 2021), and CA (Sarker et al., 2025). Beyond water vapor, resistance to liquid water is a critical aspect of barrier performance of starch-based film, which is commonly evaluated by water solubility and swelling capacity. The cross-linked network not only reduces WVP but also profoundly enhances the film's integrity upon direct contact with liquid water. By restricting chain mobility and forming a robust matrix, cross-linking significantly lowers the water solubility and suppresses the swelling of the film, thereby preventing structural disintegration and maintaining functionality in high-humidity environments or when in direct contact with moist food products (Xu, Du, et al., 2025).

Similarly, cross-linking significantly improves the resistance of starch-based films to oxygen transmission (Wang, Wang, Ma, Li, Cao, Xu, et al., 2025). The formation of a tight, cross-linked network reduces the free volume and increases the tortuosity within the polymer matrix. This forces oxygen molecules to follow a longer, more convoluted path to diffuse through the film, thereby lowering the oxygen permeability (OP) (Ren et al., 2024). This enhanced oxygen barrier is crucial for preventing the oxidative rancidity of lipids and the degradation of sensitive nutrients and pigments in food products (Yıldırım-Yalçın et al., 2021).

The introduction of a cross-linked structure transforms the loose and hydrophilic starch matrix into a denser and more organized network. By effectively reducing WVP, OP, water solubility, and swelling, cross-linking directly addresses the major limitations of native starch films, making them considerably more competitive for practical food packaging applications where comprehensive protection against moisture and oxygen is critical.

9.3.3 Thermal stability

Thermal stability is a critical attribute of starch-based films, shaping their processability and potential applications, particularly during heat-sealing steps or when packaging warm products. Native starch films show limited thermal resistance, mainly because water acts as a plasticizer and the matrix features weak intermolecular interactions (hydrogen bonds) that break down at modest temperatures. Cross-linking creates a sturdy network, substantially increasing the material's thermal endurance (Kumar et al., 2025). The fundamental mechanism behind this improvement lies in the formation of covalent bonds that are thermally more stable than the native hydrogen bonds and physical entanglements. This cross-linked network restricts the mobility of starch molecular chains, thereby increasing the energy required for thermal degradation (Miskeen et al., 2021).

This effect is quantitatively characterized by techniques such as thermogravimetric analysis (TGA) and differential scanning calorimetry (DSC) (Souza et al., 2021). TGA results typically show that the onset temperature of the main decomposition stage is shifted to a higher value in cross-linked films. This indicates that a greater input of thermal energy is required to initiate the breakdown of the polymer backbone, as the cross-links must be cleaved before the chains can volatilize (Zhao et al., 2025). DSC analysis often reveals an increase in the Tg of the starch matrix after cross-linking. A higher Tg signifies that the film transitions from a glassy to a rubbery state at a higher temperature, meaning it can maintain its rigid, glassy structure and mechanical integrity under greater thermal stress (Sharma et al., 2021). Cross-linked films frequently yield a higher amount of solid char residue at high temperatures (e.g., above 500 °C). The stable, cross-linked network favors the formation of carbonaceous char over the production of volatile fragments, reflecting improved thermal stability.

In summary, the three-dimensional network formed by cross-linking acts as a thermal stabilizer. It effectively impedes the movement of polymer segments and increases the overall cohesion of the film, leading to superior resistance to heat-induced softening and decomposition. This enhanced thermal stability broadens the processing window and practical application scope of starch-based packaging materials.

9.3.4 Biodegradability

Biodegradability is a cornerstone of the environmental appeal of starch-based packaging, representing a critical advantage over persistent petroleum-based plastics (Cui et al., 2021). However, the introduction of cross-linking modifications presents a complex and dual-faceted impact on this essential property. While cross-linking is employed to enhance durability and performance during use, it inherently alters the material's susceptibility to microbial and enzymatic attack, which is the basis of biodegradation.

The primary mechanism of starch biodegradation involves the enzymatic hydrolysis of glycosidic bonds in the polymer backbone by microorganisms such as bacteria and fungi (Lv, He, Xu, Jin, Zhao, Yuan, et al., 2025; Su et al., 2023). Cross-linking directly impedes this process by creating a dense three-dimensional network that restricts the access and diffusion of extracellular enzymes into the starch matrix (Mylkie et al., 2025). Furthermore, the covalent bonds (e.g., ester, phosphate diester linkages) themselves are often more recalcitrant to enzymatic cleavage than the native glycosidic bonds and hydrogen bonds in starch. Consequently, a higher degree of cross-linking typically leads to a reduced biodegradation rate in the initial stages (Picchio et al., 2018). Studies utilizing soil burial or enzymatic degradation tests often show that cross-linked starch films maintain their integrity and weight for a longer period compared to their native counterparts (Mylkie et al., 2025). Despite this initial retardation, it is crucial to emphasize that cross-linked starch films remain fundamentally biodegradable. The cross-links, while stable, are not indestructible and can be slowly broken down over time by specific microbial consortia or under appropriate environmental conditions (e.g., in industrial composting facilities with higher temperatures and microbial diversity) (Yu et al., 2021).

9.3.5 Antimicrobial/antioxidant

A distinctive advantage of using specific cross-linkers in starch-based active packaging lies in their intrinsic bioactivity. Beyond their primary role of forming a three-dimensional network, these multifunctional agents directly contribute antimicrobial or antioxidant properties to the film, eliminating the need for additional active compounds in some applications (Cheng et al., 2025; Ren et al., 2024; Yuan, Du, et al., 2025a).

Certain cross-linkers function by creating an environment that is inherently hostile to microorganisms. For example, as a widely used bio-based cross-linker, CA forms ester bonds with starch hydroxyl groups. Upon incorporation into the film, its free carboxyl groups can create a localized acidic microenvironment (lower pH) at the film surface, which effectively inhibits the growth of a broad spectrum of bacteria and molds (Zhang et al., 2023c). When utilized as a co-cross-linking component or in blend systems, chitosan provides a dual function. Its reactive amino and hydroxyl groups can form networks with starch, while its inherent cationic nature enables it to interact with and disrupt the negatively charged cell membranes of microorganisms, leading to cell lysis and death (Suzuki et al., 2007).

Another class of cross-linkers directly scavenges free radicals, thereby retarding the oxidative spoilage of packaged foods (Kurabetta et al., 2023; Menzel, 2020). These natural compounds are ideal multifunctional agents. They can cross-link starch chains through oxidative coupling catalyzed by laccase or via esterification. Crucially, their phenolic hydroxyl groups are potent hydrogen donors, enabling them to neutralize free radicals and terminate oxidative chain reactions, thus imparting significant antioxidant capacity to the film matrix (Shogren & Biswas, 2013).

9.3.6 Controlled release

A pivotal function of active packaging is the sustained and controlled release of incorporated bioactive compounds, such as antimicrobials or antioxidants, to maintain product quality and extend shelf-life (Almasi et al., 2021; Kuai et al., 2021). The incorporation of these agents into the packaging matrix alone is insufficient; the release kinetics must be meticulously tailored to match the spoilage kinetics of the specific product (Fan, Wang, Wen, Li, Fang, Richel, et al., 2023). Cross-linking modification plays a decisive and dual-faceted role in governing this release profile, primarily by acting as a macromolecular "gatekeeper" within the starch matrix (Xu, Du, et al., 2025). The fundamental mechanism through which cross-linking modulates release is by altering the diffusivity and creating specific interactions with the active compound (Wei et al., 2024b). The formation of a three-dimensional network increases the tortuosity of the diffusion path that the active agent must traverse to migrate from the film into the food or

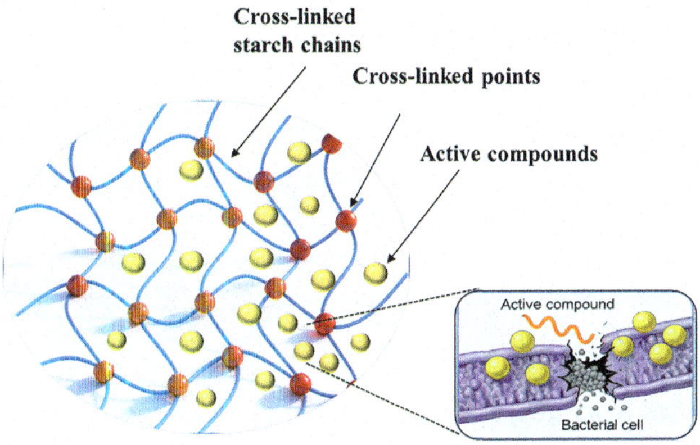

Figure 9.3 Mechanisms of active compound release and antimicrobial in cross-linked starch films.

headspace (Fig. 9.3). A higher cross-linking density results in a denser matrix with smaller mesh sizes, thereby physically hindering the diffusion of active molecules and slowing down their release rate. This often leads to a more sustained, long-term release profile, which is highly desirable for long-shelf-life products (Zhang et al., 2023c).

Beyond the physical barrier effect, chemical interactions between the cross-linker, the starch, and the active agent can significantly influence release. Cross-linking reduces the swelling of the starch film in high-humidity environments (Xu, Du, et al., 2025). Since swelling typically creates larger pores and facilitates the diffusion of active compounds, a cross-linked film that maintains its structure will exhibit a slower, more controlled release, especially in moist conditions. The chemical nature of the cross-linker can directly affect the active agent. For instance, a cross-linker like citric acid can create a more hydrophilic environment, which may slow the release of hydrophobic actives (Zhang et al., 2023c). Conversely, if the active compound (e.g., a polyphenol) possesses functional groups that can interact with the cross-linked network (via hydrogen bonding or covalent grafting), its release can be further retarded, leading to a prolonged efficacy (Cheng et al., 2025). However, a critical challenge lies in optimizing the cross-linking density. An excessively dense network may overly restrict diffusion, preventing the initial release of the active compound from reaching the minimum effective concentration (MEC) required to inhibit microbial growth or oxidation, thereby rendering the active packaging ineffective at the critical early stage.

9.4 Active packaging applications

The integration of cross-linking technology with active compounds has propelled starch-based films from simple protective barriers to functional materials that actively participate in preserving food quality. The enhanced mechanical robustness, superior barrier properties, and tunable release kinetics imparted by cross-linking are crucial for these practical applications, enabling the films to effectively extend the shelf-life of various perishable food products (Table 9.3).

9.4.1 Fruits and vegetables

The preservation of fresh produce is a major application area. Fruits and vegetables undergo rapid quality deterioration due to moisture loss, microbial growth, and physiological senescence (Zhang & Jiang, 2023). Cross-linked starch-based films incorporated with antimicrobial agents (e.g., essential oils from thyme or oregano) or antioxidants (e.g., tannic acid) can create a modified atmosphere around the produce and provide a sustained release of active compounds. For instance, a composite film was developed from corn starch (CS) and konjac glucomannan (KGM) cross-linked with CA and incorporated with thymol microcapsules (TMs) for the preservation of cherry tomatoes (Li et al., 2025). The cross-linking with CA significantly enhanced the film's barrier properties by reducing its water vapor permeability and swelling rate. Consequently, the improved water resistance prevented the film from becoming soggy and disintegrating in the high-humidity environment of fresh produce packaging. Furthermore, the film significantly suppressed the growth of *E. coli* and *S. aureus*. When applied to cherry tomatoes, this active packaging effectively minimized weight loss, maintained firmness, and enhanced the retention of total phenolic and flavonoid contents. In another study, native cassava starch was oxidized to dialdehyde starch (DAS), in which cross-linking occurred between molecules via acetalization reactions. Polyvinyl alcohol (PVA) and cellulose nanofibrils (CNF) were subsequently incorporated into the DAS matrix to form a compact composite film (Wang et al., 2025). The optimized DAS-6c-12p film (containing 6 wt% CNF and 12 wt% PVA relative to DAS) demonstrated remarkable enhancements in mechanical and barrier properties compared to the control CS film: tensile strength surged from 15.28 ± 1.11 MPa to 65.67 ± 2.50 MPa, tensile strain increased from $4.67 \pm 0.68\%$ to $6.21 \pm 0.13\%$, while the water vapor, oxygen, and edible

Table 9.3 Performance of cross-linked starch films in food preservation.

Food category	Film matrix	Cross-linker	Active agent	Key improvements in film properties	Salient results	Reference
Cherry tomatoes	Starch/konjac glucomannan	Citric acid	Thymol micro-capsules	The WVP of composite films was reduced 38%–47%	Reduced population of *E. coli* and *S. aureus*. Delayed weight loss and firmness loss, while promoted the accumulation of phenolic and flavonoid compounds of cherry tomatoes.	Li et al. (2025)
Strawberries	Dialdehyde starch/polyvinyl alcohol (PVA)/cellulose nanofibrils	Dialdehyde starch	-	Tensile strength increasing over fourfold and permeability coefficients for water vapor, oxygen, and oil decreasing by approximately 73%, 99%, and 90%	Effectively inhibiting respiration and rot of strawberries.	Wang et al. (2025)
Strawberries	Starch/polyvinyl alcohol	Glutaraldehyde	Silver sodium zirconium phosphate	Elongation at break and tensile strength of result films were significantly improved as compared with pure starch nanofibrous films. Hydrophobicity of cross-linked films was increased.	Reduced population of *E. coli* and *S. aureus*. Effectively prolonged the shelf life of strawberries.	Liu, Chen, Dong, Wang, Zhang, He, et al. (2022)

(*Continued*)

Table 9.3 Performance of cross-linked starch films in food preservation.—cont'd

Food category	Film matrix	Cross-linker	Active agent	Key improvements in film properties	Salient results	Reference
Mushroom	Corn starch/secalin	Citric acid	Citric acid	Significantly reduced water vapor permeability. The addition of an optimal content of cross-linker simultaneously improved the film's tensile strength and elongation at break.	Exhibited strong antioxidant/antimicrobial activities. Improved preservation for mushroom by reducing weight loss and visual spoilage.	Dodange and Shekarchizadeh (2025)
Ladyfinger	Cationic starch/chitosan	Tannic acid	-	Exhibited excellent barrier properties against moisture and water, supported by increased water contact angle values (86.97 ± 0.29 degrees)	Showed concurring impact against food borne pathogens like *E. coli, S. aureus and C. albicans.* Effectively prolonged the shelf life of ladyfingers.	Kurabetta et al. (2023)
Fresh meat	Amylose starch/2-hydroxypropyl-trimethyl-ammonium chloride chitosan	Glutaraldehyde	2-hydroxypropyl-trimethyl-ammonium chloride chitosan	Tensile strength and elongation at break of the optimal films reached to 16.13 MPa (an increase of 206.65%) and 53.86% (an increase of 109.49%)	Reduced population of *E. coli* (ATCC 25922) and *S. aureus* (ATCC 29213). Prolong the shelf life of fresh meat	Deng, Chen, Li, Liu, Zhou, Qin, et al. (2024)

Chicken breast fillet	Maize starch	Sodium trimeta-phosphate	Grape juice	-	Efficiently restricted the lipid oxidation and the total aerobic mesophilic and psychrophilic and Enterobacteriaceae counts in chicken breast fillets.	Yıldırım-Yalçın et al. (2021)
Minas cheese	Corn starch/ whey	transglu-taminase	broccoli leaf extract	Improved mechanical strength and reduced water solubility.	Significant efficacy in preserving Minas cheese for over 60 days by stabilizing water activity and suppressing microbial growth.	Rosseto et al. (2025)

oil permeability coefficients were drastically reduced to 3.04 ± 0.19 g mm/(m^2·d), 0.097 ± 0.003 cm^3 mm/(m^2·d·0.1 MPa), and 0.50 ± 0.02 g mm/(m^2·d), respectively. This robust barrier performance translated directly into superior strawberry preservation, effectively inhibiting respiration and rot, and surpassing the efficacy of both pure starch and commercial PE films.

9.4.2 Meat and seafood products

The high protein and moisture content in meat and seafood products make them highly susceptible to microbial spoilage and oxidative rancidity. Active packaging plays a vital role here. Cross-linked starch films carrying natural antimicrobials like nisin or organic acids can effectively inhibit the growth of pathogens such as *E. coli* on meat surfaces (Deng et al., 2024). developed an antibacterial film for meat preservation using amylose as the matrix and 2-hydroxypropyl-trimethylammonium chloride chitosan (HTCC) as the antimicrobial agent, with glutaraldehyde (GA) serving as the cross-linker. The macromolecular chains of amylose and HTCC first entangled to form a complex, followed by a Schiff base reaction between the amino groups of HTCC and the aldehyde groups of GA, which created a stable interface between HTCC and amylose. This cross-linked network resulted in remarkable mechanical enhancement, increasing the tensile strength and elongation at break by 206.65% and 109.49%, respectively. Furthermore, the film effectively inhibited the growth of *E. coli* (ATCC 25,922) and *S. aureus* (ATCC 29,213), significantly extending the shelf life of fresh meat. In another study, an edible film was developed from maize starch cross-linked with sodium trimetaphosphate and incorporated with grape juice as an active component, plasticized with glycerol. When applied as a coating on chicken fillets, this maize starch-based film markedly retarded lipid oxidation and suppressed microbial growth throughout cold storage.

9.4.3 Cheese and dairy products

Cheese requires packaging that controls moisture loss while preventing surface mold growth. Cross-linked starch films offer a biodegradable solution with added functionality (Rosseto et al., 2025). developed biodegradable coating films based on corn starch (3.5%) and whey, combined with transglutaminase (2.5% of whey protein) and broccoli leaf extract (7.5% of the film-forming solution), aimed at preserving Minas cheese. The evaluated formulations demonstrated that the combination of transglutaminase and broccoli extract effectively improved the structural, mechanical, and antioxidant properties of the films. Edible biodegradable coatings demonstrated significant efficacy in preserving Minas cheese for over 60 days by stabilizing water activity and suppressing microbial growth.

9.5 Challenges and future perspectives

Despite significant progress in the development of cross-linked starch-based active packaging, several formidable challenges remain to be addressed before widespread commercial adoption can be realized. Furthermore, the field presents exciting opportunities for future research to overcome these limitations and create next-generation smart and sustainable packaging solutions.

9.5.1 Current challenges

While chemical cross-linkers like glutaraldehyde are highly effective, their potential toxicity and consumer skepticism limit their use in food contact materials. Although natural and enzymatic cross-linkers offer superior safety, they often come with higher costs and, in some cases, lower reaction efficiency, presenting a significant challenge for large-scale, cost-sensitive packaging applications. Furthermore, a critical challenge in starch-based active packaging lies in achieving optimal release kinetics for antimicrobial or antioxidant agents. A highly dense cross-linked network can severely restrict the diffusion of these active compounds, compromising their efficacy during the critical initial period. Conversely, an excessively loose network may result in a rapid and uncontrolled burst release, prematurely exhausting the active capacity and shortening the functional shelf-life. Consequently, precisely tailoring the cross-linking density to control the release profile remains a significant challenge.

Many cross-linking strategies demonstrated in the laboratory are difficult to scale up. Issues such as the homogeneity of cross-linker distribution in large batches, the extended reaction times required for some natural or enzymatic agents, and the energy consumption of certain processes pose significant hurdles for industrial manufacturing.

The long-term stability of the cross-linked starch-based films' mechanical and barrier properties under real-world storage conditions requires further validation. Moreover, while starch is biodegradable, the impact of cross-linking on the biodegradation rate and the environmental fate of the cross-linker itself (e.g., borax, certain synthetic agents) needs thorough investigation to ensure a truly sustainable life cycle.

9.5.2 Future perspectives

Future research should focus on addressing these challenges through innovative approaches: (i) Development of Novel, Multifunctional Cross-linkers: A promising direction is the design of "green" cross-linkers that are not only non-toxic and efficient but also possess inherent active functions. For instance, cross-linkers with inherent antimicrobial (e.g., chitosan derivatives) or antioxidant properties could be synthesized, simplifying the film formulation and enhancing performance. (ii) Advanced Structuring and Hybrid Systems: Employing novel processing techniques such as electrospinning and 3D printing to create hierarchically structured materials, combined with the development of hybrid cross-linking systems (e.g., a combination of a rapid physical gelation and a slow covalent cross-linking), could offer superior control over film properties and active release. (iii) Stimuli-Responsive and Intelligent Packaging: The future lies in smart packaging. Research can be directed toward developing cross-linked starch films that are stimuli-responsive. For example, cross-links that loosen in response to specific triggers like pH changes (e.g., spoilage) or microbial enzymes could enable an "on-demand" release of active compounds, maximizing efficiency and shelf-life. (iv) Life Cycle Assessment (LCA) and Standardization: To convincingly demonstrate the environmental advantages, comprehensive LCA studies from raw material extraction to end-of-life are essential for cross-linked starch films. Furthermore, establishing standardized testing protocols for their mechanical, barrier, and active properties will be crucial for comparing data and building industry confidence.

In conclusion, while cross-linking stands as a cornerstone technology for advancing starch-based active packaging, its journey from the laboratory to the marketplace requires a multidisciplinary effort. By focusing on green chemistry, controlled release mechanisms, scalable processes, and a demonstrably sustainable profile, cross-linked starch films hold the immense potential to become a mainstream, high-performance packaging material that meets the dual demands of functionality and environmental responsibility.

References

Ahmed, R., Hira, N.ul ain, Wang, M., Iqbal, S., Yi, J., & Hemar, Y. (2024). Genipin, a natural blue colorant precursor: Source, extraction, properties, and applications. *Food Chemistry, 434*, 137498. https://doi.org/10.1016/j.foodchem.2023.137498

AL-Hassan, A. A., & Norziah, M. H. (2017). Effect of transglutaminase induced crosslinking on the properties of starch/gelatin films. *Food Packaging and Shelf Life, 13*, 15–19. https://doi.org/10.1016/j.fpsl.2017.04.006

Alizadeh Sani, M., Priyadarshi, R., Zhang, W., Khezerlou, A., & Rhim, J.-W. (2024). Innovative application of laccase enzyme in food packaging. *Trends in Food Science & Technology, 151*, s.

Almasi, H., Jahanbakhsh Oskouie, M., & Saleh, A. (2021). A review on techniques utilized for design of controlled release food active packaging. *Critical Reviews in Food Science and Nutrition, 61*(15), 2601–2621. https://doi.org/10.1080/10408398.2020.1783199

Cheng, H., Chen, L., McClements, D. J., Yang, T., Zhang, Z., Ren, F., Miao, M., Tian, Y., & Jin, Z. (2021). Starch-based biodegradable packaging materials: A review of their preparation, characterization and diverse applications in the food industry. *Trends in Food Science and Technology, 114*, 70–82. https://doi.org/10.1016/j.tifs.2021.05.017

Cheng, H., McClements, D. J., Xu, H., Zhang, Z., Zhang, R., Jin, Z., & Chen, L. (2025). Effect of various natural cross-linking agents on the properties of corn starch/soy protein isolate composite films. *Carbohydrate Polymers, 366*, 123850.

Chi, Y., Maitland, E., & Pascall, M. A. (2024). The effect of citric acid concentrations on the mechanical, thermal, and structural properties of starch edible films. *International Journal of Food Science and Technology, 59*(3), 1801–1813. https://doi.org/10.1111/ijfs.16933

Cho, J.-H., Ryu, K.-H., Kim, H.-J., & Back, J.-H. (2024). Fabrication and enhanced flexibility of starch-based cross-linked films. *Biomacromolecules, 25*(12), 7894–7903.

Cui, C., Ji, N., Wang, Y., Xiong, L., & Sun, Q. (2021). Bioactive and intelligent starch-based films: A review. *Trends in Food Science & Technology, 116*, 854–869. https://doi.org/10.1016/j.tifs.2021.08.024

Dai, L., Yu, H., Zhang, J., & Cheng, F. (2021). Preparation and characterization of cross-linked starch nanocrystals and self-reinforced starch-based nanocomposite films. *International Journal of Biological Macromolecules, 181*, 868–876.

Deng, B., Chen, J., Li, S., Liu, J., Zhou, Z., Qin, Z., Wang, H., Su, M., Li, L., & Bai, Z. (2024). An antibacterial packaging film based on amylose starch with quaternary ammonium salt chitosan and its application for meat preservation. *International Journal of Biological Macromolecules, 261*. https://doi.org/10.1016/j.ijbiomac.2024.129706

Dodange, S., & Shekarchizadeh, H. (2025). Enhanced performance of an eco-friendly and sustainable secalin-corn starch packaging film via citric acid cross-linking. *Current Research in Food Science*, 101231.

Duarte, G. A., Bezerra, M. C., Bettini, S. H. P., & Lucas, A. A. (2023). Real-time monitoring of the starch cross-linking with citric acid by chemorheological analysis. *Carbohydrate Polymers, 311*. https://doi.org/10.1016/j.carbpol.2023.120733

Fan, S., Wang, D., Wen, X., Li, X., Fang, F., Richel, A., Xiao, N., Fauconnier, M.-L., Hou, C., & Zhang, D. (2023). Incorporation of cinnamon essential oil-loaded Pickering emulsion for improving antimicrobial properties and control release of chitosan/gelatin films. *Food Hydrocolloids, 138*, 108438.

Gonçalves, I., Hernández, D., Cruz, C., Lopes, J., Barra, A., Nusnes, C., da Silva, J. A. L., Ferreira, P., & Coimbra, M. A. (2021). Relevance of genipin networking on rheological, physical, and mechanical properties of starch-based formulations. *Carbohydrate Polymers, 254*, 117236.

Gonenc, I., & Us, F. (2019). Effect of glutaraldehyde crosslinking on degree of substitution, thermal, structural, and physicochemical properties of corn starch. *Starch - Stärke, 71* (3–4). https://doi.org/10.1002/star.201800046

Han, Z., Shi, R., & Sun, D.-W. (2020). Effects of novel physical processing techniques on the multi-structures of starch. *Trends in Food Science & Technology, 97*, 126–135.

Jiang, S., Qiao, C., Liu, R., Liu, Q., Xu, J., & Yao, J. (2023). Structure and properties of citric acid cross-linked chitosan/poly(vinyl alcohol) composite films for food packaging applications. *Carbohydrate Polymers, 312*, 120842. https://doi.org/10.1016/j.carbpol.2023.120842

Karmakar, B., Sarkar, S., Chakraborty, R., Saha, S. P., Thirugnanam, A., Roy, P. K., & Roy, S. (2024). Starch-based biodegradable films amended with nano-starch and tannic acid-coated nano-starch exhibit enhanced mechanical and functional attributes with antimicrobial activity. *Carbohydrate Polymers, 341*, 122321.

Khadsai, S., Janmanee, R., Sam-Ang, P., Nuanchawee, Y., Rakitikul, W., Mankhong, W., Likittrakulwong, W., & Ninjiaranai, P. (2024). Influence of crosslinking concentration on the properties of biodegradable modified cassava starch-based films for packaging applications. *Polymers, 16*.

Kuai, L., Liu, F., Chiou, B.-S., Avena-Bustillos, R. J., McHugh, T. H., & Zhong, F. (2021). Controlled release of antioxidants from active food packaging: A review. *Food Hydrocolloids, 120*, 106992.

Kumar, Y., Singh, S., & Saxena, D. C. (2025). A comprehensive review on methods, mechanisms, properties, and emerging applications of crosslinked starches. *International Journal of Biological Macromolecules, 306*(Pt 2), 141526. https://doi.org/10.1016/j.ijbiomac.2025.141526

Kurabetta, L. K., Masti, S. P., Eelager, M. P., Gunaki, M. N., Madihalli, S., Hunashyal, A. A., Chougale, R. B., Kumar S.K., P., & Kadapure, A. J. (2023). Physicochemical and antioxidant properties of tannic acid crosslinked cationic starch/chitosan based active films for ladyfinger packaging application. *International Journal of Biological Macromolecules, 253*. https://doi.org/10.1016/j.ijbiomac.2023.127552

Li, F., Mao, S., Xu, Y., Pan, L., Wang, W., Zhang, T., & Lu, C. (2025). Crosslinked starch/konjac glucomannan active films containing encapsulated thymol for cherry tomatoes preservation. *Food Chemistry, 481*, 144086. https://doi.org/10.1016/j.foodchem.2025.144086

Lin, L., Cheng, H., Cheng, Q., Li, C., Al-Asmari, F., Sameeh, M. Y., & Cui, H. (2024). Schiff base linkage of citral to zinc-casein hydrolysate chelates for preparing starch-based active films against L. monocytogenes on ready-to-eat foods. *International Journal of Biological Macromolecules, 263*, 130401.

Liu, X., Chen, L., Dong, Q., Wang, Z., Zhang, D., He, J., Ye, Y., Zhou, J., Zhu, W., Hu, Z., Din, Z.-u., Ma, T., Ding, W., & Cai, J. (2022). Emerging starch composite nanofibrous films for food packaging: Facile construction, hydrophobic property, and antibacterial activity enhancement. *International Journal of Biological Macromolecules, 222*, 868–879.

Liu, Z., Chen, L., Qu, L., Zhang, R., Qin, Z., Zhang, H., Wei, J., Xu, J., & Hou, Z. (2024a). Cross-linked poly(ester urethane)/starch composite films with high starch content as sustainable food-packaging materials: Influence of cross-link density. *International Journal of Biological Macromolecules, 256*, 128441.

Lu, K., Folkersma, R., Voet, V. S. D., & Loos, K. (2024). Effects of the amylose/amylopectin ratio of starch on borax-crosslinked hydrogels. *Polymers, 16*(16). https://doi.org/10.3390/polym16162237

Lu, H., Ji, N., Li, M., Wang, Y., Xiong, L., Zhou, L., Qiu, L., Bian, X., Sun, C., & Sun, Q. (2019). Preparation of borax cross-linked starch nanoparticles for improvement of mechanical properties of maize starch films. *Journal of Agricultural and Food Chemistry, 67*(10), 2916–2925.

Lv, H., He, D., Xu, E., Jin, Z., Zhao, H., Yuan, C., Zhao, M., Gao, W., Li, J., Liu, Y., Wu, Z., Liu, P., & Cui, B. (2025). Starch-based biodegradable active intelligent packaging with color superimposition via emulsion electrospinning nanofiber for meat freshness monitoring and shelf-life extension. *Carbohydrate Polymers, 367*, 123940.

Matheus, J. R. V., Dalsasso, R. R., Rebelatto, E. A., Andrade, K. S., Andrade, L. M.d., Andrade, C. J.d., Monteiro, A. R., & Fai, A. E. C. (2023). Biopolymers as green-based food packaging materials: A focus on modified and unmodified starch-based films. *Comprehensive Reviews in Food Science and Food Safety, 22*(2), 1148–1183. https://doi.org/10.1111/1541-4337.13107

Mendes, A. C., & Pedersen, G. A. (2021). Perspectives on sustainable food packaging:– is bio-based plastics a solution? *Trends in Food Science and Technology, 112*, 839–846. https://doi.org/10.1016/j.tifs.2021.03.049

Menzel, C. (2020). Improvement of starch films for food packaging through a three-principle approach: Antioxidants, cross-linking and reinforcement. *Carbohydrate Polymers, 250*, 116828. https://doi.org/10.1016/j.carbpol.2020.116828

Miskeen, S., Hong, J. S., Choi, H.-D., & Kim, J.-Y. (2021). Fabrication of citric acid-modified starch nanoparticles to improve their thermal stability and hydrophobicity. *Carbohydrate Polymers, 253*, 117242.

Mylkie, K., Chełminiak-Dudkiewicz, D., Młynarczyk, D. T., Smolarkiewicz-Wyczachowski, A., & Ziegler-Borowska, M. (2025). Starch films cross-linked with boronated chitosan enriched with cannabis oil as food packaging material. *Polymer Testing, 150*, 108910. https://doi.org/10.1016/j.polymertesting.2025.108910

Picchio, M. L., Linck, Y. G., Monti, G. A., Gugliotta, L. M., Minari, R. J., & Alvarez Igarzabal, C. I. (2018). Casein films crosslinked by tannic acid for food packaging applications. *Food Hydrocolloids, 84*, 424–434. https://doi.org/10.1016/j.foodhyd.2018.06.028

Ren, J., Cao, Y., Zhang, Y., Xiao, X., Zhang, Q., Yang, Q., & Liu, F. (2024). Dual cross-linked network of the active starch film by incorporating palmitic acid and geraniol: A comprehensive evaluation of the hydrophobic mechanism and antimicrobial activity. *International Journal of Biological Macromolecules, 280*(Pt 2), 135830. https://doi.org/10.1016/j.ijbiomac.2024.135830

Rosseto, M., Toniciolli Rigueto, C. V., Ferreira Menezes, J., Santos Nunes, L., Mello da Silveira, S., Gottschalk de Abreu, G. C., Vortmann, A. V., Dettmer, A., & Pereira dos Santos Richards, N. S. (2025). Corn starch and whey-based films with transglutaminase and broccoli extract: A bioactive strategy for Minas cheese preservation. *Food Bioscience, 71*. https://doi.org/10.1016/j.fbio.2025.107285

Santhosh, R., Ahmed, J., Thakur, R., & Sarkar, P. (2024). Starch-based edible packaging: Rheological, thermal, mechanical, microstructural, and barrier properties – A review. *Sustainable Food Technology, 2*(2), 307–330. https://doi.org/10.1039/d3fb00211j

Sarker, A., Rennie, T., Shaheb, Md R., Matak, K., & Jaczynski, J. (2025). Optimization of the properties of underutilized buckwheat starch films through different modification approaches. *Food Packaging and Shelf Life, 49*, 101513. https://doi.org/10.1016/j.fpsl.2025.101513

Sharma, V., Kaur, M., Singh Sandhu, K., Kaur, S., & Nehra, M. (2021). Barnyard millet starch cross-linked at varying levels by sodium trimetaphosphate (STMP): Film forming, physico-chemical, pasting and thermal properties. *Carbohydrate Polymer Technologies and Applications, 2*, 100161. https://doi.org/10.1016/j.carpta.2021.100161

Shogren, R. L., & Biswas, A. (2013). Preparation of starch-sodium lignosulfonate graft copolymers via laccase catalysis and characterization of antioxidant activity. *Carbohydrate Polymers, 91*(2), 581–585. https://doi.org/10.1016/j.carbpol.2012.08.079

Skov, K. B., Portillo-Perez, G. A., Franco, M., & Martinez, M. M. (2025). Impact of citric acid-mediated starch crosslinking on the properties of films derived from stale bread particles: A comparative study of solvent casting and melt-mixing/hot pressing processing. *International Journal of Biological Macromolecules, 310*. https://doi.org/10.1016/j.ijbiomac.2025.143263

Souza, A. G., Ferreira, R. R., Paula, L. C., Mitra, S. K., & Rosa, D. S. (2021). Starch-based films enriched with nanocellulose-stabilized Pickering emulsions containing different essential oils for possible applications in food packaging. *Food Packaging and Shelf Life, 27*. https://doi.org/10.1016/j.fpsl.2020.100615

Srinivasan, S., Ge, C., Lewis, C. L., Begum, S. S., Samui, A. B., & Noyes, H. F. (2025). Effect of chemical modification of corn starch for the development of films for packaging applications: Impact of glutaraldehyde and organically modified montmorillonite incorporation. *Carbohydrate Polymers, 367*. https://doi.org/10.1016/j.carbpol.2025.124019

Su, C.-y., Li, D., Wang, L.-j., & Wang, Y. (2023). Biodegradation behavior and digestive properties of starch-based film for food packaging – A review. *Critical Reviews in Food Science and Nutrition, 63*(24), 6923–6945.

Suzuki, S., Ying, B., Yamane, H., Tachi, H., Shimahashi, K., Ogawa, K., & Kitamura, S. (2007). Surface structure of chitosan and hybrid chitosan-amylose films—restoration of the antibacterial properties of chitosan in the amylose film. *Carbohydrate Research, 342*(16), 2490–2493.

Tavassoli, M., Abedi-Firoozjah, R., Bahramian, B., Hashemi, M., Noori, S. M. A., Oladzadabbasabadi, N., Nagdalian, A., & Jafari, S. M. (2025). Glutaraldehyde cross-linking for improving the techno-functional properties of biopolymeric food packaging films; a comprehensive review. *Food Chemistry, 478*, 143740.

Velasquez, S. T. R., Hu, Q., Kramm, J., Santin, V. C., Völker, C., & Wurm, F. R. (2025). Plastics of the future? An interdisciplinary review on biobased and biodegradable polymers: Progress in chemistry, societal views, and environmental implications. *Angewandte Chemie - International Edition, 64*(23). https://doi.org/10.1002/anie.202423406

Wang, W., Jin, X., Zhu, Y., Zhu, C., Yang, J., Wang, H., & Lin, T. (2016). Effect of vapor-phase glutaraldehyde crosslinking on electrospun starch fibers. *Carbohydrate Polymers, 140*, 356–361.

Wang, B., Wang, X., Ma, H., Li, J., Cao, D., Xu, J., Zeng, J., Gao, W., & Chen, K. (2025). Enhancing the mechanical and barrier properties of starch-based green packaging films via acetalization reaction and hydrogen bond dual crosslinking. *Carbohydrate Polymers, 364*, 123787.

Wei, H.-N., Liu, X.-Y., Wang, C.-C., Feng, R., & Zhang, B. (2024a). Characteristics of corn starch/polyvinyl alcohol composite film with improved flexibility and UV shielding ability by novel approach combining chemical cross-linking and physical blending. *Food Chemistry, 456*, 140051.

Wei, M., Shan, M., Zhang, L., Chen, N., Tie, H., Xiao, Y., & Li, Z. (2024b). Preparation of gelatin/κ-carrageenan active films through procyanidins crosslinking: Physicochemical, structural, antioxidant and controlled release properties. *Food Hydrocolloids, 153*, 110023.

Wei, W., Wu, M., Shang, N., Mu, R., Zhang, T., & Xu, H. (2023). Low temperature preparation of high-performance citric acid crosslinked starch gels with adjustable properties based on cationic esterification synergistic crosslinking. *Food Hydrocolloids, 145*, 109098. https://doi.org/10.1016/j.foodhyd.2023.109098

Wu, F., Misra, M., & Mohanty, A. K. (2021). Challenges and new opportunities on barrier performance of biodegradable polymers for sustainable packaging. *Progress in Polymer Science, 117*, 101395. https://doi.org/10.1016/j.progpolymsci.2021.101395

Xu, Z., Du, J., Shan, P., Wang, Y., Li, L., Tao, Y., Hu, J., Fu, C., Lv, Y., Jia, Z., Song, X., & Wang, H. (2025a). Multi-mechanism cross-linking driven chain spacing shortening to construct polysaccharide-based active packaging films with high water resistance and intelligent controlled release. *Chemical Engineering Journal, 520*, 165750.

Xu, J., Li, Q., Yuan, J., Fan, X., & Gao, W. (2025b). Magnetic field-induced alignment of starch molecules enhances the mechanical strength of starch films. *Carbohydrate Polymers, 366*, 123834. https://doi.org/10.1016/j.carbpol.2025.123834

Yıldırım-Yalçın, M., Sadıkoğlu, H., & Şeker, M. (2021). Characterization of edible film based on grape juice and cross-linked maize starch and its effects on the storage quality of chicken breast fillets. *LWT, 142*, 111012. https://doi.org/10.1016/j.lwt.2021.111012

Yang, Y., Tang, Q., Deng, Z., Xiao, H., Lu, L., Lin, Q., Zhou, W., Li, J., Wang, L., Liu, G.-Q., & Luo, F. (2025). Effect of transglutaminase on the interaction of protein and rice starch. *International Journal of Biological Macromolecules, 288*, 138585.

Yang, Z., Zhang, D., & Cai, J. (2024). Robust design of starch composite nanofibrous films for active food packaging: Towards improved mechanical, antioxidant, and antibacterial properties. *International Journal of Biological Macromolecules, 260*(Pt 1), 129329. https://doi.org/10.1016/j.ijbiomac.2024.129329

Yao, S., Wang, B.-J., & Weng, Y.-M. (2022). Preparation and characterization of mung bean starch edible films using citric acid as cross-linking agent. *Food Packaging and Shelf Life, 32*, 100845.

Yates, J., Deeney, M., Rolker, H. B., White, H., Kalamatianou, S., & Kadiyala, S. (2021). A systematic scoping review of environmental, food security and health impacts of food system plastics. *Nature Food, 2*(2), 80–87. https://doi.org/10.1038/s43016-021-00221-z

Yoğurtçu, H., & Gürler, N. (2024). Evaluation of effect of boric acid on thermoplastic starch: Morphological, mechanical, barrier, and optical properties. *Polymer Engineering & Science, 64*(5), 2230–2240. https://doi.org/10.1002/pen.26688

Yu, X., Chen, L., Jin, Z., & Jiao, A. (2021). Research progress of starch-based biodegradable materials: A review. *Journal of Materials Science, 56*(19), 11187–11208. https://doi.org/10.1007/s10853-021-06063-1

Yuan, X., Du, G., Wang, Z., Zheng, T., Shi, Y., Huang, R., Fu, L., Huang, B., Xu, C., Shi, X., & Lin, B. (2025a). Dual grafting strategy via schiff base cross-linking for durable antibacterial and uvioresistant starch-based degradable packaging film. *International Journal of Biological Macromolecules, 330*, 148052. https://doi.org/10.1016/j.ijbiomac.2025.148052

Yuan, X., Wei, F., Wang, Z., Li, X., Deng, Y., Shi, X., & Lin, B. (2025b). Transparent, mechanically robust and antibacterial food packaging film based on starch/PVA/citric acid/carboxycellulose nanofibers by one-pot green process. *International Journal of Biological Macromolecules, 302*, 140573. https://doi.org/10.1016/j.ijbiomac.2025.140573

Zhang, S., Guo, H., Li, H., Qiao, X., Wu, P., Li, Y., Ai, S., Shan, Y., & Ding, S. (2025). A starch-based composite film with dual-modality essential oil controlled-release for perishable fruit preservation. *Food Packaging and Shelf Life, 52*, 101606.

Zhang, Y., & Jiang, W. (2023). Effective strategies to enhance ultraviolet barrier ability in biodegradable polymer-based films/coatings for fruit and vegetable packaging. *Trends in Food Science & Technology, 139*, 104139.

Zhang, S., Pan, X., Zhao, J., Li, J., Yu, X., Peng, Y., & Wu, J. (2023a). Characterization of ionically crosslinked mango peel pectin-based films: Effect of different cations on the improved properties of film. *Food Packaging and Shelf Life, 38*, 101131.

Zhang, W., Roy, S., Assadpour, E., Cong, X., & Jafari, S. M. (2023c). Cross-linked biopolymeric films by citric acid for food packaging and preservation. *Advances in Colloid and Interface Science, 314*, 102886. https://doi.org/10.1016/j.cis.2023.102886

Zhang, W., Roy, S., Ezati, P., Yang, D.-P., & Rhim, J.-W. (2023b). Tannic acid: A green crosslinker for biopolymer-based food packaging films. *Trends in Food Science & Technology, 136*, 11–23.

Zhang, X., Zhao, Y., Wang, S., & Jing, X. (2021b). Cross-linked polymers based on B–O bonds: Synthesis, structure and properties. *Materials Chemistry Frontiers, 5*(15), 5534–5548. https://doi.org/10.1039/d1qm00514f

Zhao, X., Chen, Z., Zhuo, H., Zhong, C., Shi, G., Liu, T., Huang, X., Zhong, L., & Peng, X. (2024b). Ultra-strong and transparent biomimetic nanocomposite through orientation effects and in situ biomineralization. *Advanced Functional Materials, 34*(1). https://doi.org/10.1002/adfm.202310094

Zhao, Y., Tian, R., Zhang, Q., Jiang, L., Wang, J., Zhang, Y., & Sui, X. (2024a). Enhancing the properties of soy protein isolate and dialdehyde starch films for food packaging applications through tannic acid crosslinking. *Carbohydrate Polymers, 332*, 121903.

Zhao, Y., Yao, G., Li, K., Ye, J., Chen, J., & Zhang, J. (2025). Preparation, characterization, and antibacterial application of cross-linked nanoparticles composite films. *Food Chemistry X, 25*, 102057. https://doi.org/10.1016/j.fochx.2024.102057

Zhao, Y., Zheng, Z., Zhao, Y., Chen, J., & Tang, S. (2024c). Cross-linked modification of tapioca starch by sodium Trimetaphosphate: An influence on its structure. *Food Chemistry X, 23*, 101670. https://doi.org/10.1016/j.fochx.2024.101670

Zhu, Y., Tao, H., Janaswamy, S., Zou, F., Cui, B., & Guo, L. (2021). The functionality of laccase- or peroxidase-treated potato flour: Role of interactions between protein and protein/starch. *Food Chemistry, 341*(Pt 2), 128082. https://doi.org/10.1016/j.foodchem.2020.128082

Zhu, W., Zhang, D., Liu, X., Ma, T., He, J., Dong, Q., Din, Z.-u., Zhou, J., Chen, L., Hu, Z., & Cai, J. (2022). Improving the hydrophobicity and mechanical properties of starch nanofibrous films by electrospinning and cross-linking for food packaging applications. *Lwt, 169*, 114005.

Further reading

Liu, W., Wen, G., Wang, J., Gao, J., Wang, H., & Yu, W. (2024b). Preparation of starch film with high guanidine chemically coupling efficiency based on crosslinking effect of glycidyl methacrylate. *Industrial Crops and Products, 222*, 119647. https://doi.org/10.1016/j.indcrop.2024.119647

Zhang, D., Chen, L., Cai, J., Dong, Q., Din, Z.-u., Hu, Z.-Z., Wang, G.-Z., Ding, W.-P., He, J.-R., & Cheng, S.-Y. (2021a). Starch/tea polyphenols nanofibrous films for food packaging application: From facile construction to enhance mechanical, antioxidant and hydrophobic properties. *Food Chemistry, 360*, 129922.

CHAPTER TEN

Starch-based packaging material added with natural colorant

Manisha Joshi[1], Ruchi D. Jena[1], Kanika Sharma[2], Deblina Biswas[3]
[1]School of Bioengineering and Food Technology, Shoolini University, Solan, Himachal Pradesh, India
[2]Department of Biotechnology, University Institute of Biotechnology, Chandigarh University, Mohali, Punjab, India
[3]Department of Instrumentation and Control Engineering, Dr. B R Ambedkar National Institute of Technology, Jalandhar, Punjab, India

10.1 Introduction

Conventional food packaging technology has been designed to act as passive packaging, mainly to protect the food items from the environment. It mostly aims at food protection, presentation, and packaging, with less interaction of the packaging materials with the packaged food product. Notably, the rising needs of the consumers to consume safe, healthy, and less processed foods, as well as the increased globalization of food distribution chains, have influenced the need to change food packaging technology from passive packaging to active packaging, which can detect information about the processed food product. Modern food packaging technology has, therefore, moved from passive packaging to more advanced active packaging technology that can detect information about the processed food product (Han et al., 2018). Contemporary innovative packaging solutions are purposely designed to act by interacting with the packaged food products, as well as the environment around them, as means of optimizing protection, as well as prolonging the shelf life of the products by reducing wastages. However, the overall aim of any food packaging solution has always been to ensure the preservation of product freshness, integrity, and quality, thereby ensuring extended storage and consumption safety (Majid et al., 2018). Within this overall context, the development of smart packaging solutions can thus be seen as an innovative milestone in the field of food packaging. Smart packaging solutions involve the integration of various functionality features that respond to changes in the environment surrounding the packaged food products. Thereby, depending on their modes of operation, there exist two types of smart packaging solutions: Active Packages and Intelligent Packages.

Starches in Smart Packaging
ISBN: 978-0-443-45396-0
https://doi.org/10.1016/B978-0-443-45396-0.00006-2

Active packaging systems involve the use of functional materials, which are added into the packaging material with the view of achieving a change in the packaged environment (Kadirvel et al., 2024). The materials added in active packaging include antioxidants, antimicrobial compounds, oxygen absorbers, carbon dioxide releasers, moisture absorbers, and ethylene absorbers, among others, whose combined effect contributes to the retardation of spoilage reactions and the maintenance of food quality (Deshmukh & Gaikwad, 2024). On the contrary, intelligent packaging is mainly intended to detect changes in the food products and the surrounding environment, albeit indirectly (Yousefi et al., 2019). They consist of incorporating common packaging materials with indicators, sensors, and data storage elements with the capacity to detect changes in parameters such as pH, gas composition, microbial growth, temperature, and integrity, followed by the transmission of data to the supplier, retailers, and consumers (Yousefi et al., 2019). Both active and intelligent packaging systems, in spite of their differences in functionality, can be combined to arrive at a combination technique-also called *smart* packaging-whose functionality is the conservation of food quality along with the continuous monitoring of their properties (Drago et al., 2020).

The packages are also incorporating a range of sensing devices such as chemical sensors and biosensors with possible future uses in the evaluation of the quality and safety of the foods in their respective supply chains. These systems are capable of sensing indicators of freshness, microbial contamination, oxygen and carbon dioxide, pH shifts, and temperature variation associated with the degradation of the foods (Roy, 2026). The smart packaging system makes it possible to monitor and track the entire process of production and supply of the products, as well as receive updates on the current condition of the products, hence enabling much better decision-making, reduction of losses, and greater confidence of consumers (Khan et al., 2024). Despite these technological advancements, environmental concerns are actually hindrances to their adoption because of nonbiodegradable plastics made out of petroleum, which contribute immensely to plastic pollution. There has been a clear indication of a need to develop biodegradable and eco-friendly materials in packing. A promising area of interest has emerged in terms of materials made out of biopolymers because of their renewability, biocompatibility, and eco-friendliness. Ideally, there is a clear indication that these materials will be an efficient barrier since they will control gas, moisture, aromas, and lipid transfer of foods along with their corresponding antimicrobial and antioxidant properties, which serve to preserve food (Moeini et al., 2022). Biopolymers, which include polysaccharides, proteins, aliphatic polyesters, and lipids, have been greatly explored for sustainable food packing (Marasinghe et al., 2024).

Furthermore, among various biopolymers, starch has attracted much interest in recent years due to its ease of availability, affordability, biodegradability, and generally recognized as safe (GRAS) status (Shoukat et al., 2025). Starch exists in large amounts in plants as their main carbohydrate food reserve. It can also be found in large amounts in maize, potatoes, cassava, wheat, as well as in various types of cereals (Wang & Guo, 2020). Starch consists of two types of polysaccharides, that is, amylose, which has a linear main chain of α-1,4-linked glucose, and amylopectin, which has strong branching with α-1,4- and α-1,6-glycosidic bonds (Nath & Dutta, 2024). The proportion of amylose to amylopectin differs according to botanical sources. It affects the physicochemical properties of starch, including its ability to form films. Nonetheless, there exist various inherent disadvantages in native starch, which make it unsuitable for direct use in food packing. These include high humidity sensitivity, lack of water-resistance, poor mechanical strength, as well as instability in humid environments. The high hydroxyl content in starch, which works against their hydrophilic nature, causes poor barrier properties, as well as lower mechanical strength compared with conventional plastics (Santhosh et al., 2024). As a result, it is important to focus on modification techniques such as plasticization, compounding, chemical modification, and the use of functional additives, as depicted in Fig. 10.1, to improve the mechanical properties and barrier properties of starch packaging materials. Furthermore, with the growing need for extended shelf life and quality of foods, in a bid to improve the overall functionality of packaged products, the demands on packaging technologies have heightened (Jia et al., 2023).

Therefore, active and intelligent packaging technologies using biodegradable materials have emerged as effective solutions for enhancing food safety and quality while addressing sustainability issues. In the case of starch-based systems, active packaging functionalities are obtained by incorporating functional agents such as organic acids, enzymes, bacteriocins, and bioactive compounds from plants, capable of migration into foods or their surrounding atmosphere, which would retard spoilage reactions. On the other hand, intelligent packaging combines indicators and sensors, such as pH indicators and time temperature indicators, which give visual or measurable information related to freshness and storage history of foods (Luo et al., 2022). Currently, particular interest has been paid to

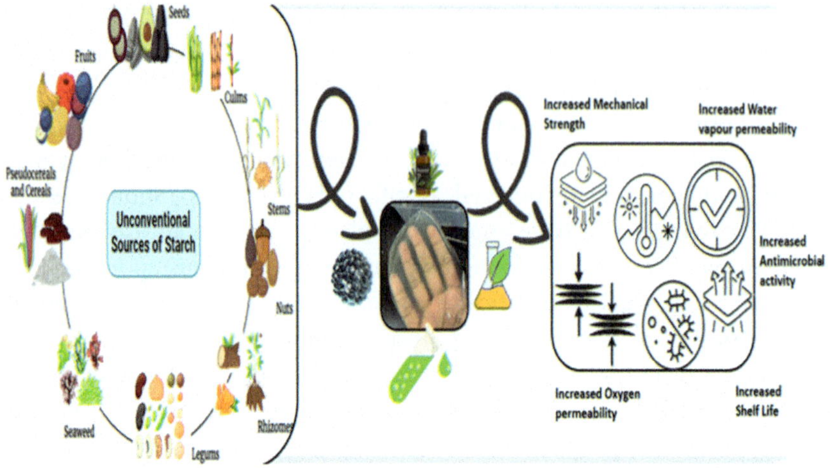

Figure 10.1 Unconventional starch sources and their functionalities in bioactive active packaging systems.

incorporating natural colorants into starch-based intelligent packaging systems. Anthocyanins, curcumin, and betalains are some natural pigments that change their color in a distinct and reversible manner with environmental stimuli comprising pH variation, gas release, and microbial metabolism occurring during food spoilage (Shaik et al., 2022). The ability of these colorants to develop colorimetric indicators that give real-time user-friendly visual information on freshness and food quality is enabled when they are embedded within starch matrices. Starch is an ideal carrier of natural colorant-based intelligent systems due to its transparency, film-forming ability, and compatibility with bioactive compounds.

Though there have been many studies conducted in the area of starch sources, methods of filming, or modification, there has been limited work done in terms of evaluating starch-based packaging materials in the context of intelligent packaging with natural colorants (Yu et al., 2021). As this forms a relevant background of this research, it has become essential for this chapter to give a comprehensive insight into starch-based intelligent packaging materials with natural colorants. Additionally, this chapter explores intelligent packaging, natural food colorants, methods of preparing starch-based intelligent films, as well as their uses in various food systems. Additionally, safe, stable, and scalable solutions are discussed, concerning starch-based intelligent packaging films, in terms of their applications in ensuring sustainable advanced food management.

10.2 Starch

Starch is a primary reserve carbohydrate in plants and a major dietary source of energy. It is a common ingredient in grains, root crops, tubers, and rhizomes, and is accumulated as granules within amyloplasts (Dereje, 2021). While starch granules differ in size and morphology depending on their botanical source, their functionality as materials in contact with food is determined more by molecular structure than by granule size. Starch consists of two polysaccharides, i.e., amylose, mainly a linear polymer of α-(1–4)-linked α-D-glucose units, and amylopectin, a highly branched polymer containing α-(1–4) links and α-(1–6) branch points (Chakraborty et al., 2022). The ratio of amylose to amylopectin, which depends upon the source of the starch, has profound effects on gelatinization behavior, film formation, and mechanical properties. Starch gelatinizes upon heating in excess moisture, where hydrogen bonds are broken and amylose leaches into the aqueous phase to form a continuous network that is required for film casting (Chakraborty et al., 2022). Native starch is very hydrophilic because of its high proportion of hydroxyl groups, hence it is sensitive to moisture, with inherent brittle film properties that preclude its use directly in packaging applications. It is widely observed that botanical origin is a key determinant of the physicochemical and functional properties of starch. Conventional starch sources are cereal grains, such as maize, wheat, and rice, as well as tubers such as potato and cassava, which are highly exploited due to their high yield and well-established processing methods (Thakur et al., 2019). These starches usually consist of 20%–30% amylose and show different crystalline structures; cereal starches exhibit A-type crystallinity, while tuber starches exhibit B-type crystallinity (El Halal et al., 2019; Svihus et al., 2005). These structural differences affect thermal stability, water absorption, retrogradation, transparency, and barrier performance of starch-based films (Bangar et al., 2024). Maize starch, for example, is often used for its balanced mechanical properties, while cassava starch is preferred for its high transparency and better compatibility with plasticizers and functional additives (Obadi et al., 2023).

In addition to traditional sources, interest has also been increasingly directed toward nonconventional starches sourced from fruit seeds, peels, pseudocereals, legumes, and even marine algae and agro-industrial coproducts for reasons of sustainability and valorization (Ayetigbo et al., 2018; Waisundara & Grichar, 2024). Their molecular arrangement and thermal properties showed higher variability, which might confer more specific properties to certain packaging uses. Further, the improved gelatinization patterns, better compatibility with active additives, and residual bioactive compounds such as phenolics or pigments have been mentioned for conferring antioxidant or functional properties to films for packaging in several studies (Carvalho et al., 2024; Fu et al., 2015; Matignon & Tecante, 2017; Tosif et al., 2023; Zabot et al., 2019). Starch-based films boast high transparency and are odorless and tasteless; thus, they are of great interest for edible and intelligent packaging applications (Frangopoulos et al., 2023). It is possible to improve transparency by gelatinization, blending, or incorporating nanofillers, which offers the opportunity to clearly view foods and, when necessary, to include colorimetric indicators (Jimenez et al., 2012; Yu et al., 2023). However, the impact of moisture sensitivity and brittleness is highly probable to affect the unfavorable tactile and barrier properties in humid conditions (Santhosh et al., 2024). Recently, substantial mechanical flexibility and moisture resistance were achieved for starch-based films through strategies related to plasticization, cross-linking, polysaccharide blending, and nanocomposites (Sadeghizadeh-Yazdi et al., 2019; Vieira et al., 2011). Overall, the molecular structure of starch, its varied sources, and its positive sensory properties make this material a promising biopolymer in developing functional, active, and intelligent biodegradable packaging systems, which is also concluded in Table 10.1.

10.3 Smart packaging

Smart packaging is an advanced form of food-packaging systems with sensing, acquisition, and communication capabilities to monitor, control, and communicate the changes occurring in food products and their immediate surroundings (Vanderroost et al., 2014). Unlike traditional packaging, considered to be a purely passive barrier, smart packaging is intended to be actively involved in maintaining the quality and freshness of foods during processing, storage, transportation, and distribution. By providing continuous or real-time evaluations of critical factors affecting the quality and safety of foods, smart packaging has contributed to better shelf-life management and reduced food losses along the supply chain. Conditions relating to smart packaging are most relevant to highly perishable fresh fruits, vegetables, meat, and seafood, whose slight deviations in storage conditions trigger rapid deterioration (Alam et al., 2021). Capable of timely detection and communication about variations in the environment or related to quality, smart packaging allows the best

Table 10.1 Sources of starch and its properties in film formation.

Film composition/ Starch type	Plasticizer/Additive	Thickness range (mm)	Moisture level (%)	Strength range (MPa)	Elongation at Break (%)	Reference
Cassava–Mungbean starch	Glycerol	0.090–0.103	19.22–22.11	2.85–9.34	18.82–21.37	Vu and Lumdubwong (2016)
	Sorbitol	0.101–0.113	8.84–9.43	6.77–19.20	10.84–14.86	
Wheat starch	Glycerol	0.074	44.5	3.29	15.21	Basiak et al. (2017)
Corn starch	Glycerol	0.112	36.7	3.72	19.13	
Potato starch	Glycerol	0.055	31.6	6.56	5.67	
Polyvinyl alcohol–pinto bean starch	-	0.061–0.070	-	27.5–52.6	108.1–241.8	Ali (2021)
Native starch	-	0.064	13.06	3.49	19.21	Shaikh et al. (2018)
Cassava starch	Metakaolin (5–15 wt%) + glycerol	-	23.1–33.2	5.5–8.1	45.5	Namory et al. (2022)
Rice starch	Sorbitol	-	-	10.75	7.56	Pornchai et al. (2022)
	NaOH (10%–50%) + sorbitol	-	2.75–9.87	11.36–53.03	-	
Octenyl succinate starch	Glycerol	0.087	29.54	9.60	32.41	Leon-Bejarano (2020)
	PSE (0.025%–0.100%) + glycerol	0.090–0.091	29.22–29.62	7.56–8.62	23.99–30.98	
Octenyl succinate starch	HSE (0.025%–0.100%) + glycerol	0.090–0.091	29.74–30.03	7.31–10.58	29.58–31.65	

decisions by manufacturers, retailers, and consumers to ensure quality and transparency throughout the supply chain. On the basis of the functional principles, there are two major types of smart packaging techniques, namely active packaging, where the packaging comes into contact with the food environment to prolong the shelf life, and intelligent packaging, where the data about the food quality and storage information is communicated.

10.3.1 Active packaging

Active food packaging represents a new approach to packaging, aside from packaging's main function of preserving foods for a certain period of time. According to European Regulation (EC) No. 450/2009, active materials are materials intentionally designed for releasing or absorbing substances into or from packaged foods or environments. Unlike passive packaging materials, active packaging materials have the ability to prevent microbiological and chemical degradation, as well as prevent depletion of sensory and visual qualities (Kumar & Gaikwad, 2025). Active agents function as absorbers, emitters, or controlled releasers of substances fundamental for maintaining food quality.

10.3.1.1 Moisture regulators

Moisture regulators are used to control the extra water in packaged foods. This extra water may be caused due to the sealing process, temperature variations, respiration in fresh produce products, or the inability of the packaging material to pass water vapor (Reinas et al., 2016). However, the presence of extra water causes rapid microbial growth and spoilage. This is evident in drip loss in meats, fish products, or condensation on fruits and vegetables (Divanshi et al., 2024). Moisture absorbers are used to reduce the water activity of the packaged foods using the properties of the absorbing components. These components can be inorganic or organic. Moisture absorbers function as relative humidity controllers or as moisture eliminators. They come in different forms, namely sachet pouches or pads, sheets, trays, or sticky labels with polymer films embedded with inorganic materials such as silica gel powder, bentonites, or calcium salts or with organic materials such as cellulose, fructose molecules, sorbitol molecules, or modified starch molecules (Kalita et al., 2024; Yadav & Dutta, 2024).

10.3.1.2 Ethylene adsorbents and scavengers

It promotes ripening and senescence in climacteric fruits and vegetables. Adsorbents physically trap ethylene molecules, reducing their concentration, and are made up of materials including silica, zeolite, montmorillonite, activated carbon, or Oya clay from Japan, all of which have high ethylene-binding capacities (Chan et al., 2017). These can be put in films or sachets. Particle modifications, for example alkali-treated halloysite nanotubes, can improve adsorbing capacities (Shu et al., 2016). On the other hand, ethylene scavengers can convert ethylene into inert materials. These include potassium permanganate, which can exist in alumina or silica forms, commonly put in sachets or films, thereby preventing direct exposure of foods (Álvarez-Hernández et al., 2019). Materials touted as nanomaterials, including TiO_2, Ag, or ZnO, Cu, Pd, have ethylene-oxidizing capacities. These have functionalities as thick films of chitosan-TiO_2, which could offer multifunctionality featuring ethylene absorption and (Saritala et al., 2025; Tabari & Farmanzadeh, 2020).

10.3.1.3 Carbon dioxide scavengers

High levels of CO_2 from microbial growth or metabolism of produce can cause deformation of packages, as well as deterioration in food quality (Lee, 2016). CO_2 scavengers, in the form of powder or sachets, assist in retaining package integrity. While chemical scavengers, including hydrated calcium oxide, convert CO_2 into calcium carbonate, physical adsorbents, including zeolites or activated carbon, remove CO_2 from the pouch headspace (Jaisan et al., 2018; Jimoh et al., 2018; Mourad et al., 2021) This regulation prevents package deformation and preserves product quality.

10.3.1.4 Antimicrobial packaging

Antimicrobial packaging improves food safety and reduces food waste by reducing microbial proliferation, hence extending the shelf life. The CO_2-emitting systems (ferrous carbonate, metal halides, sodium bicarbonate with citric/ascorbic acids) maintain controlled levels to prevent microbial proliferation and avoid deforming of packages (Khaneghah et al., 2018). Ethanol, delivered through sachets or films, targets bacteria, yeasts, and molds; the release rate depends on the film properties and the water activity of the food (Gómez et al., 2014).

Volatile agents such as chlorine dioxide and sulfur dioxide, and nonvolatile weak organic acids (acetate, sorbate, benzoate, propionate) disrupt microbial membranes (Sofos, 2024). Inorganic nanoparticles incorporated into films or pads elicit their antimicrobial activity through ion release and ROS generation: Ag, Cu, Au, Pt, Se; TiO_2, ZnO, MgO, CuO. Treatment options also exist with enhanced efficacy through bimetallic and trimetallic

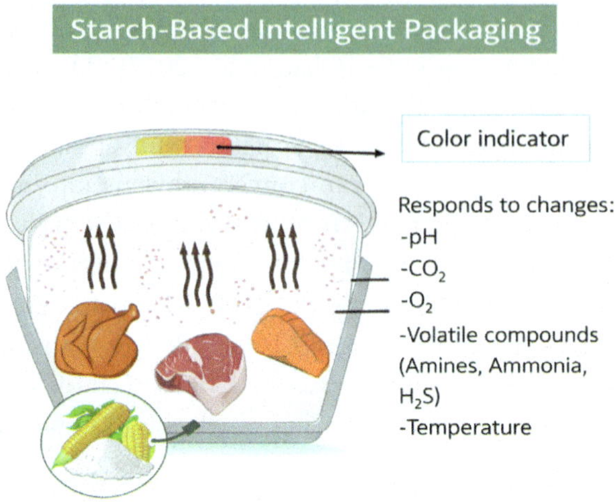

Figure 10.2 Illustration of starch-derived intelligent packaging systems.

nanoparticles, which has also improved mechanical/barrier properties (Basavegowda & Baek, 2021; Spirescu et al., 2021). Application requires consideration of food type, packaging, and potential nanoparticle toxicity.

10.3.2 Intelligent packaging

Intelligent packaging is defined as sophisticated packaging that is able to track, indicate, or react to changes within the condition or safety of packaged foods as demonstrated in Fig. 10.2. Through the combination of sensors, indicators, and data carriers, these intelligent packaging systems have the ability to provide the consumer with real-time information concerning the quality of the packaged goods (Vanderroost et al., 2014).

10.3.2.1 Time temperature indicators (TTIs)

Time Temperature Indicators (TTIs) trace cumulative exposure to temperature, which plays a critical role in the stability and shelf life of perishable foods. TTIs react to temperature changes through chemical, enzymatic, and physicochemical reactions that offer visual and measurable results corresponding to the extent of degradation. TTIs are usable for chilled, frozen, and heat-processed foods and are cost-effective and dynamic shelf-life trackers (Liu et al., 2023; Yang et al., 2025). TTIs, therefore, offer a useful tool for dynamic shelf life assessment, especially temperature-sensitive foods such as meat, because they fill a critical gap remaining in static date marking (Pandian et al., 2021). Further, accurate interpretation of TTI responses and the development of robust decision-support models are essential for effective practical application with respect to quality assessment and reduce unnecessary food waste.

10.3.2.2 Freshness indicators

Freshness indicators identify spoilage biomarkers such as volatile amines, hydrogen sulfide, carbon dioxide, or oxygen. The freshness indicators rely on pH-sensitive dyes that can easily be noticed for their color transformations upon interaction with microbes, or after a chemical reaction occurs in the food products. Both artificial dyes and natural dyes such as anthocyanins, curcumin, betalains, grape skin, and red cabbage juice have been used for colorimetric detection (Luo et al., 2022; Yang, Fu, et al., 2024; Yang, Yuan, et al., 2024). Natural dyes are preferred for their nontoxic, biodegradable, and affinity for starch-Based Films, which lays a foundation for sustainable visual intelligent packaging. Some common examples include colorimetric markers named FreshTag specifically for fish products, bio-based pH indicators composed of chitosan, and indicators based on whey protein isolate or chitosan to sense carbon dioxide production in meat, chicken, and poultry products based on transparency or color alterations caused by pH changes (Chen et al., 2025).

10.3.2.3 Biosensors and gas sensors

Biosensors and Gas Sensors are a combination of a biorecognition element with a transducer that identify specific analytes such as allergens, sugars, lipids, and microbial metabolites (Bao et al., 2024; Naresh & Lee, 2021). Sensors track levels of oxygen and carbon dioxide, using pH-responsive natural colorants, to instantly determine food respiration, microorganisms, and packaging (Heo & Lim, 2024). They are usually employed in label, tablet, coating, and film forms, combined with starch film matrices, and allow a transparent, biodegradable, and visually sensitive solution when combined with natural colorants, connecting sensors with sustainable packaging, providing an innovative way to instantly track food freshness in a variety of food categories, ranging from meats, fish, dairy, fruits, and vegetables.

10.4 Starch-based films

Starch-based films are a flexible and ecocompatible packaging material with a wide range of applications based on various industrial sectors such as food and beverages, cosmetics, pharmaceuticals, and consumer products. Starch-based packaging materials could be rationally tailored to meet various packaging requirements with tunable variables such as strength, thickness, permeability, and compatibility with the packaged product Maran et al., 2013). Besides, starch-based packaging materials exhibit excellent printability and sealing abilities with effective labeling, branding, and information representation along with product conservation and safety assurance. Starch-based packaging materials could be further functionalized for the integration of bioactive agents such as antioxidants, antimicrobial agents, and nutraceuticals. Bioactive agent-incorporated starch packaging materials could be further rationally designed for enhanced multifunctionality, hence showing active packaging materials with immense applications regarding food conservation and safety by reducing postharvest losses due to spoilage. The incorporation of bioactive agents into packaging materials ensures enhanced product shelf life along with improvement in the quality parameters of packaged foods. Through complementarity between the starch materials and the bioactive agents, the edible and fully biodegradable packaging materials could form a potpourri for ecocompatibility, with high-end implications for product safety, viability, and conservation, while at the same time having considerable repercussions on environmental stewardship (Wilfer et al., 2021). On the whole, starch-based films are a robust and avant-garde solution for food packaging according to principles of sustainability, ensuring an effective balance between material performance, functional innovation, and ecological responsibility. By strategic use of the intrinsic properties of starch and emerging technologies in packaging, such systems ensure the shift toward a sustainable packaging framework that meets the present needs without compromising the ability of future generations to meet their own needs.

10.4.1 Film formation techniques

10.4.1.1 Solvent casting

Solvent casting involves casting a starch solution and other agents, followed by drying to produce films (Majeed et al., 2023). This method is flexible in terms of creating the final product thickness and chemical composition, allowing the possibility of blending BIO, plasticizers, and other agents to enhance the mechanical properties of films. Drying conditions and solvents influence the properties of the films.

10.4.1.2 Extrusion

"Extrusion is a continuous process where a thermoplastic starch matrix is extruded, in other words, transported under heat pressure through a die, to produce films of uniform thickness" (Fitch-Vargas et al., 2016). This method facilitates the uniform dispersion of all the additives, bioactive, and fillers, while microstructure control depends on factors such as its tensile strength, extensibility, and puncture resistance. Functional starch films on a large scale are produced through this method.

10.4.1.3 Compression molding

Compression molding entails the use of starch mixed with other compounds and pressed into heated molds to form a film (Berti et al., 2024). This method allows for control over thickness, density, and the surface topography, a key aspect that enables the production of starch-based films that require desired mechanical properties such as biodegradability (Wu et al., 2023).

10.4.1.4 Electrospinning

Electrospinning can produce ultrafine nanofibers at a stiff cost in terms of starch solution concentrations in the presence of an electric field (Fonseca et al., 2019). These films have strong surface area and improved mechanical barriers. Nanofibers have the ability to release bioactive compounds; however, there is ineffectiveness at an industrial scale (Uygun et al., 2020).

10.5 Functions of starch-based smart packaging

10.5.1 Starch-based active films

Starch-based packaging offers structural support and environmental protection but inherently does not delay deterioration. Good preservation is achieved by incorporating antioxidants and/or antimicrobials into the starch matrix; this way, controlled release of bioactive compounds enables enhancing food safety, quality, and shelf life, also represented in Fig. 10.3. The release kinetics are influenced by the source of bioactive, extraction method, concentration, interactions with starch matrix, and microstructural characteristics of the film (Mahović Poljaček et al., 2024).

10.5.1.1 Antioxidant active starch-based films

Antioxidant-active packaging films reduce the oxidation of food by releasing substances in the packaging environment to slow-down

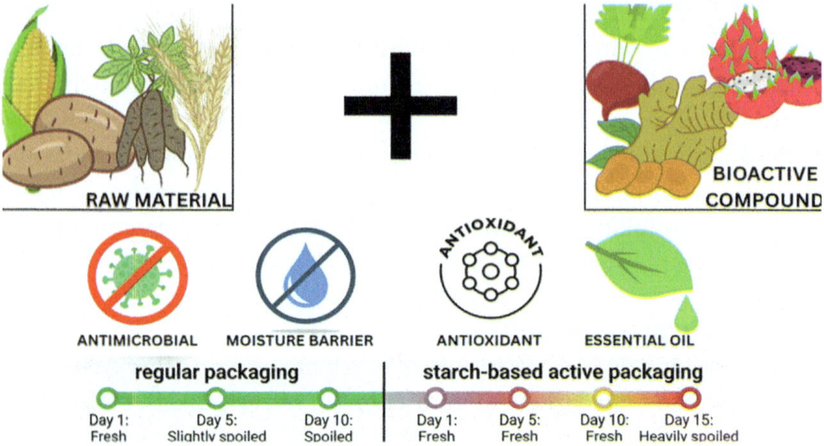

Figure 10.3 Starch-based active packaging incorporating bioactive compounds for enhanced food preservation.

oxidation (Piñeros-Hernandez et al., 2017). Natural antioxidants, such as polyphenols, tocopherols, essential oils, and plant extracts, are more in demand than synthetic ones owing to their nontoxicity and versatility (Deshmukh & Gaikwad, 2024). For instance, a pineapple peel extract was mixed with a polyvinyl alcohol and corn starch film to provide a DPPH radical scavenging ability of about 42% (Kumar et al., 2024). Additionally, potato starch films with oxidized samples containing sodium ascorbate demonstrated a seven times higher resistance to oxidation compared with the control (Kowalczyk et al., 2018). Essential oils, especially those in nanoemulsions, can readily diffuse into the packaging atmosphere to prolong the shelf life of meat (Amiri et al., 2019).

10.5.1.2 Antibacterial active starch-based films

Starch films have only limited intrinsic antibacterial activity, and functionality is improved with the addition of natural, organic, or inorganic antimicrobial agents (Syafiq et al., 2020; Zhu et al., 2023). The natural agents include essential oils, chitosan, and antimicrobial peptides, whereas inorganic agents involve metal-based or photocatalytic nanoparticles such as Ag, Cu, TiO_2, and ZnO (Babaei-Ghazvini et al., 2021). Rice starch films that contain oregano essential oil reduced the bacterial growth in frozen fish (Martins et al., 2021), and starch-CMC films with $AgNO_3$ inhibited both *S. aureus* and *E. coli* (Xing et al., 2019).

10.5.1.3 Controlled release starch-based active films

Based on the controlled release of active substances, the antioxidant and antibacterial properties of films are effective. However, too slow release may be inadequate for preservatives, while the rate of rapid release might be too quick for interaction with other compounds in the foods. Methods that enhance the rate include the modulation of the matrix, the multilayer packaging films method, the cross-linking process, microencapsulation, and the incorporation of MOFs (Almasi et al., 2021; Tian et al., 2022). Also, the rate of diffusion is influenced by the percentage of moisture in the foods. For example, corn starch films with zein-rutin nanoparticles released rutin between 27% and 37% in 12 h (Zhang & Haiyan, 2017). Also, pea starch films with quercetin particles were designed for continuous diffusion for 1 week (Farrag et al., 2018).

10.5.2 Starch-based natural colorant intelligent packaging

Intelligent starch food packaging is an innovative solution that not only has monitoring functions but also preserves food actively. These food packages are supposed to evaluate food quality inside the packing, indicating changes in food quality or within its surroundings. Real-time data about food quality, freshness, storage, and spoilage is directly provided. Natural bioactive agents in food packing films possess pharmacology, specifically bactericides, and antioxidants. It has been observed that these food packs possess both passive and active functionalities at a single time. A major point of distinction between starch food packing films and most of the biodegradable polymers is their transparency, which allows a clear indication of any change in food packing, making a clear visual observation feasible without any distortion by food matrices (Chiu & Yang, 2024). The aim of introducing starch food packing is to provide these food packs with visual indicators, which reveal semiquantitative/quantitative data about food quality. Examples of these food packing films include food freshness indicators, which react positively to volatile food molecules, and Time Temperature Indicators (TTIs), which measure food exposure to temperature changes. Additionally, using natural color molecules as sensors can add to these food packing functionalities. Pigments isolated from various materials such as red cabbage, turmeric, beetroot, blueberry, sweet potato could also vary in terms of color due to pH, spoilage materials, or even temperature changes, as depicted for other materials in Fig. 10.4. For example, pH-sensitive materials could be used to create a visual alarm system for tracking transitions from a 'fresh'

Figure 10.4 Natural pigments as indicators in intelligent packaging.

state to a 'spoiled' state, whereas intelligent labels could assist consumers to get notifications about food spoilage status even on packaging materials directly. In addition, intelligent starch-based materials could serve as a complementary component for tracking several factors such as status of spoilage of foods such as meats, temperature, and notifications regarding times and quality, among others.

10.5.2.1 Freshness indicator

This film-forming property of starch makes it a great matrix for intelligent film-based colorimetric sensors. pH-sensitive starch film sensors are popular as food spoilage sensors because of the change in local pH caused by bacterial growth. Organic pH indicators such as anthocyanins, curcumin, and carotenoids are always preferred for their nontoxic and environmentally friendly properties. Among these, the large range of colors displayed by anthocyanins makes it easier for the common man to identify the freshness of the food. Colorimetric sensors work on the basis of the change in the colors of the natural dyes because of the chemical change caused by the spoilage of food, such as the formation of ammonia, dimethylamine, and trimethylamine (Balbinot-Alfaro et al., 2019). For instance, a mixture of agar and potato starch Films containing sweet potato anthocyanins change red to green on detecting the spoilage in pork packaging (Choi et al., 2017). Chitosan, corn starch, and purple cabbage extract-based films were employed as sensors for the spoilage of fish (Silva-Pereira et al., 2015). Electrochemical sensors give better results, but the main advantage of the colorimetric sensor is that it has better sensitivities (Subramanian et al., 2022).

10.5.2.2 Time-temperature sensor

Temperate changes are very important to the quality of perishable food. Starch-derived TTIs provide eco-friendly substitutes for nonbiodegradable traditional synthetic indicators, which are based on the hydrophilic property, optical clarity, and film-forming property of starch, which are useful for inducing regulated molecular mobility or reaction rates (Zhang et al., 2023). Starch carriers can induce temperature-dependent visual changes through the chemical, enzymatic, or physicochemical change. Natural pigment, protein, or bio-derived materials can provide nontoxic, fully biodegradable TTIs designed for direct food contact. Myoglobin extract, nitrite, mixed into thermoplastic cassava starch, respectively, provided a protein thermochromic substance TCI with visible visual changes as well as measurable changes to various instruments at various temperatures (Santos et al., 2020). Blackberry powder at −127 °C, freeze-dried, also exhibited a very pronounced visual-changeable property from red to brown colors upon sterilization (Nogueira et al., 2019). Starch films act as ideal carriers for active or intelligent substances, which can improve shelf life, quality, as well as food security, while providing visual cues to monitor the spoilage (Arruda et al., 2025).

10.5.2.3 Fabrication of starch-based intelligent packaging using anthocyanin

Starch-based intelligent packaging films that include anthocyanin have been produced by solution casting, whereby the starch extracts are combined in solution form, including pigment extracts, and poured to form thin films. In these, anthocyanin, which is derived from red cabbage or black rice, has desirable pH-sensitive properties that allow visual freshness indication in food. For instance, the blend of cabbage anthocyanin and sago starch undergoes various colored transitions as it alters from acidic to basic conditions, while esterified starch blended with anthocyanins obtained from black rice undergoes colored transitions from red, then to pink, purple, and eventually green (Cheng, 2022; Song et al., 2024). Enhancement of mechanical properties can also come from combining anthocyanin-derived materials and binding them to other sugar-derived materials, including PVA and chitosan nanocomposites. Another approach involves encapsulating anthocyanin in starch/amylopectin nanoparticles. In this case, the materials have been successfully used in freshness indication of seafood, meat, and other perishable foods (Nguyen et al., 2025).

10.5.2.4 Fabrication of starch-based intelligent packaging using curcumin

Curcumin, the active polyphenolic compound of Curcuma longa, shows pH-sensitive chromatic transitions, along with the ability to act as an antioxidant and antimicrobial agent, qualifying it as an optimal bio-indicator for smart starch-based food packaging (Devi et al., 2025). It displayed a yellow-to-orange-red or reddish-brown transition when exposed to an alkaline environment, allowing easy recognition of the pH change associated with the spoilage process. Curcumin can easily be introduced into starch-based bio-indicators through established processing techniques such as solvent casting or emulsion technique. Thus, it can be introduced into Pickering emulsions before mixing it with starch/polyvinyl alcohol-based films as an optimal strategy for improving the stability of the bio-indicators. This type of film, for example, displayed a yellow-to-red transition during the spoilage process of fish products, clearly demonstrating its optimal role as an indicator for the freshness of such products (Mali & Pandey, 2024). Furthermore, recent works clearly verify the adaptability of this approach, where waste starch, Artocarpus lakoocha, combined with 2.5% curcumin, displayed distinctive pH-sensitive bio-films that act as efficient antioxidants and antimicrobial agents, detecting the real-time freshness of chicken fillets (Liu et al., 2021). Moreover, starch nanofibers obtained by electrospinning, coloaded with curcumin and other natural pigments, for instance, anthocyanins, presented rapid color transitions, with improved resistance to moisture and increased delay of spoilage when applied to packaging for shrimp or pork (Lv et al., 2025).

10.5.2.5 Fabrication of starch-based intelligent packaging using betalains

The water-soluble, nitrogen-containing natural pigments are betacyanins (red–violet) and betaxanthins (yellow–orange), generally obtained from beetroot and the peels of red pitaya. Because of the strong coloration, antioxidant activity, and sharp pH sensitivity of betalains, recently they have gained wide interest in uses as natural colorimetric probes in starch-based smart packaging systems. Betalains, during fabrication, were generally added into gelatinized starch matrixes by techniques involving solution casting or blending with compatible polymer matrices that afforded them homogeneous immobilization of pigments with retained film integrity. Cassava starch films containing betalains from agro-industrial coproducts have been obtained as intelligent labels, exhibiting distinct and reversible color transitions from red–violet to yellowish hues under alkaline conditions given off by food deterioration, while offering mechanical and barrier properties within the acceptable range (Otalora Gonzalez et al., 2022). On the other hand, Khanzada et al., 2025 have proposed the development of starch films incorporating betalains as pH-sensitive sensors, where stable pigment–starch interactions facilitated visible color responses in a wide pH range relevant for protein-rich food deteriorations. However, a further step forward was shown by Qin et al. (2020) with the fabrication of starch/polyvinyl alcohol bioactive composite films immobilized with betalains extracted from red pitaya fruit skin. Such bioactive composite films showed intelligence in terms of sensing capabilities, which included the visible distinction in color changes that discriminated the freshness of shrimp, and exhibited enhanced capabilities in terms of antioxidants, antimicrobial agents, and UV blocking. Overall, it is observed that bioactive composite films made up of starch and immobilized betalains can serve as biodegradable and nontoxic intelligent packaging systems that are color-sensitive and can detect perishable goods at site.

10.6 Applications in the food industry

The introduction of starch-based intelligent films has been applied to food packaging, which shows great promise for improving food safety and monitoring freshness and shelf life in different kinds of food systems. These natural colorants include anthocyanins, curcumin, betalains, and others, which are added to these intelligent films and indicate the quality change of foods through visual change while undertaking the antimicrobial and antioxidant functions of the film. Color changes to the films reflect chemical or microbial changes to the food; thus, spoilage or environmental deviation is monitored in real time and noninvasively (D'Almeida & de Albuquerque, 2024; García-Guzmán, 2022). Application of starch-based packaging is presented in Table 10.2.

Table 10.2 Applications of starch-based functional films in food preservation and quality monitoring.

Starch matrix source	Functional additives incorporated	Food product	Observed functional outcome	Reference
Yam starch	Eugenol	Pork	Incorporation of 3% eugenol resulted in more than a 50% extension of pork shelf life	Cheng et al. (2019)
Job's tears starch	Clove bud essential oil	Pork belly	Addition of 0.5% essential oil significantly suppressed lipid oxidation	Kang and Song (2019)
Potato starch	Carrot-derived anthocyanins	Raw milk	Film functioned as a freshness indicator by signaling spoilage-related changes	Goodarzi et al. (2020)
Cassava starch	Gelatin and casein	Guava	Shelf life of guava fruit increased by approximately 2 days	Silva et al. (2021)
Maize starch	Grape juice extract	Chicken breast fillets	Lipid oxidation and microbial growth were effectively delayed	Meral et al. (2021)
Brazilian pine seed starch	Citric pectin and functionalized additives	Grapes and bread	Product quality was maintained for up to 30 days of storage	de Oliveira et al. (2021)
Hydroxypropyl distarch phosphate	ε-Polylysine and gelatin	Fresh bread	Microbial spoilage was significantly delayed	Yue et al. (2022)
Corn starch	Carboxymethyl cellulose	Food simulant	Strong antimicrobial activity observed against *E. coli*	Ni et al. (2018)
Potato starch	Betacyanin	Fish	Color shift from pink to yellow correlated with increasing TVB-N levels	Shahab et al. (2021)
Corn starch	Curcumin-loaded Pickering emulsion	Fish	Film color changed from yellow to red during spoilage progression	Xinxin et al. (2021)
Cassava starch	*Lycium ruthenicum* anthocyanin nano-complexes	*Micropterus salmoides* fillets	Pronounced color change occurred as fish deterioration advanced	Yan et al. (2021)

10.6.1 Smart packaging technology applications in various food sectors

10.6.1.1 Fruits and vegetables

Fruits and vegetables exhibit high respiration rates, transpiration, and microbial growth. Intelligent starch films assess the crucial parameters of pH change, ethylene production, and microbial byproducts, leading to measurable visual transitions. For example, starch films high in anthocyanin change from red to blue as the pH increases as a result of microbial growth, indicating immediate freshness (Vasuki et al., 2023; Balbinot-Alfaro et al., 2019). Ethylene removal by added scavengers in packages delays ripening, thus averting spoilage (Wei et al., 2021). Time temperature indicators (TTIs) enable tracking of temperature abuse, important in respect to respiration-sensitive produce (Yahia & Carrillo-Lopez, 2018; Ghoshal, 2018).

10.6.1.2 Dairy products

Milk, cheese, and yogurt have high water activity and nutrient values, making it favorable for growth. Intelligent films based on starch and combined with oxygen scavengers, antimicrobial compounds, and natural dyes offer visual signs of decay while preventing oxidation and growth. For instance, films respond to an indicator change and microbial by-products, with visual changes in colors expressing a loss of freshness (Contò et al., 2017; Haghighi-Manesh et al., 2017). Controlled oxygen scavenging avoids off-flavors and preserves nutrient values, hence shelf life (Vasuki et al., 2023).

10.6.1.3 Meat products

Meat products have a tendency toward lipid oxidation and microbiological contamination. Smart starch films, with antioxidants and antimicrobials, prevent lipid oxidation and microbial growth, while pH indicators help determine product quality visually (Jayasena & Cheorun, 2013; Kerry & Tyuftin, 2017). The films show a change in color along with protein breakdown and pH change, making it easy for consumers to determine product freshness. Smart packaging increases shelf life by preventing lipid, as well as myoglobin, oxidation, dehydration, and microbiological growth (Umaraw et al., 2020).

10.6.1.4 Seafood products

Fish and other fish products are known to have a quality shelf life that deteriorates quickly because of their high water content and the presence of microorganisms. The smart packaging formulation for starch-based packaging can indicate the onset of spoilage in a colorimetric manner in response to the volatile nitrogenous compounds (ammonia and biogenic amines) and microorganisms (Laorenza et al., 2022; Rodrigues et al., 2021). The active ingredients of the packaging include the presence of oxygen and carbon dioxide, as well as the use of barrier packaging.

10.6.2 Edible and compostable packaging innovations

Starch-based films, which are edible, coupled with biocolors and antimicrobial agents, offer environment-friendly options that also allow direct monitoring of food quality. These films retain food characteristics, thereby extending shelf life for perishable foods, including fruits, vegetables, dairy, meat, and seafood, without chemical preservatives, thereby allowing consumer consumption of these perishable foods (Kumar et al., 2025). Composting systems, which are also compostable, provide options for attaining circular economy principles with minimal postconsumer packaging (Kędzia et al., 2022).

10.7 Challenges and limitations

With the rising environmentally hazardous conditions posed by plastic wastes, there has been a growing focus on producing environment-compatible counterparts of plastic materials, especially those used on a disposable, short-term consumption principle. Among the emerging classes of materials suitable as environment-compatible plastic materials, starch-based materials have attracted significant attention because of their potential to be utilized as cost-effective, green, and environment-compatible plastic materials. Starch, being a natural, bio-based, and generally recognized as safe (GRAS) polysaccharide material, contains multiple components, including starch, protein, and lipids, obtained from natural sources such as corn, potatoes, and cassava roots. Presently, starch-based materials possess vast applications in producing materials and products developed from petro-based plastic materials, including simple nonedible products such as packing materials, through appropriate modifications to enhance their cost, efficacy, and value as commodities. However, application of starch-based materials, especially in the form of colorimetric indicators, possess significant limitations, prime among them being their poor water-resistance property, relatively high moisture sensitivity, possible negative impacts on the structural integrity of the material, and the color-producing molecules they comprise, as well as their brittle constitution, and gas barrier properties. Leaching of natural colorants blended with starch matrices also remains an important challenge for their stability. Natural colorants such as anthocyanin, curcumin, and betalains exhibit high reactivity toward environmental factors, such as light, temperature, atmospheric oxygen, and low or high pH environments. Natural food colorants, when blended with a starch matrix, tend to degrade, fade, or show irreversible color changes, contributing toward generating ambiguous or false positive results concerning freshness detection. Moreover, blending of a starch polymer matrix with natural colorants can affect their reactivity toward specific stimuli, showing an increasing tendency toward color change at a threshold level deviated from real food-spoilage conditions. However, in an attempt to resolve such issues, researcher communities have devoted considerable effort toward material modifications for improving performance along with reliability of the indicator system. Various methods, including starch bio-nanocomposites, blending with other complementary bio-polymers, addition of nanofiller materials, and multilayer films, have been considered for improving mechanical strength, water resistance, and controlled release of colorants. Encapsulation of natural dyes using nano/microcarriers also represents an emerging paradigm for protecting dyes from deteriorative environmental effects while retaining responsiveness toward stimuli of food-spoilage. Although these advancements have significantly improved the functionality of starch-based colorimetric indicators, challenges related to long-term stability, scalability, and standardization remain and must be addressed before widespread commercial adoption can be realized.

10.8 Future prospects and research directions

In the future, research related to intelligent packaging based on starch will likely focus more on optimizing the stability, precision, and reliability of natural colorant-based indicators while still being fully biodegradable. Further research related to optimizing the interaction of starch matrices with natural dyes such as anthocyanins, curcumin, or betalains will be highly important, especially to resist moisture, pigment migration, or unwanted color development at higher humidities. Modification of materials using nanofillers, biopolymer blending, or nanoencapsulation technology will likely play an important role in optimizing the stiffness, barrier performance, or stability of colors without impairing food safety. In the meantime, hybrid materials consisting of colorimetric indicators with rudimentary digital technology, such as QR-code or NFC-enabled reader technology, could potentially increase the precision or traceability of indicators while still providing nonexperts with visual information. Moreover, it will be crucial to optimize the development of fully biodegradable indicator systems, from the pigment carriers to the sensors, to prevent unwanted environmental pollution. From the viewpoint of circular economy, future research will likely focus more on the utilization of starch from agronomic by-products, life cycle assessment for the determination of the sustainability advantages, or biodegradability criteria to secure commercial approval. Addressing these challenges will be pivotal for translating starch-based natural colorant intelligent packaging from laboratory concepts to scalable, reliable, and environmentally responsible food-packaging solutions.

10.9 Conclusion

There has been a growing demand for environmentally compatible, safe, and convenient food packaging. This has fueled considerable interest in biodegradable packaging materials that can substitute conventional nonrenewable plastic packaging. Among the biodegradable packaging alternatives, starch has been found to be an attractive bio-based product catering to the requirements of intelligent packaging. This particular bio-based product, along with the prospects of utilizing starch-based packaging as intelligent packaging alternatives, has been comprehensively reviewed in this chapter. Starch-based packaging as an intelligent packaging alternative, along with its bio-based applications, has been detailed. Starch, based on its bio-based properties, shows promising outcomes as a packaging alternative. Apart from the parameters detailed, the packaging alternative based on starch has the advantage of transparency, film-formability, and biodegradability, along with its apt compatibility. However, the biodegradable packaging alternative based on starch faces a number of drawbacks, such as moisture, fragility, and the lack of a barrier. Proper processing, including plasticization, biopolymer blends, nanocomposites, or the encapsulation of dyes, indicates greater improvements for the biodegradable packaging alternative. The use of food-grade natural colorant-based intelligent packaging in different food systems, such as meat, fish, dairy, and fresh produce, demonstrates practical feasibility by offering an efficient way to monitor quality and reduce food waste. However, pigment instability, interference due to moisture, scalability of production, lack of harmonization in regulations, and cost issues still remain a challenge. Overcoming these limitations through targeted material design, standardized safety evaluation, and life-cycle assessment is crucial for successful commercialization.

References

Álvarez-Hernández, M. H., Martínez-Hernández, G. B., Avalos-Belmontes, F., Castillo-Campohermoso, M. A., Contreras-Esquivel, J. C., & Artés-Hernández, F. (2019). Potassium permanganate-based ethylene scavengers for fresh horticultural produce as an active packaging. *Food Engineering Reviews, 11*(3), 159–183.

Alam, A. U., Rathi, P., Beshai, H., Sarabha, G. K., & Deen, M. J. (2021). Fruit quality monitoring with smart packaging. *Sensors, 21*(4), 1509.

Ali, K. (2021). Evaluation of physical, mechanical and antibacterial properties of pinto bean starch-polyvinyl alcohol biodegradable films reinforced with cinnamon essential oil. *Polymers, 13*, 2778.

Almasi, H., Jahanbakhsh Oskouie, M., & Saleh, A. (2021). A review on techniques utilized for design of controlled release food active packaging. *Critical Reviews in Food Science and Nutrition, 61*(15), 2601.

Amiri, E. (2019). Combined antioxidant and sensory effects of corn starch films with nanoemulsion of *Zataria multiflora* essential oil fortified with cinnamaldehyde on fresh ground beef patties. *Meat Science, 153*, 66–74.

Arruda, T. R., Machado, G. D. O., Marques, C. S., Souza, A. L. D., Pelissari, F. M., Oliveira, T. V. D., & Silva, R. R. A. (2025). An overview of starch-based materials for sustainable food packaging: Recent advances, limitations, and perspectives. *Macrom.*

Ayetigbo, O., Latif, S., Abass, A., & Müller, J. (2018). Comparing characteristics of root, flour, and starch of biofortified yellow-flesh and white-flesh cassava variants, and sustainability considerations: A review. *Sustainability*.

Babaei-Ghazvini, A., Acharya, B., & Korber, D. R. (2021). Antimicrobial biodegradable food packaging based on chitosan and metal/m.

Balbinot-Alfaro, E., et al. (2019). Intelligent packaging with pH indicator potential. *Food Engineering Reviews, 11*(4), 235–244.

Bangar, S. P.; Balakrishnan, G.; Navaf, M.; Sunooj, K. V. (2024) "Recent Advancements on Barnyard Millet Starch: A Sustainable Alternative to Conventional Starch.

Bao, F., Liang, Z., Deng, J., Lin, Q., Li, W., Peng, Q., & Fang, Y. (2024). Towards intelligent food packaging of biosensor and film substrate for foodborne microorganisms monitoring: A review of recent progress. *Critical Reviews in Food Science and Nutrition, 64*(12), 3920–3931.

Basavegowda, N., & Baek, K. H. (2021). Multimetallic nanoparticles as alternative antimicrobial agents: challenges and perspectives.

Basiak, E., Andrzej, L., & Frédéric, D. (2017). Effect of starch type on the physico-chemical properties of edible films. *International Journal of Biological Macromolecules, 98*, 348–356.

Berti, S., Jagus, R. J., Flores, S. K., & Gonzalez-Martinez, C. (2024). Antimicrobial edible starch films obtained by casting and thermo compression techniques. *Food and Bioprocess Technology*.

Carvalho, H. J. M., Barcia, M. T., & Schmiele, M. (2024). Non-Conventional Starches: Properties and Potential Applications in Food and Non-Food Products.

Chakraborty, I., Pooja, N., Mal, S. S., Paul, U. C., Rahman, M. H., & Mazumder, N. (2022). An insight into the gelatinization properties influencing the modified starches used in food industry: A review. *Food and Bioprocess Technology, 15*(6).

Chan, W. H., Mazlee, M. N., Ahmad, Z. A., Ishak, M. A. M., & Shamsul, J. B. (2017). The development of low cost adsorbents from clay and waste materials: A review. *Journal of Material Cycles and Waste Management, 19*(1), 1–14.

Chen, Y., Zhou, W., Li, X., Liu, D., & Zhong, Y. (2025). Chitosan-based natural colorant intelligent freshness indicator: Recent advances, properties, novel techniques, and applications. *Food and Bioprocess Technology*.

Cheng, M., et al. (2022). Characterization and release kinetics study of active packaging films based on modified starch and red cabbage anthocyanin extract. *Polymers, 14*(6), 1214.

Cheng, J., Wang, H., Kang, S., Xia, L., Jiang, S., Chen, M., & Jiang, S. (2019). Active packaging film of yam starch eugenol and its application to pork products. *Food Hydrocolloids, 96*.

Chiu, I., & Yang, T. (2024). Biopolymer-based intelligent packaging integrated with natural colourimetric sensors for food safety and sustainability. *Analytical Science Advances, 5* (5–6).

Choi, I., et al. (2017). Intelligent pH indicator film composed of agar/potato starch and anthocyanin extracts from purple sweet potato. *Food Chemistry, 218*, 122–128.

Contò, F., et al. (2017). *Advances in dairy products*.

D'Almeida, A. P., & de Albuquerque, T. L. (2024). Innovations in food packaging: From bio-based materials to smart packaging systems. *Processes, 12*(10), 2085.

de Oliveira, A., et al. (2021). Bioactivity screening of pinhão (*Araucaria Angustifolia* (Bertol.) Kuntze) seed extracts: The inhibition of cholinesterases and α-amylases, and cytotoxic and anti-inflammatory activities. *Food and Function, 12*(20), 9820–9828.

Dereje, B. (2021). Composition, morphology and physicochemical properties of starches derived from indigenous Ethiopian tuber crops: A review. *International Journal of Biological Macromolecules, 187*, 911–921.

Deshmukh, R. K., & Gaikwad, K. K. (2024). Natural antimicrobial and antioxidant compounds for active food packaging applications. *Biomass Conversion and Biorefinery*.

Devi, L. S., Krishnan, R., Begum, S., Nath, D., Mohanty, A., Misra, M., & Kumar, S. (2025). Curcumin as bioactive agent in active and intelligent food packaging systems: A comprehensive review. *Comprehensive Reviews in Food Science and Food Safety, 24**(4), e702.

Divanshi, M., Das, A., & Hv, R. (2024). Quality control and risk assessment of food storage and packaging. In *Food safety* (pp. 25–40). CRC Press.

Drago, E., Campardelli, R., Pettinato, M., & Perego, P. (2020). Innovations in smart packaging concepts for food: An extensive review. *Foods*.

El Halal, S. L. M., Kringel, D. H., Zavareze, E. D. R., & Dias, A. R. G. (2019). Methods for extracting cereal starches from different sources: A review. *Starch*.

Fitch-Vargas, P. R., Aguilar-Palazuelos, E., de Jesús Zazueta-Morales, J., Vega-García, M. O., Valdez-Morales, J. E., Martínez-Bustos, F., & Jacobo-Valenzuela, N. (2016). Physicochemical and microstructural characterization of corn starch edible films obtained by a combination ofFonseca,.

Farrag, Y., et al. (2018). Starch films loaded with donut-shaped starch-quercetin microparticles: Characterization and release kinetics. *International Journal of Biological Macromolecules, 118*, 2201–2207.

Fonseca, L. M., et al. (2019). Development of antimicrobial and antioxidant electrospun soluble potato starch nanofibers loaded with carvacrol. *International Journal of Biological Macromolecules, 139*(2019), 1182–1190.

Frangopoulos, T., Marinopoulou, A., Goulas, A., Likotrafiti, E., Rhoades, J., Petridis, D., & Karageorgiou, V. (2023). Enhancing the functional properties of starch-based biodegradable films. *Foods, 12*(14), 2812.

Fu, Z., Chen, J., Luo, S. J., Liu, C. M., & Liu, W. (2015). Effect of food additives on starch retrogradation: A review. *Starch-Stärke, 67*(1-2), 69–78.

García-Guzmán, L., et al. (2022). Progress in starch-based materials for food packaging applications. *Polysaccharides, 3*(1), 136–177.

Ghoshal, G., et al. (2018). Emerging food processing technologies. *Food processing for increased quality and consumption* (pp. 29–65). Academic Press.

Gómez, J., Gil, M. L. A., De la Rosa-Fox, N., & Alguacil, M. (2014). Diatomite releases silica during spirit filtration. *Food.*

Goodarzi, M. M., et al. (2020). Development of an easy-to-use colorimetric pH label with starch and carrot anthocyanins for milk shelf life assessment. *International Journal of Biological Macromolecules, 153*, 240–247.

Haghighi-Manesh, Soroush, & Azizi, M. H. (2017). Active packaging systems with emphasis on its applications in dairy products. *Journal of Food Process Engineering, 40*(5), e12542.

Han, J. W., Ruiz-Garcia, L., Qian, J. P., & Yang, X. T. (2018). Food packaging: A comprehensive review and future trends. *Comprehensive Reviews in Food Science and Food Safety, 17*(4), 860–877.

Heo, W., & Lim, S. (2024). A review on gas indicators and sensors for smart food packaging. *Foods, 13*(19), 3047.

Jaisan, C., An, D. S., & Lee, D. S. (2018). Application of physical gas absorbers in manipulating the CO2 pressure of kimchi package. *Journal of food science, 83*(12), 3002–3008.

Jayasena, D. D., & Cheorun, J. (2013). Essential oils as potential antimicrobial agents in meat and meat products: A review. *Trends in Food Science & Technology, 34*(2), 96–108.

Jia, R., Cui, C., Gao, L., Qin, Y., Ji, N., Dai, L., & Sun, Q. (2023). A review of starch swelling behavior: Its mechanism, determination methods, influencing factors, and impact on food quality. *Carbohydrate Polymers.*

Jimenez, A., Fabra, M. J., Talens, P., & Chiralt, A. (2012). Edible and biodegradable starch films: a review. Food and Biop.

Jimoh, O. A., Ariffin, K. S., Hussin, H. B., & Temitope, A. E. (2018). Synthesis of precipitated calcium carbonate: a review. Carbonates and.

Kędzia, G., et al. (2022). Social innovations for improving compostable packaging waste management in CE: A multi-solution perspective. *Energies, 15*, 23.

Kadirvel, V., Palanisamy, Y., & Ganesan, N. D. (2024). Active packaging system—An overview of recent advances for enhanced food quality and safety. *Packaging Technology and Science, 38*(2), 145–162. https://doi.org/10.1002/pts.2863

Kalita, S., Mohanty, A. K., Misra, M., Mukherjee, A., & Kumar, S. (2024). Introduction to active food packaging System. *Smart Food Packaging Systems: Innovations and Technology Applications.*

Kang, J., & Song, K. B. (2019). Characterization of Job's tears (Coix lachryma-jobi L.) starch films incorporated with clove bud essential oil and their antioxidant effects on pork belly during storage. *LWT, 111*, 711–718.

Kerry, J. P., & Tyuftin, A. A. (2017). Storage and preservation of raw meat and muscle-based food products: IV storage and packaging. *Lawrie's Meat Science* (pp. 297–327). Woodhead Publishing.

Khan, S., Monteiro, J. K., Prasad, A., Filipe, C. D., Li, Y., & Didar, T. F. (2024). Material breakthroughs in smart food monitoring: Intelligent packaging and on-site testing technologies for spoilage and contamination detection. *Advanced Materials, 36*(1), 230.

Khaneghah, A. M., Hashemi, S. M. B., & Limbo, S. (2018). Antimicrobial agents and packaging systems in antimicrobial active food packaging: An overview of approaches and interactions. *Food and Bioproducts Processing, 111*, 1.

Khanzada, B., Shaikh, M., Shahid, Z., Majid, M., Kanwal, M., Aziz, M., & Cheema, N. (2025). Development of Betalains Incorporated Starch Films.

Kowalczyk, D., et al. (2018). Ascorbic acid-and sodium ascorbate-loaded oxidized potato starch films: Comparative evaluation of physicochemical and antioxidant properties. *Carbohydrate Polymers, 181*, 317–326.

Kumar, L., et al. (2025). Innovations in edible packaging films, coatings, and antimicrobial agents for applications in food industry. *Comprehensive Reviews in Food Science and Food Safety, 24*(4).

Kumar, L., & Gaikwad, K. K. (2025). Emerging Technologies in Food Packaging. Concepts of Food Packaging,.

Kumar, L., Deshmukh, R. K., & Gaikwad, K. K. (2024). Functional agents from agro-waste for active and intelligent food packaging. *Agro-Waste Derived Biopolymers and Biocomposites: Innovations and Sustainability in Food Packaging* (pp. 205–239).

Laorenza, Y., et al. (2022). Polymeric packaging applications for seafood products: Packaging-deterioration relevance, technology and trends. *Polymers, 14*(18), 3706.

Lee, D. S. (2016). Carbon dioxide absorbers for food packaging applications. *Trends in Food Science & Technology, 57*, 146–155.

Leon-Bejarano, M., et al. (2020). Physical, barrier, mechanical, and biodegradability properties of modified starch films with nut by-products extracts. *Foods, 9*(2), 226.

Liu, D., Dang, S., Zhang, L., Munsop, K., & Li, X. (2021). Corn starch/polyvinyl alcohol based films incorporated with curcumin-loaded Pickering emulsion for application in intelligent packaging. *International Journal of Biological Macromolecules, 188*.

Liu, Y., Li, L., Yu, Z., Ye, C., Pan, L., & Song, Y. (2023). Principle, development and application of time–temperature indicators for packaging. *Packaging Technology and Science, 36* (10), 833.

Luo, X., Zaitoon, A., & Lim, L. T. (2022). A review on colorimetric indicators for monitoring product freshness in intelligent food packaging: Indicator dyes, preparation methods, and applications. *Comprehensive Reviews in Food Science and Food Safety, 21* (3), 2489–2519.

Lv, H., He, D., Xu, E., Jin, Z., Zhao, H., Yuan, C., & Cui, B. (2025). Starch-based biodegradable active intelligent packaging with color superimposition via emulsion electrospinning nanofiber for meat freshness monitoring and shelf-life extension. *Carbohydrate Polymers*.

Mahović Poljaček, S., Tomašegović, T., Strižić Jakovljević, M., Jamnicki Hanzer, S., Murković Steinberg, I., Žuvić, I.,. & Karlovits, I. (2024). Starch-Based Functional Films Enhanced with Bacterial Nanocellulose.

Majeed, T., Dar, A. H., Pandey, V. K., Dash, K. K., Srivastava, S., Shams, R., … Pandiselvam, R. (2023). Role of additives in starch-based edible films and coating: A review with current knowledge. *Progress in Organic Coatings, 181*, 107597.

Majid, I., Nayik, G. A., Dar, S. M., & Nanda, V. (2018). Novel food packaging technologies: Innovations and future prospective. *Journal of the Saudi Society of Agricultural Sciences, 17*(4), 454.

Mali, S. N., & Pandey, A. (2024). Development of curcumin integrated smart pH indicator, antibacterial, and antioxidant waste derived Artocarpus lakoocha starch-based packaging film. *International Journal of Biological Macromolecules, 275*, 133827.

Maran, J. P., Sivakumar, V., Sridhar, R., & Immanuel, V. P. (2013). Development of model for mechanical properties of tapioca starch based edible films. *Industrial crops and products, 42*, 159.

Marasinghe, W. N., Jayathunge, K. G. L. R., Dassanayake, R. S., Liyanage, R., Bandara, P. C., Rajapaksha, S. M., & Gunathilake, C. (2024). Structure, properties, and recent developments in polysaccharide.

Martins, P. C., Daniela, C. B., & Martins, V. G. (2021). Oregano essential oil addition in rice starch films and its effects on the chilled fish storage. *Journal of Food Science and Technology, 58*(4), 1562–1573.

Matignon, A., & Tecante, A. (2017). Starch retrogradation: From starch components to cereal products. *Food Hydrocolloids*.

Meral, Y., Hasan, S., & Mahmut, şS. (2021). Characterization of edible film based on grape juice and cross-linked maize starch and its effects on the storage quality of chicken breast fillets. *LWT, 142*, 111012.

Moeini, A., Pedram, P., Fattahi, E., Cerruti, P., Gabriella, S., et al. (2022). Edible polymers and secondary bioactive compounds for food packaging applications: Antimicrobial, mechanical, and gas barrier properties. *Polymers, 14*(12), 2395. https://doi.org/10.3390/polym14122395

Mourad, A. A. H., Mohammad, A. F., Altarawneh, M., Al-Marzouqi, A. H., El-Naas, M. H., & Al-Marzouqi, M. H. (2021). Effects of potassium hydroxide and aluminum oxide on the performance of a modified solvay process for CO_2 capture: A comparative study. *International Journal of Energy Research.*

Namory, M., Koffi, K. L., Tohoué, T. M., & Samuel, O. (2022). Effect of metakaolin content on mechanical and water barrier properties of cassava starch films. *South African Journal of Chemical Engineering.*

Naresh, V., & Lee, N. (2021). A review on biosensors and recent development of nanostructured materials-enabled biosensors. *Sensors, 21*(4).

Nath, R., & Dutta, K. (2024). Sources of starch and its morphology: A review. Interdisciplinary research in technology and management, 402-411).

Nguyen, T. T. T., Truong, N. L. Q., Chiem, T. N. Q., Truong, L. D. P., Nguyen, T. T. T., & Van Tran, T. (2025). Development of intelligent films based on jackfruit seed starch/polyvinyl alcohol incorporated with purple sweet potato peel anthocynanin as an indicator. *International Journal of Biological Macromolecules*, 145436.

Ni, S., Jiao, L., Zhang, H., Zhang, Y., Fang, G., Xiao, H., & Dai, H. (2018). Enhancing hydrophobicity, strength and UV shielding capacity of starch film via novel co-cross-linking in neutral conditions. *Royal Society Open Science, 5*, 181.

Nogueira, G. F., Farayde, M. F., & de Oliveira, R. A. (2019). Incorporation of spray dried and freeze dried blackberry particles in edible films: Morphology, stability to pH, sterilization and biodegradation. *Food Packaging and Shelf Life, 20*, 100313.

Obadi, M., Qi, Y., & Xu, B. (2023). High-amylose maize starch: Structure, properties, modifications and industrial applications. *Carbohydrate Pol.*

Otalora Gonzalez, C. M., et al. (2022). Cassava starch films with anthocyanins and betalains from agroindustrial by-products: Their use for intelligent label development. *Foods, 11* (21), 3361.

Pandian, A. T., Chaturvedi, S., & Chakraborty, S. (2021). Applications of enzymatic time–temperature indicator (TTI) devices in quality monitoring and shelf-life evaluation of food products. *Journal of Food Measurement and Characterization, 15*(2).

Piñeros-Hernandez, D., Medina-Jaramillo, C., Lopez-Córdoba, A., & Goyanes, S. (2017). Edible cassava starch films loaded with rosemary antioxidant extracts as active food packaging. *Food Hydrocol.*

Pornchai, R., Sarinthip, T., Auras, A. R., Nareekan, C., Kittisak, J., Pensak, J., Yuthana, P., Phisit, S., Noppol, L., Thanongsak, C., et al. (2022). Morphology, mechanical, and water barrier properties of carboxymethyl rice starch films: sodium hydroxide effect. *Molecules.*

Qin, Y., et al. (2020). Development of active and intelligent packaging by incorporating betalains from red pitaya (*Hylocereus polyrhizus*) peel into starch/polyvinyl alcohol films. *Food Hydrocolloids, 100*, 105410.

Reinas, I., Oliveira, J., Pereira, J., Mahajan, P., & Poças, F. (2016). A quantitative approach to assess the contribution of seals to the permeability of water vapour and oxygen in thermosealed packages. *Food Packaging and Shelf Life*, 7, 34–40.

Rodrigues, C., et al. (2021). Bio-based sensors for smart food packaging—current applications and future trends. *Sensors, 21*, 6.

Roy, L. (2026). Integration of sensor technology and artificial intelligence for the real-time measurement of food quality factors. In *Artificial intelligence in food science* (pp. 447–468). Academic Press.

Sadeghizadeh-Yazdi, J., Habibi, M., Kamali, A. A., & Banaei, M. (2019). Application of edible and biodegradable starch-based films in food packaging: A systematic review and meta-analysis. *Current research in nutrition and food science journal,* 7(3), 624–637.

Santhosh, Ahmed, J., Thakur, R., & Sarkar, P. (2024). Starch-based edible packaging: Rheological, thermal, mechanical, microstructural, and barrier properties - A review. *Sustainable Food Technology, 2*(2), 307–330.

Santos, C. T., et al. (2020). Protein time–temperature sensor for intelligent starch polymers. *Journal of Food Processing and Preservation, 44*(5), e14428.

Saritala, M. A., Muzammil, M., Quddus, M. R., Razzak, S. A., & Hossain, M. M. (2025). A review on production of ethylene oxide from epoxidation of ethylene: Catalysis, mechanism and kinetics. *Catalysts, 15*(6), 560.

Shahab, N., Masoud, R., & Mehdi, A. (2021). A starch-based pH-sensing and ammonia detector film containing betacyanin of paperflower for application in intelligent packaging of fish. *International Journal of Biological Macromolecules, 191*, 161–170.

Shaik, M. I., Azhari, M. F., & Sarbon, N. M. (2022). Gelatin-based film as a color indicator in food-spoilage observation: A review. *Foods.*

Shaikh, M., Haider, S., Ali, T. M., & Hasnain, A. (2018). Physical, thermal, mechanical, and barrier properties of pearl millet starch films influenced by the extent of acetylation and hydroxypropylation. *International Journal of Biological Macromolecules, 124*, 209–219.

Shoukat, R., et al. (2025). Rice starch chemistry, functional properties, and industrial applications: A review. *Polymers, 17*(1), 110.

Shu, Z., Chen, Y., Zhou, J., Li, T., Sheng, Z., Tao, C., & Wang, Y. (2016). Preparation of halloysite-derived mesoporous silica nanotube with enlarged specific surface area for enhanced dye adsorption. *Applied Clay Science, 132*, 114.

Silva, O. A., et al. (2021). Effects of a native cassava starch, chitosan, and gelatin-based edible coating over guavas (*Psidium guajava* L.). *ACS Food Science & Technology, 1*(7), 1247–1253.

Silva-Pereira, M. C., et al. (2015). Chitosan/corn starch blend films with extract from *Brassica oleraceae* (red cabbage) as a visual indicator of fish deterioration. *LWT-Food Science and Technology, 61*(1), 258–262.

Sofos, J. N. (2024). Antimicrobial agents. In *Food additive toxicology* (pp. 501–529). CRC Press.

Song, W., Wu, N., He, Y., Zhao, H., Xu, J., & Ren, L. (2024). Intelligent color changing packaging film based on esterified starch and black rice anthocyanins. *Food Chemistry X, 24*, 101930.

Spirescu, V. A., Chircov, C., Grumezescu, A. M., Vasile, B.Ș., & Andronescu, E. (2021). Inorganic nanoparticles and composite films for antimicrobial therapies. *International Journal of Molecular Sciences, 22*(9), 4595.

Subramanian, K., Logaraj, H., Ramesh, V., Mani, M., Balakrishnan, K., Selvaraj, H., & Aruni, W. (2022). Intelligent pH indicative film from plant-based extract for active biodegradable smart food packing. *Journal of Nanomaterials, 2022*(1), 4482114.

Svihus, B., Uhlen, A. K., & Harstad, O. M. (2005). Effect of starch granule structure, associated components and processing on nutritive value of cereal starch: A review. *Animal Feed Science and Technology, 122*(3–4), 303–320.

Syafiq, R., Sapuan, S. M., Zuhri, M. Y. M., Ilyas, R. A., Nazrin, A., Sherwani, S. F. K., & Khalina, A. (2020). Antimicrobial activities of starch-based biopolymers and biocomposites incorporated with plant essential oils: A review. *Polymers, 12*(10), 2403.

Tabari, L., & Farmanzadeh, D. (2020). Yttrium doped graphene oxide as a new adsorbent for H_2O, CO, and ethylene molecules: Dispersion-corrected DFT calculations. *Applied Surface Science, 500*, 144029.

Thakur, R., Pristijono, P., Scarlett, C. J., Bowyer, M., Singh, S. P., & Vuong, Q. V. (2019). Starch-based films: Major factors affecting their properties. *International Journal of Biological Macromolecules, 132*, 1079–1089.

Tian, S., Xue, X. A., Wang, X., & Chen, Z. (2022). Preparation of starch-based functional food nano-microcapsule delivery system and its controlled release characteristics. *Frontiers in Nutrition, 9*, 982370.

Tosif, M. M., Bains, A., Sadh, P. K., Sarangi, P. K., Kaushik, R., Burla, S. V. S., & Sridhar, K. (2023). Loquat seed starch—Emerging source of non-conventional starch: Structure, properties, and novel applications. *International Journal of Biological Macromolecules, 244*, 125230.

Umaraw, P., et al. (2020). Edible films/coating with tailored properties for active packaging of meat, fish and derived products. *Trends in Food Science & Technology, 98*, 10–24.

Uygun, E., Yildiz, E., Sumnu, G., & Sahin, S. (2020). Microwave pretreatment for the improvement of physicochemical properties of carob flour and rice starch–based electrospun nanofilms. *Food and Bioprocess Technology, 13*(5), 838–850.

Vanderroost, M., Ragaert, P., Devlieghere, F., & De Meulenaer, B. (2014). Intelligent food packaging: The next generation. *Trends in food science & technology, 39*(1), 47–62.

Vasuki, T., Mithulesh, Vijayasri Kadirvel, & Gururaj, P. N. (2023). Smart packaging—An overview of concepts and applications in various food industries. *Food Bioengineering, 2*(1), 25–41.

Vieira, M. G. A., Da Silva, M. A., Dos Santos, L. O., & Beppu, M. M. (2011). Natural-based plasticizers and biopolymer films: A review. *European Polymer Journal, 47*(3), 254–263.

Vu, H. P. N., & Lumdubwong, N. (2016). Starch behaviors and mechanical properties of starch blend films with different plasticizers. *Carbohydrate Polymers, 154*, 112–120.

Waisundara, V. Y., & Grichar, W. J. (2024). Pseudocereals: recent advances and new perspectives.

Wang, S., & Guo, P. (2020). Botanical sources of starch. In *Starch structure, functionality and application in foods* (pp. 9–27). Singapore: Springer Singapore.

Wei, H., et al. (2021). Ethylene scavengers for the preservation of fruits and vegetables: A review. *Food Chemistry, 337*, 127750.

Wilfer, P. B., Giridaran, G., Jeevahan, J. J., Joseph, G. B., Kumar, G. S., & Thykattuserry, N. J. (2021). Effect of starch type on the film properties of native starch based edible films. *Materials Today: Proceedings, 44*, 3903–3907.

Wu, H., Li, T., Peng, L., Wang, J., Lei, Y., Li, S., & Zhang, Z. (2023). Development and characterization of antioxidant composite films based on starch and gelatin incorporating resveratrol fabricated by extrusion compression moulding. *Food Hydrocolloids, 139*, 108509.

Xing, Y., et al. (2019). Antimicrobial nanoparticles incorporated in edible coatings and films for the preservation of fruits and vegetables. *Molecules, 24*(9), 1695.

Xinxin, L., Di, L., Shuai, D., Ling, Z., & Kang, M. (2021). Corn starch/polyvinyl alcohol based films incorporated with curcumin-loaded Pickering emulsion for application in intelligent packaging. *International Journal of Biological Macromolecules, 188*, 974–982.

Yadav, S., & Dutta, P. K. (2024). Moisture-absorbent food packaging systems and the role of chitosan. *Smart Food Packaging Systems: Innovations and Technology Applications*, 169–193.

Yahia, E. M., & Carrillo-Lopez, A. (Eds.). (2018). *Postharvest physiology and biochemistry of fruits and vegetables.* Woodhead publishing.

Yan, Q., Dawei, Y., Fengfeng, X., Dan, C., Juan, K., & Jun, L. (2021). Smart packaging films based on starch/polyvinyl alcohol and Lycium ruthenicum anthocyanins-loaded nano-complexes: Functionality, stability and application. *Food Hydrocolloids, 119*, 106850.

Yang, Y., Fu, J., Duan, Q., Xie, H., Dong, X., & Yu, L. (2024). Strategies and methodologies for improving toughness of starch films. *Foods, 13*(24), 4036.
Yang, L., Yuan, Q., Li, T. T., Lou, C. W., Hung, C. Y., & Lin, J. H. (2025). Exploration and application prospect of the advanced technology of time–temperature indicators. *Textile Research Journal, 95*(3–4), 404–428.
Yang, L., Yuan, Q. Y., Lou, C. W., Lin, J. H., & Li, T. T. (2024). Recent advances on pH-responsive polymers integrated with nature colorants: From preparation to applications. *Textile Research Journal, 94*(19–20), 2316–2329.
Yousefi, H., Su, H. M., Imani, S. M., Alkhaldi, K., M. Filipe, C. D., & Didar, T. F. (2019). Intelligent food packaging: A review of smart sensing technologies for monitoring food quality. *ACS Sensors, 4*(4), 808–821.
Yu, Z., Boyarkina, V., Liao, Z., Lin, M., Zeng, W., & Lu, X. (2023). Boosting food system sustainability through intelligent packaging: Application of biodegradable freshness indicators. *ACS Food Science & Technology, 3*(1), 199–212.
Yu, X., Chen, L., Jin, Z., & Jiao, A. (2021). Research progress of starch-based biodegradable materials: A review. *Journal of Materials Science, 56*(19), 11187–11208.
Yue, C., Shan, G., Wen-tao, W., Han-xue, H., & Loong-Tak, L. (2022). Low temperature extrusion blown ε-polylysine hydrochloride-loaded starch/gelatin edible antimicrobial films. *Carbohydrate Polymers, 278*, 118990.
Zabot, G. L., Silva, E. K., Emerick, L. B., Felisberto, M. H. F., Clerici, M. T. P. S., & Meireles, M. A. A. (2019). Physicochemical, morphological, thermal and pasting properties of a novel native starch obtained from annatto seeds. *Food Hydrocolloids, 89*, 321–329.
Zhang, M., Ahmed, A., & Xu, L. (2023). Electrospun nanofibers for functional food packaging application. *Materials, 16*(17), 5937.
Zhang, S., & Haiyan, Z. (2017). Preparation and properties of zein–rutin composite nanoparticle/corn starch films. *Carbohydrate Polymers, 169*, 385–392.
Zhu, L., Luo, H., Shi, Z. W., Lin, C. Q., & Chen, J. (2023). Preparation, characterization, and antibacterial effect of bio-based modified starch films. *Food Chemistry X, 17*, 100602.

CHAPTER ELEVEN

Application of starch film and coating in improving shelf life of fresh and processed food

Sadaf Mohajjel Sadeghi[1, 2], Reza Abedi-Firoozjah[3], Arezou Khezerlou[4], Mahmood Alizadeh Sani[5]
[1]Student Research Committee, Tabriz University of Medical Sciences, Tabriz, Iran
[2]Department of Food Science and Technology, Faculty of Nutrition and Food Sciences, Tabriz University of Medical Sciences, Tabriz, Iran
[3]Student Research Committee, Kermanshah University of Medical Sciences, Kermanshah, Iran
[4]Department of Food Safety and Hygiene, School of Health, Gonabad University of Medical Sciences, Gonabad, Iran
[5]Department of Food Science and Technology, School of Nutritional Sciences and Dietetics, Tehran University of Medical Sciences, Tehran, Iran

11.1 Introduction

To prevent significant quality loss caused by oxidation and microbes, effective preservation strategies are vital for raw materials and food from farm to table. Ensuring food integrity throughout the supply chain is essential for maintaining freshness and safety (Liu et al., 2022). Packaging plays a vital role in the food industry by acting as a barrier against microbial growth, oxidation, spoilage, and contamination, thereby preserving product quality, freshness, and safety (Putri et al., 2023). Due to the environmental impact of nonbiodegradable plastics, which account for a large share of the 300 million tons produced annually, 40% of which is used for packaging, interest has surged in eco-friendly alternatives. In recent years, synthetic plastics have dominated food packaging worldwide, but the need for sustainable options is now urgent. Ultimately, developing biodegradable replacements is essential to reduce pollution and promote environmental health (Liu et al., 2022). Studies indicate that active packaging is the most effective approach for enhancing food safety and extending Shelf Life (S_{life}) (Sadeghi et al., 2025). Starch-based films, among various biopolymers, are odorless, colorless, and exhibit sufficient resistance to oxygen permeation (Sanches et al., 2021). Since pure starch films have weaker mechanical strength and poor moisture barriers due to their hydrophilicity, improving these properties through modification or blending is crucial. Unlike synthetic

Starches in Smart Packaging
ISBN: 978-0-443-45396-0
https://doi.org/10.1016/B978-0-443-45396-0.00015-3

plastics, their limitations highlight the need for enhancement to meet packaging demands. Ultimately, such improvements are vital for expanding their practical applications (Liu et al., 2022). In this chapter of the book, we intend to discuss the application of starch-based packaging to increase the S_{life} of fresh and processed food products.

11.2 Application in fresh fruits and vegetables

Fresh produce requires specific storage conditions that vary by type, not only in temperature but also in light exposure, humidity, and whether it is stored near other foods (Schifferstein, 2024). Rich in nutrients, fruits and vegetables are vital components of a healthy diet. However, their high susceptibility to microbial spoilage and short S_{life} pose serious challenges to food safety, sustainability, and economic viability (Sahu et al., 2024; Sequino et al., 2022). Postharvest losses in fruits and vegetables are caused by moisture loss, physical damage, microbial growth, and adverse climatic conditions (Basumatary et al., 2022). Regarding postharvest storage of fruits, modern technologies such as low temperature, ethylene inhibitors, and packaging are essential to maintain fruit quality (Sganzerla et al., 2020).

Among various biopolymers, starch is considered as a packaging material for fruits and vegetables. However, as revealed, starch in its natural form has poor barrier properties and mechanical stability (Sivapragasam et al., 2024). To improve the limited functional properties of starch-based films, various techniques such as chemical modification and nanoparticle integration have been developed (Yudhistira et al., 2024). The following are studies on the application of starch-based active packaging in the preservation of fruits and vegetables. A number of studies are also mentioned in Table 11.1.

Bananas are widely consumed worldwide. They contain high levels of nutrients and exhibit antimicrobial and medicinal properties. Due to their high moisture, bananas are highly perishable, and enzymatic browning from polyphenol oxidase darkens their appearance, lowering quality and value. To improve shelf life and quality, multiple preservation strategies have been suggested (Chen et al., 2023; Martí et al., 2024; Mohapatra et al., 2011). In one study, poly (butylene adipate-co-terephthalate) (PBAT) and thermoplastic cassava starch (TPS) containing TiO_2 nanoparticles (0%, 3%, and 5% (w/w)) were developed as functional active packaging for banana stored at ambient conditions for 12 h. For 7 minutes, the films were exposed to UV light. The banana control group indicated

Table 11.1 Application of starch-based films for extending shelf life of different fresh and processed food products.

Film base	Additives	Method preparation	Food sample/ stored condition	Food application	References
[A] Fresh fruits and vegetables					
Cassava starch/ PBAT	Modified zeolite/ Bromothymol blue and methyl red (ripeness indicators)	Blown extrusion	Banana/Stored at 25 °C for 7 days	Zeolite failed to sufficiently scavenge ethylene in PFZ and NPFZ to slow color change, and nonperforated treatments (NPCF, NPFZ) did not establish a modified atmosphere. Packaging effectively reduced banana weight loss. Packaged bananas exhibited lower firmness loss rates (K) than NPB	Bilck et al. (2021)
WSS	RPPAEs (0%, 3.5%, 7.0%, and 10.5% (w/w, WSS basis))	Coating	"Guoguang" apples/Stored at 25 ± 2 °C for 25 days	Coating treatments effectively reduced weight loss, preserved the firmness of apples during storage, preserved soluble solids, and maintained total phenolic content and antioxidant capacities in "Guoguang" apples	Jiang et al. (2024)

(*Continued*)

Table 11.1 Application of starch-based films for extending shelf life of different fresh and processed food products.—cont'd

Potato starch	VC-loaded nanoliposomes	Casting	Wrap apples and green beans/ Storage at different pH (4 and 7) And temperatures (20, 25, and 37°C)	The VC content in packed apples and beans firstly increased for approximately 4 and 3 h, respectively. Following this initial rise, a gradual decrease in VC levels was observed in both packed samples. In beans and apples, a maximum of up to 60% and 26% of VC release was observed at 37°C, respectively.	Khuntia et al. (2022)
[B] Fresh meat and poultry					
Esterified starch	Tartary buckwheat bran extract and chitosan	Casting	Mutton meats/ Stored at 4 ± 1°C for 11 days	Reduced microbial growth, minimized pH changes, and suppressed protein/lipid oxidation; S_{life} extended by 2 days versus control.	Tian et al. (2021)
TPS/PBAT	Oregano, clove and rosemary oils	Extrusion	Beef/Stored under refrigerated conditions for 15 days	Packages containing cloves showed the highest total phenolic and antioxidant capacity ($P < .05$) and had the greatest effect against lipid oxidation when used in beef.	Cestari et al. (2021)

Potato starch/guar gum	BLP (2%, 4%, and 8% of the total weight of potato starch and guar gum)	Casting	Chicken meat/ Stored at 2–4 °C for 12 days	The extract-embedded film efficiently preserved the quality of chicken meat for up to 12 days under refrigerated storage conditions.	Nandi et al. (2024)
Arrowroot starch/ nanocellulose suspension	Neem oil	Casting	Chicken meat/ Stored at 4 °C for about 12 days	Decreasing lipid oxidation	Mukkadan et al. (2024)
Starch/gellan gum (90:10)/polyester blend films (PLA: PHV, 75:25)	Ferulic, *p*-coumaric or protocatechuic acid at 2% (w/w)	Melt-blending and compression molding	Pork meat/Stored at 5 °C and 48% RH for 15 days	Improved pork shelf life by decreasing lipid oxidation and microbial growth	Hernández-Garcí et al. (2022)
Cassava starch/ gelatin	Quercetin/TBHQ	Casting	Fresh pork/Stored at 4 ± 1 °C for about 12 days	Delayed lard oxidation (>35 days) and postponed pork redness discoloration	Tongdeesoontorn et al. (2021)
[C] Seafoods					
PLA/PBAT/TPS	ScPS-CEO	Industrial melt extrusion	Fresh salmon/ Stored at 4 ± 1 °C for 12 days	Prolonged salmon S_{life} by ~100% through suppression of spoilage mechanisms	Gui et al. (2024)
Corn starch	FA	Coating	Fresh silver pomfret/Stored at 4 ± 0.2°C for 15 days	Prolonged S_{life} of silver pomfret steaks to 15 days by synergistically inhibiting microbial growth and lipid oxidation	Remya et al. (2022)

(*Continued*)

Table 11.1 Application of starch-based films for extending shelf life of different fresh and processed food products.—cont'd

Starch/PVA	CIN	Casting	Large yellow croakers/Stored at −23°C for 30 days	High-humidity starch/PVA/CIN film best preserved frozen large yellow croaker quality	Zong et al. (2022)
Corn starch/Fish gelatin	CTEO (0%, 8%, 14%, 20%, 25%, and 30% w:w)	Casting	Snakehead fish/Stored at 4 °C for 12 days	Reduced spoilage indicators	Deng et al. (2024b)
Starch/PVA	FA/AN	Casting	Largemouth bass/Stored at 4 °C for 10 days	Successfully extended the S_{life} of bass fillets	Qin et al. (2021)
Cornstarch	Chl (1, 3 and 5 based on CS weight)	Casting	Shrimp/Stored in 4 °C for 5 days	Shrimp wrapped in CS films containing 5% Chl and irradiated showed acceptable levels of nitrogen and bacterial counts for up to 4 days.	Ukwatta et al. (2024)
Starch/PVA	Cu-MOF (0, 3, 6, and 9 wt%)	Casting	Shrimp/Stored at 4 °C for 8 days	Efficiently extending the S_{life} of the shrimp	Kang et al. (2024)
[D] Dairy products					
PVA/starch	Propolis extract (0.5%, 2%, 5%, 10% and 20%)/anthocyanin	Casting	Pasteurized milk/Stored for 48h	Film with 20% propolis extract has the maximum antibacterial activity with No spilt milk and color change	Mustafa et al. (2021)

Potato starch/CNC	*Moringa oleifera* extract and gelatin	Casting	Paneer, Khoa, peda/Stored at 37 ± 1 °C and 4 ± 1 °C °C for 18 days	Prolonging product S_{life}	Nair et al. (2025)
PVA/starch	LAE and mulberry anthocyanin	Casting	Fresh milk/Stored at room temperature for 7 days	Slowed down the spoilage process	Sharma et al. (2022)
Starch/cellulose	TDCEOs	Airgel	Koopeh cheese/ Stored 7 °C for 21 days	During 21 days of storage, treated cheese experienced a notable 3-log reduction in psychrophilic bacteria	Mirmoeini et al. (2023a)
Starch	Clove oil	Casting	Cheese/Stored at refrigerator temperature for 15 days	Enhanced the quality properties of soft cheese	Zaiden et al. (2025)
[E] Nuts					
Cassava starch/ CMC	Tocopherol	Coating	Brazil nuts/Stored at 25°C for 120 days	Delaying browning and oxidative changes, with the browning index increasing only 2% over 120 days	da Costa et al. (2024)
[F] Processed meats/sausage					
Cassava starch	GTE	Casting	Sliced calabrese sausage/Stored at 4 °C for 18 days	pH declines and acidification, maintaining sensory quality with scores around 4.7–5.2	Hamann et al. (2024)

(Continued)

Table 11.1 Application of starch-based films for extending shelf life of different fresh and processed food products.—cont'd

Starch	ZAEO	Casting	Beef sauce/Stored at room temperature for 12 days	Extending S_{life} by 4–6 days	Liu et al. (2023)
Starch	MEO	Extrusion	Pork sausage/ Stored at 25°C for 72 h	Microbial count of the pork sausages remained below the safety threshold for 40 h	Nadon et al. (2025a)

BLP, polyphenolic extract of waste petioles of betel leaf; *Chl*, chlorophyllin; *CMC*, carboxymethyl cellulose; *CNC*, rice straw nanocellulose; *CTEO*, cinnamon and thyme essential oils; *FA*, fumaric acid, CIN; cinnamaldehyde; *FA /AN*, free Lycium ruthenicum anthocyanins/anthocyanins-loaded nanocomplexes; *GTE*, green tea extract; *LAE*, ethyl lauroyl arginate; *MEO*, Makwaen essential oil; *NPB*, nonpacked bananas; *NPCF*, nonperforated control films; *NPFZ*, nonperforated films with zeolite; *PBAT*, poly butylene adipate-co-terephthalate; *PFZ*, perforated films with zeolite; *RPPAEs*, red pear peel anthocyanin extracts; *ScPS-CEO*, sodium trimetaphosphate-crosslinked porous starch-clove essential oil; *TDCEOs*, Thymus daenensis Celak essential oil; *TNC*, total number of colonies; *TPS*, thermoplastic starch; *VC*, Vitamin C; *WSS*, wampee seed starch; *ZAEO*, *Zanthoxylum armatum* DC. essential oil.

faster discoloration and decay. The degree of yellowing reduced with increasing of TiO_2 amounts. On day 9, banana control group displayed black spots, showing consumer refusal, while banana packed with PBAT/TPS/5% TiO_2 had the best look without darkening Fig. 11.1a. This study

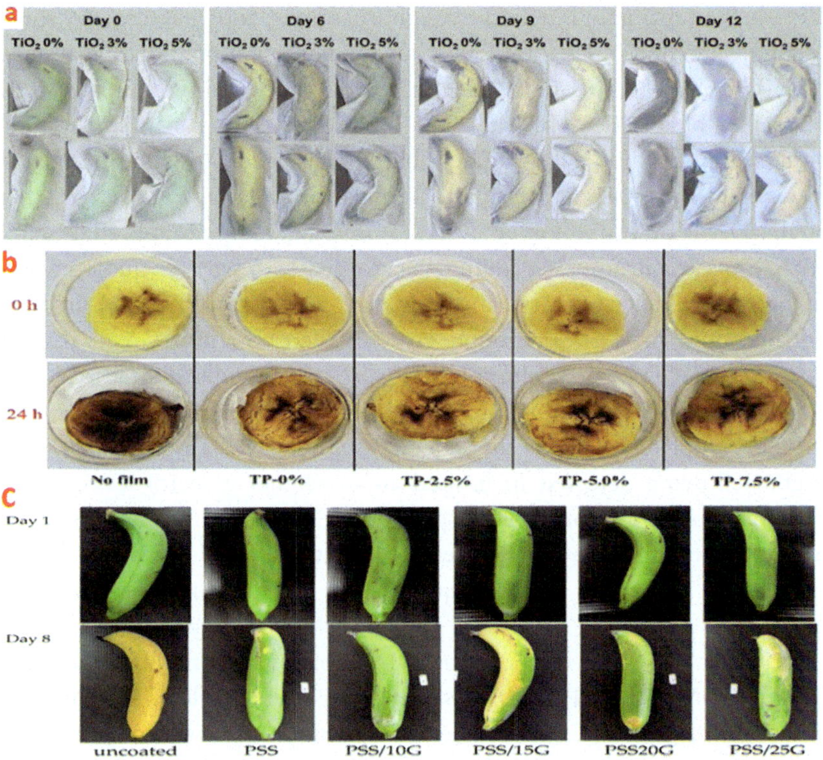

Figure 11.1 (a) Visual changes in bananas packaged in TiO2 @PBAT/TPS-based films. (b) Effect of film containing different concentrations of TP on the color of freshly cut bananas for 24 hour. (c) Photographs of coated and control bananas during storage. *(a) Photos reproduced from Phothisarattana, D. (2021). Biodegradable poly (butylene adipate-co-terephthalate) and thermoplastic starch-blended TiO2 nanocomposite blown films as functional active packaging of fresh fruit. Polymers, 13(23) MDPI, open access. (b) Photos reproduced from Chen, Nan, Gao, Hao-Xiang, He, Qiang & Zeng, Wei-Cai. (2023). Potato Starch-Based Film Incorporated with Tea Polyphenols and Its Application in Fruit Packaging. Polymers, 15(3). https://doi.org/10.3390/polym15030588 MDPI, open access. (c) Photos reproduced from Bumrungnok, Krongkarn, Threepopnatkul, Poonsub, Amornsakchai, Taweechai, Chia, Chin Hua, Wongsagonsup, Rungtiwa & Smith, Siwaporn Meejoo. (2023). Toward a Circular Bioeconomy: Exploring Pineapple Stem Starch Film as Protective Coating for Fruits and Vegetables. Polymers, 15(11) MDPI, open access. https://doi.org/10.3390/polym15112493.*

confirmed that bioplastic food packaging shows enhanced functional performance in prolonging the S_{life} of fresh produce (Phothisarattana et al., 2021). Another packaging film based on Potato Starch (PS)/tea polyphenols (TP, 2.5%, 5.0% and 7.5%) was developed for fruit preservation by Chen et al. (2023). Freshly cut bananas and blueberries were considered as model foods. Throughout storage, starch films—particularly those with TP—helped maintain blueberry weight, hardness, and chewiness by minimizing water and oxygen transfer, thus reducing transpiration and respiration. Ultimately, these films effectively preserved quality compared with uncoated blueberries. In the case of freshly cut bananas, the TP-containing film showed a stronger ability to prevent browning because of its significant oxygen barrier and antioxidant properties Fig. 11.1b (Chen et al., 2023). Bumrungnok et al. (2023) investigated pineapple stem starch (PSS) films rich in amylose, adding glycerol at 10–25 wt% (excluding water) to produce plasticized variants. Ultimately, these modifications aimed to enhance film properties through glycerol incorporation. Green bananas coated with the prepared film solution were stored under laboratory conditions and observed daily. In contrast to the control bananas, which had turned completely yellow, the coated samples stayed largely green, with slight yellowing detected in limited areas. Higher glycerol levels in the coating resulted in more yellow spots on the pineapple, likely due to uneven ripening caused by inconsistent coating thickness. Ultimately, coating uniformity influences ripening quality Fig. 11.1c (Bumrungnok et al., 2023).

In 2023, the effects of quercetin (Q) and organically modified montmorillonite (OMMT) on PBAT/TPS active films were explored using bananas and blueberries to assess their preservation ability at room temperature. Ultimately, the study highlighted the potential of these films for fruit protection. Fig. 11.2a displays bananas at the start and after 6 days of storage, revealing that all films minimized oxidation compared with unwrapped and LDPE-wrapped samples. The PBAT/TPS/Q/OMMT-2.61 vol% film exhibited the strongest antioxidant activity, demonstrating that adding Q and OMMT improves the packaging's antioxidant properties. Blueberries served as a model to illustrate how WVTR influences fruit storage duration. Fig. 11.2b reveals that blueberries in PE foil mold and decay after 6 days, whereas those in PBAT/TPS/Q/OMMT film remained fresh and intact, highlighting its antioxidant and barrier benefits for extending fruit shelf life (Yang et al., 2022). A new bilayer film made of chitosan, tannic acid, and corn starch (Ch/TA/CS) was created, effectively doubling banana

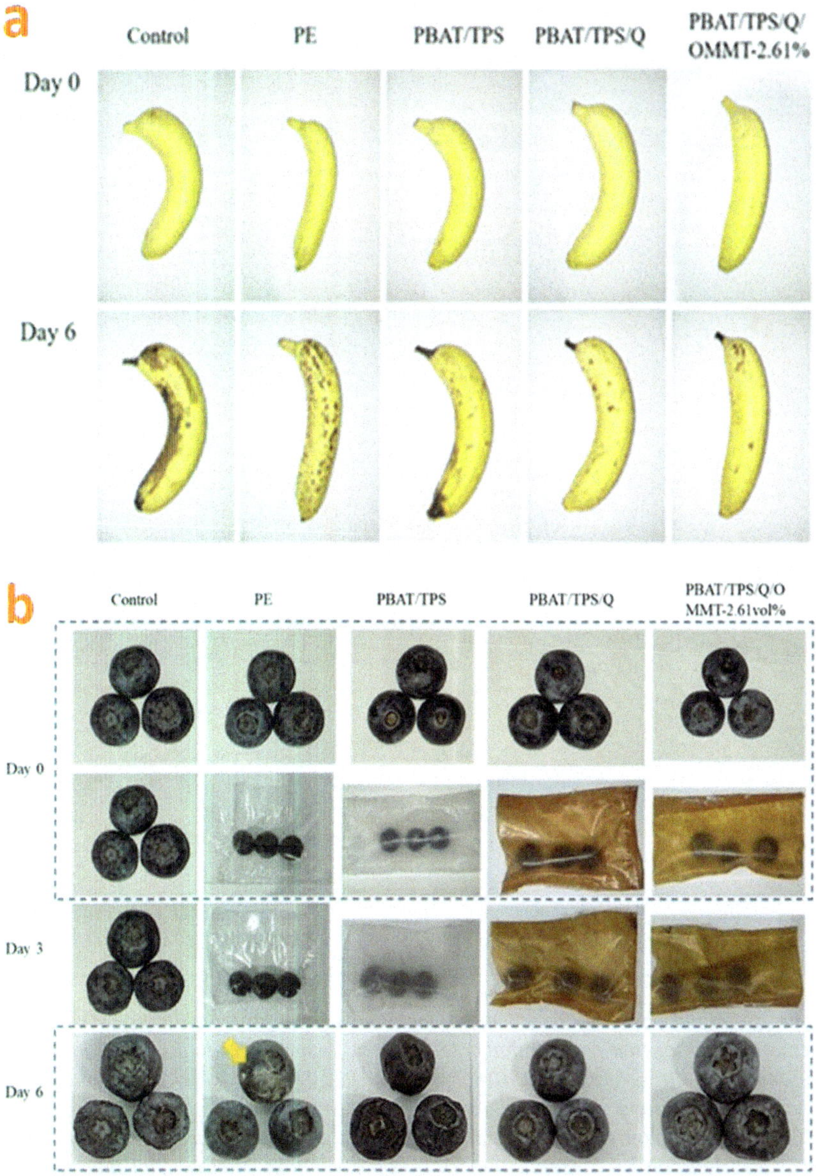

Figure 11.2 (a and b) Photograph of bananas and blueberries packaged in packaging films during storage for 6 days. (c) The effect of Ch/TA/CS on changes in appearance, and (d) weight loss of bananas during 6 days of storage. (e) Images of bananas after different storage times in different groups. *(a and b) Photos reproduced from Yang & et al. (n.d.). Effects of quercetin and organically modified montmorillonite on the properties of poly (butylene adipate-co-terephthalate)/thermoplastic starch active packaging films. ACS Omega, 8(1), 663–672 ACS, open access. (c) From Zhao, Shuangshuang, Jia, Ruoyu, Yang, Jie, Dai, Lei, Ji, Na, Xiong, Liu & Sun, Qingjie. (2022). Development of*

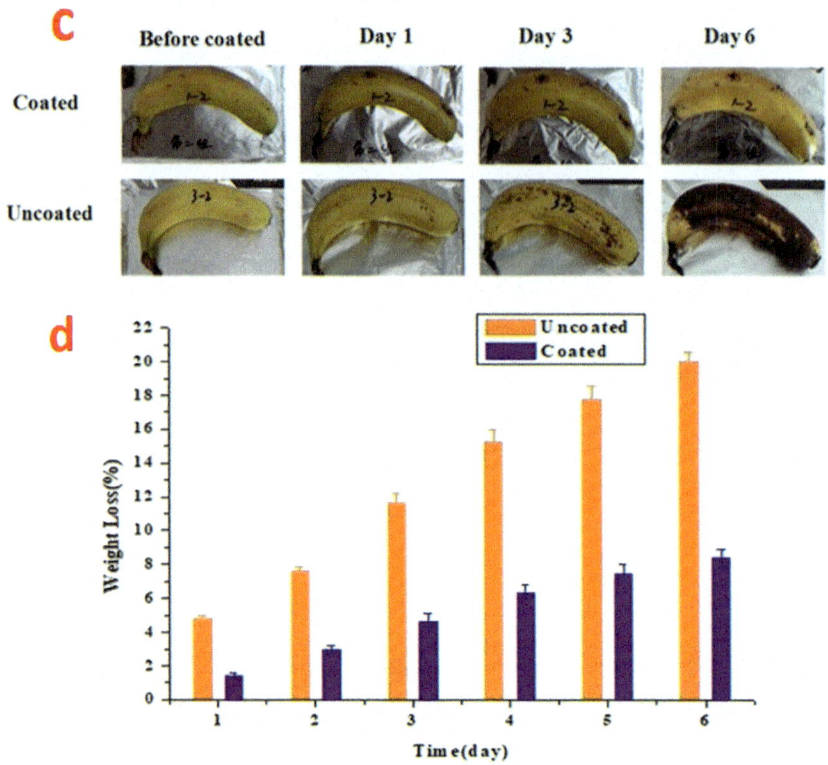

Figure 11.2 (*continued*).

chitosan/tannic acid/corn starch multifunctional bilayer smart films as pH-responsive actuators and for fruit preservation. International Journal of Biological Macromolecules, 205, 419–429. https://doi.org/10.1016/j.ijbiomac.2022.02.101 . (d) Photos reproduced from Zhao, Shuangshuang, Jia, Ruoyu, Yang, Jie, Dai, Lei, Ji, Na, Xiong, Liu & Sun, Qingjie. (2022). Development of chitosan/tannic acid/corn starch multifunctional bilayer smart films as pH-responsive actuators and for fruit preservation. International Journal of Biological Macromolecules, 205, 419–429. https://doi.org/10.1016/j.ijbiomac.2022.02.101 with permission. (e) Photos reproduced from Gong, Dezhuang, Zhang, Xinhua, Li, Jiaxu, Li, Yingying, Guo, Jialiang, Zhang, Xiuling & Zhang, Wentao. (2024). Carbon dot/g-C3N4-mediated self-activated antimicrobial nanocomposite films for active packaging applications. Food Chemistry, 438. https://doi.org/10.1016/j.foodchem.2023.137939 with permission.

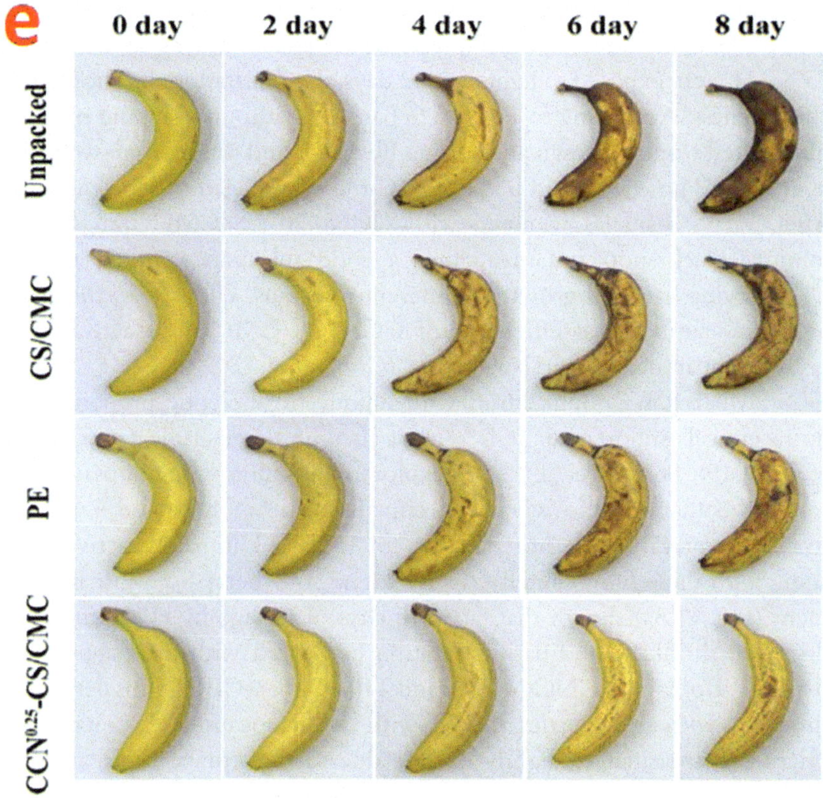

Figure 11.2 (*continued*).

shelf life from 3 to 6 days through its protective properties. As shown in Fig. 11.2c, uncoated bananas showed obvious enzymatic browning on the third day, whereas coated bananas retained their fresh appearance. On the sixth day, the exterior of uncoated bananas showed severe enzymatic browning and decay, accompanied by a weight loss of around 22%. In contrast, coated bananas remained fresh throughout the storage period, with weight loss consistently below 8% (Fig. 11.2d). These results demonstrate that the Ch/TA/CS coating effectively preserves banana freshness and delays quality deterioration. The effect of the coating on mangoes was also evident. Uncoated mangoes developed small spots by day 8 of storage; by day 11, these had expanded into distinctly large black spots, suggesting pulp deterioration. In contrast, coated mangoes remained fresh throughout the storage period (Zhao et al., 2022). Gong et al. (2024) created CS/CMC films with photoactive nanofiller CCN (0–0.25 mg/mL), selecting the CCN0.20 variant for optimal properties. Bananas stored at 20°C and 75% RH were exposed to light every 2 days, demonstrating the packaging's effectiveness. According to Fig. 11.2e bananas without wrapping or in PE film started turning brown and rotting by day 4, and the spoilage worsened over time. Those wrapped in CS/CMC film also began to spoil by day 6. After 6 days, bananas wrapped with CCN0.20-CS/CMC film showed minimal black spots and maintained a good appearance, unlike samples with varying bacterial growth at different positions. Ultimately, this highlights the antibacterial efficiency of CCN-CS/CMC films activated by visible light in prolonging banana S_{life} (Gong et al., 2024).

Strawberry is a popular fruit with unique organoleptic properties and nutraceutical importance. It is one of the most perishable fruits (Azam et al., 2019). Ferreira et al. (2021) explored TPS films with chitosan nanocapsules containing Ho wood and cinnamon oils at 1, 3, and 5 wt% for strawberry preservation, stored at 4°C for 15 days. Ultimately, active packaging with these films can effectively extend strawberry S_{life}. Visual assessment (Fig. 11.3a) shown that, after 15 days of storage in TPS films, fruits displayed shrinkage and dark coloration compared with their appearance on day 1. Both TPS/CNCs-HEOs and TPS/CNCs-CEOs films performed similarly, with strawberries treated with 1% nanocapsules maintaining a bright red appearance, slightly opaque due to the storage time and signs of dehydration. Adding 3 wt% nanocapsules gave strawberries that looked similar overall, but they lost more water and turned darker. This is probably because the capsules created microchannels that let more gas in and out. Both nanocapsule levels effectively preserved strawberries by preventing fungal growth and reducing weight loss, demonstrating film efficacy. However, 5 wt% nanocapsules, particularly TPS/CNCs-CEOs 5%, caused severe decay and fungal contamination. This may be due to the films' higher water vapor permeability, which promotes microbial growth on fruit surfaces (Ferreira et al., 2021). Cui et al. (2023) developed an intelligent PS-based active film incorporating MCM-41 as a THY carrier, responsive to storage conditions. Strawberries wrapped in this film stored at 25°C and 40%–50% RH for 5 days demonstrated its effectiveness. According to Fig. 11.3b.a strawberries began to spoil after only 2 days of storage; by day 5, most exhibited extensive microbial contamination and severe deterioration. In contrast, packaged strawberries remained fresher for a longer period, owing to the film's effective barrier properties. Throughout storage, strawberries wrapped in the THY-MCM-41/PS film preserved their color

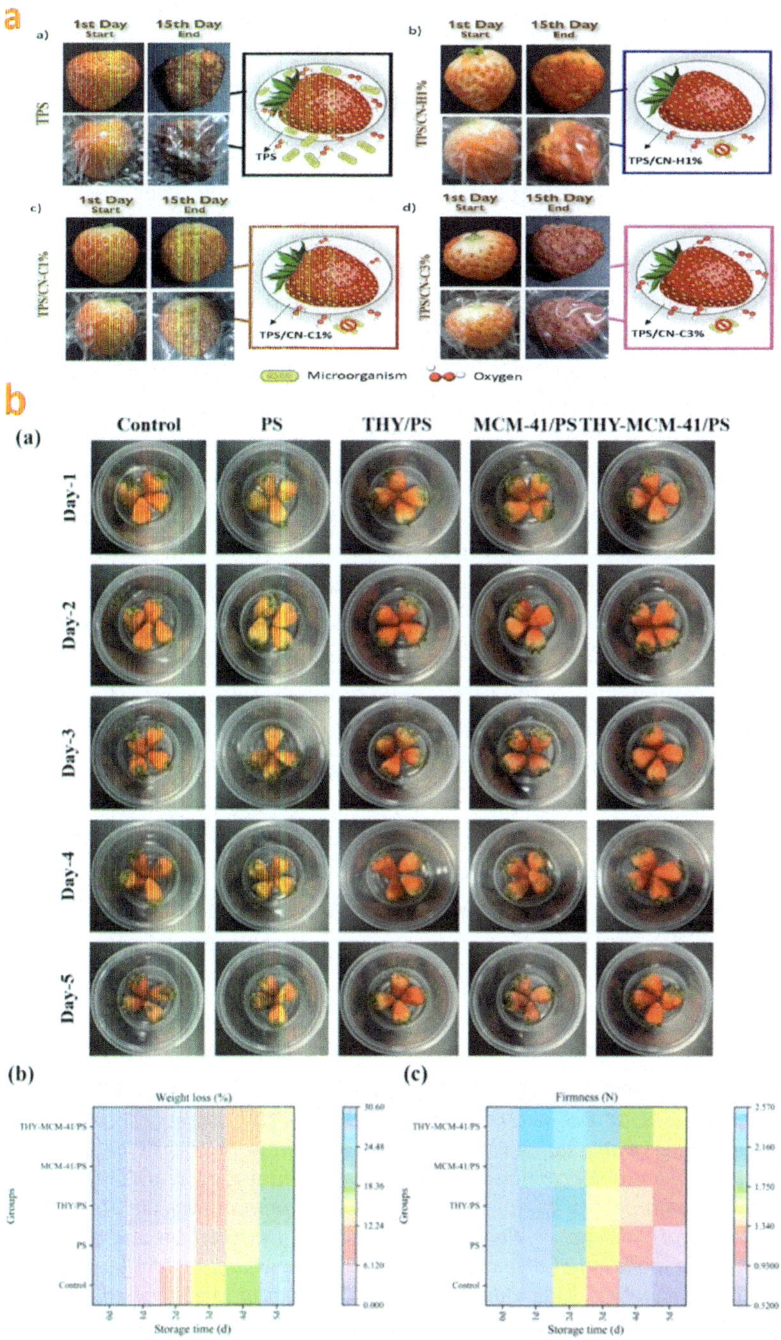

Figure 11.3 a) Visual evidence of the protection process in strawberries coated with a) TPS; b) TPS/CNCs -HEOs 1%; c) TPS/CNCs- CEOs 1%; and d) TPS/CNCs- CEOs 3%. b) Evaluation of the effect of different packaging groups on (a) appearance of strawberries,

and quality better than other methods. This preservation effect is due to two key factors: the film's excellent barrier properties, and its ability to release THY in response to increased humidity generated by strawberry respiration and transpiration during storage, which effectively suppresses microbial growth and proliferation. The packaged strawberries exhibited significantly slower weight loss compared with the control. Notably, after 5 days of storage, strawberries wrapped in the THY-MCM-41/PS film lost only 13.44% of their initial weight ($P < .05$) (Fig. 11.3b.b). The film's superior barrier properties minimized moisture loss, while the THY-MCM-41/PS group also demonstrated the highest stiffness retention at 1.52 N, indicating enhanced preservation (Fig. 11.3b.c). According to the results obtained, THY-MCM-41/PS film has the potential to be used as an active food packaging for food storage (Cui et al., 2024).

Other studies have also been conducted on the use of starch-based active packaging for the preservation of various other fruits and vegetables. One study was conducted to optimize the production of chitosan (0.75%, 2% w/w)-corn starch (CH-CS) films with ratios of 2:1, 1:1 and 1:2, using grapefruit seed extract (GFSE). Sensory evaluation by 20 semitrained judges showed that grapes in CH-CS/GFSE packaging received the highest scores, outperforming CH-CS and unwrapped controls. Ultimately, the packaging with CH-CS/GFSE provided the best sensory quality (Fig. 11.4a). The sensory analysis indicated that GFSE and CH synergistically delayed ripening, dehydration, and browning while preserving grape appearance without harming aroma or taste, thereby reducing postharvest fungal decay and maintaining 'Sultana' grape quality (Ghoshal et al., 2024). In 2023, a study was conducted to develop PVA/starch films containing basil leaf water extract (5, 10, and 15 wt%) as an antimicrobial agent (B5, B10 and B15

(b) weight loss, and (c) firmness. *(a) Photos reproduced from Ferreira, R.R., Souza, A.G., Quispe, Y.M. & Rosa, D.S. (2021). Essential oils loaded-chitosan nanocapsules incorporation in biodegradable starch films: A strategy to improve fruits shelf life. International Journal of Biological Macromolecules, 188, 628–638. https://doi.org/10.1016/j.ijbiomac.2021.08.046 with permission. (b) Photos reproduced from Cui, Yingjun, Zhang, Rongfei, Cheng, Meng, Li, Mengge & Wang, Xiangyou. (2024). Development and application of mathematical modeling of thymol release from environmental-responsive potato starch active packaging films. International Journal of Biological Macromolecules, 271(Pt 1). https://doi.org/10.1016/j.ijbiomac.2024.132353 with permission.*

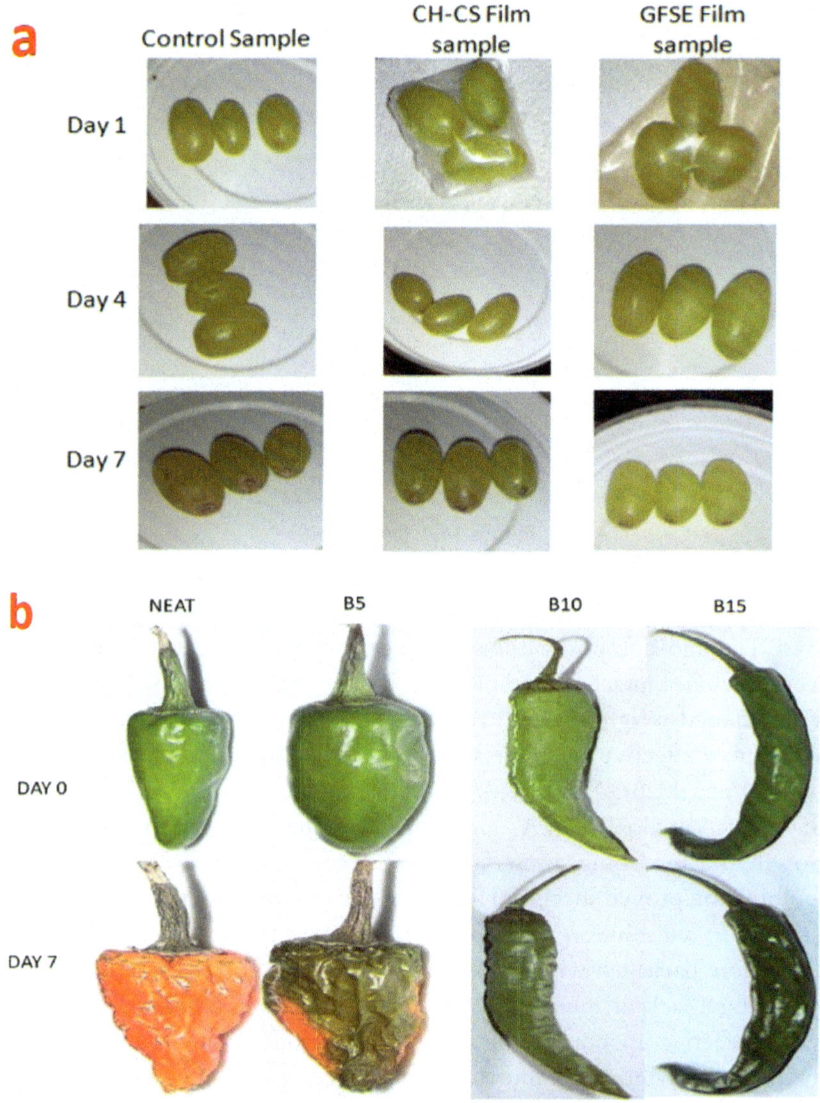

Figure 11.4 (a) Sensory analysis of grapes packaged in CH-CS, CH-CS/GFSE film and control. (b) Photo of green chili wrapped in different types of films before and after 7 days. (c) Preservation of fresh apples using prepared films for 24 hour. *(a) Photos reproduced from Ghoshal, Gargi & Singh, Jagseer. (2024). Study of coating effectiveness of grape fruit seed extract incorporated Chitosan/cornstarch-based active packaging film on grapes. Food Chemistry Advances, 4. https://doi.org/10.1016/j.focha.2024.100651 Elsevier, open access. (b) Photos reproduced from Varghese, S.A., Pulikkalparambil, H., Rangappa, S.M., Parameswaranpillai, J. & Siengchin, S. (2023). Antimicrobial active packaging based on PVA/Starch films incorporating basil leaf extracts. In Materials Today: Proceedings (Vol. 72, pp. 3056–3062). Elsevier Ltd. https://doi.org/10.1016/j.matpr.2022.09.062 with permission. (c) Photos reproduced from Xi, Wanting, Liu, Peng, Ling, Jiandi, Xian, Dongni, Wu, Linlin, Yuan, Yang, Zhang, Jianguo & Xie, Fengwei. (2024). Pre-gelatinized high-amylose starch enables easy preparation of flexible and antimicrobial composite films for fresh fruit preservation. International Journal of Biological Macromolecules, 254(Pt 2). https://doi.org/10.1016/j.ijbiomac.2023.127938 Elsevier, open access.*

Figure 11.4 (***continued***).

respectively). The tightly sealed bags were stored at room temperature for 7 days. The bio-composite films with basil extract maintained humidity, keeping chilies intact and edible after 7 days, unlike pure film packaging, which caused weight loss and color change (Fig. 11.4b). Ultimately, these films prove effective for preserving delicate produce during storage (Varghese et al., 2023). Xi et al. (2024) developed composite films with PG-G50 (61% amylose), PVA, and TA that effectively inhibited bacterial growth and browning, preserving fresh apple slices. Ultimately, this film combination proved successful for maintaining apple quality. After applying 1 mL of *E. coli* solution to fresh apple slices and storing them at 30°C for 24 h, those packaged with PG G50/PVA showed visible bacterial growth, indicating a lack of inherent antibacterial properties. Ultimately, this film was ineffective in controlling bacterial contamination. Conversely, the absence of bacterial colonies on apple slices packaged with PG-G50/PVA/TA and NCS/PVA/TA films demonstrated their potent antibacterial properties (Fig. 11.4c) (Xi et al., 2024).

A casting method produced an active PVA/CS/WMRE (10% v/v) thermoplastic film, which was tested for antimicrobial activity on fresh purple cabbage stored at 4°C for 3 days. Ultimately, this film demonstrated effective preservation, highlighting its potential for extending produce S_{life}. While microbial growth on fresh cabbage increased over time in PET containers and PVA/CS wraps, the PVA/CS/WMRE film slightly reduced microbial count by day 3. Ultimately, this film effectively slows microbial spoilage, potentially extending the shelf life of fresh vegetables (Todhanakasem et al., 2022).

Marzano-Barreda et al. (2021) developed active films based on cassava starch/PBAT using extrusion blow molding, incorporating zeolite as an ethylene adsorbent. In this study, broccoli was packaged and stored for 7 days at 12 °C. Overall, the active biodegradable packaging slowed metabolic activity in fresh broccoli florets at 12 °C, maintaining their color and vitamin C over 7 days (Marzano-Barreda et al., 2021).

According to the above results, it can be concluded that starch-based active packaging can be used in fruits and vegetables and can increase the S_{life} and quality of products.

11.3 Application in fresh meat and poultry

Meat has received considerable attention due to its high content of essential proteins, vitamins, minerals, and micronutrients, and it constitutes a major component of the global diet. However, lipid and protein oxidation, along with microbial contamination, pose significant challenges to its safety and quality. These attributes are strongly influenced by the choice of packaging materials and technologies (Fang et al., 2017; Song et al., 2021). Additionally, due to its high perishability and short S_{life}, raw poultry meat requires advanced strategies to meet modern consumer expectations for safety and quality. In this context, studies have been conducted on the application of starch-based packaging for meat and poultry preservation. A number of studies in this field are mentioned below, and other studies are also listed in Table 11.1.

Sreekanth et al. (2024) created chitosan/starch films with CEOs and nanocellulose, which effectively reduced bacterial growth in meat stored at 4 °C for 5 days. While C–S and polyethylene-packaged meat showed higher microbial levels, those in C-S-CEO and C-S-NC-CEO films remained significantly cleaner. Ultimately, these films proved capable of controlling microbial proliferation during refrigerated storage. The combined effect of chitosan in the starch/chitosan/cellulose matrix and the incorporated CEO likely explains this property (Fig. 11.5a) (Sreekanth et al., 2024). Fan et al. (2023) created CS/starch films with different *Portulaca oleracea* extract (POE) concentrations (0.15%, 0.30%, 0.45%), which were tested on pork stored at 4°C. Ultimately, these films demonstrated potential for effective meat preservation through active packaging.

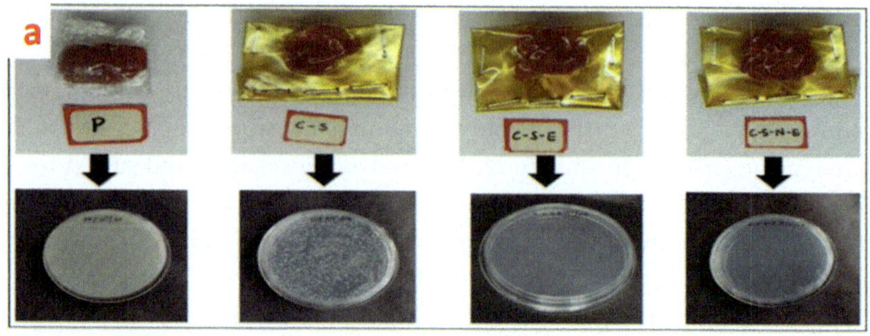

Figure 11.5 a) Evaluation of packaged meat samples in prepared films. b) Evaluation of the effects of CS/POE films on chilled meat. (a) Chilled meat after 6 days of storage, (b) color changes, (c) TBARS, and (d) TVB-N changes. c) Appearance of meat packaged in developed films. *(a) Photos reproduced from K., Sreekanth, K.P., Sharath, C.D., Midhun Dominic, Mathew, Divya & E.K., Radhakrishnan. (2024). Microbial load reduction in stored raw beef meat using chitosan/starch-based active packaging films incorporated with cellulose nanofibers and cinnamon essential oil. Meat Science, 216. https://doi.org/10.1016/j.meatsci.2024.109552 with permission. (b) Photos reproduced from Fan, Xiaojing, Zhang, Bin, Zhang, Xing, Ma, Zhiqing & Feng, Xianchao. (2023). Incorporating Portulaca oleracea extract endows the chitosan-starch film with antioxidant capacity for chilled meat preservation. Food Chemistry: X, 18. https://doi.org/10.1016/j.fochx.2023.100662 Elsevier open access. (c) Photos reproduced from Katekhong, Wattinee, Wongphan, Phanwipa, Klinmalai, Phatthranit & Harnkarnsujarit, Nathdanai. (2022). Thermoplastic starch blown films functionalized by plasticized nitrite blended with PBAT for superior oxygen barrier and active biodegradable meat packaging. Food Chemistry, 374. https://doi.org/10.1016/j.foodchem.2021.131709 with permission.*

Throughout storage, L* values declined; however, all CS/POE treatments retained higher brightness compared with polyethylene and CS film groups. Collectively, the findings demonstrated that CS/POE films effectively preserved meat color and suppressed the appearance alteration of meat. At all sampling times, TBARS values for CS/POE (0.30%) and CS/POE (0.45%) films were significantly lower ($P < .05$), demonstrating their superior ability to inhibit lipid oxidation. By the end of storage, TVB-N values for CS/POE (0.30%) and CS/POE (0.45%) films remained below the threshold, measuring 14.89 mg/100 g and 14.57 mg/100 g respectively. These results confirm that POE is a potent natural antioxidant with significant promise for use in oxidative-protective active packaging for chilled meat (Fig. 11.5b) (Fan et al., 2023). Blown-film extrusion produced functional bioplastic packaging from TPS with 1%–5% nitrite and PBAT (30/70 and 40/60 ratios). Pork sirloin steaks were vacuum-packed between PBAT/TPS films in OPA/LLDPE bags at −980 mbar and stored at 4°C.

b

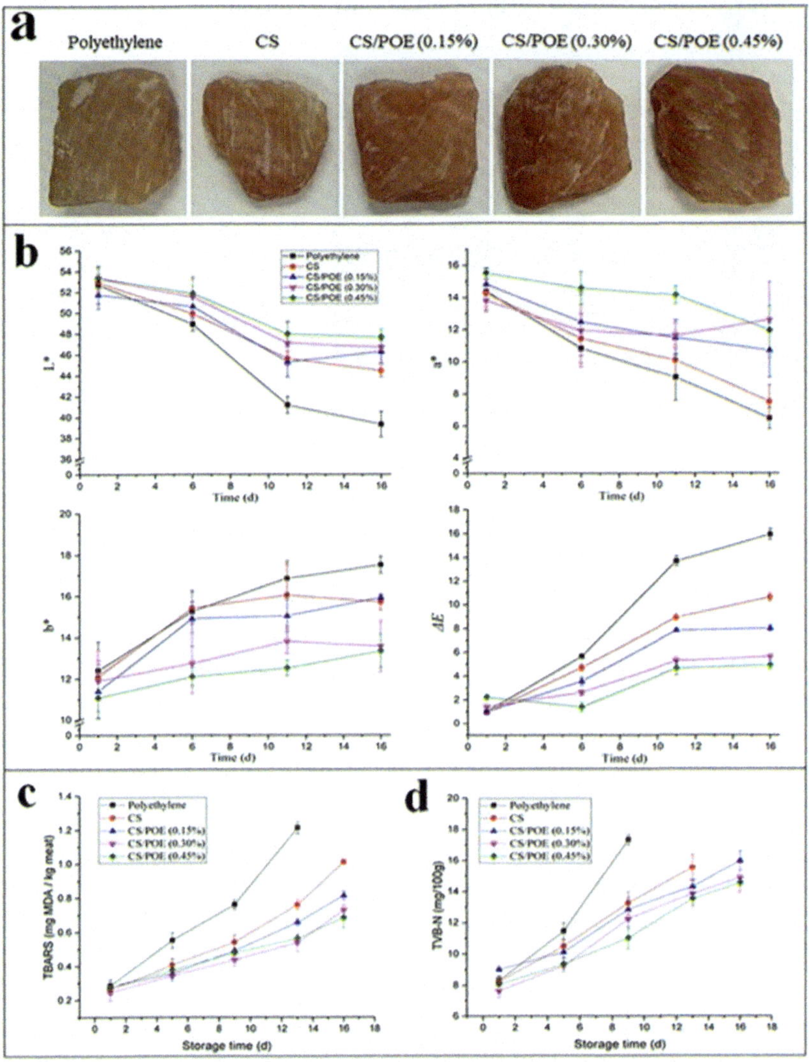

Figure 11.5 (***continued***).

Ultimately, this method showcased effective packaging for meat preservation. During the initial 3 days, nitrosyl myoglobin formed in the pork and remained stable for the entire 12-day storage period. Fig. 11.5c illustrates that pork in control films without nitrite lost its pink hue and

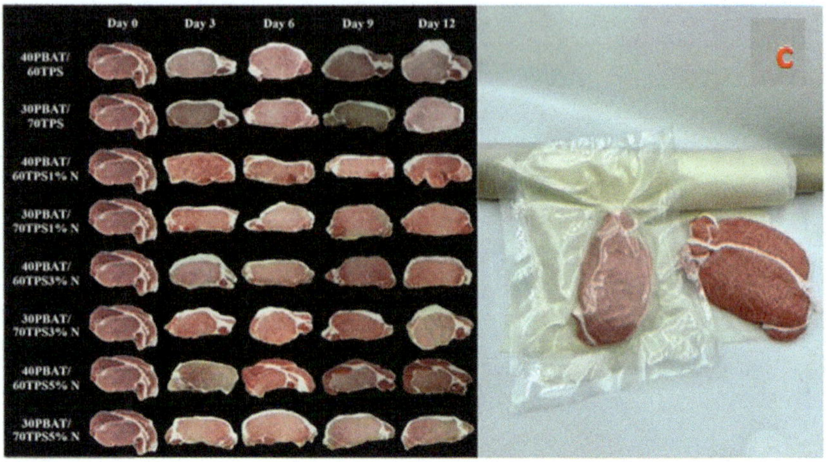

Figure 11.5 (*continued*).

discolorated, while meat in nitrite-infused films maintained its color effectively over 12 days (Katekhong et al., 2022).

Sanches et al. (2021) evaluated starch films containing red cabbage extract and sweet whey, with ground beef stored at around 4 °C for 4 days. The T2 film (64.18% RCE, 4.36% sweet whey) showed the least color change, indicating its effectiveness. Additionally, T2 was the treatment that best preserved the original color of the packaged ground beef. In comparison to other treatments, ground beef with T2 film showed the lowest pH, while T7 (50% RCE, 0% sweet whey) and T10 (50% RCE, 15% sweet whey) had similar pH levels. Nonetheless, both were significantly lower than the un-packaged control sample, which faced the most oxidative stress (Sanches et al., 2021). A novel antibacterial edible film based on potato starch, incorporating thyme oil microemulsions or microcapsules (1%–20%), was developed and used to package pork stored at 4 °C. The meat in films with 5% microcapsules lasted up to 14 days, outperforming the control (6 days) and PE film (10 days), demonstrating enhanced preservation (Yuan et al., 2021). Chandrasekar et al. (2024) created starch-based films containing FeO and ZnO nanoparticles, which were used to package chicken and mutton samples stored at cold condition for S_{life} assessment. According to the results, the double-layer Nano-Biothermoplastic packaging system significantly reduced TVC, *Pseudomonads*, and *Lactobacilli* and Enterobacteriaceae in both chicken and mutton, prolonging their S_{life} to 20 and 15 days, respectively. It also successfully reduced lipid and protein oxidation, indicating strong potential for controlling both microbial and chemical-physical deterioration during storage (Chandrasekar et al., 2024).

A study was conducted to create a gelatin and corn starch-based film incorporating corn stigma extract (CSE) to preserve the quality of chilled beef stored at chilled condition. Bioactive films including CSE significantly decries lipid oxidation in rump steaks during 7 days of refrigerated storage, with 25% CSE display a 60% lower TBARS value compared with the control, so extending S_{life} further than conventional packaging. This antioxidant effect is attributed to phenolic and bioactive compounds in CSE. Moreover, treatments containing 15% and 25% CSE maintained the lowest pH values, which corresponded to the lowest microbial counts in samples packaged with CSE (Boeira et al., 2021). In another study, biodegradable PLA-TPS bilayer films containing rice straw (RS) extract (incorporated into the PLA layer) and cellulose fibers (CF) (incorporated into the TPS layer) were developed. Pork samples packaged with these films were stored for 16 days at 4 °C. The RS extract revealed significant antioxidant activity, and overall, the results exhibited that the bilayer film effectively decreased lipid oxidation, microbial growth, and pH increase in fresh pork throughout the storage period (Freitas et al., 2023).

Adame et al. (2024) formulated a multifunctional film with Litsea cubeba essential oil (LC-EO) at 1%–7% concentrations, using pullulan and tapioca starch as carriers. Beef cubes wrapped in PTS3% and control films were stored at 4°C, showing the film's potential for meat preservation. Strong antibacterial activity was observed in the PTS3% film, with inhibition zones of 18.59 mm for *S. aureus* and 17.32 mm for *E. coli*, demonstrating its effectiveness. Overall, the $PTS_{3\%}$ film effectively preserved the quality of beef meat during refrigerated storage by maintaining texture, color, pH, and TBARS levels within acceptable ranges, outperforming conventional low-density polyethylene (LDPE) bags in preventing spoilage and increasing S_{life} (Adame et al., 2024).

Deng et al. created antibacterial films from amylose starch and chitosan 2-hydroxypropyl-trimethylammonium chloride (HTCC) for meat preservation, blending them at ratios of 1:2, 1:4, 1:10, and 1:20 to evaluate their effectiveness. For the assessment of HTCC/amylose films in meat preservation, sterile Petri dishes were each loaded with a meat sample. Samples were inoculated with 50 μL of *E. coli* (~10^3 CFU/mL), covered with film, and incubated at 37 °C and 50% RH for 48 h. After 48 h meat wrapped in plastic or F0 (amylose starch film) film exhibited substantial bacterial growth and spoilage, whereas HTCC/amylose F1–F4 and Ag^+ antibacterial adhesive film reduced colonies, with HTCC/amylose films outperforming even the Ag^+ adhesive film (Fig. 11.6a and b) (Deng et al., 2024a). A study produced TPS-LLDPE composite films with 0.25%–2% nitrite, assessing their effectiveness as active packaging for pork steaks stored at 4 °C over 12 days. As shown in Fig. 11.6d, LLDPE/TPS films with at least 0.25% nitrite improved pork redness, reaching fourfold enhancement at 2%, with nitrite migration under 1.5 ppm, below safety limits. These films notably inhibited microbial growth and maintained tenderness over storage. These findings approve that Nitrite-containing LLDPE/TPS films serve as effective active packaging systems for preserving the quality of red meat (Chatkitanan et al., 2021).

Another study explored UV-C treatment (15–120 min) to enhance starch-based films for minced meat packaging, stored at 4 °C for 6 days, with 60-minute exposure producing optimal functional properties. Throughout storage, minced meat wrapped in UV-C-treated films, especially those exposed for 60 minutes, showed significantly reduced thiobarbituric acid values and total aerobic bacteria compared with untreated films. These findings underscore the promising application of UV-C-modified starch films as efficient packaging for meat preservation (Uyarcan et al., 2024). Guo and colleagues studied how ultracold storage at −1.3°C, combined with potato starch films containing 0%–3% SSF, affected chilled beef quality. The findings revealed that 3% SSF packaging significantly extended beef shelf life, demonstrating its effectiveness for preservation under superchilled conditions. Beef samples packaged in SSF-3% films retained higher lightness (L*) and redness (a*) values compared with SSF-0% during storage. Based on TBARS, pH, TVB-N, microbial counts, and sensory data, SSF-3% effectively slowed quality loss. In conclusion, starch film with SBP is an optimal packaging choice for enhancing chilled beef quality and extending shelf life under superchilled conditions (Guo et al., 2021).

Marzlan et al. (2022) created an edible starch film infused with torch ginger (Etlingera elatior Jack) inflorescence essential oil (TGIEO). Chicken meat samples wrapped with this film were stored at 3 °C ± 1 and 85% ± 2 RH for up to 6 days, with periodic quality evaluations. Chicken samples wrapped in the TF film exhibited the lowest coliform count (4.98 ± 0.07 log CFU/g) and the lowest TBARS content (0.212 ± 0.031 mg MDA/kg), indicating reduced microbial contamination and lipid oxidation. The sensory analysis showed no significant differences ($P < .05$) among chicken samples, indicating TGIEO did not harm organoleptic qualities. Consequently, the TF film exhibited strong antibacterial and antioxidant

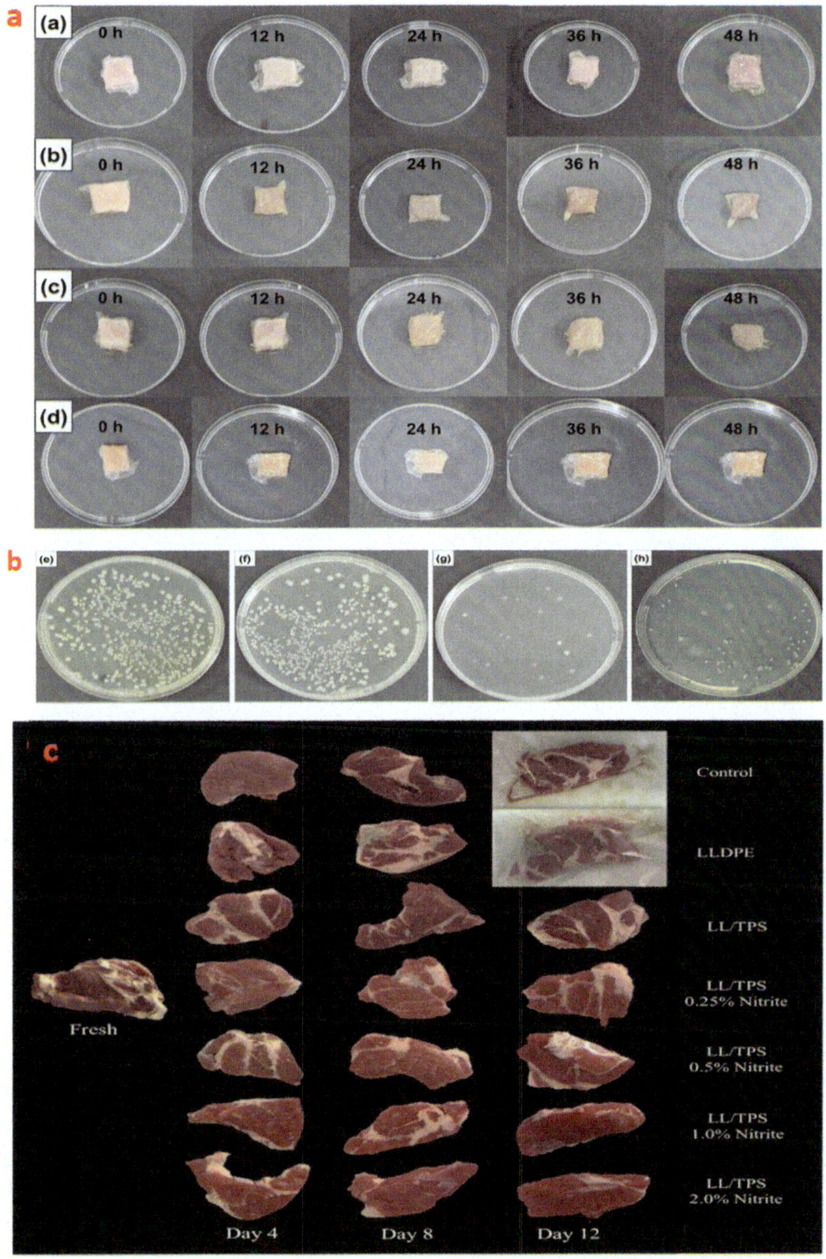

Figure 11.6 a) Photographs of samples packaged in different films and stored for 48 h and the (b) photo of the plate count results with the same dilution: (a, e) plastic cling film, (b, f) F0 film, (c, g) HTCC/amylose F1-4 film, (d, h) Ag + antibacterial cling films. c) Photo of the packaged pork in the prepared films and d) the degree of its redness changes. *(a and b) Photos reproduced from Deng, Bin, Chen, Jingwen, Li, Shaobo, Liu,*

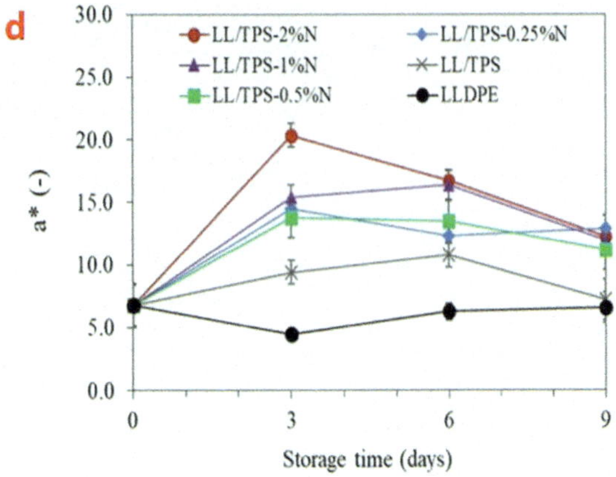

Figure 11.6 (*continued*).

effects, promising enhanced quality and shelf life during refrigeration (Marzlan et al., 2022). Mathew et al. (2025) developed arrowroot starch-based composite films reinforced with CEO nanoemulsions for active, biocompatible packaging. Their antimicrobial effectiveness was tested on vacuum-sealed raw chicken stored at 4 °C for 5 days, demonstrating promising preservation capabilities. The ARS film alone moderately inhibited microbial growth compared with polyethylene packaging, likely due to its semipermeable barrier properties. Incorporating CEO into the film enhanced its antimicrobial activity in a dose-dependent manner, showing a direct link between CEO concentration and inhibition strength (Mathew et al., 2025).

A study evaluated starch-based composite foam with 2% nisin (SCFN) for packaging fresh chicken meat stored at 2°C over 8 days. According to the results, SCF and SCFN exhibited a higher liquid retention capacity compared with expanded polystyrene (EPS) when used for storing breast and thigh meat. The inclusion of 2% nisin in the starch-based foam reduced microbial growth and preserved the freshness of chicken meat during

Jing, Zhou, Zhongkai, Qin, Zhu, Wang, Huixing, Su, Mengxiang, Li, Li & Bai, Zongchun. (2024). An antibacterial packaging film based on amylose starch with quaternary ammonium salt chitosan and its application for meat preservation. International Journal of Biological Macromolecules, 261(Pt 2). https://doi.org/10.1016/j.ijbiomac.2024.129706 with permission. (c) Photos reproduced from Chatkitanan, Tanaporn & Harnkarnsujarit, Nathdanai. (2021). Effects of nitrite incorporated active films on quality of pork. Meat Science, 172. https://doi.org/10.1016/j.meatsci.2020.108367 with permission.

storage. The enhanced redness of the breast meat is likely due to nisin's antioxidant properties, which inhibit myoglobin oxidation. Overall, SCFN shows promise for extending poultry shelf life by combining antimicrobial and antioxidant effects (Velasco et al., 2022).

Considering the above, it can be concluded that starch-based films can be considered as a proper technique to extend the S_{life} of fresh meat and poultry. This will not only increase the safety and S_{life} of the products but also reduce waste and meet the growing needs of consumers.

11.4 Application in seafood products

Seafood rapidly loses freshness due to its high perishability, posing risks of spoilage and potential health hazards if not properly stored. Consequently, seafood must be processed or packaged promptly to maintain safety and quality, yet consumers increasingly favor fresh, minimally processed products that preserve their natural sensory and nutritional qualities. To reconcile these demands, several innovative preservation and packaging technologies have been developed (Kontominas et al., 2021; Siddiqui et al., 2024). Active packaging offers a hopeful approach to decreasing spoilage in fresh seafood. A number of studies are mentioned below and a number of studies are also listed in Table 11.1.

Wang et al. (2022) created starch/polyvinyl alcohol (STA/PVA) films infused with oregano essential oil (OEO) or OEO-loaded microporous starch (OEO-MS). Sea bass wrapped in these films was vacuum-packed in polyethylene bags and stored at 4 °C to assess preservation efficacy. The TVC results exhibited that films containing OEO, particularly the STA/PVA/OEO-MS film, more efficiently inhibited microbial growth and prolonged the S_{life} of sea bass, due to their slower OEO release and higher oxygen barrier properties. Moreover, these OEO-containing films, especially STA/PVA/OEO-MS, repressed lipid oxidation during storage. Additionally, the STA/PVA/OEO-MS film effectively delayed protein degradation, as indicated by the lowest TVB-N values, which can be attributed to its oxygen barrier performance and the persistent, slow release of OEO (Wang et al., 2022).

In a study assessing the application of edible films composed of cassava starch and fish gelatin incorporated with 1%, 5%, and 10% cinnamon essential oil nanoemulsion (CEON), designated as SGCEON1, SGCEON2, and SGCEON3, three treatment groups were tested to assess their efficiency in prolonging the S_{life} of fish fillets during 10 days of storage. The control group (Group 1) remained untreated, while Group 2 was dipped in a cassava starch and fish gelatin bioactive film for 1 minute before being stored at 4 °C for 30 min. In contrast, Group 3 was wrapped with a similar edible film infused with CEON, highlighting different treatment approaches for preservation. The results showed that fillets treated with the SGCEON3 film significantly suppressed the production of volatile nitrogenous bases caused by microbial activity throughout the 10 day storage period. By day 10, MDA levels had increased to 0.30 ± 0.07 mg/kg in the control group, compared with 0.10 ± 0.005 mg/kg in dipped fillets and 0.11 ± 0.005 mg/kg in wrapped fillets, showing antioxidant protection in the treated samples. Moisture loss was rapid in wrapped fillets, whereas dipped ones maintained better moisture levels. Over time, the control exhibited a gradual reduction in moisture content, highlighting the effectiveness of dipping versus wrapping. In terms of Sensory attributes, evaluated based on texture, odor, color, and gaping, both dipped and wrapped fillets using the CEON enriched bio-polymer-based active packaging film remained acceptable for consumption up to 8 days, whereas control fillets spoiled by day 6 (Arul Raj et al., 2024).

A study assessed rice starch-based films infused with oregano essential oil (OEO) at 0%, 4%, 4.5%, 5%, 6%, and 7% (based on solid content) to evaluate their effectiveness as packaging for fish fillets. Two film types were tested: a control with 3% rice starch and 30% glycerol, and an active version adding 4.5% OEO. Fish fillets wrapped in these films were stored at 4 °C to assess preservation effects, and shelf-life study were done in triplicate on days 0, 2, 4, and 6 of refrigerated storage. By day 6, fillets packaged in the OEO-containing film showed higher mechanical strength (13.4 N), lesser lipid oxidation (1.65 mg malondialdehyde/kg sample), and reduced microbial growth (10^7 CFU/g) compared with those packaged in the control film (Martins et al., 2021).

Wang et al. (2024) created starch/chitosan films loaded with radish-derived carbon dots at different concentrations (0.0%–1.5%), and their ability to preserve salmon fillets stored at 4 °C was evaluated. The fish quality slowly declined, as verified by darkening, the fading of white lines, and increased exudation of water and oil from the surface. While treated salmon only showed deterioration by day 10, both films significantly delayed spoilage, ultimately extending S_{life} up to 4 days through inhibition of bacteria, protein degradation, and lipid oxidation (Fig. 11.7a and b). This extension of S_{life}, by 4 days compared with the control, can be attributed to the antibacterial and antioxidant properties of the R-CDs-loaded films (Wang et al., 2024).

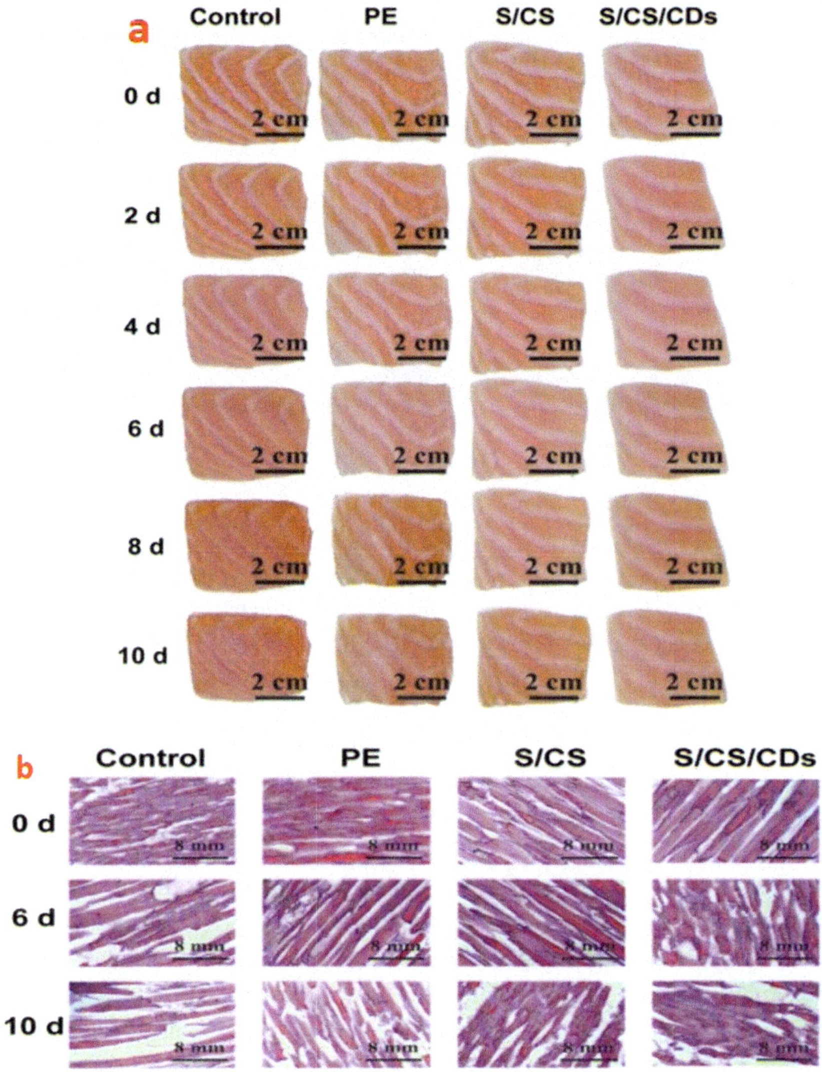

Figure 11.7 -Evaluating the performance of different groups in storing salmon fillets a) Morphological changes of salmon fillets packaged in synthetic, control and PE (polyethylene) films b) Alterations in the muscle tissues of salmon fillets. *(a and b) Photos reproduced from Wang, Dangfeng, Wang, Xin, Zhou, Shanshan, Ren, Likun, Meng, Yuqiong, Ma, Rui, Wang, Shulin, Liu, Zhiteng, Alamri, Abdulhakeem S., Alhomrani, Majid, Zhang, Zihui, Cui, Fangchao, Li, Tingting & Li, Jianrong. (2024). Radish residue carbon dots-based novel starch/chitosan film with high antioxidant, biocompatibility, and antibacterial activities for salmon fillets' active packaging. International Journal of Biological Macromolecules, 273(Pt 2). https://doi.org/10.1016/j.ijbiomac.2024.133107 with permission.*

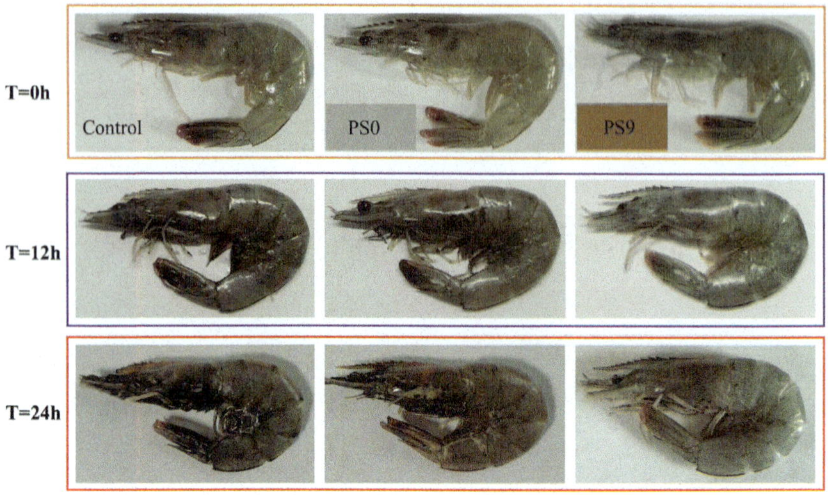

Figure 11.8 Photo of shrimp packed in different groups for 24 hour. *Adapted from Kang, Yongfeng, Cai, Duhong, Zhao, Dandan, Li, Xiying, Hu, Jie, Chen, Siwen, Chen, Weiqi & Li, Li. (2025). A novel dual-signal response Zn-MOF intelligent active label prepared from polyvinyl alcohol/starch for real-time visual monitoring shrimp freshness. International Journal of Biological Macromolecules, 311(Pt 2). https://doi.org/10.1016/j.ijbiomac.2025.143538 with permission.*

In another study, Zn-MOF at varying concentrations was incorporated into a PVA-starch matrix, leading to the preparation of PS@Zn-MOF composite films. To evaluate preservation, shrimp covered with the film were kept at 25 °C and 75% RH, revealing significant spoilage in control and PS0 samples, characterized by blackened bodies and discolored heads (Fig. 11.8). Also, the results show that the PS9 composite film remarkably delayed microbial activation and kept the TVB-N and TVC levels below the safety threshold. These findings show that PS@Zn-MOF films are effective active packaging materials for delaying the spoilage of seafood (Kang et al., 2025).

In general, incorporating antimicrobial and antioxidant substances into starch-based active films can prolong seafood shelf life and fulfill consumer requirements.

11.5 Application in dairy products

Milk and dairy products are packaged using various materials tailored to their properties, processing, storage, and use. Packaging primarily acts as a barrier against damage and contamination, ensuring product quality. Extending S_{life} depends on packaging's ability to minimize weight and nutrient loss (Mirza Alizadeh et al., 2021). Effective packaging ensures convenient handling, storage, and transport of defined product units, while also providing legal labeling that informs consumers and regulators about ingredients, as required by law (Mirza Alizadeh et al., 2021; Rejeesh et al., 2023). Key factors in choosing packaging materials for dairy products include safety and compatibility, impact resistance, sanitation, odor and light protection, tamper-proof features, appropriate shape, size, and weight, visual appeal, printability, and cost-effectiveness (Biegańska, 2022; Rejeesh et al., 2023). While traditional packaging was inert and did not interact with dairy products, modern trends include active, smart, or intelligent packaging designed to extend S_{life} (Abedi-Firoozjah et al., 2025a; Rejeesh et al., 2023).

Starch-based packaging materials can be ideal for dairy and cheese products, but they frequently need enhancements because of lower barrier properties and moisture resistance compared with traditional plastics (Bahramian et al., 2024; Mustafa et al., 2021). While they offer biodegradability benefits, it's important to ensure they meet the safety standards of dairy products and provide adequate protection through the product's S_{life}, as they may not be ideal for long-term storage without modifications (Li et al., 2022; Nair et al., 2025). A number of studies in this area are mentioned below and some are included in Table 11.1.

Addressing the need for eco-friendly packaging, Mustafa et al. (2021) developed a PVA/starch-based biodegradable film, enhanced through glutaraldehyde cross-linking for better strength and barrier performance. The film incorporated propolis extract (PE) as an antimicrobial agent, with formulations containing 20% PE exhibiting the highest tensile strength (15.9 MPa) and optimal inhibition zones of 25 mm against *E. coli* and MRSA. The PVA/starch matrix facilitated controlled release of PE's polyphenols and flavonoids, which disrupted microbial membranes and scavenged free radicals, delaying milk spoilage. The film effectively reduced microbial growth by over 3 log CFU in 7 days, maintained peroxide values below 15 meq/kg for 12 days, and minimized weight loss and lipid oxidation. Driven by the need for sustainable packaging, biodegradable films, such as potato starch reinforced with rice straw nanocellulose (CNC), *Moringa oleifera* extract, and gelatin, have been developed (Nair et al., 2025). In a recent research by Nair et al. (2025), the films were developed through the acid hydrolysis of rice straw to extract CNC and subsequent ultrasonication of *Moringa* extracts, combined with starch and gelatin. Incorporating 2% CNC significantly enhanced tensile strength from 0.83 to 4.48 MPa, while 1.5% *Moringa* extract further increased it to 7.63 MPa and improved elongation, demonstrating effective antimicrobial activity against pathogens and reducing microbial counts in food products. For instance, in khoa coated with *Moringa* films, yeast and mold counts remained around 6 log CFU/g by day 18, compared with over 6 log CFU/g for starch-coated samples, showcasing the extract's antifungal efficacy. Similarly, a 15-day trial with peda showed *Moringa*-coated samples maintaining counts of about 3.6 log CFU/g against S peda's 4 log CFU/g. The incorporation of bioactive compounds, such as flavonoids in *Moringa*, proved effective for food preservation, while the films degraded at a rate of 63.9% after 30 days in soil, emphasizing their environmental benefits. In a another study, a novel antibacterial and pH-sensitive film has been developed using PVA/starch, enhanced with ethyl lauroyl arginate (LAE) and mulberry anthocyanin for preserving milk (Sharma et al., 2022). This composite film demonstrated a porous, heterogeneous structure that facilitated the immobilization of the active agents. The mechanical properties were largely unaffected with up to 20% mulberry anthocyanin, while incorporating LAE significantly increased antibacterial efficacy against *E. coli* (inhibition zones; 23 mm) and *S. aureus* (inhibition zones; 19 mm). Importantly, spoilage tests revealed that films containing 5% of both LAE and anthocyanin extended the S_{life} of milk, maintaining microbial counts below 3.5 log CFU/g over 15 days.

Cheese is a globally popular, nutritious dairy product rich in fats, minerals, and proteins with health benefits such as antihypertensive and antioxidant effects. Its consumption varies culturally, but due to its nutrient density, cheese is susceptible to microbial contamination and spoilage (El-Sayed et al., 2024; Mirmoeini et al., 2023b). Raw milk is the main contamination source in cheese, with poor hygiene and environmental factors such as air, water, soil, and contact surfaces also contributing. Microorganisms can rapidly grow on cheese surfaces due to favorable pH and moisture levels, increasing contamination risk during manufacturing and storage (Biegańska, 2022; El-Sayed et al., 2024; Rejeesh et al., 2023). Cheese packaging is essential for quality, safety, and marketability, considering type, S_{life}, and regulations. Edible, eco-friendly films/coatings are emerging as sustainable alternatives, promising increased adoption through future innovations (Mirmoeini et al., 2023b; Zaiden et al., 2025).

In a study, Zaiden and Al-Hadidy (2025) evaluated the effectiveness of starch-based edible films incorporated with clove oil (*Dianthus Caryophyllus* L.) for preserving soft cheese through antimicrobial action and improved physicochemical properties (Zaiden et al., 2025). The films demonstrated significant antimicrobial effectiveness against *Klebsiella pneumonia* and *Candida albicans*, with the highest inhibitory effect of 27 mm at a 4% clove oil concentration. Different treatments of stored cheese (uncoated and films with varying clove oil concentrations) showed significant differences in pH, moisture content, and free fatty acids, particularly at a 95% confidence level. After 15 days at 4°C, uncoated cheese's pH dropped to 5.1, while the pH for coated cheese with 4% clove oil remained higher at 6.3, indicative of reduced microbial growth. Moisture percentages decreased with time, reaching 58.22% for uncoated cheese versus 60.46% for the films, revealing better moisture retention in coated samples. Free fatty acids levels also differed significantly: while uncoated cheese saw a major increase (3.995%), the films maintained free fatty acids below 1.1%, showing the antioxidant properties of clove oil, which inhibited lipase activity. Focusing on preservation, Lopes et al. (2024) developed edible films using LBMD (locust bean milling dust) and potato starch to enhance the shelf life of cheese and oat cookies (Lopes et al., 2024). These films effectively immobilize active agents and enhance food S_{life} through specific interactions among their components. LBMD and potato starch increase the mechanical strength of the films, yielding an improved Young's modulus of approximately 48 MPa compared with 12 MPa for LBMD alone. The films displayed enhanced barrier properties, with a water contact angle of 87 degrees, indicating reduced moisture absorption and increased rigidity. Cheese packaged in LBMD/starch films showed remarkable preservation qualities. Over 72 hour at 25°C, the pH level increased from 5.58 to 6.51, reflecting microbial activity. The expressible moisture content (EMC) of cheese increased significantly, with treated films maintaining lower EMC values (0.62%–8.37%) versus control films (0.89%–19.79%). TBARS values for the active films were 0.28–5.54 mg MDA/kg, demonstrating reduced lipid oxidation compared with control samples (0.30–5.89 mg MDA/kg). Further, the total microbial count in treated samples remained below the safety threshold for 40 hour, showcasing effective spoilage prevention.

With their low density, high surface area, and sorption abilities, aerogels are innovative porous materials formed by extracting liquid from gels, offering a promising avenue for creating advanced antimicrobial-emitting systems (Zheng et al., 2020). Thanks to their micro- and nano-structure, aerogels can effectively carry and release antimicrobials. Bio-based matrices such as polysaccharides, proteins, and seed mucilages are commonly used to create these aerogels for antimicrobial applications (Abedi-Firoozjah et al., 2025b; Mirmoeini et al., 2023a). Accordingly, Mirmoeini et al. (2023) focused on developing starch/cellulose aerogels infused with *Thymus daenensis Celak* essential oil (TDCEO) for the preservation of Koopeh cheese (Mirmoeini et al., 2023a). A formulation of 1% cellulose (extracted from sunflower stalks) and 5% starch was optimized for antimicrobial packaging. The airgel's-controlled release mechanism, aided by its low porosity, ensures a steady delivery of TDCEO, which exhibits antimicrobial action due to its lipophilic nature, effectively disrupting microbial cell membranes. The MID of TDCEO vapor was established at 256 μL/L, and cheese wrapped with SC-TDCEO50 MID exhibited a 3-log decrease in psychrophiles over 21 days, with *E. coli* becoming undetectable after 14 days. Yeast-mold populations decreased by 1 log in the SC-TDCEO50 MID group. Sensory analysis indicated that cheese treated with SC-TDCEO25 MID maintained better sensory attributes compared with the control, with a notable preference for its color and odor. Ma et al. (2024) examined the use of biodegradable films for preserving Mongolian cheese, various compositions were analyzed, focusing on Konjac Glucomannan (KG), cassava starch, Maltodextrin (M), and Gelatin (G). The films demonstrated lower L* values, particularly 21.14 ± 2.14 for MSA films, 32.47 ± 2.8 for starch-CD films, and 18.34 ± 1 for MSA + starch-CD films, indicating increased color density relative to unmodified groups. These observations suggest that the interactions among anthocyanins and film components significantly influenced lightness and color stability. Highlighting their food application, films containing these components successfully reduced weight loss and microbial proliferation in cheese during storage, showing their preservation potential. After 30 days, weight loss measurements showed 73.49% for starch, 50.60% for KG, and 54.22% for CSKG films compared with control samples. Total viable counts (TVC) and total yeast and mold counts (TYMC) also revealed enhanced preservation, with control samples exhibiting values significantly higher than those treated with the various films.

Starch-based packaging systems are an eco-friendly alternative for dairy products. These materials provide a sustainable option for packaging dairy items, helping reduce plastic waste while maintaining protection, freshness, and safety during storage and transportation. However, it may require enhancements to ensure it meets the necessary performance standards.

11.6 Application in nuts

Nuts are nutritious, rich in healthy fats, proteins, vitamins, and antioxidants, making them popular as snacks and in health foods. While essential, processing methods such as roasting and drying are susceptible to rancidity due to high unsaturated fat content, which negatively impacts quality and S_{life}. Traditional preservation with synthetic packaging and antioxidants raises health and environmental concerns. As a sustainable alternative, biodegradable packaging and edible coatings offer eco-friendly solutions for maintaining nut quality (Mukwevho et al., 2025; Zhou et al., 2025). da Costa et al. (2024) proposed an edible starch/CMC packaging system infused with tocopherol, applied to Brazil nuts to prevent oxidation (da Costa et al., 2024). Coated nuts showed significantly less mass loss (around 4%) after 120 days compared with uncoated nuts, which lost over 10%. The coating delayed browning and oxidative changes, with the browning index increasing only 2% over 120 days in coated samples. Oxidative markers such as peroxide and conjugated dienes remained lower in coated nuts, with peroxide values staying within safe limits (<15 meq/kg). The coatings extended the oil's induction time, coating with soy lecithin and no tocopherol mix reaching up to 130.74 hour in accelerated tests, indicating improved stability. The interaction between cassava starch, CMC, and tocopherol created an effective barrier and controlled release system, enhancing nut preservation. Babapour and his research team developed a potato starch-based film reinforced with nano zinc oxide (Nano-ZnO) and infused with fennel essential oil (FEO) to preserve pistachios (Babapour et al., 2022). The starch's hydroxyl groups formed hydrogen bonds with Nano-ZnO and FEO, immobilizing active compounds and controlling their release. The film's structure improved mechanical strength and reduced oxygen permeability, slowing oxidation and mold growth. During 15 days of refrigerated storage, coated pistachios showed only 4% weight loss versus over 10% in controls, with moisture content maintained at 6.55%, compared with 5.07%. The coating delayed lipid oxidation, with peroxide values staying below 15 meq/kg, and reduced aflatoxin production; the lowest aflatoxin level (2.80 ppb) was observed in samples with 5% Nano-ZnO and 3% FEO. The active agents disrupted microbial membranes, preventing mold and mycotoxin formation.

According to studies, it can be said that starch-based packaging can be effective in packaging these products. Additionally a number of studies in this field are also included in Table 11.1.

11.7 Application in processed and packaged foods

Processed and packaged foods require specialized packaging to ensure safety, freshness, and S_{life}. Starch-based active packaging materials are suitable for processed and packaged food products. They are biodegradable, environmentally friendly, and can incorporate active agents such as antimicrobials or antioxidants to enhance food safety and S_{life}, making them a promising choice for various processed foods (Hamann et al., 2024; Nadon et al., 2025b).

Oxidation significantly degrades processed meats such as calabrese sausage, affecting flavor, color, texture, and nutrition through lipid oxidation, enzyme activity, and microbial growth (Fell et al., 2024; Hamann et al., 2024). Natural and synthetic antioxidants are used to slow oxidation by breaking peroxides, disrupting chain reactions, reducing oxygen, and inhibiting catalysts. Biodegradable films with natural antioxidants provide an eco-friendly alternative, with starch-based polymers such as cassava starch being prominent because of their excellent properties, affordability, and film-forming ability. In an interesting work, Hamann et al. (2024) developed cassava starch-based biocompatible active packaging system containing green tea extract (GTE; 1%, 2.5%, and 5%) to preserve sliced calabrese sausage (Hamann et al., 2024). The starch' hydroxyl groups formed hydrogen bonds with GTE's catechins, stabilizing active compounds. The 1% GTE films showed the best antioxidant activity (IC_{50}; 0.012 mg/mL) over 60 days, effectively inhibiting lipid oxidation (based on the TBARS values) for 18 days, compared with untreated samples. During storage, coated sausages-maintained moisture (6.55%) versus 5.07% in control groups, and lipid oxidation markers (peroxide and TBARS) remained lower, showing delayed spoilage. The films also slowed pH decline and acidification, maintaining sensory quality with scores around 4.7–5.2 over 15 days. The active agents' phenolic compounds scavenged free radicals and inhibited microbial activity, preventing mold and oxidation. Liu et al. (2023) found that a starch-based edible film loaded with *Zanthoxylum armatum* essential oil (ZAEO) effectively preserved beef sauce (Liu et al., 2023). The starch's interaction with ZAEO bioactive compounds enhanced its barrier and antioxidant properties, slowing lipid oxidation, microbial growth, and spoilage indicators. During 15 days of storage, beef sauce packaged in the ZAEO-infused film showed only 4% weight loss versus over 10% in controls, with peroxide values remaining below 15 meq/kg and TBARS inhibited for at least 12 days. The film also delayed pH decline, reduced TVB-N levels, and lowered aerobic plate counts by 45%, extending S_{life} by 4–6 days.

Generally, the limited physicochemical and mechanical properties of starch-based films hinder their industrial use, prompting modifications such as plasma technology. This method effectively alters surface characteristics, improving adhesion, transparency, strength, and gas barrier properties of natural polysaccharide films (Abedi-Firoozjah et al., 2021; Nadon et al., 2025a). Nadon et al. (2025a) explored the potential of cassava starch-based edible films, incorporating Makwaen essential oil (MEO), to enhance the preservation of pork sausages (Nadon et al., 2025a). The film was produced using a blend of cassava starch and polybutylene succinate (80:20 w/w), subjected to Argon cold plasma (Ar-CP) treatment to improve film characteristics. The optimal treatment conditions were 120 W for 15 seconds, leading to a tensile strength of 19.84 MPa and an elongation of 244%. The antimicrobial activity of the films was attributed to MEO's major components, such as sabinene and D-limonene, which effectively inhibited microbial growth. In practical food applications, pork sausages stored at 25°C for 72 h showed significantly reduced spoilage. The pH of sausage samples increased from 5.58 to 6.51, indicating microbial activity. Notably, the expressible moisture content (EMC) rose, with the treated film maintaining lower EMC (8.37%) compared with the control (19.79%). TBARS values were also lower in the Ar-CP treated film (5.54 mg MDA/kg) than in the control (5.89 mg MDA/kg), underscoring reduced lipid oxidation. The total microbial count of the pork sausages remained below the safety threshold for 40 hour with the active film, compared with 36 h for the control. Cruz-Gálvez et al. designed potato starch films incorporating acetonic and methanolic extracts of Hibiscus sabdariffa to preserve sausages (Cruz-Gá et al., 2018). The films' physicochemical properties were altered by extract addition: methanolic extracts increased fracture tension and decreased permeability, while acetonic extracts enhanced antioxidant activity. In vitro, acetonic extracts showed higher antimicrobial efficacy, especially against *L. monocytogenes*. When used as a coating, the films significantly reduced microbial growth, with *L. monocytogenes* decreasing by over 3 log CFU after 7 days, and delayed lipid oxidation, maintaining peroxide and TBARS levels below critical limits for 12 days. The films also slowed pH decline and water vapor loss, extending S_{life} by 6–8 days.

The results of the studies generally show the effectiveness of starch-based packaging in these products.

11.7.1 Application in bread

As a staple food worldwide, especially in developing countries, bread quickly deteriorates in freshness due to staling and microbial contamination. To extend S_{life}, key strategies include moisture retention through proper packaging, reducing staling by adjusting formulations and processing, and controlling microbial growth (Chandrasekar et al., 2023; Khezerlou et al., 2023). Traditional packaging often falls short in safeguarding food quality, safety, and hygiene, especially for baked goods, where sensory qualities deteriorate over time (Araú et al., 2023; Noshirvani et al., 2025). Starch-based active packaging that extends S_{life} and reduces preservatives is a major advancement. Incorporating antimicrobial and antioxidant additives with biodegradable materials such as starch offers an eco-friendly, cost-effective solution.

Araújo et al. (2023) suggested biodegradable corn starch films loaded with pectin extract and *Turmeric Longa Linn* essential oil (TLEOs) for bread preservation (Araú et al., 2023). Corn starch's rich amylose and amylopectin content forms a strong matrix, where TLEO's active compounds such as Turmerone (36.47%) exert antimicrobial effects, inhibiting fungi such as *Aspergillus niger*. Here, TLEOs diffusing slowly into bread, preventing microbial growth and mold for up to 9 weeks, significantly extending S_{life} compared with controls and cellophane (which showed browning from week 6 (Fig. 11.9a)). The interaction between starch and TLEOs stabilizes active compounds and enhances barrier properties, reducing water permeability and controlling moisture. The pectin extract adds antioxidant activity, further inhibiting microbial spoilage. Films effectively maintained bread quality, delaying mold growth, and preserving texture and color. In an interesting recent study, Chang et al. (2025) proposed tapioca starch-based films infused with clove essential oil (CEOs) at various concentrations (0%–1.25%) for bread preservation (Chang et al., 2025). Tapioca starch's high amylose content forms a strong, film-forming matrix, while CEO's active compounds, such as eugenol, exhibit potent antimicrobial effects, especially against bacteria such as *S. aureus* and *E. coli*. The films showed enhanced microbial inhibition, extending bread's S_{life} from 6 days (control) to 9 days without mold growth (Fig. 11.9b). The film's mechanism involves CEO's gradual release, disrupting microbial activity, while tapioca starch's chemical interactions stabilize the active compounds and improve barrier performance.

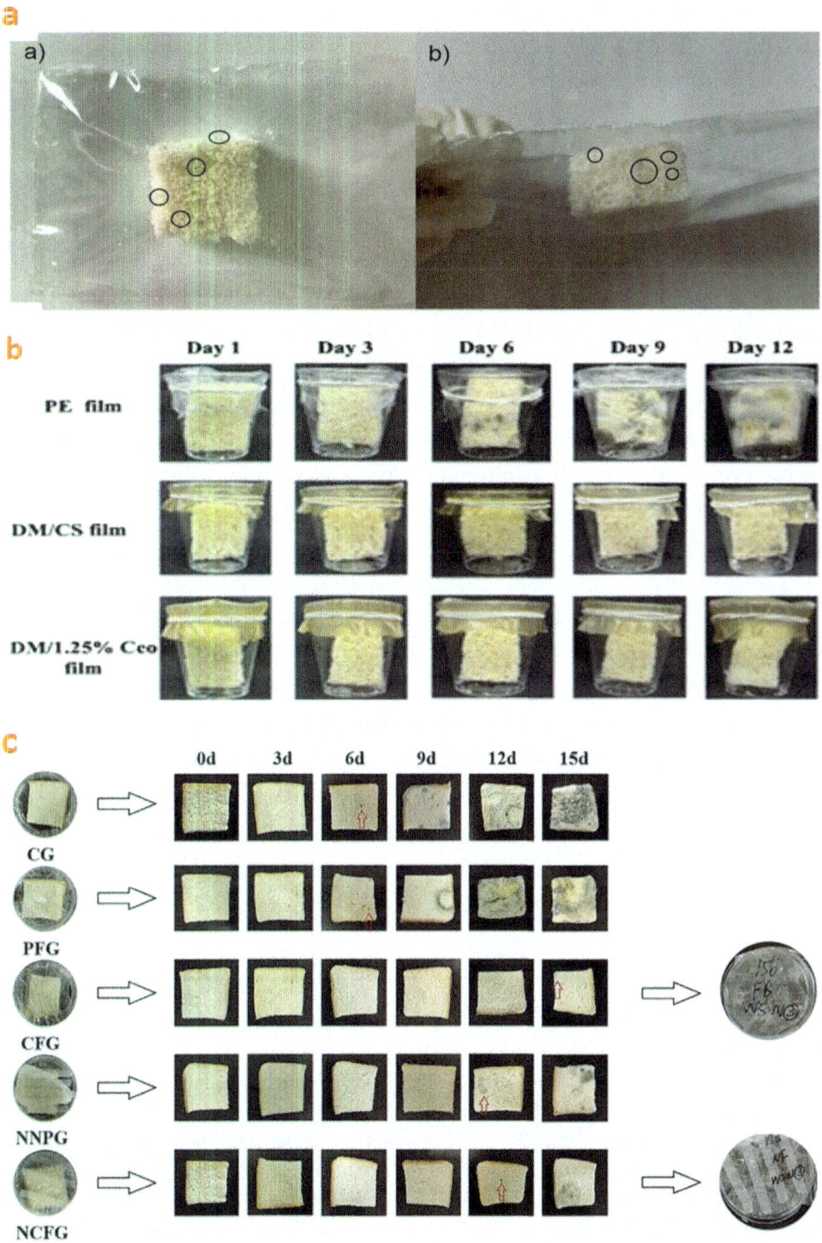

Figure 11.9 -Improving bread fresh-keeping quality during storage; a) corn starch films loaded with pectin extract and TLEOs, b) tapioca starch-based films infused with clove essential oil (CEOs; 0%–1.25%) and c) corn starch-based active film with nanoemulsified clove essential oil (NCEOs; 10%, 20%, and 40%). *(a) Photos reproduced from Araújo, M.N.P., Grisi, C.V.B., Duarte, C.R., de Almeida, Y.M.B. & Vinhas, G.M.*

Encapsulating essential oils in nanoemulsions helps preserve its properties and reduce flavor impact. Nanoemulsions offer stability, controlled release, and enhanced antimicrobial activity. In a study by Fan et al. (2024), the authors developed a corn starch-based active film with nanoemulsified clove essential oil (NCEOs) through casting methods, tested at at 10%, 20%, and 40% concentrations (Fan et al., 2024). As NCEOs increased, films became thinner, rougher, and more yellow, with decreased water content (25.80 ± 2.05 degrees→91.57 ± 5.14 degrees) and permeability ($10.24 \pm 0.19 \rightarrow 5.32 \pm 0.06$), but higher tensile strength (7.19 ± 2.23 MPs→ 13.9 ± 3.44 MPs), antioxidant (7.14% →99.46%), and antimicrobial activity (2.18 log CFU/g→7.00 log CFU/g). The 40% NCEOs film notably extended bread S_{life} to 15 days, preventing mold and spoilage, unlike controls, which showed mold after 6–9 days (Fig. 11.9c). NCEO's active compounds, such as eugenol, diffusing slowly, inhibiting microbial growth and oxidation, while starch's chemical interactions stabilize NCEOs within the matrix.

Fonseca et al. (2021) demonstrated that starch/carvacrol nanofibers effectively inhibit fungal growth in bread, extending S_{life} (Fonseca et al., 2021). The nanofibers, produced through electrospinning, encapsulate carvacrol (30% concentration), which enhances thermal stability and controls volatile release, providing sustained antifungal activity. The interaction between starch and carvacrol stabilizes the active compound within nanofibers, facilitating slow, prolonged release that inhibits fungi such as *Penicillium* and *Aspergillus* during storage. The nanofibers' chemical structure improves barrier properties, reducing oxygen permeability (7.31 g/m^2·s) and water vapor permeability (1.35 g/m^2·s). In vitro MICs for nanofibers

(2023). Active packaging of corn starch with pectin extract and essential oil of Turmeric Longa Linn: Preparation, characterization and application in sliced bread. International Journal of Biological Macromolecules, 226, 1352–1359. https://doi.org/10.1016/j.ijbiomac.2022.11.248 with permission. (b) Photos reproduced from Chang, Hui, Zhao, Ying, Zhang, Jie, Chen, Jian & Yang, Tao. (2025). Slow-release antibacterial film loaded with clove essential oil based on tapioca starch used for bread preservation. Food Chemistry: X, 29. https://doi.org/10.1016/j.fochx.2025.102677 Elsevier open access. (c) Photos reproduced from Fan, Saifeng, Yin, Xiaoyu, Liu, Xingxun, Wang, Guangyu & Qiu, Weifen. (2024). Enhancing bread preservation through non-contact application of starch-based composite film infused with clove essential oil nanoemulsion. International Journal of Biological Macromolecules, 263(Pt 1). https://doi.org/10.1016/j.ijbiomac.2024.130297 with permission.

with 30% carvacrol were 0.098 mg/mL against *Penicillium* and 9.8 mg/mL against *A. flavus*; the nanofibers exerted fungicidal effects and maintained activity at baking temperatures (160 °C). Bread treated with these nanofibers showed no fungal growth after 7 days, surpassing controls, which exhibited mold from day 3. The biodegradable nanofibers degraded 73.5% in 20 days, far exceeding traditional plastics' degradation rates (<2%). Chandrasekar et al. (2023) developed a cellulose-reinforced starch-based bioactive films from recycled banana peel waste, rich in amylopectin (88.55%) and thermally stable (~295 °C). Extracts contained bioactive compounds such as β-sitosterol, offering antioxidant (74.43%) and antimicrobial effects, enhancing film properties (Chandrasekar et al., 2023). Nanocellulose fibers improved mechanical strength six-fold and reduced oxygen permeability by 40%. The films exhibited ~98% UV protection, effectively extending bread S_{life} by 10 days at 4 °C, compared with 5 days in conventional PE packaging. The active compounds inhibited microbial growth, lowering initial microbial counts significantly ($P < .05$). Moisture content decreased more in starch-based films, but antimicrobial action prevented mold development longer. The films' mechanism involves starch's chemical interactions with bioactives and nanocellulose, creating a barrier that controls moisture, oxygen, and microbial activity. An interesting study developed an eco-friendly active film using rice husk fiber (RH) and starch, incorporating peppermint essential oil (PPEO) at 1%–6% levels (Srivastava et al., 2024). The film's properties improved with 1% PPEO and RH 10%, achieving maximum tensile strength (2.25 MPa) and elastic modulus (57.4 MPa). PPEO's active compounds, mainly menthol, disrupted microbial cell membranes, preventing bacterial and fungal growth. The starch matrix stabilized PPEO, enabling sustained release and enhancing barrier properties. The interactions between starch, RH, and PPEO created a durable, antimicrobial film with reduced water solubility and moisture content, and increased optical barrier. When used in bread packaging, the biocomposite extended S_{life} beyond 15 days, compared with 3–4 days for untreated bread, with no visible mold growth. The film created a microenvironment that inhibited microbial proliferation by disrupting cell membranes, thus delaying spoilage.

11.7.2 Application in cake

Cake packaging is vital for preserving quality, safety, and S_{life}. Traditional materials such as plastic and foil are common but lack sustainability. Eco-friendly options such as biopolymer films (like starch) can include antimicrobials, extending freshness. In a recent study, Lemos et al. (2025) produced bean starch-based edible active films infused with orange peel essential oil (OPEO; 5% and 10% concentrations) to preserve cake (Lemos et al., 2025). The starch's hydroxyl groups facilitated strong interactions with OPEO's lipophilic compounds, mainly D-limonene, which disrupt fungal cell membranes by increasing permeability and causing leakage of cellular contents. Here, the modification of starch improved film stability, morphology, and barrier properties, with a tensile strength of 5.44 MPa and elongation at break of 52%. During application, films with 10% OPEO completely prevented fungal growth on cakes for up to 9 days, significantly extending S_{life} compared with controls, which showed mold after 3 days. The volatile compounds from OPEO acted through rapid diffusion, inhibiting fungi such as *Penicillium* and *Aspergillus*.

Glutinous rice cake, valued in East Asia for its viscoelastic texture, is prone to spoilage due to high water content, leading to nutrient loss and contamination. Promhuad et al. (2025) fabricated a biodegradable film combining 40% PBAT with 60% TPS, incorporating maltol and ethyl maltol at 1%–5% (Promhuad et al., 2025). Prepared through extrusion, these films combined polymer components with natural antimicrobials, offering sustainable, effective packaging solutions. The films showed smoother morphology, better homogeneity, and increased crystallinity due to hydrogen bonding between TPS and PBAT. Volatile compounds such as maltol and ethyl maltol effectively inhibited Staphylococcus aureus and significantly delayed mold growth on cake, extending S_{life} beyond 16 days, compared with 8 days in control samples (Fig. 11.10a). The release of volatile maltol and ethyl maltol disrupted fungal cell membranes, causing leakage and cell death, while maintaining film integrity. The interaction between TPS and PBAT stabilized maltol and ethyl maltol, providing controlled release and enhancing antimicrobial efficacy. Films with $\geq$3% active agents completely prevented mold during storage. Glycerol-hydrogenated rosin (GEHR), from natural rosin, is used in foods such as gum and soft drinks. It is soluble in hydrocarbons and oils, providing antioxidant benefits and improving preservation (Luo et al., 2022). Luo et al. (2022) suggested an edible active packaging film combining corn starch and rock bean protein (RP) with embedded D-limonene-loaded GEHR microparticles, produced through electrostatic spray (Luo et al., 2022). The film's structure was stabilized by hydrogen bonding between films components, enhancing thermal stability (179.2°C) and mechanical

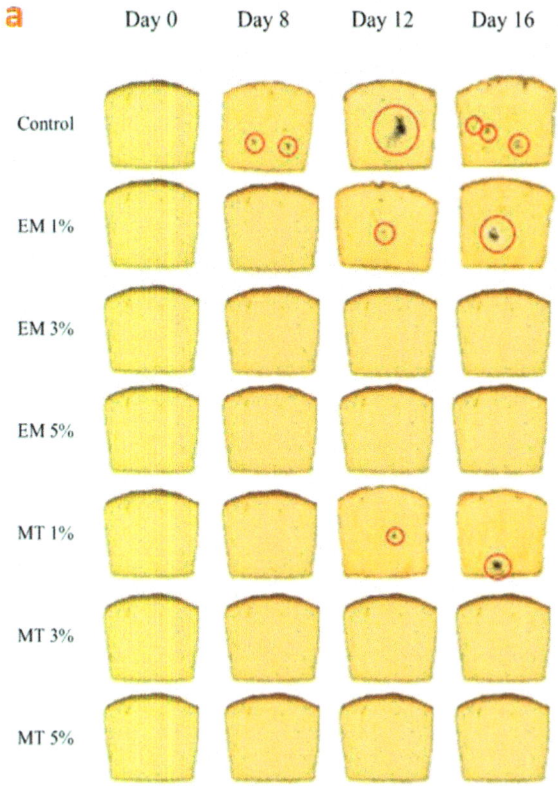

Figure 11.10 Improving cake fresh-keeping quality after treatment with starch-based packaging systems; a) PBAT/TPS incorporating maltol and ethyl maltol at 1%–5% and b) corn starch/rock bean protein (RP) with embedded D-limonene-loaded GEHR microparticles. *(a) Photos reproduced from Promhuad, Khwanchat, Ebel, Lucie & Harnkarnsujarit, Nathdanai. (2025). Thermoplastic starch/poly(butylene adipate-co-terephthalate) blown film with maltol and ethyl maltol preserving cake quality: Morphology and antimicrobial function. Food Chemistry, 464(Pt 1). https://doi.org/10.1016/j.foodchem.2024.141646 with permission. (b) Photos reproduced from Luo, Songming, Chen, Jundong, He, Jing, Li, Haisong, Jia, Qi, Hossen, Md Alomgir, Dai, Jianwu, Qin, Wen & Liu, Yaowen. (2022). Preparation of corn starch/rock bean protein edible film loaded with d-limonene particles and their application in glutinous rice cake preservation. International Journal of Biological Macromolecules, 206, 313–324. https://doi.org/10.1016/j.ijbiomac.2022.02.139 with permission.*

strength (TS; 0.88 MPa, EAB; 54.36%). The active agents, mainly D-limonene, disrupted microbial cell membranes, exerting strong antifungal and antibacterial potential. When applied to cake preservation, the film extended S_{life} by 2–4 days, with mold appearing after 7 days in control

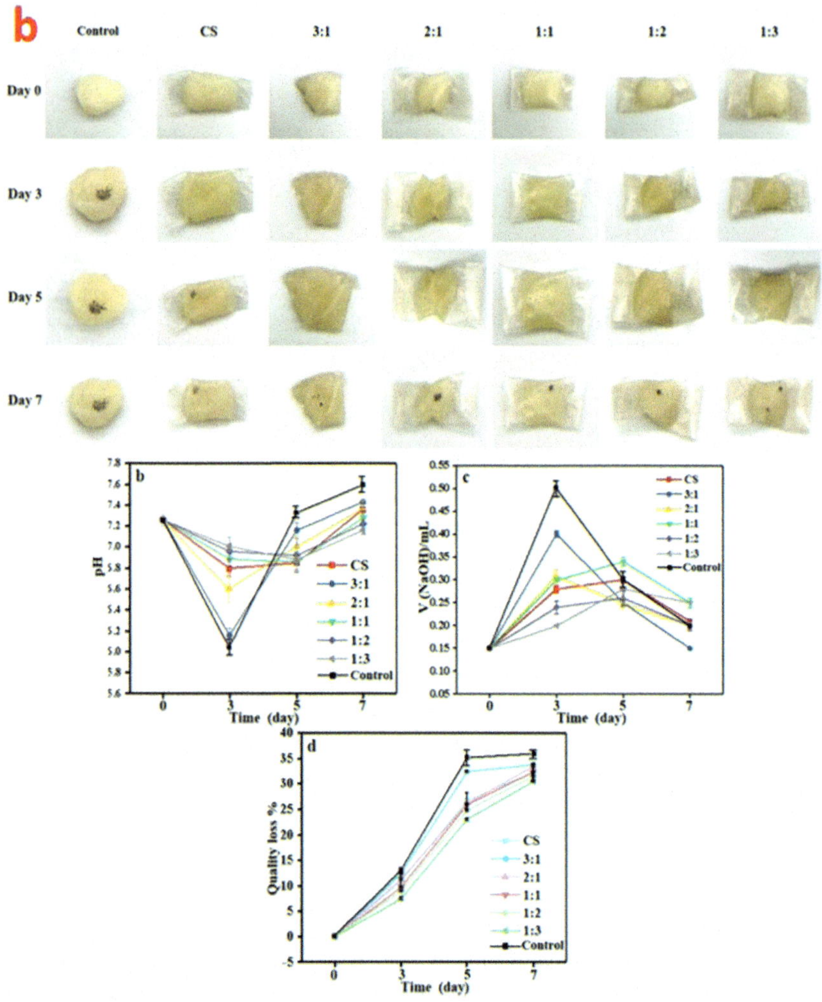

Figure 11.10 (*continued*).

groups but delayed up to 11 days in treated samples. Apart from antimicrobial properties, the film's barrier properties prevented excessive moisture loss, maintaining cake freshness (Fig. 11.10b).

Starch-based coatings for cakes are eco-friendly options that improve moisture retention, texture, and shelf life. These biodegradable films form a protective layer, helping to preserve freshness and prevent contamination, making them a sustainable alternative to synthetic coatings (Khezerlou et al., 2023). Eom et al. synthesized an edible coating using waxy corn starch

(WCS) combined with gellan gum (GG) and sodium alginate (SA) to preserve rice cakes, effectively reducing starch retrogradation (Eom et al., 2018). The WCS/GG blend (1g WCS + 1.25 g GG/100 mL) formed a thick, smooth, and homogeneous film that enhanced water retention and barrier properties. During 4 days of storage at 25 °C and 45% RH, coated rice cakes lost only 1.08% moisture, compared with 3.36% in control samples, delaying moisture loss and hardening. The film's structure, stabilized by hydrogen bonding, slowed retrogradation, with a 9.73% reduction in retrogradation rate (k = 0.408 vs. 0.452), and maintained softness longer. The coating also slowed hardness increase, delaying it by up to 18 h. The film's low water vapor permeability helped retain cake moisture, and its strong water-holding capacity improved texture.

11.8 Conclusion

With the potential to advance eco-friendly packaging, starch-based materials serve as a promising substitute for conventional plastics that significantly harm the environment. Also numerous studies have confirmed that starch can serve as an effective base for active packaging when combined with antioxidant and/or antimicrobial agents. Such active starch-based packaging can prolong the S_{life} of a wide variety of foods, thereby helping to reduce food waste, which is especially crucial given the world's growing population. However the current application of starch-based materials in packaging is primarily limited by their relatively poor mechanical and barrier properties, which stem from the inherent characteristics of starch itself. To address these challenges, strategies such as physical and chemical starch modifications, as well as blending with other biopolymers and incorporating functional additives, are under investigation. While some of these approaches have successfully improved the performance of starch-based films further research is still required to develop starch-based packaging materials whose mechanical and barrier properties can match those of traditional petroleum-based plastics.

AI disclosure

During the preparation of this work, the author(s) used chat.qwen.ai and deep AI in order to to make grammatical corrections, summarize content, find synonyms, Text improvement, etc. After using this tool/service, the author(s) reviewed and edited the content as needed and take(s) full responsibility for the content of the publication.

References

Abedi-Firoozjah, R., et al. (2021). Non-thermal techniques: A new approach to removing pesticide residues from fresh products and water. *Toxin Reviews, 40*(4), 562–575.

Abedi-Firoozjah, R., et al. (2025a). Comparison of the pH-and NH3-sensitivity of chitosan/polyvinyl alcohol smart films containing anthocyanins or betacyanins for monitoring fish freshness. *Carbohydrate Polymer Technologies and Applications, 9*, Article 100639.

Abedi-Firoozjah, R., et al. (2025b). Unlocking the potential of cyclodextrin/metal–organic frameworks in food packaging systems: An exclusive review on the latest trends. *Food and Bioprocess Technology, 18*(6), 5010–5037.

Adame, M. Y., et al. (2024). Fabrication and characterization of pullulan/tapioca starch-based antibacterial films incorporated with Litsea cubeba essential oil for meat preservation. *International Journal of Biological Macromolecules, 268*, Article 131775.

Araújo, M. N. P., et al. (2023). Active packaging of corn starch with pectin extract and essential oil of Turmeric Longa Linn: Preparation, characterization and application in sliced bread. *International Journal of Biological Macromolecules, 226*, 1352–1359.

Arul Raj, J. S., et al. (2024). Green packaging solutions for extending the shelf life of fish fillet: Development and evaluation of cinnamon essential oil-infused cassava starch and fish gelatin edible films. *ACS Omega, 9*(46), 45898–45910.

Azam, M., et al. (2019). Postharvest quality management of strawberries. In *Strawberry-pre- and post-harvest management techniques for higher fruit quality*.

Babapour, H., et al. (2022). Effects of active packaging based on potato starch/nano zinc oxide/fennel (Foeniculum vulgare miller) essential oil on fresh pistachio during cold storage. *Journal of Nuts, 13*(2), 105–123.

Bahramian, B., et al. (2024). Resveratrol-biopolymer materials: A sustainable approach to food packaging. *Trends in Food Science & Technology, 154*, Article 104761.

Basumatary, I. B., et al. (2022). Biopolymer-based nanocomposite films and coatings: Recent advances in shelf-life improvement of fruits and vegetables. *Critical Reviews in Food Science and Nutrition, 62*(7), 1912–1935.

Biegańska, M. (2022). Packaging of dairy products: Emerging strategies. In *Food packaging: The smarter way* (pp. 127–164). Springer.

Bilck, A. P., Yamashita, F., & Marzano-Barreda, L. A. (2021). Characterization and application of starch/polyester packaging produced by blown extrusion. *Carbohydrate Polymer Technologies and Applications, 2*, Article 100088.

Boeira, C. P., et al. (2021). Antioxidant and antimicrobial effect of an innovative active film containing corn stigma residue extract for refrigerated meat conservation. *Journal of Food Processing and Preservation, 45*(9), Article e15721.

Bumrungnok, K., et al. (2023). Toward a circular bioeconomy: Exploring pineapple stem starch film as protective coating for fruits and vegetables. *Polymers, 15*(11), 2493.

Cestari, L. A., et al. (2021). Production, antioxidant characterization and application of active starch-based films containing essential oils for beef packaging. *Research, Society and Development, 10*(8).

Chandrasekar, C. M., et al. (2023). Development and characterization of starch-based bioactive thermoplastic packaging films derived from banana peels. *Carbohydrate Polymer Technologies and Applications, 5*, Article 100328.

Chandrasekar, C. M., et al. (2024). Fabrication of double layer nanoparticle infused starch-based thermoplastic food packaging system for meat preservation. *International Journal of Biological Macromolecules, 254*, Article 127689.

Chang, H., et al. (2025). Slow-release antibacterial film loaded with clove essential oil based on tapioca starch used for bread preservation. *Food Chemistry X, 29*, Article 102677.

Chatkitanan, T., & Harnkarnsujarit, N. (2021). Effects of nitrite incorporated active films on quality of pork. *Meat Science, 172*, Article 108367.

Chen, N., et al. (2023). Potato starch-based film incorporated with tea polyphenols and its application in fruit packaging. *Polymers, 15*(3), 588.
Cruz-Gálvez, A. M., et al. (2018). Antimicrobial activity and physicochemical characterization of a potato starch-based film containing acetonic and methanolic extracts of Hibiscus sabdariffa for use in sausage. *LWT, 93*, 300–305.
Cui, Y., et al. (2024). Development and application of mathematical modeling of thymol release from environmental-responsive potato starch active packaging films. *International Journal of Biological Macromolecules, 271*, Article 132353.
da Costa, D. S., et al. (2024). Cassava starch/carboxymethyl cellulose edible coating added of tocopherol: A strategy to preserve the oxidative stability of Brazil nuts. *Foods, 13*(17), 2732.
Deng, B., et al. (2024a). An antibacterial packaging film based on amylose starch with quaternary ammonium salt chitosan and its application for meat preservation. *International Journal of Biological Macromolecules, 261*, Article 129706.
Deng, N., et al. (2024b). Characterization, antibacterial and snakehead fish (Channa argus) preservation effects of active films infused with compounded cinnamon and thyme essential oils. *Food Packaging and Shelf Life, 46*, Article 101353.
El-Sayed, S. M., & Youssef, A. M. (2024). Emergence of cheese packaging by edible coatings for enhancing its shelf-life. *Journal of Food Measurement and Characterization, 18*(7), 5265–5280.
Eom, H., et al. (2018). Development of a starch/gum-based edible coating for rice cakes to retard retrogradation during storage. *Lwt, 97*, 516–522.
Fan, X., et al. (2023). Incorporating Portulaca oleracea extract endows the chitosan-starch film with antioxidant capacity for chilled meat preservation. *Food Chemistry X, 18*, Article 100662.
Fan, S., et al. (2024). Enhancing bread preservation through non-contact application of starch-based composite film infused with clove essential oil nanoemulsion. *International Journal of Biological Macromolecules, 263*, Article 130297.
Fang, Z., et al. (2017). Active and intelligent packaging in meat industry. *Trends in Food Science & Technology, 61*, 60–71.
Fell, E. L., et al. (2024). Nisin or organic acid salt mixtures for the calabrese-type sausages in industrial-scale production. *Brazilian Journal of Food Technology, 27*, e2023061.
Ferreira, R. R., et al. (2021). Essential oils loaded-chitosan nanocapsules incorporation in biodegradable starch films: A strategy to improve fruits shelf life. *International Journal of Biological Macromolecules, 188*, 628–638.
Fonseca, L. M., et al. (2021). Suitability of starch/carvacrol nanofibers as biopreservatives for minimizing the fungal spoilage of bread. *Carbohydrate Polymers, 252*, Article 117166.
Freitas, P. A., González-Martínez, C., & Chiralt, A. (2023). Using rice straw fractions to develop reinforced, active PLA-starch bilayers for meat preservation. *Food Chemistry, 405*, Article 134990.
Ghoshal, G., & Singh, J. (2024). Study of coating effectiveness of grape fruit seed extract incorporated chitosan/cornstarch based active packaging film on grapes. *Food Chemistry Advances, 4*, Article 100651.
Gong, D., et al. (2024). Carbon dot/g-C3N4-mediated self-activated antimicrobial nanocomposite films for active packaging applications. *Food Chemistry, 438*, Article 137939.
Gui, H., et al. (2024). Sustained release, antimicrobial, and antioxidant properties of modified porous starch-based biodegradable polylactic acid/polybutylene adipate-co-terephthalate/thermoplastic starch active packaging film. *International Journal of Biological Macromolecules, 267*, Article 131657.
Guo, Z., et al. (2021). Changes in chilled beef packaged in starch film containing sea buckthorn pomace extract and quality changes in the film during super-chilled storage. *Meat Science, 182*, Article 108620.

Hamann, D., et al. (2024). Starch-based and green tea extract films: Characterisation and evaluation of the shelf life of interleaved sliced calabrese sausage. *International Journal of Food Science and Technology, 59*(11), 8899–8907.
Hernández-García, E., Vargas, M., & Chiralt, A. (2022). Starch-polyester bilayer films with phenolic acids for pork meat preservation. *Food Chemistry, 385*, Article 132650.
Jiang, H., et al. (2024). Multifunctional food packaging based on wampee seed starch and red pear peel extracts for simultaneous freshness preservation and visual freshness detection. *Food Packaging and Shelf Life, 43*, Article 101293.
Kang, Y., et al. (2024). Novel copper-based metal–organic skeleton smart tags that respond to ammonia for real-time visual freshness monitoring of shrimp. *Chemical Engineering Journal, 495*, Article 153388.
Kang, Y., et al. (2025). A novel dual-signal response Zn-MOF intelligent active label prepared from polyvinyl alcohol/starch for real-time visual monitoring shrimp freshness. *International Journal of Biological Macromolecules*, Article 143538.
Katekhong, W., et al. (2022). Thermoplastic starch blown films functionalized by plasticized nitrite blended with PBAT for superior oxygen barrier and active biodegradable meat packaging. *Food Chemistry, 374*, Article 131709.
Khezerlou, A., et al. (2023). Combining non-thermal processing techniques with edible coating materials: An innovative approach to food preservation. *Coatings, 13*(5), 830.
Khuntia, A., et al. (2022). Release behavior of vitamin C nanoliposomes from starch–vitamin C active packaging films. *Journal of Food Process Engineering, 45*(9), Article e14075.
Kontominas, M. G., et al. (2021). Recent developments in seafood packaging technologies. *Foods, 10*(5), 940.
Lemos, G. S., et al. (2025). Active food packages for cake conservation: Antifungal potential of bean starch biodegradable films with orange peel essential oil. *International Journal of Biological Macromolecules, 310*, Article 143441.
Li, N., et al. (2022). Development of pH-indicative and antimicrobial films based on polyvinyl alcohol/starch incorporated with ethyl lauroyl arginate and mulberry anthocyanin for active packaging. *Coatings, 12*(10), 1392.
Liu, D., et al. (2022). Recent advances and applications in starch for intelligent active food packaging: A review. *Foods, 11*(18), 2879.
Liu, J., et al. (2023). Starch-based film functionalized with Zanthoxylum armatum essential oil improved the shelf life of beef sauce. *LWT, 183*, Article 114930.
Lopes, J., et al. (2024). Heat-sealable bioplastic films of blended locust bean and potato byproducts for active packaging of fatty foods: Cheese and oat cookies as case studies. *Food Hydrocolloids, 147*, Article 109322.
Luo, S., et al. (2022). Preparation of corn starch/rock bean protein edible film loaded with d-limonene particles and their application in glutinous rice cake preservation. *International Journal of Biological Macromolecules, 206*, 313–324.
Martínez, S., et al. (2024). Banana drying: A review on methods and advances. *Food Reviews International, 40*(8), 2188–2226.
Martins, P. C., Bagatini, D. C., & Martins, V. G. (2021). Oregano essential oil addition in rice starch films and its effects on the chilled fish storage. *Journal of Food Science and Technology, 58*(4), 1562–1573.
Marzano-Barreda, L. A., Yamashita, F., & Bilck, A. P. (2021). Effect of biodegradable active packaging with zeolites on fresh broccoli florets. *Journal of Food Science and Technology, 58*(1), 197–204.
Marzlan, A. A., et al. (2022). Incorporating torch ginger (Etlingera elatior Jack) inflorescence essential oil onto starch-based edible film towards sustainable active packaging for chicken meat. *Industrial Crops and Products, 184*, Article 115058.

Mathew, D., Sudheep, N., & Thomas, B. (2025). Cinnamon essential oil-enhanced arrowroot starch-based biodegradable films for sustainable packaging and microbial preservation of chicken meat. *Results in Surfaces and Interfaces*, Article 100617.
Mirmoeini, S. S., et al. (2023a). Essential oil-loaded starch/cellulose aerogel: Preparation, characterization and application in cheese packaging. *International Journal of Biological Macromolecules, 244*, Article 125356.
Mirmoeini, S. S., et al. (2023b). Essential oil-loaded starch/cellulose aerogel: Preparation, characterization and application in cheese packaging. *International Journal of Biological Macromolecules, 244*, Article 125356.
Mirza Alizadeh, A., et al. (2021). Trends and applications of intelligent packaging in dairy products: A review. *Critical Reviews in Food Science and Nutrition, 62*(2), 383–397.
Mohapatra, D., et al. (2011). Post-harvest processing of banana: Opportunities and challenges. *Food and Bioprocess Technology, 4*(3), 327–339.
Mukkadan, M. J., Preetha, R., & Sreejit, V. (2024). Development of active packaging film reinforced with nano cellulose extracted from Palmyra palm (Borassus flabellifer) residues for prolonging the shelf life of chicken meat. *International Journal of Food Science and Technology, 59*(3), 2033–2041.
Mukwevho, P. L., Kaseke, T., & Fawole, O. A. (2025). Innovations in biodegradable packaging and edible coating of shelled temperate nuts. *Food and Bioprocess Technology*, 1–32.
Mustafa, P., et al. (2021). Improving functional properties of PVA/starch-based films as active and intelligent food packaging by incorporating propolis and anthocyanin. *Polymers and Polymer Composites, 29*(9), 1472–1484.
Nadon, S., et al. (2025a). Application of argon cold plasma on active polybutylene succinate and thermoplastic cassava starch film with Makwaen essential oil to improve shelf-life of pork sausage. *Food Packaging and Shelf Life, 49*, Article 101470.
Nadon, S., et al. (2025b). Application of argon cold plasma on active polybutylene succinate and thermoplastic cassava starch film with Makwaen essential oil to improve shelf-life of pork sausage. *Food Packaging and Shelf Life, 49*, Article 101470.
Nair, S., et al. (2025). Improved mechanical characteristics and shelf life of sustainable biodegradable food packaging films via integration of rice straw nanocellulose, Moringa oleifera extract and gelatin. *International Journal of Biological Macromolecules, 321*, Article 145977.
Nandi, S., & Guha, P. (2024). Development, characterization and application of starch-based film containing polyphenols of piper betle L. waste in chicken meat storage. *Food Chemistry, 431*, Article 137103.
Noshirvani, N., & Abolghasemi Fakhri, L. (2025). Advances in extending the microbial shelf-life of bread and bakery products using different technologies: A review. *Food Reviews International, 41*(1), 87–112.
Phothisarattana, D., et al. (2021). Biodegradable poly (butylene adipate-co-terephthalate) and thermoplastic starch-blended TiO2 nanocomposite blown films as functional active packaging of fresh fruit. *Polymers, 13*(23), 4192.
Promhuad, K., Ebel, L., & Harnkarnsujarit, N. (2025). Thermoplastic starch/poly(butylene adipate-co-terephthalate) blown film with maltol and ethyl maltol preserving cake quality: Morphology and antimicrobial function. *Food Chemistry, 464*, Article 141646.
Putri, T. R., et al. (2023). Effect of different starch on the characteristics of edible film as functional packaging in fresh meat or meat products: A review. *Materials Today: Proceedings, 87*, 192–199.
Qin, Y., et al. (2021). Smart packaging films based on starch/polyvinyl alcohol and Lycium ruthenicum anthocyanins-loaded nano-complexes: Functionality, stability and application. *Food Hydrocolloids, 119*, Article 106850.
Rejeesh, C., & Anto, T. (2023). Packaging of milk and dairy products: Approaches to sustainable packaging. *Materials Today: Proceedings, 72*, 2946–2951.

Remya, S., et al. (2022). Influence of corn starch based bio-active edible coating containing fumaric acid on the lipid quality and microbial shelf life of silver pomfret fish steaks stored at 4 C. *Journal of Food Science and Technology, 59*(9), 3387–3398.

Sadeghi, S. M., et al. (2025). Orange peel–derived carbon dots and H. sabdariffa L. anthocyanin-integrated gelatin and pectin pH-responsive films for detection of meat freshness. *International Journal of Biological Macromolecules*, Article 146915.

Sahu, M., & Shrivas, A. (2024). Exploring microbial spoilage in fruits: Causes, detection methods, and preventive measures. *International Journal of Innovative Research in Technology and Science, 12*(2), 470–477.

Sanches, M. A. R., et al. (2021). Active packaging with starch, red cabbage extract and sweet whey: Characterization and application in meat. *Lwt, 135*, Article 110275.

Schifferstein, H. N. (2024). Changes in appearance during the spoilage process of fruits and vegetables: Implications for consumer use and disposal. *Cleaner and Responsible Consumption, 12*, Article 100184.

Sequino, G., et al. (2022). Specific microbial communities are selected in minimally-processed fruit and vegetables according to the type of product. *Foods, 11*(14), 2164.

Sganzerla, W. G., et al. (2020). Bioactive food packaging based on starch, citric pectin and functionalized with Acca sellowiana waste by-product: Characterization and application in the postharvest conservation of apple. *International Journal of Biological Macromolecules, 147*, 295–303.

Sharma, V., et al. (2022). Agro-industrial food waste as a low-cost substrate for sustainable production of industrial enzymes: A critical review. *Catalysts, 12*(11), 1373.

Siddiqui, S. A., et al. (2024). Applying innovative technological interventions in the preservation and packaging of fresh seafood products to minimize spoilage-A systematic review and meta-analysis. *Heliyon, 10*(8).

Sivapragasam, N., Maqsood, S., & Rupasinghe, H. V. (2024). Berry bioactive compounds immobilized in starch matrix for active and intelligent packaging: A review. *Future Foods, 10*, Article 100397.

Song, D.-H., et al. (2021). Edible films on meat and meat products. *Coatings, 11*(11), 1344.

Sreekanth, K., et al. (2024). Microbial load reduction in stored raw beef meat using chitosan/starch-based active packaging films incorporated with cellulose nanofibers and cinnamon essential oil. *Meat Science, 216*, Article 109552.

Srivastava, V., Singh, S., & Das, D. (2024). Development and characterization of peppermint essential oil/rice husk fibre/corn starch active biocomposite film and its performance on bread preservation. *Industrial Crops and Products, 208*, Article 117765.

Tian, Z., et al. (2021). An active packaging film based on esterified starch with Tartary buckwheat bran extract and chitosan and its application for mutton preservation. *Journal of Food Processing and Preservation, 45*(12), Article e16000.

Todhanakasem, T., et al. (2022). Active thermoplastic starch film with watermelon rind extract for future biodegradable food packaging. *Polymers, 14*(16), 3232.

Tongdeesoontorn, W., et al. (2021). Antioxidant films from cassava starch/gelatin biocomposite fortified with quercetin and TBHQ and their applications in food models. *Polymers, 13*(7), 1117.

Ukwatta, R. H., et al. (2024). The characterisation of cornstarch-based chlorophyllin composite film for preservation of shrimp under photodynamic irradiation. *International Journal of Food Science and Technology, 59*(6), 3950–3966.

Uyarcan, M., & Güngör, S. C. (2024). Improving functional properties of starch-based films by ultraviolet (UV-C) technology: Characterization and application on minced meat packaging. *International Journal of Biological Macromolecules, 282*, Article 137085.

Varghese, S. A., et al. (2023). Antimicrobial active packaging based on PVA/Starch films incorporating basil leaf extracts. *Materials Today: Proceedings, 72*, 3056–3062.

Velasco, V., et al. (2022). Starch-based composite foam for chicken meat packaging. *Journal of Food Science and Technology, 59*(12), 4594–4602.
Wang, J., Chen, C., & Xie, J. (2022). Loading oregano essential oil into microporous starch to develop starch/polyvinyl alcohol slow-release film towards sustainable active packaging for sea bass (Lateolabrax japonicus). *Industrial Crops and Products, 188*, Article 115679.
Wang, D., et al. (2024). Radish residue carbon dots-based novel starch/chitosan film with high antioxidant, biocompatibility, and antibacterial activities for salmon fillets' active packaging. *International Journal of Biological Macromolecules, 273*, Article 133107.
Xi, W., et al. (2024). Pre-gelatinized high-amylose starch enables easy preparation of flexible and antimicrobial composite films for fresh fruit preservation. *International Journal of Biological Macromolecules, 254*, Article 127938.
Yang, F., et al. (2022). Effects of quercetin and organically modified montmorillonite on the properties of poly (butylene adipate-co-terephthalate)/thermoplastic starch active packaging films. *ACS Omega, 8*(1), 663–672.
Yuan, L., et al. (2021). Effect of potato starch-based antibacterial composite films with thyme oil microemulsion or microcapsule on shelf life of chilled meat. *Lwt, 139*, Article 110462.
Yudhistira, B., et al. (2024). Progress in the application of emerging technology for the improvement of starch-based active packaging properties: A review. *ACS Food Science & Technology, 4*(9), 1997–2012.
Zaiden, I., & Al-Hadidy, Y. (2025). The effectiveness of edible film from corn starch supported with clove oil (Dianthus carphyllus L.) on the microbial, physiochemical and sensory evaluation properties of soft cheese. In *IOP conference series: Earth and environmental science*. IOP Publishing.
Zhao, S., et al. (2022). Development of chitosan/tannic acid/corn starch multifunctional bilayer smart films as pH-responsive actuators and for fruit preservation. *International Journal of Biological Macromolecules, 205*, 419–429.
Zheng, Q., et al. (2020). Fabrication and application of starch-based aerogel: Technical strategies. *Trends in Food Science & Technology, 99*, 608–620.
Zhou, J., et al. (2025). Nuts: An overview on oxidation, affecting factors, inhibiting measures, and prospects. *Journal of Food Composition and Analysis, 139*, Article 107134.
Zong, L., et al. (2022). Effects of starch/polyvinyl alcohol active film containing cinnamaldehyde on the quality of large yellow croaker (Pseudosciaena crocea) proteins during frozen storage. *Food Chemistry, 389*, Article 133065.

CHAPTER TWELVE

Application of starch-based packaging in monitoring quality of fresh and processed food

Neha Chatterjee[1, 2], **Oshin Sahni**[3]
[1]Department of Food Science, University of Wisconsin–Madison, Madison, WI, United States
[2]Glanbia, Twin Falls, ID, United States
[3]Department of Food Science, University of Tennessee, Knoxville, TN, United States

12.1 Introduction

The global food supply chain faces constant challenges related to food safety, quality assurance, and waste reduction. Traditional food packaging has historically served the primary functions of containment and protection, acting as a passive barrier against external environmental factors such as moisture, light, and microbial contamination (Biji et al., 2015). However, in response to increasing consumer demand for safer, higher-quality foods with extended shelf life and a growing imperative to minimize food loss, the packaging industry has evolved beyond this passive role (Boukid, 2022). This evolution has given rise to advanced packaging concepts, broadly categorized as smart packaging, which are designed to perform functions beyond simple containment, thereby enhancing food preservation and providing critical information about product status (Biji et al., 2015). This chapter explores the application of starch-based polymers in the development of intelligent packaging systems for real-time monitoring of food quality.

12.1.1 Overview of smart and intelligent packaging

The term smart packaging generally refers to two related but different concepts: active packaging and intelligent packaging. Active packaging is designed to interact directly with the food or the environment around it. Here, certain compounds or functional materials are included in the packaging so they can absorb or release substances in order to maintain product quality or extend shelf life. For example, oxygen scavengers slow down oxidation, moisture absorbers help manage water activity, and antimicrobial

Starches in Smart Packaging
ISBN: 978-0-443-45396-0
https://doi.org/10.1016/B978-0-443-45396-0.00018-9

films can slowly release preservative agents to limit microbial growth. In this case, the packaging is doing more than just acting as a barrier — it's actively helping to protect the food.

On the other hand, intelligent packaging focuses on monitoring rather than interacting. These systems are used to sense changes in the condition of the food or the internal atmosphere of the package and then communicate that information. This could be through a visible color change, a label that records temperature exposure, or even sensor data connected to a digital platform. The key idea is that intelligent packaging provides real-time insights into freshness, safety, or storage conditions, making it easier for manufacturers, retailers, and consumers to make informed decisions (Biji et al., 2015). Unlike active systems, intelligent packaging does not directly interfere with the food and its primary function is informational (Dodero et al., 2021). This is achieved through the integration of indicators or sensors that can detect and signal changes in various quality-related parameters. These systems provide real-time data on factors such as package integrity, temperature history, headspace gas composition, pH changes, and the presence of microbial metabolites. The information is typically conveyed through a visually perceptible change, such as a shift in color, or via an electronic signal (Khodaei et al., 2022; Mkhari et al., 2025).

Active packaging acts upon the food system to preserve it, raising direct considerations regarding the safety and migration of its components into the food. Intelligent packaging reports on the state of the food system, with its primary challenges revolving around the accuracy, reliability, and interpretability of the information it provides. The development of both fields has seen exponential growth in research interest, driven by the need to deliver safer and higher-quality food products with improved shelf-life management (Boukid, 2022).

12.1.2 The need for real-time food quality monitoring

The conventional approach to shelf-life management relies on static date labels, such as "best-before" or "use-by" dates. Temperature is widely recognized as the most significant environmental factor affecting the rate of chemical, physical, and microbial degradation in food products (Khodaei et al., 2022). Deviations from the recommended temperature profile during transportation and storage which is a common occurrence in complex supply chains can drastically accelerate spoilage, rendering a date label inaccurate and potentially misleading (Sejani & Dhamsaniya, 2021).

This discrepancy between predicted and actual shelf life contributes to two major problems: food waste and foodborne illness. Consumers and retailers may discard perfectly safe food based on a conservative date label, while food that has been temperature-abused may be consumed before visible signs of spoilage appear, posing a health risk (Mkhari et al., 2025). Furthermore, traditional laboratory methods for quality assessment, such as microbial plating or chemical analysis, are destructive, time-consuming, and unsuitable for providing real-time information on individual packages throughout the supply chain (Zeng et al., 2022).

Intelligent packaging helps address the shortcomings of traditional shelf-life dates by basing quality on the product's actual condition rather than on a preset estimate. For example, Time–Temperature Indicators (TTIs) show how much heat a product has been exposed to over its lifetime, which often reflects its remaining shelf life better than a printed date. Freshness indicators that detect compounds such as ammonia or carbon dioxide provide a straightforward signal of spoilage as microbial activity increases. With this kind of information, manufacturers, retailers, and consumers can make more informed choices—leading to better food safety, stronger traceability, and less unnecessary waste (Biji et al., 2015; Khodaei et al., 2022; Mkhari et al., 2025).

12.1.3 Starch as a sustainable polymer for food packaging

The search for sustainable alternatives to petroleum-derived plastics has become a global priority, driven by concerns over plastic pollution, fossil fuel depletion, and climate change (Raghunathan et al., 2025). In this context, starch has emerged as one of the most promising biopolymers for developing environmentally friendly food packaging materials (Liu et al., 2022). As the second most abundant natural polysaccharide after cellulose, starch offers numerous advantages: it is derived from renewable agricultural resources, fully biodegradable, non-toxic, globally available, and has a low cost. It can be extracted from a wide array of sources, including major crops like corn, potato, and wheat, as well as agricultural by-products, contributing to a circular economy (Arruda et al., 2025).

From a material science perspective, starch possesses intrinsic properties that make it highly suitable as a matrix for food packaging films. Starch-based films are typically clear, colorless, and tasteless, ensuring that they do not interfere with the sensory properties of the packaged food (Karnwal et al., 2025). Furthermore, the dense, hydrogen-bonded network formed by starch polymer chains provides a moderate to good barrier against gases such as oxygen and carbon dioxide, which is beneficial for preventing oxidative degradation and inhibiting the growth of aerobic microorganisms (Arruda et al., 2025).

However, the same chemical features that give starch its useful properties also create some major limitations. Because the polymer contains so many hydroxyl groups, it tends to absorb water easily, which makes it highly hydrophilic. As a result, starch-based materials usually show poor moisture resistance and allow water vapor to pass through more readily than desired (Liu et al., 2022). This moisture sensitivity can compromise the mechanical integrity of the film in humid environments. Additionally, the strong intermolecular hydrogen bonding makes native starch films brittle and lacking in the flexibility required for many packaging applications (Arruda et al., 2025).

12.2 Fundamentals of starch-based films

The potential of starch for intelligent packaging depends on its material properties, which are shaped by its chemical structure, botanical source, and processing method. This section explores how starch type, film formation, additives, and biodegradability influence the performance of starch-based films.

12.2.1 Sources and properties of starch

Starch is a biopolymer synthesized by plants for energy storage and is composed of two primary glucose polymers: amylose and amylopectin (Kusuma et al., 2024). Amylose is a predominantly linear macromolecule consisting of α-1,4-linked D-glucose units, which allows it to adopt a helical conformation and align into ordered, crystalline structures (Karnwal et al., 2025). Amylopectin is a much larger, highly branched polymer containing both α-1,4 and **α**-1,6 glycosidic linkages, which disrupt crystalline packing and create more amorphous regions. The ratio of amylose to amylopectin is the single most important intrinsic variable governing the properties of starch and the films derived from it (Arruda et al., 2025). This ratio varies significantly depending on the botanical source of the starch (Borges et al., 2015). Starches with a higher amylose content tend to form stronger intermolecular hydrogen bonds upon film formation, leading to films that are stronger (higher tensile strength), more crystalline, and less permeable to gases and water vapor (Molavi et al., 2018). However, this increased order and rigidity also result in films that are more brittle and less flexible (Kusuma et al., 2024). Conversely, starches with a higher amylopectin content produce films that are more flexible and transparent but have weaker mechanical and barrier properties (Figueroa-Lopez et al., 2024).

This structure-property relationship means that the selection of the starch source is a foundational step in designing a film for a specific packaging application. For example, wheat starch, which typically has a high amylose content (around 25%), forms strong, less permeable films, making it an excellent candidate for applications where barrier performance is critical (Raghunathan et al., 2025). Potato starch is also known for its high molecular weight and relatively high amylose content, yielding films with good barrier properties but potentially lower mechanical strength (Figueroa-Lopez et al., 2024). Corn starch is commonly used due to its good film-forming ability, attributed to a favorable amylose content (Dodero et al., 2021). In contrast, starches like cassava (tapioca) have a lower amylose content (around 17%), which results in films with higher water solubility and flexibility, making them suitable for different types of applications where these properties are desired (Nath, 2024). This shows that there is no single "best" starch; the ideal choice depends on the specific requirements of the packaging application. A comparative summary of properties from different starch sources is provided in Table 12.1.

12.2.2 Film formation and characteristics

Starch naturally comes as semi-crystalline granules that don't dissolve in cold water. To turn it into a continuous film, the granules need to be broken down through a process called gelatinization. This is done by heating a starch–water mixture (usually 3%–12% w/v) above its gelatinization temperature, which depends on the starch source—around 70 °C for cassava and 90 °C for yam. As the granules heat up, water penetrates and breaks the hydrogen bonds in the crystalline regions, making the granules swell permanently. At the same time, linear amylose chains leak out and the branched amylopectin molecules start to unwind (Rodrigues et al., 2021). The result is a viscous, solubilized solution or paste where the polymer chains are dispersed. When this solution is cast and dried, the solvent evaporates, and the polymer chains re-associate through hydrogen bonding to form a solid, continuous matrix, i.e., the film (Bangar et al., 2021). An important process that occurs as starch films cool and age is retrogradation. This happens when the dispersed amylose and amylopectin chains,

Table 12.1 Comparative properties of starches from various sources for film production.

Starch Source	Typical amylose content (%)	Key granule characteristics	Resulting film properties	References
Wheat	25–28	Small, lenticular granules	High tensile strength, low water vapor permeability, crystalline, less flexible. Excellent barrier performance.	Raghunathan et al. (2025)
Corn	25–28	Polygonal granules	Good film-forming ability, good mechanical properties, moderate barrier performance.	Dodero et al. (2021)
Potato	20–23	Large, oval granules	Good barrier to oxygen and water vapor, high transparency, but mechanically weaker and more moisture sensitive.	Figueroa-Lopez et al. (2024)
Cassava (tapioca)	16–20	Round, truncated granules	High water solubility, good flexibility (tough and rigid), higher elongation. Weaker barrier properties.	Nath (2024)
Yam	18–25	Variable shape (oval, polygonal)	Good flexibility, high elongation at rupture, moderate barrier properties.	Rodrigues et al. (2021)
Rice	15–25	Small, polygonal granules	Flexible, low water solubility, suitable for food coatings.	Borges et al. (2015)

especially the linear amylose realign and recrystallize into a more ordered structure. Some crystallinity is beneficial, improving film strength and barrier properties. However, too much retrogradation can make the film brittle, cloudy, and prone to syneresis (water release), which can reduce its shelf life and overall performance (Rodrigues et al., 2021).

The intrinsic characteristics of films made from native starch are a direct consequence of this chemistry. The dense network of hydrogen bonds formed between the abundant hydroxyl groups on the glucose units creates a structure that is a good barrier to nonpolar molecules like oxygen and carbon dioxide (Arruda et al., 2025). However, these same hydroxyl groups have a strong affinity for water, making the films highly hydrophilic. This results in poor barrier properties against water vapor, as well as a tendency to absorb moisture from the environment, which plasticizes the film and reduces its mechanical strength in wet conditions (Liu et al., 2022). Furthermore, the strong intermolecular forces and semi-crystalline nature of the starch matrix make unmodified films rigid and brittle, with low flexibility and elongation at break (Arruda et al., 2025).

12.2.3 Plasticizers and additives in starch films

Native starch is naturally brittle and difficult to process, so adding plasticizers is essential. Plasticizers are small, low-molecular-weight molecules that insert themselves between the starch chains. By doing so, they weaken the strong hydrogen bonds holding the chains together and increase the space between them, making the material more flexible and easier to work with (Bangar et al., 2021). This reduces the film's internal cohesion, increases the mobility of the polymer chains, and lowers the glass transition temperature, resulting in a material that is significantly more flexible and less brittle. The most commonly used plasticizers for starch films are polyols like glycerol and sorbitol, which are non-toxic and safe for food contact. Both the type and amount of plasticizer strongly influence the final film properties. For example, glycerol's small size and high affinity for water make it an effective plasticizer, easily integrating into the starch matrix and greatly increasing flexibility and elongation at break (Sirbu et al., 2024). However, its hygroscopic nature can also increase the film's moisture sensitivity and water vapor permeability, and it may reduce tensile strength and thermal stability (Wang et al., 2022). In contrast, sorbitol, a larger molecule, is a less efficient plasticizer, resulting in films that are more rigid and less flexible (Li & Huneault, 2011). Sorbitol, on the other hand, interacts more strongly with starch chains and attracts less water than glycerol, which can improve a film's thermal stability and water resistance. The differing effects of glycerol and sorbitol illustrate a key trade-off in choosing plasticizers. This has prompted research into combining plasticizers to create a synergistic effect, producing films with a balanced mix of flexibility, strength, and stability that a single plasticizer alone cannot achieve (Sirbu et al., 2024).

Beyond plasticizers, a variety of other additives are incorporated to further enhance the performance of starch films. Reinforcing agents are used to improve mechanical strength and barrier properties. These include natural polymers like chitosan and cellulose (in the form of nanofibers or nanocrystals), which can form strong hydrogen bonds with the starch matrix, creating a denser and more robust network structure. Inorganic nanoparticles, such as montmorillonite nanoclay and zinc oxide (ZnO), can also be dispersed within the starch matrix. These nanoparticles create a more tortuous path for diffusing gas and water molecules, significantly improving barrier properties, and can also act as reinforcing fillers to enhance mechanical strength (Liu et al., 2022).

12.2.4 Biodegradability and environmental impact

A primary driver for the development of starch-based packaging is its superior environmental profile, particularly its biodegradability. Unlike conventional plastics that persist in the environment for centuries, starch-based materials are designed to be completely degraded by microorganisms. The process of biodegradation involves microorganisms such as bacteria, fungi, and algae secreting enzymes (like amylases) that break down the complex starch polymer into smaller molecules, eventually mineralizing it into carbon dioxide (CO_2), water, and biomass. The rate and extent of biodegradation are highly dependent on the environmental conditions. Key influencing factors include the type and population of microorganisms present, temperature, humidity, and oxygen availability. Studies have shown that degradation is significantly faster in biologically active environments like compost soil compared to standard planting soil, due to a higher concentration of microbes. Quantitative studies have demonstrated rapid weight loss of starch films under various conditions. For example, one study reported that starch/chitosan nanoparticle films lost a significant portion of their weight within 8 days of soil burial (Othman et al., 2023; Sander, 2019). Another investigation found that films made from taro peel starch completely degraded within 5 days in both simulated river water and

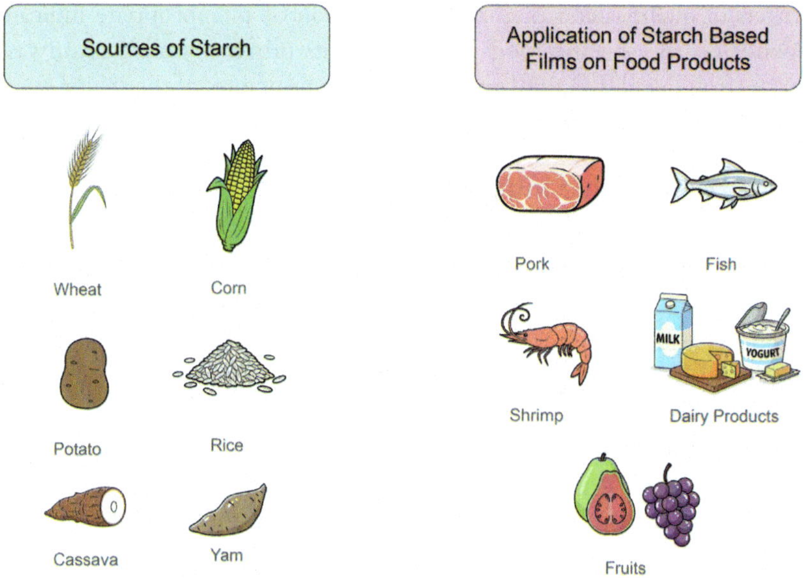

Figure 12.1 Sources and applications of starch-based films.

composting soil environments (Bidari et al., 2023). In laboratory tests using specific amylolytic bacteria, films made from potato waste starch showed degradation levels as high as 78.86% within just 24 h (Karki et al., 2020).

This capacity for rapid and complete biodegradation distinguishes modern starch-based materials from early-generation "degradable" plastics. Those earlier materials often consisted of a synthetic polymer blended with starch; upon disposal, only the starch component would degrade, leaving behind a matrix of microplastic fragments that still posed an environmental hazard (Liu et al., 2022). In contrast, fully bio-based starch films offer a true "end-of-life" solution, returning their constituent elements to the natural carbon cycle and helping to mitigate the accumulation of plastic waste (Othman et al., 2023). Depicts various sources and applications of starch based films (Fig. 12.1).

12.3 Principles of food spoilage indicators

Intelligent packaging derives its functionality from the ability to detect and signal changes associated with food spoilage. This is achieved by incorporating indicators that respond to specific chemical or physical markers of quality deterioration. The selection of an appropriate indicator depends on the specific food product and its primary spoilage pathways. This section details the fundamental working principles of the major classes of indicators relevant to starch-based packaging systems.

12.3.1 pH-sensitive indicators

Changes in pH are a common and reliable marker for spoilage in a wide variety of food products. For instance, the microbial decomposition of proteins in meat and fish generates volatile alkaline compounds like ammonia, leading to an increase in pH. Conversely, the fermentation of sugars in dairy products or fruit juices produces organic acids (e.g., lactic acid), causing a decrease in pH. pH-sensitive indicators, also known as halochromic indicators, are chemical compounds that change color in response to variations in pH, providing a direct visual signal of these spoilage processes (Che Hamzah et al., 2022; Neodini Remedio & Parada Quinayá, 2024).

Among the various options, anthocyanins have emerged as ideal candidates for use in intelligent food packaging. Anthocyanins are a class of natural, water-soluble pigments responsible for the red, purple, and blue colors found in many fruits, vegetables, and flowers. They are considered safe for consumption and can be extracted from many different edible sources, such as red cabbage, black rice, butterfly pea flowers, and various berries. The pH-sensing mechanism of anthocyanins is based on structural transformations of the core flavylium cation molecule. The specific structure, and thus the color, is dependent on the proton concentration of the surrounding medium. In strongly acidic conditions ($pH < 3$), the flavylium cation form predominates, which is typically red or pink. As the pH increases toward neutral, the molecule undergoes hydration and deprotonation reactions to form the colorless carbinol pseudo-base and the purple or blue quinonoidal base structures. Under alkaline conditions i.e., $pH > 7$, these structures can further transform into the yellow chalcone form. This series of pH-dependent equilibria results in a distinct and often dramatic spectrum of colors across the pH range, making anthocyanins excellent visual indicators. When these pigments are immobilized within a starch film matrix, they can react with volatile metabolites released by the food product. For example, ammonia gas from spoiling fish will increase the pH in the package headspace, causing an anthocyanin-based film to shift from its initial color (e.g., purple) to blue or green, providing a clear and intuitive signal of spoilage (Liu, Zhang, et al., 2022; Zeng et al., 2022; Zhao et al., 2022). A significant advantage of anthocyanins is their multifunctionality and in addition to their role as pH indicators, they are also known to possess potent antioxidant and antimicrobial properties (Neodini Remedio & Parada Quinayá, 2024).

12.3.2 Gas-sensitive indicators

The composition of the headspace gas within a food package is a dynamic environment that provides crucial information about product quality and package integrity. Gas-sensitive indicators are designed to detect changes in the concentration of specific gases that serve as markers for spoilage or packaging failure (Mani et al., 2025).

The spoilage of protein-rich foods like meat, poultry, and fish is characterized by the enzymatic and microbial breakdown of amino acids, which produces a mixture of volatile compounds known as total volatile basic nitrogen (TVBN) (Bhowmik et al., 2024). Ammonia (NH_3) is a primary and significant component of TVBN and serves as a key indicator of protein decomposition (Heo & Lim, 2024). As a volatile base, ammonia released from the food product increases the pH of the package headspace. Therefore, ammonia indicators typically operate on a halochromic (pH-sensitive) principle. Natural dyes such as anthocyanins and curcumin, when embedded in a polymer film, will undergo a distinct color change upon exposure to ammonia gas, signaling the onset of spoilage. For example, a curcumin-based indicator changes from yellow to reddish-brown in the presence of ammonia, while anthocyanin-based systems can shift from red/purple to blue/green (Fadhila et al., 2024).

Carbon dioxide (CO_2) is another critical gas to monitor in food packaging. An elevated CO_2 concentration inside a package can indicate microbial spoilage, as CO_2 is a major byproduct of microbial respiration (Biji et al., 2015). Conversely, in modified atmosphere packaging (MAP), where high levels of CO_2 are often used to inhibit microbial growth, a decrease in CO_2 concentration can signal a package leak (Puligundla et al., 2012). Most colorimetric CO_2 indicators work by detecting the drop in pH that occurs when CO_2 dissolves in water to form carbonic acid (H_2CO_3) (Heo et al., 2024). A pH-sensitive dye is incorporated into a hydrophilic matrix; as CO_2 levels rise, the pH of the matrix decreases, triggering a color change (Zhang & Lim, 2017). Other systems utilize materials whose physical properties are pH-dependent. For example, indicators based on whey protein isolate (WPI) or chitosan can change from transparent to opaque as the pH drops near their isoelectric point, providing an irreversible signal of CO_2 exposure (Biji et al., 2015).

Oxygen (O_2) indicators are primarily used to verify the integrity of anaerobic packaging systems, such as MAP and vacuum packaging (Khodaei et al., 2022). The presence of oxygen inside such a package is a clear sign of a leak, which can lead to rapid oxidative degradation and the growth of aerobic spoilage microorganisms (Heo et al., 2024). The most common working principle for oxygen indicators is an irreversible redox reaction (Mills, 2005). These systems typically consist of a redox dye (e.g., methylene blue), a reducing agent (e.g., glucose or an iron-based compound), and an alkaline compound, all encapsulated within a polymer matrix. In the absence of oxygen, the reducing agent converts the dye to its reduced, colorless form. If oxygen enters the package, it oxidizes the dye back to its original colored state (e.g., blue for methylene blue), providing a clear, irreversible visual warning that the package integrity has been compromised (Heo & Lim, 2024).

12.3.3 Indicators for microbial contaminants

While pH and gas indicators are effective for signaling general spoilage, they do not typically identify the specific microorganisms responsible. For applications where the detection of specific foodborne pathogens such as *Salmonella*, *Listeria monocytogenes*, *Escherichia coli*, is critical, biosensors offer a highly specific and sensitive solution. Microbes such as *L. monocytogenes* can even grow at refrigeration temperatures making it a very dangerous foodborne pathogen and is found in foods such as deli meats and dairy products made with raw milk (Chatterjee, 2023). Most Salmonella infections originate from seven food categories—chicken, fruits, pork, seeded vegetables, other produce (e.g., nuts), beef, and turkey and more than 80% of E. coli O157 illnesses are associated with vegetable row crops (particularly leafy greens) and beef (Centers for Disease Control and Prevention [CDC], 2021). A biosensor is an analytical device that combines a biological recognition element with a physicochemical transducer. The biorecognition element is a biological component that selectively binds to a specific target analyte. Common biorecognition elements include antibodies (for immunosensors), which bind to specific antigens on a pathogen's surface; nucleic acid probes (DNA or RNA), which hybridize with complementary genetic sequences of a microorganism; aptamers, which are single-stranded nucleic acids that fold into specific structures to bind targets; and enzymes, which catalyze specific reactions. The transducer, in turn, converts the biochemical binding event at the biorecognition element into a measurable physical signal. Transduction mechanisms can be electrochemical (measuring changes in current, potential, or impedance), optical (measuring changes in light absorbance, fluorescence, or color), or piezoelectric (measuring changes in mass on a crystal surface) (Inês & Cosme, 2025). For packaging applications, these biosensor components can be immobilized onto a solid support, such as a starch-based film. When the packaged food becomes contaminated with the target pathogen, the pathogen or its specific toxins diffuse to the sensor surface and bind to the biorecognition elements. This binding event triggers the transducer to produce a signal—often a visible color change—indicating the presence of the contaminant. The development of polymer-based biosensors is a promising area of research, as biopolymers like starch can provide a biocompatible and stable matrix for immobilizing the sensitive biological components (Inês & Cosme, 2025; Wang et al., 2023).

12.3.4 Time-Temperature Indicators (TTIs)

Time-Temperature Indicators (TTIs) are simple and cost-effective devices that monitor the cumulative exposure of a product to temperature over time. Unlike other indicators that detect the byproducts of spoilage, TTIs monitor the primary environmental condition—temperature history—that governs the rate of nearly all deteriorative reactions in food (Khodaei et al., 2022). The response of a TTI is a distinct, irreversible visual change (e.g., color development or movement) whose rate is temperature-dependent, increasing at higher temperatures and decreasing at lower temperatures. For a TTI to be effective, its temperature-dependent kinetics, characterized by its activation energy (Ea), must closely match the kinetics of the quality loss reaction in the food product it is monitoring. TTIs can be classified based on their working principle, with enzymatic and microbiological systems being particularly relevant as they can closely mimic biological spoilage processes (Pavelková, 2013; Vaikousi et al., 2008).

Enzymatic TTIs are based on a temperature-dependent enzymatic reaction. A common design involves the hydrolysis of a lipid substrate by a lipase enzyme. This reaction releases fatty acids, causing a gradual decrease in pH. A pH indicator dye co-immobilized in the system then undergoes a progressive color change (e.g., from green to yellow or red) as the pH drops.

The rate of the enzymatic reaction, and thus the rate of color change, is directly controlled by temperature. By adjusting the concentrations of the enzyme and substrate, the TTI can be calibrated to match the shelf life of various food products (Talukder et al., 2018).

Microbiological TTIs are often considered the closest analog to actual microbial spoilage because they rely on the real growth and metabolism of a non-pathogenic microorganism. Typically, these indicators use a selected strain—commonly a lactic acid bacterium such as *Lactobacillus sakei*—suspended in a small volume of growth medium that also contains a pH-sensitive dye. As the microorganism grows, it produces acidic metabolites like lactic acid, which gradually lowers the pH of the medium and causes a visible color shift in the indicator. Since the growth rate of the microorganism follows the same time–temperature kinetics that drive spoilage in the food itself, microbiological TTIs are able to closely mirror actual quality deterioration. Additionally, the endpoint of the indicator can be finely controlled by adjusting the starting concentration of the microbial culture, allowing the TTI's response time to align with the expected shelf life of the specific product (Vaikousi et al., 2008).

The different types of indicators represent different levels of specificity. TTIs provide an indirect but powerful measure of the overall potential for quality loss based on environmental conditions. Gas and pH indicators offer a more direct, though still general, signal that spoilage reactions are actively occurring. Biosensors provide the highest level of specificity, confirming the presence of a particular hazardous agent. The selection of the most appropriate indicator technology is therefore a strategic decision based on the specific food product, its primary spoilage mechanisms, and the critical quality or safety attributes that need to be monitored. An overview of these indicator systems is provided in Table 12.2.

12.4 Fabrication of starch-based indicator labels and films

The successful translation of the principles of food spoilage indicators into functional packaging materials requires robust and scalable fabrication methods. This section details the practical aspects of creating starch-based intelligent films, from the incorporation of sensing components to the primary manufacturing techniques and the essential characterization methods used to validate their performance.

Table 12.2 Overview of food spoilage indicator systems for starch-based packaging.

Indicator type	Principle of operation	Target analyte(s)	Example sensing materials/reagents	Typical signal change	References
pH-sensitive (halochromic)	Structural change of dye molecule in response to H^+ concentration.	pH, volatile bases (e.g., ammonia), organic acids (e.g., lactic acid)	Anthocyanins (from red cabbage, black rice, etc.), curcumin, Alizarin	Color change across a spectrum (e.g., red → purple → green)	Zeng et al. (2022)
Gas-sensitive (oxygen)	Irreversible redox reaction.	Oxygen(O_2)	Methylene blue + reducing agent (e.g., glucose, iron)	Colorless to blue	Heo and Lim (2024)
Gas-sensitive (carbon dioxide)	pH decreases due to carbonic acid formation.	Carbon dioxide (CO_2)	pH indicator dyes (e.g., bromothymol blue), whey protein isolate, chitosan	Color change (e.g., blue → yellow) or transparency change (clear → opaque)	Biji et al. (2015)
Gas-sensitive (ammonia)	pH increases due to volatile base.	Ammonia (NH_3), TVBN	pH indicator dyes (anthocyanins, curcumin, polyaniline)	Color change (e.g., purple → green, yellow → red)	Fadhila et al. (2024)
Biosensor	Specific binding of target analyte to a biological recognition element.	Specific pathogens (*Salmonella*, *E. coli*), toxins, allergens	Antibodies, nucleic acid probes, aptamers, enzymes	Electrochemical signal or optical signal (e.g., color change, fluorescence)	Inês and Cosme (2025)
Time-temperature indicator (enzymatic)	Temperature-dependent rate of an enzymatic reaction causing a pH change.	Cumulative time-temperature exposure	Lipase + lipid substrate + pH indicator	Progressive color change (e.g., green → yellow → red)	Talukder et al. (2018)
Time-temperature indicator (microbiological)	Temperature-dependent rate of microbial growth and metabolism causing a pH change.	Cumulative time-temperature exposure	*Lactobacillus sakei* + growth medium + pH indicator	Progressive color change (e.g., red → yellow)	Vaikousi et al. (2008)

12.4.1 Incorporation of natural and synthetic dyes

The integration of indicator dyes into the starch matrix is a critical step in fabricating colorimetric intelligent films. This is typically achieved by adding the dye directly to the aqueous film-forming solution prior to the main processing step. For intelligent packaging, natural dyes are strongly preferred over synthetic ones due to their perceived safety, consumer acceptance, and often, their origin from edible sources, which simplifies regulatory approval for food contact applications. Anthocyanins, extracted from sources like butterfly pea flower, red cabbage, black raspberry, or roselle, are commonly used for this purpose. The extraction is often performed using an acid-alcohol mixture to stabilize the pigment. Once the dye is added to the starch suspension, it becomes physically entrapped and immobilized within the polymer network as the film dries and solidifies. The interactions between the dye molecules and the starch polymer chains, primarily through hydrogen bonding, help to stabilize the dye and ensure it is retained within the film matrix, minimizing leaching into the food product. The incorporation of these dyes is not always a passive process. The presence of dye molecules can influence the physicochemical properties of the final film. For example, studies have shown that the addition of anthocyanins can affect the film's mechanical properties, water vapor permeability, and thermal stability. In some cases, the dye can even act as a compatibilizer in polymer blends, such as between starch and polyvinyl alcohol (PVA), improving the homogeneity of the film matrix. Therefore, the formulation must be carefully optimized to ensure that the addition of the indicator enhances, or at least does not compromise, the essential barrier and mechanical functions of the packaging film (Che Hamzah et al., 2022; Mary et al., 2020; Zhai et al., 2017). Other natural pigments like melanin are also added in active packaging films for its antimicrobial, antioxidant and UV light barrier properties and is known to enhance the shelf life of various food products (Reji et al., 2026).

12.4.2 Solution casting and extrusion methods

Two primary methods are employed for the fabrication of starch-based films: solution casting and extrusion. The choice of method has significant implications for the film's properties, scalability, and commercial viability.

Solution casting is the most prevalent method for laboratory-scale research and development due to its simplicity and low cost. The process involves three main steps: solubilization, where starch is dispersed in a solvent with plasticizers and dye and heated to form a homogeneous solution; casting, where the solution is poured onto a smooth surface; and drying, where the solvent is evaporated over a period of 12–24 h to form the solid film. The main advantages of solution casting are its ease of implementation and the excellent optical quality of the resulting films. However, it is a slow, batch-wise process with high energy consumption, making it difficult and costly to scale up for mass production. Extrusion is another method which is the standard, continuous process used in commercial production and represents the most viable path for large-scale manufacturing. In this process, starch powder is mixed with plasticizers and additives and fed into an extruder, where intense heat and mechanical shear melt the polymer into a thermoplastic starch (TPS) melt, which is then forced through a die to form a continuous film. Extrusion offers high throughput, lower cost, and better process control, often resulting in enhanced mechanical properties compared to cast films (Bangar et al., 2021). However, the high temperatures and shear stresses can degrade thermally sensitive components like natural dyes, and the resulting films are often thicker and more opaque (Ezati & Rhim, 2021). This discrepancy creates a significant "lab-to-fab" gap, where a formulation successful in the lab may fail or yield different properties on an industrial extrusion line. Bridging this gap is a critical area of research (Calderón-Castro et al., 2018).

12.4.3 Characterization of intelligent films

Once fabricated, the intelligent films must undergo rigorous characterization to validate their performance and suitability for food packaging applications. This involves a suite of analytical techniques to assess their mechanical integrity, optical characteristics, and, most importantly, their sensing capabilities. The mechanical properties are very important for ensuring the packaging can withstand physical stress. These are typically evaluated using tensile testing to determine key parameters such as Tensile Strength (TS), which indicates the film's overall strength; Elongation at Break (EB), which measures its flexibility; and Young's Modulus, which represents its stiffness or rigidity (Kusuma et al., 2024).

Optical properties are important for consumer acceptance and are characterized to assess transparency, gloss, and color (Kusuma et al., 2024). Transparency and opacity are measured using a UV-Visible spectrophotometer, which also assesses the film's ability to block UV light (Menchaca-Rivera et al., 2019). The color of the film, both initially and after exposure to an analyte, is quantified using a colorimeter and expressed in the CIE L*a*b* color space, where L* represents lightness, a* the red-green axis, and b* the yellow-blue axis (Guggilla et al., 2016).

The sensing properties are the most critical characterization for an intelligent film. For pH-sensitive films, this involves exposing samples to buffer solutions of known pH and measuring the resulting color change (Mary et al., 2020). For gas-sensitive films, samples are exposed to controlled concentrations of the target gas in a sealed chamber (Ameri et al., 2024). The colorimetric response is quantified by calculating the Total Color Difference (ΔE) from the L*, a*, and b* values before and after exposure. A **ΔE** value greater than 3.0 is generally considered a color difference easily perceptible to the human eye, confirming the film's utility as a visual indicator (Guggilla et al., 2016).

12.5 Applications for fresh food products

The principles and fabrication methods described in the preceding sections culminate in the development of intelligent packaging systems tailored for specific food products. Fresh foods, with their active metabolism and high susceptibility to microbial spoilage, represent a primary application area where real-time quality monitoring can provide significant benefits in terms of safety and waste reduction. Starch-based intelligent packaging systems have attracted increasing interest for monitoring the quality of fresh and processed foods due to their biodegradability, film-forming ability, low cost, and compatibility with natural sensing agents. These systems function by responding to physicochemical changes occurring during food deterioration, including pH variation, accumulation of volatile nitrogenous compounds, moisture migration, and ethylene release. As a result, starch-based indicators have been widely investigated for real-time quality monitoring in meat, seafood, fruits, vegetables, and ready-to-eat food products. This section presents case studies on the application of starch-based intelligent films for monitoring the freshness of meat, seafood, and fruits and vegetables.

12.5.1 Monitoring freshness of meat and poultry

The spoilage of fresh meat and poultry is a complex process driven primarily by the growth of microorganisms that metabolize proteins and amino acids. This metabolic activity leads to the production of various volatile compounds, most notably alkaline substances such as ammonia and other amines, which collectively form total volatile basic nitrogen (TVBN) (Khodaei et al., 2022). The accumulation of these compounds causes a progressive increase in the pH of the meat surface and the surrounding headspace, a reliable indicator of spoilage (Liu et al., 2022).

Starch-based intelligent films functionalized with pH-sensitive dyes are exceptionally well-suited for monitoring this process. By incorporating natural anthocyanin extracts into a starch-based matrix, researchers have developed indicator labels that provide a direct, visual assessment of meat freshness. For instance, in one study, an intelligent label was fabricated by blending pea starch with κ-carragenan and incorporating anthocyanin-rich black raspberry extract. When applied to packaged pork stored at 4 °C, the label exhibited a distinct and gradual color change that correlated with the progression of spoilage, as measured by pH and TVB-N levels, providing a clear, non-destructive method for consumers to visually assess the freshness of the product (Jiang et al., 2024). Another investigation utilized a film made from agar, potato starch, and anthocyanins extracted from sweet potatoes. When used to package pork, the film's color shifted from red to green as the meat deteriorated and its pH increased, effectively signaling the transition from fresh to spoiled (Liu et al., 2022). Furthermore, a more advanced system was developed using a starch film containing both carbon dots (as a reinforcing agent) and anthocyanins from *Clitoria ternatea* (butterfly pea flower). This film demonstrated a clear visual color change from purple to green as packaged pork spoiled over time, providing a low-cost and intuitive indicator for consumers (Koshy et al., 2021). These examples demonstrate the practical utility of starch-based pH indicators. They translate a complex biochemical process (protein decomposition) into a simple, easy-to-interpret visual cue, empowering consumers and retailers to make safer and more informed decisions about product quality without needing to open the package (Jiang et al., 2024).

12.5.2 Quality assessment of fish and seafood

Fish and seafood are among the most perishable food products due to their high water activity, neutral pH, and the presence of endogenous enzymes that accelerate autolysis. Similar to meat, their spoilage is characterized by the rapid production of TVB-N, making them ideal candidates for monitoring with pH-sensitive intelligent packaging (Abelti et al., 2022). Several studies have successfully applied starch-based indicator films for this purpose. A smart film was developed using starch extracted from jackfruit seeds and incorporated with anthocyanins from black grapes. When used to cap containers holding tilapia fish and shrimp, the film changed color in response to the volatile amines produced during spoilage, and the indicator function was deemed positive, providing a convenient, non-destructive method to monitor freshness (Costa et al., 2020). Researchers have also developed starch/polyvinyl alcohol (SPVA) films containing roselle anthocyanins. When used to monitor the freshness of silver carp at refrigeration temperatures, the films showed visible color changes over time that correlated well with the increase in TVB-N levels, confirming their ability to act as real-time freshness indicators for fish (Zhai et al., 2017). Similarly, Shikonin, another natural pH-sensitive pigment, has been integrated into both starch-based and agar-based films for monitoring shrimp freshness. The films exhibited a color change from red to blue as the pH increased during spoilage, providing a clear visual signal of quality deterioration (Ezati & Rhim, 2021). The application of these intelligent films in seafood packaging offers a significant advantage over traditional sensory evaluation (smell, texture), which is subjective and can only be performed after the package is opened. These indicators provide an objective, continuous record of quality, enhancing safety and reducing waste in a highly perishable food category (Abelti et al., 2022).

12.5.3 Tracking ripeness and spoilage in fruits and vegetables

The monitoring needs for fresh fruits and vegetables are often different from those for protein-based foods. While microbial spoilage is a concern, a primary quality attribute for many fruits is ripeness, a complex developmental process regulated by plant hormones. For climacteric fruits—such as apples, bananas, tomatoes, and mangoes—the ripening process is initiated and coordinated by the gaseous hormone ethylene (C_2H_4) (M. Liu et al., 2015). Ethylene production triggers a cascade of changes, including softening, color development, and aroma production. Monitoring the concentration of ethylene in the package headspace can therefore provide a direct indication of the fruit's ripeness stage and help predict its remaining shelf life (Janssen et al., 2014).

Intelligent indicators for fruit ripeness can be designed to detect ethylene. One such concept involves a colorimetric sensor based on the reducing properties of ethylene gas. An indicator was developed using molybdenum compounds, where the ethylene reduces Mo(VI) to Mo(V), causing a distinct color change from white/light yellow to blue. The rate and intensity of this color change are proportional to the ethylene concentration, allowing for the visual tracking of the ripening process in apples (Lang & Hübert, 2011). In addition to ethylene, carbon dioxide (CO_2) is another key metabolite, as its production rate is directly linked to the fruit's respiration rate, which often peaks during ripening. An intelligent indicator label for monitoring mango ripeness was developed using a methylcellulose-based film (a polysaccharide derivative similar in principle to starch) containing pH-sensitive dyes. As the mango ripened, its respiration rate increased, elevating the CO_2 concentration inside the package. This dissolved CO_2 lowered the pH of the indicator film, causing a visible color change from blue (unripe), through green (half-ripe), to yellow (fully ripe). This color change correlated well with traditional ripeness indices like firmness and soluble solid content, providing a simple, naked-eye assessment of the fruit's maturity (Noiwan et al., 2022). A summary of these and other application case studies is presented in Table 12.3.

12.6 Applications for processed food products

While fresh foods present a major application area for intelligent packaging, processed foods also benefit significantly from real-time quality monitoring. For these products, quality assurance often centers on maintaining the integrity of the packaging system itself or tracking spoilage pathways that differ from those of raw agricultural commodities. Starch-based indicators can be adapted to meet these specific needs for products such as ready-to-eat meals and dairy.

12.6.1 Package integrity and leak detection

For a vast range of processed foods, particularly those preserved using Modified Atmosphere Packaging (MAP), maintaining package integrity is the most critical factor for ensuring shelf life and safety. MAP involves altering the gaseous environment inside a package—typically by reducing oxygen and increasing carbon dioxide and/or nitrogen—to inhibit microbial growth and slow down oxidative reactions (Berk, 2009). However, even a microscopic leak in the package seal or material can compromise this protective atmosphere, allowing external oxygen to enter and the protective gases to escape (Khodaei et al., 2022). This ingress of oxygen can fuel the growth of aerobic spoilage organisms and accelerate quality degradation, rendering the product unsafe long before its printed expiration date (Heo et al., 2024).

Table 12.3 Application case studies of starch-based and related intelligent packaging systems.

Food product	Polymer matrix	Indicator agent	Target analyte	Observed color change	Correlation with spoilage metric	References
Pork	Pea Starch/ κ-Carrageenan	Black raspberry anthocyanins	pH/TVB-N	Not specified, but distinct color change observed	Correlated with pH and TVB-N increase	Jiang et al. (2024)
Pork	Starch/Carbon dots	Clitoria ternatea anthocyanins	pH/TVB-N	Purple → green	Correlated with increased storage time and spoilage	Koshy et al. (2021)
Pork	Agar/Potato Starch	Sweet potato anthocyanins	pH	Red → green	Correlated with pH change and sample deterioration	Liu et al. (2022)
Fish (tilapia)	Jackfruit seed Starch	Grape anthocyanins	pH/Volatile amines	Initial color → blue	Correlated with spoilage after 146 h	Costa et al. (2020)
Fish (silver carp)	Starch/PVA	Roselle anthocyanins	TVB-N	Visible color changes over time	Correlated with increase in TVB-N	Zhai et al. (2017)
Shrimp	Starch	Shikonin	pH/TVB-N	Red → blue	Correlated with spoilage over 36 h	Ezati and Rhim (2021)
Guava	Cassava starch + casein + gelatin	-	Moisture retention/ ripening delay	Coated fruit stayed green longer	Shelf life increased by ~2 days; uncoated fruit lost green color by day 3, coated remained greenish up to day 9	Pellá et al. (2020)
Grapes	Brazilian pine seed starch + citric pectin + feijoa extract)	Antioxidants, phenolic compounds, bioactive compounds	Moisture retention, firmness maintenance, inhibition of senescence and biochemical degradation, antioxidant protection	N/A	Film and coating packaging maintained grape firmness and weight, reducing water loss, Maintained pH, total soluble solids/acid ratio, and antioxidant activity over 30 days, Delayed senescence and postharvest deterioration compared to control	Sganzerla et al. (2021)

Milk	Gelatin/Starch	Anthocyannins	pH	Color changed with acidification	Color change corresponded with decreasing pH upon spoilage	Neodini Remedio and Parada Quinayá (2024)
Shrimp	κ-Carrageenan/PVA	Mexican petunia anthocyanins	TVB-N/pH	Pale brown → green → brownish-yellow	Color shift matched increasing TVB-N and shrimp deterioration (fresh → spoiled)	Kanphai et al. (2025)
Chicken	Sodium alginate/almond gum	Curcumin + betacyanin blend	TVB-N/pH	Varies with spoilage (e.g., darkening)	Indicator correlated with TVB-N increase and chicken quality loss	Thiyagarajan et al. (2025)

Intelligent gas indicators integrated into the packaging serve as reliable, continuous leak detectors throughout the distribution chain. Oxygen indicators are the most direct method for detecting leaks in anaerobic or low-oxygen MAP systems. These indicators are typically based on a redox dye that remains in a colorless, reduced state within the intended oxygen-free environment. If a leak occurs and oxygen enters the package, the dye is oxidized, triggering a distinct and irreversible color change (e.g., to blue), which provides an unambiguous visual signal that the package integrity has been breached (Mills, 2005). Starch-based films can serve as the matrix for immobilizing these redox dye systems, creating a biodegradable and food-safe indicator label (Mills, 2006). For packages flushed with high concentrations of CO_2, a carbon dioxide indicator can also function as a leak detector. In this scenario, the indicator is designed to signal a *decrease* in CO_2 concentration. A leak would cause the internal CO_2 to diffuse out and be replaced by ambient air, leading to a change in the indicator's color or transparency, thereby alerting to the packaging failure (Puligundla et al., 2012). These indicators provide a level of quality assurance that cannot be achieved by conventional, intermittent quality control checks, which typically involve the destructive testing of a small sample of packages from a production batch. An integrated indicator provides 100% monitoring for every single package, from the point of sealing to the moment of consumption (Mills, 2005).

12.6.2 Monitoring dairy product spoilage

The primary spoilage mechanism for many dairy products, such as milk and yogurt, involves the microbial fermentation of lactose into lactic acid. This process is driven by lactic acid bacteria and results in a progressive decrease in the product's pH, leading to souring and coagulation. Therefore, pH-sensitive indicators incorporated into starch-based films can be effectively used to monitor the freshness of dairy products. An intelligent label for milk packaging could be designed using a starch matrix containing a pH indicator dye or a natural extract like anthocyanins (Kalita et al., 2026). The indicator would be calibrated to undergo a visible color change when the pH of the milk drops below a critical threshold that correlates with the onset of spoilage such as $pH < 6.2$ for fresh milk (Che Hamzah et al., 2022). For example, a prototype indicator based on the diffusion of lactic acid has been developed that changes color from green to red as the pH decreases, demonstrating the feasibility of this approach (Guggilla et al., 2016). Such a system would provide consumers with a clear signal to "use" or "discard" the product, which is more reliable than relying solely on smell or a fixed expiration date (Yousefi et al., 2019).

Beyond general spoilage, starch-based pH indicators have also been developed for more specific safety applications in the dairy industry. One study demonstrated the creation of a starch-PVA film containing the pH indicator bromocresol purple to detect penicillin residues in raw milk. The enzymatic hydrolysis of penicillin produces penicilloic acid, which lowers the pH. The indicator film was able to detect this pH change, shifting color from dark violet to light violet in the presence of penicillin, offering a rapid method for screening antibiotic contamination at the farm level (Saengdee et al., 2022).

12.6.3 Shelf-life indication for ready-to-eat meals

Ready-to-eat (RTE) meals and other chilled prepared foods represent a growing market segment characterized by convenience and a relatively short, temperature-dependent shelf life (Dodero et al., 2021). For these products, maintaining the cold chain is absolutely critical to ensure both quality and microbiological safety. A break in the cold chain can lead to rapid bacterial growth and render the product unsafe for consumption, even if it is still within its printed "use-by" date (Sejani & Dhamsaniya, 2021).

TTIs are the ideal intelligent packaging technology for this application category. A TTI label affixed to the packaging of an RTE meal provides a continuous, cumulative record of its thermal history. The irreversible visual change of the TTI progresses at a temperature-dependent rate that is designed to mimic the degradation rate of the food product itself (Pavelková, 2013). This allows the TTI to provide a dynamic, real-time indication of the product's remaining shelf life and quality status (Sejani & Dhamsaniya, 2021). For example, a TTI might show a green color when fresh, progressively changing to yellow and then red as its shelf life is consumed by exposure to time and temperature. Starch can play a direct role in the functionality of certain TTI systems. For instance, enzymatic TTIs can be designed based on the reaction between amylase and starch. The temperature-dependent rate of starch hydrolysis by the enzyme can be coupled to a colorimetric reaction, creating a TTI where the starch itself is a key component of the time-temperature sensing mechanism (Pavelková, 2013).

12.7 Advantages, challenges, and limitations

The development of starch-based intelligent packaging represents a significant technological advancement with the potential to transform the food industry. However, a balanced and critical assessment is necessary to understand not only its benefits but also the substantial scientific, economic, and regulatory challenges that must be overcome for widespread adoption.

12.7.1 Benefits of starch-based systems

The most compelling advantages of using starch as the foundational polymer for smart packaging are rooted in sustainability and economics. In terms of sustainability and environmental impact, starch is a fully renewable resource derived from a wide range of agricultural crops, which helps to decouple the industry from its reliance on finite fossil fuels (Raghunathan et al., 2025). Most importantly, starch-based materials are completely biodegradable, designed to break down into harmless natural components like carbon dioxide, water, and biomass, offering a powerful solution to the global problem of plastic pollution (Arruda et al., 2025). Regarding cost and availability, starch is one of the most abundant and inexpensive biopolymers available globally, and it can be sourced from agricultural and industrial by-products, further reducing costs and promoting a circular economy (Liu et al., 2022). Finally, from a food safety perspective, starch is an edible, food-grade material generally recognized as safe (GRAS) for food contact, which simplifies the regulatory approval process (Karnwal et al., 2025).

12.7.2 Challenges

Despite its advantages, native starch has several inherent limitations that present significant technical challenges. The main hurdle is its moisture sensitivity and poor barrier properties due to its hydrophilic nature, which leads to high water vapor permeability and reduced mechanical strength in humid conditions. Unmodified starch films also have poor mechanical properties, being typically brittle and lacking in strength and flexibility compared to conventional plastics (Liu et al., 2022). While plasticizers can improve flexibility, they often compromise tensile strength and can worsen moisture sensitivity (Wang et al., 2022). Another challenge is indicator stability; natural dyes like anthocyanins are sensitive to degradation from light, heat, and oxygen, which can cause the indicator to fail or provide a false reading (Zhao et al., 2022). Lastly, component migration is a paramount safety concern. All components, including plasticizers, dyes, and nanoparticles, must be rigorously tested to ensure they do not migrate from the packaging into the food in quantities that exceed legally mandated safety limits, adding complexity and cost to the development process (Dodero et al., 2021).

12.7.3 Consumer acceptance and regulatory hurdles

Beyond technical challenges, successful commercialization depends on navigating complex social and regulatory landscapes. Consumer acceptance is a key factor; while there is a positive perception of the benefits of intelligent packaging, such as enhanced safety and reduced waste, consumer awareness and understanding remain low. Acceptance is influenced by the perceived usefulness and ease of interpretation of the indicator, trust in its reliability, and concerns about potential health risks (Jayasinghe et al., 2024). Cost is a significant barrier, as consumers may have a limited willingness to pay a premium for these benefits (Boukid, 2022). On the regulatory front, food contact materials are subject to stringent safety regulations. In the European Union, for example, materials must comply with Regulation (EC) No. 1935/2004, and active and intelligent materials are further governed by Regulation (EC) No. 450/2009 (AGES - Food contact materials, n.d.). Manufacturers must perform extensive migration testing and provide a formal Declaration of Compliance (DoC) to verify safety (Food Contact Material Compliance - An Overview, 2023). This rigorous, time-consuming, and expensive regulatory process represents a significant barrier to entry for new and innovative materials (Venkatesan et al., 2025).

12.8 Future trends and commercialization

Despite the existing challenges, the field of starch-based intelligent packaging is dynamic and continues to advance, driven by innovations in material science, electronics, and data technology. The future of this field lies in creating more sophisticated, functional, and commercially viable systems that can be seamlessly integrated into the global food supply chain. This section explores key future trends and outlines the critical steps on the path to industrial scalability and market entry.

12.8.1 Integration with RFID and NFC technologies and development of multifunctional intelligent packaging

While colorimetric indicators provide a simple visual interface, their utility can be significantly enhanced by integrating them with digital communication technologies like Radio Frequency Identification (RFID) and Near Field Communication (NFC). This integration bridges the gap between the physical state of the food and the digital information ecosystem. An RFID or NFC tag embedded in the packaging label can store a unique product identifier and receive input from the intelligent sensor. This enables enhanced supply chain management, as RFID readers can automatically scan packages at checkpoints, providing real-time, item-level data on product location and status for highly accurate inventory management and traceability. It also facilitates improved consumer interaction; by tapping a smartphone on the package, consumers can access a wealth of information via NFC, including real-time freshness status, product provenance, and nutritional details, thereby building brand loyalty and increasing confidence in product quality and safety (Sobhan et al., 2025).

The future direction of smart packaging is moving away from single-function systems toward the development of multifunctional materials that combine both active and intelligent properties within a single, integrated platform. Starch-based systems are particularly well-suited for this evolution, especially when combined with natural bioactive compounds that possess multiple functionalities. As previously discussed, anthocyanins are a prime example. When incorporated into a starch film, they can act as an intelligent pH indicator to signal spoilage, function as an active antioxidant to slow oxidative degradation, and provide active antimicrobial effects to inhibit microbial growth (Neodini Remedio & Parada Quinayá, 2024). Similarly, a packaging film could be designed to incorporate an antioxidant like clove oil to actively preserve the food, while also featuring a printed, irreversible oxygen indicator to intelligently monitor for package leaks (Karnwal et al., 2025). The development of such multifunctional films, where different active and intelligent components work synergistically within a biodegradable starch matrix, represents a key frontier in packaging research. These systems promise to deliver enhanced performance and greater value by addressing multiple aspects of food preservation and quality monitoring concurrently.

12.8.2 Industrial scalability and market entry

For starch-based intelligent packaging to move from the laboratory to the supermarket shelf, several critical commercialization hurdles must be addressed. A primary challenge is overcoming the "lab-to-fab" gap by developing formulations and indicator systems robust enough to withstand industrial-scale processing methods like extrusion (Bangar et al., 2021). Another key factor is cost reduction and economic viability; achieving cost-competitiveness with conventional packaging is essential for market adoption and can be pursued by optimizing formulations, improving manufacturing efficiency, and utilizing low-cost agricultural by-products (Boukid, 2022). Navigating regulatory approval is a non-negotiable and often lengthy process that requires early engagement with regulatory bodies and comprehensive safety and migration testing to ensure full compliance with standards like the EU's Framework Regulation (EC) No. 1935/2004 (Venkatesan et al., 2025). Finally, building consumer and industry trust is crucial and will require a concerted effort to educate stakeholders about the technology's benefits and reliability, using clear, standardized visual indicators for consumers and demonstrating the return on investment for retailers through reduced food waste (Chirilli et al., 2022).

12.9 Conclusion

Research on starch-based intelligent packaging shows that starch is a strong sustainable packaging option because it's renewable, biodegradable, and low-cost. However, it absorbs moisture and can be brittle, so it usually needs plasticizers or fillers to improve flexibility and strength. The starch source also matters, since the amylose-to-amylopectin ratio affects film properties.

These films become "intelligent" when indicators are added. Natural pigments like anthocyanins are especially useful — they shift color as food spoils and often have antioxidant or antimicrobial activity. Gas indicators (for oxygen, CO_2, or ammonia) help detect leaks or microbial growth, while biosensors can target specific pathogens. Time–temperature indicators track thermal history, which is important for cold-chain foods.

The main challenge is scaling production. Most lab work uses solution casting, which doesn't translate well to industry. Industrial extrusion is scalable but can damage heat-sensitive indicators. Although there are successful case studies for monitoring meat, seafood, and fruit ripeness, commercialization is slowed by performance, cost, and regulatory requirements.

Smart starch packaging can reduce both plastic waste and food waste by giving real-time freshness information instead of relying on fixed expiration dates. Future progress will likely focus on multifunctional films and connecting physical indicators with digital systems like RFID or NFC. As manufacturing improves and demand for sustainability grows, starch-based smart packaging is well-positioned to move from research into real-world use.

References

Abelti, A., Abera Teka, T., Forsido, S., Tamiru, M., Bultosa, G., Alkhtib, A., & Burton, E. (2022). Bio-based smart materials for fish product packaging: A review. *International Journal of Food Properties, 25*, 857–871. https://doi.org/10.1080/10942912.2022.2066121

AGES - Food contact materials. (n.d.). AGES. Retrieved November 15, 2025, from https://www.ages.at/en/human/nutrition-food/food-information/food-contact-materials.

Ameri, M., Ajji, A., & Kessler, S. (2024). Characterization of a food-safe colorimetric indicator based on black rice anthocyanin/PET films for visual analysis of fish spoilage. *Packaging Technology and Science, 37*(8), 769–780. https://doi.org/10.1002/pts.2824

Arruda, T. R., Machado, G. de O., Marques, C. S., Souza, A. L. de, Pelissari, F. M., Oliveira, T. V. de, & Silva, R. R. A. (2025). An overview of starch-based materials for sustainable food packaging: Recent advances, limitations, and perspectives. *Macromol, 5*(2), 19. https://doi.org/10.3390/macromol5020019

Bangar, S. P., Purewal, S. S., Trif, M., Maqsood, S., Kumar, M., Manjunatha, V., & Rusu, A. V. (2021). Functionality and applicability of starch-based films: An eco-friendly approach. *Foods, 10*(9), 2181. https://doi.org/10.3390/foods10092181

Berk, Z. (2009). Food packaging. In *Food process engineering and technology* (pp. 545–559). Elsevier. https://doi.org/10.1016/B978-0-12-373660-4.00026-0

Bhowmik, S., Agyei, D., & Ali, M. (2024). Smart chitosan films as intelligent food packaging: An approach to monitoring food freshness and biomarkers. *Food Packaging and Shelf Life, 46*, 101370. https://doi.org/10.1016/j.fpsl.2024.101370

Bidari, R., Abdillah, A. A., Ponce, R. A. B., Charles, A. L., Bidari, R., Abdillah, A. A., Ponce, R. A. B., & Charles, A. L. (2023). Characterization of biodegradable films made from taro peel (colocasia esculenta) starch. *Polymers, 15*(2). https://doi.org/10.3390/polym15020338

Biji, K. B., Ravishankar, C. N., Mohan, C. O., & Srinivasa Gopal, T. K. (2015). Smart packaging systems for food applications: A review. *Journal of Food Science and Technology, 52* (10), 6125–6135. https://doi.org/10.1007/s13197-015-1766-7

Borges, J. A., Romani, V., Cortez-Vega, W., & Martins, V. (2015). Influence of different starch sources and plasticizers on properties of biodegradable films. 22, 2346–2351.

Boukid, F. (2022). Smart food packaging: An umbrella review of scientific publications. *Coatings, 12*(12), 1949. https://doi.org/10.3390/coatings12121949

Calderón-Castro, A., Vega-García, M. O., de Jesús Zazueta-Morales, J., Fitch-Vargas, P. R., Carrillo-López, A., Gutiérrez-Dorado, R., Limón-Valenzuela, V., & Aguilar-Palazuelos, E. (2018). Effect of extrusion process on the functional properties of high amylose corn starch edible films and its application in mango (*Mangifera indica* L.) cv. Tommy Atkins. *Journal of Food Science and Technology, 55*(3), 905–914. https://doi.org/10.1007/s13197-017-2997-6

Centers for Disease Control and Prevention [CDC]. (2021). Interagency Food Safety Analytics Collaboration. Foodborne illness source attribution estimates for 2021 for Salmonella, Escherichia coli O157, and Listeria monocytogenes using multi-year outbreak surveillance data, United States. GA and D.C.: U.S. Department of Health and Human Services, Centers for Disease Control and Prevention, Food and Drug Administration, U.S. Department of Agriculture's Food Safety and Inspection Service. 2023. https://www.cdc.gov/ifsac/media/pdfs/P19-2021-report-TriAgency-508.pdf.

Chatterjee, N. (2023). Regulation of the c-di-AMP phosphodiesterase PgpH in Listeria monocytogenes Thesis. https://minds.wisconsin.edu/handle/1793/95590.

Che Hamzah, N. H., Khairuddin, N., Muhamad, I. I., Hassan, M. A., Ngaini, Z., & Sarbini, S. R. (2022). Characterisation and colour response of smart sago starch-based packaging films incorporated with brassica oleracea anthocyanin. *Membranes, 12*(10), 913. https://doi.org/10.3390/membranes12100913

Chirilli, C., Molino, M., & Torri, L. (2022). Consumers' awareness, behavior and expectations for food packaging environmental sustainability: Influence of socio-demographic characteristics. *Foods, 11*(16), 2388. https://doi.org/10.3390/foods11162388

Costa, L., Diogenes, I., Oliveira, M., Ribeiro, S., Furtado, R., Bastos, M., Silva, M., & Benevides, S. (2020). Smart film of jackfruit seed starch as a potential indicator of fish freshness. *Food Science and Technology, 41*. https://doi.org/10.1590/fst.06420

Dodero, A., Escher, A., Bertucci, S., Castellano, M., & Lova, P. (2021). Intelligent packaging for real-time monitoring of food-quality: Current and future developments. *Applied Sciences, 11*(8), 3532. https://doi.org/10.3390/app11083532

Ezati, P., & Rhim, J.-W. (2021). Starch and agar-based color-indicator films integrated with shikonin for smart packaging application of shrimp. *ACS Food Science & Technology, 1* (10), 1963–1969. https://doi.org/10.1021/acsfoodscitech.1c00292

Fadhila, A. N., Warsiki, E., & Yuliani, S. (2024). Halochromic polyaniline as smart detector packaging for fresh food: A review. *IOP Conference Series: Earth and Environmental Science, 1358*(1), 012013. https://doi.org/10.1088/1755-1315/1358/1/012013

Figueroa-Lopez, K. J., Villabona-Ortíz, Á., & Ortega-Toro, R. (2024). Sustainable starch-based films from cereals and tubers: A comparative study on cherry tomato preservation. *Polymers, 16*(20), 2913. https://doi.org/10.3390/polym16202913

Food Contact Material Compliance—An Overview. (2023, July 31). Institute of Food Science and Technology. https://www.ifst.org/resources/information-statements/food-contact-material-compliance-overview.

Guggilla, M., Rajeshwar, B., & Matche, S. (2016). Lactic Acid Based Spoilage Indicator for Milk.

Heo, W., & Lim, S. (2024). A review on gas indicators and sensors for smart food packaging. *Foods, 13*(19), 3047. https://doi.org/10.3390/foods13193047

Heo, W., Lim, S., Heo, W., & Lim, S. (2024). A review on gas indicators and sensors for smart food packaging. *Foods, 13*(19). https://doi.org/10.3390/foods13193047

Inês, A., & Cosme, F. (2025). Biosensors for detecting food contaminants—an overview. *Processes, 13*(2), 380. https://doi.org/10.3390/pr13020380

Janssen, S., Schmitt, K., Blanke, M., Bauersfeld, M. L., Wöllenstein, J., & Lang, W. (2014). Ethylene detection in fruit supply chains. *Philosophical Transactions. Series A, Mathematical, Physical, and Engineering Sciences, 372*(2017), 20130311. https://doi.org/10.1098/rsta.2013.0311

Jayasinghe, C. V. L., Jayatilake, S., Rajapakse, H. U. K. D. Z., Gunarathne, H. M. N. R., & Wanigasinghe, H. G. (2024). Consumer perceptions of smart packaging technologies for food. In *Intelligent packaging* (pp. 365–382). Elsevier. https://doi.org/10.1016/B978-0-443-15388-4.00014-6

Jiang, C., Liu, G., Zhang, Q., Wang, S., & Zou, Y. (2024). Preparation and characterization of intelligent packaging labels based on pea starch, κ-carrageenan and black raspberry extract for monitoring freshness of pork. *Food Chemistry X, 24*, 101850. https://doi.org/10.1016/j.fochx.2024.101850

Kalita, P., Sahni, O., Chatterjee, N., Sharma, N., & Roy, S. (2026). Innovations in intelligent packaging sensors for real-time Milk freshness monitoring. *Food Chemistry, 514*. https://doi.org/10.1016/j.foodchem.2026.149095

Kanphai, K., Athipornchai, A., Sirion, U., Suksai, C., & Trakulsujaritchok, T. (2025). Green smart packaging film based on kappa-carrageenan/poly(Vinyl Alcohol) enriched with mexican petunia anthocyanins for visual detection of food freshness. *ACS Omega, 10*(32), 36104–36114. https://doi.org/10.1021/acsomega.5c03712

Karki, D., Kc, Y., Khanal, H., Bhattarai, P., Koirala, B., & Khatri, S. (2020). Analysis of biodegradable films of starch from potato waste. *Asian Food Science Journal*, 28–40. https://doi.org/10.9734/afsj/2020/v14i330132

Karnwal, A., Rauf, A., Jassim, A. Y., Selvaraj, M., Al-Tawaha, A. R. M. S., Kashyap, P., Kumar, D., & Malik, T. (2025). Advanced starch-based films for food packaging: Innovations in sustainability and functional properties. *Food Chemistry X, 29*, 102662. https://doi.org/10.1016/j.fochx.2025.102662

Khodaei, S. M., Gholami-Ahangaran, M., Karimi Sani, I., Esfandiari, Z., & Eghbaljoo, H. (2022). Application of intelligent packaging for meat products: A systematic review. *Veterinary Medicine and Science, 9*(1), 481–493. https://doi.org/10.1002/vms3.1017

Koshy, R. R., Koshy, J. T., Mary, S. K., Sadanandan, S., Jisha, S., & Pothan, L. A. (2021). Preparation of pH sensitive film based on starch/carbon nano dots incorporating anthocyanin for monitoring spoilage of pork. *Food Control, 126*, 108039. https://doi.org/10.1016/j.foodcont.2021.108039

Kusuma, I., Ma'arif, A., & Rahmadhia, S. N. (2024). Starch based for intelligent food packaging: A review. *BIO Web of Conferences, 146*, 01099. https://doi.org/10.1051/bioconf/202414601099

Lang, C., & Hübert, T. (2011). A colour ripeness indicator for Apples. *Food and Bioprocess Technology, 5*. https://doi.org/10.1007/s11947-011-0694-4

Li, H., & Huneault, M. (2011). Comparison of sorbitol and glycerol as plasticizers for thermoplastic starch in TPS/PLA blends. *Journal of Applied Polymer Science, 119*. https://doi.org/10.1002/app.32956

Liu, M., Pirrello, J., Chervin, C., Roustan, J.-P., & Bouzayen, M. (2015). Ethylene control of fruit ripening: Revisiting the complex network of transcriptional regulation1. *Plant Physiology, 169*(4), 2380–2390. https://doi.org/10.1104/pp.15.01361

Liu, D., Zhang, C., Pu, Y., Chen, S., Liu, L., Cui, Z., & Zhong, Y. (2022). Recent advances in pH-responsive freshness indicators using natural food colorants to monitor food freshness. *Foods, 11*(13), 1884. https://doi.org/10.3390/foods11131884

Mani, R., Jothi, V. K., Balakrishnan, M., Ramalakshmi, A., & Rhim, J.-W. (2025). Smart sensors in food packaging: Sensor technology for real-time food safety and quality monitoring. *Journal of Food Process Engineering, 48*. https://doi.org/10.1111/jfpe.70120

Mary, S. K., Koshy, R. R., Daniel, J., Koshy, J. T., Pothen, L. A., & Thomas, S. (2020). Development of starch based intelligent films by incorporating anthocyanins of butterfly pea flower and TiO_2 and their applicability as freshness sensors for prawns during storage. *RSC Advances, 10*(65), 39822–39830. https://doi.org/10.1039/D0RA05986B

Menchaca-Rivera, J. A., Gonzalez-Reyna, M. A., Avilés-Arellano, L. M., Fernández-Loyola, R., Morales-Sánchez, E., & Pérez Robles, J. F. (2019). Determination of optical properties of a corn starch biofilm. *Journal of Applied Polymer Science, 136*(9), 47111. https://doi.org/10.1002/app.47111

Mills, A. (2005). Oxygen indicators and intelligent inks for packaging food. *Chemical Society Reviews, 34*(12), 1003. https://doi.org/10.1039/b503997p

Mills, A. (2006). Oxygen indicators and intelligent inks for packaging food. *Chemical Society Reviews, 34*, 1003–1011. https://doi.org/10.1039/b503997p

Mkhari, T., Adeyemi, J. O., & Fawole, O. A. (2025). Recent advances in the fabrication of intelligent packaging for food preservation: A review. *Processes, 13*(2), 539. https://doi.org/10.3390/pr13020539

Molavi, H., Behfar, S., Shariati, M. A., Kaviani, M., & Atarod, S. (2018). A review on biodegradable starch based film. *Journal of Microbiology, Biotechnology and Food Sciences*. https://doi.org/10.15414/jmbfs.2015.4.5.456-461

Nath, S. (2024). Advancements in food quality monitoring: Integrating biosensors for precision detection. *Sustainable Food Technology, 2*(4), 976–992. https://doi.org/10.1039/D4FB00094C

Neodini Remedio, L., & Parada Quinayá, D. (2024). Intelligent packaging systems with anthocyanin: Influence of different polymers and storage conditions. *Polymers, 16*, 2886. https://doi.org/10.3390/polym16202886

Noiwan, D., Suppakul, P., & Rachtanapun, P. (2022). Preparation of methylcellulose film-based CO2 indicator for monitoring the ripeness quality of mango fruit cv. nam dok mai si thong. *Polymers, 14*(17), 3616. https://doi.org/10.3390/polym14173616

Othman, S. H., Ronzi, N. D. A., Shapi'i, R. A., Dun, M., Ariffin, S. H., & Mohammed, M. A. P. (2023). Biodegradability of starch nanocomposite films containing different concentrations of chitosan nanoparticles in compost and planting soils. *Coatings, 13*(4), 777. https://doi.org/10.3390/coatings13040777

Pavelková, A. (2013). Time temperature indicators as devices intelligent packaging. *Acta Universitatis Agriculturae et Silviculturae Mendelianae Brunensis, 61*(1), 245–251. https://doi.org/10.11118/actaun201361010245

Pellá, M. C. G., Silva, O. A., Pellá, M. G., Beneton, A. G., Caetano, J., Simões, M. R., & Dragunski, D. C. (2020). Effect of gelatin and casein additions on starch edible biodegradable films for fruit surface coating. *Food Chemistry, 309*, 125764. https://doi.org/10.1016/j.foodchem.2019.125764

Puligundla, P., Jung, J., & Ko, S. (2012). Carbon dioxide sensors for intelligent food packaging applications. *Food Control, 25*. https://doi.org/10.1016/j.foodcont.2011.10.043

Raghunathan, R., Jeevarathinam, G., Vadukkumchery Prasannan, M., Rustagi, S., & Asdaq, S. M. B. (2025). Starch-based polymer composites for sustainable food packaging: Advances in bioplastic development. *Starch - Stärke,* 77(9), e70092. https://doi.org/10.1002/star.70092

Reji, R. E., Boruah, T., Chatterjee, N., Nayik, G. A., Kalita, P., & Roy, S. (2026). Recent advances in natural melanin added food packaging. *Advances in Food and Nutrition Research*. Elsevier. In press.

Rodrigues, A., Costa, R., Ferreira, L., Silva, S., Britto, D., & Lima, M. (2021). Properties and characterization of biodegradable films obtained from different starch sources. *Food Science and Technology, 41*. https://doi.org/10.1590/fst.28520

Saengdee, P., Nuanthong, T., Chamnan, P., Pattamang, P., Thongsook, O., Meananeatra, R., Ranron, N., Pankong, K., Uahchinkul, W., Jeamsaksiri, W., Vongkamjan, K., Chokboribal, J., & Atthi, N. (2022). Development of starch-polyvinyl alcohol films-based pH indicator for detection of penicillin g residue in raw milk. *Journal of Physics: Conference Series, 2175*(1), 012040. https://doi.org/10.1088/1742-6596/2175/1/012040

Sander, M. (2019). Biodegradation of polymeric mulch films in agricultural soils: Concepts, knowledge gaps, and future research directions. *Environmental Science & Technology, 53*(5), 2304–2315. https://doi.org/10.1021/acs.est.8b05208

Sejani, V., & Dhamsaniya, N. (2021). Intelligent Packaging for Food: A Review.

Sganzerla, W. G., da Rosa, C. G., da Silva, A. P. G., Ferrareze, J. P., Azevedo, M. S., Forster-Carneiro, T., Nunes, M. R., & de Lima Veeck, A. P. (2021). Application *in situ* of biodegradable films produced with starch, citric pectin and functionalized with feijoa (*Acca sellowiana* (Berg) Burret) extracts: An effective proposal for food conservation. *International Journal of Biological Macromolecules, 189*, 544–553. https://doi.org/10.1016/j.ijbiomac.2021.08.146

Sirbu, E.-E., Dinita, A., Tănase, M., Portoacă, A.-I., Bondarev, A., Enascuta, C.-E., & Calin, C. (2024). Influence of plasticizers concentration on thermal, mechanical, and physicochemical properties on starch films. *Processes, 12*(9), 2021. https://doi.org/10.3390/pr12092021

Sobhan, A., Hossain, A., Wei, L., Muthukumarappan, K., & Ahmed, M. (2025). IoT-enabled biosensors in food packaging: A breakthrough in food safety for monitoring risks in real time. *Foods, 14*(8), 1403. https://doi.org/10.3390/foods14081403

Talukder, S., Soni, A., Jaiswal, R., Mendiratta, S., & Saini, B. (2018). Enzymatic time temperature indicators: A review. *Innovation Journal, 7*, 643–647.

Thiyagarajan, A., Subramaniyam, N., Thiyagarajan, V., Pushparaj, C., Kandasamy, S., Rajendran K, S., & Subramani, R. (2025). Betacyanin–curcumin smart films for detecting fresh chicken quality in real time. *Sustainable Food Technology, 3*(6), 1880–1889. https://doi.org/10.1039/D5FB00180C

Vaikousi, H., Biliaderis, C. G., & Koutsoumanis, K. P. (2008). Development of a microbial time/temperature indicator prototype for monitoring the microbiological quality of chilled foods. *Applied and Environmental Microbiology, 74*(10), 3242–3250. https://doi.org/10.1128/AEM.02717-07

Venkatesan, R., Alrashed, M. M., Vetcher, A. A., & Kim, S.-C. (2025). Next-generation food packaging: Progress and challenges of biopolymer-based materials. *Polymers, 17* (17), 2299. https://doi.org/10.3390/polym17172299

Wang, B., Huang, D., & Weng, Z. (2023). Recent advances in polymer-based biosensors for food safety detection. *Polymers, 15*(15), 3253. https://doi.org/10.3390/polym15153253

Wang, B., Xu, X., Fang, Y., Yan, S., Cui, B., & Abd El-Aty, A. M. (2022). Effect of different ratios of glycerol and erythritol on properties of corn starch-based films. *Frontiers in Nutrition, 9*. https://doi.org/10.3389/fnut.2022.882682

Yousefi, H., Su, H.-M., Imani, S. M., Alkhaldi, K., M. Filipe, C. D., & Didar, T. F. (2019). Intelligent food packaging: A review of smart sensing technologies for monitoring food quality. *ACS Sensors, 4*(4), 808–821. https://doi.org/10.1021/acssensors.9b00440

Zeng, F., Ye, Y., Liu, J., & Fei, P. (2022). Intelligent pH indicator composite film based on pectin/chitosan incorporated with black rice anthocyanins for meat freshness monitoring. *Food Chemistry X, 17*, 100531. https://doi.org/10.1016/j.fochx.2022.100531

Zhai, X., Shi, J., Zou, X., Wang, S., Jiang, C., Zhang, J., Huang, X., Zhang, W., & Holmes, M. (2017). Novel colorimetric films based on starch/polyvinyl alcohol incorporated with roselle anthocyanins for fish freshness monitoring. *Food Hydrocolloids, 69*, 308–317. https://doi.org/10.1016/j.foodhyd.2017.02.014

Zhang, Y., & Lim, L.-T. (2017). Colorimetric array indicator for NH_3 and CO_2 detection. *Sensors and Actuators B: Chemical, 255*. https://doi.org/10.1016/j.snb.2017.09.148

Zhao, L., Liu, Y., Zhao, L., & Wang, Y. (2022). Anthocyanin-based pH-sensitive smart packaging films for monitoring food freshness. *Journal of Agriculture and Food Research, 9*, 100340. https://doi.org/10.1016/j.jafr.2022.100340

CHAPTER THIRTEEN

Industrialization application of starch-based smart packaging

Łukasz Łopusiewicz[1, 2], Danila Merino[3, 4]
[1]Vizja University, School of Medical & Health Sciences, Warsaw, Poland
[2]University of Greifswald, Department Pharmaceutical Biology, Institute of Pharmacy, Greifswald, Germany
[3]University of the Basque Country (UPV/EHU), POLYMAT Institute, San Sebastian, Spain
[4]Basque Foundation for Science, Ikerbasque, Bilbao, Spain

13.1 Introduction

The growing problem of environmental pollution caused by plastic waste is one of the key challenges facing the modern materials economy. Global plastic waste production is estimated at hundreds of millions of tons per year and shows a steady upward trend, which is directly related to the widespread use of synthetic polymers in short-term applications, particularly in the packaging sector (Abdullah & Dong, 2018; Rocha et al., 2025). The dominance of these materials is due to their favorable performance properties, such as good mechanical strength, thermal stability, barrier properties, and low processing cost, but at the same time, they generate a significant environmental burden at the end of the product life cycle (Abdullah & Dong, 2018; Lauer & Smith, 2020). Packaging plays a fundamental role in the functioning of modern food systems, being an integral part of the process of food processing, transport, storage, distribution, and consumption (D'Almeida & de Albuquerque, 2024). Its task is not only to physically protect the product, but also to limit the microbiological, chemical, and physical processes that lead to the loss of sensory quality and nutritional value (Liu et al., 2022; Onyeaka et al., 2022). The scale of packaging material use in the food sector is significant – it is estimated that over 40% of all plastics produced globally are used in this industry, making it one of the main sources of environmental pressure (Rocha et al., 2025; Lauer & Smith, 2020).

Despite their key protective function, conventional petrochemical plastics have limited recyclability and high resistance to environmental degradation. Repeated processing leads to a significant loss of mechanical properties, which in practice forces the constant consumption of fossil

Starches in Smart Packaging
ISBN: 978-0-443-45396-0
https://doi.org/10.1016/B978-0-443-45396-0.00014-1

raw materials and perpetuates the linear model of material management (Lauer & Smith, 2020). The consequence of this approach is the accumulation of plastic waste in the environment, including the formation of microplastics, which pose a threat to terrestrial and marine organisms and the stability of ecosystems (Onyeaka et al., 2022; Chia et al., 2022). Despite growing awareness of these problems, plastic packaging still accounts for a significant proportion of total plastic demand, which further highlights the urgency of finding alternative material solutions (Rocha et al., 2025; Onyeaka et al., 2022). In response to these challenges, there has been intensive development of packaging materials based on bio-based polymers. This interest stems both from the need to reduce negative environmental impact and from the desire to use renewable raw materials and biodegradable materials (Abdullah & Dong, 2018; Arruda et al., 2025). Particular attention is being paid to natural biopolymers, including polysaccharides and proteins, which combine raw material availability with a favorable environmental profile and the potential for designing functional packaging structures (Arruda et al., 2025; Cheng et al., 2022). This direction remains consistent with global sustainable development goals, including responsible production, waste reduction, and climate action (Rocha et al., 2025).

Among the available biopolymers, starch stands out as a material with particularly high application potential. Its wide availability, low cost, renewability, non-toxicity, and chemical and physical modification opportunities make it one of the most intensively researched raw materials in the context of biodegradable packaging (Chia et al., 2022; Amaraweera et al., 2021). Market data indicate that starch-based materials account for a significant portion of global demand for biopolymers, which highlights their importance not only in the laboratory but also in industry (Amaraweera et al., 2021). At the same time, the use of starch in disposable packaging requires meeting the requirements for durability, functionality, and economic viability on an industrial scale (Amaraweera et al., 2021; Garavito et al., 2024). Parallel to the development of biodegradable materials, the importance of active and intelligent packaging capable of interacting with the packaged product or its environment is growing. Smart packaging allows for the monitoring of environmental changes and food condition, while active systems involve the deliberate introduction of functional ingredients into the material matrix or packaging space to improve product quality and safety (Onyeaka et al., 2022; Cheng et al., 2022). The integration of "smart" functions with biodegradable polymer matrices, including starch-based ones, is currently one of the key areas of research into modern packaging materials, responding to simultaneous environmental, technological, and industrial requirements (Rocha et al., 2025; Chia et al., 2022).

The chapter presents starch as a key biopolymer for sustainable packaging systems, discussing its raw material availability, material properties, and potential for integrating active and smart functions in the context of industrial applications.

13.2 Starch as an sustainable biopolymer

Starch is one of the most common plant-based polysaccharides and, at the same time, one of the key biopolymers considered as an alternative to conventional packaging materials (Fig. 13.1). Its complete renewability, biodegradability, and compostability, combined with its low cost of acquisition, make starch a consistent focus of research into sustainable polymer materials (Abdullah & Dong, 2018; Arruda et al., 2025; Amaraweera et al., 2021). As the end product of photosynthesis, it is the primary form of energy storage in plants, occurring as semi-crystalline, water-insoluble granules located in various plant organs, which further facilitates its isolation and processing (Abdullah & Dong, 2018). From the perspective of the circular economy, it is particularly important that starch can be obtained both from conventional agricultural raw materials and from low-quality commercial materials and waste streams from the agri-food industry. The use of waste and by-products, such as low-grade fruit, vegetables, or post-process biomass, reduces raw material losses and alleviates the environmental pressure associated with their disposal (Arruda et al., 2025; Fatima et al., 2024). This approach strengthens the arguments in favor of starch

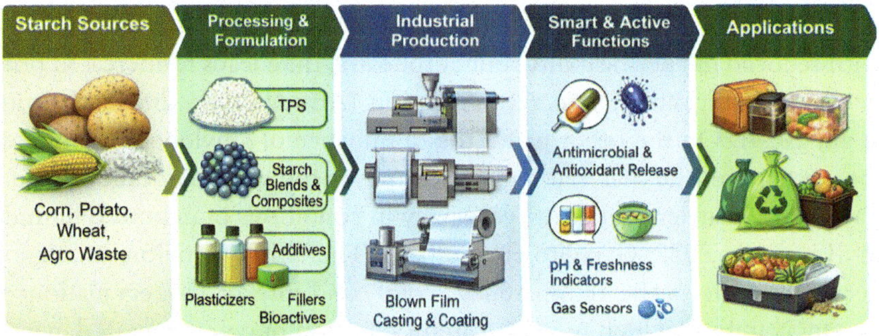

Figure 13.1 From starch to industrial smart packaging: Schematic representation of feedstocks, TPS processing, functionalization strategies, and industrial applications.

as a raw material compatible with the principles of sustainable production and industrial-scale implementation.

Industrially, starch is obtained from a wide range of botanical sources, including cereals, tubers, roots, and legumes. The most commonly used raw materials include corn, wheat, rice, potatoes, and cassava, with alternative sources such as bananas, mangoes, pineapples, and oats also attracting increasing interest (Abdullah & Dong, 2018; Onyeaka et al., 2022). Global starch production is growing steadily, driven by increasing demand from the food, chemical, and materials sectors; the largest production capacities are concentrated in North America and East Asia, with Europe also playing a significant role (Onyeaka et al., 2022). These data clearly indicate that starch is not only an easily accessible raw material, but also one with a well-established infrastructure, which is crucial from the point of view of the industrialization of packaging materials.

At the molecular level, starch is a heterogeneous material composed mainly of two polysaccharides: amylose and amylopectin, whose ratio depends on the botanical origin (Arruda et al., 2025; Oyekunle et al., 2024). In most natural starches, amylose accounts for about 15%–30% and amylopectin for 70%–80% of the polymer composition, although there are also varieties with almost 100% amylopectin content (waxy starches) or increased amylose content (Abdullah & Dong, 2018; Amaraweera et al., 2021). Amylose has a essentially linear structure based on α-(1 $\rightarrow$ 4) bonds, while amylopectin is characterized by a highly branched architecture resulting from the presence of α-(1 $\rightarrow$ 4) and α-(1 $\rightarrow$ 6) bonds, which leads to the formation of a semi-crystalline organization of starch granules (Arruda et al., 2025; Yu et al., 2021).

The physicochemical and mechanical properties of starch are a direct consequence of its molecular composition and degree of crystallinity. A higher amylose content promotes the formation of more compact structures with reduced gas permeability, but at the same time leads to increased brittleness and susceptibility to retrogradation (Arruda et al., 2025; Garavito et al., 2024). On the other hand, the dominance of amylopectin increases the elasticity of the material, but this comes at the expense of dimensional stability and barrier properties, especially in conditions of increased humidity (Liu et al., 2022; Garavito et al., 2024). The ratio of amylose to amylopectin also affects the transparency, color, and rate of biodegradation of starchy materials, which requires careful selection of raw materials depending on the planned application (Garavito et al., 2024; Oyekunle et al., 2024).

However, in its native state, starch has significant material limitations that hinder its direct application in packaging engineering. Its high hydrophilicity, resulting from the presence of numerous hydroxyl groups, leads to intense water absorption, increased water vapor permeability, and degradation of mechanical properties in a humid environment (Liu et al., 2022; Fatima et al., 2024). Starch materials are also brittle and prone to cracking, and their thermal stability is limited (D'Almeida & de Albuquerque, 2024; Arruda et al., 2025). A particular challenge remains the phenomenon of retrogradation, which involves the reassociation of amylose and amylopectin chains after gelatinization, leading to increased crystallinity, loss of elasticity, and deterioration of barrier properties over time (Arruda et al., 2025).

In order to improve processability, starch is most often converted into a thermoplastic form through the addition of plasticizers and the application of heat and shear, which leads to the destruction of the crystalline structure of the granules and the formation of a homogeneous polymer phase (Abdullah & Dong, 2018; Rocha et al., 2025). Thermoplastic starch (TPS) can be processed using methods typical for the plastics industry, but even in this form, its mechanical and barrier properties remain inferior to those of petrochemical polymers (Rocha et al., 2025; D'Almeida & de Albuquerque, 2024). As a result, starch currently serves primarily as a base matrix, requiring further modification to obtain packaging materials with parameters that meet application and industrial requirements.

13.3 The importance of smart and active packaging in modern food packaging systems

The current development of packaging technologies is directly related to the limitations of classic, passive packaging systems, which, despite their effective protective function, do not offer the possibility of assessing the actual condition of the product during its distribution and storage (D'Almeida & de Albuquerque, 2024; Abdullah et al., 2022). Traditional packaging only provides information about the expected shelf life, without taking into account dynamic environmental changes or the actual degradation processes occurring in food. As a result, this leads both to the premature disposal of products that are still fit for consumption and to the risk of consuming food of reduced quality or safety (Rocha et al., 2025; D'Almeida & de Albuquerque, 2024). Smart packaging addresses this functional gap by enabling continuous or semi-continuous monitoring of product quality and storage conditions (Abdullah et al., 2022). By definition, these are packaging systems capable of collecting, recording, and transmitting information about the condition of food or the environment surrounding the product, allowing for more informed decision-making by producers, distributors, and consumers (D'Almeida & de Albuquerque, 2024; Abdullah et al., 2022). In this context, the informational function of packaging ceases to be static and becomes a dynamic element of the quality management system.

Indicators and sensors that respond to changes in parameters critical to food freshness and safety, such as pH, temperature, humidity, and the composition of the atmosphere inside the packaging, play a key role in smart packaging systems (D'Almeida & de Albuquerque, 2024; Abdullah et al., 2022). Colorimetric indicators are particularly important because they provide direct, visual information, eliminating the need for additional measuring devices (Liu et al., 2022). Such solutions are easy to interpret and can be effectively implemented on a large scale, which is important from an industrial point of view. pH and freshness indicators exploit the fact that food spoilage processes, especially those involving animal products, lead to the accumulation of volatile nitrogenous bases such as ammonia, trimethylamine, and dimethylamine, which cause local pH changes (Liu et al., 2022; Huang, Wang, Xu, et al., 2024). These changes can be directly reflected by a change in the color of the indicator, allowing for an unambiguous assessment of the degree of product degradation. Such solutions are particularly attractive in the context of perishable foods, where quick and unambiguous information about freshness is crucial for consumer safety (Cheng et al., 2022; Huang, Wang, Hu, et al., 2024).

Freshness indicators are supplemented by time and temperature indicators (TTI), which enable the assessment of the impact of temperature history on the actual shelf life (Liu et al., 2022). Since food quality degradation is a process dependent on both time and temperature, TTI offer a more realistic assessment of product shelf life than traditional best-before dates. Their mechanisms of action are based on irreversible chemical, enzymatic, microbiological, or physical changes leading to gradual color changes correlated with the cumulative effect of temperature (Liu et al., 2022). From a cold chain logistics perspective, TTIs enable rapid detection of breaks in the cold chain and reduce losses resulting from improper storage. At the same time, sensor technologies and identification and tracking systems, such as RFID and QR codes, are being developed to enable the collection and transmission of data on storage conditions, location, and transport history of products (D'Almeida & de Albuquerque, 2024). The integration of these solutions with digital data management systems, including blockchain technologies, increases the transparency and reliability of information throughout the supply chain, which is important for both quality control and building consumer confidence (D'Almeida & de Albuquerque, 2024).

A complementary category to smart packaging is active packaging, which, instead of merely monitoring, deliberately influences the microenvironment of the packaged food (Rocha et al., 2025; Abdullah et al., 2022). Active packaging is defined as systems in which functional substances capable of controlled release or absorption of specific compounds have been introduced into the packaging material or packaging space (Abdullah et al., 2022). In practice, this includes, among others, systems that absorb oxygen, carbon dioxide, ethylene, or moisture, as well as films that release antioxidants and antimicrobial substances (Rocha et al., 2025; Liu et al., 2022; Abdullah et al., 2022). Active films containing antioxidant and antimicrobial compounds are particularly important in the context of reducing the two main mechanisms of food spoilage: oxidation and the growth of microorganisms (Liu et al., 2022; Koshy et al., 2023). The effectiveness of such systems depends not only on the type of active substance used, but also on its interaction with the polymer matrix and the kinetics of its release into the food environment (Liu et al., 2022). Too slow release may lead to insufficient product protection, while too fast release may lead to excessive concentration of the active substance and potentially affect the sensory properties of food (Liu et al., 2022).

In response to these challenges, controlled release systems are being intensively developed, in which the diffusion mechanism of the active substance is closely linked to the environmental conditions inside the packaging, such as humidity or water activity (Liu et al., 2022). In the case of starch matrices, it is particularly important that an increase in humidity—which is also a factor accelerating food spoilage—leads to increased mobility of polymer chains and intensified release of active substances, creating a functional feedback loop between the condition of the product and the response of the packaging material (Liu et al., 2022; Fatima et al., 2024).

The integration of active and intelligent functions in a single packaging system leads to advanced solutions that simultaneously extend product shelf life and provide information about its actual quality status (Koshy et al., 2023; Sivapragasam et al., 2024). This approach enables more rational food management throughout the supply chain, reduces waste, and improves consumer safety. In this context, biopolymer packaging materials, in particular starch films, are increasingly being designed as platforms for active and intelligent systems, combining smart functionality with sustainability requirements (Arruda et al., 2025).

13.4 Starch-based smart and active packaging systems

The dynamic development of intelligent and active starch-based systems stems directly from the need to overcome the limitations of traditional passive materials, which, despite their effective barrier function, do not provide information about the actual condition of the product or actively control degradation processes (Rocha et al., 2025). In this context, starch is no longer seen merely as a biodegradable substitute for petrochemical polymers, but begins to play the role of a functional matrix capable of integrating sensory, active, and informational elements into a single material system (Rocha et al., 2025; Liu et al., 2022).

Starch-based smart packaging is defined as systems basedn on starch that monitor the internal and external conditions of the packaging and provide information on freshness, spoilage, temperature, or packaging integrity without directly interfering with the microenvironment of the product (Rocha et al., 2025; Fatima et al., 2024). A key feature of these solutions is their informational function, which supports decision-making at various stages of the supply chain – from the manufacturer, through logistics, to the end consumer (Rocha et al., 2025). The most commonly studied systems include colorimetric indicators, chemical and physical sensors, and identification solutions such as RFID, which, when combined with a starch matrix, create biodegradable "smart" platforms (Rocha et al., 2025).

The exceptional usefulness of starch as a carrier of smart functions results, among other things, from its natural transparency and colorlessness, which enable unambiguous observation of color changes in indicators without interference from the color of the packaging material (Liu et al., 2022; Che Hamzah et al., 2024). This feature is fundamental in the design of visual indicators, as it allows consumers to interpret the signal directly, without the need for additional analytical tools (Liu et al., 2022). In practice, starch indicator films enable the transmission of qualitative, semiquantitative, or quantitative information in an intuitive and easily industrially scalable manner.

The most developed trend in the field of smart starch systems is pH-sensitive films, in which natural pigments, primarily anthocyanins, act as indicators (Liu et al., 2022; Arruda et al., 2025). These compounds exhibit distinct, reversible color changes in response to pH changes, making them particularly useful for monitoring the freshness of foods susceptible to rapid spoilage, such as meat, fish, and seafood (Arruda et al., 2025; Kumar et al., 2024). As protein degradation and the accumulation of volatile basic amines progress, there is a local increase in pH, which directly initiates a change in the color of the indicator, allowing for visual assessment of the condition of the product (Cheng et al., 2022; Erna et al., 2022). The incorporation of anthocyanins into the starch matrix not only imparts sensory properties, but also significantly modifies the structure of the material. Hydrogen and electrostatic interactions between pigments and starch chains promote the formation of more homogeneous and compact polymer networks, which can result in improved barrier properties and water resistance (Cheng et al., 2022; Erna et al., 2022). At the same time, the intensity of the color response, its stability, and reversibility are strongly dependent on the pigment concentration, the type of starch, and the degree of its structural modification (Mahović Poljaček et al., 2024). The stability of indicators under conditions of increased humidity remains one of the key challenges for starch-based smart films. In response to this problem, strategies are being developed to strengthen the structure of the material through the use of additives such as bacterial nanocellulose, amylopectin nanoparticles, or cross-linking protein components, which limit matrix swelling and pigment migration (Mahović Poljaček et al., 2024; Wang et al., 2023; Zheng et al., 2023). This approach allows for high indicator sensitivity while increasing the mechanical and functional stability of the film under real-world conditions.

Parallel to the development of smart systems, intensive research is being conducted on active starch-based packaging that affects the microenvironment of packaged food through the controlled release or absorption of specific substances (Gao et al., 2024). In these systems, starch acts as a biodegradable carrier of antioxidants and antimicrobial substances, enabling their gradual release during storage (Gao et al., 2024; Kılıç & Kahyaoglu, 2025). The mechanism and kinetics of release remain crucial to effectiveness, as they determine both the efficiency of food protection and its sensory properties (Gao et al., 2024). Systems combining active and intelligent functions in a single material, known as dual-function packaging, are becoming increasingly important. An example of this are thermoplastic starch films containing extracts rich in anthocyanins, which simultaneously exhibit antioxidant activity and the ability to visually monitor food spoilage (Kılıç & Kahyaoglu, 2025). Such solutions not only extend product shelf life, but also provide early warning signals, significantly increasing the functionality of packaging throughout the supply chain.

Advanced starch-based smart systems increasingly integrate hybrid components such as conductive polymers or nanocrystalline metal-organic structures, which give films the ability to selectively detect gases characteristic of food spoilage, in particular ammonia (Chia et al., 2022; Huang, Wang, Xu, et al., 2024; Xu et al., 2024). The integration of these components with the starch matrix leads to improved phase compatibility, mechanical properties, barrier properties, and colorimetric stability, while maintaining the biodegradable nature of the material (Huang, Wang, Hu, et al., 2024; Xu et al., 2024). Application studies clearly indicate that smart and active starch systems are capable of effectively monitoring the freshness of meat, poultry, and seafood under real storage conditions, with the observed color changes showing a high correlation with classic chemical indicators of spoilage (Cheng et al., 2022; Wang et al., 2023; Kılıç & Kahyaoglu, 2025). At the same time, the presence of active components contributes to slowing down oxidative and microbiological processes, confirming the potential of these materials as fully-fledged, multifunctional packaging systems of the new generation.

The avilable literature data clearly confirms that starch is a versatile and scalable platform for designing smart and active packaging that combines quality monitoring, active food protection, and compliance with the principles of sustainable development (Rocha et al., 2025; Arruda et al., 2025). The integration of these functions in a single material opens up real prospects for industrial implementation, particularly in the disposable and short-term packaging segment.

13.5 Industrial applications of thermoplastic starch

As highlighted in previous sections, thermoplastic starch (TPS) is obtained by combining native starch, commonly with low molecular weight agents such as water, glycerol, or sorbitol, and subsequently applying heat and mechanical force. This process disrupts the granular structure of starch, transforming it into a flexible, thermoplastic material that can be molded, extruded, or cast (Gutiérrez & Valencia, 2021; Quilez-Molina & Merino, 2023). The resulting TPS exhibits sufficient processability to be shaped into films, sheets, or complex three-dimensional structures, expanding its potential for industrial applications (Table 13.1).

In industrial settings, TPS is predominantly processed via extrusion, followed by secondary shaping techniques such as blow-molding, thermoforming, or injection molding (Bastioli, 1998; Panda et al., 2023). Emerging research has also explored the use of TPS filaments in additive manufacturing, enabling the production of customized, complex parts through 3D printing (Haryńska et al., 2021; Jiang et al., 2025). Despite these advances, TPS is rarely processed in its pure form due to inherent limitations: most notably its narrow processing window, high moisture sensitivity and its tendency to become brittle over time through retrogradation (Quilez-Molina & Merino, 2023; Panda et al., 2023). To overcome these challenges, TPS is often blended with other biopolymers or additives, enhancing its mechanical strength, durability, and water resistance (Quilez-Molina et al., 2022; Menossi et al., 2025). Such blends create thermoplastic materials with properties suitable for commercial use across diverse application sectors.

The relevance of TPS spans packaging (Arruda et al., 2025; Quilez-Molina et al., 2022, 2024), agriculture (Merino et al., 2018, 2019), and consumer goods (He et al., 2023; Paola Bracciale et al., 2024; Azimi et al., 2025), among others. Several companies have emerged as key players in the TPS market, developing proprietary formulations and processing technologies to mitigate the material's intrinsic weaknesses (Novamont, 2025; Beteiligungs, 2025; Kuraray, 2025; Biotec, 2025). These commercial blends are designed to meet the demands of specific applications, balancing performance, processability, and sustainability. The following sections present case studies of TPS-based products, accompanied by a comparative overview of their main properties as reported in company data sheets and relevant research papers.

13.5.1 TPS in industry: case studies

TPS has found widespread industrial applications through the development of tailored, multi-component bioplastics that overcome the intrinsic limitations of pure starch. Several companies have developed proprietary formulations that combine TPS with biodegradable polymers, plasticizers, and

Table 13.1 Commercial TPS-Based bioplastic products: Composition, properties, certifications, and applications.

COMMERCIAL product	Company	Composition/ formulation	Renewable raw material content	MECHANICAL properties	THICKNESS	END–OF–LIFE/ CERTIFICATIONS	APPLICATIONS	Reference
Mater-Bi®	Novamont (Italy)	TPS + biodegradable polymers (PLA, PHB/ PHBV, PCL, PVA, EVOH, chitosan, cellulose derivatives)	40%–60% (depending on grade)	Max tensile strength ~40 MPa; flexibility tailored via formulation	10–12 μm (thin films) – 200 μm (bags, agricultural films)	Industrial & home compostable (EN 13432, OK Compost Home)	Food packaging (bags, plates, cups, cutlery), agricultural films (mulching, raffia, clips)	Mater-Bi (2025a, 2025b, 2025c, 2025d, 2025e)
PLANTIC™/ PLANTIC™ HP	Kuraray (Australia)	High-amylose starch; reactive extrusion; optionally blended with PLA or PCL	Up to 80% (85% for HP)	Tailored via reactive extrusion; high gas barrier (O_2, CO_2); anti-static; oil resistant	<300 μm (flexible films, lidding, paper films)	Biobased and 100% biodegradable; home compostable (EN 13432)	Flexible packaging, lidding, high-barrier paper films; fresh and cooked meat, pasta, chocolates, dairy, snacks, convenience foods, seafood, bakery, petfoods	PLANTIC (2025a), Khemani et al. (2008)
Bioplast® 300	Biotec (Germany)	Starch-based TPS + epoxide-containing polymer (glycidyl (meth)acrylate copolymer); plasticized with glycerol/sorbitol	32%	Tensile strength 2–10 N/mm^2; elongation 80%–200%	10–58 μm (can be drawn down to 10 μm); max thickness OK compost INDUSTRIAL 84 μm, HOME 79 μm	Industrial & home compostable (EN 13432)	Ultra-light single-use bags, single- and multi-use bags, technical films; suitable for VFFS	哈拉尔德·施密特; 克里斯托夫·黑塞; 约翰内斯·马塔; 拉尔夫·哈克福特 (2013), Bioplast (2025)

Bioplast® 600	Biotec (Germany)	Starch-based TPS + epoxide-containing polymer (glycidyl (meth)acrylate copolymer); plasticized with glycerol/sorbitol; core layer in co-extrusion	63%	Tensile strength 2–10 N/mm^2; elongation 80%–200%	Max thickness OK compost INDUSTRIAL 94 μm, HOME 64 μm	Industrial & home compostable (EN 13432); ISCC PLUS	Core layer in co-extruded films; ultra-light single-use bags, multi-use bags, technical films	哈拉尔德·施密特; 克里斯托夫·黑塞; 约翰内斯·马塔; 拉尔夫·哈克福特 (2013), Bioplast (2025)
AMITROPLAST®	AGRANA Bioplastics	Thermoplastic starch (as additive to polymer compounds)	100%	In a compound with 40% AMITROPLAST® + 60% biodegradable polyester: ~35 MPa tensile strength, ~300% extensibility (for 20 μm film)	An example: 20 μm film (in a compound)	Biodegradable, compostable in home or industrial composting; recyclability when in compounds.	Extrusion & thermoforming: carrier bags, fruit/vegetable bags, protective packaging, non-woven fibers, technical films, forestry applications. Injection molding: agricultural clips, pots, 3D-print filament, cutlery, sanitary products, packaging, etc.	Products (2025)
AGENACOMP®	AGRANA Bioplastics	Compound of thermoplastic starch (AMITROPLAST®) + a biodegradable polymer	Up to 50%	N/A	Films <10 μm by standard film extrusion possible	Home-compostable; certified according to EN 13432	Flexible packaging films, mulch films, 3D-printing, multilayer & monolayer films, bags.	Products (2025)

reactive additives to provide flexibility, strength, and processability, while maintaining biodegradability and high renewable content. The following case studies illustrate representative TPS-based products used in packaging, agriculture, and consumer goods.

13.5.1.1 *Mater-Bi*

Mater-Bi is a commercially available family of biodegradable bioplastics based on TPS, representing one of the most studied examples of starch-based materials for industrial applications (Paola Bracciale et al., 2024). TPS in Mater-Bi is obtained by plasticizing native starch, typically amylopectin-rich, with low molecular weight plasticizers such as glycerol or sorbitol and subsequently applying heat and mechanical force to disrupt the starch granules (Bastioli et al., 1995). To overcome the intrinsic limitations of pure TPS, Mater-Bi is formulated as a multi-component blend with biodegradable polymers and this way it is processable through blow-molding, thermoforming, or injection molding (Bastioli, 1998).

The formulations of Mater-Bi generally include TPS combined with synthetic or semi-synthetic polymers, such as polyesters (polylactic acid, polyhydroxybutyrate/valerate, polycaprolactone), ethylene-vinyl alcohol copolymer, polyvinyl alcohol, and selected hydrophilic polymers such as chitosan, chitin, cellulose derivatives, and thermoplastic gums (Bastioli, Bellotti, & Montino, 1993). Some of these polymers are chemically modified, for example, through partial substitution of hydroxyl groups with oxo or alkyl carbonyl groups, to tailor hydrophilicity, mechanical strength, and barrier properties (Bastioli, Bellotti, Del, et al., 1993). By combining TPS with these polymers, Mater-Bi achieves improved flexibility, water resistance, mechanical strength, and barrier performance compared to pure starch, while maintaining biodegradability and compostability.

Mater-Bi is used across multiple sectors, including retail, food packaging, foodservice, and agriculture. Examples of applications include ultra-light fruit and vegetable bags (Mater-Bi, 2025f), reusable shopping bags (Mater-Bi, 2025a), organic waste bags (Mater-Bi, 2025g), and disposable foodservice items such as plates, cups, cutlery, and bowls (Mater, 2025). The material is suitable for microwave heating and designed to withstand typical handling stresses. In agriculture, Mater-Bi is applied in biodegradable mulching films (Mater-Bi, 2025h), and twines and clips (Mater-Bi, 2025b), providing weed control and mechanical support during crop growth while fully biodegrading in soil without the need for collection at the end of the cycle.

Commercial grades of Mater-Bi vary in the proportion of renewable raw materials, ranging from minimum 40% to over 60% depending on regulatory requirements and application needs (Mater-Bi, 2025c, 2025d). Thicknesses of products vary widely, from 10 to 12 μm for thin films to over 200 μm for reusable shopping bags, and up to 200 μm for agricultural applications (Mater-Bi, 2025a, 2025c, 2025d, 2025e). Mechanical performance is similarly tailored, with maximum tensile strength reaching approximately 40 MPa for specific formulations (Mater-Bi, 2025e).

Mater-Bi products are certified for industrial and home compostability according to standards such as EN 13432 and TÜV Austria OK Compost Home (Mater-Bi, 2025c, 2025d, 2025e). Some products also include traceability and carbon footprint monitoring systems, providing data on renewable content and environmental impact in line with ISO 14067 (Mater-Bi, 2025i). Overall, Mater-Bi serves as a representative example of how TPS-based bioplastics can be engineered to meet diverse industrial requirements while maintaining biodegradability, compostability, and a high content of renewable resources.

13.5.1.2 PLANTIC

PLANTIC is a biomass-derived thermoplastic material produced from high-amylose starch, primarily sourced from corn. The starch is obtained from naturally high-amylose varieties that have been selectively bred over multiple generations to provide enhanced functional properties (PLANTIC, 2025b). PLANTIC is commercialized since 2003 and is designed as a sustainable, high-performance alternative to conventional plastics for food packaging and other flexible applications. The material typically contains up to 80% renewable content, with formulations such as PLANTIC HP reaching 85% renewable raw material (PLANTIC, 2025a).

The material exhibits exceptional gas barrier properties, particularly for oxygen and carbon dioxide, making it suitable for packaging fresh produce and other perishable foods. These barrier characteristics extend shelf life and improve food quality during storage and transportation. PLANTIC products are also inherently anti-static, oil resistant, sealable, printable, and laser-etchable. In addition, they are designed to be biodegradable and home compostable in accordance with EN 13,432 standards, and are certified as biobased and biodegradable according to European benchmarks. The production of PLANTIC reportedly uses up to 40% less energy than conventional polymers, while offering cost-competitive performance and long-term stability due to the relatively stable price of corn compared to petrochemical feedstocks (PLANTIC, 2025a).

The processing of PLANTIC relies on reactive extrusion modification of starch and other polysaccharides. According to a patent of the company, the process involves reacting pre-gelatinized high-amylose starch in an aqueous state with modifying agents, such as cyclic esters, cyclic carbonates, epoxides, lactide, or glycolide, in the presence of optional Lewis acid catalysts (Khemani et al., 2008). This reactive extrusion allows the grafting of functional groups onto the polysaccharide backbone, yielding a thermoplastic material with improved mechanical strength, flexibility, and processability. Modified PLANTIC polymers can also be blended with other biodegradable polymers, such as polylactic acid or polycaprolactone, to further tailor functional properties for specific applications.

PLANTIC is available in a range of flexible structures, typically less than 300 μm thick, suitable for thermoforming, pouches, lidding, and high-barrier paper films. Specific grades include PLANTIC EF for high-barrier flexible packaging, PLANTIC TF for lidding applications, and PLANTIC PF for paper films (PLANTIC, 2025c). These materials have been applied in a variety of food markets, including pasta, fresh meat, cooked meat, chocolates, dairy, snack foods, convenience foods, seafood, poultry, bakery products, and pet foods.

13.5.1.3 Bioplast

Bioplast, developed by the German company Biotec, represents a family of starch-based, compostable thermoplastic materials designed for flexible industrial applications (Biotec, 2025). These materials illustrate how TPS can be engineered with plasticizers and reactive polymers to overcome the inherent limitations of pure starch, such as brittleness, moisture sensitivity, and retrogradation.

TPS in Bioplast is obtained by plasticizing native or modified starch with low-molecular-weight agents such as glycerol or sorbitol, followed by the application of thermal and mechanical energy to disrupt the granular structure. To enhance mechanical strength, flexibility, and processability, Bioplast incorporates small quantities of epoxide group-containing polymers, such as glycidyl (meth)acrylate copolymers (**哈拉尔德·施密特; 克里斯托夫·黑塞; 约翰内斯·马塔; 拉尔夫·哈克福特**, 2013). These act as compatibilizers, reacting with the starch matrix to improve interfacial adhesion and film integrity. The resulting polymer is thermoplastically processable and can be formed into films, fibers, or molded articles with controlled mechanical properties.

Bioplast formulations vary depending on application requirements. Bioplast 300, for example, is suitable for blown film extrusion in Vertical Form Film Seal (VFFS) processes, and is designed to be home-compostable even at higher thicknesses (~58 μm) (Bioplast, 2025). It contains approximately 32% bio-based carbon and is printable, sealable, and compatible with coloring using compostable masterbatches. Bioplast 600 is developed as a core layer for co-extruded films, particularly in markets with specific regulatory requirements such as France and Italy (Bioplast, 2025). This grade features 63% bio-based carbon and holds ISCC PLUS certification for renewable content. Both grades are fully biodegradable and compliant with EN 13,432 standards.

Bioplast films exhibit tensile strengths ranging from 2 to 10 MPa and elongation at break between 80% and 200%, depending on the grade and processing method (哈拉尔德·施密特; 克里斯托夫·黑塞; 约翰内斯·马塔; 拉尔夫·哈克福特, 2013). Water content is tightly controlled during processing to ensure consistent thermoplastic behavior, typically between 1 and 7 wt.% 38. The materials are compatible with conventional industrial processing techniques, including extrusion, thermoforming, and co-extrusion, enabling the production of ultra-light single-use bags, multi-use bags, technical films, and co-extruded packaging structures.

The versatility of Bioplast allows its application across multiple sectors, from consumer goods to agriculture. Home-compostable films can be used for shopping bags and waste collection, while co-extruded films provide tailored barrier properties for specialty packaging.

13.5.1.4 AMITROPLAST and AGENACOMP

AMITROPLAST and AGENACOMP are TPS products developed by AGRANA, an Austrian company with a longstanding expertise in starch and sugar production (Kuraray, 2025). AMITROPLAST is a versatile thermoplastic starch designed for use in extrusion, film blowing, injection molding, and 3D printing. It enables the incorporation of significant TPS content into bioplastic compounds, allowing manufacturers to increase the bio-based fraction of their products while maintaining compatibility with conventional polymer processing equipment. The material is 100% bio-based, biodegradable, and cost-effective, providing flexibility for a wide range of industrial applications.

Applications of AMITROPLAST include carrier and shopping bags, protective packaging, waste bags, bin liners, non-woven fibers, technical films, and forestry products. In injection molding, it is used to produce agricultural clips, plant pots, cutlery, sanitary products, and packaging items. AMITROPLAST compounds demonstrate high mechanical performance: for example, a 20 μm film composed of 40% AMITROPLAST and 60% biodegradable polyester typically achieves an extensibility of ~300% and tensile strength of 35 MPa, with tensile strength further increasing to 60 MPa under controlled stretching (Products, 2025). The material also minimizes smoke formation during film blowing and can be recycled and reintroduced into the manufacturing process.

AGENACOMP complements AMITROPLAST as a compound combining thermoplastic starch with biodegradable polymers, enhancing mechanical properties while maintaining home compostability. Designed for flexible packaging films, mulch films, and 3D printing, AGENACOMP allows extrusion of ultra-thin films (<10 μm) on standard equipment. Product variants differ in bio-based content and starch proportion, enabling tailored solutions for specific applications: F30 offers mechanical performance comparable to polyethylene, F40 contains 33% bio-based carbon, F50 contains 40%, F51 exceeds 50%, and F61 exceeds 60% bio-based carbon (Agrana, 2025).

The production of AMITROPLAST and AGENACOMP involves extrusion of thermoplastic starch with polyols (e.g., glycerol, sorbitol, erythritol, mannitol), epoxidized plant oils (soybean, linseed, sunflower, or rapeseed oils), and small quantities of acids (citric, malic, tartaric, or acetic acid) to enhance functionality. Extrusion temperatures typically range from 100 to 175 °C, with degassing applied through twin-screw extrusion to reduce moisture content (Fahrngruber et al., 2022). The resulting compounds can be processed into transparent films, molded articles, or 3D-printed objects that are fully biodegradable and suitable for both home and industrial composting.

13.6 Safety, regulatory, economic and Environmental Aspects

This section summarizes the main regulatory requirements, safety testing, biodegradability certification schemes, life-cycle evidence and economic/market factors that determine whether TPS-based materials (including the commercial blends described in Section 5) can be deployed at industrial scale, with emphasis on food-contact use, end-of-life claims and market acceptance. Each subsection highlights practical implications for formulators and manufacturers and points to key reference standards and guidance.

13.6.1 Food-contact safety and regulatory frameworks

In the EU, food-contact materials (FCMs) are regulated by a two-tier framework: the general framework Regulation (EC) No 1935/2004 (safety, inertness and documentation/traceability) (Regulation, 2021) together with specific measures for plastics (Commission Regulation (EU) No 10/2011) that set rules for plastic materials and list authorized substances, overall-migration limits (OML) and specific-migration limits (SMLs) (Regulation, 2025). Compliance for plastics is normally demonstrated by migration testing (overall and specific migration) under the worst-case time/temperature/food-simulant conditions and by a Declaration of Compliance (DoC) from the manufacturer. Recent EU updates (2025) have further revised some FCM rules and assessment procedures, so manufacturers must verify the current implementing acts before placing new formulations on the market.

With the adoption of Regulation (EU) 2025/351 (Brito et al., 2019), the EU has overhauled and tightened the regulatory framework for plastics as FCMs. The new Regulation introduces a formal "high degree of purity" requirement for any substance used in plastic food-contact materials, meaning that only minimal levels of non-intentionally added substances (NIAS) may be present, and each such NIAS must individually satisfy one of four criteria (e.g., comply with existing specifications, undergo risk assessment, or limit migration to either $\leq$0.05 mg/kg (if genotoxicity is ruled out) or $\leq$ 0.00,015 mg/kg otherwise).

For plasticizers, which are often among the most mobile additives in polymers, these changes significantly raise the bar: their identity, purity, and migration potential must now be more rigorously characterized and justified, and they are explicitly subject to NIAS thresholds and risk assessment obligations. The revised Regulation also updates migration testing rules (Annex V), requiring repeated migration tests to demonstrate that migration does not increase over successive cycles, and mandating that measurement uncertainty be considered in compliance assessments.

For plastics in the EU, the typical OML value used for compliance is 10 mg/dm^2 (or 60 mg/kg food), while SMLs apply to individual regulated substances. Migration testing follows EN/ISO standards and the Commission's rules in Regulation October 2011 (Regulation, 2025), simulating intended use (food type, temperature, time) and worst-case conditions for thermal and hydrolytic stability, important for starch-based materials. Because TPS is hydrophilic and interacts strongly with aqueous or acidic foods, migration into polar simulants and potential leaching of plasticizers (e.g., glycerol and its derivatives) must be carefully examined. TPS and TPS-based blends are therefore treated as FCMs and must comply with the same migration testing regime, with special attention to low-molecular-weight additives (plasticizers, compatibilizers, residual monomers and oligomers), which often determine SML/OML compliance risk.

13.6.2 Biodegradability, compostability and end-of-life claims

Biodegradability and compostability claims for TPS-based materials are assessed using standardized test methods that define environmental conditions, timescales, and pass/fail criteria.

EN 13,432:2001/AC:2005 (Commission Regulation, 2025) (Europe) is the key harmonized industrial compostability standard for packaging and plastics. It specifies a comprehensive test scheme under controlled composting conditions, covering.

- Disintegration: Within 12 weeks, at least 90% of the material must physically fragment into pieces smaller than 2 mm, meaning no more than 10% dry weight may remain on a 2 mm sieve.
- Biodegradation: Within 6 months (180 days), at least 90% of the organic carbon must be converted to CO_2 in aerobic composting, relative to a reference material (typically cellulose).
- Compost quality and ecotoxicity: The resulting compost must not negatively affect plant growth, and the material must comply with strict limits on heavy metals and other potentially harmful substances.

Materials that meet all these criteria are considered industrially compostable, i.e., suitable for treatment in managed composting facilities, though not necessarily under home-composting conditions. Certification bodies such as TÜV Austria, issue conformity marks including OK Compost INDUSTRIAL, and OK Compost HOME, which are often required for commercial claims (UNE, 2025).

OK Compost HOME apply less stringent and lower-temperature/time conditions (e.g., ~20–30 °C instead of 55–60°C for OK Compost INDUSTRIAL), making the test more challenging for some materials 48. Achieving home-compostable certification requires demonstrating adequate disintegration and biodegradation at these milder conditions, which is often more difficult for thick or slowly degrading TPS-based products.

Marine biodegradability standards (e.g., ASTM D6691, ISO/TS 18,830, and TÜV's OK Biodegradable MARINE) assess biodegradation in seawater environments (OK compost, 2025; OK biodegradable, 2025; ASTM, 2025). These tests are more demanding and rarely passed by starch-based blends without specific formulation strategies to enhance degradation. As a result, marine certification remains uncommon, and such claims should only be made when supported by robust laboratory and field evidence.

Many TPS blends (Mater-Bi, Bioplast, AGRANA grades) are formulated to meet industrial and/or home compostability requirements and are therefore commonly certified under EN 13,432/OK Compost; nevertheless, compostability is conditional on local collection and composting infrastructure and feedstock contamination limits. When making public claims, manufacturers must state the specific certificate and test conditions.

13.6.3 Environmental, economic aspects and market adoption

Life cycle assessment (LCA) studies indicate that TPS and TPS-based blends can, in some cases, exhibit higher carbon footprints than conventional fossil-based plastics, particularly for films, due to low starch yields, energy-intensive processing, and agricultural inputs such as fertilizers and pesticides (Forfora et al., 2024). For example, TPS films may range between 3.2 and 5.8 kg CO_2-eq/kg, exceeding LDPE or PET in some scenarios, whereas TPS foams show a more favorable profile (1.3–3.2 kg CO_2-eq/kg) compared with EPS foams (5.1–6.7 kg CO_2-eq/kg). These results highlight that environmental performance is strongly influenced by feedstock efficiency, crop management, and manufacturing technology.

A recent study evaluated pellet- and film-based products from corn starch, identifying the main hotspots in manufacturing and end-of-life treatments. While the addition of PLA and PBAT enhances thermo-mechanical properties, PBAT was identified as the primary contributor to environmental impacts across most categories. Six disposal scenarios were assessed, with incineration compared against fossil-based LDPE; in these cases, starch-based plastics demonstrated savings of 1.63–2.10 kg CO_2 and 29.6–76.1 MJ per kilogram relative to fossil counterparts (Poveda-Giraldo et al., 2025). Notably, anaerobic digestion and integrated energy recovery (IPR and AD) scenarios showed the lowest environmental impacts, providing credits through electricity generation and fertilizer substitution, and highlighting the compatibility of TPS materials with circular energy strategies. Recommendations from this work suggest that reducing PBAT content or substituting it with bio-based monomers could decrease carbon footprints by 16%–34% and reduce net greenhouse gas emissions by up to 1.2 kg CO_2 per kilogram in anaerobic digestion treatments (Poveda-Giraldo et al., 2025).

Despite the potential for higher carbon emissions, TPS materials offer critical advantages: they are derived from renewable resources and are biodegradable, reducing persistent plastic waste and mitigating pollution in marine and terrestrial ecosystems. Economically, TPS and TPS-based blends remain generally more expensive than petrochemical polymers, reflecting feedstock costs, limited economies of scale, and additional processing requirements such as compounding or certification (by, 2025). Nevertheless, TPS products could be cost-competitive in niche applications, such as compostable bags, agricultural films, or specialty packaging, especially where local feedstocks or high-volume supply chains are available.

Packaging is the dominant application segment for bioplastics (≈45% of current global bioplastics market), and TPS-based materials are widely used for films, bags and foodservice items. Market reports forecast continued growth in production capacity through 2030 (Thermoplastic Starch Market, 2025). Consumer research shows growing awareness and favorable attitudes toward biodegradability and compostability, although misunderstandings about how to recognize biodegradable packaging, proper disposal routes and concerns about performance and price remain important barriers (Bojanowska & Sulimierska, 2023; Cao et al., 2025). Clear labeling, education and compatible collection/composting infrastructure are critical to increase adoption.

13.7 Short summary and research needs

Available literature data indicate that starch-based materials, in particular thermoplastic starch formulations, have reached a level of maturity that allows them to be used in food contact packaging and compostable systems. Advances in physical, chemical, and composite modifications have significantly improved the mechanical and barrier properties of starch films, bringing them closer to conventional packaging materials. At the same time, starch has shown high potential as a platform for active and intelligent packaging, enabling both the extension of food shelf life through the controlled release of antioxidants and antimicrobial substances, and real-time freshness monitoring using colorimetric indicators and sensory systems.

Despite these achievements, the widespread industrial adoption of starch-based packaging systems remains limited by a number of barriers. A key challenge is the inherent properties of starch, in particular its high hydrophilicity, low barrier stability under conditions of elevated humidity, and susceptibility to retrogradation and variability in properties depending on the type of starch, processing technology, and storage conditions. In addition, most research to date has been conducted on a laboratory scale, making it difficult to assess the stability, safety, and repeatability of materials under actual production conditions. Priority areas for further research include the development of formulations with increased hydrolytic resistance and better barrier properties while maintaining biodegradability and compostability, as well as the search for alternative, low-cost sources of starch and the intensification of technological processes to reduce the environmental footprint. In the area of active packaging, further research is needed on the kinetics of controlled release of functional substances and their compatibility with the starch matrix to ensure effectiveness without adversely affecting food safety and sensory characteristics.

In the case of smart packaging, a significant research gap remains in the limited functionality of current systems, which are mainly based on pH indicators. Interactions between active ingredients and the product environment can lead to signal distortion and misinterpretation by the consumer. Therefore, research is needed on multifunctional, multi-reactive starch systems capable of selectively responding to various environmental stimuli and integrating sensory functions with controlled release systems. At the same time, the lack of standardized methods for assessing antimicrobial and antioxidant activity in realistic food matrices remains a challenge. The development of reliable testing procedures, control of low-molecular-weight additive migration, access to reliable LCA data, transparency of environmental declarations, and the existence of efficient end-of-life management pathways will be key to the commercialization of starch-based packaging systems.

In summary, despite existing limitations, starch remains one of the most promising raw materials for the development of active and intelligent biodegradable packaging. Its availability, structural flexibility, and compatibility with advanced functionalization strategies provide a solid foundation for further development and industrial implementation, provided that research is focused on real technological and environmental needs.

References

Abdullah, Z. W., & Dong, Y. (2018). Recent Advances and Perspectives on Starch Nanocomposites for Packaging Applications. *Journal of Materials Science, 53*(22), 15319–15339. https://doi.org/10.1007/s10853-018-2613-19

Abdullah, Cai, J., Hafeez, M. A., Wang, Q., Farooq, S., Huang, Q., Tian, W., & Xiao, J. (2022). Biopolymer-Based functional films for packaging applications: A review. *Frontiers in Nutrition, 9*, Article 1000116. https://doi.org/10.3389/fnut.2022.1000116

Agrana. Agenacomp. https://www.agrana-bioplastics.com/products/ (accessed 2025-10-01).

Amaraweera, S. M., Gunathilake, C., Gunawardene, O. H. P., Fernando, N. M. L., Wanninayaka, D. B., Dassanayake, R. S., Rajapaksha, S. M., Manamperi, A., Fernando, C. A. N., Kulatunga, A. K., & Manipura, A. (2021). Development of Starch-Based materials using Current modification techniques and their applications: A review. *Molecules, 26*(22), 6880. https://doi.org/10.3390/molecules26226880

Arruda, T. R., Machado, G. D., Marques, C. S., Souza, A. L., Pelissari, F. M., Oliveira, T. V., & Silva, R. R. (2025). An overview of starch-based materials for sustainable food packaging: Recent advances, limitations, and perspectives. *Macromol, 5* (2), 19. https://doi.org/10.3390/macromol5020019

ASTM D6691 Aerobic Biodegradation of Plastic Materials in the Marine Environment by a Defined Microbial Consortium Lab In US - Infinita Lab. https://infinitalab.com/astm/astm-d6691-aerobic-biodegradation-of-plastic-materials-in-the-marine-environment-by-a-defined-microbial-consortium/?srsltid=AfmBOorsdnpJAnnaEHJYA8W8zAdDLr0Z_pQqbg0swDd5GfoWLl83qI32 (accessed 2025-10-01).

Azimi, N. R., Sonchaeng, U., Leelaphiwat, P., Tongdeesoontorn, W., & Bumbudsanpharoke, N. (2025). Comparative Study of packaging performance: Thermoplastic starch blends and polyester blends for biodegradable personal care packaging. *International Journal of Biological Macromolecules, 320*, Article 145739. https://doi.org/10.1016/j.ijbiomac.2025.145739

Bastioli, C. (1998). Properties and applications of Mater-Bi starch-based materials. *Polymer Degradation and Stability, 59*(1), 263–272. https://doi.org/10.1016/S0141-3910(97)00156-0

Bastioli, C.; Bellotti, V.; Del, G. L.; Lombi, R. Polymer Compositions for the Production of Articles of Biodegradable Plastics Material and Methods of Their Preparation. US5258430A, November 2, 1993.

Bastioli, C.; Bellotti, V.; Del, T. G.; Montino, A.; Ponti, R. Film-Forming, Starchy, Polymeric Composition and Shaped Articles, Particularly Films and Sheets, Which Can Be Produced from the Composition and Have a Good Barrier Effect, and a Method of Producing the Articles. US5462980A, October 31, 1995.

Bastioli, C.; Bellotti, V. V. M. E. G.; Montino, A. Biodegradable Polymeric Compositions Based on Starch and Thermoplastic Polymers. EP0575349A1, December 29, 1993.

Beteiligungs-AG, A. AGRANA. https://www.agrana.com/ (accessed 2025-09-19).

Bioplast 300 – Biotec. https://www.biotec.de/products/bioplast-300/ (accessed 2025a-09-19).

Bioplast 600 – Biotec. https://www.biotec.de/products/bioplast-600/ (accessed 2025b-09-19).

Biotec. BIOTEC - Advanced biopolymers for building a better tomorrow. https://www.biotec.de/ (accessed 2025-09-19).

Bojanowska, A., & Sulimierska, A. (2023). Consumer awareness of biodegradability of food products packaging. *Sustainability, 15*(18), Article 13980. https://doi.org/10.3390/su151813980

Brito, T. B., Carrajola, J. F., Gonçalves, E. C. B. A., Martelli-Tosi, M., & Ferreira, M. S. L. (2019). Fruit and vegetable residues flours with different granulometry range as raw material for pectin-enriched biodegradable film preparation. *Food Research International, 121*, 412–421. https://doi.org/10.1016/j.foodres.2019.03.058

by, W. Cost Comparison of Biodegradable Materials: What You Should Know. https://planetpristine.com/sustainable/products/cost-comparison-of-biodegradable-materials/ (accessed 2025-10-01).

Cao, Y., Otsuka, A., Teah, H. Y., Kanematsu, Y., & Kikuchi, Y. (2025). Consumers' knowledge, perceptions, and preferences regarding Bioplastics: Evidence from a Survey in Japan. *Sustainable Production and Consumption, 55*, 1–10. https://doi.org/10.1016/j.spc.2025.02.003

Che Hamzah, N. H., Khairuddin, N., & Muhamad, I. I. (2024). Immobilisation of anthocyanin in starch-based film as a potential colorimetric oxygen indicator in modified atmosphere condition. *Adv Agric Food Res J, 5*(2). https://doi.org/10.36877/aafrj.a0000503

Cheng, M., Yan, X., Cui, Y., Han, M., Wang, X., Wang, J., & Zhang, R. (2022). An eco-friendly film of pH-Responsive indicators for smart packaging. *Journal of Food Engineering, 321*, Article 110943. https://doi.org/10.1016/j.jfoodeng.2022.110943

Chia, M.-R., Ahmad, I., & Phang, S.-W. (2022). Starch/Polyaniline biopolymer film as potential intelligent food packaging with colourimetric ammonia sensor. *Polymers, 14*(6), 1122. https://doi.org/10.3390/polym14061122

Commission Regulation (EU) 2025/351 of 21 February 2025 Amending Regulation (EU) No 10/2011 on Plastic Materials and Articles Intended to Come into Contact with Food, Amending Regulation (EU) 2022/1616 on Recycled Plastic Materials and Articles Intended to Come into Contact with Foods, and Repealing Regulation (EC) No 282/2008, and Amending Regulation (EC) No 2023/2006 on Good Manufacturing Practice for Materials and Articles Intended to Come into Contact with Food as Regards Recycled Plastic and Other Matters Related to Quality Control and Manufacturing of Plastic Materials and Articles Intended to Come into Contact with Food; 2025. http://data.europa.eu/eli/reg/2025/351/oj/eng (accessed 2025-10-01).

D'Almeida, A. P., & de Albuquerque, T. L. (2024). Innovations in food packaging: From bio-based materials to smart packaging systems. *Processes, 12*(10), 2085. https://doi.org/10.3390/pr12102085

Erna, K. H., Felicia, W. X., Vonnie, J. M., Rovina, K., Yin, K. W., & Nur'Aqilah, M. N. (2022). Synthesis and physicochemical characterization of polymer film-based anthocyanin and starch. *Biosensors, 12*(4), 211. https://doi.org/10.3390/bios12040211

Fahrngruber, B.; Wastyn, M. M.; Kozich, M. Compound or Film Containing Thermoplastic Starch and a Thermoplastic Polymer. US2022064411A1, March 3, 2022.

Fatima, S., Khan, M. R., Ahmad, I., & Sadiq, M. B. (2024). Recent advances in modified starch based biodegradable food packaging: A review. *Heliyon, 10*(6). https://doi.org/10.1016/j.heliyon.2024.e27453

Forfora, N., Azuaje, I., Kanipe, T., Gonzalez, J. A., Lendewig, M., Urdaneta, I., Venditti, R., Gonzalez, R., & Argyropoulos, D. (2024). Are starch-based materials more eco-friendly than Fossil-Based? A critical assessment. *Cleaner Environmental Systems, 13*, Article 100177. https://doi.org/10.1016/j.cesys.2024.100177

Gao, S., Li, M., Zhai, X., Wang, W., & Hou, H. (2024). Starch as a smart, cheap, and green gatekeeper for the controlled release of Propyl gallate from antioxidant biodegradable packaging films. *Food Chemistry, 453*, Article 139627. https://doi.org/10.1016/j.foodchem.2024.139627

Garavito, J., Peña-Venegas, C. P., & Castellanos, D. A. (2024). Production of starch-based flexible food packaging in developing countries: Analysis of the processes, challenges, and requirements. *Foods, 13*(24), 4096. https://doi.org/10.3390/foods13244096

Gutiérrez, T. J., & Valencia, G. A. (2021). Reactive extrusion-processed native and phosphated starch-based food packaging films governed by the hierarchical structure. *International Journal of Biological Macromolecules, 172*, 439–451. https://doi.org/10.1016/j.ijbiomac.2021.01.048

Haryńska, A., Janik, H., Sienkiewicz, M., Mikolaszek, B., & Kucińska-Lipka, J. (2021). PLA–Potato thermoplastic starch filament as a sustainable alternative to the conventional PLA filament: Processing, characterization, and FFF 3D printing. *ACS Sustainable Chemistry & Engineering, 9*(20), 6923–6938. https://doi.org/10.1021/acssuschemeng.0c09413

He, X., Zhao, S., Zhang, Z., Dai, L., Qin, Y., Ji, N., Xiong, L., Shi, R., & Sun, Q. (2023). A combined extrusion, retrogradation, and cross-linking strategy for preparing starch-based straws with desirable mechanical properties. *International Journal of Biological Macromolecules, 227*, 1089–1097. https://doi.org/10.1016/j.ijbiomac.2022.11.289

Huang, X., Wang, F., Hu, W., Zou, Z., Tang, Q., Li, H., & Xu, L. (2024). Smart packaging films based on corn Starch/Polyvinyl alcohol containing nano SIM-1 for monitoring food freshness. *International Journal of Biological Macromolecules, 256*, Article 128373. https://doi.org/10.1016/j.ijbiomac.2023.128373

Huang, K., Wang, Y., Xu, Z., Zou, Z., Tang, Q., Li, H., & Peng, D. (2024). Novel intelligent packaging films based on Starch/PVA with Cu-ICA nanocrystal as functional compatibilizer for monitoring food freshness. *International Journal of Biological Macromolecules, 271*, Article 132373. https://doi.org/10.1016/j.ijbiomac.2024.132373

Jiang, G., Zhang, W., Fan, L., Luo, C., Leng, P., & Tao, D. (2025). One-Step strategy for preparing renewable polylactic acid (PLA)/Thermoplastic starch (TPS) composite filaments for 3D printing. *Industrial Crops and Products, 233*, Article 121444. https://doi.org/10.1016/j.indcrop.2025.121444

Kılıç, N. N., & Kahyaoglu, L. N. (2025). Smart thermoplastic starch films with dual functionality: Quality preservation of tomatoes and spoilage detection in chicken. *International Journal of Biological Macromolecules, 330*, Article 147898. https://doi.org/10.1016/j.ijbiomac.2025.147898

Khemani, K. C.; Morris, B.; Moad, G.; Li, G. Reactive Extrusion Modification of Functional Polymers. WO2008074096A1, 2008. https://patents.google.com/patent/WO2008074096A1/en (accessed 2025-09-19).

Koshy, J. T., Vasudevan, D., Sangeetha, D., & Prabu, A. A. (2023). Biopolymer based multifunctional films loaded with anthocyanin rich floral extract and ZnO nano particles for smart packaging and wound healing applications. *Polymers, 15*(10), 2372. https://doi.org/10.3390/polym15102372

Kumar, Y., Bist, Y., Thakur, D., Nagar, M., & Saxena, D. C. (2024). A review on the role of pH-Sensitive natural pigments in biopolymers based intelligent food packaging films. *International Journal of Biological Macromolecules, 276*, Article 133869. https://doi.org/10.1016/j.ijbiomac.2024.133869

Kuraray. https://www.kuraray.com/global-en/ (accessed 2025-09-19).

Lauer, M. K., & Smith, R. C. (2020). Recent advances in starch-based films toward food packaging applications: Physicochemical, mechanical, and functional properties. *Comprehensive Reviews in Food Science and Food Safety, 19*(6), 3031–3083. https://doi.org/10.1111/1541-4337.12627

Liu, D., Zhao, P., Chen, J., Yan, Y., & Wu, Z. (2022). Recent advances and applications in starch for intelligent active food packaging: A review. *Foods, 11*(18), 2879. https://doi.org/10.3390/foods11182879

Mahović Poljaček, S., Tomašegović, T., Strižić Jakovljević, M., Jamnicki Hanzer, S., Murković Steinberg, I., Žuvić, I., Leskovac, M., Lavrič, G., Kavčič, U., & Karlovits, I. (2024). Starch-Based functional films enhanced with bacterial nanocellulose for smart packaging: Physicochemical properties, pH sensitivity and colorimetric response. *Polymers, 16*(16), 2259. https://doi.org/10.3390/polym16162259

Mater-. Envases de alimentos. Mater-Bi. https://materbi.com/es/aplicaciones/envases-de-alimentos/ (accessed 2025-09-19).

Mater-Bi. La rafia y clips. Mater-Bi. https://materbi.com/es/aplicaciones/la-rafia-y-clips/ (accessed 2025b-09-19).
Mater-Bi. Data Sheet - FRUIT AND VEGETABLE BAGS STRONG, BIODEGRADABLE AND COMPOSTABLE. https://materbi.com/wp-content/uploads/2024/06/scheda_saccofruttaverdura_EN.pdf (accessed 2025c-09-19).
Mater-Bi. Data Sheet - MATER-BI FOR BIOWASTE BIODEGRADABLE AND COMPOSTABLE SOLUTIONS FOR ORGANIC WASTE COLLECTION. https://materbi.com/wp-content/uploads/2024/03/scheda-waste_EN_TUV_LR-.pdf (accessed 2025d-09-19).
Mater-Bi. Data Sheet - Mater-Bi Para Acolchado Soluciones Para La Biodegradabilidad En El Suelo. https://materbi.com/wp-content/uploads/2024/10/Pacciamatura-_SPA.pdf (accessed 2025e-09-19).
Mater-Bi. Bolsa para fruta y verdura | Biodegradabili Mater-Bi. Mater-Bi. https://materbi.com/es/aplicaciones/bolsa-para-fruta-y-verdura/ (accessed 2025f-09-19).
Mater-Bi. Bolsa para residuos orgánicos. Mater-Bi. https://materbi.com/es/aplicaciones/bolsa-para-residuos-organicos/ (accessed 2025g-09-19).
Mater-Bi. Film para acolchado | Biodegradabile Mater-Bi. Mater-Bi. https://materbi.com/es/aplicaciones/film-para-acolchado/ (accessed 2025h-09-19).
Mater-Bi. Mater-Bi -Bioplásticos biodegradables y compostables. Mater-Bi. https://materbi.com/es/mater-bi/ (accessed 2025i-09-19).
Mater-Bi. Bolsa para la compra. Mater-Bi. https://materbi.com/es/aplicaciones/bolsa-para-la-compra/ (accessed 2025a-09-19).
Menossi, M., Salcedo, F., Mansilla, A. Y., Alvarez, V. A., & Ludueña, L. N. (2025). Poly (Lactic Acid)/Starch biodegradable blend with antioxidant activity: A sustainable alternative for single-use active plastic packaging. *Journal of Applied Polymer Science, 142*(21), Article e56933. https://doi.org/10.1002/app.56933
Merino, D., Gutiérrez, T. J., & Alvarez, V. A. (2019). Potential agricultural Mulch films based on native and phosphorylated corn starch with and without surface functionalization with Chitosan. *Journal of Polymers and the Environment, 27*(1), 97–105. https://doi.org/10.1007/s10924-018-1325-1
Merino, D., Gutiérrez, T. J., Mansilla, A. Y., Casalongué, C. A., & Alvarez, V. A. (2018). Critical evaluation of starch-based antibacterial nanocomposites as agricultural Mulch films: Study on their interactions with water and light. *ACS Sustainable Chemistry & Engineering, 6*(11), 15662–15672. https://doi.org/10.1021/acssuschemeng.8b04162
Novamont S.p.A. - Chimica Vivente per la Qualità della Vita. https://www.novamont.com/ (accessed 2025-09-19).
OK biodegradable. OK Certification. https://okcert.tuvaustria.com/ok-biodegradable-en/ (accessed 2025-10-01).
OK compost. TÜV Austria Iberia ES. https://es.tuvaustria.com/ok-compost/ (accessed 2025-10-01).
Onyeaka, H., Obileke, K., Makaka, G., & Nwokolo, N. (2022). Current research and applications of starch-based biodegradable films for food packaging. *Polymers, 14*(6), 1126. https://doi.org/10.3390/polym14061126
Oyekunle, D. T., Nia, M. H., & Wilson, L. D. (2024). Recent progress on the application of Chitosan, starch and Chitosan–Starch composites for meat Preservation—A mini review. *Journal of Composites Science, 8*(8), 302. https://doi.org/10.3390/jcs8080302
Panda, G., Sarkar, C., & Saha, S. (2023). Processing of biodegradable polymers. In S. Saha, & C. Sarkar (Eds.), *Biodegradable polymers and their emerging applications* (pp. 27–47). Singapore: Springer Nature Singapore. https://doi.org/10.1007/978-981-99-3307-5_2
Paola Bracciale, M., De Gioannis, G., Falzarano, M., Muntoni, A., Polettini, A., Pomi, R., Rossi, A., Sarasini, F., Tirillò, J., & Zonfa, T. (2024). Disposable Mater-Bi® bioplastic

tableware: Characterization and assessment of anaerobic biodegradability. *Fuel, 355*, Article 129361. https://doi.org/10.1016/j.fuel.2023.129361

PLANTIC. Starch Technologies. Plantic. https://plantic.com.au/technology/innovation/starch-technologies.html (accessed 2025b-09-19).

PLANTIC. Flexible films. Plantic. https://plantic.com.au/product/plantic-flexible/ (accessed 2025c-09-19).

PLANTIC. Plantic Advantages. Plantic. https://plantic.com.au/technology/plantic-advantages.html (accessed 2025a-09-19).

Poveda-Giraldo, J. A., Yoo, S., Yoo, S., Kim, I., Lan, K., Venditti, R., & Park, S. (2025). Environmental life cycle assessment of renewable starch alternative plastic: A comparative analysis of end-of-life scenarios in the Republic of Korea. *Journal of Cleaner Production, 520*, Article 146178. https://doi.org/10.1016/j.jclepro.2025.146178

Products – Agrana Bioplastics. https://www.agrana-bioplastics.com/products/ (accessed 2025-09-19).

Quilez-Molina, A. I., Chandra Paul, U., Merino, D., & Athanassiou, A. (2022). Composites of thermoplastic starch and lignin-rich agricultural waste for the packaging of fatty foods. *ACS Sustainable Chemistry amp; Engineering, 10*(47), 15402–15413. https://doi.org/10.1021/acssuschemeng.2c04326

Quilez-Molina, A. I., & Merino, D. (2023). Chapter 3 - From waste to resource: Methods for vegetable waste transformation into sustainable plant-based bioplastics. In S. Ahmed, & Annu (Eds.), *Advanced applications of biobased materials* (pp. 61–110). Elsevier. https://doi.org/10.1016/B978-0-323-91677-6.00023-4

Quilez-Molina, A. I., Merino, D., & Dumon, M. (2024). Porous starch embedded with Anthocyanins-CMC coating as bifunctional packaging with seafood freshness monitoring properties. *Food Hydrocolloids, 154*, Article 110114. https://doi.org/10.1016/j.foodhyd.2024.110114

Regulation (EC) No 1935/2004 of the European Parliament and of the Council of 27 October 2004 on Materials and Articles Intended to Come into Contact with Food and Repealing Directives 80/590/EEC and 89/109/EEC; 2021. http://data.europa.eu/eli/reg/2004/1935/2021-03-27/eng (accessed 2025-10-01).

Regulation - 10/2011 - EN - EUR-Lex. https://eur-lex.europa.eu/legal-content/EN/TXT/?uri=CELEX:32011R0010 (accessed 2025-10-01).

Rocha, A. C. S., Muniz, J. D. B., Silvino, A. C., da Silva, E. O., & de Menezes, L. R. (2025). Smart food packaging with biodegradable polymers: Progress and prospects of colorimetric indicators. *Polymers for Advanced Technologies, 36*(11), Article e70411. https://doi.org/10.1002/pat.70411

Sivapragasam, N., Maqsood, S., & Rupasinghe, H. P. V. (2024). Berry bioactive compounds immobilized in starch matrix for active and intelligent packaging: A review. *Future Foods, 10*, Article 100397. https://doi.org/10.1016/j.fufo.2024.100397

Thermoplastic Starch Market - TPS Size, Share & Industry Analysis. Mordor Intelligence. https://www.mordorintelligence.com/industry-reports/global-thermoplastic-starch-market (accessed 2025-10-01).

UNE-EN 13432:2001/AC:2005 Envases y embalajes. Requisitos de l. https://www.une.org/encuentra-tu-norma/busca-tu-norma/norma?c=N0034238 (accessed 2025-10-01).

Wang, L., Yang, C., Deng, X., Peng, J., Zhou, J., Xia, G., Zhou, C., Shen, Y., & Yang, H. (2023). A pH-Sensitive intelligent packaging film harnessing Dioscorea Zingiberensis starch and anthocyanin for meat freshness monitoring. *International Journal of Biological Macromolecules, 245*, Article 125485. https://doi.org/10.1016/j.ijbiomac.2023.125485

Xu, R., Xia, L., Tang, Q., Tang, F., Pang, S., Li, H., & Zou, Z. (2024). High-Performance carboxymethyl Starch/PVA based intelligent packaging films engineered with Cu-Trp nanocrystal as functional compatibilizer. *Food Chemistry, 454*, Article 139696. https://doi.org/10.1016/j.foodchem.2024.139696

Yu, X., Chen, L., Jin, Z., & Jiao, A. (2021). Research progress of starch-based biodegradable materials: A review. *Journal of Materials Science, 56*(19), 11187–11208. https://doi.org/10.1007/s10853-021-06063-1

Zheng, L., Liu, L., Yu, J., Farag, M. A., & Shao, P. (2023). Intelligent Starch/Chitosan-Based film incorporated by anthocyanin-encapsulated amylopectin nanoparticles with high stability for food freshness monitoring. *Food Control, 151*, Article 109798. https://doi.org/10.1016/j.foodcont.2023.109798

哈拉尔德·施密特; 克里斯托夫·黑塞; 约翰内斯·马塔; 拉尔夫·哈克福特. Polymer Material and Method for the Production Thereof. CN101835836B, 2013. https://patents.google.com/patent/CN101835836B/en (accessed 2025-09-19).

CHAPTER FOURTEEN

Biodegradability, compostability, environmental fate, and relevant standards in starch-based packaging

Olaniyi Amos Fawole[1], Anthony Temitope Idowu[2]
[1]University of Johannesburg, South African Chair in Sustainable Postharvest and Agroprocessing Research, Department of Botany and Plant Biotechnology, Johannesburg, South Africa
[2]University of Limerick, Department of Biological Science, Limerick, Ireland

14.1 Introduction

Plastics are polymers that are made up of long-chain organic compounds with desirable properties (Wang et al., 2022) and numerous advantages when compared to metals, ceramics, and woods. For example, plastics are low-cost, lightweight, adaptable, and chemically inert, easily shaped, pliable, and multifunctional (Nizamuddin & Chen, 2024). Plastic materials are rapidly used in our day-to-day activities, and their usage in electronics, textiles, healthcare products, toys, and packaging applications is inevitable (Hottle et al., 2013). It is expected that global plastic production will surpass 800 million tons before 2050 (Sardon & Dove, 2018).

While plastics are produced in large amounts for numerous applications, about 40% is only attributed to use for products with short shelf life (i.e., packaging), and 60% account for their use for products with long shelf life (Nizamuddin & Chen, 2024). At the end of life, high-value plastics are rendered into waste, negating their economic value due to end-of-life disposal or unpopular reuse methods, and only about (<9%) of the waste plastics are recycled (Stubbins et al., 2021). Statistically, more than 80% of plastics become futile and are littered in the environment (oceans, rivers, and lands) at the end of their lives, leading to heavy economic loss of around 80–100 billion USD per year (Zhang et al., 2021) and a serious environmental issue (MacLeod et al., 2021).

Plastics undergo a steady degradation process due to the presence of high molecular weight covalent bonds, which form static networks resulting in unwanted material proliferations (Webber & Tibbitt, 2022). Likewise,

Starches in Smart Packaging
ISBN: 978-0-443-45396-0
https://doi.org/10.1016/B978-0-443-45396-0.00001-3

plastic waste discarded on land or water degrades into micro or nanoparticles (Guan et al., 2021), which contaminate lives through ingestion or entanglement and harm nutrients (Rahimi & García, 2017) and transfer toxins to food chains (Seltenrich, 2015). To mitigate these challenges, it was pertinent to develop materials having biodegradability with or without compostability properties, which can degrade rapidly (Christensen et al., 2019) and can be explored as a replacement for traditional polymers (Hillmyer, 2017; Song et al., 2009).

Starch has been regarded as a viable option to petroleum-derived plastics, which have numerous environmental limitations. Starch is a starting material for a lot of green material products. A diverse source of starch exists, and the chief among them is tubers, seeds such as corn, and roots of plants (Asaf Kleopas, 2008). Starch-based materials have numerous intrinsic biodegradability and biocompatibility and are used in edible coatings and for the development of films (packaging material) in the food industries (Bemiller & Whistler, 2009). Starch can be derived from diverse botanical sources with varying proportions and molecular weight of its major components (amylose and amylopectin). This impact on its chemical and physical properties as well as processing and its potential applications (Castano et al., 2017; Kim et al., 2015; Srichuwong et al., 2005). Nevertheless, the structure of starch poses numerous challenges in its application as a packaging material. For instance, starch films are brittle, which is a significant weakness that impacts their mechanical properties, leading to inefficiency in maintaining film structural integrity and resisting deformation damage during subsequent processing (Yang et al., 2024). Advancements in research in the last decade indicated that a solution exists in overcoming the limitations geared toward the applicability of starch-based materials as a potential packaging raw material. Therefore, this chapter aims to present an overview of the biodegradability, compostability, and environmental impact of starch-based packaging. This is vital when considering using starch polymer for the development of packaging items in the industries.

14.2 Introduction to biodegradability and compostability of plastics

Degradation of plastics can occur via different mechanisms: chemical, thermal, biological, and photodegradation (Alshehrei, 2017). Plastics are biodegraded aerobically in compost and soil. Water and carbon dioxide are produced during aerobic biodegradation (Alshehrei, 2017). On the contrary, plastics can be anaerobically biodegraded in sediments and landfills, which can lead to the production of water, carbon dioxide, and methane (Ishigaki et al., 2004). The terms "biodegradability" and "compostability" of plastics have received a lot of attention, and the terminologies are often used interchangeably when describing plastic waste, but these terms are not the same. To provide clarity, biodegradable plastics can be degraded by the action of naturally occurring microbes such as bacteria, actinomycetes, algae, and fungi (Goel et al., 2021). Varieties of biodegradable plastics existed that originated from petroleum sources, such as aliphatic polyesters and copolyesters, e.g., poly(ε-caprolactone), poly(vinyl alcohol) (PVA), polybutylene succinate (PBS), and polycaprolactone (PCL), polyesteramides (PEA), and their blends (Nizamuddin & Chen, 2024). The largest market share of biodegradable plastics emanated from thermoplastic starch (TPS), polylactic acid (PLA), and polyhydroxyalkanoates (PHAs) in comparison to others (PCL and PBS) due to their biobased and specific mechanical characteristics. Bioplastic market share is estimated to be around 1% of the 370 million tons of the accumulated global plastic production (Coppola et al., 2021). Bioplastic materials signify a true alternative to conventional plastic and its application. The term "compostability" refers to the ability of plastics to break down and integrate into the compost system, which is a distinct type of biodegradation (Wojtowicz, 2010). However, compostability requires controlled conditions such as moisture, heat, and microorganisms to completely break down compostable plastics into water, carbon dioxide, and new microbial biomass (Wojtowicz, 2010). It is expected that the breakdown of compostable plastics should not leave any trace of harmful chemical fragments or residues (Dey et al., 2024).

14.2.1 Assessments on biodegradability and compostability of plastics

It is well documented that studies regarding the biodegradability and compostability of plastics have been reported. Some of the studies are presented herein to give an overview of the biodegradability and compostability of plastics. For instance, Koutny et al. (2006) assessed the biodegradability of low-density polyethylene film (LDPE) and high-density polyethylene film (HDPE) consisting of antioxidants and prooxidants. Both polymers were tested against microbial strains following an abiotic pretreatment consisting of photooxidation and artificial high thermo-oxidation (60°C), which was aimed at mimicking around 3 years of outdoor weathering.

The samples were inoculated, then incubated up to 200 days (27 °C), and their metabolic activities were analyzed. Fast microbial growth was observed at the initial phase, suggesting the likely utilization of low molecular weight compounds, followed by a long stabilization of metabolic activity. Incubation at the end showed that biodegradation likely impacted only the surface layer of the material, according to the study. In a separate study, Mohee and Unmar (2007) investigated the biodegradability of two varying plastics (identified as plastic A and B) under controlled and natural composting conditions. Plastic A consisted of PDQ-H additive, which is a plastic additive that breaks down plastics by photo and oxidative methods. This was used with either polyethylene or polypropylene. Plastic B consisted of the addition of 3% mass of the totally degradable plastic additive called totally degradable plastic additive (TDPA) (used to make plastic degrade) to polyethylene. The outcome revealed that the cumulative CO_2 evolution for plastic A was higher than for plastic B. This indicated that plastic A showed a higher level of biodegradation in terms of CO_2 evolution than plastic B under controlled composting conditions. In the same study, plastic A and B were compared to a reference plastic, namely, a compostable plastic bag (Mater-Bi product) denoted plastic C but made from a starch resource, and they were all exposed to a natural composting condition. After 55 days of composting, it was observed that plastic C was completely degraded, while plastic A and B did not undergo any tangible degradation. The study concluded that naturally based plastic material from starch can degrade completely in a short time frame of around 60 days, while plastics consisting of biodegradable additives would require an extended time. Elsewhere, Napper and Thompson (2019) assessed the possible biodegradation ability of some plastic carrier bags called biodegradable, compostable, and HDPE carrier bags (labeled as a conventional carrier bag). The outcome of the study revealed that some level of deterioration of the plastic carrier bag occurred when exposed to the soil, marine, and control environment over a 3-year duration. However, an enhanced level of deterioration occurs in the biodegradable bags compared to the conventional polyethylene. In a separate study, Mörtl et al. (2024) assessed the biodegradation of bioplastic bags under an industrial composting environment. In the study, the bioplastic bags were composted over a period of 12 weeks. The outcome revealed that 95% of the bags disintegrated by 12 weeks. Therefore, confirming the possible degradation of bioplastics over synthetic plastics. Afshar et al. (2025) investigated the breakdown of 10 commercially available biodegradable plastic products under simulated industrial composting conditions. The products assessed included polymer compositions of either polybutylene adipate terephthalate (PBAT)/starch, PLA, and PHA covering rigid and flexible plastics. These products consisted of waste bag drawstrings, flexible plastics, rigid plastics, and many more. Out of this, nine were marked as compostable. The outcome showed that disintegration degrees ranged from 75% to 100%, with five products attaining complete disintegration. Furthermore, microplastics were detected in compost undersieves for two of the 10 biodegradable plastic products, while no microplastics were detected for conventional plastics. Overall, various studies emphasize the importance of selecting natural polymers over synthetic ones when considering the development of plastics for various purposes.

14.2.2 Enzymatic and microbial disintegration of biodegradable plastics

Biodegradable polymers are known to contain hydrolyzable bonds such as carbonates, glycosides, ortho-esters, esters, amides, urea, urethanes, anhydrides, and many more (Göpferich, 1996). Polymers possessing strong covalent bonds in the backbone (like C–C) and with no hydrolyzable groups require an extended time of degradation (Hasırcı et al., 2001). Hydrolysis reaction may be catalyzed by enzymes called hydrolases, which include esterases, proteases, glycosidases, phosphatases, lipases, etc. These enzymes comprise cell-derived proteins that are accountable for the catalysis of several reactions in the human body (Shah et al., 2014). Other hydrolases like cellulases, pectinases, and xylanases are responsible for the hydrolysis of cellulose, xylan, and pectin, respectively. In this context, it is expected that some of these enzymes may contribute to the degradation of biomaterials through hydrolysis. Most of the biodegradable plastics are polyesters, i.e., their degradation may be catalyzed by esterolytic enzymes such as esterases, proteases, and lipases with some variation in substrate preference or interfacial activation (Bornscheuer, 2002). All enzymes tested for polyester degradation have so-called catalytic triad in their active site, which is peculiar to all serine hydrolases. However, the form, depth, and surroundings of the catalytic center seem to vary significantly (Mueller, 2007).

The contribution of microorganisms to the degradation of plastics cannot be underestimated. Microorganisms use the mechanism of enzymatic degradation (hydrolysis) by secreting extracellular enzymes to foster possible disintegration of biodegradable plastics (Shah et al., 2014). Different microorganisms, such as actinomycetes, fungi, and bacteria, have been

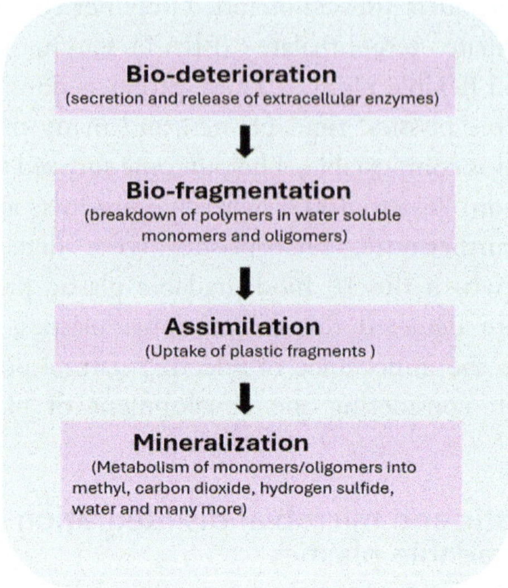

Figure 14.1 The process of biodegradation by microorganisms.

identified to possess biodegradable properties (Fig. 14.1) (Li et al., 2008; Nakajima-Kambe et al., 2009; Song et al., 2009). For instance, bacteria such as *Roseateles depolymerans* strain TB-87, isolated from freshwater, could disintegrate different types of aliphatic polyester biodegradable plastics, such as polybutylene succinate adipate (PBSA), PBS, PCL, and polyethersulfone (PES), excluding PLA (Shah et al., 2013). Novel esterases with wide substrate specificity and high activity against aliphatic copolyester, both in solid and liquid states, have been reported from the bacterial strain (TB-87) (Shah et al., 2013). Elsewhere, a bacterial strain, *Pseudomonas* sp. Strain AKS2 from soil reportedly degrades aliphatic polyesters (Tribedi et al., 2012). Aliphatic polyesters are biodegradable polymers that can be formed by polycondensation reactions or ring-opening polymerization of lactones (Lecomte & Jérôme, 2011). The disintegration of PHAs by different bacterial and *Actinomycetes* strains, such as *Bacillus* sp. AF3 (Shah et al., 2007), *Pseudomonas fluorescens* (Schirmer et al., 1993), *Comamonas* sp. (Molitoris et al., 1996), *Streptomyces* sp. *SNG9* (Mabrouk & Sabry, 2001), *Actinomadura* sp. AF-555 (Shah et al., 2010), and *Streptomyces venezuelae* (Santos et al., 2013), and many more have been reported from diverse natural environments such as composts, marine water, and soils.

Biological recycling of polyesters through composting is a promising technique; some polyester-degrading thermophilic bacteria and their enzymes have been reported from the compost environment. For instance, two kinds of thermo-active esterases have been reported from the genus Thermobifida, one from *Thermobifida fusca* strain 43793 (Chen et al., 2008; Kleeberg et al., 2005) and the other from *Tardiphaga alba* strain Est119 (Hu et al., 2010). *Thermobifida fusca* is a thermophilic soil bacterium known for its ability to disintegrate plant biomass to generate thermostable enzymes like cellulases and polyethylene terephthalate (PET)-degrading enzymes (Luzics et al., 2025). *Tardiphaga alba* is a heavy-metal-tolerant bacterium that could be obtained from garden soil (Bao et al., 2024). It has been documented that these microbes could depolymerize aliphatic aromatic copolyesters. Aliphatic–aromatic copolyesters belong to the class of biodegradable plastics that integrate the flexibility of aliphatic (carbon-chain) polyesters with the strength and thermal stability of aromatic (benzene-ring) polyesters (Ge et al., 2025).

Microbial degradation could be through aerobic (in the presence of oxygen) or anaerobic (in the absence of oxygen), which is dependent on the microbial strain and the environmental factors (Gu, 2003). A brief schematic expression of how microorganisms break down biodegradable plastics is shown in Fig. 14.1. Microorganisms initiate the attack of the biodegradable plastics by secreting enzymes that could disintegrate the long polymer chains into smaller units called the monomers or oligomers (low molecular weight polymers consisting of a small number of repeat units) (Mandal et al., 2024); this process is referred to as depolymerization (Alshehrei, 2017). Summarily, the mechanism of biodegradation can be presented sequentially to involve bio-deterioration, bio-fragmentation, assimilation, and mineralization (Khan & Majeed, 2019). In this step, the microorganism attaches to the plastic surface, alters its chemical and physical properties, and then polymer disintegration occurs (bio-fragmentation) through enzymatic cleavage by different hydrolytic enzymes (e.g., protease and lipase), which break down different bonds of the polymer (Loredo-Treviño et al., 2012). Through assimilation, the fragmented polymer is then taken up by microbes to be mineralized into methyl, carbon dioxide, hydrogen sulfide, and water (Singh & Susan, 2018; Tokiwa et al., 2009).

14.2.3 Factors influencing biodegradability and compostability of plastics

Several factors have been pinpointed to be responsible for the biodegradability and compostability of plastics, as illustrated in Fig. 14.2. As indicated, the characteristics of the polymer as well as the exposure (environmental) conditions both influenced the biodegradation process (Lucas et al., 2008); some of these factors are further discussed in this section.

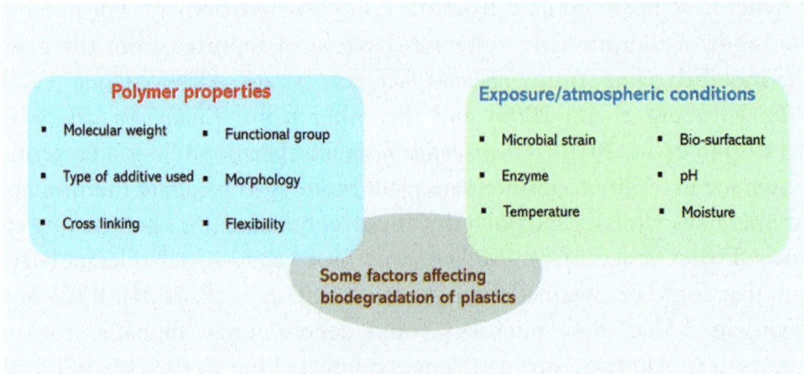

Figure 14.2 Factors that influence the biodegradation of plastics.

14.2.4 Polymer properties

14.2.4.1 Chemical structures and functional groups

It is notable to state that polymer degradation is influenced by the chemical structures of the polymers, particularly the presence of functional groups and hydrophobicity–hydrophilicity balance (Deshayes & Kasko, 2013). The process of disintegration varies depending on the polymer type. In an instance, the ester linkages in polyester allow for easy breaking, but the C–C bonds in polyolefins limit easy disintegration (Kumari et al., 2025). Hence, differences in molecular compositions affect its adsorption and polymer degradability (Taniguchi et al., 2019).

The presence of functional groups and high resistance to breaking chemical bonds, such as ester and C–C bonds, is a significant factor influencing the rate of plastic disintegration. For instance, a greater specific surface area will disintegrate more quickly since most plastics break down near the surface first. Nevertheless, increased adsorption from a higher specific surface area influences disintegration by adsorbing other chemicals, which is a function of the hydrophobic–hydrophilic nature of the polymer. Plastic adsorption strength signified the degree to which the polymer adheres to bacteria. According to Kumari and Chaudhary (2020), the disintegration of plastics made from polyethylene, polypropylene, polyvinyl chloride, and polystyrene is impacted by their highly stable C–C bonds, the structure's molecular weight, and surface hydrophobicity. Photocatalysis appears to be an alternative to disintegrating polyethylene material (Kumari et al., 2025; Nabi et al., 2020).

14.2.4.2 Morphological characteristics and molecular weight

The morphological characteristics, such as crystallinity, are an important factor influencing polymer degradation (Chen et al., 2008). Polymer crystallinities indicate the extent to which polymer chains are organized into a three-dimensional structure. It has a major influence on the polymer's properties, such as melting point, strength, and hardness (Venkatram et al., 2023). According to Shah et al. (2014), the process of degradation is influenced by the temperature of the medium because the rate of enzymatic hydrolysis of polyesters increases with the enhancement of the temperature. Enzyme and microbial accessibility are limited in more organized crystals, making them challenging to degrade. Amorphous crystals are irregularly arranged, and it is usually the origin of polymer disintegration (Kumari et al., 2025). Polypropylene barely disintegrates in aquatic environments owing to its high surface roughness, hydrophobicity, and high molecular weight. It depends primarily on photo-oxidative degradation. Its hydrophilic feature is improved by generating esters, acids, polar carbonyl groups, and ketones when exposed to oxygen and ultraviolet light (Khoironi et al., 2020).

14.2.4.3 Additive

Additives are chemical substances that are added to the production process of a material to optimize its performance when molded (Sanchez-Rexach et al., 2020). Additives can also be added during plastic formation when considering changing the properties of a polymer to get a specific performance out of the plastic. Sanchez-Rexach et al. (2020) highlighted that pigment impurities and fillers can affect the disintegration resistance of polyolefins. Likewise, increasing lignocellulosic filler loading can slightly lower thermal stability and increase ash content (Kumari et al., 2025). Therefore, interfacial dispersion and adhesion between the thermoplastic polymer and filler are key features that should be considered. Metals such as manganese portray prooxidant properties in polyolefins, making them susceptible to thermal-oxidation disintegration (Kumari et al., 2025). Prooxidants are substances that cause oxidative stress by generating reactive oxygen species (ROS) (Ozougwu, 2016). Prooxidants produce free radicals on the polyethylene chain when activated by heat in the presence of oxygen, thereby altering polymers' physical properties (Gurer-Orhan et al., 2004).

14.2.5 Environmental factors

As per environmental factors, low pH introduces hydrogen ions, which enhance plastic disintegration and then facilitate better colloidal interaction with nanoparticles (Kumari et al., 2025). Meanwhile, low- temperature fragments of microplastics increase the surface area and interaction of nitrogen–carbon–titanium dioxide (Ariza-Tarazona et al., 2020). According to Xiong et al. (2020), pH levels influence the electrostatic interactions between polymers and pollutants, modifying adsorption behavior. Elsewhere, it was stated that the disintegration of polyethylene plastics by *Pseudomonas aeruginosa* is inhibited in an acidic environment, but this limitation can be mitigated by low-temperature conditions (Tamnou et al., 2021). Other environmental factors, such as temperature, moisture, oxygen, microbial presence, and strains, can influence polymer disintegration. A warm climate, assisted with relative humidity, increases the disintegration rate because of microbes (Kumari et al., 2025). Humidity has a major effect on the process of degradation, especially in dry and hot climates, where these processes tend to occur at a slower pace (Nagai et al., 1999). An increase in humidity and temperature can speed up hydrolytic disintegration, as photodisintegration has a notable effect due to moisture and temperature (Andrady et al., 2003). Excess humidity can reduce the efficacy of light stabilizers, leading to disintegration. High temperature and solar UV radiation accelerate plastic disintegration, which results in rapid weathering at elevated temperatures (Fig. 14.2).

14.3 An overview of starch composition and starch-based packaging

14.3.1 Composition

Starch chiefly consists of amylose and amylopectin in high proportions, and then lipids, proteins, and minerals in small fractions (Alcázar-Alay & Meireles, 2015). Glucose is the basic unit in starch, contributing to the formation of the amylose and amylopectin polymers (Kong, 2020). The amount of amylose and amylopectin differs based on the starch source (Bertolini, 2009). Molecular weight of amylose and amylopectin are within 105–107 and 107–109 Da, respectively. Starch is widely used as a biopolymer for the production of coatings and films (Gabor & Tita, 2012) due to its biodegradable and renewable properties (Gadhave et al., 2018; Wróblewska-Krepsztul et al., 2018). The conventional content of amylose, amylopectin, and overall starch composition of some edible crops is presented in Table 14.1.

Native (i.e., unprocessed) starch cannot be used directly in plastics; its granular structure needs to be disintegrated with water, heat, and also plasticizers such as glycerol (Avérous, 2004) to generate TPS. Thereafter, TPS can be processed as in the case of many other plastics, e.g., injection molding or extrusion. The limitation of pure TPS is that it has a poor mechanical characteristic and is prone to water, thus restraining its possible product applications. Consequently, TPS is often reactively compounded with other

Table 14.1 Conventional content of amylose, amylopectin, and overall starch composition of some edible crops.

Crops	Amylose content (%)	Amylopectin (%)	Starch (%)	References
Rice	15–35	65–85	81–92	Omar et al. (2016), Marichelvam et al. (2019), Zakaria et al. (2017)
Maize	15–20	80–85	70–72	García-Guzmán et al. (2022), Ma et al. (2020)
Potato	25–31	69–75	60–80	Robertson et al. (2018), Marichelvam et al. (2019), Zakaria et al. (2017)
Wheat	17–34	75–80	60–70	Kim and Kim (2021), Marichelvam et al. (2019), Zakaria et al. (2017)
Pulses	36–42	58–64	50–60	García-Guzmán et al. (2022), Robinson et al. (2019)
Cassava	19–22	28–81	20–32	Surtono et al. (2019), Marichelvam et al. (2019), Zakaria et al. (2017)
Arrowshoot	20.5	79.5	88–89	Rodrigues et al. (2018), Phukan and Nongkhlaw (2022), Marichelvam et al. (2019)
Yam	9–15	85–91	16–20	Li et al. (2019), Zakaria et al. (2017)

polymers, such as aliphatic polyesters, to enhance mechanical properties. Since hydrophilic starch and hydrophobic polyesters are immiscible, it was imperative to add compatible additives to allow good adhesion between components to enhance technical performance (Gunawardene et al., 2021). Varying these components used during compounding (copolymers, native starch, and additives) assists in generating starch plastics with a wide range of technical properties.

Starch plastics, a mixture of starch with other polymers, are currently being developed for diverse uses. Starch plastic production capacity equaled to around 430 metric kilotons in 2016, constituting about 10.3% of the worldwide capacity of biobased plastics (Aeschelmann et al., 2017). Currently, mercantile starch plastics are produced for injection molding (e.g., disposable tableware, flowerpots), films (e.g., bags, biodegradable packaging, agricultural mulching films), and foam applications (e.g., loose fill packaging).

14.4 Current state of starch-based film or packaging

Packaging of consumable material is regarded as a broad-spectrum tool that pinpoints manufacturing, nutrient details, and expiry information, making it suitable for customers. Food products are available in many forms, such as solid, liquid, and semisolid textures, therefore necessitating that they are packaged appropriately. Efficient packaging warrants protection of foods from microbial contamination, dust, and undesirable pollutants (Samsudin & Hani, 2017). Hence, the packaging of food needs extra protection against detrimental factors such as environmental, chemical, and microbial contaminations, among others, that may affect consumers. It is known that each packaging material possesses varying attributes (such as gas barrier, moisture barrier, high impact strength, transparency, and oil resistance). Therefore, it is critical to select suitable packaging materials for maintaining the quality and safety of food products to extend the shelf life.

The use of starch-based material for the development of film or packaging material has been reported in the literature. For instance, edible film was developed from native potato starch with plasticizer (glycerol) according to a study by Sessini et al. (2016). It was further reinforced with catechin and starch nanocrystals (SNCs) that were generated by acidic hydrolysis from waxy maize starch granules. The mechanical performance was also improved in the ternary bio-nanocomposite edible films. There was an improvement in mechanical performance in the ternary bio-nanocomposite edible films. The developed edible films fully disintegrate in compost conditions, indicating their possible use as biodegradable edible films. In another study, Basiak et al. (2017) investigated the effect of starch type on the physicochemical properties of edible films. In the study, potato, wheat, or corn was used as a starch, while glycerol was used as a plasticizer for the development of film. It was observed that films developed from potato starch constituted a greater barrier for water and oxygen, though they have weaker mechanical properties than corn and wheat starch films. It was observed that the starch origin affected the optical properties and thickness. Starch with more amylose led to opalescent and thicker films, while starch with less amylose had transparent and thinner films. Elsewhere, Luchese et al. (2017) developed starch-based films from corn and cassava by casting assisted with glycerol and xanthan gum. The outcome revealed that higher biopolymer amounts enhanced the thickness and reduced water vapor permeability of starch-based films. An increase in starch content of films promoted improved mechanical properties. The mechanical properties of the starch-based films, such as tensile strength, maximum elongation, and elastic modulus, were compared to low-density polyethylene-based films used in food packaging. It was observed that the cassava and corn were more promising and could serve as an alternative when considering packaging material development. In another study, Wang et al. (2017) investigated the possibility of improving the properties of collagen films from high amylose starch, normal starch, and waxy maize starch. Collagen film is popularly used as an artificial sausage casing in the food industry (Wang et al., 2015). The outcome showed that adding all the starches enhanced the thermal stability of the collagen film. Film solubility in water decreased with the addition of starch, depending on the amylose content and starch concentration. Elsewhere, Nawab et al. (2017) developed a composite film from mango kernel starch, guar, and xanthan gum. Irrespective of the gum type, the tensile strength of the composite films increased as gum concentration increased. Elongation at break showed a decrease as the gum concentration increased. The addition of guar and xanthan gum enhanced the water vapor permeability and solubility of the composite film. Song et al. developed edible films from wheat and corn starch and lemon essential oil, as well as surfactants (Tween 80 and Span 80). Film containing only Tween 80 showed higher solubility, water content, and water vapor permeability. The addition of lemon essential oil improved the antimicrobial properties of the film. In another study, Silva et al. (2019) synthesized edible films from cassava, chitosan, and gelatin by casting that can be used as a fruit coating. The developed film possessed satisfactory barrier and mechanical properties. It was observed that lower concentrations of starch produced less opaque, more flexible, and soluble films. Furthermore, the combination of a higher proportion of starch and chitosan in film development was responsible for the lowered water vapor transmission rate observed in the film. In a separate study, Dai et al. (2019) developed an edible film from botanical starch (such as cassava, waxy corn, sweet potato, corn, wheat) and different commercial modified cassava starches (oxidized cassava starch, cross-linked cassava starch, and esterified cassava starch) by the casting method. The outcome revealed that transparency and color were a function of the plant sources and the method of modification of the starch. It was also observed that cassava starch film had the best comprehensive properties among the native starch. The modified cassava starch had a better comprehensive property among the crosslinked cassava starch films. In another study, Hirpara et al. (2021) developed biodegradable packaging film from potato starch assisted with glycerol as a plasticizer. The film was developed by casting technique using a film-forming solution. The physicochemical properties of the developed film included water vapor permeability (0.0058 g mm/m^2 kPa), moisture content (23.1%), water absorption capacity (190%), and transparency (69.54%). The outcome showed a promising possibility to develop starch film from potatoes. Fonseca-García et al. (2021) reported the possibility of developing a starch-based film from corn starch, chitosan, and Pluronic F127 (a copolymer). Chitosan is a polymer obtained from shrimp shells and crabs and can be used to produce biodegradable materials (Adeyemi & Ojoawo, 2023). It was observed that the surface of starch–chitosan composite films obtained was more homogeneous and smoother when Pluronic F127 was increased. Overall, these studies highlighted the possibility of developing packaging material from starch-based sources.

14.5 Strategies for producing starch-based films

Different techniques have been explored as a replacement when considering the development of starch-based packaging. This includes physical, chemical, and enzymatic modification of native starch (Lin et al., 2024). In addition, mixing starches with varying crystalline properties or inducing partial retrogradation enables the fine-tuning of films' permeability, mechanical, and biodegradability features, which smoothen the development of customized bio-based packaging solutions.

One standout technique for tailoring the properties of starch is by cross-linking. This method is induced by the formation of covalent bonds between and within starch molecules. It can greatly alter the properties of the material, including increasing its hydrophobicity. Common reagents used to crosslink starch are epichlorohydrin, glyoxal, and sodium-trimetaphosphate. Nevertheless, most of these reagents are expensive, toxic, or inefficient, which lowers their application in foods (Lin et al., 2024). Some alternate reagents used for starch cross-linking are urea (Min et al., 2023) and citric acid (Itkor et al., 2024). It is noteworthy to state that thermal degradation of starch occurs at 275°C, when a significant loss of macromolecules begins to occur, distinguishing it as a material with good thermal resistance (Arruda et al., 2025). This allows the application of starch in technologies that need thermal stability and preservation of structural properties, e.g., production of films and other innovative materials (Wang et al., 2025). For instance, the casting method is a "wet" technique that is widely used in the production of starch-based films, where the starch is gelatinized in an aqueous medium when undergoing heating and stirring, which is vital for the formation of a cohesive and homogenous polymeric matrix. Thereafter, it is poured on a flat surface and then subjected to drying to obtain a film (Chaudhary et al., 2024). A graphical illustration of starch film solution development by casting is shown in Fig. 14.3. Overall, adequate optimization of the raw material mix may be necessary for the production of a suitable film.

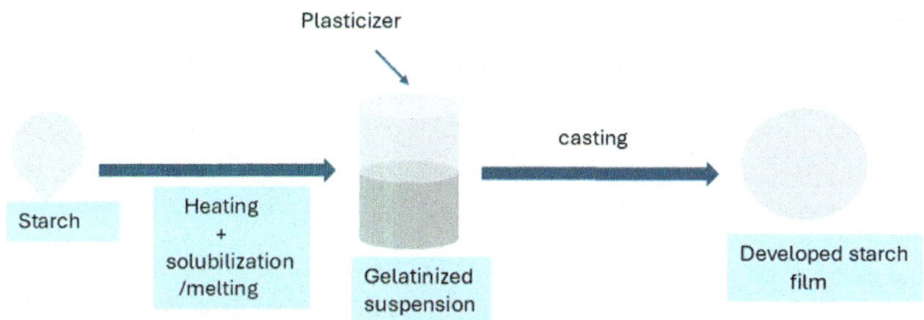

Figure 14.3 Graphical illustration of starch film solution development by casting.

14.6 Environmental fate and standards in starch-based packaging

Starch packaging's environmental fate is generally positive. This is due to the promising ability of renewable resources (potato, corn, wheat, tapioca, and rice) to biodegrade into water and carbon dioxide within months under proper conditions (microbes, heat, and moisture), unlike synthetic plastics, which are excellent for reducing landfill waste. Elsewhere, Onyeaka et al. (2022) highlighted the unique advantage of starch-based packaging to be environmental benefits such as biodegradability and ease of accessing renewable sources, as well as the ability to decompose faster than conventional plastics. Nevertheless, its fate depends heavily on the additives added as well as regulatory standards. Nonbiodegradable components (e.g., synthetic fibers, copolymers) added as additives can potentially slow down degradation and can constitute moderate environmental risks, while additives significantly affect greenhouse gas emissions (Andreeßen & Steinbüchel, 2019).

Regarding standards, EN 13432 and ISO 18606 guide compostability, requiring proper waste streams (anaerobic digestion, industrial/home composting) for optimal breakdown, with additives influencing degradation impact and time, needing clear labeling for effective end-of-life management. EN 13432 is the European standard defining the requirements for packaging to be considered biodegradable and compostable in industrial composting facilities. ISO 18606 describes the requirements and procedures for packaging to be deemed industrially compostable, specifically in aerobic environments (De Wilde et al., 2013; Wojtowicz, 2009). It aims to ensure breakdown within a timeframe, leaves no toxic residue, disintegrates, and doesn't harm the quality of the resulting compost (Avella et al., 2001). Standards such as ISO 17088 harmonize specifications for compostable plastics globally (De Wilde et al., 2013). In terms of compliance, manufacturers use specific captions or logos to show consumers that their products, such as food packaging and bioplastics, are admittedly sustainable alternatives to traditional plastics.

14.7 Future recommendations on starch-based packaging

There is no doubt that starch-based films are an emerging ideal solution to the menace of plastic waste that can be used for a variety of packaging applications. Major investment in starch-based packaging will help to ameliorate the challenges associated with synthetic/crude oil-based packaging produced and used. It is recommended that further research should be conducted to understand how the genotype of starch influences the properties of the final film material. Likewise, the combination of other polymers with starch raw material and how it can improve the barrier and mechanical properties of packaging should be investigated. Additives such as glycerol are often added to starch film to act as a plasticizer, reduce brittleness, and boost flexibility. It is necessary to consider the augmentation of additives with other additives to obtain a starch film with superior packaging properties. Additives such as essential oils, consisting of several bioactivities (antimicrobial and antioxidant), are also used in many cases when considering starch film development. More investigation should be carried out to identify diverse essential oils from plants and their efficiencies when considering starch film packaging. Further investigation should be carried out to resolve the challenges of brittleness, heat sensitivity, and retrogradation (recrystallization) of starch-based films, making them less flexible and opaque in nature.

14.8 Conclusion

The continuous shift from synthetic plastics owing to the numerous biodegradation limitations is expected to continue. Likewise, enzymatic degradation of plastics has both opportunities and challenges when considering sustainable plastic management. Challenges such as slow degradation rate make them resistant to enzymatic attack. In addition, the high cost of enzyme production and scalability challenges appear to impede their application. However, opportunities exist for plant material consisting of starch to be considered for the development of starch-based films or packaging material.

References

Adeyemi, A. O., & Ojoawo, S. O. (2023). Production and characterization of chitosan of crustacean shells. *Materials Today: Proceedings, 88*, 128–134. https://doi.org/10.1016/j.matpr.2023.05.729

Aeschelmann, F., Carus, M., Baltus, W., Carrez, D., Guzman, D., Käb, H. & Ravenstijn, J. (2017). Bio-based building blocks and polymers: global capacities and trends 2016–2021. Nova Institute GmbH and Europeans Bioplastics Association.

Afshar, S. V., Boldrin, A., Christensen, T. H., Corami, F., Daugaard, A. E., Rosso, B., & Hartmann, N. B. (2025). Disintegration of commercial biodegradable plastic products under simulated industrial composting conditions. *Scientific Reports, 15*(1). https://doi.org/10.1038/s41598-025-91647-z

Alcázar-Alay, S. C., & Meireles, M. A. A. (2015). Physicochemical properties, modifications and applications of starches from different botanical sources. *Food Science and Technology (Brazil), 35*(2), 215–236. https://doi.org/10.1590/1678-457X.6749

Alshehrei, F. (2017). Biodegradation of synthetic and natural plastic by microorganisms. *Journal of Applied & Environmental Microbiology, 5*, 8–19.

Andrady, A. L., Hamid, H. S., & Torikai, A. (2003). Effects of climate change and UV-B on materials. *Photochemical & Photobiological Sciences, 2*, 68–72.

Andreeßen, C., & Steinbüchel, A. (2019). Recent developments in non-biodegradable biopolymers: Precursors, production processes, and future perspectives. *Applied Microbiology and Biotechnology, 103*, 143–157.

Ariza-Tarazona, M. C., Villarreal-Chiu, J. F., Hernández-López, J. M., Rivera De la Rosa, J., Barbieri, V., Siligardi, C., & Cedillo-González, E. I. (2020). Microplastic pollution reduction by a carbon and nitrogen-doped TiO2: Effect of pH and temperature in the photocatalytic degradation process. *Journal of Hazardous Materials, 395*. https://doi.org/10.1016/j.jhazmat.2020.122632

Arruda, T. R., Machado, G.d. O., Marques, C. S., Souza, A. L.d., Pelissari, F. M., Oliveira, T. V.d., & Silva, R. R. A. (2025). An overview of starch-based materials for sustainable food packaging: Recent advances, limitations, and perspectives. *Macromol, 5*(2). https://doi.org/10.3390/macromol5020019

Asaf Kleopas, S. (2008). *Synthesis and properties of starch based bio-materials*. Groningen: University of Groningen.

Avella, M., Bonadies, E., Martuscelli, E., & Rimedio, R. (2001). European current standardization for plastic packaging recoverable through composting and biodegradation. *Polymer Testing, 20*, 517–521.

Avérous, L. (2004). Biodegradable multiphase systems based on plasticized starch: A review. *Journal of Macromolecular Science: Part C: Polymer Reviews, 44*(3), 231–274. https://doi.org/10.1081/mc-200029326

Bao, Z., Wang, C., Cao, J., Zhang, T., Guo, Y., Sato, Y., Nishizawa, T., & Ohta, H. (2024). Tardiphaga alba sp. Nov., a heavy-metal-tolerant bacterium isolated from garden soil. *International Journal of Systematic and Evolutionary Microbiology, 74*(1). https://doi.org/10.1099/ijsem.0.006238

Basiak, E., Lenart, A., & Debeaufort, F. (2017). Effect of starch type on the physico-chemical properties of edible films. *International Journal of Biological Macromolecules, 98*, 348–356.

Bemiller, J. N., & Whistler, R. L. (2009). *Starch: Chemistry and technology*. Academic Press.

Bertolini, A. C. (2009). *Starches: Characterization, properties, and applications starches: Characterization, Properties, and applications* (pp. 1–270). Brazil: CRC Press.

Bornscheuer, U. (2002). Microbial carboxyl esterases: Classification, properties and application in biocatalysis. *FEMS Microbiology Reviews, 26*(1), 73–81. https://doi.org/10.1016/s0168-6445(01)00075-4

Castano, J., Rodríguez-Llamazares, S., Sepúlveda, E., Giraldo, D., Bouza, R., & Pozo, C. (2017). Morphological and structural changes of starch during processing by melt blending. *Starch-Stärke, 69*, Article 1600247.

Chaudhary, S., Kour, M., & Kumar, R. (2024). Bioplastic films from starch of *Colocasia esculenta* and its waste: A smart template for sensing applications. *International Journal of Biological Macromolecules, 281*, 136218.

Chen, Y., Tan, L., Chen, L., Yang, Y., & Wang, X. (2008). Study on biodegradable aromatic/aliphatic copolyesters. *Brazilian Journal of Chemical Engineering, 25*(2), 321–335. https://doi.org/10.1590/s0104-66322008000200011

Christensen, P. R., Scheuermann, A. M., Loeffler, K. E., & Helms, B. A. (2019). Closed-loop recycling of plastics enabled by dynamic covalent diketoenamine bonds. *Nature Chemistry, 11*(5), 442–448. https://doi.org/10.1038/s41557-019-0249-2

Coppola, G., Gaudio, M. T., Lopresto, C. G., Calabro, V., Curcio, S., & Chakraborty, S. (2021). Bioplastic from renewable biomass: A facile solution for a greener environment. *Earth Systems and Environment, 5*(2), 231–251. https://doi.org/10.1007/s41748-021-00208-7

Dai, L., Zhang, J., & Cheng, F. (2019). Effects of starches from different botanical sources and modification methods on physicochemical properties of starch-based edible films. *International Journal of Biological Macromolecules, 132*, 897–905.

De Wilde, B., Mortier, N., Verstichel, S., Briassoulis, D., Babou, M., Mistriotis, A., & Hiskakis, M. (2013). Report on current relevant biodegradation and ecotoxicity standards. *Knowledge Based Bio-based Products Pre-Standardization, 1*, 14–20.

Deshayes, S., & Kasko, A. M. (2013). Polymeric biomaterials with engineered degradation. *Journal of Polymer Science, Part A: Polymer Chemistry, 51*(17), 3531–3566. https://doi.org/10.1002/pola.26765

Dey, S., Veerendra, G. T. N., Anjaneya Babu, P. S. S., Phani Manoj, A. V., & Nagarjuna, K. (2024). Degradation of plastics waste and its effects on biological ecosystems: A scientific analysis and comprehensive review. *Biomedical Materials & Devices, 2*(1), 70–112. https://doi.org/10.1007/s44174-023-00085-w

Göpferich, A. (1996). *Mechanisms of polymer degradation and erosion1 the Biomaterials: Silver Jubilee Compendium* (pp. 117–128). Germany: Elsevier Ltd. https://doi.org/10.1016/B978-008045154-1.50016-2

Fonseca-García, A., Jiménez-Regalado, E. J., & Aguirre-Loredo, R. Y. (2021). Preparation of a novel biodegradable packaging film based on corn starch-chitosan and poloxamers. *Carbohydrate Polymers, 251*, 117009.

Gabor, D., & Tita, O. (2012). Biopolymers used in food packaging: A review. *Acta Universitatis Cibiniensis. Series E: Food Technology, 16*(2), 3–19.

Gadhave, R. V., Das, A., Mahanwar, P. A., & Gadekar, P. T. (2018). Starch based bioplastics: The future of sustainable packaging. *Open Journal of Polymer Chemistry, 08* (02), 21–33. https://doi.org/10.4236/ojpchem.2018.82003

García-Guzmán, L., Cabrera-Barjas, G., Soria-Hernández, C. G., Castaño, J., Guadarrama-Lezama, A. Y., & Rodríguez Llamazares, S. (2022). Progress in starch-based materials for food packaging applications. *Polysaccharides, 3*(1), 136–177. https://doi.org/10.3390/polysaccharides3010007

Ge, T., Zhao, D., Liu, X., Yang, Y., He, X., Qi, Y., Liu, W., & Sun, Q. (2025). Synthesis and properties of biodegradable copolyester with phenyl side groups. *Macromolecular Chemistry and Physics, 226*(6). https://doi.org/10.1002/macp.202400442

Goel, V., Luthra, P., Kapur, G. S., & Ramakumar, S. S. V. (2021). Biodegradable/Bioplastics: Myths and realities. *Journal of Polymers and the Environment, 29*(10), 3079–3104. https://doi.org/10.1007/s10924-021-02099-1

Gu, J. D. (2003). Microbiological deterioration and degradation of synthetic polymeric materials: Recent research advances. *International Biodeterioration & Biodegradation, 52*(2), 69–91. https://doi.org/10.1016/S0964-8305(02)00177-4

Guan, Q. F., Yang, H. B., Zhao, Y. X., Han, Z. M., Ling, Z. C., Yang, K. P., Yin, C. H., & Yu, S. H. (2021). Microplastics release from victuals packaging materials during daily usage. *EcoMat, 3*(3). https://doi.org/10.1002/eom2.12107

Gunawardene, O. H. P., Gunathilake, C., Amaraweera, S. M., Fernando, N. M. L., Wanninayaka, D. B., Manamperi, A., Kulatunga, A. K., Rajapaksha, S. M., Dassanayake, R. S., Fernando, C. A. N., & Manipura, A. (2021). Compatibilization of starch/synthetic biodegradable polymer blends for packaging applications: A review. *Journal of Composites Science, 5*(11). https://doi.org/10.3390/jcs5110300

Gurer-Orhan, H., Sabır, H. U., & Özgüneş, H. (2004). Correlation between clinical indicators of lead poisoning and oxidative stress parameters in controls and lead-exposed workers. *Toxicology, 195*, 147–154.

Hasırcı, V., Lewandrowski, K., Gresser, J. D., Wise, D. L., & Trantolo, D. J. (2001). Versatility of biodegradable biopolymers: Degradability and an in vivo application. *Journal of Biotechnology, 86*(2), 135–150. https://doi.org/10.1016/s0168-1656(00)00409-0

Hillmyer, M. A. (2017). The promise of plastics from plants. *Science, 358*(6365), 868–870. https://doi.org/10.1126/science.aao6711

Hirpara, N. J., Dabhi, M., & Rathod, P. (2021). Development of potato starch based biodegradable packaging film. *Journal of Food Processing and Technology, 12*, 529–541.

Hottle, T. A., Bilec, M. M., & Landis, A. E. (2013). Sustainability assessments of bio-based polymers. *Polymer Degradation and Stability, 98*, 1898–1907.

Hu, X., Thumarat, U., Zhang, X., Tang, M., & Kawai, F. (2010). Diversity of polyester-degrading bacteria in compost and molecular analysis of a thermoactive esterase from Thermobifida alba AHK119. *Applied Microbiology and Biotechnology, 87*(2), 771–779. https://doi.org/10.1007/s00253-010-2555-x

Ishigaki, T., Sugano, W., Nakanishi, A., Tateda, M., Ike, M., & Fujita, M. (2004). The degradability of biodegradable plastics in aerobic and anaerobic waste landfill model reactors. *Chemosphere, 54*(3), 225–233. https://doi.org/10.1016/s0045-6535(03)00750-1

Itkor, P., Singh, A. K., Lee, M., Boonsiriwit, A., & Lee, Y. S. (2024). Effects of starch–citric acid cross-linking on the fibrous composites using waste paper pulp material for eco-friendly packaging. *Biomass Conversion and Biorefinery, 14*(13), 14693–14705. https://doi.org/10.1007/s13399-022-03564-0

Khan, A. K., & Majeed, T. (2019). Biodegradation of synthetic and natural plastics by microorganisms: A mini review. *Journal of Natural and Applied Sciences Pakistan, 1*, 180–184.

Khoironi, A., Hadiyanto, H., Anggoro, S., & Sudarno, S. (2020). Evaluation of polypropylene plastic degradation and microplastic identification in sediments at Tambak Lorok coastal area, Semarang, Indonesia. *Marine Pollution Bulletin, 151*. https://doi.org/10.1016/j.marpolbul.2019.110868

Kim, S. R. B., Choi, Y.-G., Kim, J.-Y., & Lim, S.-T. (2015). Improvement of water solubility and humidity stability of tapioca starch film by incorporating various gums. *LWT-Food Science and Technology, 64*, 475–482.

Kim, K. H., & Kim, J. Y. (2021). Understanding wheat starch metabolism in properties, environmental stress condition, and molecular approaches for value-added utilization. *Plants, 10*(11). https://doi.org/10.3390/plants10112282

Kleeberg, I., Welzel, K., VandenHeuvel, J., Müller, R.-J., & Deckwer, W.-D. (2005). Characterization of a new extracellular hydrolase from Thermobifida fusca degrading aliphatic–aromatic copolyesters. *Biomacromolecules, 6*(1), 262–270. https://doi.org/10.1021/bm049582t

Kong, X. (2020). *Fine structure of amylose and amylopectin starch structure, functionality and application in foods* (pp. 29–39). China: Springer Singapore. https://doi.org/10.1007/978-981-15-0622-2_3

Koutny, M., Sancelme, M., Dabin, C., Pichon, N., Delort, A. M., & Lemaire, J. (2006). Acquired biodegradability of polyethylenes containing pro-oxidant additives. *Polymer Degradation and Stability, 91*(7), 1495–1503. https://doi.org/10.1016/j.polymdegradstab.2005.10.007

Kumari, A., & Chaudhary, D. R. (2020). *Engineered microbes and evolving plastic bioremediation technology Bioremediation of Pollutants: From Genetic Engineering to Genome Engineering* (pp. 417–443). India: Elsevier. https://doi.org/10.1016/B978-0-12-819025-8.00021-1

Kumari, S., Yadav, D., Yadav, S., Selvaraj, M., Sharma, G., Karnwal, A., & Yadav, S. (2025). From macro to micro: The key parameters influencing the degradation mechanism and the toxicity of microplastics in the environment. *Polymer Degradation and Stability, 233*, Article 111174. https://doi.org/10.1016/j.polymdegradstab.2025.111174

Lecomte, P., & Jérôme, C. (2011). *Recent developments in ring-opening polymerization of lactones. 245* (pp. 173–217). Springer Science and Business Media LLC. https://doi.org/10.1007/12_2011_144

Li, H., Gidley, M. J., & Dhital, S. (2019). High-amylose starches to bridge the "Fiber Gap": Development, structure, and nutritional functionality. *Comprehensive Reviews in Food Science and Food Safety, 18*, 362–379.

Li, F., Wang, S., Liu, W., & Chen, G. (2008). Purification and characterization of poly(l-lactic acid)-degrading enzymes from Amycolatopsis orientalis ssp. orientalis. *FEMS Microbiology Letters, 282*(1), 52–58. https://doi.org/10.1111/j.1574-6968.2008.01109.x

Lin, Z., Cheng, H., He, K., McClements, D. J., Jin, Z., Xu, Z., Meng, M., Peng, X., & Chen, L. (2024). Recent progress in the hydrophobic modification of starch-based films. *Food Hydrocolloids, 151*, Article 109860. https://doi.org/10.1016/j.foodhyd.2024.109860

Loredo-Treviño, A., Gutiérrez-Sánchez, G., Rodríguez-Herrera, R., & Aguilar, C. N. (2012). Microbial enzymes involved in polyurethane biodegradation: A review. *Journal of Polymers and the Environment, 20*(1), 258–265. https://doi.org/10.1007/s10924-011-0390-5

Lucas, N., Bienaime, C., Belloy, C., Queneudec, M., Silvestre, F., & Nava-Saucedo, J. E. (2008). Polymer biodegradation: Mechanisms and estimation techniques - A review. *Chemosphere, 73*(4), 429–442. https://doi.org/10.1016/j.chemosphere.2008.06.064

Luzics, S., Baka, E., Otto, M., Kosztik, J., Szalontai, H., Bata-Vidács, I., Nagy, I., Tóth, Á., Táncsics, A., Pápai, M., Nagy, I., Orsini, M., & Kukolya, J. (2025). High-quality de novo genome assembly and functional genomic insights into Thermobifida alba DSM43795T, a mesophilic actinobacterium isolated from garden soil. *Biologia Futura, 76*(4), 541–558. https://doi.org/10.1007/s42977-025-00283-1

Mörtl, M., Damak, M., Gulyás, M., Varga, Z. I., Fekete, G., Kurusta, T., Rácz, Á., Székács, A., & Aleksza, L. (2024). Biodegradation assessment of bioplastic carrier bags under industrial-scale composting conditions. *Polymers, 16*(24). https://doi.org/10.3390/polym16243450

Ma, X., Liu, Y., Liu, J., Zhang, J., & Liu, R. (2020). Changes in starch structures and in vitro digestion characteristics during maize (Zea mays L.) germination. *Food Science and Nutrition, 8*(3), 1700–1708. https://doi.org/10.1002/fsn3.1457

Mabrouk, M. M., & Sabry, S. A. (2001). Degradation of poly (3-hydroxybutyrate) and its copolymer poly (3-hydroxybutyrate-co-3-hydroxyvalerate) by a marine Streptomyces sp. SNG9. *Microbiological Research, 156*(4), 323–335. https://doi.org/10.1078/0944-5013-00115

MacLeod, M., Arp, H. P. H., Tekman, M. B., & Jahnke, A. (2021). The global threat from plastic pollution. *Science, 373*(6550), 61–65. https://doi.org/10.1126/science.abg5433

Mandal, M., Roy, A., Popek, R., & Sarkar, A. (2024). Micro- and nano-plastic degradation by bacterial enzymes: A solution to 'white pollution. *Microbe Magazine, 3*, Article 100072.

Marichelvam, M., Jawaid, M., & Asim, M. (2019). Corn and rice starch-based bio-plastics as alternative packaging materials. *Fibers, 7*, 32.

Min, Y., Yi, J., Dai, R., Liu, W., & Chen, H. (2023). A novel efficient wet process for preparing cross-linked starch: Impact of urea on cross-linking performance. *Carbohydrate Polymers, 320*, Article 121247. https://doi.org/10.1016/j.carbpol.2023.121247

Mohee, R., & Unmar, G. (2007). Determining biodegradability of plastic materials under controlled and natural composting environments. *Waste Management, 27*(11), 1486–1493. https://doi.org/10.1016/j.wasman.2006.07.023

Molitoris, H. P., Moss, S. T., de Koning, G. J. M., & Jendrossek, D. (1996). Scanning electron microscopy of polyhydroxyalkanoate degradation by bacteria. *Applied Microbiology and Biotechnology, 46*(5–6), 570–579. https://doi.org/10.1007/s002530050863

Mueller, R. (2007). Mechanism of polyester degradation by enzymes. *Polymer Preprints, 48*.

Nabi, I., Bacha, A. U. R., Li, K., Cheng, H., Wang, T., Liu, Y., Ajmal, S., Yang, Y., Feng, Y., & Zhang, L. (2020). Complete photocatalytic mineralization of microplastic on TiO2 nanoparticle film. *iScience, 23*(7). https://doi.org/10.1016/j.isci.2020.101326

Nagai, Y., Ogawa, T., Nishimoto, Y., & Ohishi, F. (1999). Analysis of weathering of a thermoplastic polyester elastomer II. Factors affecting weathering of a polyether–polyester elastomer. *Polymer Degradation and Stability, 65*(2), 217–224. https://doi.org/10.1016/s0141-3910(99)00007-5

Nakajima-Kambe, T., Ichihashi, F., Matsuzoe, R., Kato, S., & Shintani, N. (2009). Degradation of aliphatic–aromatic copolyesters by bacteria that can degrade aliphatic polyesters. *Polymer Degradation and Stability, 94*(11), 1901–1905. https://doi.org/10.1016/j.polymdegradstab.2009.08.006

Napper, I. E., & Thompson, R. C. (2019). Environmental deterioration of biodegradable, Oxo-biodegradable, compostable, and conventional plastic carrier bags in the Sea, soil, and open-air over a 3-year period. *Environmental Science and Technology, 53*(9), 4775–4783. https://doi.org/10.1021/acs.est.8b06984

Nawab, A., Alam, F., Haq, M. A., Lutfi, Z., & Hasnain, A. (2017). Mango kernel starch-gum composite films: Physical, mechanical and barrier properties. *International Journal of Biological Macromolecules, 98*, 869–876. https://doi.org/10.1016/j.ijbiomac.2017.02.054

Nizamuddin, S., & Chen, C. (2024). Biobased, biodegradable and compostable plastics: Chemical nature, biodegradation pathways and environmental strategy. *Environmental Science and Pollution Research, 31*(6), 8387–8399. https://doi.org/10.1007/s11356-023-31689-w

Omar, K. A., Salih, B. M., Abdulla, N. Y., Hussin, B. H., & Rassul, S. M. (2016). Evaluation of starch and sugar content of different rice samples and study their physical properties. *Indian Journal of Natural Sciences, 6*, 11084–11088.

Ozougwu, J. C. (2016). The role of reactive oxygen species and antioxidants in oxidative stress. *International Journal of Research, 1*, 1–8.

Phukan, M., & Nongkhlaw, F. T. (2022). Starch: An overview of its conventional and unconventional sources, pharmaceutical and industrial applications. *Curr Trends Pharm Res, 9*, 140–161.

Rahimi, A. R., & García, J. M. (2017). Chemical recycling of waste plastics for new materials production. *Nature Reviews Chemistry, 1*(6). https://doi.org/10.1038/s41570-017-0046

Robertson, T. M., Alzaabi, A. Z., Robertson, M. D., & Fielding, B. A. (2018). Starchy carbohydrates in a healthy diet: The role of the humble potato. *Nutrients, 10*(11). https://doi.org/10.3390/nu10111764

Robinson, G., Balk, J., & Domoney, C. (2019). Improving pulse crops as a source of protein, starch and micronutrients. *Nutrition Bulletin, 44*, 202–215.

Rodrigues, L. B. O., Veloso, C. M., Bonomo, R. C. F., Rodrigues, L. B., Minim, L. A., & Costa, R. A. S. (2018). Rheological and textural studies of arrowroot (Maranta arundinacea) starch–sucrose–whey protein concentrate gels. *Journal of Food Science and Technology, 55*(8), 2974–2984. https://doi.org/10.1007/s13197-018-3215-x

Samsudin, H., & Hani, N. M. (2017). *Use of starch in food packaging. Starch-based materials in food packaging*. Elsevier.

Sanchez-Rexach, E., Johnston, T. G., Jehanno, C., Sardon, H., & Nelson, A. (2020). Sustainable materials and chemical processes for additive manufacturing. *Chemistry of Materials, 32*(17), 7105–7119. https://doi.org/10.1021/acs.chemmater.0c02008

Santos, M., Gangoiti, J., Keul, H., Möller, M., Serra, J. L., & Llama, M. J. (2013). Polyester hydrolytic and synthetic activity catalyzed by the medium-chain-length poly(3-hydroxyalkanoate) depolymerase from Streptomyces venezuelae SO1. *Applied Microbiology and Biotechnology, 97*(1), 211–222. https://doi.org/10.1007/s00253-012-4210-1

Sardon, H., & Dove, A. P. (2018). Plastics recycling with a difference. *Science, 360*(6387), 380–381. https://doi.org/10.1126/science.aat4997

Schirmer, A., Jendrossek, D., & Schlegel, H. G. (1993). Degradation of poly (3-hydroxyoctanoic acid)[P (3HO)] by bacteria: purification and properties of a P (3HO) depolymerase from *Pseudomonas fluorescens* GK13. *Applied and Environmental Microbiology, 59*, 1220–1227.

Seltenrich, N. (2015). New link in the food chain? Marine plastic pollution and seafood safety. *Environmental Health Perspectives, 123*(2). https://doi.org/10.1289/ehp.123-a34

Sessini, V., Arrieta, M. P., Kenny, J. M., & Peponi, L. (2016). Processing of edible films based on nanoreinforced gelatinized starch. *Polymer Degradation and Stability, 132*, 157–168. https://doi.org/10.1016/j.polymdegradstab.2016.02.026

Shah, A. A., Eguchi, T., Mayumi, D., Kato, S., Shintani, N., Kamini, N. R., & Nakajima-Kambe, T. (2013). Purification and properties of novel aliphatic-aromatic co-polyesters degrading enzymes from newly isolated roseateles depolymerans strain TB-87. *Polymer Degradation and Stability, 98*(2), 609–618. https://doi.org/10.1016/j.polymdegradstab.2012.11.013

Shah, A. A., Hasan, F., & Hameed, A. (2010). Degradation of poly (3-hydroxybutyrate-co-3-hydroxyvalerate) by a newly isolated *Actinomadura* sp. AF-555, from soil. *International Biodeterioration & Biodegradation, 64*, 281–285.

Shah, A. A., Hasan, F., Hameed, A., & Ahmed, S. (2007). Isolation and characterization of poly(3-hydroxybutyrate-co-3-hydroxyvalerate) degrading bacteria and purification of PHBV depolymerase from newly isolated Bacillus sp. AF3. *International Biodeterioration & Biodegradation, 60*(2), 109–115. https://doi.org/10.1016/j.ibiod.2007.01.004

Shah, A. A., Kato, S., Shintani, N., Kamini, N. R., & Nakajima-Kambe, T. (2014). Microbial degradation of aliphatic and aliphatic-aromatic co-polyesters. *Applied Microbiology and Biotechnology, 98*(8), 3437–3447. https://doi.org/10.1007/s00253-014-5558-1

Silva, O. A., Pellá, M. G., Pellá, M. G., Caetano, J., Simões, M. R., Bittencourt, P. R. S., & Dragunski, D. C. (2019). Synthesis and characterization of a low solubility edible film based on native cassava starch. *International Journal of Biological Macromolecules, 128*, 290–296. https://doi.org/10.1016/j.ijbiomac.2019.01.132

Singh, N., & Susan, A. (2018). Polymer nanocomposites for water treatments. In *Polymer-based nanocomposites for energy and environmental applications*. Elsevier.

Song, J. H., Murphy, R. J., Narayan, R., & Davies, G. B. H. (2009). Biodegradable and compostable alternatives to conventional plastics. *Philosophical Transactions of the Royal Society B: Biological Sciences, 364*(1526), 2127–2139. https://doi.org/10.1098/rstb.2008.0289

Srichuwong, S., Sunarti, T. C., Mishima, T., Isono, N., & Hisamatsu, M. (2005). Starches from different botanical sources I: Contribution of amylopectin fine structure to thermal properties and enzyme digestibility. *Carbohydrate Polymers, 60*(4), 529–538. https://doi.org/10.1016/j.carbpol.2005.03.004

Stubbins, A., Law, K. L., Muñoz, S. E., Bianchi, T. S., & Zhu, L. (2021). Plastics in the Earth system. *Science, 373*(6550), 51–55. https://doi.org/10.1126/science.abb0354

Surtono, A., Aprilliana, P., Supriyanto, A., Pauzi, G. A., & Junaidi, S. W. (2019). Suciyati, warsito, measuring of cassava starch content by using strain gauge. *Journal of Physics: Conference Series, 1338*(1), Article 012019. https://doi.org/10.1088/1742-6596/1338/1/012019

Tamnou, E. B. M., Arfao, A. T., Nougang, M. E., Metsopkeng, C. S., Ewoti, O. V. N., Moungang, L. M., Nana, P. A., Takang-Etta, L.-R. A., Perrière, F., & Sime-Ngando, T. (2021). Biodegradation of polyethylene by the bacterium Pseudomonas aeruginosa in acidic aquatic microcosm and effect of the environmental temperature. *Environmental Challenges, 3*, Article 100056.

Taniguchi, I., Yoshida, S., Hiraga, K., Miyamoto, K., Kimura, Y., & Oda, K. (2019). Biodegradation of PET: Current status and application aspects. *ACS Catalysis, 9*(5), 4089–4105. https://doi.org/10.1021/acscatal.8b05171

Tokiwa, Y., Calabia, B. P., Ugwu, C. U., & Aiba, S. (2009). Biodegradability of plastics. *International Journal of Molecular Sciences, 10*(9), 3722–3742. https://doi.org/10.3390/ijms10093722Japan

Tribedi, P., Sarkar, S., Mukherjee, K., & Sil, A. K. (2012). Isolation of a novel Pseudomonas sp from soil that can efficiently degrade polyethylene succinate. *Environmental Science and Pollution Research, 19*(6), 2115–2124. https://doi.org/10.1007/s11356-011-0711-1

Venkatram, S., McCollum, J., Stingelin, N., & Brettmann, B. (2023). A close look at polymer degree of crystallinity versus polymer crystalline quality. *Polymer International, 72*(10), 855–860. https://doi.org/10.1002/pi.6508

Wang, C., Han, H., Wu, Y., & Astruc, D. (2022). Nanocatalyzed upcycling of the plastic wastes for a circular economy. *Coordination Chemistry Reviews, 458*, Article 214422. https://doi.org/10.1016/j.ccr.2022.214422

Wang, K., Wang, W., Ye, R., Liu, A., Xiao, J., Liu, Y., & Zhao, Y. (2017). Mechanical properties and solubility in water of corn starch-collagen composite films: Effect of starch type and concentrations. *Food Chemistry, 216*, 209–216. https://doi.org/10.1016/j.foodchem.2016.08.048

Wang, Z., Zhang, X., Fu, J., Huang, J., & Wang, L. (2025). Fish collagen mediated alteration of wheat starch thermal properties during multi-species co-fermentation. *International Journal of Biological Macromolecules, 295*, Article 139987. https://doi.org/10.1016/j.ijbiomac.2025.139987

Wang, W., Zhang, Y., Ye, R., & Ni, Y. (2015). Physical crosslinkings of edible collagen casing. *International Journal of Biological Macromolecules, 81*, 920–925. https://doi.org/10.1016/j.ijbiomac.2015.09.032

Webber, M. J., & Tibbitt, M. W. (2022). Dynamic and reconfigurable materials from reversible network interactions. *Nature Reviews Materials, 7*(7), 541–556. https://doi.org/10.1038/s41578-021-00412-x

Wojtowicz, A. (2009). Biodegradability and compostability of biopolymers. In *Thermoplastic starch: A green material for various industries* (pp. 55–76).

Wojtowicz, A. (2010). *Biodegradability and compostability of biopolymers thermoplastic starch: A green material for various industries* (pp. 55–76). Poland: Wiley-VCH. https://doi.org/10.1002/9783527628216.ch3

Wróblewska-Krepsztul, J., Rydzkowski, T., Borowski, G., Szczypiński, M., Klepka, T., & Thakur, V. K. (2018). Recent progress in biodegradable polymers and nanocomposite-based packaging materials for sustainable environment. *International Journal of Polymer Analysis and Characterization, 23*(4), 383–395. https://doi.org/10.1080/1023666X.2018.1455382

Xiong, Y., Zhao, J., Li, L., Wang, Y., Dai, X., Yu, F., & Ma, J. (2020). Interfacial interaction between micro/nanoplastics and typical PPCPs and nanoplastics removal via electrosorption from an aqueous solution. *Water Research, 184*, Article 116100. https://doi.org/10.1016/j.watres.2020.116100

Yang, Y., Fu, J., Duan, Q., Xie, H., Dong, X., & Yu, L. (2024). Strategies and methodologies for improving toughness of starch films. *Foods, 13*(24), 4036. https://doi.org/10.3390/foods13244036

Zakaria, N., Muhammad, N., & Abdullah, M. (2017). Potential of starch nanocomposites for biomedical applications. In *IOP conference series: Materials science and engineering* (p. 012087). IOP Publishing.

Zhang, F., Zhao, Y., Wang, D., Yan, M., Zhang, J., Zhang, P., Ding, T., Chen, L., & Chen, C. (2021). Current technologies for plastic waste treatment: A review. *Journal of Cleaner Production, 282*, Article 124523. https://doi.org/10.1016/j.jclepro.2020.124523

CHAPTER FIFTEEN

Regulatory frameworks and consumer acceptance of starch-based packaging

Seyed Alireza Banihashemi[1], Saber Amiri[1], Masoud Aman Mohammadi[2]
[1]Department of Food Science and Technology, Faculty of Agriculture, Urmia University, Urmia, Iran
[2]Food and Beverages Safety Research Center, Urmia University of Medical Sciences, Urmia, Iran

15.1 Introduction

The growing environmental concerns surrounding traditional plastic packaging have significantly driven the demand for sustainable alternatives. Among these, starch-based packaging (SBP) has emerged as a promising solution due to its biodegradability and potential to reduce reliance on petroleum-based plastics. Starch, a renewable and abundant resource, offers a flexible foundation for the development of packaging materials that can meet the rising demand for eco-friendly products in various industries, especially food packaging. This transition is increasingly vital in the face of the global push toward reducing plastic waste and enhancing sustainability in product packaging (Branca et al., 2024; Commission, 2022; Khezerlou et al., 2021).

This chapter explores the evolving regulatory frameworks governing starch-based packaging across various global contexts, with a specific focus on the European Union (EU), the United States (US), and international standards. The intersection of regulatory policy, scientific advancements, and consumer acceptance plays a pivotal role in shaping the market dynamics of SBP. In particular, this chapter delves into the regulatory landscape in the EU, where stringent food-contact material (FCM) laws such as Regulation (EC) 1935/2004 and Regulation (EU) October 2011 set the foundation for safety and migration limits of materials in contact with food (Commission, 2023). These regulations ensure that materials, including starch-based composites, do not endanger human health or cause undesirable changes to food properties. The chapter also discusses how the US FDA, employing a risk-based approach, evaluates food-contact materials and provides an alternative framework that emphasizes exposure-based safety assessments (EFSA Panel on Food Contact

Starches in Smart Packaging
ISBN: 978-0-443-45396-0
https://doi.org/10.1016/B978-0-443-45396-0.00016-5

Materials, Enzymes, Flavourings and Processing Aids, 2016; Sharma et al., 2023).

Moreover, as SBP continues to gain traction, its safety and long-term viability are scrutinized through scientific methodologies, particularly in the realms of migration science and the potential environmental risks associated with biodegradable microplastics. These concerns are integral to understanding the full lifecycle of starch-based materials and their integration into a circular economy (Miralles et al., 2025). The increasing complexity of migration studies, particularly those focusing on non-intentionally added substances (NIAS), demands further attention to ensure compliance with regulatory standards. Such scientific advancements contribute to the evolving regulatory criteria for bioplastics and their inclusion in food-contact material safety evaluations (Zellmer et al., 2024).

Consumer acceptance of SBP plays a crucial role in its widespread adoption. Research on consumer willingness to pay (WTP), the effectiveness of eco-labeling, and the challenges in understanding terms like "biodegradable" and "compostable" illustrate that while there is a general shift toward environmentally friendly packaging, practical barriers still hinder its mainstream market acceptance. These barriers include misconceptions about the end-of-life disposal of bioplastics and the lack of standardized labeling. This chapter highlights the importance of addressing these concerns through regulatory clarity and consumer education, ensuring that SBP can be integrated effectively into the circular economy (Allison et al., 2024; Zwicker et al., 2021).

In conclusion, the transition to starch-based packaging presents a tangible opportunity to reduce the environmental footprint of single-use plastics. However, the successful integration of SBP into the market relies heavily on the alignment of scientific innovation with robust regulatory frameworks, alongside the cultivation of consumer trust through transparent, credible communication. It is through these combined efforts that the potential of SBP to contribute to a sustainable future in packaging can be fully realized (Branca et al., 2024; Halmans, 2024).

15.2 Regulatory landscapes

15.2.1 European Union: food-contact safety and circularity rules

15.2.1.1 Food-contact safety

In the European Union, the foundational regulatory framework for food-contact materials is Regulation (EC) 1935/2004, which mandates that any material in contact with food must not endanger human health or cause unacceptable changes to food composition or organoleptic properties. When starch-based materials fall within the definition of plastics, either as blends, coatings, or multilayer systems, they are subject to the Plastics Regulation (EU) No 10/2011, which introduces a positive list of authorized substances and imposes overall and specific migration limits (OML/SMLs) (Commission, 2004, 2023). The overall migration limit of 10 mg/dm^2 (or 60 mg/kg food) acts as a toxicological safeguard, ensuring that total migrant quantities remain within acceptable exposure thresholds. The European Food Safety Authority (EFSA) plays a central role in assessing the safety of new monomers, additives, and other constituents used in FCMs. Its risk assessment principles, recently reaffirmed by the CEP Panel, emphasize migration-driven exposure evaluation and toxicological threshold testing for both intentionally and non-intentionally added substances (NIAS) (Muncke et al., 2017; Peters et al., 2019). EFSA's current focus on natural-origin mixtures, plasticizers, and recycled inputs is particularly relevant to SBP systems that incorporate biofillers, plasticizers, or recycled layers.

15.2.1.2 Circularity and packaging policy

The European regulatory environment is also being transformed by the replacement of the Packaging and Packaging Waste Directive (94/62/EC) with the Packaging and Packaging Waste Regulation (EU) 2025/40 (PPWR), adopted in December 2024 and entering application from mid-2026 (Commission, 2025). As a directly applicable regulation, the PPWR introduces harmonized measures on design-for-recycling, reuse quotas, recycled content, and labeling, reinforcing circular economy principles. By tightening definitions of "compostable" and "recyclable", the PPWR discourages vague environmental claims and aligns product labeling with actual waste-management capabilities. Policy analyses in 2024 highlighted that the PPWR's waste-reduction targets and selective format bans, especially on unnecessary single-use packaging, are expected to create favorable conditions for high-performance compostable and biobased alternatives (Financial Times). SBP, when verifiably compostable under EN 13432, becomes not only an environmentally favorable material but also a compliance lever within an evolving regulatory landscape.

15.2.2 United States: FDA food-contact pathways

In the United States, regulatory oversight of food-contact materials is governed by the Federal Food, Drug, and Cosmetic Act (FFDCA) and administered by the Food and Drug Administration (FDA). The primary pathways include Food Contact Notifications (FCNs), food additive regulations, Generally Recognized as Safe (GRAS) determinations, and Threshold of Regulation exemptions. The FCN program, introduced in 2000, has streamlined market entry by enabling case-specific notifications for new substances, provided that exposure estimates remain below defined thresholds (Sharma et al., 2023). PLA and related biopolymer systems already appear in multiple FCNs, signifying prior acceptance of migration and toxicology data under the FDA's risk-based framework. Although the US model is less prescriptive than the EU's positive-list approach, FDA nonetheless expects proportionality between exposure potential and toxicological data (EFSA Panel on Food Contact Materials, Enzymes, Flavourings and Processing Aids, 2016; Miralles et al., 2025). Historical migration studies on PLA and starch-based composites confirm low migration under non-fatty food conditions and moderate temperatures, which provides a precedent for future SBP submissions.

15.2.3 International standards and harmonization

At the international level, the drive for harmonization is led by standards organizations such as ISO, CEN, and ASTM. The EU's EN 13432 and EN 14995 define compostability and biodegradability parameters, including $\geq$90% disintegration within 12 weeks and $\geq$90% biodegradation within 6 months, along with ecotoxicity and heavy-metal limits ((CEN), 2000, 2006; Lavagnolo et al., 2024). Their North American counterparts, ASTM D6400 (plastics) and ASTM D6868 (coated papers), serve analogous roles, establishing test comparability across regions. Meanwhile, Codex Alimentarius (CXC 1–1969) provides horizontal hygiene and labeling requirements and has begun scoping guidance for recycled materials in packaging (International, 2021a, 2021b; Ye et al., 2025). Such developments are instrumental in bridging global policy fragmentation by establishing consistent safety baselines for materials combining recycled and compostable components, a key consideration for multilayer SBP systems (Table 15.1).

15.3 Safety assessment and migration science

15.3.1 Migration test design and performance of SBP

Migration testing remains the primary means of verifying food-contact safety. The EU's standardized test conditions (OM1-OM6) and simulants (A/B/C/D1/D2) ensure reproducibility and comparability across laboratories (Standardization), 2022). Numerous studies on starch-based

Table 15.1 Comparison of regulatory frameworks: EU versus FDA versus Codex.

Criterion	EU	FDA	Codex
Regulatory approach	Positive list system	Exposure-based system	Hygiene-based framework
Approval pathway	EFSA evaluation + regulation (EU) No 10/2011	FCN/GRAS/Threshold of regulation (ToR)	General guidance (non-binding)
Migration limits	OML/SML requirements	Exposure-driven limits	No strict SML values
NIAS requirements	Mandatory assessment	Case-by-case evaluation	Decentralized/not explicitly defined
Compostability standards	EN 13432	ASTM D6400	Recommended only, not mandatory

composites, ranging from LDPE-starch blends to modified starch derivatives, report overall migration values far below the 10 mg/dm^2 limit, often in the range of 3–4 mg/dm^2 under standard aqueous and ethanol simulants (Onyeaka et al., 2022; Wang et al., 2025a). Such results underscore the intrinsic inertness of many starch-based matrices when properly plasticized and processed. The 2022 revision of EN 1186 further refines these methodologies by clarifying analytical parameters and time-temperature combinations (Diyana et al., 2021; Standardization, 2022). Compliance certificates increasingly include chain-of-custody reporting and simulator combinations, supporting traceability and demonstrating suitability for "all food types".

15.3.2 NIAS (non-intentionally added substances) in biobased systems

A critical evolution in migration science concerns NIAS, chemical entities formed during production or degradation that are not intentionally added. Modern analytical workflows now rely on untargeted liquid chromatography-high-resolution mass spectrometry (LC-HRMS), gas chromatography-mass spectrometry (GC-MS), and bioassays to screen for NIAS (Miralles et al., 2025; Zellmer et al., 2024). These methods, originally developed for petroleum-based plastics, are being adapted to complex biobased matrices where degradation intermediates and natural-origin compounds introduce analytical challenges. By establishing comprehensive

NIAS profiles, regulatory agencies and industry can better apply threshold of toxicological concern (TTC) approaches and prioritize risk management strategies. For SBP, NIAS scrutiny is particularly important given the heterogeneity of starch sources, additives, and microbial degradation pathways.

15.3.3 Health risk considerations: microplastics and particles

Recent discourse on microplastics has extended to biodegradable materials, prompting debates about "biodegradable microplastics". Biopolymers such as PLA, PBAT, or starch blends can fragment into microscopic particles before complete mineralization, raising concerns about persistence and bioavailability. Current toxicological understanding indicates that the risk depends on particle size, surface chemistry, and exposure dose (Zhao et al., 2020). EFSA's present approach emphasizes migration-based exposure over particle-based metrics until standardized assays for micro and nanoplastics hazards are validated. Research on starch-derived microfragments suggests low bioaccumulation potential, yet comprehensive dose-response data remain limited. Future risk frameworks will likely need to integrate polymer degradation kinetics with toxicokinetic modeling to accurately assess SBP's long-term safety.

15.4 Functional performance: from "edible films" to advanced packaging

15.4.1 Mechanical and barrier upgrades via nanocomposites and hybrids

Material performance has historically constrained the broader adoption of SBP. Starch films exhibit high hydrophilicity, leading to poor water-vapor and oxygen barrier properties. Nanocomposite engineering has emerged as a transformative approach, with montmorillonite (MMT) nanoclays, nanocellulose, ZnO, TiO_2, and layered silicates imparting tortuosity and improved load transfer within the starch matrix (Dhiman et al., 2025; Sapuan et al., 2024; Singh et al., 2023). Properly dispersed nanoscale fillers can achieve tensile strength and modulus increases of over 50%, alongside notable reductions in water-vapor permeability. Processing parameters, including extrusion shear, plasticizer concentration, and ultrasound pretreatment, play a decisive role in determining nanoparticle dispersion (Aguirre-Loredo et al., 2023). The "rheological memory" of starch-nanofiller systems underscores the need for strict manufacturing control to ensure consistent performance. Starch-cellulose hybrids, often combined with polyvinyl alcohol (PVA) or PLA layers, exhibit synergistic mechanical enhancement, confirmed by microscopy studies showing well-bonded interfaces (Liu et al., 2009; Soni et al., 2020).

15.4.2 Active packaging: antimicrobial and antioxidant systems

Active packaging leverages bioactive compounds to extend shelf life and improve food safety. Chitosan and ε-polylysine are among the most studied actives incorporated into starch matrices due to their biocompatibility and antimicrobial spectrum against *Staphylococcus aureus* and *Escherichia coli* (Momtaz et al., 2024; Sha et al., 2025). These systems often rely on synergistic mechanisms combining ionic interactions, film barrier improvement, and controlled release. Plant-derived polyphenolic extracts, such as basil, berry, and tropical fruit extracts, as well as bioactive compounds of animal origin, have been reported to introduce antioxidant activity, contributing to lipid oxidation delay in food products (Banihashemi et al., 2020; Li et al., 2026; Periyasamy et al., 2025). However, these additions necessitate careful risk assessment of migration levels and potential NIAS formation. Both EU and FDA frameworks apply existing migration and exposure limits to such actives, underscoring the regulatory expectation that "active" packaging must meet the same safety benchmarks as inert systems (Silva et al., 2019).

15.4.3 Intelligent packaging: indicators and sensors in starch matrices

Intelligent or "smart" packaging integrates sensor elements that monitor food freshness or quality. Anthocyanins derived from red cabbage, roselle, or berries are widely used as natural pH indicators embedded in starch films, changing color in response to spoilage-related pH shifts (Roy & Rhim, 2021; Wu et al., 2024; Zhao et al., 2022). These innovations reflect a broader movement toward biodegradable sensor systems compatible with compostable matrices. However, challenges remain concerning stability, humidity sensitivity, and potential migration of indicator compounds. Strategies such as crosslinking, encapsulation, and multilayer design are being pursued to maintain readability and regulatory compliance (Wang et al., 2025b). Emerging "smart ecosystems" integrate such indicators with data carriers or gas scavengers, envisioning fully biodegradable sensor platforms for perishable goods (Mafe et al., 2025).

15.5 Consumer acceptance and market dynamics

15.5.1 Willingness to pay, information effects, and heterogeneity

Consumer studies consistently indicate a positive willingness to pay (WTP) for biodegradable and biobased packaging (Jaiswal et al., 2026; Zwicker et al., 2021). However, these preferences are highly context-dependent and often moderated by information exposure, eco-label presence, and perceived credibility. Experiments employing discrete-choice and contingent-valuation methodologies reveal that while many consumers express pro-environmental attitudes, their actual purchase behaviors are moderated by cost, convenience, and familiarity, exemplifying the "attitude-behavior gap" (Alenazi, 2025; Jaiswal et al., 2026). Socio-demographic analysis reveals that younger, urban, and environmentally conscious consumers exhibit higher WTP, though cultural and economic variations persist across markets (Jati & Gravitiani, 2025; Nguyen et al., 2021). This heterogeneity suggests that successful market adoption of SBP requires segmentation-based marketing, aligning product narratives with local value systems and waste management realities.

15.5.2 Label clarity and claim credibility

Terminological confusion between "biodegradable", "compostable", and "recyclable" remains a significant barrier to adoption. Misleading or unverified claims can erode consumer trust and provoke regulatory scrutiny. The PPWR's forthcoming labeling provisions are designed to mitigate such issues by linking environmental claims to standardized testing (EN/ASTM) and demonstrable end-of-life performance (Allison et al., 2024; Bojanowska & Sulimierska, 2023; Mhaddolkar et al., 2024). In this context, SBP producers must ensure alignment between laboratory-tested compostability and the practical realities of municipal composting infrastructures.

15.5.3 Market signals and applications

Commercially, SBP adoption is gaining traction in sectors such as fresh produce, bakery packaging, snack wrapping, and coated paper products, particularly where active or intelligent features can demonstrably extend shelf life (Fatima et al., 2024). The alignment of supply chains, from material converters to municipal waste processors, is critical to ensuring that compostable materials are effectively diverted from landfill and do not contaminate recycling streams. The PPWR's harmonized labeling and collection provisions are expected to enhance such coordination, improving consumer participation and cross-border regulatory consistency (Cristóbal et al., 2023; Raźniewska, 2022).

15.6 Technical roadmap for starch-based packaging

15.6.1 Formulation strategies

The development of SBP follows a systematic formulation strategy beginning with the selection and modification of starch sources (e.g., corn, potato, cassava). Plasticizer systems such as glycerol or sorbitol confer thermoplasticity, while nanoclay and cellulose nanofiber reinforcements improve mechanical and barrier properties (Fatima et al., 2024). Multilayer architectures, such as PLA/starch/PLA, combine stiffness, moisture resistance, and biodegradability in a single structure. Active formulations integrate food-approved bioactives with validated release kinetics, while intelligent designs use immobilized anthocyanin indicators optimized for optical readability and moisture stability (Bayer, 2021; Falcão et al., 2022; Qin et al., 2021; Wu et al., 2024). Process optimization, particularly extrusion temperature and plasticizer balance, remains essential for industrial scalability and reproducibility.

15.6.2 Safety and compliance workflow

The regulatory compliance process begins with defining the intended use (food type, time, and temperature) followed by migration testing under EN 1186 or equivalent FDA protocols. Next, SML and NIAS screenings using LC-HRMS and GC-MS identify potential migrants, and toxicological triage via TTC or QSAR modeling assesses their safety relevance. Compostability must be demonstrated through EN 13432 or ASTM D6400 tests, including disintegration, biodegradation, and ecotoxicity verification. Finally, labeling and claims must align with PPWR and regional state laws (EFSA Panel on Food Contact Materials, Enzymes, Flavourings and Processing Aids, 2016; Koster et al., 2015).

15.7 Outlook: policy, technology, and market interaction

The interplay between regulation, technology, and market acceptance defines the trajectory of starch-based packaging. The PPWR's harmonized rules, coupled with growing global awareness of plastic pollution, provide significant policy tailwinds for SBP adoption. However, these opportunities are accompanied by heightened expectations for scientific substantiation, life-cycle transparency, and verifiable end-of-life performance (Commission, 2022; Halmans, 2024). Research and development efforts should prioritize enhanced mechanical reinforcement through nanoclay and cellulose composites, multilayer architectures combining biopolymers with complementary properties, and early-stage NIAS and migration assessment to streamline regulatory approval (Calambas et al., 2021; Singh et al., 2023). Bridging the attitude-behavior gap in consumer adoption will require transparent labeling, product-specific life-cycle analyses, and collaboration between retailers, municipalities, and converters to ensure that environmental claims correspond to actual collection and processing outcomes (Branca et al., 2024; Wijesinghe, 2025). Overall, starch-based packaging stands at the intersection of material innovation, regulatory transformation, and consumer behavior. Its future success will depend on aligning scientific credibility with policy compliance and market trust, a synthesis that embodies the very principles of the circular economy.

References

Aguirre-Loredo, R. Y., Fonseca-García, A., Calambas, H. L., Salazar-Arango, A., & Caicedo, C. (2023). Improvements of thermal and mechanical properties of achira starch/chitosan/clay nanocomposite films. *Heliyon, 9*(6).

Alenazi, S. A. (2025). Sustainability awareness, price sensitivity, and willingness to pay for eco-friendly packaging: A discrete choice and valuation study in the Saudi retail sector. *Sustainability, 17*(16), 7287.

Allison, A. L., Băitanu, A., Purkiss, D., Lorencatto, F., Michie, S., & Miodownik, M. (2024). Enabling desired disposal of compostable plastic packaging: An evaluation of disposal instruction labels. *Frontiers in Sustainability, 5*, Article 1376519.

Banihashemi, S. A., Nikoo, M., Ghasempour, Z., & Ehsani, A. (2020). Bioactive peptides fractions from traditional Iranian Koopeh cheese; lactic fermentation products. *Biocatalysis and Agricultural Biotechnology, 29*, Article 101798.

Bayer, I. S. (2021). Biopolymers in multilayer films for long-lasting protective food packaging: A review. *Sustainable food packaging technology*, 395–426.

Bojanowska, A., & Sulimierska, A. (2023). Consumer awareness of biodegradability of food products packaging. *Sustainability, 15*(18), Article 13980.

Branca, G., Resciniti, R., & Babin, B. J. (2024). Sustainable packaging design and the consumer perspective: A systematic literature review. *Italian Journal of Marketing, 2024*(1), 77–111.

Calambas, H. L., Fonseca, A., Adames, D., Aguirre-Loredo, Y., & Caicedo, C. (2021). Physical-mechanical behavior and water-barrier properties of biopolymers-clay nanocomposites. *Molecules, 26*(21), 6734.

(CEN), E. C. f. S. (2000). EN 13432: Requirements for packaging recoverable through composting and biodegradation — Test scheme and evaluation criteria for the final acceptance of packaging (EN 13432). C. E. C. f. Standardization).

(CEN), E. C. f. S. (2006). EN 14995: Plastics — Evaluation of compostability — Test scheme and specifications (EN 14995). E. C. f. S. (CEN).
Commission, E. (2004). Regulation (EC) No 1935/2004 of the European Parliament and of the Council of 27 October 2004 on materials and articles intended to come into contact with food (Regulation (EC) No 1935/2004). O. J. o. t. E. Union.
Commission, E. (2022). EU Policy Framework on Biobased, Biodegradable and Compostable Plastics.
Commission Regulation (EU) No 10/2011 on plastic materials and articles intended to come into contact with food, (2023). https://eur-lex.europa.eu.
Commission, E. (2025). Regulation (EU) 2025/40 of the European Parliament and of the Council of 19 December 2024 on packaging and packaging waste. O. J. o. t. E. Union.
Cristóbal, J., Albizzati, P. F., Giavini, M., Caro, D., Manfredi, S., & Tonini, D. (2023). Management practices for compostable plastic packaging waste: Impacts, challenges and recommendations. *Waste Management, 170*, 166–176.
Dhiman, S., Kumari, A., Kumari, S., & Sharma, R. (2025). Advanced biodegradable starch-based nanocomposite films incorporating zinc oxide nanoparticles: Synthesis, characterization, and efficacy in antibacterial food packaging applications. *Journal of Environmental Chemical Engineering, 13*(3), Article 116296.
Diyana, Z., Jumaidin, R., Selamat, M. Z., Ghazali, I., Julmohammad, N., Huda, N., & Ilyas, R. (2021). Physical properties of thermoplastic starch derived from natural resources and its blends: A review. *Polymers, 13*(9), 1396.
EFSA Panel on Food Contact Materials, Enzymes, Flavourings and Processing Aids. (2016). Recent developments in the risk assessment of chemicals in food and their potential impact on the safety assessment of substances used in food contact materials. *EFSA Journal, 14*(1), 4357.
Falcão, L.d. S., Coelho, D. B., Veggi, P. C., Campelo, P. H., Albuquerque, P. M., & de Moraes, M. A. (2022). Starch as a matrix for incorporation and release of bioactive compounds: Fundamentals and applications. *Polymers, 14*(12), 2361.
Fatima, S., Khan, M. R., Ahmad, I., & Sadiq, M. B. (2024). Recent advances in modified starch based biodegradable food packaging: A review. *Heliyon, 10*(6).
Halmans, J. (2024). Reaching a Sustainable Consumer Packaging Industry.
International, A. (2021a). ASTM D6400: Standard Specification for Labeling of Plastics Designed to be Aerobically Composted in Municipal or Industrial Facilities (ASTM D6400). A. International.
International, A. (2021b). ASTM D6868: Standard Specification for Labeling of End Items that Incorporate Plastics and Polymers as Coatings or Additives with Paper and Other Substrates Designed to be Aerobically Composted (ASTM D6868). A. International.
Jaiswal, S., Schutz, A., & Jaiswal, A. K. (2026). Consumer behavior and willingness to pay for sustainable food packaging. In *Safety and sustainability aspects of food packaging* (pp. 417–441). Elsevier.
Jati, B., & Gravitiani, E. (2025). The benefit of biodegradable plastics for supporting sustainable development: A case study of willingness to pay in surakarta city, Indonesia. *Nature Environment and Pollution Technology, 24*(1).
Khezerlou, A., Zolfaghari, H., Banihashemi, S. A., Forghani, S., & Ehsani, A. (2021). Plant gums as the functional compounds for edible films and coatings in the food industry: A review. *Polymers for Advanced Technologies, 32*(6), 2306–2326.
Koster, S., Bani-Estivals, M., Bonuomo, M., Bradley, E., Chagnon, M.-C., Garcia, M., Godts, F., Gude, T., Helling, R., & Paseiro-Losada, P. (2015). Guidance on best practices on the risk assessment of non intentionally added substances (NIAS) in food contact materials and articles.

Lavagnolo, M. C., Poli, V., Zampini, A. M., & Grossule, V. (2024). Biodegradability of bioplastics in different aquatic environments: A systematic review. *Journal of Environmental Sciences, 142*, 169–181.

Li, B., Li, F., Cai, S., Tang, T., Chen, H., Feng, K., & Bai, Y. (2026). Tea polyphenol–loaded chitosan nanoparticles incorporated into carboxymethyl cellulose/potato starch to develop sustained-release bioactive films for pork preservation. *Packaging Technology and Science, 39*(2), 161–180.

Liu, H., Xie, F., Yu, L., Chen, L., & Li, L. (2009). Thermal processing of starch-based polymers. *Progress in Polymer Science, 34*(12), 1348–1368.

Mafe, A. N., Edo, G. I., Ali, A. B., Akpoghelie, P. O., Yousif, E., Isoje, E. F., Igbuku, U. A., Opiti, R. A., Ajiduku, L. A., & Owheruo, J. O. (2025). Next-generation biopolymers for sustainable food packaging: Innovations in material science, circular economy, and smart technologies. *Food and Bioprocess Technology*, 1–57.

Mhaddolkar, N., Tischberger-Aldrian, A., Astrup, T. F., & Vollprecht, D. (2024). Consumers confused 'where to dispose biodegradable plastics?': A study of three waste streams. *Waste Management & Research, 42*(9), 776–787.

Miralles, P., Fuentes-Ferragud, E., Socas-Hernández, C., & Coscollà, C. (2025). Recent trends and challenges on the non-targeted analysis and risk assessment of migrant non-intentionally added substances from plastic food contact materials. *Toxics, 13*(7), 543.

Momtaz, F., Momtaz, E., Mehrgardi, M. A., Momtaz, M., Narimani, T., & Poursina, F. (2024). Enhanced antibacterial properties of polyvinyl alcohol/starch/chitosan films with NiO–CuO nanoparticles for food packaging. *Scientific Reports, 14*(1), 7356.

Muncke, J., Backhaus, T., Geueke, B., Maffini, M. V., Martin, O. V., Myers, J. P., Soto, A. M., Trasande, L., Trier, X., & Scheringer, M. (2017). Scientific challenges in the risk assessment of food contact materials. *Environmental Health Perspectives, 125* (9), Article 095001.

Nguyen, A. T., Yến-Khanh, N., & Thuan, N. H. (2021). Consumers' purchase intention and willingness to pay for eco-friendly packaging in Vietnam. In *Sustainable packaging* (pp. 289–323). Springer.

Onyeaka, H., Obileke, K., Makaka, G., & Nwokolo, N. (2022). Current research and applications of starch-based biodegradable films for food packaging. *Polymers, 14*(6), 1126.

Periyasamy, T., Asrafali, S. P., & Lee, J. (2025). Recent advances in functional biopolymer films with antimicrobial and antioxidant properties for enhanced food packaging. *Polymers, 17*(9), 1257.

Peters, R. J., Groeneveld, I., Sanchez, P. L., Gebbink, W., Gersen, A., de Nijs, M., & van Leeuwen, S. P. (2019). Review of analytical approaches for the identification of non-intentionally added substances in paper and board food contact materials. *Trends in Food Science & Technology, 85*, 44–54.

Qin, Y., Yun, D., Xu, F., Chen, D., Kan, J., & Liu, J. (2021). Smart packaging films based on starch/polyvinyl alcohol and Lycium ruthenicum anthocyanins-loaded nano-complexes: Functionality, stability and application. *Food Hydrocolloids, 119*, Article 106850.

Raźniewska, M. (2022). Compostable packaging waste management—Main barriers, reasons, and the potential directions for development. *Sustainability, 14*(7), 3748.

Roy, S., & Rhim, J.-W. (2021). Anthocyanin food colorant and its application in pH-responsive color change indicator films. *Critical Reviews in Food Science and Nutrition, 61*(14), 2297–2325.

Sapuan, S. M., Harussani, M. M., Ismail, A. H., Zularifin Soh, N. S., Mohamad Azwardi, M. I., & Siddiqui, V. U. (2024). Development of nanocellulose fiber reinforced starch biopolymer composites: A review. *Physical Sciences Reviews, 9*(3), 1171–1211.

Sha, J., Zhu, Z., Lin, M., Zhu, L., Wang, Z., & Xiao, K. (2025). Chitosan/MXene/ε-polylysine based biodegradable packaging films with synergistic photothermal antimicrobial properties. *Food Packaging and Shelf Life, 50*, Article 101544.

Sharma, S., Jaiswal, A. K., Duffy, B., & Jaiswal, S. (2023). Food contact surfaces: Challenges, legislation and solutions. *Food Reviews International, 39*(2), 1086–1109.

Silva, F., Becerril, R., & Nerín, C. (2019). Safety assessment of active food packaging: Role of known and unknown substances. In *Advances in the determination of xenobiotics in foods* (pp. 1–41). Bentham Science Publishers.

Singh, P., Kaur, G., Singh, A., & Kaur, P. (2023). Starch based bio-nanocomposite films reinforced with montmorillonite and lemongrass oil nanoemulsion: Development, characterization and biodegradability. *Journal of Food Measurement and Characterization, 17*(1), 527–545.

Soni, R., Asoh, T.-A., & Uyama, H. (2020). Cellulose nanofiber reinforced starch membrane with high mechanical strength and durability in water. *Carbohydrate polymers, 238*, Article 116203.

Standardization, C. E. C.f. (2022). *EN 1186: Materials and articles in contact with foodstuffs — Plastics — Test methods for overall migration. EN 1186.* Brussels: CEN (European Committee for Standardization).

Wang, Y., Ju, J., Diao, Y., Zhao, F., & Yang, Q. (2025a). The application of starch-based edible film in food preservation: A comprehensive review. *Critical Reviews in Food Science and Nutrition, 65*(14), 2731–2764.

Wang, Y., Zhong, J., McClements, D. J., Zhang, Z., Zhang, R., Jin, Z., & Chen, L. (2025b). Stability enhancement methods for natural pigments in intelligent packaging: A review. *Critical Reviews in Food Science and Nutrition, 65*(29), 6233–6248.

Wijesinghe, C. (2025). Sustainable Food Packaging and Consumer Purchase Intentions: A Systematic Literature Review from a Theory of Planned Behaviour Perspective.

Wu, X., Yan, X., Zhang, J., Wu, X., Luan, M., & Zhang, Q. (2024). Preparation and characterization of pH-sensitive intelligent packaging films based on cassava starch/polyvinyl alcohol matrices containing Aronia melanocarpa anthocyanins. *Lwt, 194*, Article 115818.

Ye, H., Li, Q., Li, J., Li, D., & Ao, Z. (2025). Review on the abiotic degradation of biodegradable plastic poly (butylene adipate-terephthalate): Mechanisms and main factors of the degradation. *Chinese Chemical Letters, 36*(1), Article 109861.

Zellmer, S., Brüschweiler, B. J., Ebner, I., John, L., Kreuzer, K., Kühne, F., von Leliwa, S., Lorenz, C., Tietz, T., & Luch, A. (2024). Non-intentionally added substances (NIAS) in food contact material (FCM)-evaluation of consumer safety.

Zhao, L., Liu, Y., Zhao, L., & Wang, Y. (2022). Anthocyanin-based pH-sensitive smart packaging films for monitoring food freshness. *Journal of Agriculture and Food Research, 9*, Article 100340.

Zhao, L., Rong, L., Zhao, L., Yang, J., Wang, L., & Sun, H. (2020). Plastics of the future? The impact of biodegradable polymers on the environment. In *Microplastics in terrestrial environments: Emerging contaminants and major challenges* (pp. 423–445). Springer.

Zwicker, M. V., Brick, C., Gruter, G.-J. M., & van Harreveld, F. (2021). (Not) doing the right things for the wrong reasons: An investigation of consumer attitudes, perceptions, and willingness to pay for bio-based plastics. *Sustainability, 13*(12), 6819.

Further reading

Regulation (EC) No 1935/2004 on materials and articles intended to come into contact with food, L 338, 334–317 (2023). https://eur-lex.europa.eu.

CHAPTER SIXTEEN

Life cycle assessment and sustainability of starch-based packaging

Rituparna Borah[1, 2], Rabin Debnath[1], Akash Sharma[1], Tapoban Bordoloi[3], Raja Chakraborty[1]
[1]Institute of Pharmacy, Assam Don Bosco University, Guwahati, Assam, India
[2]School of Pharmacy, Girijananda Chowdhury University, Guwahati, Assam
[3]NEPEDS College of Pharmaceutical Sciences, Sonapur, Assam, India

16.1 Introduction

In recent years, the demand for sustainable materials has increased significantly in response to the environmental challenges associated with synthetic petroleum-based plastics. The accumulation of plastic waste, depletion of natural resources, and rising greenhouse gas emissions have necessitated the development of biodegradable and environmentally friendly alternatives. Among these, starch-based packaging has emerged as a promising solution in the field of biopolymers and green materials engineering. Starch is derived from abundant and renewable sources such as maize, potatoes, cassava, and wheat, and serves as a biodegradable matrix that can substantially reduce the environmental impact of conventional packaging systems (Avérous and Halley, 2014; Peelman et al., 2013). Despite the growing interest in starch-based packaging, a comprehensive evaluation of its environmental sustainability across the entire life cycle remains limited. Most studies emphasize material properties and biodegradability, while fewer systematically assess environmental impacts using life cycle assessment (LCA). Therefore, this chapter aims to evaluate the sustainability of starch-based packaging through an LCA perspective, identifying key environmental hotspots and potential strategies for improvement.

16.1.1 Definition and characteristics

The term "starch-based packaging" describes materials where starch is a primary ingredient, either unmodified or in its natural state, for the production of films, coatings, or molded goods used in packaging applications. The two

Starches in Smart Packaging
ISBN: 978-0-443-45396-0
https://doi.org/10.1016/B978-0-443-45396-0.00008-6

glucose polymers, amylose and amylopectin, make up the majority of starch, a polysaccharide, and their relative amounts affect its mechanical and physicochemical properties (Singh et al., 2007). Starch can be thermoplasticized (usually with glycerol, sorbitol, or other polyols) to form a thermoplastic polymer, which can then be processed using standard plastic production methods such as extrusion, injection molding, and casting (Avérous, 2004).

Renewability, biodegradability, film-forming ability, and relative abundance are important properties of starch-based materials. Still, starch may not be able to serve as an effective moisture barrier or retain its structural integrity in humid conditions due to its natural hydrophilia. To solve these problems, a variety of approaches have been developed, including chemical alteration (e.g., esterification or oxidation), physical blending with other biodegradable polymers (e.g., polylactic acid or polycaprolactone), and the addition of nanofillers (Thakur et al., 2019). The above techniques seek to improve the material's mechanical, thermal, and barrier features in order to satisfy the demands of contemporary packaging.

16.1.2 Importance of sustainable packaging solutions

The use of sustainable materials, reducing waste production, and minimizing environmental impacts are becoming increasingly critical considerations in packaging design. In this paradigm, starch-based packaging is essential because it provides a compostable and renewable substitute for plastics made from petroleum. Starch promotes resource efficiency and supports a circular bioeconomy because it can be derived from agricultural feedstocks and even agro-industrial byproducts (Muller et al., 2017). The long-term environmental challenges associated with persistent plastics are also lessened by the biodegradability of starch-based products, which guarantees their safe reintegration into natural ecosystems through microbial decomposition (Jafarzadeh et al., 2020).

Utilizing packaging composed of starch additionally supports a number of various global sustainability initiatives, such as the Sustainable Development Goals (SDGs) of the UN, especially those that deal with climate action (SDG 13) and responsible consumption and production (SDG 12). According to an industrial point of view, the creation of packaging materials based on starch helps to diversify raw material sources, lower carbon emissions, and promote innovation in the bio-based manufacturing industries (Siracusa et al., 2021).

16.1.3 Measuring sustainability: The role of life cycle assessment (LCA)

A detailed understanding of the consequences of starch-based packaging during its life cycle is necessary in order to assess its environmental advantages objectively. In accordance with ISO 14040 and ISO 14044, LCA is a systematic approach for evaluating the possible environmental effects of a product at every step of its life, from the extraction and manufacturing of raw materials to its use, disposal, or recycling. Researchers and industry stakeholders can measure factors including energy demand, greenhouse gas emissions, water use, and waste generation in comparison to conventional plastic systems by using LCA to starch-based products (Yates & Barlow, 2013).

Finding environmental "hotspots" in the production chain and improving material formulations, processing parameters, and end-of-life plans are made feasible by LCA. This strategy guarantees that starch-based packaging offers measurable environmental advantages over the course of its lifecycle in addition to seeming sustainable in theory. As a result, LCA is an essential tool for decision-making that directs the creation of policies, industry adoption, and ongoing development of packaging solutions that are genuinely sustainable (Kumar et al., 2020).

16.2 Sources of starch and environmental impacts

16.2.1 Primary starch sources

Starch, one of the most widespread biopolymers, is the major reserve of energy for many plants such as cereals, tubers, roots, fruits and seeds. It is the second largest biomass of earth and is considered a natural, renewable, and biodegradable polymer. Cereal grains (namely corn, wheat and rice) and tuberous vegetal organs (as potato or sweet potato), with cassava being a notable exception are the main commercial sources of starch. These included corn, potato, cassava and wheat starches in their native as well as modified forms. Approximately 99% of all the starch-based materials used in animal feed are derived from these four plant sources. Starch is a major component in cereal grains, the carbohydrate portion of which accounts for 60%–75% by weight and contributes approximately 70%–80% of calorific intake in human diet globally. The structure, granule morphology and composition of starch differ according to botanical source, resulting in differences in physicochemical characteristics and functionality. For instance, granules from corn and wheat are polyhedral in shape while those from potato and tapioca are ovate or truncated. This difference has implications for industry uses, including food processing, bioethanol production and biodegradable plastic manufacturing. The knowledge of these properties is essential for the choice of starch feedstock in the development approach to sustainable and high-performing biopolymers. As the main sources of primary starch ingredients, potato, corn and cassava possess their own molecular structures of comparison and performance features that are related to their suitability for polymer packaging applications (Gul et al., 2021).

Starch, a naturally occurring polysaccharide, is the primary storage carbohydrate of tubers, roots and cereal grains. Starch is produced mainly from tubers (potato, cassava, sweet potato, and mashua) and grains (maize, rice, wheat, barley, sorghum, and quinoa). Various botanical sources yield starches with distinctive physicochemical characteristics owing to its particular amylose/amylopectin content. Potato starch is composed of 20%–30% amylose and 70%–80% amylopectin and small amounts of lipids, proteins, and phosphorus, which affect its swelling capacity and gelatinization property. Similarly, cassava starch, with 15%–26% amylose and 74%–85% amylopectin, is valued for its complete gelatinization at 60–70 °C, making it ideal for biodegradable coatings and biofilms. Grains such as maize (25%–28% amylose) and rice (15%–25% amylose) have starches that are well adapted for film-forming as well as hypoallergenic uses for both edible and pharmaceutical applications. These sources differ in their granule size, morphology and composition, which influences their prospective application in the industry for biopolymer production, especially for bio-plastics, bio-coatings and biodegradable packaging. Therefore, tubers and grains offer renewable, economical, and versatile starch sources that are critical to sustainable biopolymer technologies (Culqui-Arce et al., 2025). Starch is one of the most abundant biopolymers found in nature and it is the main form of energy reserve in various plant types, which include cereals as well as tubers and roots (Culqui-Arce et al., 2025; Gul et al., 2021).

16.2.1.1 Potato

Potato (*Solanum tuberosum* L.) is one of the most cultivated tuber crops and ranks third after maize and wheat in terms of starch production. It comprises approximately 70%–85% starch by dry weight and yields more calories per hectare than most cereals, thus offering an effective carbohydrate supply for biopolymer manufacture. Potato starch structurally comprises two major polymers that are amylose (18%–21%) and amylopectin (79%–82%), with the latter containing glucose phosphate monoester substituents (0.2%–0.4% of glucose units) at C-6 and less frequently at position C-3. These phosphate groups provide a strong negative charge, and contribute to water absorption, swelling and viscosity properties by which specification of potato as opposed to cereal starches may rely (Reyniers et al., 2020).

The granules are sizable (10–110 μm), ovoid or irregular, with a B-type crystal in structure containing 36 molecules of water per unit cell and the hexagonal lattice. The structure combined with low lipid and protein contents the potato starch remarkable paste clarity, gel strength and a lower gelatinization temperature (58–68 °C). Environmental factors, especially temperature, affect biosynthesis, lower temperatures increase amylose production and higher temperatures facilitate sugar conversion (Tong et al., 2023). In addition, genetic variation between wild Solanum species has also been utilized through molecular breeding and marker-assisted selection to enhance starch content, phosphate content and processing quality. Potato is valued as a source of starch because it has a high gelatinization temperature, swelling power, viscosity, and good film forming property. Indeed, while tuber starch (e.g., potato) possesses specific gel strength and hydration characteristics, cereal starch (e.g., corn) delivers greater yield as well as processing efficiency all of which are details considered in the next section (Bradeen & Kole, 2016).

16.2.1.2 Corn

Corn (*Zea mays* L.), commonly known as maize, is the primary cereal and most important industrial starch source in the world, accounting for approximately 35% of worldwide starch production. Corn kernels consist of about 72% carbohydrates, including 9% protein and 4.5% lipids (mostly starch). Starch, which comprises 75% of the kernel's dry mass, is located in the endosperm and is dominated by amylopectin (≈75%) and amylose (≈25%) from a starch perspective, with their ratio determining its physicochemical and processing (Rouf Shah et al., 2016).

Corn starch is mostly produced by the wet milling process, requiring approximately 1.5 m^3 of water per tonne of grain in the separation of its three fractions of starch, protein and oil. The novel technologies, e.g., enzymatic and ultrasound pretreatment, wet-milling, enhance 10% recovery of starch with a decrease in sulfur dioxide emissions and energy consumption up to 5% (Yu & Moon, 2021). Both high-amylose (up to 94% amylose) and waxy corn (≈100% amylopectin) are special genetic types selected for industrial purposes. High-amylose starch has greater film strength, better thermal stability, and excellent biodegradability for environmentally friendly packing and amylopectin-rich waxy corn has superior adhesion and covering qualities. Developed genome editing tools (CRISPR–Cas9) and marker-assisted selection have additionally increased yield and quality of starch with limited environmental impact (Ostrander, 2014). Also, zein from maize, an alcohol-soluble protein, gives a shiny hydrophobic film, which is serving as coatings of foods and for biodegradable packages. Its high productivity, renewable and extensive distribution characteristic of diverse climates where the crop grows. Corn is therefore the main cereal starch used for industrial purposes, particularly in the global north. Nevertheless, in tropical areas, cassava has been identified as an alternative renewable starch source yielding the highest under marginal conditions, and it presents very promising perspectives for the biodegradable packaging issue.

16.2.1.3 Cassava

As one of the most important tropical root crops, cassava (*Manihot esculenta* Crantz) represents a primary source of renewable starch, notably in Africa, South America and Southeast Asia. With a total production in the world of more than 270 million tons per year, cassava represents about 11% of global starch production (Amelework et al., 2021). Tuberous cassava roots have 17%–35% fresh-weight basis (up to 80% dry weight) starch, which is made up of amylose and amylopectin. The starch granules have oval or truncated form and present the A or C type crystallinity (25%–34% relative crystallinity) with average diameters ranging from 2 to 30 μm Due to high fat content, amylose, and hydrophilic nature of cassava starch, it has higher swelling power compared with other food-based biopolymers, which are important for the formation of a biodegradable film (Chisenga et al., 2019).

Cassava starch gelatinizing temperature (63–72°C), as well as its good film-forming capability, gives way to flexible, transparent and edible packaging materials. Cassava starch-based films have tensile strengths reaching 8.8 MPa and water permeability close to low polyethene (LDPE), showing good mechanical and barrier properties (Luchese et al., 2017). However, the natural cassava starch possesses restrictions with low shear and thermal stability, making necessary chemical treatments primarily cross-linking with reagents such as sodium acetate, ammonium phosphate, and acid in order to get stability, solubility and resistance to acidic medium. These adaptations improve further food wrapping, adhesives and biodegradable coating of starch. The flexibility, high yield per acre, and tailorable nature of cassava starch make it a promising and sustainable cost-effective option for starch-based packaging and biofilm applications (Akpa & Dagde, 2012).

16.2.2 Use of agricultural residues and waste biomass

Agricultural residues and waste biomass utilization play a key, LCA-relevant role in the establishment of sustainable starch-based packaging through valorization of ubiquitous lignocellulosic and starch-rich wastes into polymeric precursors, functional blends and energy generators. Agro-industrial residues such as sugar palm peel, pineapple peel, rice straw, copra meal, brewer's spent grain, spent tea and coffee wastes, maize cobs and stover, rice husks wheat bran banana pseudostems cassava and potato peels bagasse groundnut shells and coffee husks exhibited high contents of cellulose (50%–85%), hemicellulose (10%–65%), lignin (2%–35%) along with presence of starch in most cases ranging from 2% to 70%, indicating the low cost renewable nature that can be used to produce bio-polymers cellulosic films nanocelluloses reinforcements and biodegradable composites (Aruwajoye et al., 2020; Chimtong et al., 2016; Souza et al., 2022; Xu et al., 2023). From an LCA standpoint, valorization of such residues decreases fossil-dependency and greenhouse gas intensity by replacing petrochemical polymers with the formation of energy in coproduct streams (e.g., bioethanol, biohydrogen or carbonized briquettes). Nevertheless, the overall environmental benefit is sensitive to process parameters such as pretreatment intensity, enzyme dosage, logistics and conversion yields. At the biochemical conversion stage, nonstarch polysaccharide (NSP)-degrading enzyme cocktails (e.g., Pentozyme comprising xylanase, amylase, β-glucanase, cellulose, mannanase and pectinase), hydrolyze complex plant polymers to oligosaccharides and reducing sugars we have experimental evidence that tea leaves produce particularly high concentrations of this sugar and these oligosaccharides are prebiotic and antibacterial in nature, which presents them with multifunctionality over their use as carbon feedstocks for bioplastic formation (Chimtong et al., 2016). Other next-generation processes, including Simultaneous Hydrolysis Enzymatic Synthesis and Fermentation (SHESF) process produce artificial starch (amylose and amylopectin) with cellulose digestibility of 89%, with titers approximating 3.75 g/L starch and 3.23 g/L protein from pretreated corn stover cellulose. This strategy decreases enzyme expenses and the demand of external energy, illustrating how integrated biorefinery operations improve resource utilization efficiency and environmental sustainability (Xu et al., 2023). Moreover, some of these green extraction techniques (subcritical water, supercritical CO_2, deep eutectic solvents) can be used for selective recovery of polysaccharide fractions and derivatives (pectin, inulin, chitin-like materials), which contribute biocompatibility, gelation ability, barrier properties and antimicrobial activity to packaging material while enhancing feedstock efficiency (Souza et al., 2022). However, several obstacles still lie ahead strong pretreatment conditions can form inhibitory byproducts and demand flow of energy, costs for enzymes as well as product inhibition are cost-effective, the feedstock's heterogeneity and seasonal availability, together with its logistical requirements influence emissions of some residues, such as rice husks containing high amounts of silica, affect ash amount and behavior during combustion. Empirically, well-managed valorization systems have reported material and energy savings (e.g., mixed starch-based by-products having been used for integrated bioethanol and biohydrogen to enhance resource recovery and minimize effluents), or carbonized residues in combination with binders such as starch yielding briquettes of up to ~23 MJ/kg showing both economic and environmental positives under the tech-term waste-to-resource paradigm (Aruwajoye et al., 2020; Lubwama et al., 2022). Finally, agricultural residues and waste biomass are potentially useful multifunctional feedstocks for the production of sustainable starch-based packaging. Optimal LCA performance will be achieved through the integration of low-impact pretreatment and extraction, enzyme and energy inputs to a minimum, logistic optimization, and coproduct valorization leading to minimized fossil-based input materials, lower greenhouse gas emissions and support for circular bioeconomy principles.

16.2.3 Resource inputs in cultivation

16.2.3.1 Land use

Land is a key natural resource in the cycle of growing (cultivation) starch-feedstock crops, including potato, corn and cassava. When natural ecosystems forests, grasslands, wetlands are turned into croplands, there come many environmental burdens come with that change, loss of biodiversity and carbon stores. A recent review reveals that over 90% of global land-use related impacts toward biodiversity derive from agriculture, with crop farming accounting for some 72% (Cabernard et al., 2024). Additionally, land-use change usually leads to soil organic carbon and biomass-carbon stock reductions, which compromise ecosystem resistance and impair carbon sequestration capacity (Marques et al., 2019).

Land use is a critical issue from the perspective of starch-based packaging feedstocks how many hectares are required for the production of a tonne of starch whether expansion entails high-carbon or high-biodiversity landscapes, and how intensification can offset land footprint while introducing other burdens.

16.2.3.2 Irrigation water

Another critical input for starch crops is water, notably in irrigated systems. Worldwide, agriculture represents close to 70% of freshwater withdrawal indicating the extent of water use in cropping systems (Amelework et al., 2021). In several poor countries where water is scarce, agriculture can account for more than 90% of the freshwater withdrawals, as a result of this overreliance on irrigation (Fujs and Kashiwase, 2024).

When investigating the water input for starch-feedstock production, it needs to be analyzed in several aspects: in volume (m three per hectare or per tonne of starch), source (surface vs. groundwater), irrigation efficacy (drip vs. flood) and context of water stress. Ecological pressures such as aquifer depletion, streamflow reduction and soil salinization can aggregate due to overirrigation or dependence on nonsustainable water sources (Amelework et al., 2021). Therefore, when assessing the potential of starch as a feedstock for packaging, it is important to consider water-use intensity per unit of starch output and sustainability of the irrigation regime in the assessed region, rather than yield per area.

16.2.3.3 Agrochemical inputs

Most crops cultivated for extraction of starch also require fertilizer (the two common fertilizers are nitrogen and phosphorous) applied chemically to the soil, as well as pesticides and herbicides. These inputs are costly to produce (energy and raw materials) and apply in the field (application, emissions, runoff). For instance, artificial nitrogen fertilizers account for about 8.3% of farm-gate greenhouse-gas (GHG) emissions worldwide (Menegat et al., 2022). Supplementary Nitrogen is often lost to the environment, only a fraction of applied Nitrogen is absorbed by the crop, and large portions may volatilize as N_2O (a potent greenhouse gas), leach out in form of nitrate into groundwaters, or be drained into aquatic systems (Bhatt et al., 2025).

Moreover, eutrophication of fresh and marine waters resulting in algal blooms, hypoxia and loss of biodiversity in water systems has also been associated with excessive application of Nitrogen and Phosphorous fertilizers (Jwaideh et al., 2022).

On the pesticide/herbicide sector, the use of agrochemicals may result in soil degradation, adverse effects on nontarget organisms (e.g., beneficial insects and soil microfauna), as well as water body pollution, both originating from runoff or spray drift. Although product-derived starch data are currently less balanced, the generic agricultural challenge applies, higher input systems raise upstream burdens and may cancel out possible environmental benefits associated with starch feedstocks Farmerline. (n.d.). From a starch-based packaging feedstock perspective, the level of agrochemical inputs per tonne of starch serves as a key indicator that high levels could potentially outweigh the land or water benefits by contributing significant upstream burdens in energy use, emissions and ecosystem effects. Accordingly, starch sourced from lower-input or more efficient systems (e.g., better fertilization and integrated pest management) serves to enhance the sustainability profile.

16.2.4 Environmental impacts of cultivation

16.2.4.1 Greenhouse gas emissions

Starch-based feedstock crops, such as maize, wheat and cassava are grown in harsher environments that affects the carbon and water cycles of the ecosystem due to greenhouse gas (GHG) emissions from soil process and fertilizer use for grain cultivation. Nearly 12% of total anthropogenic GHG emissions come from agriculture, with carbon dioxide (CO_2), methane (CH_4), and nitrous oxide (N_2O) predominantly emitted from soils after fertilization and waterlogging (Quadros, 2023). The global-warming potential of CH_4 and N_2O is 21 and 310 times that of CO_2, respectively, even though they are present in lower amounts. The overuse of nitrogen fertilizer promotes N_2O fluxes, mainly through the microbial nitrification and denitrification processes. Flooded rice (and other starch crops under anaerobic conditions) releases large amounts of CH_4 as a result of methanogenesis. The manufacturing of fertilizers itself uses up about 1.2% of the planet's energy and generates a similarly large share of global GHGs. In addition, application of manure, compost and bio-solid amendments lead to higher CO_2 and CH_4 emissions when they are not properly managed. Overall, this means that farming for starch feedstocks does not only fuel direct soil-related emissions but also carries upstream energy expenses. Precision nutrient management, conservation tillage practices and water-saving irrigation technologies can greatly reduce these emissions to improve the life-cycle sustainability for starch-based packaging materials.

Cassava the main bioplastic packaging starch crop is primarily responsible for greenhouse gas (GHG) emissions in its production chain. According to Trakulvichean et al., cassava planting is responsible for around 52% of the overall carbon footprint in cassava starch production processes, which is about 203–281 kg CO_2 eq per tonne of starch. These emissions are mainly generated from the manufacture and application of fertilizers, soil nitrogen transformations contributing to nitrous oxide (N_2O) emissions, as well as diesel for land preparation and harvesting. Variations in farming systems, such as fertilizer application rates, irrigation and harvesting time, influence the overall emissions intensity. The study emphasizes that increased starch yield and extraction efficiency may diminish the raw material demand with a subsequent reduction in GHG emissions (for cultivation-bound GHGs). In contrast, starch processing accounts for 36% of electricity use and transport only 8% of the total emissions, suitably highlighting the importance of farming in an environmental context. Recommended mitigation measures would include precision nutrient management, planting of disease-resistant and high starch content cultivars, and modification in harvesting timings. These efficiencies not only reduce fertilizer-derived N_2O emissions, but also increase the yield improvement efficiency that ultimately brings the cassava starch sector in line with Thailand's 2050 carbon-neutrality goal (Trakulvichean et al., 2025).

Emissions of GHGs associated with cultivation of starch crop species such as maize, cassava, wheat and potato etc provide a major contribution to the agricultural carbon emissions. Veiga et al., noted that agriculture, excluding land-use change, contributes about 25% of Brazil's CO_2 emissions overall. In the crop sector, direct emissions account for 64%, mostly originating from fertilizer, pesticide and diesel use, whereas indirect emissions comprise of 36% related to input production and transport. The average emissions of GHG from major Brazilian crops is equal to 2728.78 kg CO_2-eq ha^{-1}, while the starch crops, such as cassava and maize produced emission intensities that are relatively lower than those of other commodities in terms of tonnes of dry matter (195.5–230.4 kg CO_2-eq Mg^{-1}), thanks to their efficiency on nutrients use and the increase in yield. Nitrous oxide (N_2O) from the usage of fertilizers continues to be the most significant source of GHG because of its high global warming potential. On the other hand, crops with biological N_2 fixation (some leguminous species) have lower emissions. CO_2 emissions produced by diesel combustion and the production of agrochemicals are also elevated due to mechanization and chemical inputs. Smart farming, precision agriculture and low-input cultivation practices are therefore critical strategies to minimize GHG footprint of starch-based raw materials for sustainable packaging (Veiga et al., 2024).

Cassava, a major starchy food crop as well as feedstock for bioplastics production, serves a double purpose in GHG accounting, acting not only as a carbon store in the accumulated biomass but also an emitter of GHGs through soil, fertilizer and postharvest pathways. In a recent work Molina-Botero et al., showed that when cassava replaces energy from grain-based concentrates in livestock diets, cassava contribution to attenuating GHG emissions is considerable. Supplementation of 30% cassava roots or leaves in the diets of dairy cows reduced the global warming potential (GWP) from milk production by 6%–8%, with a CH_4 intensity ranging between 1.30 and 1.54 kg CO_2-eq per kg fat-protein-corrected milk. Daily CH_4 production ranged from 194 to 234 g CH_4 per animal, while replacing concentrate with cassava led to a reduction in units of digested protein by about 25%. The study also highlighted that on-farm agricultural CH_4 comprised 32%–39% of the total GHG while nitrous oxide from mineral fertilizer application was in (28%–31%) and emissions release (26%–28%) categories. The findings suggest that a sustainable cassava production process integrating optimal fertilizer management and residue use could lead to reduced methane and nitrous oxide emission, thereby enhancing the life-cycle sustainability of starch-based packaging-grade raw material (Molina-Botero et al., 2024).

16.3 Manufacturing starch-based packaging

16.3.1 Conversion of native starch to thermoplastic starch (TPS)

The conversion of native starch to thermoplastic starch (TPS) involves transforming starch into a thermoplastic material by applying heat and mechanical forces in the presence of plasticizers. This process allows starch to behave like conventional thermoplastics, making it suitable for various applications, including biodegradable packaging. The conversion process is crucial because it addresses the inherent limitations of native starch such as brittleness and water sensitivity by enhancing its flexibility and processability.

16.3.1.1 Role of plasticizers (glycerol, sorbitol)

Plasticization Mechanism: Both glycerol and sorbitol act by inserting themselves between starch molecules, reducing intermolecular forces and increasing the mobility of the polymer chains. This process is essential for converting the rigid structure of native starch into a flexible thermoplastic material (Carvalho, 2016; Gómez-Caturla et al., 2023).

Mechanical Properties: Sorbitol generally provides higher tensile strength and modulus compared with glycerol. This is attributed to sorbitol's higher molecular weight and ability to form stronger hydrogen bonds with starch, resulting in a more robust material. Glycerol, on the other hand, tends to increase the elongation at break, making the material more flexible but less strong (Esmaeili et al., 2017; Gómez-Caturla et al., 2023; Li & Huneault, 2011).

Thermal Stability: The thermal stability of TPS is enhanced with the use of sorbitol, as it forms more stable hydrogen bonds with starch. This results in a material that is more resistant to thermal degradation. A higher sorbitol-to-glycerol ratio has been shown to improve the thermal stability of TPS (Esmaeili et al., 2017).

Water Sensitivity: Both glycerol and sorbitol influence the hydrophilicity of TPS. At lower concentrations, they can exhibit antiplasticizing effects, reduce water absorption and increase the material's moisture resistance. However, at higher concentrations, they enhance plasticization, increasing water uptake and flexibility (Shimazu et al., 2007).

Morphological Effects: The presence of glycerol and sorbitol affects the microstructure of TPS. Sorbitol tends to produce a finer morphology and better dispersion of starch particles, thereby improving mechanical properties. Glycerol, while enhancing flexibility, may lead to a coarser morphology (Gómez-López et al., 2025; Li & Huneault, 2011).

Coplasticization: Combining glycerol and sorbitol can optimize the properties of TPS. A balanced ratio of these plasticizers can enhance mechanical properties, thermal stability, and resistance to retrogradation. For instance, a sorbitol-to-glycerol ratio of 2:1 has been found to yield TPS with superior properties (Esmaeili et al., 2017; González-Torres et al., 2021).

Environmental Impact: The use of biodegradable plasticizers such as glycerol and sorbitol aligns with producing environmentally friendly materials. These plasticizers are derived from renewable resources and contribute to the biodegradability of TPS, making it a sustainable alternative to conventional plastics (González-Torres et al., 2021; Zhang et al., 2014).

While glycerol and sorbitol are effective plasticizers for TPS, their impact on the material's properties can vary significantly depending on concentration and specific application requirements. The choice between glycerol and sorbitol, or their combination, should be guided by the desired balance between flexibility, strength, and environmental considerations. Further research into alternative plasticizers and coplasticization strategies could enhance the performance and applicability of TPS in various industries.

16.3.2 Blending with other biopolymers

Blending polymers such as Polylactic Acid (PLA) and Polycaprolactone (PCL) with native starch to create TPS is a significant advancement in the development of biodegradable materials. This process enhances the mechanical, thermal, and physicochemical properties of TPS, making it a viable alternative to conventional plastics. The blending of these polymers addresses the inherent limitations of each component such as PLA's brittleness and TPS's limited mechanical properties by leveraging the complementary characteristics of each material. This synthesis results in a biodegradable product that is both cost-effective and environmentally friendly.

16.3.2.1 PLA

Mechanical Enhancement: PLA is known for its desirable mechanical properties, which include high tensile strength and rigidity. When blended with TPS, PLA improves the overall mechanical performance of the material, making it suitable for applications such as packaging and biomedicine (Villadiego et al., 2022; Jikui et al., 2013).

Thermal Properties: PLA contributes to the thermal stability of the blend. It helps maintain the structural integrity of TPS at higher temperatures, which is crucial for processing and applications across various environments (Mina et al., 2013).

Compatibility Challenges: Despite its benefits, PLA is hydrophobic, which leads to poor interfacial adhesion with the hydrophilic starch. This immiscibility is a challenge that researchers address by using compatibilizers and plasticizers to improve phase interactions and blend homogeneity (Gómez-López et al., 2025; Villadiego et al., 2022).

16.3.2.2 PCL

Flexibility and Processability: PCL is a flexible polymer that enhances the ductility and processability of TPS. It acts as a plasticizer, improving the blend's flexibility and making it easier to process through extrusion and injection molding (Manoe et al., 2022).

Thermal Stability: PCL contributes to the thermal stability of the blend, increasing the onset temperature of degradation. This property is beneficial for applications requiring thermal resistance (Manoe et al., 2022).

Moisture Absorption: The incorporation of PCL can lead to increased moisture absorption in the blend, which may affect the material's mechanical properties over time. This aspect needs to be managed depending on the intended application (Mina et al., 2013).

16.3.3 Processing and compatibility

Blending Techniques: The blending of TPS with PLA and PCL typically involves extrusion and thermo-compression techniques. These methods ensure a uniform distribution of components and enhance the mechanical and thermal properties of the final product (Debiagi et al., 2017; Manoe et al., 2022).

Use of Compatibilizers: To address the immiscibility between PLA and TPS, compatibilizers and plasticizers such as glycerol and isosorbide are used. These agents improve the interfacial adhesion and phase dispersion, resulting in a more homogeneous blend (Gómez-López et al., 2025).

Environmental Impact: The resulting blends are fully biodegradable, offering a sustainable alternative to traditional plastics. The biodegradability of these materials ensures that they do not contribute to long-term environmental pollution (Bulatović et al., 2021; Loercks et al., 1997).

While blending PLA and PCL with TPS offers numerous advantages, it is essential to balance mechanical properties and biodegradability. The choice of blending ratios, processing conditions, and the use of compatibilizers can significantly influence the performance of the final product.

16.3.4 Enhancing mechanical and barrier properties

Starch-based packaging materials are increasingly being explored as sustainable alternatives to traditional plastic packaging due to their biodegradability and environmental benefits. However, the mechanical and barrier properties of these materials are critical for their practical application in food packaging. Several factors influence these properties, including the source and type of starch, processing methods, and the incorporation of additives or other polymers.

Starch Source and Composition: The source of starch significantly affects its mechanical and barrier properties. Different starches, such as corn, potato, and cassava, have varying amylose-to-amylopectin ratios, which influence film-forming properties and mechanical strength (Lauer & Smith, 2020). The amylose content is particularly important as higher amylose levels generally enhance film strength and reduce solubility, improving barrier properties (Radfar et al., 2020).

Processing Techniques: The method of starch extraction and film formation plays a crucial role. Techniques such as casting, extrusion, and thermoplastic processing can impact the film's mechanical properties and barrier performance (Jiménez et al., 2012). TPS is often produced by plasticizing starch with glycerol, which enhances flexibility but may reduce barrier properties. Balancing plasticizer content is key to optimizing performance (Lopez-Gil et al., 2014).

Additives and Blending: Incorporating additives, such as plasticizers, cross-linkers, and reinforcing agents, can significantly improve the mechanical and barrier properties of starch films. For instance, the addition of natural fibers or nanoparticles can enhance strength and reduce permeability (Lopez-Gil et al., 2014; Singh et al., 2022). Blending starch with other biopolymers, such as PLA or low-density polyethylene (LDPE), can improve compatibility and mechanical properties. However, challenges with immiscibility must be addressed (Radfar et al., 2020).

Functional Additives: Active ingredients, such as antimicrobial agents and antioxidants, can be integrated into starch films to extend the shelf life of packaged foods and provide additional functionality (Dutta & Sit, 2024; Kusuma et al., 2024). These additives not only enhance the functional properties of the films but also improve barrier properties by reducing oxygen and moisture permeability (Singh et al., 2022).

While starch-based packaging materials offer a promising sustainable alternative, their mechanical and barrier properties must be carefully optimized to meet the demands of food packaging applications. The integration of advanced materials and processing techniques continues to enhance the performance of these biodegradable films, paving the way for broader adoption in the packaging industry.

16.3.5 Energy use and emissions during manufacturing

Energy Sources and Consumption: The production of starch-based packaging materials involves substantial energy inputs, primarily from natural gas and electricity. For instance, in a potato starch factory, 77.4% of the energy is utilized, with 46.4% as steam heat and 17.2% as heating gas, indicating a high energy demand for processes such as starch drying and modification (Meuser et al., 1981). Similarly, a corn starch factory uses 69.3% natural gas and 30.7% electricity, with significant energy allocated to starch extraction and drying (German & Meuser, 1986).

Energy Intensity: The energy intensity varies across different starch products. For example, the production of extruded starch requires 17,840 MJ/t TS, while dried native starch requires 12,120 MJ/t TS (Meuser et al., 1981). These figures highlight the energy-intensive nature of starch processing, a critical consideration in the Life Cycle Assessment of starch-based packaging.

Greenhouse Gas Emissions: Starch-based packaging materials, such as those made from PLA and starch blends, exhibit lower greenhouse gas emissions compared with conventional plastics such as polyethylene terephthalate (PET). For instance, PLA/starch trays have 51.38% lower GHG emissions than PET trays, primarily due to the lower weight and different resin production processes (Suwanmanee et al., 2012). However, emissions remain significant, particularly when considering the entire life cycle from cradle to grave.

Environmental Impact: Despite lower carbon emissions, starch-based products can have a broader environmental impact, including eutrophication and acidification, due to agricultural inputs such as fertilizers and pesticides used in biomass production (Forfora et al., 2024). Additionally, land-use change (LUC) associated with renewable feedstocks such as corn and cassava contributes to the environmental footprint of PLA and PLA/starch products (Suwanmanee et al., 2013).

16.3.5.1 Challenges and considerations

Process Optimization: Enhancing the mechanical properties and reducing the energy consumption of starch-based packaging can be achieved through process optimization. Techniques such as plasticizer optimization and response surface method (RSM) have been proposed to improve the mechanical properties and reduce the crystallinity of thermoplastic starch, thereby potentially lowering energy requirements (Wu et al., 2023).

Comparative Environmental Impact: While starch-based packaging offers a biodegradable alternative, its environmental impact is not uniformly lower than that of fossil-based plastics. For instance, when clean energy sources are used, the environmental impact of PLA and PLA/starch boxes can be higher than that of polystyrene boxes (Suwanmanee et al., 2013). This underscores the importance of considering the entire life cycle and energy sources in environmental assessments.

16.4 End-of-life scenarios and environmental outcomes

16.4.1 Disposal pathways

End-of-life (EoL) is a critical element in determining the environmental sustainability and circularity of starch-based, as well as other, bioplastics packaging systems. Bioplastics, such as PLA, TPS and Bio-PET, have been developed as sustainable alternative to fossil-based polymers because they are based on renewable resources and may also be degradable. But the sustainability merits of these materials largely depend on how they are handled after use (recycled, composted, incinerated or landfilled). In reality, as observed in the real world, such sustainable waste loops deviate widely from what is assumed within policy frameworks and (life cycle) models. Lack of waste infrastructure, low public awareness and incoherent labeling are challenges restricting efficient implementation of the environmental benefits by EoL bioplastics.

The end-of-life routes for bioplastics (PLA and Bio-PET) include recycling, incineration at waste-to-energy plants, and in some cases dumping into the environment. While the PLA you find is sold as compostable, its degradation does only take place under industrial level required temperature and humidity conditions in the form of Mature Wastes. Accordingly, the majority of PLA waste is sent to municipal incineration as opposed to composting. Incineration of PLA produces about 1.35 kg CO_2 eq/kg of material (slightly mitigated by the energy recovery). Recycling is preferred for its environmental benefits but has not been as widely adopted due to the need for specific infrastructure and limited market demand of recycled PLA. Its potential for negative emissions is significant; being able to reach a climate impact from recycling of -0.70 kg CO_2 eq per kg clearly distinguishes it as a superior material with regard to emission reduction. Whereas Bio-PET, is chemically identical to PET, and has achieved more significant integration into established recycling streams reaching collection rates of up to 82%. Yet, unrecycled Bio-PET is mainly incinerated, which releases about 4 kg CO_2 eq per kg. Furthermore, poorly managed or disposed-of PLA and Bio-PET will also contribute to the environmental concerns from microplastic accumulation; PLA has a slow degradation rate under Swedish terrestrial and marine conditions, while Bio-PET might last for several decades. As a result, the overall environmental benefits of these starch-based and biobased polymers at their end-of-life are minimized due to inadequate disposability infrastructure as well as public confusion about biodegradability (Hansson, 2019).

Building on this work (Molina-Besch, 2022), stressed that the environmental performance of bioplastics in the EoL process are dominated by waste management systems and the methodological footprints used for life cycle assessments (LCAs). Typical forms of disposal are industrial composting, landfill, incineration upgraded to combined heat and power or more recently through chemical or mechanical recycling. Although composting and anaerobic digestion should be on the plus side for polymers such as PLA and TPS since they are theoretically biodegradable, in reality these advantages can be implemented only where there is appropriate composting or digestion. In areas without such processes, bioplastics are frequently disposed of in landfills or by incineration. Landfilling has the potential of methane formation under anaerobic conditions and incineration causes biogenic CO_2 to emit but recovery of energy achieved. Recycling likely remains the best alternative for both and energy recovery in terms of circular economy, but it is still limited because of contamination issues and lack of infrastructure. Furthermore, LCA results have shown there to be a large variation in the impact between regions depending on waste management performance and energy mix assumptions. Therefore, the sustainable disposal of starch-based packaging requires not only technical and infrastructural enhancements but also scenario-specific simulation techniques and consumer involvement to avoid environmental burden shifting between life cycle stages (Molina-Besch, 2022).

A further analysis of the EoL pathways of TPS packaging highlights different environmental impacts depending on the disposal scenario. A complete LCA analysis of bio-degradable dry packaging considering six disposal options for 1 kg TPS: mechanical recycling (MR), industrial composting (IC), incidentally anaerobic digestion (AD), direct fuel substitution (DFS) at industrial facilities, municipal sol id waste incineration with energy recovery (MSWI) and landfilling (LF) was performed. These were modeled for European circumstances, including dynamic greenhouse gas emissions and 100 years long-term degradation. The results showed that mechanical recycling was the most environmentally friendly option of replacing 0.74 kg virgin TPS, when Global warming potential and water use could be reduced through material recovery and avoiding production. DFS was found to be the next favorable pathway, where TPS's high HHV (18.6 MJ/kg) would allow for replacing heavy fuel oil as well as efficient energy recovery with low direct emissions. Anaerobic degradation yields were acceptable (with 90% of carbon degraded and 380 g/kg methane produced) with the composting in postupgrading could result in up to 10% may be released as methane. It generated 0.744 kWh/kg of electricity and 7.77 MJ/kg of heat, for energy recovery despite direct emissions. In contrast, composting and landfilling both showed relatively poor performances; 80% of TPS was degraded during composting with minimal improvement in the quality of resulting compost or energy benefits gained from the degradation process, while landfilling led to almost complete (99%) degradation over a century with significant methane leakages and low energy recovery (22%). This work also found that recycling and energy recovery were preferable to options for composting and landfilling uncontaminated dry TPS packaging, undermining dogmatic interpretations of the waste hierarchy and advocating case-specific LCA modeling (Rossi et al., 2015).

The EoL phase is also important for the environmental impact of starch-based bio-composites such as in the case of starch–Brazil nut fiber materials. A study showed environmental impacts of these biocomposites are dependent on the provenance of starch, plasticizer type and filler materials. Starch-based materials are biodegradable, so its disposal at the end-of-life can be managed by composting or natural degradation, which makes it less dependent on landfill as compared with synthetic counterparts such as PP or PLA composites. It's usually incinerated (sometimes in local landfills), emitting CO_2 and other pollutants, because there is insufficient capacity to process it through composting or recycling. However, bio-based origin of starches and including issues on conversion to biodegradable polymers with natural plasticizers such as glycerol results in lower their environmental impact than the composites based on PLA or PP. By using agricultural residues such as Brazil nut fibers for natural fillers, we can additionally practice a circular material use and waste valorization leading to improved sustainability. Landfilling or incineration resulting effects are still high in some areas, but recycling and composting always present lower global warming potential and resource depletion impacts, confirming these processes environmental superiority in LCA methodologies (Rojas-Bringas et al., 2021).

Similarly, more comprehensive LCA investigations stress the significant contribution of the EoL stage to sustainability performances in different starch-based bio-packaging systems. TPS shows natural biodegradability and compostability, it can be degraded by microorganisms to CO_2 and water, which will also form the basis of cyclical carbon closed loop through followed photosynthesis uptake. Widespread end-of-life pathways include composting, incineration with energy recovery, landfilling, mechanical recycling and open dumping in a few instances. It can reduce environmental loads, when composted under controlled conditions, but unmanaged is also prone to methane emissions. Between 15.9 and 36.2 MJ/kg of energy can be recovered from incineration, depending on the efficiency of the system, thus offsetting fossil fuel consumption but generating CO_2 and N_2O emissions. It is reported that incineration with energy recovery can be particularly advantageous for some bio-based packaging solutions such as the mango cardboard boxes because life cycle impacts could be reduced by two%–14% when compared with alternatives. Mechanical recycle also benefits the sustainability in terms of the ultimate conservation of material and energy; in particular, remilling and extruding HIPS trays can be processed using low energy intensive (0.398 kWh/kg) with low emission technology. In contrast, landfills have higher ecological footprint due to methane emissions, being composted or recycled can contribute to reduce global warming and resource impacts. Notably, for all three aspects of human health, ecosystem quality and resource use the optimized EoL strategies (including energy recovery incineration or mechanical recycling) generate the lowest environmental footprints, underlining their importance in sustainable waste treatment (Salwa et al., 2021).

In general, the EoL treatment of starch-based and bioplastic packaging also remains a key factor in determining their overall environmental sustainability. Despite their obvious merits intrinsically (biodegradability, renewable source and potential circularity), the benefits depend greatly on infrastructural support, consumer behavior and policy-motivated waste management systems. Recycling and energy recovery seem to be the better options, although composting and landfilling are more variable according to the specific context. Successful achievement of sustainability benefits is thus dependent on integrated aerobic waste management systems, regional life cycle modeling and public awareness to help close the disconnect with theoretical biodegradability and measurable environmental performance. This is the only way for starch and other bioplastics to exploit the potential as sustainable replacement materials for conventional petroleum-based packaging.

16.4.2 Degradation rates and influencing factors

The degradation properties of the starch-based and other biodegradable plastics are a crucial aspect among the environmental considerations, especially in relation to LCA and sustainability assessment. The ability of these materials to degrade in certain environments differs them from traditional petroleum-based polymers, which remain in the environment for hundreds of years (Molina-Besch, 2022). But its efficiency and how much it breaks down is influenced by a combination of structural, environmental and operating factors that dictate the rate at which it decomposes. These values are key parameters to enable evaluation of the EoL performance of starch-based packaging and facilitate prediction of its environmental life cycle in practice (Äkräs et al., 2024).

According to Moshood et al., provide decomposition half-times of biodegradable plastics, such as starch-based polymers, polylactic acid (PLA), and polyhydroxyalkanoates (PHAs), in the range from 20 to 45 days under ideal composting conditions, which include proper humidity, oxygen, and microbial activity. Such conditions are often found in industrial composting installations or aerobic landfill environments, which usually have a high level of microbial populations and controlled environmental parameters. Conventional plastics, on the other hand, do not have hydrolyzable bonds and are highly crystalline, which can take a few hundred to even thousands of years yet still contribute immensely to our worldwide plastic pollution. The present work also emphasizes that a variety of native material properties, notably including intrinsic molecular features (e.g., polymer molecular structure, crystallinity, molecular weight and degree of monomer distribution), very significantly modulate rates of degradation by controlling microbial access to the polymeric matrix and hydrolytic cleavage. Greater crystallinity and closely packed molecular structure help to protect water penetration and enzymatic degradation, which results in slower degradation. In contrast, amorphous and hydrophilic materials, such as starch-based polymers, induce rapid microbial colonization and hydrolysis. Indeed, temperature, O_2 incrustation and moisture contents are 'environmental catalyzers' for these reactions. High humidity and aerated conditions foster microbial activity and the action of enzymes, while dry or anoxic environments (such as marine sediments or poorly designed landfills) can slow down, even block, degradation. Naturally occurring plasticizers such as glycerol and organic fillers (lignocellulosic fibers) enhance the flexibility of the polymer and its hydrophilicity, which in turn contributes to increased biodegradability. But synthetic stabilizers, or insufficient composting systems can offset these impacts resulting in longer degradation times. Hence, the degradation properties of biodegradable plastics are directly influenced by both material composition and disposal situation, highlighting a demand for efficient composting approaches along with well-balanced polymer compositions for environmentally friendly products (Moshood et al., 2022).

Similarly, Moshood et al., stress that environmental and material-specific factors both determine how quickly biodegradable plastics degrade in the environment. Under the most ideal composting conditions, decomposition will occur in a matter of weeks, but if left to nature and anoxic environment, it may considerably prolong. For example, in aquatic or anaerobic environments with anoxic conditions, the lack of cells and limited exposure to oxygen may retard degradative enzyme activity by several orders of magnitude. The structure of the polymer, such as covalent structure and topology, molecular weight and morphology (e.g., crystalline vs. amorphous) still remains a crucial factor in determining their biodegradability. Materials with high crystallinity are more resistant to microbial attack, whereas amorphous regions are more susceptible to enzymatic hydrolysis. In contrast, starch-based polymers, due to their hydrophilic and predominantly amorphous nature, generally degrade more rapidly than more crystalline biopolymers such as polylactic acid (PLA). Microbial metabolism and rates of biochemical reaction are further controlled by environmental factors, such as temperature, water content, and accessibility of oxygen. Moreover, natural additives such as glycerol promote the polymer-water interaction, which helps in the microbial colonization and degradation. Thus, a smart material engineering of starch-based packaging materials as well as the availability of suitable waste management infrastructure is needed in order to obtain the desired degradation rate. It was found that by combining system optimization with the development of innovative materials a significant closure in the gap between theoretical biodegradability and in situ behavior can be achieved (Moshood et al., 2022).

Barring starch-based plastics, the biodegradability of cellulose nanofibril (CNF)-reinforced chitosan composite films provides an alternate route for sustainable packaging applications especially when compared with fossil fuel originated polymers, such as LDPE and PLA 333. Although the discussed LCA concerns a cradle-to-gate scope and does not consider specific end-of-life scenarios, it underscores the inherent compostability and biodegradability of natural biopolymers such as chitosan—obtained from shrimp shell waste and CNF sourced from wood pulp. They do not only degrade naturally but are also both nontoxic and less persistent than synthetic plastics. Their biocompatibility and nontoxic degradation products make them less hazardous to be assimilated in the environment. Nevertheless, the degradation of the composites may be retarded using cross-linking agents such as citric acid to enhance film stability due to its potential adverse effect on microbial decomposition. Furthermore, end-of-life emissions for shrimp shell-based materials are included into landfill disposal municipal solid waste (MSW) category, highlighting the impact of the waste management system on the LCA results. This emphasizes the need for including end-of-life modelling in future LCAs to characterize degradation efficiency and confirm the superiority of the sustainable performance of biopolymer-based composite compared with conventional fossil film technology (Ponnusamy & Mani, 2022).

In a study according the efficiency of degradation depends on the physical and chemical nature of polymer such as its crystallinity, molecular weight, or functional additives. Starch-based plastics, containing predominantly hydrophilic polysaccharides degrades more easily than fossil-based polymers during aerobic composting due to the high microbial enzymatic activities. Environmental factors such as temperature, oxygen level, relative humidity, and microbial population are also significant factors for which market composting plants accomplish total biodegradation within weeks. By comparison, such materials decompose much less rapidly in natural soil or aquatic environments, which are subject to variable or suboptimal conditions. The report also emphasizes the significance of region-specific modelling in LCAs to represent local waste management activities and climate. Ignoring such parameters may cause over- or underestimation of environmental benefits. Consequently, the extent of degradation of starch-based packaging is determined by not only its polymer nature but also its environmental and infrastructure settings for disposal (Nessi et al., 2021).

Biodegradable starch-based packaging made from renewable agricultural sources, including cornstarch, potato and sugarcane, shows immense potential as a responsible way to achieve reduction of plastic pollution and carbon footprint on the environment. Such materials, as well as other bioplastics (such as PLA), clearly represents an advantage over the classical plastics because of their ability to bio-degrade under composting or natural conditions leading to reduced long term build-up in terrestrial and marine ecosystems. However, the rate of degradation is very sensitive to the disposal condition and wastes management method. Starch-based materials disintegrate comparably quickly in controlled landfill or composting the environments, but when deposited along with nondisintegrable waste such as plastic bottles into airless drifting landfills, their decomposition may slow down dramatically or even stop. PLA, while often advertised as compostable, commonly needs specialized commercial facilities with controlled temperature, moisture and oxygen to fully biodegrade into natural elements, which may be scarce. Moreover, there are still economic limitations to the large-scale use, focusing on the higher costs of production and scarce concentrated processing capabilities. Mechanical durability of natural starch packaging is a challenge in current research where the general trend is to increase its strength without losing the biodegradable nature. The study suggests that it is green if these alternatives are made of starch with a significantly lower dependency on fossil fuel and faster disintegration (depending on where they end up), providing there is proper infrastructure for their disposal, phased recycling in the system and change in material design to avoid hitting condition specific barriers for degradation (Gupta et al., 2024).

16.4.3 Environmental burdens of biopolymer blends

One of the most important aspects in the overall sustainability of biobased packaging materials is certainly associated to the environmental impacts of biopolymer blends (in particular, those made up by starch-based polymers, such as TPS and PLA) (Jozinović et al., 2024). Blends of biopolymers are generally developed to take advantage of the renewable origin and biodegradability typical of the nature polymers, and the mechanical performance exhibited by synthetic or semisynthetic ones. Although these mixtures result in decreased reliance on fossil fuel and subsequent reduction of greenhouse gases, their life cycles are complex from agricultural resource harvest to polymer production, processing use, and disposal at EoL leading to complicated environmental trade-offs. LCA analyses have demonstrated that the overall sustainability of starch-based and other biopolymer blends is strongly determined by feedstock sourcing, processing efficiency, degradation pattern, and end-of-life routes for waste. Such dependencies are crucial for the consideration of a shift to bioplastics in relation to circular economy objectives and environmental impact goals (Yadav & Nikalje, 2024).

According to LCA studies of biopolymer blends demonstrate renewable resource application decreases fossil depletion and global warming potential (GWP), however, these benefits are frequently counteracted with greater eutrophication and acidification as a consequence of increased agricultural inputs of fertilizers and pesticides. Starch-based and PLA materials, which are derived from crops such as corn and sugarcane, sequester biogenic carbon as they grow, so their net CO_2 emissions are lower relative to traditional plastics. The catch is that growing them requires a lot of space, irrigation and nitrogen fertilizers—this leads to nitrate running off into rivers, soil erosion and water pollution in rivers. Furthermore, the degradation behavior of blend compostable polymers such as PLA and TPS is complex when biodegraded. In anaerobic landfill conditions the leachate formed from these mixtures can produce large amounts of methane (CH_4), which is a potent greenhouse gas if the waste fails to be managed effectively. This highlights the extent to which disposal of biopolymers is crucial for environmental improvements. Composting offers a more managed and eco-friendly solution route for the degradation of polymers to CO_2, water, and biomass in aerobic conditions. However, also the composting process itself involves numerous environmental loads such as diesel use for hauling and (turning of) waste, water for maintaining moisture content in waste after landfill mining/landfill mining enhancement and operational CO_2 emissions. Recycling is the environmentally preferred end-of-life option with lowest GWP and quantities of energy used, yet it only applies to a relatively narrow range of biopolymer blends who are resistant to contamination or due to lack of dedicated recycling infrastructure. Thus, sustainability of starch/PLA and their blends rely on optimal utilization of renewable resources, suitable degradation conditions for optimization effects at the EoL disposal practices that avoid undesirable environmental impacts (Hottle et al., 2017).

Moreover, LCAs that focus on individual starch-based materials such as MaterBi blends confirm degradation rates and related environmental consequences greatly differ per disposal location. MaterBi is generally composted under industrial conditions Independently replicated studies have shown that approximately 60% mass of MaterBi was degraded during this process, with 95 wt% converted into CO_2 and 5 wt% into CH_4 The scenario scores poor in LCA evaluations mostly due to high energy and carbon demand and small credits for the quality of compost. In contrast, anaerobic digestion (AD) is a more sustainable EoL process achieving 85% decomposition and producing methane suitable for energy recovery with an efficiency of 36% at a 95% gas capture rate. This makes AD potent for the reduction of GWP and NREU. In contrast, the range of landfill degradation rates is quite large, from 0% to 85%, and is strongly influenced by temperature, moisture and extent of gas capture. GWP of landfill methane emissions can be substantially increased by leakage from the landfill if landfills do not have adequate containment. Domestic composting generates approximately the same energy from 11% to 22% of the gross calorific value of material to useful energy such as incineration with energy recovery but it does not apply controlled microbial and temperature condition, caused inhomogeneous degradation. Composting temperature (50–60°C, in industry), oxygen concentration and microbial activity rate under anaerobic versus aerobic conditions markedly influence the overall degradation rate. Material composition is also a determining factor, since the higher starch presents better biodegradability, while PLA content hinders this act because of supreme crystallinity. The technical maturity of treatment infrastructure for waste also adds uncertainty to LCA results and the need for efficient processing and end-of-life system of bio-packaging as opposed to traditional polyethylene-based ones, in order to gain realisticity sustainability benefits (Hottle et al., 2017).

The environmental consequences of starch-based biopolymer composites are not only limited to the end-of-life treatment, they also have upstream impacts from raw material and production. However, in some cases the overall environmental performance of such materials is composition-dependent and not so obvious if it comes to degradation pathways (Sinha, 2024). Combining with other biopolymers can improve mechanical strength, flexibility and barrier properties but make their degradation and recycling more complex. For instance, fossil-fuel dependency and greenhouse gas (GHG) emissions of 30%–80% less than traditional plastics can be achieved with TPS–PLA blends. But the presence of PLA (having semicrystalline structure and low hydrolysis rate) can limit the biodegradation under some environmental conditions. Highly humid and biologically active composting conditions result in relatively rapid decomposition of such mixtures, while under anaerobic landfill conditions, decomposition is often incomplete and methane formed. In addition, land use pressure, loss of biodiversity and fertilizer-induced eutrophication as consequences of the feedstock culture for starch and sugarcane partly mitigates the CO_2 -savings gained from biogenic feedstocks. The LCAs always show that biopolymer blends have lower carbon footprints and energy requirements than petrochemical plastics, yet impact indicators depend on additive addition. Ecotoxicity and resource depletion could be increased by processing agents, nanofillers or chemical cross-linkers used in performance improvement. Thus, achieving minimal ecological impact of lifecycle in the case of starch blends still requires tuning of polymer composition, wetting out biodegradation conditions and improve recycling and energy-recover infrastructure (Sinha, 2024).

These results are in agreement with those of Mafe et al., stressed that these environmental benefits of starch-based, PLA, and polyhydroxyalkanoate (PHA) packaging materials are dependent on end-of-life treatment and the efficiency of their biodegradation as compared with those offered by conventional plastics, e.g., PET and HDPE. Replacing the fossil-based feedstocks with renewable biomass can decrease CO_2 emissions, dependence on fossil sources and the build-up of long-term plastic waste. Nevertheless, degradation rates of biopolymer are highly dependent on polymer chemistry and on the environment. Bioplastic films such as starch-based and PLA composites can be completely broken down within a few months in aerobic composting but are likely to last for years in anaerobic, or marine habitats where the microbial populations and the oxygen concentration are not enough. It is also the case that agricultural feedstock production (starch and sugarcane) requires a high amount of water and fertilizer, leading to eutrophication or acidification resulting in GWP impacts during biodegradation. LCA reveal that biopolymers may release 50%–60% less CO_2 than petroleum-derived analogs, but the advantages are often offset by energy for processing polymers and lack of upscaling industrial composting facilities. Cost aspects also limit large-scale use, particularly in developing countries where composting or biogas recovery facilities are lacking. Therefore, ROC of biopolymer blends and promotion of EoL treatment systems such as anaerobic digestion, energy recovery, and selective recycling are essential to reduce environmental burden and achieve the sustainable circularity in starch-based packaging (Mafe et al., 2024). The environmental performance of starch-based and other biopolymer blends depends on feedstock cultivation, energy-intensive production, and end-of-life degradation processes. While these materials reduce fossil fuel use and carbon emissions, they still contribute to eutrophication and land-use impacts during agriculture. Composting and anaerobic digestion support energy recovery but add secondary energy burdens, whereas landfilling can release methane. Therefore, sustainability relies not on material composition alone but on efficient waste separation, circular recycling, and region-specific end-of-life management strategies that balance performance with environmental integrity.

16.4.4 Recycling potential and challenges

The EoL treatment of starch-based (and other bio-based) plastics is one of the key factors affecting their sustainability and LCA performance. With the transition of global economy from a linear "take–make–dispose" to circular model, recycling becomes one of the key instruments to mitigate greenhouse gas (GHG) emissions, preserve resources and close material cycles. From several routes to disposal, including composting, AD and incineration, recycling provides the greatest potential to maintain material value as well as reducing production of virgin polymers. The recycling of starch-based materials The ability to recycle starch-based plastics will depend upon a number of factors including polymer composition, blend compatibility, levels of contamination and the availability of infrastructure. However, because they are produced from natural materials, such as wood and cotton linters, technical difficulties as well as logistic problems may limit their industrial recycling. Understanding these nuances through LCA facilitates more informed decisions in optimizing sustainability aspects and incorporating starch products packaging into the circular bioeconomy (Mafe et al., 2024; Rosenboom et al., 2022).

According to Spierling et al., mechanical and chemical recycling are promising EoL options for bioplastics such as polylactic acid (PLA), thermoplastic starch (TPS) and bio-polyethylene (Bio-PE). Mechanical recycling transforms end-of-life consumer or industrial plastic waste into new products, and has a potential to save up to 94 million tons of CO_2 equivalent emissions per year relative to polymer production. But this relies on the quality and homogeneity of waste streams: If additives or residues from food stuff degrade recycle, the approach could be compromised. Chemical recycling schemes (depolymerization or solvent recovery) will return monomers identical to virgin materials, but based upon energies (and costs), they still are not practical. A major hurdle when recycling biodegradable polymers, such as PLA and TPS into traditional recycling schemes is the difference in their thermal and chemical behaviors may compromise fossil-based polymers from PET or PP, thus dedicated recycling streams, upgraded sorting techniques, and standardized labeling are necessary to enhance material recovery efficiency and environmental burdens over their life cycle for food packaging materials based on starch (Spierling et al., 2020).

Outside recycling, composting and anaerobic digestion (AD) are valuable EoL scenarios that return carbon and nutrients to the biological cycles. Fermentable starch-based products are consumed by microorganisms, which produce a biologically rich compost and improve the fertility of the soil– long gone are the days of building up a big pile of waste that never breaks down or turning plastic bottles into microplastics. One ton of starch-based packaging may produce 200–400 kg of compost in 75–135 days, depending on moisture, particle size and compost mixture. Alike, AD based on starch derived materials can render 800–1200 m^3 of biogas per ton while up to 1144–1716 kWh of usable electricity as well as a possible CO_2 capture about approximately 1299.5 kg per ton. These natural processes are the foundation for cradle-to-cradle sustainability, assuming that substances used in the products are nontoxic and free from disruptive agents. But it may work successfully based on feasible infrastructure requirement, reverse logistics and consumer's involvement for retrieving back the materials (Casarejos et al., 2018).

In the face of these environmentally friendly alternatives to fossil-based materials, some important limitations hamper the recyclability potential from starch-based packaging. According to a report the market share of bioplastics such as for example starch-based polymers (both in packaging and film applications), PLA, and TPS is limited by recyclability due low market share, lackluster segregation, and contaminations. Traditional recycling processes developed for PET or PP are frequently not adapted to bioplastics due to their distinct thermal stability and chemical structure. If BBPs become accidently incorporated into fossil-based plastics, they reduce the quality of the recyclate and can cause process disruptions. As a result, most LCAs cover composting and AD instead of recycling, even though recycling is, in theory, the option with the lowest global warming potential (GWP) if applied correctly. Some investment in advanced sorting systems, polymer compatibility research and regulatory frameworks to support dedicated bioplastic recycling streams will be needed to overcome current barriers (Kakadellis & Harris, 2020).

From an LCA point of view, however, the sustainability of starch-based materials also relies on the origin of feedstock, composition's design and addition's use (Broeren et al., 2017). revealed that virgin and reclaimed starch-based plastics releases up to 80% fewer GHGs, consuming 60% less nonrenewable energy (NREU) compared with petroleum plastic such as polypropylene (PP) pp and LDPE. But the addition of petrochemical contents such as PBAT, or PBS; compatibilizing additives could be compensated these advantages, which may even add up to 40% amount of overall GHG. Savings in eutrophication potential are 40% and land use is reduced by 60%, underlining the relevance of resource recovery for improved circularity. However, multi component mixtures and complexity in the additives hinder recycling and separation. To reduce this, researchers suggest decreasing the level of artificial fillers, enhancing polymer compatibility and incorporating the closed-loop recycling loop, particularly for starch composites (Broeren et al., 2017).

16.5 Life cycle assessment: Starch-based versus petroleum-based plastics

16.5.1 LCA framework and methodology

The LCA framework, following ISO 14,040 and 14,044 standards, evaluates the environmental performance of starch-based bioplastics compared with fossil polymers such as LDPE, PP, and PET. While petroplastics are nondegradable and carbon-intensive, starch-based materials offer renewable, CO_2-sequestering, and biodegradable alternatives that can save up to 1.63–2.10 kg CO_2 equivalent and 29.6–76.1 MJ equivalent of nonrenewable energy per functional unit. The LCA process includes four key phase's goal and scope definition, life cycle inventory (LCI), life cycle impact assessment (LCIA), and interpretation covering cradle-to-grave stages from raw material cultivation to end-of-life. Using industrial and Ecoinvent v3.10 data, impacts such as global warming potential, eutrophication, land use, and energy demand are quantified. Overall, LCAs consistently show starch-based bioplastics outperform fossil-based plastics in most environmental categories.

According to Poveda-Giraldo et al., starch-derived biopolymers exhibit a significant decrease in carbon emissions and fossil energy consumption, which can be attributed to the renewable feedstocks and sequestration of bio-based carbon through plant cultivation. On the other hand, LCAs also show that agricultural stages may cause eutrophication and land use impacts associated with fertilizer application and energy requirements for cultivation (which could counteract some of the environmental benefits). Pinpointing such "hotspots" and giving indications on how sustainability can be improved (e.g., use of low impact agriculture tools, energy efficiency during polymer synthesis, end-of-life recovery system) would then become part of the interpretation phase in LCA (Poveda-Giraldo et al., 2025). The LCA methodology, used for bioplastic packaging systems, distances from materials comparisons examining more complex end-of-life scenarios in a real context. For instance, biodegradation tests performed in soil and composting environments according to standardized procedures such as ASTM D5988 and ISO 17,088 guarantee that the degradation profile of a product matches with well-established environmental safety benchmarks. These procedure document the rate and extent of biodegradation by the measurement of CO_2 evolution, thereby indicating that starch-containing materials are indeed capable of being microbiologically assimilated in toto under appropriate conditions. This integration of biodegradation data in life cycle impact modeling opens up the possibility for evaluating environmental performance considering the entire product life time (Malinconico, 2017). However, fossil-based PE foils are characterized by outstanding mechanical and chemical stability but as a consequence these are not degradable in short-term causing enhanced life cycle environmental persistence and GHG emissions. Indeed, they are made out of nonrecyclable materials and largely contribute to the pile up of waste and microplastic contaminations. In all LCA results, it is evident that starch polymers result in a significantly lower overall impact especially considering energy recovery (in the form of composting or anaerobic digestion) and biogenic carbon accounting within system boundaries. However, young bioplastic production may consume a greater process energy and the lack of available universal composting facilities compromise their best performance in some regions (Malinconico, 2017; Piemonte et al., 2013). In general, the LCA tool and methodology offers a scientific basis for assessment and improvement of environmental sustainability of starch-based packaging. The life cycle analysis, from cradle to grave, that is based on the interaction between feedstock cultivation, energy consumption, emissions and degradation processes provides information for the optimization of technology especially in respect of policy development. With the harmonization of the impact categories, based on real biodegradation data and energy demand evaluations, LCA analyses have continuously demonstrated that starch-based bioplastics are the most viable solution to replace fossil fuel-based plastics resulting in a circular and low-carbon economy.

16.5.2 Comparative environmental impacts

Starch-based bioplastics are evaluated and compared with conventional petroleum-based products in terms of environmental aspects and their life cycle performance including carbon footprint, energy and resource consumption. Starch-based packaging materials manufactured from renewable staple crops such as corn, cassava, potato, or wheat starch are a sustainable substitute for nonbiodegradable plastics, which can be used in place of usual oil-derived plastics polyethylene (PE), polypropylene (PP), and polyethylene terephthalate (PET). Numerous LCA studies show that starch-based polymers lead to dramatically lower greenhouse gas (GHG) emissions and lower nonrenewable energy input with higher sustainability. Nonetheless, these advantages depend on various factors such as the agricultural methods to raise raw materials, the energy and type of materials used for production, transportation distances, and the applied EoL management systems (Yadav & Nikalje, 2024).

16.5.2.1 Carbon footprint reductions (30%–40%)

When compared with petroleum-based polymers, starch-based bioplastics can yield significant environmental benefit. Some studies report life-cycle GHG emissions of starch-based films and biopolymer blends (e.g., thermoplastic starch (TPS) or starch–PLA composites) in the range of ~1.5–2.5 kg CO_2-eq per kg of polymer, whereas typical fossil feed-stock polymers such as PP or LDPE can be in the ~3.0–4.0 kg CO_2-eq per kg range translating into a 30%–40% reduction under favorable conditions (Poveda-Giraldo et al., 2025). As starch-based materials engage in a biogenic carbon cycle (CO_2 is captured during crop growth and released during degradation), they contrast with fossil-based plastics, which release long-sequestered fossil carbon. However, the actual benefit depends greatly on avoiding landfill disposal (which may generate methane) and instead utilizing industrial composting or anaerobic digestion. Additionally, agricultural practices (fertilizer, irrigation, land-use change) may increase emissions by 15%–20% unless improved management is adopted (Atiwesh et al., 2021).

Comparison of the biogenic carbon cycle in starch-based plastics with the linear carbon release of fossil-based polymers have been demonstrated in Fig. 16.1.

16.5.2.2 Energy use and resource efficiency

Beyond carbon, energy input and resource efficiency are critical to comparing starch-based with petro-plastics. Life-cycle inventory studies indicate that 1 kg of PE or PP may require ~70–85 MJ of cumulative energy demand (CED) from nonrenewable sources, whereas starch-based polymers can require ~40–60 MJ/kg under improved production conditions. This corresponds to a potential reduction of up to ~50% in nonrenewable energy demand when renewable feedstocks and energy sources are used (Yadav & Nikalje, 2024). Still, trade-offs remain: starch crops demand land, water and fertilizers raising risks of eutrophication and soil degradation if not properly managed. Using nonfood feedstocks or integrating food-industry residues can enhance resource-efficiency. End-of-life routes (combustion, anaerobic digestion) offer energy recovery (e.g., ~15–25 MJ/kg) but must be managed to avoid negative impacts. In summary, starch-based bioplastics can achieve ~30–40% lower carbon footprint and ~40–50% improved

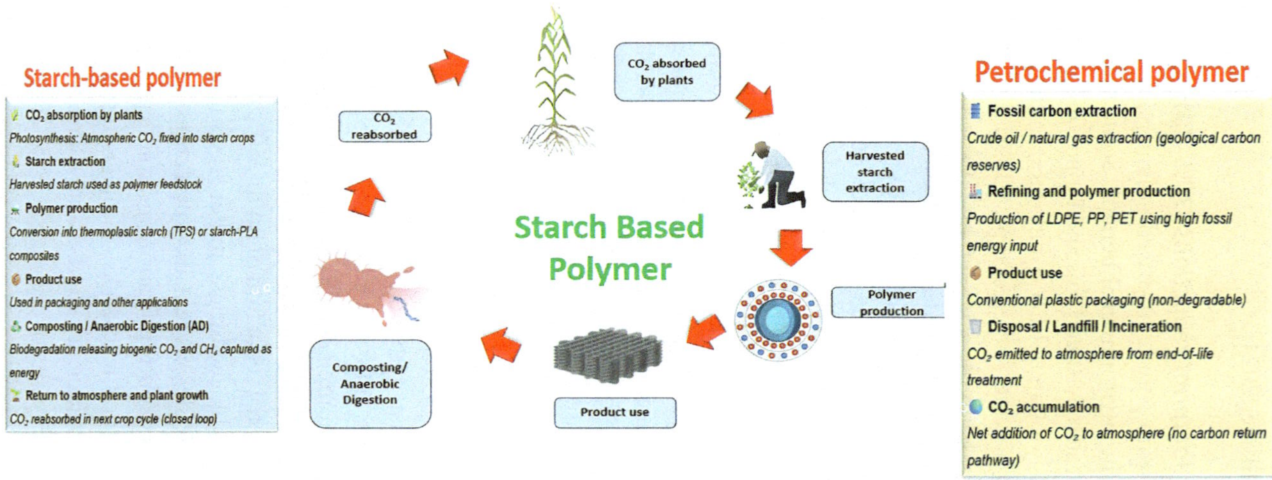

Figure 16.1 Starch-based polymer versus petrochemical polymer.

energy/resource efficiency versus petro-plastics provided that agricultural feedstocks, processing and EoL systems are optimized for circular economy integration (Synani et al., 2024).

16.5.3 Limitations and uncertainties in LCA studies

LCA is a comprehensive and standardized method for the quantification of environmental impacts due to products along their whole life cycle, beginning with the extraction of resources and ending with product recycling or disposal. It is now recognized as a valuable tool to compare the sustainability of starch-based bioplastics against oil-based plastics. Nevertheless, LCA is associated with methodological and data uncertainties that limit the accuracy and comparability of results, even though it offers valuable tools for gauging environmental performance. These deficiencies arise mainly from an absence of consistency with regard to system boundaries, data quality, geographical coverage and adopted impact assessment methods that may then result in inconsistent (and even contradictory) findings between studies. Ramesh and Vinodh in a recent study consider that one of the most important drawbacks of polymer LCA studies lies in partial or incomplete inclusion of system boundaries, and many analyses contribute to a "cradle-to-gate" whilst neglecting key EoL activities such as recycling, incineration or composting. This incomplete evaluation does not take into account emissions and energy recovery potential, leading to an insufficient environmental profile. In addition, variation in functional unit definition, LCA tool (e.g., SimaPro) and database used (such as Ecoinvent) contribute to uncertainty, whist regional differences in feedstock growing conditions and energy source serve to exacerbate the situation. E.g. Starch-based bioplastics produced from cassava or corn does, however, differ dramatically in their land use, fertilizer demand and emission profiles depending on climate zone and agricultural practice, making global comparisons more complex. The further highlights that even though the GHG emissions of biopolymers are lower compared with fossil plastics, different modeling assumptions lead to differing environmental results (Ramesh & Vinodh, 2020). Similarly, Hottle et al. (2013) emphasize that numerous LCAs of starch-based materials consider only the production of resins or granules, and do not account for downstream life cycle stages including use, distribution and end-of-life treatment, which can reduce reported overall GHG emissions and energy demand. Although the addition of EoL pathways (e.g., incineration, anaerobic digestion or composting), which provide a more realistic view on available recycling options still involves a series of additional uncertainties given limitations in empirical data available for waste management and biopolymer degradation rates. In addition, dissonances in LCI datasets due to regional agricultural practices or differences between energy grids can strongly perturb results. In the majority of these studies, environmental concerns such as GWPs and fossil energy are paid more attention to than other important CO_2 -based resources, including eutrophication, water scarcity, and land use change (Hottle et al., 2013).

From a comparative LCA perspective, pectin–corn starch biocomposite films were compared with LDPE film by Günkaya and Banar (2016) using the CML-IA method and Monte Carlo simulations. Their results indicated that system boundary constraints such as the lack of inclusion of use and disposal stages resulted in underestimation of environmental impacts. Additionally, LCA parameterization tools did not allow users to input biogenic CO2 emissions due to degradation rendering end-of-life characterization less complete. It was also found that the electricity consumption, which occurred during starch modification and extrusion, is the key uncertainty affecting multiple impact categories such as GWP and human toxicity. Monte Carlo analysis also demonstrated spread in ecotoxicity, abiotic depletion and photochemical oxidation (such that small changes in inputs would lead to a different rank order of environmental impacts. The lack of large-scale industrial operational data, poor biodegradation modeling capabilities and process-specific energy uncertainties were identified by the authors as factors affecting the ability to predict LCA for starch-based packaging materials (Günkaya & Banar, 2016).

So, despite the strength of LCA as a tool for assessing environmental sustainability, it is worth exercising caution in interpreting its results for starch-based and petra-based plastics. These limitations need to be addressed by standardized methods, transparent reporting, region-specific data, and more accurate modeling of end-of-life processes. By accounting for these details, LCA may become a more precise and uniform tool that can be used to support sustainable material design, industrial innovation, and policy making in the bioplastics industry.

16.6 Emerging trends and innovations

16.6.1 Use of waste biomass as feedstock

In the context of sustainable packaging, a shift toward the utilization of waste biomass as a renewable source to substitute petrol-based polymers has been emerging. Relationship to other studies in recent years, a variety of

LCA works have highlighted the environmental benefits of using agricultural residues and food industry by-products for starch-based packaging. This strategy responds directly to two pressing global challenges: plastic pollution and agri-food waste valorization, while contributing to the targets of the circular bioeconomy. According to Versino et al., agri-food waste, including fruit peels, cereal husks and lignocellulosic residues can be efficiently valorized into starch- and cellulose-based biopolymers, facilitating dual benefits, including less generation of waste streams and reduced reliance on virgin raw materials. Moreover, biopolymers produced from waste also show increased mechanical properties, oxygen barrier and hydrophobicity even in the presence of adding nanofillers or natural fillers such as essential oils that upgrade their use for active and intelligent packaging (Versino et al., 2023).

Surendren et al., also extended this principle by showing that starch coplasticization by applying organic acids produced from citrus waste or lignocellulosic residues (e.g., citric acid and succinic anhydride) leads to similar lower environmental impacts compared with traditional glycerol or urea-based compositions. Their comparative LCA demonstrated how the incorporation of coplasticizers produced from waste can significantly reduce fossil fuel depletion by up to 33% with no significant loss in mechanical and functional performance. In contrast, the fossil-energy demand for the synthesis pathways of urea-based systems meant that this system had higher global-warming potential and ecotoxicity issued. These results further support that the valorization of biowaste intermediates along starch plasticization pathways not only reduces carbon footprints, but also enhances the biodegradability and end-of-life properties of starch films (Surendren et al., 2024). Senila et al., performed a cradle-to-gate LCA of bioplastic manufacturing from orchard residues, apple and plum prunings, being volumetrically converted to PLA and poly(3-hydroxybutyrate) (PHB). In that respect, the results of the present study showed that lignocellulosic waste valorization may reduce CO_2 emissions by approximately 33% compared with conventional PLA synthesis (still much less regarding petroleum-based plastics). Energy for biomass pretreatment and fermentation remained the energy intensive stages, although largely reversible by inclusion of renewable energy and process optimization. Crucially, repurposing these waste streams for production of high-value polymers can contribute to the resilience of rural bioeconomies by converting materials historically burned or sent to landfills into valuable packaging feedstocks (Senila et al., 2024).

Recent studies also emphasize that the environmental impact of waste derived feedstock does not stop at material replacement. Hemachandra et al., noted that packaging LCAs are more and more likely to include indirect benefits in climate change, notably when waste-based materials lower food wasting through improved active and modified-atmosphere packaging. Indeed, given that food loss contributes some 10% to global greenhouse gas emissions on its own, the incorporation of biodegradable starch coatings or films from agro-waste may compensate for higher production inputs by facilitating food preservation and offering a longer shelf life. This systemic viewpoint recasts packaging as not only a waste-management issue, but also a mitigation strategy in the larger context of food-system sustainability. The authors highlight that consequent LCA studies will need to explicitly assess these indirect benefits, in order to support evidence-based design and policy-making for systems of circular packaging (Hemachandra et al., 2024).

In another study Sobeih et al., described recent progress on the use of nonconventional waste (i.e., banana peel, potato remnants, cassava pulp) to produce starch for large-scale bioplastics. These waste-derived starches showed similar or superior films properties as compared with the natural fiber films (bagasse and kenaf) when reinforced with nanoparticles. The resulting hybrid composites not only improved strength and moisture resistance, but also realized large reductions in life-cycle greenhouse gas emissions relative to those for fossil-based plastics. In addition, while harnessing the power of renewable nanoclays together with essential oils allowed elongation of shelf life and preserving the materials' antimicrobial properties, which is yet another step in the direction toward zero-waste and circular-economy paradigms (Sobeih et al., 2025).

These studies support that using waste biomass as a feedstock may be one of the most promising development directions for sustainability in starch-based packaging. Furthermore, LCAs are more carbon-efficient, resource efficient and biodegradable. This literature describes remaining obstacles such as fluctuations in feedstock composition, weak hierarchy among these facilities, and the energy required for pretreatment and fermentation. Ongoing development efforts aim to tackle these hurdles by incorporating renewable energy feedstocks, green catalysis, and scaling modular biorefineries' capacities to process various waste streams into homogeneous biopolymer precursors.

This evidence confirms that the choice of waste biomass for strategic incorporation in starch-based packaging (e.g., lignocellulosic by-products, orchard prunings, and food-processing subproducts) decisively improves its environmental and functional attributes. Transition from linear to circular business models is demonstrated here using multicriteria LCA assessments and it paves the way for carbon neutral, fully biodegradable, and resource-efficient packaging solutions, which reconcile green pack integrity with economic flourish.

16.6.2 Active and intelligent packaging systems

Intelligent and active food packaging systems have emerged as a fundamental element of the new generation of sustainable food packaging, combining developments in material science, nanotechnology and artificial intelligence with improved food preservation, product traceability capabilities and environmental performance. There is more to active packaging than passive containment, instead of resting inert, it responds actively to the food or its external environment emitting or absorbing components to enhance shelf life and enhance safety. Contrarily, smart packaging acts as a reactive system that can perceive, check, and transmit the state status of products in real time. These are developments that facilitate better-quality and safer food, according to LCA method principles, through the reduction of waste, optimized use of resources and aid in driving toward a circular economy. Active packaging generally involves controlled release or uptake processes. There is an incorporation of natural antimicrobial and antioxidant additives, such as essential oils, aldehydes and phenolic compounds in biodegradable carriers (example-starch, alginate and chitosan). These are compounds that act under environmental triggers such as pH, temperature, or moisture changes to reduce microbial growth and oxidative spoilage. In a study Hou et al., noted that "smart" active films could respond to surrounding factors, such as enzymes, humidity and UV light by releasing bioactive compounds in a controlled way correlated with microbial growth. For instance, pH-sensitive chitosan films are degraded under acidic environment to release aldehydes, humidity-responsive hydrogels, and light-driven nanocapsules has been introduced as essential oils or antimicrobials delivery systems at predetermined points, reducing reliance on synthetic preservatives (Hou et al., 2023). Such adaptive systems also prolong food shelf-life and promote sustainability by reducing waste and chemical residues.

Extending these principles, the introduction of artificial intelligence (AI) has introduced a new era of smart packaging systems capable of autonomously collecting data, recognizing patterns, and making predictions. Packaging with AI (intelligent packaging) contains sensors, convolutional networks and computer vision to spot spoilage from color, gas or odor differences in real time. Li et al., state that such systems are capable of determining the freshness level among fruits, meats and dairy products by visual and volatile compound analysis with error rates that cannot be achieved through conventional way. AI combined with radio-frequency identification (RFID) and augmented reality (AR) can provide a better traceability and consumer transparency on food were written across the field of the food supply. It becomes possible to track inventories with precision, minimize spoil due to mishandling in storage and enhance customer confidence by displaying visible fresh data on smart labels. Chipless RFID sensors, which do not require embedded microchips or batteries, with the reduction in cost and recyclability also aspired to lead further toward environmental as well as economic sustainability (Li et al., 2023). Recent developments have also pursued material synthesis and sensor downsizing to achieve maximum performance with the least impact on biodegradability. According to Mkhari et al., nanotechnology has been an essential aspect to enhance the mechanical, barrier and antimicrobial characteristics of intelligent packaging. Nanoclays, zinc oxide (ZnO), titanium dioxide (TiO_2) and silver nanoparticles have been widely added in starch/chitosan-based films to improve barrier properties and resistance against oxidative degradation and maintaining biodegradability. The nanoparticles strengthen the films and serve as antimicrobial barriers against pathogens. Furthermore, the availability of nanosensors for O_2, CO_2 and volatile amines has opened opportunities to noninvasively perform accurate monitoring of food freshness. These nanosized sensors can identify very small biochemical alterations in the product, where data are transmitted by wireless to monitoring systems for real-time quality control of the package.

The manufacturing of these systems is being revolutionized by 3D printing and microfluidics, which permit the creation of personalized and multifunctional packages. Bio-pigments, including berried anthocyanins and curcumin, can be integrated with the biodegradable polymer films by 3D printing as an example, a noticeable color change in response to spoilage metabolites acts as a visual pH or gas indicator. These printed films are light and flexible, and allow both preservation and sensor tasks to be incorporated in the same layer, with a gain in material economy, enhancing eco-efficiency. Microfluidic platforms meanwhile, enable multianalyte, high-throughput detection of contaminants and pathogens with minimal reagent usage. For example, droplet-based sensor microfluidics can detect heavy metals such as lead and mercury and biological hazards such as *Listeria monocytogenes* at unimaginably low levels for safer food regulations (Mkhari et al., 2025).

Recent electromagnetic sensors, such as radio frequency identification (RFID) and chipless wireless communication systems, have continued to change the food logistics by monitoring the temperature, gas composition, and humidity in various supply chains globally. The incorporation of these systems with AI-driven predictive models generates dynamic shelf-life estimation and automated quality control, resulting in a further cut in food losses throughout the distribution networks (Li et al., 2023; Mkhari et al., 2025).

Sustainable development is still at the forefront of these technological advancements. Biopolymer materials produced from starch, cellulose, chitosan, etc., have acquired a place for themselves in the field of commodity plastics, replacing petrochemical products for their biodegradability, compostability and lower carbon footprint. When these renewable polymers are further reinforced by nanocomposites or functionalized using 3D printing technology, the plastics ensure the required mechanical properties and adaptability for intelligent applications in packaging. By overlaying Life Cycle Assessment criteria at the design level, scientists guarantee that packaging materials don't exchange low energy requirements and minimal environmental impact for anything over its durations of use and disposal (Hou et al., 2023).

Altogether, the merging of nanotechnology, artificial intelligence, and green material design is shifting a passive barrier into an interactive, on-demand, controlling and sustainable system in a rethought package. Active and smart packaging not only improves the safety and shelf life of products but also expands global waste minimization and resource management efforts. With enhanced sensing, controlled release and digital traceability applied in these systems, it can be a significant advance towards sustainable food systems that combine with environmental consciousness and technology innovation.

16.6.2.1 Antioxidant and antimicrobial additives

Integration of antioxidant and antimicrobial agents in packaging material based on starch is considered a notable progress in active and intelligent food packaging sector. These reinforcing agents, which are bioactives such as lubricants, can be employed to enhance performance properties of biopolymer films, such as the resistance to oxidative degradation, inhibition of growth of microorganisms and extension of the shelf life of food products, while maintaining the degradability or biodegradability and portions at least equal to environmental acceptation. The sustainability of these starch-based systems would be further enhanced when such additives are derived from renewable, natural or waste-origin sources following LCA and circular economy guidelines (Ali et al., 2023; Muñoz-Gimena et al., 2023; Nikmanesh et al., 2023; Sun et al., 2024).

Herbal extracts that are a rich source of phytochemicals are known as potent bio-additives. Ali et al., prepared cornstarch films incorporated with *Aconitum heterophyllum*, *Artemisia annua* and *Thymus serpyllum* extracts, having up to 69% DPPH radical scavenging activity owing to their phenolic and flavonoid levels. These films also exhibited antimicrobial activity against *Staphylococcus aureus* and *Salmonella* spp., in conjunction with enhanced barrier properties and crystallinity (Ali et al., 2023).

Nano-composite reinforcements have allowed the creation of active multifunctional films. Muñoz-Gimena et al., found that inclusion of nanoparticles such as Ag, ZnO, TiO_2, CuO, and MgO into thermoplastic starch improves mechanical and antimicrobial behavior. ZnO and TiO_2 provide UV protection and ROS-induced bacterial inactivation, whereas Ag nanoparticles disrupt microbial membrane integrity and the DNA (Muñoz-Gimena et al., 2023). In addition, bio-conjugate systems exhibit enhanced additive stability. Sun et al., prepared soy protein isolate (SPI) films incorporated with dialdehyde starch–quercetin (DAS–QR) conjugate, which showed 79.8% DPPH scavenging and 92.7% antibacterial inhibition against *S. aureus*. This methodology decreased the oxidation and microbial spoilage in food at storage, therefore proving potential for the system of controlled release as smart packaging (Sun et al., 2024). Nikmanesh et al., also developed this approach using *Viola odorata* extract in potato starch films to produce a phenolic content of 16.65 mg GAE/g and powerful antibacterial activity with inhibition zones that can reach 34 mm for *S. aureus*. They observed that these films were more flexible and oxygen impermeable than the PEF film, supporting the additive effect of phenolic–polymer cross-linking (Nikmanesh et al., 2023).

These advances show the synergy between natural and renewable sources of antioxidants, nanostructures and conjugates to develop high-performance materials. Their reduced reliance on synthetic stabilizers implies lower overall carbon footprint and waste generation throughout the packaging life cycle and encourages the move to circular bioeconomy solutions.

16.6.2.2 Freshness indicators

The freshness indicators form the intelligent part of the smart packaging since they allow visual, real-time detection of food spoilage based on measurable biochemical reactions. The detected indicators have a response to TVB-N, biogenic amines, and organic acids formed during microbial decomposition, which could provide nondestructive quality measuring solutions, reduce waste, and improve safety (Devi et al., 2025; Liu et al., 2022; Shukla et al., 2015; Tan et al., 2024; Zhai et al., 2025). Chemical dyes that are harmful have been replaced by eco-friendly freshness indicators using natural pigments. A recent study have demonstrated that starch or chitosan films incorporated with anthocyanins, curcumin, and betalains also showed pH-responsive color changes with reversibility, red in acid and green in base conditions (Liu et al., 2022).

Zhai et al., extended this idea by describing colorimetric food freshness indicators (CFFIs), which use several chromophores such as porphyrins, anthocyanins, and alizarin. These hybrids enhance gas selectivity and response time by means of electrospun and 3D-printed nanostructures, while dealing with humidity as well as leaching issues due to the inclusion of metal organic frameworks (MOFs) (Zhai et al., 2025).

Tan et al., developed a dual-pigment indicator with purple potato anthocyanin (PA) and alizarin in a soy protein–locust bean gum matrix, it showed 90% pigment stability as well as accurate pH response for shrimp and chicken spoilage tests. Similarly (Devi et al., 2025), have said that, apart from their pH-sensitive color transitions, curcumin-based indicators embedded in biopolymers, such as starch or gelatin can act as antioxidants and antimicrobials and successfully stabilize volatile amines in (hydrophobic effect) and inhibit oxidation. Finally, *Shukla* et al., constructed an inexpensive bromophenol blue sensor in which the color change (from yellow to blue) was related to bacterial growth and TVB-N accumulation of buffalo meat ($r = 0.914$). This sensor can fit well in the starch-based biodegradable films due to its solvent-free and scalable approach (Tan et al., 2024).

Therefore, these platforms constitute a complete shift from the traditional pH-sensing toward multifunctional and biodegradable indicators by integrating both natural pigments and hybrid biopolymers to nanotechnology.

16.7 Future prospects

Starch-based packaging is gaining popularity as a sustainable alternative to traditional plastics because it is biodegradable and compostable, and is made from renewable resources such as corn, potatoes, and cassava. This helps reduce environmental impact and carbon footprint. Advances in materials science, such as nanotechnology, composite blending, and the use of natural plasticizers, have enhanced mechanical and barrier properties, making these packages more effective for various applications, including food, medical, and agricultural uses. The global market is growing quickly as regulations encourage businesses to adopt eco-friendly packaging. However, high costs and difficulties with large-scale production, moisture resistance, and shelf life remain barriers to broader adoption. By utilizing nonfood starch sources and agricultural by-products, starch-based packaging fosters a circular economy and strengthens local economies, particularly in developing countries with limited recycling infrastructure. Ongoing innovation in manufacturing methods, government policy support, and the incorporation of new features such as active and intelligent packaging are anticipated to drive further growth and sustainability in this field over the next decade.

16.8 Conclusion

Starch-based packaging stands out as an eco-friendly alternative to traditional plastics by using renewable resources such as corn, potatoes, and cassava. This shift significantly reduces our reliance on fossil fuels and lowers greenhouse gas emissions associated with packaging. Besides being biodegradable and compostable, starch-based materials help tackle the growing plastic pollution problem by breaking down naturally, reducing waste in landfills and oceans. However, their sustainability depends on several factors, including responsible farming practices, energy-efficient processing, and proper end-of-life management systems such as recycling and composting. Challenges such as moisture sensitivity, costs, and infrastructure limitations need to be addressed for these materials to reach their full potential. Advances in material formulation and better waste management can improve performance and environmental outcomes. With continued innovation, supportive policies, and investment in infrastructure, starch-based packaging offers a promising path toward more sustainable packaging solutions that balance environmental benefits with functional needs, contributing meaningfully to reducing plastic waste and climate change impacts.

References

Äkräs, L., Silvenius, F., Baniasadi, H., Vahvaselkä, M., Ilvesniemi, H., & Seppälä, J. (2024). A cradle-to-gate life cycle assessment of polyamide-starch biocomposites: Carbon footprint as an indicator of sustainability. *Clean Technologies and Environmental Policy, 26*(10), 3297–3312. https://doi.org/10.1007/s10098-024-02884-1

Akpa, J. G., & Dagde, K. K. (2012). Modification of cassava starch for industrial uses. *International Journal of Engineering and Technology, 2*(6), 913–919.

Ali, A., Basit, A., Hussain, A., Sammi, S., Wali, A., Goksen, G., Muhammad, A., Faiz, F., Trif, M., Rusu, A., & Manzoor, M. F. (2023). Starch-based environment friendly, edible and antimicrobial films reinforced with medicinal plants. *Frontiers in Nutrition, 9*. https://doi.org/10.3389/fnut.2022.1066337

Amelework, A. B., Bairu, M. W., Maema, O., Venter, S. L., & Laing, M. (2021). Adoption and promotion of resilient crops for climate risk mitigation and import substitution: A case analysis of cassava for South African agriculture. *Frontiers in Sustainable Food Systems, 5*, Article 617783. https://doi.org/10.3389/fsufs.2021.617783

Aruwajoye, G. S., Kassim, A., Saha, A. K., & Gueguim Kana, E. B. (2020). Prospects for the improvement of bioethanol and biohydrogen production from mixed starch-based agricultural wastes. *Energies, 13*(24). https://doi.org/10.3390/en13246609

Atiwesh, G., Mikhael, A., Parrish, C. C., Banoub, J., & Le, T. A. T. (2021). Environmental impact of bioplastic use: A review. *Heliyon, 7*(9). https://doi.org/10.1016/j.heliyon.2021.e07918

Avérous, L. (2004). Biodegradable multiphase systems based on plasticized starch: A review. *Journal of Macromolecular Science: Part C: Polymer Reviews, 44*(3), 231–274. https://doi.org/10.1081/mc-200029326

Avérous, L., & Halley, P. J. (2014). Biocomposites based on plasticized starch. In *Biodegradable polymers* (pp. 13–40). Springer.

Bhatt, R., Moulick, D., Bárek, V., Brestic, M., Gaber, A., Skalicky, M., & Hossain, A. (2025). Sustainable strategies to limit nitrogen loss in agriculture through improving its use efficiency—aiming to reduce environmental pollution. *Journal of Agriculture and Food Research, 22*, Article 101957.

Bradeen, J. M., & Kole, C. (2016). *Genetics, genomics and breeding of potato*. CRC Press.

Broeren, M. L. M., Kuling, L., Worrell, E., & Shen, L. (2017). Environmental impact assessment of six starch plastics focusing on wastewater-derived starch and additives. *Resources, Conservation and Recycling, 127*, 246–255. https://doi.org/10.1016/j.resconrec.2017.09.001

Bulatović, V. O., Mandić, V., Kučić Grgić, D., & Ivančić, A. (2021). Biodegradable polymer blends based on thermoplastic starch. *Journal of Polymers and the Environment, 29*(2), 492–508. https://doi.org/10.1007/s10924-020-01874-w

Cabernard, L., Pfister, S., & Hellweg, S. (2024). Biodiversity impacts of recent land-use change driven by increases in agri-food imports. *Nature Sustainability, 7*(11), 1512–1524. https://doi.org/10.1038/s41893-024-01433-4

Carvalho, A. J. F. (2016). Chapter 6: Chemical modification of thermoplastic starch. *RSC Green Chemistry*, 217–235. https://doi.org/10.1039/9781782622796-00217

Casarejos, F., Bastos, C. R., Rufin, C., & Frota, M. N. (2018). Rethinking packaging production and consumption vis-à-vis circular economy: A case study of compostable cassava starch-based material. *Journal of Cleaner Production, 201*, 1019–1028. https://doi.org/10.1016/j.jclepro.2018.08.114

Chimtong, S., Saenphoom, P., Karageat, N., & Somtua, S. (2016). Oligosaccharide production from agricultural residues by non-starch polysaccharide degrading enzymes and their prebiotic properties. *Agriculture and Agricultural Science Procedia, 11*, 131–136. https://doi.org/10.1016/j.aaspro.2016.12.022

Chisenga, S. M., Workneh, T. S., Bultosa, G., & Alimi, B. A. (2019). Progress in research and applications of cassava flour and starch: A review. *Journal of Food Science and Technology, 56*(6), 2799–2813. https://doi.org/10.1007/s13197-019-03814-6

Culqui-Arce, C., Paucar-Menacho, L. M., Schmiele, M., Vito-Villa, J., Camacho-León, J., Mori-Mestanza, D., Auquiñivin-Silva, E. A., Cayo-Colca, I. S., Castro-Alayo, E. M., & Balcázar-Zumaeta, C. R. (2025). Unveiling the potential of starch from tubers and grains as primary source of biopolymers. *Discover Food, 5*(1). https://doi.org/10.1007/s44187-025-00598-x

Debiagi, F., Mello, L. R. P. F., & Mali, S. (2017). Thermoplastic starch-based blends: Processing, structural, and final properties. *Starch-based materials in food packaging: Processing, characterization and applications*. Elsevier Inc. https://doi.org/10.1016/B978-0-12-809439-6.00006-6

Devi, L. S., Krishnan, R., Begum, S., Nath, D., Mohanty, A., Misra, M., & Kumar, S. (2025). Curcumin as bioactive agent in active and intelligent food packaging systems: A Comprehenscve Review. *Comprehensive Reviews in Food Science and Food Safety, 24* (4), Article 15414337. https://doi.org/10.1111/1541-4337.70228

Dutta, D., & Sit, N. (2024). Comprehensive review on developments in starch-based films along with active ingredients for sustainable food packaging. *Sustainable Chemistry and Pharmacy, 39*, Article 101534. https://doi.org/10.1016/j.scp.2024.101534

Esmaeili, M., Pircheraghi, G., & Bagheri, R. (2017). Optimizing the mechanical and physical properties of thermoplastic starch via tuning the molecular microstructure through co-plasticization by sorbitol and glycerol. *Polymer International, 66*(6), 809–819. https://doi.org/10.1002/PI.5319

Farmerline. The impact of fertilizers on the environment: Inorganic vs. organic. https://farmerline.co/the-impact-of-fertilizers-on-the-environment-inorganic-vs-organic/.

Fujs, T., & Kashiwase, H. (2024). *Strains on freshwater resources: The impact of food production on water consumption*. World Bank Blogs. https://blogs.worldbank.org/en/opendata/strains-freshwater-resources-impact-food-production-water-consumption.

Forfora, N., Azuaje, I., Kanipe, T., Gonzalez, J. A., Lendewig, M., Urdaneta, I., Venditti, R., Gonzalez, R., & Argyropoulos, D. (2024). Are starch-based materials more eco-friendly than fossil-based? A critical assessment. *Cleaner Environmental Systems, 13*. https://doi.org/10.1016/j.cesys.2024.100177

Gómez-Caturla, J., Ivorra-Martinez, J, García-García, D., Balart, R., Dominici, F., Puglia, D., & Torre, L. (2023). Thermoplastic starch (TPS) obtained from mango Kernel flour by means of extrusion process with different plasticizers. *Key Engineering Materials, 957*, 21–27. https://doi.org/10.4028/p-w62C6Y

Günkaya, Z., & Banar, M. (2016). An environmental comparison of biocomposite film based on orange peel-derived pectin jelly-corn starch and LDPE film: LCA and biodegradability. *International Journal of Life Cycle Assessment, 21*(4), 465–475. https://doi.org/10.1007/s11367-016-1042-8

German, H., & Meuser, F. (1986). Energieverbrauch und Energiefluß einer Maisstärkefabrik. *Starch - Stärke, 38*(4), 109–114. https://doi.org/10.1002/star.19860380402

Gómez-López, R. A., Montilla-Buitrago, C. E., Villada-Castillo, H. S., & Serna-Cock, L. (2025). Multidimensional study of starch co-plasticized with glycerol and isosorbide in TPS/PLA blends: Morphological, mechanical, thermal, and structural properties. *Journal of Applied Polymer Science, 142*(5), Article e56439. https://doi.org/10.1002/app.56439

González-Torres, B., Robles-García, M.Á., Gutiérrez-Lomelí, M., Padilla-Frausto, J. J., Navarro-Villarruel, C. L., Del-Toro-sánchez, C. L., Rodríguez-Félix, F., Barrera-Rodríguez, A., Reyna-Villela, M. Z., Avila-Novoa, M. G., & Reynoso-Marín, F. J. (2021). Combination of sorbitol and glycerol, as plasticizers, and oxidized starch improves the physicochemical characteristics of films for food preservation. *Polymers, 13* (19). https://doi.org/10.3390/polym13193356

Gul, K., Mir, N. A., Yousuf, B., Allai, F. M., & Sharma, S. (2021). In *Starch: An overview food biopolymers: Structural, functional and nutraceutical properties* (pp. 3–17). https://doi.org/10.1007/978-3-030-27061-2_1

Gupta, S., Bagdwal, S., & Agarwal, S. (2024). A comparative study between biodegradable and plastic waste packaging material. *Proceedings of the 3rd International Conference on Optimization Techniques in the Field of Engineering (ICOFE)-2024*. https://doi.org/10.2139/ssrn.5086897

Hansson, T. (2019). *End-of-life scenarios for bioplastic food and drinking packages*.

Hemachandra, S., Hadjikakou, M., & Pettigrew, S. (2024). A scoping review of food packaging life cycle assessments that account for packaging-related food waste. *International Journal of Life Cycle Assessment, 29*(10), 1899–1915. https://doi.org/10.1007/s11367-024-02349-z

Hottle, T. A., Bilec, M. M., & Landis, A. E. (2013). Sustainability assessments of bio-based polymers. *Polymer Degradation and Stability, 98*(9), 1898–1907.

Hottle, T. A., Bilec, M. M., & Landis, A. E. (2017). Biopolymer production and end of life comparisons using life cycle assessment. *Resources, Conservation and Recycling, 122*, 295–306. https://doi.org/10.1016/j.resconrec.2017.03.002

Hou, T., Ma, S., Wang, F., & Wang, L. (2023). A comprehensive review of intelligent controlled release antimicrobial packaging in food preservation. *Food Science and Biotechnology, 32*(11), 1459–1478. https://doi.org/10.1007/s10068-023-01344-8

Jafarzadeh, S., Alias, A. K., Ariffin, F., Mahmud, S., & Najafi, A. (2020). Characterization of active biodegradable films based on κ-carrageenan and starch incorporated with cinnamon essential oil and nanoclay. *International Journal of Biological Macromolecules, 152*, 1112–1120.

Jikui, W., Yanchang, L., Si, P., Yuming, W., Kai, W., Xu, Z., Diqi, Z., & Weihong, G. (2013). *All-degradable thermoplastic starch/polylactic acid blend material and preparation method thereof*.

Jiménez, A., Fabra, M. J., Talens, P., & Chiralt, A. (2012). Edible and biodegradable starch films: A review. *Food and Bioprocess Technology, 5*(6), 2058–2076. https://doi.org/10.1007/s11947-012-0835-4

Jozinović, A., Kovač, M., Ocelić Bulatović, V., Kučić Grgić, D., Miloloža, M., Šubarić, D., & Ačkar, D. (2024). Biopolymeric blends of thermoplastic starch and polylactide as sustainable packaging materials. *Polymers, 16*(9), Article 1268. https://doi.org/10.3390/polym16091268

Jwaideh, M. A., Sutanudjaja, E. H., & Dalin, C. (2022). Global impacts of nitrogen and phosphorus fertiliser use for major crops on aquatic biodiversity. *The International Journal of Life Cycle Assessment, 27*(8), 1058–1080. https://doi.org/10.1007/s11367-022-02078-1

Kakadellis, S., & Harris, Z. M. (2020). Don't scrap the waste: The need for broader system boundaries in bioplastic food packaging life-cycle assessment – A critical review. *Journal of Cleaner Production, 274*, Article 122831. https://doi.org/10.1016/j.jclepro.2020.122831

Kumar, N. M., Chopra, S. S., & Rajput, P. (2020). Life cycle assessment and environmental impacts of solar PV systems. *Photovoltaic Solar Energy Conversion*, 391–411. https://doi.org/10.1016/B978-0-12-819610-6.00012-0

Kusuma, I., Ma'arif, A., & Rahmadhia, S. N. (2024). Starch based for intelligent food packaging: A review. *BIO Web of Conferences, 146*, Article 01099. https://doi.org/10.1051/bioconf/202414601099

Lauer, M. K., & Smith, R. C. (2020). Recent advances in starch-based films toward food packaging applications: Physicochemical, mechanical, and functional properties. *Food Science and Food Safety, 19*(6), 3031–3083. https://doi.org/10.1111/1541-4337.12627

Li, H., & Huneault, M. A. (2011). Comparison of sorbitol and glycerol as plasticizers for thermoplastic starch in TPS/PLA blends. *Journal of Applied Polymer Science, 119*(4), 2439–2448. https://doi.org/10.1002/app.32956

Li, X., Liu, D., Pu, Y., & Zhong, Y. (2023). Recent advance of intelligent packaging aided by artificial intelligence for monitoring food freshness. *Foods, 12*(15), 2976. https://doi.org/10.3390/foods12152976

Liu, D., Zhao, P., Chen, J., Yan, Y., & Wu, Z. (2022). Recent advances and applications in starch for intelligent active food packaging: A review. *Foods, 11*(18). https://doi.org/10.3390/foods11182879

Loercks, J., Winfried, P., Schmidt, H., Timmermann, R., Grigat, E., & Wolfgang, S.-S. (1997). *Biodegradable polymeric mixtures based on thermoplastic starch.*

Lopez-Gil, A., Rodriguez-Perez, M. A., De Saja, J. A., Bellucci, F. S., & Ardanuy, M. (2014). Strategies to improve the mechanical properties of starch-based materials: Plasticization and natural fibers reinforcement. *Polímeros Ciência e Tecnologia*, 36–42. https://doi.org/10.4322/polimeros.2014.054

Lubwama, M., Yiga, V. A., & Lubwama, H. N. (2022). Effects and interactions of agricultural waste residues. *Biomass Conversion and Biorefinery, 12*(11), 4979–4999. https://doi.org/10.1007/s13399-020-01001-8

Luchese, C. L., Spada, J. C., & Tessaro, I. C. (2017). Starch content affects physicochemical properties of corn and cassava starch-based films. *Industrial Crops and Products, 109*, 619–626. https://doi.org/10.1016/j.indcrop.2017.09.020

Mafe, A. N., Edo, G. I., Othuke Akpoghelie, P., Joshua, O. A., Isoje, E. F., Igbuku, U. A., & Essaghah, A. E. A. (2024). Comparative analysis of the environmental impact of biopolymer-based and conventional plastic packaging in food engineering applications. *Al-Mustaqbal Journal of Sustainability in Engineering Sciences, 2*(2). https://doi.org/10.62723/2959-5932.1017

Malinconico, M. (2017). *Soil degradable bioplastics for a sustainable modern agriculture.* Springer.

Manoe, A. F, Claro, P. I, Mattoso, L. H, Marconcini, J. M, & Mantovani, G. L (2022). Biodegradable blends of thermoplastic waxy starch and poly(ε-caprolactone) obtained by high shear extrusion: Rheological, mechanical, morphological and thermal properties. *Journal of Polymer Research, 29*(7), Article 256. https://doi.org/10.21203/rs.3.rs-884067/v1

Marques, A., Martins, I. S., Kastner, T., Plutzar, C., Theurl, M. C., Eisenmenger, N., Huijbregts, M. A. J., Wood, R., Stadler, K., Bruckner, M., Canelas, J., Hilbers, J. P., Tukker, A., Erb, K., & Pereira, H. M. (2019). Increasing impacts of land use on biodiversity and carbon sequestration driven by population and economic growth. *Nature Ecology and Evolution, 3*(4), 628–637. https://doi.org/10.1038/s41559-019-0824-3

Menegat, S., Ledo, A., & Tirado, R. (2022). Greenhouse gas emissions from global production and use of nitrogen synthetic fertilisers in agriculture. *Scientific Reports, 12*(1), Article 14490. https://doi.org/10.1038/s41598-022-18773-w

Meuser, F., German, H., & Knobloch, G. (1981). Energiebedarf und Energiefluß in einer Kartoffelstärkefabrik. *Starch–Stärke, 33*(10), 332–337. https://doi.org/10.1002/STAR.19810331003

Mina, J. H., Valadez-González, A., Herrera-Franco, P., Zuluaga, F., & Delvasto, S. (2013). Preparación y caracterización físico-química y mecánica de mezclas ternarias de polilactida (PLLA), policaprolactona (PCL) y almidón termoplástico (TPS). *Revista Latinoamericana de Metalurgia y Materiales, 33*(1), 82–91.

Mkhari, T., Adeyemi, J. O., & Fawole, O. A. (2025). Recent advances in the fabrication of intelligent packaging for food preservation: A review. *Processes, 13*(2), Article 539. https://doi.org/10.3390/pr13020539

Molina-Besch, K. (2022). Use phase and end-of-life modeling of biobased biodegradable plastics in life cycle assessment: A review. *Clean Technologies and Environmental Policy, 24*(10), 3253–3272. https://doi.org/10.1007/s10098-022-02373-3

Molina-Botero, I. C., Gaviria-Uribe, X., Rios-Betancur, J. P., Medina-Campuzano, M., Toro-Trujillo, M., González-Quintero, R., Ospina, B., & Arango, J. (2024). Methane emission, carbon footprint and productivity of specialized dairy cows supplemented with bitter cassava (Manihot esculenta crantz). *Animals, 14*(1). https://doi.org/10.3390/ani14010019

Moshood, T. D., Nawanir, G., Mahmud, F., Mohamad, F., Ahmad, M. H., & Abdulghani, A. (2022). A literature review on sustainability of bio-based and biodegradable plastics: Challenges and opportunities. *Energy Engineering: Journal of the Association of Energy Engineering, 119*(4), 1611–1647. https://doi.org/10.32604/ee.2022.019028

Muller, J., González-Martínez, C., & Chiralt, A. (2017). Combination of poly (lactic) acid and starch for biodegradable food packaging. *Materials, 10*(8), Article 952. https://doi.org/10.3390/ma10080952

Muñoz-Gimena, P. F., Oliver-Cuenca, V., Peponi, L., & López, D. (2023). A review on reinforcements and additives in starch-based composites for food packaging. *Polymers, 15*(13), 2972. https://doi.org/10.3390/polym15132972

Nessi, S., Sinkko, T., Bulgheroni, C., Garcia-Gutierrez, P., Giuntoli, J., Konti, A., Sanye-Mengual, E., Tonini, D., Pant, R., Marelli, L., & Ardente, F. (2021). *Life cycle assessment (LCA) of alternative feedstocks for plastics production.* Publications Office of the European Union.

Nikmanesh, A., et al. (2023). Development and characterization of antioxidant and antibacterial films based on potato starch incorporating Viola odorata extract. *Foods, 12*(15), 2955.

Ostrander, B. M. (2014). Maize starch for industrial applications. *Industrial Crops: Breeding for BioEnergy and Bioproducts*, 171–189. https://doi.org/10.1007/978-1-4939-1447-0_9

Peelman, N., Ragaert, P., De Meulenaer, B., Adons, D., Roos, P., Cardon, L., Van Impe, F., & Devlieghere, F. (2013). Application of bioplastics for food packaging. *Trends in Food Science and Technology, 32*(2), 128–141. https://doi.org/10.1016/j.tifs.2013.06.003

Piemonte, V., De Falco, M., & Basile, A. (2013). Inside the bioplastics world: An alternative to petroleum-based plastics sustainable development. In *Chemical engineering innovative technologies* (pp. 181–198). Wiley. https://doi.org/10.1002/9781118629703.ch8

Ponnusamy, P. G., & Mani, S. (2022). Life cycle assessment of manufacturing cellulose nanofibril-reinforced chitosan composite films for packaging applications. *International Journal of Life Cycle Assessment, 27*(3), 380–394. https://doi.org/10.1007/s11367-022-02035-y

Poveda-Giraldo, J. A., Yoo, S., Yoo, S., Kim, I., Lan, K., Venditti, R., & Park, S. (2025). Environmental life cycle assessment of renewable starch alternative plastic: A comparative analysis of end-of-life scenarios in the Republic of Korea. *Journal of Cleaner Production, 520*, Article 146178. https://doi.org/10.1016/j.jclepro.2025.146178

Quadros, J., Epoxidation of vegetable oils and their use in plasticized polyvinyl chloride (PVC) applications BioProducts: Green materials for an emerging circular and sustainable economy. De Gruyter, Brazil De Gruyter, Brazil (2023), 293–306. doi: 10.1515/9783110791228-016.

Radfar, R., Hosseini, H., Farhoodia, M., Średnicka-Toberd, D., Ghasemie, I., Fathif, M., & Mousavi Khaneghah, A. (2020). Factors affecting mechanical properties of low-density polyethylene (LDPE)/starch blends: A review. *CRPJ Food Science Journal*, 6–26. https://doi.org/10.34302/crpjfst/2020.12.5.1

Ramesh, P., & Vinodh, S. (2020). State of art review on life cycle assessment of polymers. *International Journal of Sustainable Engineering, 13*(6), 411–422. https://doi.org/10.1080/19397038.2020.1802623

Reyniers, S., Ooms, N., Gomand, S. V., & Delcour, J. A. (2020). What makes starch from potato (Solanum tuberosum L.) tubers unique: a review. *Comprehensive Reviews in Food Science and Food Safety, 19*(5), 2588–2612. https://doi.org/10.1111/1541-4337.12596

Rojas-Bringas, P. M., De-la-Torre, G. E., & Torres, F. G. (2021). Influence of the source of starch and plasticizers on the environmental burden of starch-Brazil nut fiber biocomposite production: A life cycle assessment approach. *Science of the Total Environment, 769*, Article 144869. https://doi.org/10.1016/j.scitotenv.2020.144869

Rosenboom, J. G., Langer, R., & Traverso, G. (2022). Bioplastics for a circular economy. *Nature Reviews Materials,* 7(2), 117–137. https://doi.org/10.1038/s41578-021-00407-8

Rossi, V., Cleeve-Edwards, N., Lundquist, L., Schenker, U., Dubois, C., Humbert, S., & Jolliet, O. (2015). Life cycle assessment of end-of-life options for two biodegradable packaging materials: Sound application of the European waste hierarchy. *Journal of Cleaner Production, 86*, 132–145. https://doi.org/10.1016/j.jclepro.2014.08.049

Rouf Shah, T., Prasad, K., & Kumar, P. (2016). Maize—A potential source of human nutrition and health: A review. *Cogent Food & Agriculture, 2*(1), Article 1166995. https://doi.org/10.1080/23311932.2016.1166995

Salwa, H. N., Sapuan, S. M., Mastura, M. T., Zuhri, M. Y. M., & Ilyas, R. A. (2021). Life cycle assessment of bio-based packaging products. *Bio-based Packaging: Material, Environmental and Economic Aspects*, 381–411. https://doi.org/10.1002/9781119381228.ch22

Senila, L., Kovacs, E., Resz, M.-A., Senila, M., Becze, A., & Roman, C. (2024). Life cycle assessment (LCA) of bioplastics production from Lignocellulosic waste (study case: PLA and PHB). *Polymers, 16*(23), 3330. https://doi.org/10.3390/polym16233330

Shimazu, A. A., Mali, S., & Grossmann, M. V. E. (2007). Plasticizing and antiplasticizing effects of glycerol and sorbitol on biodegradable cassava starch films. *Semina: Ciências Agrárias, 28*(1), 79–88. https://doi.org/10.5433/1679-0359.2007v28n1p79

Shukla, V., Kandeepan, G., & Vishnuraj, M. R. (2015). Development of on package indicator sensor for real-time monitoring of meat quality. *Veterinary World, 8*(3), 393–397. https://doi.org/10.14202/vetworld.2015.393-397

Singh, G. P., Bangar, S. P., Yang, T., Trif, M., Kumar, V., & Kumar, D. (2022). Effect on the properties of edible starch-based films by the incorporation of additives: A review. *Polymers, 14*(10). https://doi.org/10.3390/polym14101987

Singh, S., Gamlath, S., & Wakeling, L. (2007). Nutritional aspects of food extrusion: A review. *International Journal of Food Science and Technology, 42*(8), 916–929. https://doi.org/10.1111/j.1365-2621.2006.01309.x

Sinha, S. (2024). An overview of biopolymer-derived packaging material. *Polymers from Renewable Resources, 15*(2), 193–209. https://doi.org/10.1177/20412479241226884

Siracusa, V., Blanco, I., Romani, S., & Scampicchio, M. (2021). Biobased packaging materials: New perspectives and challenges. *Foods, 10*(12), 2791.

Sobeih, M. O., Sawalha, S., Hamed, R., Ali, F., & Kim, M. P. (2025). Starch-derived bioplastics: Pioneering sustainable solutions for industrial use. *Materials, 18*(8), Article 19961944. https://doi.org/10.3390/ma18081762

Souza, M. A.d., Vilas-Boas, I. T., Leite-da-Silva, J. M., Abrahão, P.d. N., Teixeira-Costa, B. E., & Veiga-Junior, V. F. (2022). Polysaccharides in agro-industrial biomass residues. *Polysaccharides, 3*(1), 95–120. https://doi.org/10.3390/polysaccharides3010005

Spierling, S., Venkatachalam, V., Mudersbach, M., Becker, N., Herrmann, C., & Endres, H. J. (2020). End-of-life options for bio-based plastics in a circular economy-status quo and potential from a life cycle assessment perspective. *Resources, 9*(7). https://doi.org/10.3390/RESOURCES9070090

Sun, Y., Ju, Y., Xie, Q., Tao, R., Wang, L., Fan, B., & Wang, F. (2024). Active packaging film developed by incorporating starch aldehyde–quercetin conjugate into SPI matrix. *Antioxidants, 13*(7), 810. https://doi.org/10.3390/antiox13070810

Surendren, A., Hasan, Y., Mohanty, A. K., Abbassi, B., & Misra, M. (2024). Comparative environmental life cycle assessment on corn starch plasticization and co-plasticization processes. *Sustainability, 16*(17), 7406. https://doi.org/10.3390/su16177406

Suwanmanee, U., Leejarkpai, T., & Mungcharoen, T. (2012). Assessment of greenhouse gas (GHG) emissions of polylactic acid (pla)/starch and polyethylene terephthalate (PET) trays. *Advanced Materials Research, 518*, 1468–1474. https://doi.org/10.4028/www.scientific.net/AMR.518-523.1468

Suwanmanee, U., Varabuntoonvit, V., Chaiwutthinan, P., Tajan, M., Mungcharoen, T., & Leejarkpai, T. (2013). Life cycle assessment of single use thermoform boxes made from polystyrene (PS), polylactic acid, (PLA), and PLA/starch: Cradle to consumer gate. *International Journal of Life Cycle Assessment, 18*(2), 401–417. https://doi.org/10.1007/s11367-012-0479-7

Synani, K., Abeliotis, K., Kelly, V., Maragkaki, A., Manios, T., & Lasaridi, K. (2024). Environmental impact and sustainability of bioplastic production from food waste. *Sustainability, 16*(13), 5529. https://doi.org/10.3390/su16135529

Tan, Y., Liu, X., Cheng, Z., Zhan, Q., & Zhao, L. (2024). Preparation and characterization of bio-based freshness indicator labels loaded with natural pigments with high stability and sensitivity. *Foods, 13*(24), 4049. https://doi.org/10.3390/foods13244049

Thakur, R., Pristijono, P., Scarlett, C. J., Bowyer, M., Singh, S. P., & Vuong, Q. V. (2019). Starch-based films: Major factors affecting their properties. *International Journal of Biological Macromolecules, 132*, 1079–1089. https://doi.org/10.1016/j.ijbiomac.2019.03.190

Tong, C., Ma, Z., Chen, H., & Gao, H. (2023). Toward an understanding of potato starch structure, function, biosynthesis, and applications. *Food Frontiers, 4*(3), 980–1000. https://doi.org/10.1002/fft2.223

Trakulvichean, S., Lerdlattaporn, R., Lueangwattanapong, K., Chokmongkol, N., Saengchan, K., & Songkasiri, W. (2025). Carbon footprint of cassava starch in Thailand: A comprehensive study of production scale and production chain. *Cleaner and Responsible Consumption, 18*, Article 100293. https://doi.org/10.1016/j.clrc.2025.100293

Veiga, J. P. S., Popin, G. V., Cerri, C. E. P., & Romanelli, T. L. (2024). Carbon footprint of Brazilian agriculture based on field operations. *Agronomy, 14*(7). https://doi.org/10.3390/agronomy14071343

Versino, F., Ortega, F., Monroy, Y., Rivero, S., López, O. V., & García, M. A. (2023). Sustainable and bio-based food packaging: A review on past and current design innovations. *Foods, 12*(5). https://doi.org/10.3390/foods12051057

Villadiego, K. M., Tapia, M. J. A., Useche, J., & Macías, D. E. (2022). Thermoplastic starch (TPS)/polylactic acid (PLA) blending methodologies: A review. *Journal of Polymers and the Environment, 30*(1), 75–91. https://doi.org/10.1007/s10924-021-02207-1

Wu, Y., Tang, R., Guo, A., Tao, X., Hu, Y., Sheng, X., Qu, P., Wang, S., Li, J., & Li, F. (2023). Enhancing starch-based packaging materials: Optimization of plasticizers and process parameters. *Materials*. https://doi.org/10.2139/ssrn.4490110

Xu, X., Zhang, W., You, C., Fan, C., Ji, W., Park, J. T., Kwak, J., Chen, H., Zhang, Y. H. P. J., & Ma, Y. (2023). Biosynthesis of artificial starch and microbial protein from agricultural residue. *Science Bulletin, 68*(2), 214–223. https://doi.org/10.1016/j.scib.2023.01.006

Yadav, K., & Nikalje, G. C. (2024). Comprehensive analysis of bioplastics: Life cycle assessment, waste management, biodiversity impact, and sustainable mitigation strategies. *PeerJ, 12*(9), Article e18013. https://doi.org/10.7717/peerj.18013

Yates, M. R., & Barlow, C. Y. (2013). Life cycle assessments of biodegradable, commercial biopolymers - A critical review. *Resources, Conservation and Recycling, 78*, 54–66. https://doi.org/10.1016/j.resconrec.2013.06.010

Yu, J. K., & Moon, Y. S. (2021). Corn starch: Quality and quantity improvement for industrial uses. *Plants, 11*(1), Article 92. https://doi.org/10.3390/plants11010092

Zhai, X., Xue, Y., Sun, Y., Ma, X., Ban, W., Marappan, G., Tahir, H. E., Huang, X., Wu, K., Chen, Z., Zou, W., Liu, B., Zhang, L., Yang, Z., & Katona, J. (2025). Colorimetric food freshness indicators for intelligent packaging: Progress, shortcomings, and promising solutions. *Foods, 14*(16), 2813. https://doi.org/10.3390/foods14162813

Zhang, Y., Rempel, C., & McLaren, D. (2014). Thermoplastic starch. *Innovations in food packaging* (pp. 391–412). https://doi.org/10.1016/B978-0-12-394601-0.00016-3

CHAPTER SEVENTEEN

Safety aspects, challenges, and future outlook in starch-based packaging

Abhishek Parasar[1], Himangshu Deka[2], Pratap Kalita[1], Abdul Baquee Ahmed[3]

[1]Pratiksha Institute of Pharmaceutical Sciences, Guwahati, Assam, India
[2]Faculty of Pharmaceutical Science, Assam down town University, Guwahati, Assam, India
[3]School of Pharmaceutical Sciences, Girijananda Chowdhury University, Tezpur Campus, Guwahati, Assam, India

17.1 Introduction

Global plastic use has risen to critical levels, with annual production exceeding 368 million metric tons, making it a major source of environmental damage (Ayassamy, 2025). Nearly 79% of discarded plastic is accumulated in landfills or released into natural ecosystems, causing serious ecological impacts such as ocean contamination and threats to wildlife (Jambeck et al., 2015). The urgency for biopolymer-based packaging solutions has intensified as sustainability becomes a critical focus for both consumers and corporations. Among biobased polymers, starch is regarded as one of the most promising options for future material development. It is the second most abundant natural polysaccharide after cellulose (Tagliapietra et al., 2021) and offers several benefits, including good compatibility with other materials, ease of processing, and moderate-to-good barrier properties against oxygen and carbon dioxide (Iamareerat et al., 2018; Nagy et al., 2017).

As a widely available biopolymer, starch has been extensively explored as a sustainable alternative to petroleum-based plastics that contribute to environmental problems. Owing to its natural biodegradability and biocompatibility, starch-based materials present viable substitutes for conventional polymers and are also applied as edible coatings in food packaging (BeMiller & Whistler, 2009). Starch-based materials are promising alternatives to conventional polymers because of their natural biodegradability and biocompatibility, and they are already applied as edible coatings in food packaging. In addition, starch can be extracted from a wide range of plant

Starches in Smart Packaging
ISBN: 978-0-443-45396-0
https://doi.org/10.1016/B978-0-443-45396-0.00004-9

sources, each with varying ratios and molecular weights of amylose and amylopectin. These differences strongly influence its physicochemical properties, processability, and end-use applications. However, the inherent structure of starch also limits its industrial use as a replacement for traditional plastics. In particular, the presence of three hydroxyl groups per D-glycosyl unit makes starch-based films highly hydrophilic, which restricts their performance (Arruda et al., 2025). Furthermore, the brittleness of starch films is a significant weakness, as their mechanical properties are usually inadequate to maintain the film's structural integrity and resist deformation and damage during subsequent processing (Yang et al., 2024).

The three ways starch can be used in producing biodegradable films are as follows: First, small quantities of starch may be blended with conventional plastics to enhance the biodegradability of petroleum-based materials. Second, starch can be used to form composites in which it constitutes more than half of the total mass. Third, biodegradable films can be produced through extrusion processes that involve blends containing granular starch (Flieger et al., 2003).

The growing adoption of starch-based biodegradable materials in food packaging offers several benefits, including lower dependence on petroleum resources, improved environmental sustainability, and safe and reliable food protection (Liu et al., 2019). However, despite these advantages, biodegradable plastics still represent less than 1% of global plastic production. Their limited use is mainly due to higher production costs and inferior mechanical and barrier properties compared with conventional packaging materials. As performance demands for packaging films continue to increase, developing starch-based films with enhanced properties, such as antibacterial, antioxidant, and other multifunctional functions, holds strong potential for future applications (Liu et al., 2022).

Recent research on starch-based films has mainly focused on their sources, processing methods, and the associated opportunities and limitations of starch-based materials. However, there is still a lack of comprehensive reviews addressing the current status and applications of emerging technologies used to modify active starch-based packaging. Figure 17.1 presents a schematic diagram showing the sources and processing of starch into a film, and highlights its key advantages. This book chapter aims to present a comprehensive overview of the advancements, highlighting key strategies used to improve starch functionality, summarizing recent research developments, and discussing future perspectives for starch-based packaging. By focusing on environmentally friendly and safe modification approaches,

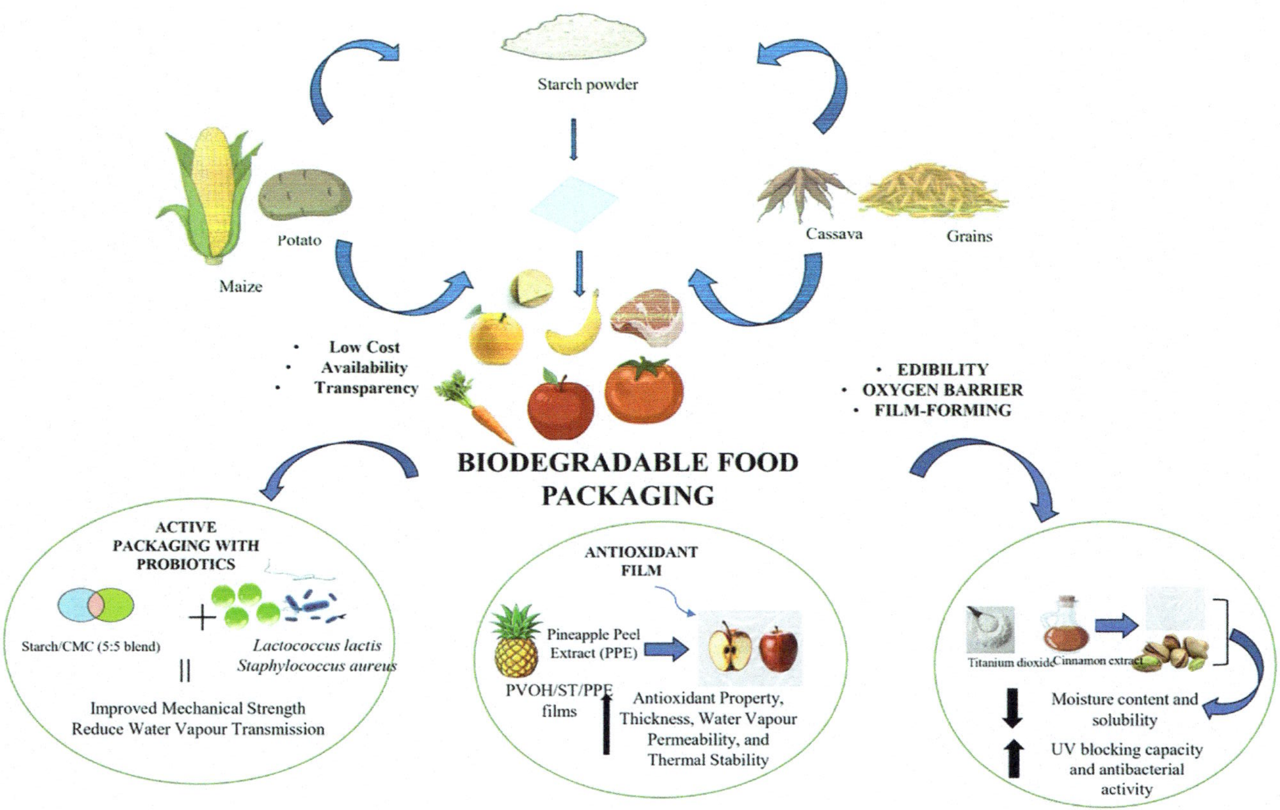

Figure 17.1 Development of starch based packaging.

this chapter underscores the growing potential of starch as a sustainable alternative to conventional plastics in food packaging applications.

17.2 Advancement of starch-based packaging

In recent times, there has been a growing interest in eco-friendly materials that has encouraged the development of biodegradable polymers from renewable sources, with starch emerging as a strong candidate. Modifying starch is a practical way to adjust its physical and mechanical behavior, which supports its use in active packaging, smart packaging, and edible film applications (Sohrabi et al., 2025; Pooja et al., 2024). Shabazi et al. reported that several physical and chemical modification methods have been used to address the limitations of starch. These approaches improve its functional performance while preserving its biodegradable and sustainable nature. As a result. The modified starch shows better thermal stability, stronger mechanical properties, lower water solubility, and improved ability to interact with other compounds (Shahbazi et al., 2018). Table 17.1 summarizes some recent studies on starch-based packaging.

In a recent study done by Du et al., gelatinization was highlighted as one of the common physical modification methods. In his study, when starch is heated with water, the granules take up moisture, expand, and lose their crystalline structure, which in turn forms a thick paste (Du et al., 2025). This transformation produces more uniform starch films and improves interactions between polymer chains, which leads to better flexibility (Wang et al., 2024).

To enhance starch performance and broaden its uses, chemical modifications such as esterification can be applied. In this method, ester groups are added to the starch structure, which alters its behavior and increases its hydrophobic nature (Sun et al., 2025).

Yudhistira et al. highlighted how emerging technologies such as cold plasma, high-pressure processing (HPP), ultrasound, and pulsed electric fields (PEF) can improve the performance of starch-based active packaging since they can reduce the carbon footprint of the product regardless of whether their application needs to be optimized (Yudhistir et al., 2024). Cold plasma can modify starch through processes such as reticulation, cross-linking, etching, or adding functional groups. It leaves no chemical residues and is suitable for heat-sensitive materials (Sifuentes-Nieves et al., 2019).

Cold plasma contains reactive species such as electrons, ions, and free radicals. Its effect on starch depends on the plasma type and the starch's

Table 17.1 Recent advancements in starch-based packaging materials.

Sl. No.	Strategy/Approach	Modification/Additives	Key improvement observed	Packaging application	Reference
1	Physical modification	Gelatinization	Improved film uniformity, flexibility, and polymer chain interaction	Active, edible films	Du et al. (2025), Wang et al. (2024)
2	Chemical modification	Esterification	Increased hydrophobicity and altered starch behavior	Active packaging	Sun et al. (2025)
3	Emerging nonthermal technologies	Cold plasma	Enhanced mechanical properties, reduced solubility, increased hydrophobicity, improved chemical and morphological features	Active packaging films	Sifuentes-Nieves et al. (2019) Goiana et al. (2024)
4	Emerging nonthermal technologies	Pulsed electric fields (PEF)	Strengthened cross-links, improved biopolymer interactions without chemicals	Biobased composite films	Giteru et al. (2021)
5	Physical modification	Ultrasound	Disrupted polymer networks, improved physicochemical properties	Starch-based films	Raghunathan et al. (2021)

(*Continued*)

Table 17.1 Recent advancements in starch-based packaging materials.—cont'd

Sl. No.	Strategy/Approach	Modification/Additives	Key improvement observed	Packaging application	Reference
6	Nonthermal processing	High pressure processing (HPP)	Modified structural and physicochemical properties via cavitation	Packaging films	Raghunathan et al. (2021)
7	Active packaging agents	Essential oils (cinnamon, clove, potassium sorbate) with ZnO nanoparticles	Enhanced antimicrobial, UV barrier, and mechanical properties; reduced tensile strength at higher concentrations	Antimicrobial food packaging	Raigond et al. (2019)
8	Pickering emulsions	Nanocellulose-stabilized essential oils (ho wood, cardamom, cinnamon)	Improved mechanical strength, thermal stability, reduced crystallinity, controlled release	Biodegradable active packaging	Souza et al. (2021)
9	Bioactive fillers	Mango puree, mango peel, pineapple pomace	Increased antioxidant, antimicrobial activity, phenolic content, and biodegradability	Edible packaging	Susmitha et al. (2021)
10	Agro-industrial nanofillers	Carbon dots from pineapple leaf waste	Improved mechanical strength, UV barrier, antioxidant, and antimicrobial properties; extended shelf life	Active meat packaging	Kuchaiyaphum et al. (2024)

11	Nanocomposite reinforcement	Zinc oxide nanoparticles (1%–3%)	Improved mechanical strength, antibacterial activity, UV, and water vapor barrier performance	Active packaging films	Dhiman et al. (2025)
12	Smart packaging systems	Freshness indicators and pH-sensitive dyes	Visual color change in response to spoilage metabolites	Intelligent packaging	Sharaby et al. (2024), Asikkutlu and Yildirim-Yalcin (2025)
13	Intelligent bioactive films	Quaternary ammonium chitosan and anthocyanins	pH-responsive color change, UV blocking, antioxidant, and antibacterial	Intelligent food packaging	Xie et al. (2025)

own properties. It can modify starch through cross-linking, depolymerization, surface erosion, or the addition of new functional groups. These changes improve the structure and performance of starch granules, making them more suitable for packaging applications (Ma & Jiang, 2024). In a study by Goiana et al., cold plasma was applied at 100, 200, and 300 Hz for 0, 10, and 20 min to three types of starch-based films: films made from plasma-treated starch, films in which the film-forming solution was treated, and films treated after formation. The treatment produced marked changes in the physical, chemical, and morphological properties of the samples. The strongest chemical effects occurred when starch was treated before film formation, while films made from untreated starch showed mostly physical changes. Plasma-treated films also became less soluble and more hydrophobic (Goiana et al., 2024).

PEF treatment can also be used to produce composite biobased films by changing biomacromolecule interactions, including molecular polarization and electrochemical reactions that generate free radicals, which strengthen cross-links and promote bioconjugations (Giteru et al., 2021). PEF is a low-temperature, chemical-free approach for altering biomacromolecules.

Ultrasound is considered one of the most effective physical methods for modifying starch-based films. It relies on sound waves and can serve as an alternative to chemical treatments. HPP is another nonthermal technique used for liquids or solids. Through cavitation, it disrupts starch polymer networks and changes their physicochemical and structural properties (Raghunathan et al., 2021).

Considerable research has focused on overcoming the drawbacks of starch-based packaging, particularly its poor resistance to external factors and limited mechanical strength, which commonly limit its application in food packaging. Barrier performance is a critical aspect of packaging effectiveness, especially for protecting food and other sensitive products. Therefore there is a clear need to develop approaches that enhance the durability, resistance, and overall functionality of starch-based materials (Yang et al., 2025; Zhang et al., 2025).

In recent years, adding functional agents such as antioxidants and antimicrobial compounds to starch-based films has become an important approach to enhance their performance. These active packaging materials act as semipermeable barriers, helping to prolong shelf life by limiting moisture transfer, reducing solute loss associated with fruit respiration, and slowing oxidative reactions. Essential oils are a prominent example of such active components, as their antioxidant and antimicrobial properties contribute significantly to improving food quality and safety (Susmitha et al., 2021; Yao et al., 2021).

In a study by Raigond et al., biodegradable and antimicrobial nanocomposite films were developed from potato starch using a fixed concentration of zinc oxide nanoparticles through the casting technique. Three antimicrobial agents, cinnamon oil, clove oil, and potassium sorbate, were incorporated into the films, and their activity was evaluated against *S. aureus*, *E. coli*, *Salmonella typhi*, and *Campylobacter jejuni*. Films containing clove oil showed the highest inhibition against *S. aureus* (22%–100%), while cinnamon oil-based films were most effective against *C. jejuni* (19%–22%). The strongest inhibition of *E. coli* growth (33%–40%) was observed in films containing potassium sorbate. Complete suppression of *C. jejuni* and *E. coli*, however, required higher levels of cinnamon oil and potassium sorbate. An increase in antimicrobial concentration led to a reduction in tensile strength, with decreases of up to 13% for cinnamon oil films, 23% for clove oil films, and 34% for potassium sorbate films. Overall, cinnamon oil proved to be the most suitable antimicrobial agent, providing broad antibacterial activity while having the least negative impact on film tensile strength (Raigond et al., 2019).

Also, Souza et al. prepared thermoplastic starch films containing nanocellulose-stabilized pickering emulsions of hawthorn, cardamom, and cinnamon essential oils by the casting method and evaluated their suitability as packaging materials. The films were characterized for chemical interactions, mechanical, crystallinity, thermal, and water vapor barrier properties. Ho wood-based films exhibited strong molecular interactions, resulting in improved mechanical strength, whereas cardamom and cinnamon films showed lower compatibility and reduced tensile strength. All films displayed reduced crystallinity compared to neat TPS, indicating changes in starch chain organization. Enhanced thermal stability was observed due to strong interactions between starch and the emulsions, supporting their safe use in food packaging. In addition, reduced water vapor transmission suggests improved controlled release of active compounds. Overall, ho wood essential oil-based films showed the greatest potential as` biodegradable active packaging materials (Souza et al., 2021).

Also, Susmitha et al. evaluated corn starch (CS)–gelatin films enriched with mango puree, mango puree with peel, and pineapple pomace (5%–15% w/v) for active packaging applications. The additives enhanced physicochemical properties and increased antioxidant and antimicrobial activity, and total phenolic content, while mechanical strength was highest in the control films. FTIR confirmed good component compatibility, TGA showed thermal behavior similar to the control, and all films achieved over 50% biodegradability within 15 days. These fruit-based biodegradable films show promising potential for edible food packaging (Susmitha et al., 2021).

Using agroindustrial by-products as sources of nanofillers offers an effective approach for developing starch-based nanocomposite films. For example, pineapple leaf waste can serve as a novel precursor for synthesizing carbon dots through a simple hydrothermal process. Kuchaiyaphum et al. incorporated these carbon dots with an average size of 2.36 ± 0.33 nm into pineapple stem starch-based active packaging films, resulting in significant improvements in mechanical strength and UV barrier performance, along with enhanced antioxidant and antimicrobial activities. These improvements contributed to extending the shelf life of fresh pork (Kuchaiyaphum et al., 2024).

Additional nanocomponents can be incorporated into starch-based matrices to enhance performance and impart bioactive functions. In this context, Dhiman et al. reported the development of biodegradable starch-based nanocomposite films containing zinc oxide nanoparticles. When ZnO nanoparticles were added at concentrations of 1%, 2%, and 3%, the resulting films exhibited improved mechanical strength, antibacterial activity, and UV barrier properties. A marked improvement in water vapor barrier performance was also observed, with values increasing from 1.16 g mm/m^2 day kPa for neat starch films to 2.25 g mm/m^2 day kPa after nanoparticle incorporation (Dhiman et al., 2025).

One of the key applications of starch-based smart packaging is the use of freshness indicators that undergo color changes in response to metabolites released during food spoilage, such as volatile amines in meat (Sharaby et al., 2024). In addition, starch-based films can incorporate pH-sensitive colorimetric sensors that visually signal the deterioration of food products (Asikkutlu & Yildirim-Yalcin, 2025).

For example, Xie et al. developed a starch-based composite film containing quaternary ammonium chitosan and *Lycium ruthenicum* anthocyanins by a simple casting method for intelligent food packaging. The optimized film exhibited good mechanical strength, reduced water sensitivity, strong UV blocking, antioxidant and antibacterial activity, and clear pH-responsive color changes from pink to green (pH 2–12). When applied to shrimp, the film extended shelf life by 16 h while enabling real-time freshness monitoring, demonstrating its potential as a multifunctional intelligent packaging material (Xie et al., 2025).

17.3 Case studies on starch-based food packaging

Starch is a polysaccharide composed of amylose and amylopectin and is widely sourced from maize, potatoes, cassava, and grains. It is one of the promising biodegradable polymers for food packaging because of its low cost, availability, edibility, transparency, and excellent oxygen barrier and film-forming properties (Perera et al., 2023). Recent progress in starch-based materials for food packaging comes from new preparation methods, improved equipment, and the use of nanoscale modifications. These developments have made it possible to create materials with unique features and reliable performance.

Research done by Lan et al. found that starch/carboxymethyl cellulose (NS/CMC) composite films incorporated with *Lactococcus lactis* offer promising active packaging solutions. A 5:5 NS/CMC blend with 1.5% *Lactococcus lactis* yielded films of superior performance, marked by enhanced strength, which reduced water vapor transmission and stable probiotic viability for 30 days at 4°C. The system released up to 3.35 mg/mL nisin, which effectively inhibits *Staphylococcus aureus*. SEM and FTIR confirmed uniform bacterial incorporation via hydrogen bonding. Despite minor reductions in strength at higher *L. lactis* concentrations, the films remained suitable for packaging dry foods, highlighting their potential as sustainable, antimicrobial packaging materials (Lan et al., 2021).

Kumar et al. examine the fabrication of polyvinyl alcohol (PVOH) and corn starch (ST) films enriched with pineapple peel extract (PPE) as a natural antioxidant agent, which is an abundant by-product from the food processing industry, via the casting method. PPE concentrations from 5% to 20% were tested for their effects on antioxidant activity, appearance, thermal behavior, mechanical strength, and barrier properties. Higher PPE levels slightly increased film thickness, water vapor permeability, and flexibility, while also improving thermal stability and antioxidant capacity. As a result, PVOH/ST/PPE films showed good potential as active biodegradable packaging for food prone to oxidation (Kumar et al., 2021)

Another study done by Arezoo et al. demonstrated that incorporating cinnamon essential oil (CEO) and nano titanium dioxide (TiO_2–N) into sago starch films significantly improved their antimicrobial and functional properties. While the CEO enhanced flexibility, TiO_2–N improved mechanical strength and reduced water and oxygen permeability. The combined additives decreased the moisture content and solubility, which in turn improved UV-blocking capacity and provided strong antibacterial activity against *E. coli, S. aureus*, and *S. typhimurium*. These synergistic effects make the film suitable for active, biodegradable food packaging, especially for fresh pistachio packaging (Arezoo et al., 2020).

Song et al. examined corn-wheat starch films containing lemon essential oil (LO) and the surfactants Span 80 or Tween 80. LO reduced water content, transparency, whiteness, permeability, solubility, and tensile strength, while increasing elongation and haze. Higher LO levels improved antibacterial activity and altered film structure. Surfactants further boosted antimicrobial effects and solubility but lowered strength and transparency. Tween 80 produced films with higher moisture, solubility, and WVP than Span 80. As a result, these formulations show potential for food packaging (Song et al., 2018). In another study, Lozano-Navarro et al. examined polysaccharide composite films to assess their physical, optical, antimicrobial, and thermal properties after adding a natural antioxidant such as anthocyanins, betalains, resveratrol, thymol, and carvacrol. Only slight FTIR changes occurred because of the low extract levels, but spectroscopic tests and Folin-Denis analysis confirmed the antioxidants' presence. SEM showed no major changes in film morphology, while TGA indicated that extract type affected weight loss. Overall, incorporating natural antioxidants improved key characteristics of chitosan—starch films and supports their use in food packaging (Lozano-Navarro et al., 2018). Also, Uttami et al. tested tapioca starch–glycerol films containing 0%–2% cinnamon bark essential oil to see how they performed and how well they preserved fresh beef. The oil did not change film thickness but reduced water vapor transmission, increased tensile strength up to 1.5%, lowered elongation, and improved antibacterial activity. The 2% oil film was most effective, slowing microbial growth and lipid oxidation and keeping beef fresh for up to 15 days (Utami et al., 2019). Aitboulahsen et al. examined how different plasticizers influence the physicochemical and mechanical properties of films made from gelatin, pectin, starch, and their blends (gelatin/pectin, gelatin/starch, starch/pectin), prepared via casting. Among the six formulations, the gelatin/pectin film plasticized with sorbitol exhibited superior performance, showing tensile strength of 52.46 MPa, elongation at break of 19.80%, moisture content of 19.96%, water vapor permeability of 3.24×10^{7} $g\,s^{-1}m^{-1}Pa^{-1}$, and water resistance of 19.96%. Incorporation of *Mentha pulegium* and *Lavandula angustifolia* essential oils into this film enhanced antibacterial activity and reduced water vapor permeability, though mechanical strength decreased (Aitboulahsen et al., 2020).

de Souza et al. tested thermoplastic starch films containing a hybrid of carvacrol essential oil and montmorillonite at 4.5, 9, and 15 wt%. FTIR and Raman showed new hydrogen bonds with starch, altering thermal behavior. XRD indicated slight crystallinity changes, with the 15% film becoming more amorphous and allowing better essential oil diffusion. All hybrid films inhibited *E. coli*, with the strongest effect in the 15% sample because of improved oil release and greater disruption of bacterial cell membranes. These results show the hybrid compound is a promising antimicrobial additive for starch films used in fresh food packaging (de Souza et al., 2020). In another study, de Oliveira Filho et al., found that, Arrowroot starch (AA)-based films incorporated with carnauba wax nanoemulsion (CWN), cellulose nanocrystal (CNC) and essential oil (Eos) from Mentha spicata (MEO) and Cymbopogon martini (CEO) were produce using the casting technique evaluated for barrier, mechanical, thermal, optical, and microstructural properties, as well as antifungal activity. Adding CNCs reduced moisture content and water vapor permeability, while adding essential oils lowered transparency and altered film structure. Both oils improved thermal stability and provided strong protection against Rhizopus stolonifer and Botrytis cinerea. With effective moisture, light, and fungal resistance, these films show good potential for active food packaging or coating applications (de Oliveira Filho et al., 2021). Also, Cai et al. demonstrated in research that thyme essential oil microcapsules (TEO-M) were incorporated into starch films via casting to enhance mango preservation. SEM showed intact microcapsules, while FTIR indicated hydrogen bonding with starch. TEO-M films had lower thermal stability but inhibited *Botryodiplodia theobromae* and *Colletotrichum gloeosporioides*. They increased opacity, thickness, tensile strength, and water solubility, while reducing elongation, moisture, and water vapor permeability. Mangoes wrapped in TEO-M films had a shelf life of 10 days at 25 °C, demonstrating their potential for fruit preservation (Cai et al., 2020).

In a recent study done by Chandrasekar et al., starch-based bioactive thermoplastic films were developed using FeO and ZnO nanoparticles synthesized via clove extract, exhibiting strong oxygen scavenging and antimicrobial properties, respectively. A double-layer nano-biothermoplastic system was formed, with nanoparticle distribution and diffusion influenced by starch's amorphous structure and film swelling. The packaging effectively maintained microbial, chemical, and color quality of mutton and chicken at 4 °C, demonstrating its potential for extending food shelf life (Chandrasekar et al., 2024). Also, Deng et al. developed a biodegradable antibacterial film from amylose starch and HTCC for meat preservation. Molecular analysis revealed intermolecular hydrogen bonding between amylose and HTCC, cross-linked via glutaraldehyde, enhancing mechanical properties with tensile strength and elongation increased by 206.7% and 109.5%, respectively. The films exhibited strong antibacterial activity, low cytotoxicity, and reduced UV transmittance, effectively improving meat shelf life. These characteristics highlight HTCC/amylose films as promising candidates for sustainable antibacterial food packaging (Deng et al., 2024). Li et al. developed CS-based bioactive films with 0%–30% *Rosa roxburghii* Tratt fruit pomace (RRTP) as antioxidant fillers using extrusion compression molding. Incorporation of RRTP significantly enhanced the films' barrier, mechanical, and antioxidant properties. The 30% RRTP composite showed the lowest water vapor and oxygen permeability (2.94×10^{-10} g m m^{-2}·s^{-1} Pa^{-1} and 0.87×10^{-14} kg m m^{-2}·s^{-1} Pa^{-1}), highest tensile strength (3.41 MPa), and strongest DPPH scavenging activity (88.59%). LC-MS confirmed abundant phenolics in RRTP, while FTIR, XRD, and SEM analyses indicated good compatibility and hydrogen-bond formation with starch. Packaging mushrooms in the CS/30% RRTP film maintained freshness for 12 days at 4 °C, highlighting its potential as antioxidant active packaging (Li et al., 2025). Also, Aswathy et al. developed a triple-extract cassava starch-based film (SPFPOMC) incorporating passion fruit peel, pomegranate peel, and cinnamon bark to overcome the limitations of single-extract systems. The optimized film exhibited enhanced tensile strength (6.64 MPa), moderate elongation (13.78%), low water vapor permeability (8.66×10^{-12} g/m Pa s), high contact angle (98 degrees), sustained antioxidant activity, UV protection, and improved antibacterial and antifungal efficacy. SPFPOMC demonstrated thermal stability up to 70 °C, high biocompatibility (93.25% cell viability), rapid biodegradability, and effectively extended the shelf life of strawberries, offering a sustainable alternative to conventional plastic packaging while valorizing fruit waste (Aswathy et al., 2025). In another study, Orhotohwo et al. developed a potato starch (PS) and κ-carrageenan (KCa) composite film with broccoli by-product powder (BBPP) and extract (BBPE) for active food packaging. Structural analysis confirmed strong polymer compatibility and intermolecular interactions. BBPP improved tensile strength (43.16%) and Young's modulus (50.13%), while BBPE enhanced flexibility (58.27%) and antimicrobial activity, reducing *E. coli* and *S. aureus* counts. Applied to fresh-cut kiwis, BBPE-based coatings minimized weight loss, preserved soluble solids, and maintained bacterial counts below 4 Log CFU/g over 11 days at 4 °C, demonstrating their potential as sustainable, active packaging materials (Orhotohwo et al., 2025). Also, Yassaroh et al. developed a composite film from sago starch and microcrystalline cellulose (MCC), with glycerol as a plasticizer and ginger extract as an antioxidant. Ultrasonication reduced MCC particle size and improved dispersion, enhancing mechanical and water vapor barrier properties. The antioxidant films effectively inhibited browning and shrinkage in sliced apples stored at 4 °C for 10 days. The films were fully biodegradable within 2 weeks, demonstrating their potential as sustainable, food-grade packaging materials. Table 17.2 represented the summary of research on starch-based films for food packaging.

Tan et al. developed a pH-sensitive intelligent film using sodium carboxymethylcellulose, sodium alginate, and cassava starch, incorporating *Lycium ruthenicum* anthocyanins (LRA) and tea polyphenols (TP) to monitor pork freshness and enhance shelf life. LRA provided strong pH responsiveness across a wide range, while the addition of LRA and TP increased film thickness, reduced water content and water vapor permeability, and improved light stability. FTIR confirmed good polymer compatibility and successful incorporation of bioactive compounds. TP significantly boosted antioxidant capacity, and the films produced clear color changes in response to volatile basic nitrogen during pork spoilage. The formulation containing 0.2% TP effectively slowed lipid oxidation and extended pork freshness. These findings highlight the potential of CSC/LRA/TP films as antioxidant, freshness-indicating materials for intelligent pork packaging (Tan et al., 2025).

Cryogels made from starch extracted from Brazilian purple yam were developed by Rosas et al., with and without citric acid to serve as active packaging components for fresh cheese. The addition of citric acid increased cryogel density and reduced porosity, while all formulations showed high water absorption capacity and biodegradability. When applied to cheese contaminated with *Listeria monocytogenes* and *Escherichia coli*, the cryogels fully absorbed exudate and exhibited antimicrobial activity, completely inhibiting *E. coli* and partially inhibiting *L. monocytogenes*. These findings demonstrate the potential of starch-based cryogels as a sustainable and effective approach for improving the safety and quality of fresh cheeses (Rosas et al., 2025).

Di Matteo et al. developed starch-based thermoplastic biopolymer films by incorporating thymol and carvacrol (5% and 8% w/w) to create biodegradable active packaging with antioxidant and antimicrobial properties. The functionalized films retained their key physicochemical and thermal

Table 17.2 Summary of research on starch-based films for food packaging.

Sl. No.	Material/Additives	Key functional improvements	Antimicrobial/Antioxidant effect	Application	Ref.
1	Starch/CMC film with *Lactococcus lactis*	Better strength, lower WVP, stable probiotics	Released 3.35 mg/mL nisin; inhibited *S. aureus*	Suitable for dry-food packaging; uniform dispersion confirmed by SEM/FTIR	Lan et al. (2021)
2	PVOH/corn starch with pineapple peel extract	Higher thermal stability, increased flexibility	Strong antioxidant activity	Active biodegradable films for oxidation-prone foods	Kumar et al. (2021)
3	Sago starch with cinnamon oil and nano-TiO_2	Lower WVP/O_2 permeability, stronger mechanics	Strong activity versus *E. coli*, *S. aureus*, and *S. typhimurium*	Effective packaging for fresh pistachios	Arezoo et al. (2020)
4	Corn–wheat starch with lemon oil + Span/Tween	Lower transparency, strength; higher elongation	Strong antibacterial activity enhanced by surfactants	Potential for general food packaging	Song et al. (2018)
5	Chitosan–starch with natural antioxidants	Minor FTIR changes; improved thermal behavior	Antioxidant extracts confirmed via spectroscopic tests	Enhanced functional properties for food packaging	Lozano-Navarro et al. (2018)
6	Tapioca starch–glycerol with cinnamon oil	Lower WVP, higher tensile strength (1.5%)	Better inhibition of spoilage bacteria	Preserved beef quality for 15 days	Utami et al. (2019)
7	Gelatin–pectin/starch blends + essential oils	Sorbitol-plasticized gelatin/pectin showed best mechanics	Mentha and Lavandula oils improved antibacterial activity	Suitable for biodegradable antimicrobial films	Aitboulahsen et al. (2020)
8	Thermoplastic starch + carvacrol–montmorillonite hybrid	Modified crystallinity; improved thermal behavior	All films inhibited *E. coli*, strongest at 15% loading	Promising antimicrobial additive for fresh foods	de Souza et al. (2020)

9	Arrowroot starch + CWN, CNC, MEO/CEO	Lower WVP, higher stability; reduced transparency	Strong antifungal action versus *R. stolonifer*, *B. cinerea*	Candidate for coatings and active packaging	de Oliveira Filho et al. (2021)
10	Starch + thyme oil microcapsules	Higher strength, opacity; lower WVP	Inhibited major mango pathogens	Extended mango shelf life to 10 days	Cai et al. (2020)
11	Starch thermoplastics with FeO and ZnO NPs	Strong oxygen-scavenging; enhanced diffusion	Broad antimicrobial effect	Maintained the quality of mutton/chicken at 4°C	Chandrasekar et al. (2024)
12	Amylose + HTCC cross-linked with glutaraldehyde	Strength and elongation increased >100%	Strong antibacterial activity; low cytotoxicity	Improved preservation of meat products	Deng et al. (2024)
13	Corn starch + 0–30% RRTP	Lowest WVP/O_2 permeability at 30%; strongest tensile strength	High DPPH activity (88.59%)	Preserved mushrooms for 12 days at 4°C	Li et al. (2025)
14	Cassava starch with triple fruit-waste extracts	High TS (6.64 MPa), low WVP, good UV barrier	Strong antibacterial and antifungal activity	Extended strawberry shelf life; highly biodegradable	Aswathy et al. (2025)
15	Potato starch + κ-carrageenan + broccoli by-products	BBPP ↑ strength; BBPE ↑ flexibility	BBPE reduced *E. coli* and *S. aureus*	Preserved fresh-cut kiwi for 11 days	Orhotohwo et al. (2025)
16	Sago starch + MCC + ginger extract	Better mechanics and barrier properties	Antioxidant activity delayed browning	Preserved apple slices for 10 days; fully biodegradable	Yassaroh et al. (2025)
17	Cassava starch–CMC–alginate + LRA + tea polyphenols	Lower WVP; better light stability	Strong antioxidant action; pH-based color shift	Intelligent packaging for pork freshness	Tan et al. (2025)
18	Purple yam starch cryogels ± citric acid	Higher density, strong exudate absorption	Complete inhibition of *E. coli*	Improved safety of fresh cheese	Rosas et al. (2025)
19	Thermoplastic starch + thymol/carvacrol	Lower WVP; preserved thermal stability	>99.9% pathogen reduction; strong radical scavenging	Extended shelf life of Brie cheese	Di Matteo et al. (2026)

characteristics, with the main change being reduced water vapor permeability, particularly in films containing 5% and 8% thymol. Both compounds delivered strong radical-scavenging activity, exceeding 50% for $DPPH^{\bullet}$ and 90% for $ABTS^{\bullet+}$. Films with 8% carvacrol or thymol achieved more than 99.9% reduction in major foodborne pathogens and spoilage organisms, while *Pseudomonas aeruginosa* showed a 90%–99% reduction. When applied to Brie cheese, the 8% active films maintained pronounced antimicrobial effects after 7 and 10 days of storage, without inhibiting lactobacilli. These results highlight their potential as sustainable active packaging materials for extending the shelf life of high-moisture foods (Di Matteo et al., 2026).

17.4 Safety measurement for starch-based packaging

Biodegradable starch-based packaging materials, made from renewable resources such as corn, potato, tapioca, wheat, and rice, offer an alternative to petroleum-based plastics, reducing microplastic pollution and waste disposal in landfills (Kumari et al., 2022). Safety testing is first of all migration testing to prevent the transfer of toxicants into food, followed by a number of tests for mechanical resistance, barrier properties, biodegradability under controlled conditions, and, finally, full toxicity testing; this according to strict regulations, including FDA 21 CFR 176/177, EU Regulations 1935/2004 and 10/2011, EN 13432 for composability, and the latest ASTM D6400 and arising standards relating to microplastics in 2025. These methods neutralize starch's inherent hydrophilic nature, which naturally promotes water absorption, as well as the possibilities of leaching with the addition of other additives such as glycerol, sorbitol, citric acid, plasticizers, or reinforcements such as CNC, MMT clay, or ZnO nanoparticles (Gupta et al., 2024).

17.4.1 Migration testing protocols

Migration testing is employed for the quantification of the additives and other degradative materials that diffuse from the starch films into food simulants under stimulatory conditions that mimic the storage and abuse scenarios (Mandliya et al., 2024). Standard protocol requires the immersion of the film samples, usually pieces of 10 cm^2 dipped into 10 mL simulants that includes for aqueous food deionized water, for acidic foods 3% acetic acid, 10% for low-alcohol item is ethanol, and for fatty foods) 95% ethanol. According to the EU Regulation October 2011, the film piece is immersed for a period of up to 10 days, maintaining temperatures from 5°C to 60°C.

The overall Migration Limit for foods is set at 60 mg/kg food, complemented by Specific Migration Limits for key plasticizers, such as glycerol at 30 mg/kg (Wongphan et al., 2025).

After completion of the immersion period, the starch films undergo several analytical procedures, such as evaporation of the solvent, weighing of the residue, and, for quantification of the additives and other materials, advanced analytical techniques like high-performance liquid chromatography, gas chromatography coupled to mass spectrometry, or Fourier-transform infrared spectroscopy are used. Glycerol and sorbitol were found to be the major migrants in TPS owing to their hydrophilic properties. They migrated from 20% to 50% of their initial content, while the migration values were still below the threshold limit of the regulatory bodies. In addition, no heavy metals or phthalates were reportedly detected. Important factors are film thickness, which has an inverse relationship with migration; temperature, upon which migration increases exponentially according to Arrhenius kinetics; pH, because films swell in acidic simulants; and prolonged contact time. The addition of nanocomposite reinforcement—subsequently, clay or zinc oxide nanoparticles—further reduces the permeability, increasing barrier properties while still maintaining safety. Native components underpin the low-risk profiles, since starch has a Generally Recognized as Safe (GRAS) status from the FDA (Garavito et al., 2024; Kumar et al., 2024).

17.4.2 Cytotoxicity and toxicological evaluations

In vitro cytotoxicity studies validate the biological safety of migrated substances using cell models relevant to oral exposure pathways. Human Caco-2 intestinal epithelial cells, Huh7 hepatocytes, and IHH immortalized human hepatocytes receive DMSO-solubilized extracts at concentrations of 0.04%–0.4% v/v or direct film dissolutions at 2.5%–10% v/v for 72 h, with mitochondrial viability assessed through resazurin reduction assays yielding over 90% survival relative to low-density polyethylene controls. Certain starch extracts, particularly from carboxymethyl cellulose-starch blends, even induce subtle cell proliferation between 105% and 120%, attributable to benign oligosaccharide release mimicking nutrient effects (Tan et al., 2024).

17.4.3 Biodegradation and composability standards

Biodegradation and compostability of starch-based films are evaluated against internationally harmonized standards that establish performance requirements for safe and complete environmental degradation. In the United

States, ASTM D6400 stipulates that certified compostable plastics must achieve at least 90% biodegradation within 180 days under industrial composting conditions, demonstrate substantial disintegration within 12 weeks, and comply with ecotoxicity and heavy-metal guidelines (ASTM International, 2021). Degradation behavior under various environmental conditions is further characterized by supporting test techniques, such as ASTM D5338 for aerobic composting and ASTM D5988 for soil biodegradation. Similar international standards, such as ISO 17088, ISO 14855, and ISO 20200, demand total disintegration, $\geq$90% mineralization in comparison to cellulose controls, and no phytotoxic residues (International Organization for Standardization, 2018, 2021a, 2021b). The main European standard for certifying compostable packaging is EN 13432, which has standards for chemicals, biodegradation, disintegration, and ecotoxicity (European Committee for Standardization, 2000). With regulatory oversight from the Central Pollution Control Board, India adheres to comparable standards under IS/ISO 17088:2021 (Bureau of Indian Standards, 2021). The high compostability of starch films is supported by empirical evidence. For instance, Zainuddin et al. (2022) showed that sago-starch films blended with biodegradable fillers achieved more than 90% disintegration and 78%–92% CO_2 evolution within 45 days under controlled composting, confirming rapid biodegradation consistent with ASTM and ISO thresholds. Starch films usually fulfill or surpass these requirements because of their hydrophilic, enzyme-susceptible polymer matrix, completely disintegrating in industrial composting settings in three to 6 months.

17.4.4 Tests for mechanical and barrier properties

Mechanical integrity testing prevents physical failures that might arise from contamination breaches while handling and transport. Universal Testing Machines per ASTM D882 test tensile strength in the range of 10–30 MPa, elongation at break from 20% to 50%, and Young's modulus, while dart-drop methods are used to assess puncture resistance. Thermal stability in the hot-fill process is assured by differential scanning calorimetry, delineating glass transition temperatures from -10 to 50°C and melting endotherms at 100–150°C. Barrier tests aim for oxygen transmission rates of less than 10 cc/m^2/day and water vapor transmission rates of less than 1458 g/m^2/day, as tested in ASTM E96 permeation cells, while improvements of two to five fold have been obtained with the addition of nanocellulose or chitosan fillers. The mechanical and barrier properties of the starch films can be improved by adopting various optimization techniques, such as response surface methods, which help in selecting the appropriate concentration of the plasticizer and other ingredients. The optimization process significantly improves the flexibility, migration control, and strength of starch-based films (Garavito et al., 2024; Tavares et al., 2020).

17.4.5 Regulatory compliance

The regulatory frameworks that ensure positive listings of such starch additives are the FDA 21 CFR 177, FSSAI standards in India, and Codex Alimentarius, along with Declarations of Compliance for new formulations. Persistent challenges involve scaling production of antimicrobial composites and addressing lifecycle assessment gaps. Upcoming research directions would revolve around in vivo pharmacokinetic studies and the design of formulations using artificial intelligence for enhanced translational potential (Prajapati, 2022).

17.5 Challenges and strategies for starch-based packaging

17.5.1 Material and performance challenges

Starch-based films have many issues; these materials have low water vapor barrier properties and undergo absorption and subsequent swelling due to an affinity with water. By extension, they will be more brittle compared with polyolefins. As a result, they will have low tensile strength and elongation at break compared with polyolefins. Their uses will be restricted to very demanding locations. Examples include stretch wrap and heavy-duty packaging bags.

An additional issue regards variability with regard to botanical origin and ratio of amylose and amylopectin, as well as processing history. Small changes in starch and water contents can greatly affect gelatinization, retrogradation, mechanical properties, and permeability; thus, it becomes a constraint on large-scale production. Also, there might be retrogradation, embrittlement, and shrinkage due to storage. Lastly, the barrier properties with regard to oxygen, lipids, and aroma still have limitations compared to conventional plastics. Another important problem is the variability of properties depending on botanical source, amylose/amylopectin ratio, and history of processing. Small variations in starch or moisture content may strongly influence gelatinization, retrogradation, mechanical behavior, and permeability; this is a limitation to achieve consistent large-scale processing. Thermoplastic starch may also undergo retrogradation over time, embrittlement, and dimensional instability during storage. Finally, the barrier performance for oxygen, lipids, and aroma compounds often remains lower than for conventional plastics, which limits the extension of shelf life for sensitive foods.

17.5.2 Processing and scale-up issues

The processing of starch-based bioplastics, from the point of view of a processor, demands stringent control of moisture, temperature, and shear to achieve homogeneous thermoplastic starch with minimal degradation. The window of operation is small: too little water or plasticizer results in fragile, inhomogeneous melts; too much leads to bubble formation, resulting in stickiness and poor film quality. Extrusion and blow-molding of starch films require specific screw designs, venting, and temperature profiles, which can be capital-intensive for small and medium enterprises.

Scale-up is further constrained by the need for drying, conditioning, and tight moisture control before and after processing, adding steps and energy costs compared with commodity plastics. In many regions, especially developing countries, access to industrial-grade starch with consistent quality is limited, as well as access to suitable compounding and film-forming equipment. This makes the economies of scale and productivity hard to achieve compared to those enjoyed by petrochemical plastics.

17.5.3 Limitations in economy and supply chain

From an economic perspective, starch packaging generally costs more per unit function compared with fossil-based plastics. These costs include not only raw materials but also plasticizers, compatibilizers, nanofillers, and expensive processing steps for property tuning (Kostag & Schmid, 2020; Siracusa et al., 2008). The presence of added components, either biodegradable polyesters such as PLA and PBAT biodegradable packaging materials or nanofillers, raises costs and reduces the advantage of "low-tech, low-cost" (Faiella et al., 2021). There is also a conflict between food and materials, particularly if starch is a source of food. Use of food-grade starch for packaging may put pressure on food supply chains and result in food price instability (Siracusa et al., 2008). Collection and composting or biodegradation facilities for starch packaging may be poorly developed. Employment of proper waste treatment methods for packaging made from starch would not allow realizing environmental value. In addition, market adoption would be hampered due to performance disparities, additional costs associated with packaging made from starch, as well as ambiguous labeling and standards for compostable starch packaging (Avérous & Pollet, 2012; Shah et al., 2021).

17.5.4 Trade-offs: environmental and functional

Although starch itself is biodegradable, the biodegradability rate and extent can be regulated via formulation and environmental factors. For example, higher amylose and cross-linking will delay biodegradability, and these factors will enhance stability but compromise the biodegradability rate. On the other hand, highly plasticized and biologically modified materials might have higher biodegradability rates but with questionable stability. Biodegradability can be achieved with additives involving synthetic polyesters, plasticizers, and inorganic nanoparticles, which have doubtful ecotoxicity and recycling properties (Dharmaraj et al., 2020). From a functionality perspective, it would be difficult for pure starch films to achieve simultaneously the light weight, strength, sealability, and barrier properties attainable with a multi-layer structure based on petrochemicals. As a consequence, there will be an added motivation for research on multi-layer or hybrid packaging solutions incorporating starch as just one component, thus making recycling and composting more complicated (Shah et al., 2021).

17.5.5 Formula and structural approaches

A prominent method for addressing the deficiencies related to mechanics and barrier properties is starch blending with biodegradable polymers such as PLA (Polylactic Acid), PCL (Polycaprolactone), PBAT (Polybutyl Adipate Terephthalate), or PHAs (Polyhydroxyalkanoates). The toughness, processability, and water-resistance properties of these blends could be greatly improved, but it should be noted that a compatibilizer or reactive extrusion may be required as a method for improving interfacial interactions. Apart from glycerol and sorbitol, some research on alternative plasticizers involving citrate esters, urea, and deep eutectic solvents is emerging with aims for optimizing malleability and suppressing retrogradation.

The other strategy would be based on reinforcement with micro- and nanofillers: cellulose nanofibers, nanoclays, lignin, chitosan, and finally, graphene and its derivatives. Their role would be the creation of tortuous paths for diffusion and thus improving barrier properties. Cross-linking with physical methods, such as heating and radiation, as well as with chemicals, such as citric acid, epichlorohydrin, and isocyanates, would decrease solubility, water sensitivity, and overcross-linking, causing lower biodegradability and brittleness. Modifications at the surface level via hydrophobic coatings and plasma treatment would improve water barrier properties with minimal effects on the matrix, which would remain primarily starch (Dharmaraj et al., 2020; Faiella et al., 2021).

17.5.6 Processing and technology strategies

On the processing side, optimized extrusion and thermoforming protocols represent the key to overcoming scale-up limitations. Tailoring screw design, barrel temperature profiles, and die geometry allows for better control of gelatinization and plasticization and helps in obtaining films with uniform thickness and properties. Integrated in-line moisture sensing and closed-loop control systems can stabilize product quality despite fluctuations in ambient humidity or feedstock moisture. Emerging technologies under investigation to improve filler dispersion, reduce plasticizer content, and provide lightweight foams for cushioning or trays include reactive extrusion, ultrasound, or microwave-assisted processing, and supercritical fluid-assisted foaming (Tang et al., 2021). Strategies for the resource-limited setting include modular, low-cost extruders, using locally fabricated equipment where possible, and harnessing solar or biomass-based energy to reduce operating costs. Decentralized microplants near sources of starch or agro-residue can minimize transport burdens and reinforce local circular economies. Value chain and sustainability strategies. At the value-chain level, several strategies have been put in place in order to balance material demand with food security. The preferential use of nonedible starch sources and agro-industrial residues adds value to the by-products. Cassava peels or potato processing wastes are examples, besides the underutilized tubers, while reducing competition with food uses. Design of application-specific formulations is another approach: for long shelf-life foods, a high-barrier and high-durability material would be produced, and fast-degrading, simpler formulations for short-life or secondary packaging, so that the performance and biodegradation profiles fit the purpose. By developing and harmonizing standards for compostability, biodegradability, and biobased content, in addition to clear eco-labeling, consumers and industries will be better equipped to choose appropriate products and foster investments in infrastructure. The economic feasibility of starch-based packaging can be further enhanced by introducing policies on green public procurement, tax incentives for biobased materials, and extended producer responsibility schemes. Additionally, designs based on life-cycle assessment will enable trade-offs and hot spots to be identified, which will direct research toward those formulations and processes that actually provide benefits to the environment. Putting together all of these tactics into a systemic approach, starch-based packaging is used where its benefits, such as renewability, biodegradability, and compatibility with agricultural residues, can be fully utilized while its drawbacks are reduced through smart formulation design, advanced processing technologies, and supportive policy frameworks (Cui et al., 2021; Karnwal et al., 2025; Venkatesh et al., 2020). Fig. 17.2 summarizes the major challenges associated with starch-based packaging. Fig. 17.2 represented some key challenges for starch-based packaging.

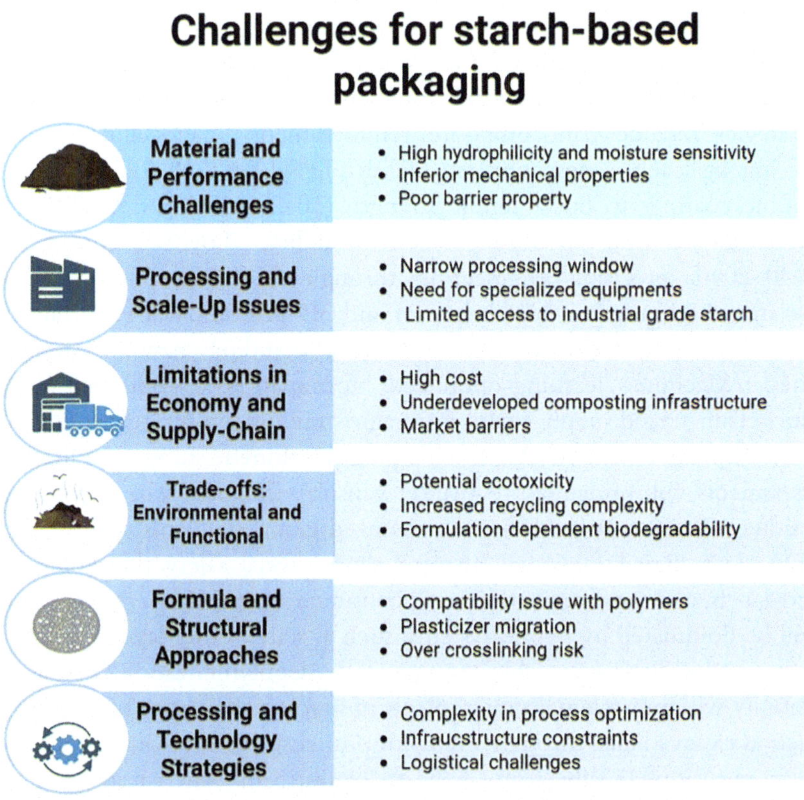

Figure 17.2 Some key challenges for starch-based packaging.

17.6 Future perspectives: safety, challenges, and perspectives in starch-based packaging

Although starch packaging offers great potential as an eco-friendly and biodegradable alternative to petroleum packaging materials, it necessitates innovative solutions for safety concerns. Currently, low toxicity for glycerol/sorbitol migrants and GRAS status for starch are validated by safety criteria that include migration testing (SML/OML as per EU 10/2011), cytotoxicity tests (Caco-2 viability rate surpassing 90%), and biodegradability standards (EN 13432). Future safety enhancements would concentrate on multiomics investigation for subtoxicity characteristics, such as endocrine disruption, brought on by novel packaging additives, as well as predictive toxicology based on AI-modeled simulations for extended exposure duration (Garavito et al., 2024). Hydrophilic brittleness, retrogradation, and variable amylose/amylopectin ratios continue to be problems that lead to moisture sensitivity (WVTR >1458 g/m^2/day), inconsistent barriers, and scale-up obstacles such as limited processing windows and increased costs. Economic tensions from food versus materials competition and underdeveloped composting infrastructure further impede adoption. Strategies include nanocomposite reinforcements (nanocellulose, ZnO <5 mg/kg leach), compatibilized blends (TPS-PBAT/PLA), and hydrophobic coatings to boost tensile strength (10–30 MPa) and OTR (<10 cc/m^2/day) while preserving compostability. Looking ahead, by 2030–2040, precision engineering through CRISPR-edited starches-variation high-amylose biofabrication and 3D printing will yield tailored films with on-demand degradation triggers, including enzyme-responsive ones. Machine learning-optimized formulations, combined with blockchain-traced supply chains from agro-residues, ensure reproducibility and compliance with GRAS. In vivo pharmacokinetic studies and lifecycle assessments will bridge lab-to-market gaps in validating zero microplastic residues. Regulatory evolution-harmonized global standards with Codex/FDA with digital DoC-will require nanomaterial safety thresholds and eco-labels, and work toward gaining consumer trust. Multilayer packaging will be dominated by hybrid systems, such as starch outer layers over PLA cores, and will be recyclable by enzymatic depolymerization. Economic viability will improve with microplants in starch-rich regions of the world, such as cassava belts, through a reduction in costs by 20%–30% because of solar extrusion. Finally, from niche to mainstream, starch packaging develops to guarantee food safety (zero genotoxicity), environmental closure

(90% biodegradation in 180 days), and circularity. An interdisciplinary convergence of nanotechnology, biotechnology, and AI places it as the cornerstone of zero-waste economies: plastic pollution cut down by 50% in food sectors by mid-century (Onyeaka et al., 2022; Jiang et al., 2020).

17.7 Conclusion

Growing environmental concerns, regulatory challenges, and the need for circular economy solutions have turned starch-based packaging materials into a possible class of sustainable substitutes for traditional petroleum-derived plastics. Starch films are one of the attractive options for food and consumer packaging applications because of their intrinsic renewability, biodegradability, and compatibility with agroindustrial residues. Considerable enhancement in their mechanical strength, barrier performance, and processability was achieved through formulation techniques like blending with biodegradable polyesters, improved plasticization, reinforcing micro- and nanofillers, and controlled cross-linking.

Despite the advantages, there are still so many challenges that need to be taken care of. Because of the requirement of sophisticated processes, instruments, and complex formulation, starch-based packing materials are still far behind conventional fossil-based plastics. Along with that limitation in the supply chain, insufficient infrastructure for composting decreases the expected environmental benefits. Balancing between durability and biodegradability often creates challenges in designing the packaging materials. Other issues, such as ecotoxicity produced by the additives and food material competition, required careful selection of the raw ingredients. In the near future, starch-based packaging can be implemented with a systemic integration of material science, advanced processing technologies, and optimization of the logistics.

References

Aitboulahsen, M., El Galiou, O., Amin, L., Bakkali, M., & Zerrouk, M. H. (2020). Effect of plasticizer type and essential oils on mechanical, physicochemical, and antimicrobial characteristics of gelatin, starch, and pectin-based films. *Journal of Food Processing and Preservation, 44*(6). https://doi.org/10.1111/jfpp.14480

Arezoo, E., Mohammadreza, E., Maryam, M., & Abdorreza, M. N. (2020). The synergistic effects of cinnamon essential oil and nano TiO2 on antimicrobial and functional properties of sago starch films. *International Journal of Biological Macromolecules, 157*, 743–751. https://doi.org/10.1016/j.ijbiomac.2019.11.244

Arruda, T. R., Machado, G.d. O., Marques, C. S., Souza, A. L.d., Pelissari, F. M., Oliveira, T. V.d., & Silva, R. R. A. (2025). An overview of starch-based materials for sustainable food packaging: Recent advances, limitations, and perspectives. *Macromol, 5*(2). https://doi.org/10.3390/macromol5020019

Asikkutlu, A. G., & Yildirim-Yalcin, M. (2025). Optimization of mechanical and water barrier properties of avocado seed starch based film and its application as smart pH indicator by adding blue butterfly pea flower extract. *Food Chemistry X, 25*, Article 102155.

ASTM International. (2021). *Astm D6400-21: Standard specification for labeling of plastics designed to be aerobically composted in municipal or industrial facilities*. ASTM International.

Aswathy, E., Elias, E., & Sreejith, L. (2025). Development and characterization of cassava starch-based eco-friendly films powered by fruit peel extracts for enhancing the shelf-life of perishable foods. *International Journal of Biological Macromolecules*, Article 147913.

Avérous, L., & Pollet, E. (2012). *Biodegradable polymers* (Vol 50, pp. 13–39). Springer Science and Business Media LLC. https://doi.org/10.1007/978-1-4471-4108-2_2

Ayassamy, P. (2025). Ocean plastic pollution: A human and biodiversity loop. *Environmental Geochemistry and Health, 47*(4). https://doi.org/10.1007/s10653-025-02373-4

BeMiller, J. N., & Whistler, R. L. (Eds.). (2009). *Starch: Chemistry and technology*. Academic press.

Bureau of Indian Standards. (2021). *IS/ISO 17088:2021—Specifications for compostable plastics*. BIS.

Cai, C., Ma, R., Duan, M., Deng, Y., Liu, T., & Lu, D. (2020). Effect of starch film containing thyme essential oil microcapsules on physicochemical activity of mango. *LWT, 131*, Article 109700. https://doi.org/10.1016/j.lwt.2020.109700

Chandrasekar, C. M., Nespoli, L., Bellesia, T., Ghaani, M., Farris, S., & Romano, D. (2024). Fabrication of double layer nanoparticle infused starch-based thermoplastic food packaging system for meat preservation. *International Journal of Biological Macromolecules, 254*(Pt 1), Article 127689. https://doi.org/10.1016/j.ijbiomac.2023.127689

Cui, C., Ji, N., Wang, Y., Xiong, L., & Sun, Q. (2021). Bioactive and intelligent starch-based films: A review. *Trends in Food Science & Technology, 116*, 854–869.

de Oliveira Filho, J. G., Albiero, B. R., Cipriano, L., de Oliveira Nobre Bezerra, C. C., Oldoni, F. C. A., Egea, M. B., de Azeredo, H. M. C., & Ferreira, M. D. (2021). Arrowroot starch-based films incorporated with a carnauba wax nanoemulsion, cellulose nanocrystals, and essential oils: A new functional material for food packaging applications. *Cellulose, 28*(10), 6499–6511. https://doi.org/10.1007/s10570-021-03945-0

de Souza, A. G., dos Santos, N. M. A., da Silva Torin, R. F., & dos Santos Rosa, D. (2020). Synergic antimicrobial properties of Carvacrol essential oil and montmorillonite in biodegradable starch films. *International Journal of Biological Macromolecules, 164*, 1737–1747. https://doi.org/10.1016/j.ijbiomac.2020.07.226

Deng, B., Chen, J., Li, S., Liu, J., Zhou, Z., Qin, Z., Wang, H., Su, M., Li, L., & Bai, Z. (2024). An antibacterial packaging film based on amylose starch with quaternary ammonium salt chitosan and its application for meat preservation. *International Journal of Biological Macromolecules, 261*(Pt 2), Article 129706. https://doi.org/10.1016/j.ijbiomac.2024.129706

Dharmaraj, N., Anandharamakrishnan, C., & Satheesh, N. (2020). Starch-based biodegradable films for food packaging: A review. *International Journal of Biological Macromolecules, 165*, 1–15. https://doi.org/10.1016/j.ijbiomac.2020.08.104

Dhiman, S., Kumari, A., Kumari, S., & Sharma, R. (2025). Advanced biodegradable starch-based nanocomposite films incorporating zinc oxide nanoparticles: Synthesis, characterization, and efficacy in antibacterial food packaging applications. *Journal of Environmental Chemical Engineering, 13*(3), Article 116296. https://doi.org/10.1016/j.jece.2025.116296

Di Matteo, A., Stanzione, M., Orlo, E., Russo, C., Lavorgna, M., Buonocore, G. G., & Isidori, M. (2026). Biodegradable films enriched with thymol and carvacrol with antioxidant and antibacterial properties to extend cheese shelf-life. *Food Control, 179*. https://doi.org/10.1016/j.foodcont.2025.111541

Du, J., Lu, Z., Cheng, K., Li, C., Wu, H., Xu, B., & Qian, J.-Y. (2025). Relationship between starch granule-associated proteins and in vitro digestibility of buckwheat starches: From the perspective of gelatinization degree. *Food Chemistry, 473*, Article 143115. https://doi.org/10.1016/j.foodchem.2025.143115

European Committee for Standardization. (2000). *EN 13432:2000—Requirements for packaging recoverable through composting and biodegradation*. CEN.

Faiella, F., La Rotonda, M. I., & Lamberti, G. (2021). Biodegradable polymer blends for packaging applications: Properties and performance. *Polymers, 13*(12), 2001. https://doi.org/10.3390/polym13122001

Flieger, M., Kantorová, M., Prell, A., Řezanka, T., & Votruba, J. (2003). Biodegradable plastics from renewable sources. *Folia Microbiologica, 48*(1), 27–44. https://doi.org/10.1007/bf02931273

Garavito, J., Peña-Venegas, C. P., & Castellanos, D. A. (2024). Production of starch-based flexible food packaging in developing countries: Analysis of the processes, challenges, and requirements. *Foods, 13*(24), 4096.

Giteru, S. G., Ali, A., & Oey, I. (2021). Understanding the relationship between rheological characteristics of pulsed electric fields treated chitosan-zein-poly (vinyl alcohol)-polyethylene glycol composite dispersions and the structure-function of their resulting thin-films. *Food Hydrocolloids, 113*, Article 106452.

Goiana, M. L., Mattos, A. L. A., Rosa, M.d. F., & Fernandes, F. A. N. (2024). Use of cold plasma as an alternative to improve corn starch-based films: Effect of the plasma application strategy. *Processes, 12*(7). https://doi.org/10.3390/pr12071429

Gupta, R. K., Pipliya, S., Karunanithi, S., Eswaran U, G. M., Kumar, S., Mandliya, S., & Kovács, B. (2024). Migration of chemical compounds from packaging materials into packaged foods: Interaction, mechanism, assessment, and regulations. *Foods, 13*(19), 3125.

Iamareerat, B., Singh, M., Sadiq, M. B., & Anal, A. K. (2018). Reinforced cassava starch based edible film incorporated with essential oil and sodium bentonite nanoclay as food packaging material. *Journal of Food Science and Technology, 55*(5), 1953–1959. https://doi.org/10.1007/s13197-018-3100-7

International Organization for Standardization. (2018). *ISO 20200:2018—Plastics: Determination of the degree of disintegration of plastic materials under simulated composting conditions in a laboratory-scale test*. ISO.

International Organization for Standardization. (2021a). ISO 17088:2021—Specifications for compostable plastics. ISO.

International Organization for Standardization. (2021b). ISO 14855-1: 2012—Determination of the ultimate aerobic biodegradability of plastic materials under controlled composting conditions. ISO.

Jambeck, J. R., Geyer, R., Wilcox, C., Siegler, T. R., Perryman, M., Andrady, A., Narayan, R., & Law, K. L. (2015). Plastic waste inputs from land into the ocean. *Science, 347*(6223), 768–771. https://doi.org/10.1126/science.1260352

Jiang, T., Duan, Q., Zhu, J., Liu, H., & Yu, L. (2020). Starch-based biodegradable materials: Challenges and opportunities. *Advanced Industrial and Engineering Polymer Research, 3*(1), 8–18.

Karnwal, A., Rauf, A., Jassim, A. Y., Selvaraj, M., Al-Tawaha, A. R. M. S., Kashyap, P., & Malik, T. (2025). Advanced starch-based films for food packaging: Innovations in sustainability and functional properties. *Food Chemistry, X*, Article 102662.

Kostag, M., & Schmid, M. (2020). Advances in starch-based films: Mechanical and barrier improvements. *Carbohydrate Polymers, 232*, Article 115805. https://doi.org/10.1016/j.carbpol.2020.115805

Kuchaiyaphum, P., Amornsakchai, T., Chotichayapong, C., Saengsuwan, N., Yordsri, V., Thanachayanont, C., Batpo, P., & Sotawong, P. (2024). Pineapple stem starch-based films incorporated with pineapple leaf carbon dots as functional filler for active food packaging applications. *International Journal of Biological Macromolecules, 282*(Pt 1), Article 137224. https://doi.org/10.1016/j.ijbiomac.2024.137224

Kumar, P., Gautam, S., Bansal, D., & Kaur, R. (2024). Starch-based antibacterial food packaging with ZnO nanoparticle. *Journal of Food Science and Technology, 61*(1), 178–191.

Kumar, P., Tanwar, R., Gupta, V., Upadhyay, A., Kumar, A., & Gaikwad, K. K. (2021). Pineapple peel extract incorporated poly(vinyl alcohol)-corn starch film for active food packaging: Preparation, characterization and antioxidant activity. *International Journal of Biological Macromolecules, 187*, 223–231. https://doi.org/10.1016/j.ijbiomac.2021.07.136

Kumari, S. V. G., Pakshirajan, K., & Pugazhenthi, G. (2022). Recent advances and future prospects of cellulose, starch, chitosan, polylactic acid and polyhydroxyalkanoates for sustainable food packaging applications. *International Journal of Biological Macromolecules, 221*, 163–182.

Lan, W., Zhang, R., Ji, T., Sameen, D. E., Saeed, A., Wen, Q., Dai, J., He, L., & Liu, Y. (2021). Improving nisin production by encapsulated Lactococcus lactis with starch/carboxymethyl cellulose edible films. *Carbohydrate Polymers, 251*, Article 117062. https://doi.org/10.1016/j.carbpol.2020.117062

Li, T., Guan, L., Zeng, M., Li, T., Lei, S., Tang, X., Pan, Y., Li, S., Zhou, M., Yuan, X., Zhang, Z., & Wu, H. (2025). Fabrication of corn starch-based composite films functionalized by Rosa roxburghii Tratt fruit pomace via extrusion compression molding for active food packaging. *International Journal of Biological Macromolecules, 284*(Pt 1), Article 138091. https://doi.org/10.1016/j.ijbiomac.2024.138091

Liu, P., Ying, L., Shang, X., & Xie, F. (2019). Starch–zinc complex and its reinforcement effect on starch-based materials. *Carbohydrate Polymers, 206*, 528–538. https://doi.org/10.1016/j.carbpol.2018.11.034

Liu, D., Zhao, P., Chen, J., Yan, Y., & Wu, Z. (2022). Recent advances and applications in starch for intelligent active food packaging: A review. *Foods, 11*(18), 2879. https://doi.org/10.3390/foods11182879

Lozano-Navarro, J. I., Díaz-Zavala, N. P., Velasco-Santos, C., Melo-Banda, J. A., Páramo-García, U., Paraguay-Delgado, F., García-Alamilla, R., Martínez-Hernández, A. L., & Zapién-Castillo, S. (2018). Chitosan-starch films with natural extracts: Physical, chemical, morphological and thermal properties. *Materials, 11*(1). https://doi.org/10.3390/ma11010120

Ma, S., & Jiang, H. (2024). The effect of cold plasma on starch: Structure and performance. *Carbohydrate Polymers, 340*, Article 122254.

Mandliya, S., Srivastav, P. P., Suthar, T., Shaikh, A. M., Harsányi, E., & Kovács, B. (2024). Migration of chemical compounds from packaging materials into packaged foods: Interaction, mechanism. *Assessment, and Regulations. Foods, 13*(19), 3125.

Nagy E.M. Coţa C. Cioica N. Gyorgy Z. Todica M. Cozar O. 2017 FT-IR investigation of starch based composite reinforced with Miscanthus fibers. AIP Conference Proceedings 15517616 American Institute of Physics Romania 1917 doi: 10.1063/1.5018291.

Onyeaka, H., Obileke, K., Makaka, G., & Nwokolo, N. (2022). Current research and applications of starch-based biodegradable films for food packaging. *Polymers, 14*(6), 1126.

Orhotohwo, O. L., Fanesi, B., Darko, H. S. O., Ismaiel, L., Lucci, P., Pacetti, D., Jaiswal, A. K., & Jaiswal, S. (2025). Development and characterisation of broccoli by-product-enriched starch-carrageenan films for extending the shelf life of fresh-cut fruits. *Food Bioscience, 69*. https://doi.org/10.1016/j.fbio.2025.106976

Perera, K. Y., Jaiswal, A. K., & Jaiswal, S. (2023). Biopolymer-based sustainable food packaging materials: Challenges, solutions, and applications. *Foods, 12*(12), 2422.

Pooja, N., Shashank, S., Singh, B. N., & Mazumder, N. (2024). Advancing sustainable bioplastics: Chemical and physical modification of starch films for enhanced thermal and barrier properties. *RSC Advances, 14*(33), 23943–23951. https://doi.org/10.1039/d4ra04263h

Prajapati, M. (2022). *Packaging edibles: New challenges and regulatory aspects edible food packaging: Applications, innovations and sustainability* (pp. 387–410). India: Springer Nature. https://doi.org/10.1007/978-981-16-2383-7_20

Raghunathan, R., Pandiselvam, R., Kothakota, A., & Mousavi Khaneghah, A. (2021). The application of emerging non-thermal technologies for the modification of cereal starches. *LWT, 138*. https://doi.org/10.1016/j.lwt.2020.110795

Raigond, P., Sood, A., Kalia, A., Joshi, A., Kaundal, B., Raigond, B., Dutt, S., Singh, B., & Chakrabarti, S. K. (2019). Antimicrobial activity of Potato starch-based active biodegradable nanocomposite films. *Potato Research, 62*(1), 69–83. https://doi.org/10.1007/s11540-018-9397-9

Rosas, A. L. G., Kleinübing, N. R., Bierhals, C. K., da Fonseca Antunes, B., Carloto, A. F., Silveira, D. R., & Meinhart, A. D. (2025). Cryogels based on Brazilian purple yam (Dioscorea trifidaLf) starch: An antimicrobial active packaging approach for microbial and moisture control in fresh cheese. *International Journal of Biological Macromolecules*, Article 149281.

Shah, A., Hasan, F., & Hameed, A. (2021). Biopolymers in sustainable packaging: Challenges and opportunities. *Environmental Chemistry Letters, 19*, 1–19. https://doi.org/10.1007/s10311-021-01245-2

Shahbazi, M., Majzoobi, M., & Farahnaky, A. (2018). Physical modification of starch by high-pressure homogenization for improving functional properties of κ-carrageenan/starch blend film. *Food Hydrocolloids, 85*, 204–214.

Sharaby, M. R., Soliman, E. A., & Khalil, R. (2024). Halochromic smart packaging film based on montmorillonite/polyvinyl alcohol-high amylose starch nanocomposite for monitoring chicken meat freshness. *International Journal of Biological Macromolecules, 258*, Article 128910.

Sifuentes-Nieves, I., Hernández-Hernández, E., Neira-Velázquez, G., Morales-Sánchez, E., Mendez-Montealvo, G., & Velazquez, G. (2019). Hexamethyldisiloxane cold plasma treatment and amylose content determine the structural, barrier and mechanical properties of starch-based films. *International Journal of Biological Macromolecules, 124*, 651–658. https://doi.org/10.1016/j.ijbiomac.2018.11.211

Siracusa, V., Rocculi, P., Romani, S., & Rosa, M. D. (2008). Biodegradable polymers for food packaging: A review. *Trends in Food Science and Technology, 19*(12), 634–643. https://doi.org/10.1016/j.tifs.2008.07.003

Sohrabi, S., Hosseini, H., Shojaee-Aliabadi, S., Hosseini, S. M., & Mirmoghtadaie, L. (2025). Effect of dual modification on physicochemical, structural properties and functional groups distribution of corn starch: A molecular approach. *International Journal of Biological Macromolecules, 304*. https://doi.org/10.1016/j.ijbiomac.2025.140792

Song, X., Zuo, G., & Chen, F. (2018). Effect of essential oil and surfactant on the physical and antimicrobial properties of corn and wheat starch films. *International Journal of Biological Macromolecules, 107*, 1302–1309.

Souza, A. G., Ferreira, R. R., Paula, L. C., Mitra, S. K., & Rosa, D. S. (2021). Starch-based films enriched with nanocellulose-stabilized Pickering emulsions containing different essential oils for possible applications in food packaging. *Food Packaging and Shelf Life, 27*. https://doi.org/10.1016/j.fpsl.2020.100615

Sun, K., Yi, J., Dai, R., & Chen, H. (2025). Highly efficient esterification of waxy maize starch in choline chloride/acetic acid acidic deep eutectic solvent system. *Carbohydrate Research, 548*, Article 109345. https://doi.org/10.1016/j.carres.2024.109345

Susmitha, A., Sasikumar, K., Devi, R., Padmakumar M, A., & Nampoothiri, K. M. (2021). Development and characterization of corn starch-gelatin based edible films incorporated with mango and pineapple for active packaging. *Food Bioscience, 41*, Article 100977. https://doi.org/10.1016/j.fbio.2021.100977

Tagliapietra, B. L., Felisberto, M. H. F., Sanches, E. A., Campelo, P. H., & Clerici, M. T. P. S. (2021). Non-conventional starch sources. *Current Opinion in Food Science, 39*, 93–102. https://doi.org/10.1016/j.cofs.2020.11.011

Tan, X., Sun, A., Cui, F., Li, Q., Wang, D., Li, X., & Li, J. (2024). The physicochemical properties of Cassava Starch/Carboxymethyl cellulose sodium edible film incorporated of Bacillus and its application in salmon fillet packaging. *Food Chemistry X, 23*, Article 101537.

Tan, S., Zhou, C., Rao, P., Tan, H., & Wang, J. (2025). Antioxidant pH-sensitive films incorporating CMC/SA/starch, anthocyanins, and tea polyphenols for monitoring freshness of pork. *Meat Science, 225*, Article 109808. https://doi.org/10.1016/j.meatsci.2025.109808

Tang, X., Alavi, S., & Herald, T. J. (2021). Effects of processing parameters on starch-based biodegradable materials. *Journal of Polymers and the Environment, 29*, 1–14. https://doi.org/10.1007/s10924-020-01893-5

Tavares, K. M., de Campos, A., Luchesi, B. R., Resende, A. A., de Oliveira, J. E., & Marconcini, J. M. (2020). Effect of carboxymethyl cellulose concentration on mechanical and water vapor barrier properties of corn starch films. *Carbohydrate polymers, 246*, Article 116521.

Utami, R., Khasanah, L. U., Manuhara, G. J., & Ayuningrum, Z. K. (2019). Effects of cinnamon bark essential oil (cinnamomum burmannii) on characteristics of edible film and quality of fresh beef. *Pertanika Journal of Tropical Agricultural Science, 42*(4), 1173–1184.

Venkatesh, G., Ahlgren, S., & Aneesh, R. (2020). Life cycle assessment of bio-based materials: Challenges and perspectives. *Renewable and Sustainable Energy Reviews, 124*, Article 109777. https://doi.org/10.1016/j.rser.2020.109777

Wang, J., Xu, X., Cui, B., Wang, B., & Abd El-Aty, A.M. (2024). Changes in the properties of the corn starch glycerol film in a time-dependent manner during gelatinization. *Food Chemistry, 458*, Article 140183.

Wongphan, P., Canellas, E., Nerín, C., Estremera, C., Harnkarnsujarit, N., & Vera, P. (2025). Screening and relative quantification of migration from novel thermoplastic Starch and PBAT blend packaging. *Foods, 14*(13), 2171.

Xie, F., Qin, Z., Luo, Y., He, Z., Chen, Q., & Cai, J. (2025). Synergistically engineered starch-based composite films: Multifunctional platforms integrating quaternary ammonium chitosan and anthocyanins for intelligent food monitoring and sustainable packaging. *Food Chemistry, 478*, Article 143560. https://doi.org/10.1016/j.foodchem.2025.143560

Yang, Y., Fu, J., Duan, Q., Xie, H., Dong, X., & Yu, L. (2024). Strategies and methodologies for improving toughness of starch films. *Foods, 4036.*

Yang, B., Wang, Z., & Fan, B. (2025). Solo water-gelatinized starch enhances the barrier properties of starch/PBAT. *International Journal of Biological Macromolecules, 302*, Article 140621.

Yao, X., Qin, Y., Zhang, M., Zhang, J., Qian, C., & Liu, J. (2021). Development of active and smart packaging films based on starch, polyvinyl alcohol and betacyanins from different plant sources. *International Journal of Biological Macromolecules, 183*, 358–368. https://doi.org/10.1016/j.ijbiomac.2021.04.152

Yassaroh, Y., Daulay, A., Prabowo, S., Wulandari, R., Yuliati, F., Yurohman, Y., Paramitasari, D., Pramana, Y. S., Putra, O. N., Suparman, S., Pudjianto, K., & Rahmawati, S. (2025). Biodegradable food packaging film derived from sago starch/microcrystalline cellulose enriched with ginger (Zingiber Officinale) extract for sliced apple preservation. *Journal of Food Science and Technology.* https://doi.org/10.1007/s13197-025-06308-w

Yudhistira, B., Husnayain, N., Punthi, F., Gavahian, M., Chang, C. K., & Hsieh, C. W. (2024). Progress in the application of emerging technology for the improvement of starch-based active packaging properties: A review. *ACS Food Science & Technology, 4* (9), 1997–2012.

Zainuddin, M., Ahmad, I., Kahar, N. A. M., & Abdullah, H. Z. (2022). Biodegradation behavior of sago starch-based biodegradable films under controlled composting conditions. *International Journal of Biological Macromolecules, 207*, 451–459. https://doi.org/10.1016/j.ijbiomac.2022.03.145

Zhang, Y., Li, Z., Zeng, J., Gao, H., & Qi, J. (2025). Barrier properties characterization and release kinetics study of citrus fibers-reinforced functional starch composites. *Food Hydrocolloids, 164*, Article 111195. https://doi.org/10.1016/j.foodhyd.2025.111195

Index

Note: Page numbers followed by "*f*" indicate figures and "*t*" indicate tables.

A

F

G

Q

R

S

Printed and bound by CPI Group (UK) Ltd, Croydon, CR0 4YY
14/09/2026
02216410-0006